CLASSICAL RECURSION THEORY

**The Theory of Functions and
Sets of Natural Numbers**

STUDIES IN LOGIC

AND

THE FOUNDATIONS OF MATHEMATICS

VOLUME 125

Editors:

J. BARWISE, *Stanford*
H.J. KEISLER, *Madison*
P. SUPPES, *Stanford*
A.S. TROELSTRA, *Amsterdam*

N·H

ELSEVIER
AMSTERDAM • LAUSANNE • NEW YORK • OXFORD • SHANNON • SINGAPORE • TOKYO

CLASSICAL RECURSION THEORY

The Theory of Functions and Sets of Natural Numbers

Piergiorgio ODIFREDDI
University of Turin
Turin, Italy

N·H

ELSEVIER
AMSTERDAM • LAUSANNE • NEW YORK • OXFORD • SHANNON • SINGAPORE • TOKYO

ELSEVIER SCIENCE B.V.
Sara Burgerhartstraat 25
P.O. Box 211, 1000 AE Amsterdam, The Netherlands

First edition 1989 / Second impression 1992

Paperback edition 1992 / Second impression 1999

Library of Congress Cataloging in Publication Data
A catalog record from the Library of Congress has been applied for.

ISBN: 0-444-87295-7 (hardbound edition) / ISBN: 0-444-89483-7 (paperback edition)

Transferred to digital printing 2005

♾ The paper used in this publication meets the requirements of ANSI/NISO Z39.48-1992 (Permanence of Paper).

To Lidia

What thou lovest well remains, the rest is dross
What thou lov'st well shall not be reft from thee
What thou lov'st well is thy true heritage.
(Pound, *Cantos*, LXXXI)

Foreword

Odifreddi has written a delightful yet scholarly treatise on recursion theory. Where else can one read about mezoic sets? His book constitutes his answer to the central question of recursion theory: what is recursion theory? His answer, I am pleased to note, is idiosyncratic. He makes numerous references to set theory, for example Baire's category theorem, the analytical hierarchy, the constructible hierarchy, and the axiom of determinateness. To my mind an understanding of recursion theory, even at the level of Turing degrees and recursively enumerable sets, is incomplete until the connection to higher levels is made via set theory.

If recursion theory is about computations, then the familiar finite case allows only a shallow view of the matter. Infinitely long computations, as in Kleene's account of finite type objects, or as in Takeuti's version of recursive functions of ordinals, permit a deeper insight into the nature of computation. This is borne out by the work of Slaman and others on fragments of arithmetic and polynomial reducibility, in which ideas from high up are applied low down.

The author's use of 'classical' in his title is partially meant, in this volume, to date the material he covers. He concentrates on the early days of recursion theory. Perhaps those were the glory days. Perhaps only the early results will survive.

The author makes the set theoretic connection but does not pursue it fully here. Let us hope he writes his next volumes on 'modern' recursion theory. His sparkling first volume proves him worthy of the task.

G.E. Sacks
Harvard University and M.I.T.
November 1987

vii

Preface

The origins of this book go back to fourteen years ago when, having done my studies in a country that, as Kreisel later remarked to me, was 'logically underdeveloped', I thought I could learn Recursion Theory by writing it. There were at the time a few textbooks, prominent among them Kleene [1952] and Rogers [1967], but I was unsatisfied with them because papers I was interested in, on progressions of formal systems, seemingly required many results that they did not cover. Thus I sat down, read a lot, and wrote a first version in Italian. Fortunately, I did not publish it.

Meanwhile, I had gotten in touch with some recursion theorists, and decided I would go to the United States to study some more. The Italian Center of Researches (C.N.R.) provided support, and in 1978 I landed in Urbana-Champaign, where the world opened up to me. I found there a very sensitive and kind teacher, Carl Jockusch, who taught me in one pleasant year more than I could have taught myself in a lifetime. And I found a friend in Dick Epstein, from whom I learned how to write mathematics. Then, having read some of their papers, I went to the U.C.L.A. people for a year, and I'm afraid I tried their patience with my many questions. There I learned what I know of Set Theory and Generalized Recursion Theory, through the teaching and help of Tony Martin, Yannis Moschovakis and John Steel. Back in Italy, I rewrote the whole book, this time in English.

In the meantime, I had grown aware of the fact that mathematics was not the universal science that I had once thought it was: not only personal, but also social and historical influences shape the work of the researchers. More specifically, I had learned that the Soviets were doing, in Recursion Theory, work that the Westerners did not know much about, and they themselves were largely unaware of what people did in the West. I found this an odd situation, and decided I would go to the Soviet Union to bridge the gap, at least in my knowledge. The Italian and Soviet State Departments provided support, and I stayed in Novosibirsk for one and a half years, in 1982-83, again learning a lot, in both mathematical and human terms. In particular, great help was provided by Marat Arslanov, Sergei Denisov, Yuri Ershov, and Victor Selivanov. Despite some difficulties there, which cost me a marriage among other things, I came back with more experience, and the book was now ready.

A final and unexpected touch was added by Anil Nerode and Richard Shore, who invited me to Cornell for a year in 1985, and in the following summers. With them I started a (for me) very fruitful collaboration, partly financed by a joint N.S.F.-C.N.R. grant. In particular, Richard Shore has relentlessly proved theorems that covered blank spots in the book. In Cornell I also met Juris Hartmanis, who changed my perspective in Complexity Theory.

In addition to all the people mentioned above, I was greatly helped by those who have read, and commented upon, substantial parts of the manuscript, or have taught me different things, including Klaus Ambos-Spies, Felice Cardone, Alexander Degtev, Leo Harrington, Georg Kreisel, Georgi Kobzev, Manny Lerman, Jim Lipton, Gabriele Lolli, Flavio Previale, Mark Simpson, and Bob Soare. Many other people have provided various kinds of help and corrections, in particular those who attended classes and seminars on various parts of the book in Torino and Siena (Italy), Urbana, U.C.L.A. and Cornell (United States), Novosibirsk and Kazan (Soviet Union). It would take too much space to mention them all, but to everybody go my sincerest thanks.

Since Gutenberg, books have usually been written to be printed. In my case this was made possible by Solomon Feferman and Richard Shore, who introduced the book to different editors. Michael Morley convinced me that I could type it myself in LaTeX, at a time when I did not even know how to turn a computer on, and he and Anil Nerode helped afterwards with the machines, in many ways. While I was preparing the typescript, the amazing Bill Gasarch and Richard Shore provided overall corrections in real time. Finally, support for typesetting was provided by the C.N.R. Thanks to all of them, too.

Different and very special thanks go to Lidia. She was there before it all, saw the book taking shape, and heard about it more than anybody else. She followed me on my pilgrimages, and I could perceive how great a toll this was taking on her only when it was too late. She is not here anymore, to see the end of it, and this is most sad. The immense amount of time stolen from her and devoted to this work is partly responsible for her absence. No doubt it was a stupid trade, but now, after fourteen years, here is the book: devoted to her as a partial, late compensation for what she deserved, and I was unable to give.

Torino - Urbana - Los Angeles
Novosibirsk - Ithaca
1974–1988

Preface to the Second Edition

After its first appearance in 1989, *Classical Recursion Theory* has seen a second printing and a paperback edition in 1992, all sold out. With the appearance of Volume II in 1999, the time has come to publish a second edition of Volume I, revised and corrected of all the misprints and mistakes that have been noticed through the years. I seize this opportunity to thank all the readers who have kindly pointed out imperfections in the text, thus helping me to make it better.

Recursion Theory has seen many exciting developments in the last decade, but most of them concern advanced areas which are treated in Volume II. Since Volume I has been left largely untouched by these developments, I have not attempted any major rewriting of the original text. Instead, I have inserted in the two volumes a number of forward and backward pointers (to pages, results and chapters) to make the whole book more interconnected, following the new Gospel of the Net.

Torino, June 1999.

Contents

Foreword by G.E. Sacks   vii

Preface   ix

Preface to the Second Edition   xi

Introduction   1
    What is 'Classical' 1
    What is in the Book 2
    Applications of Recursion Theory 5
    How to Use the Book 11
    Notations and Conventions 13

I   RECURSIVENESS AND COMPUTABILITY   17
    I.1   **Induction** 18
       Definitions by inductions 20
       Proofs by induction 20
       Recursiveness 21
       Historical roots of Recursion Theory $\star$ 22
       Formal Arithmetic $\star$ 23
       Some primitive recursive functions and predicates 24
       Codings of the plane 26
       Elimination of primitive recursion 28
    I.2   **Systems of Equations** 31
       The formalism of equations 31
       Definability by systems of equations 33
       Derivability from systems of equations 36
       A logical programming language $\star$ 38
    I.3   **Arithmetical Formal Systems** 39
       Notions of representability 39

	Formal systems representing the recursive functions	42
	Invariant definability ⋆	44
	Definability of functions ⋆	45
I.4	**Turing Machines**	46
	Variations of the Turing machine model	49
	Physical Turing machines ⋆	51
	Finite automata ⋆	52
	Turing machine computability	53
	Machine-dependent programming languages ⋆	59
I.5	**Flowcharts**	61
	Unstructured programming languages ⋆	63
	Unlimited register, random access machines ⋆	64
	Flowchart computability	65
	Structured programming languages ⋆	68
	Programs for primitive recursion ⋆	70
	Petri nets ⋆	74
I.6	**Functions as Rules**	75
	λ-calculus	76
	Other formulations of the λ-calculus ⋆	82
	λ-definability	83
	Functional programming languages ⋆	86
I.7	**Arithmetization**	87
	Historical remarks ⋆	87
	Numerical tools for arithmetization	88
	The Normal Form Theorem	90
	Equivalence of the various approaches to recursiveness	97
	The basic result of the foundations of Recursion Theory	100
I.8	**Church's Thesis** ⋆	101
	Introduction to Church's Thesis	102
	Historical remarks	105
	Computers and physics	106
	Classical mechanics	107
	Probabilistic physics	109
	Computers and thought	113
	The brain	115
	Constructivism	118
	Conclusion	122

II BASIC RECURSION THEORY   125

II.1 **Partial Recursive Functions** 126
    The notion of partial function 127
    Partial recursive functions 127
    Universal Turing machines and computers ★ 132
    Recursively enumerable sets 134
    R.e. sets as foundation of Recursion Theory ★ 143
    A programming language based on r.e. sets ★ 144

II.2 **Diagonalization** 145
    The essence of diagonalization 145
    Recursive undecidability results 146
    Limitations of mechanisms ★ 149
    Fixed-Point Theorem 152
    Limitations of formalism ★ 158
    Self-reference ★ 165
    Self-reproduction and cellular automata ★ 170

II.3 **Partial Recursive Functionals** 174
    Oracle computations and Turing degrees 175
    The notion of functional 177
    Partial recursive functionals 178
    First Recursion Theorem 181
    Recursive programs ★ 185
    Topological digression 186
    Iteration and fixed-points ★ 192
    Models of λ-calculus (part I) ★ 194
    Different notions of recursive functionals ★ 196
    Higher Types Recursion Theory ★ 199
    Computability on abstract structures ★ 202

II.4 **Effective Operations** 205
    Effective operations on partial recursive functions 205
    Effective operations on total recursive functions 208
    Effective operations in general ★ 210
    Recursive analysis ★ 213

II.5 **Indices and Enumerations** ★ 214
    Acceptable systems of indices 215
    Axiomatic Recursion Theory ★ 221
    Models of λ-calculus (part II) ★ 223
    Indices for recursive and finite sets 225
    Enumerations of classes of r.e. sets 228
    The Theory of Enumerations ★ 236

II.6 **Retraceable and Regressive Sets** ★ 238
    Retraceable versus recursive 239

Regressive versus r.e. 242
Existence theorems and nondeficiency sets 245
Regressive versus retraceable 248

III POST'S PROBLEM AND STRONG REDUCIBILITIES   251
III.1 Post's Problem 252
        Origins of Post's Problem ⋆ 253
        Turing reducibility on r.e. sets 254
III.2 Simple Sets and Many-One Degrees 256
        Many-one degrees 257
        Simple sets 259
        Effectively simple sets ⋆ 263
III.3 Hypersimple Sets and Truth-Table Degrees 267
        Truth-table degrees 269
        Hypersimple sets 272
        The permitting method ⋆ 277
III.4 Hyperhypersimple Sets and Q-Degrees 280
        Q-reducibility 281
        Hyperhypersimple sets 282
        Maximal sets ⋆ 288
III.5 A Solution to Post's Problem 294
        Semirecursive sets 294
        η-hyperhypersimple sets 299
III.6 Creative Sets and Completeness 304
        Effectively nonrecursive sets 304
        Creative sets 306
        Quasicreative sets ⋆ 311
        Subcreative sets ⋆ 314
        Effectively inseparable pairs of r.e. sets 316
III.7 Recursive Isomorphism Types 319
        Mezoic sets and 1-degrees 320
        Recursive isomorphism types 324
        Recursive equivalence types and isols ⋆ 328
III.8 Variations of Truth-Table Reducibility ⋆ 330
        Bounded truth-table degrees 331
        Weak truth-table degrees 337
        Other notions of reducibility ⋆ 339
III.9 The World of Complete Sets ⋆ 341
        Relationships among completeness notions 341
        Structural properties and completeness 348
III.10 Formal Systems and R.E. Sets ⋆ 349
        Formal systems and r.e. sets ⋆ 350

Undecidability 352
Essential undecidability 353
Independent axiomatizability 357

IV HIERARCHIES AND WEAK REDUCIBILITIES   **361**
IV.1 **The Arithmetical Hierarchy** 363
The definition of truth ⋆ 363
Truth in First-Order Arithmetic 363
The Arithmetical Hierarchy 365
The levels of the Arithmetical Hierarchy 367
Δ_2^0 sets 373
Relativizations ⋆ 374
IV.2 **The Analytical Hierarchy** 375
Truth in Second-Order Arithmetic 376
The Analytical Hierarchy 377
The levels of the Analytical Hierarchy 380
Π_1^1 sets 381
Δ_1^1 sets 387
Descriptive Set Theory ⋆ 392
Relativizations ⋆ 394
Post's Theorem in the Analytical Hierarchy ⋆ 395
IV.3 **The Set-Theoretical Hierarchy** 397
Truth in Set Theory 397
Standard structures 401
The Set-Theoretical Hierarchy 405
Δ_1^{GKP} functions 406
The levels of the Set-Theoretical Hierarchy 411
$\mathcal{HF}$ and the Arithmetical Hierarchy 414
Absoluteness and the Analytical Hierarchy 418
Admissible sets ⋆ 421
IV.4 **The Constructible Hierarchy** 422
The Constructible Hierarchy 422
The levels of the Constructible Hierarchy 424
The structure of L 425
Constructible sets of natural numbers 432
Σ_2^1 sets 437
$\mathcal{HC}$ and the Analytical Hierarchy 441
Recursion Theory on the ordinals ⋆ 443
Relativizations ⋆ 444

V TURING DEGREES   **447**

  V.1 **The Language of Degree Theory** 448
      The join operator 448
      The jump operator 450
      First properties of degrees 451
      The Axiom of Determinacy ⋆ 453

  V.2 **The Finite Extension Method** 456
      Incomparable degrees 457
      Embeddability results 459
      The splitting method 463
      Forcing the jump 467

  V.3 **Baire Category** ⋆ 471
      Topologies on total functions 472
      Comeager sets 473
      Baire Category and Degree Theory 477
      Meager sets of degrees 481
      Measure Theory and Degree Theory ⋆ 484

  V.4 **The Coinfinite Extension Method** 484
      Exact pairs and ideals 485
      Greatest lower bounds and least upper bounds 488
      Extensions of embeddings 490

  V.5 **The Tree Method** 493
      Hyperimmune-free degrees 495
      Minimal degrees 498
      Minimal upper bounds ⋆ 502
      König's Lemma and Π_1^0 classes ⋆ 505
      Complete extensions of Peano Arithmetic ⋆ 510

  V.6 **Initial Segments** ⋆ 516
      Uniform trees 516
      Minimal degrees by recursive coinfinite extensions 520
      The three-element chain 523
      The initial segments of the degrees ⋆ 528

  V.7 **Global Properties** 530
      Definability from parameters 530
      The complexity of the theory of degrees 536
      Absolute definability 540
      Homogeneity 543
      Automorphisms 546

  V.8 **Degree Theory with Jump** ⋆ 550

VI MANY-ONE AND OTHER DEGREES   **555**
  VI.1 **Distributivity**   555
    Distributive uppersemilattices   556
    Ideals of distributive uppersemilattices   558
  VI.2 **Countable Initial Segments**   561
    Finite initial segments   562
    Countable initial segments   566
  VI.3 **Uncountable Initial Segments**   569
    Strong minimal covers   569
    Uncountable linear orderings   570
    Uncountable initial segments   571
  VI.4 **Global Properties**   574
    Characterization of the structure of many-one degrees   575
    Definability, homogeneity, and automorphisms   575
    The complexity of the theory of many-one degrees   577
  VI.5 **Comparison of Degree Theories** ⋆   582
    1-degrees   582
    Truth-table degrees and weak truth-table degrees   584
    Elementary inequivalences   590
  VI.6 **Structure Inside Degrees** ⋆   591
    Cylinders   591
    Inside many-one degrees   594
    Inside truth-table degrees   598
    Inside Turing degrees   600

Bibliography   **603**

Notation Index   **643**

Subject Index   **649**

Introduction

Classical Recursion Theory is the study of real numbers or, equivalently, functions over the natural numbers. As such it has a long history, and a number of notions and results that were originally proved in different fields and for different purposes are incorporated, unified and extended in a systematic study. We are thinking here, for example, of the different equivalent definitions of real number, of Cantor's theorem that the real numbers are uncountable, of Gödel's class of constructible real numbers, and so on. All of these are now part of Recursion Theory and of our study, but the theory also provides new tools of its own, the origins of which can be traced back to Dedekind [1888]: he introduced the study of functions definable over the set ω of the natural numbers by recurrence using the well-ordered structure of ω, whence the name Recursion Theory.

The power of recursion as a tool for defining functions was analyzed in detail by Skolem [1923], Peter [1934], and Hilbert and Bernays [1934], but its limitations were also pointed out. Gradually the collective work of Post [1922], Church [1933], Gödel [1934], Kleene [1936], and Turing [1936], led to the identification of the most general form of the recursion principle and to what we now call recursive functions. In a bold philosophical abstraction Church [1936] proposed to identify the notion of 'effectively computable function' of natural numbers with that of recursive function, thus providing a feeling of absoluteness to the notion. With Post [1944] Recursion Theory became an independent branch of mathematics, studied for its own sake.

What is 'Classical'

In more recent decades Recursion Theory has been generalized in various ways to different domains: ordinals bigger than ω, functionals of higher order, abstract sets. All these subjects belong to what we call **Generalized Recursion Theory**. We use the word 'classical' to emphasize the fact that we confine our treatment to the original setting, and we will deal with notions of Generalized

1

Recursion Theory only when the theory provides results for the case we are interested in.

If we see *classical mathematics* as the study of concrete structures, like the set of natural numbers in Number Theory, or the set of functions over the real or complex numbers in Analysis (as opposed to *modern mathematics*, where the emphasis is on abstract structures, like algebraic or topological ones), then Classical Recursion Theory is part of classical mathematics, and sits between Number Theory and Analysis. This provides another reason for the word 'classical' in our subject.

Mathematics is usually formalized in well-established systems of Set Theory such as ZFC (the Zermelo-Fraenkel system, together with the Axiom of Choice). Our final use of the word 'classical' emphasizes the fact that we will be working mostly in ZFC. It is not surprising, due to the well-known independence results of Gödel [1938] and Cohen [1963], that only a part of the study of real numbers can be carried out in ZFC and we will point out the limits of our approach, together with possible extensions of ZFC suitable for Recursion Theory, at the end of the book.

What is in the Book

The basic methods of analysis of the real numbers that we are going to use are two:

Hierarchies. A hierarchy is a stratification of a class of reals built from below, starting from a subclass that is taken as primitive (either because well understood, or because already previously analyzed), and obtained by iteration of an operation of class construction.

Degrees. Degrees are equivalence classes of reals under given equivalence relations, that identify reals with similar properties. Once a class of reals has been studied and understood, degrees are usually defined by identifying reals that look the same from that class point of view. Degrees were used for the purpose of a classification of reals already in Euclid's *Book X* (w.r.t. a geometrical equivalence relation, between rational and algebraic dependence). See Knorr [1983] for a survey.

As might be imagined the two methods are complementary: first a class is analyzed in terms of intrinsic properties, for example by appropriately stratifying it in hierarchies, and then the whole structure of real numbers is studied modulo that analysis with the appropriate notion of degrees induced by the given class. The two methods also have a different flavor: the first is essentially definitional, the second essentially computational.

To give the reader an idea of what (s)he will find in the book we outline its bare skeleton, referring to the introductions of the various chapters for more detailed outlines.

The starting point of our study is the class of **recursive functions** introduced in Chapter I. The idea of its definition is simple: we try to isolate the functions over ω that are 'computable' in ways appealing both to the mathematician and to the computer scientist. Having many different approaches available, and various different intuitions of the notion of computability, we try them all, and discover that they all produce, once appropriately formalized, the same class of functions (and sets, through characteristic functions).

Chapter II considers two fundamental generalizations of the notion of recursiveness. **Partial recursive functions** are the natural formalization of algorithms: these, in the common use of the term, do not necessarily define total functions but only provide for specifications that allow the computation of values if particular conditions are satisfied. **Partial recursive functionals** take care of a different aspect of computations, namely the interactive procedure according to which a machine can be piloted, in its behavior, by a human agent. This can be formalized by the use of oracles that help the computation when requested by the machine.

A set is recursive if membership in it is effectively computable. The next level of complexity is reached when a set is effectively generated. In this case membership still can be effectively determined by waiting long enough in the generation of the set until the given element appears, but nonmembership requires waiting forever, and thus does not have effective content. Such sets are called **recursively enumerable**, and are the subject of Chapter III. But the emphasis of the study here is on the relative difficulty of computation. In other words, we identify sets which are equally difficult to compute. Then we attack the problem of whether the only relevant distinction among recursively enumerable sets, from a computational point of view, is between recursive and nonrecursive. The answer is that the world of recursively enumerable sets is a variegated one, in which different nonrecursive effectively generated sets may have different computational difficulty.

Chapter IV introduces the first hierarchies, by building on the fact that the recursively enumerable sets are exactly those definable in the language of First-Order Arithmetic with exactly one existential quantifier (coding the fact that an element is in a given recursively enumerable set if and only if there is a stage of the enumeration in which it appears). A natural hierarchy is thus obtained by looking at the **arithmetical sets** as those sets which are definable in First-Order Arithmetic, counting the number of alternations of quantifiers. Other hierarchies in the same vein are possible: counting alternations of function quantifiers in Second-Order Arithmetic which stratifies the **analytical sets**; or measuring the complexity of the definition of a set of natural numbers in the language of Set Theory in terms of previously defined sets which defines the **constructible sets of integers**.

Hierarchies are, by their nature, only partial tools of analysis. The notion of degree is instead a global one, classifying all sets modulo some equivalence relation. Chapters V and VI study the structure of the continuum with respect to two notions of relative computability, **Turing degrees** and **m-degrees**, and obtain two structural results. The first equates the complexities of the decision problem for the theories of Turing and m-degrees with that of Second-Order Arithmetic, the second gives a complete algebraic characterization of the continuum in terms of the structure of m-degrees. The Baire Category method, in both its original version and generalized forms, is the basic method of proof.

This completes Volume I, which introduces the fundamental notions and methods. Volumes II and III are a deeper and more sophisticated study of the same topics, in which the structures already introduced are revisited and analyzed more carefully and thoroughly. Volume II deals with sets of the arithmetical hierarchy, Volume III with the rest.

Chapters VII and VIII resume the analysis of the fundamental objects in Recursion Theory, the recursive sets and functions, and provide a microscopic picture of them. We start in Chapter VII with an abstract study of the complexity of computation of recursive functions. Then in Chapter VIII we will attempt to build from below the world of recursive sets and functions that was previously introduced in just one go. A number of subclasses of interest from a computational point of view are introduced and discussed, among them: the **polynomial time (or space) computable functions** which provide an upper bound for the class of feasibly computable functions (as opposed to the abstractly computable ones); the **elementary functions**, which are the smallest known class of functions closed under time (deterministic or not) and space computations; the **primitive recursive functions**, which are those computable by the 'for' instruction of programming languages like PASCAL, i.e. with a preassigned number of iterations (as opposed to the recursive functions, computable by the 'while' instruction, which permits an unlimited number of iterations).

Chapters IX and X return to the treatment of recursively enumerable sets. In Chapter III a good deal of information on their structure had been gathered, but here a systematic study of the structures of both the lattice of **recursively enumerable sets** and of the partial ordering of **recursively enumerable degrees** is undertaken. Special tools for their treatment are introduced, most prominent among them being the **priority method**, a constructive variation of the Baire Category method.

Chapter XI deals with **limit sets**, also known as Δ_2^0 sets, which are limits of recursive functions. They are a natural formalization of the notion of sets for which membership can be determined by effective trials and errors, unlike recursive sets (for which membership can be effectively determined), and recursively enumerable sets (for which membership can be determined with at most one mistake, by first guessing that an element is not in the set, and then changing opinion if it shows up during the generation of the set).

The following chapters produce an analysis of the sets introduced, and only touched upon, in Chapter IV, in particular **arithmetical, hyperarithmetical, Δ_2^1,** and **constructible sets**, and various other classes. In all these chapters the study proceeds by first analyzing the classes themselves, and then looking at the notions of degree associated with them (respectively: arithmetical degrees, hyperdegrees, Δ_2^1-degrees, constructibility degrees, as well as degrees with respect to appropriate admissible ordinals).

The final chapter deals with **nonclassical set-theoretical worlds** in order to point out the limitations of the classical approach, to exactly establish its limits, and to reach beyond it by adding appropriate axioms (prominent among them the Axiom of Projective Determinacy).

Starred subsections deal with topics related to the ones at hand thought sometimes quite far away from the immediate concern. They provide those connections of Recursion Theory to the rest of mathematics and computer science which make our subject part of a more articulate and vast scientific experience. Limitations of our knowledge and expertise in these fields make our treatment of the connections rather limited, but we feel they add important motivation and direct the reader to more detailed references.

Particular themes on which continuous commentary is made throughout the book are relationships with **computers**, **logic**, and the theory of **formal systems**, in particular the results known as Gödel's Theorems. As our development becomes more technical, connections to fields outside logic in general, and other branches of Recursion Theory in particular, become less important.

As will be clear by now, we have opted for breadth rather than depth, and have provided rudiments of many branches of Classical Recursion Theory, rather than complete and detailed expositions of a small number of topics. In this respect our book is in the tradition of Kleene [1952] and Rogers [1967], and differs from recent texts like Hinman [1978], Epstein [1979], Moschovakis [1980], Lerman [1983], and Soare [1987], which can be used as useful complements and advanced textbooks in their specialized areas.

Applications of Recursion Theory

No sound mathematical theory is self-contained or detached from the rest of mathematics or science. It takes inspiration from, and provides matter of reflection to other branches of knowledge. Recursion Theory is no exception and, despite this being a book on the pure theory, we will touch on applications and connections whenever possible. Here we give an idea of the applications that our subject can have in other branches of science some of which will be taken up again in more detail in the book.

Philosophy

If one of the main goals of Philosophy of Science is the conceptual analysis of epistemological notions, then the foundations of Recursion Theory provide some astounding successes for it. One of the original concerns of Recursion Theory had been the analysis of the notion of *effective computability* and of the related concept of algorithm. The isolation of the technical notion of recursiveness as a formal proposal intended to capture the essence of computability on natural numbers (see Chapter I) is a first success of the philosophical side of the theory, but by no means the only one. After all, computability on natural numbers is just one part of the whole story.

A great deal of work has been spent on axiomatizing the abstract notion of computability (see p. 221), and on analyzing the role of the special properties of natural numbers in computations. Decent notions of elementary computability have been proposed for abstract domains (see p. 202), and deeper properties have been shown to extend to a variety of domains more general than ω (such as admissible ordinals, see p. 443). This has required an analysis of the role of *finiteness* in computations, and an isolation of its essential properties. The familiarity of the notion involved, which is usually used unconsciously, magnifies the success obtained.

The concern of Recursion Theory with *predicativity* predates even its concern with computability (see p. 22), and it is reflected in its widespread use of hierarchies as a mean of building classes of functions from below. One of these hierarchies (the hyperarithmetical, see p. 391) has turned out to be particularly interesting and to provide for an upper bound to the notion of a predicatively defined set of natural number (Kreisel [1960]). Related work has subsequently been able to isolate a precise analogue of this notion (Feferman [1964]), thus doubling the success obtained with computability.

Computer Science

The area of Recursion Theory that deals with recursiveness is part of Theoretical Computer Science. Turing's analysis of computability in terms of machines provided the conceptual basis for the construction of *physical computers* in the late Forties: in the United States through Von Neumann, who knew Turing's work, and in the United Kingdom through Turing himself (see p. 132). Different approaches to recursiveness generate different types of *programming languages*, and we discuss (Chapter I and Section II.1) how the computational core of

PASCAL, LISP, PROLOG, and SNOBOL can easily be obtained from the appropriate versions of recursiveness. Finally, a good deal of Recursion Theory is devoted to the analysis of the *complexity of algorithms* and to a classification of recursive functions according to the tools needed to compute them. This is rapidly becoming a field of its own, called Complexity Theory, with methods and results strongly influenced by other parts of Recursion Theory (see Chapters VII and VIII).

Number Theory

The very origins of Recursion Theory place it close to Number Theory: the motivation of Dedekind [1888] was the analysis of the concept of natural number (see p. 22), while Skolem [1923] wanted to present a formulation of Arithmetic that avoided the difficulties of the common solutions to the paradoxes. But perhaps the most striking application of Recursion Theory to Number Theory is the solution of *Hilbert's Tenth Problem* (see p. 135) which asked for a decision procedure to determine the existence of solutions of given diophantine equations. Matiyasevitch [1970] proved a representation theorem, showing that the sets of (non-negative) solutions of diophantine equations are exactly the recursively enumerable sets. A negative solution to Hilbert's Tenth Problem then follows from the existence of a recursively enumerable, nonrecursive set.

Algebra

Until the second half of the last century, including the work of Lagrange, Gauss, Abel, and Galois, algebra had been developed in a strictly constructive way. The dichotomy between constructive and nonconstructive methods arose with the notion of prime ideal, which both Kronecker and Dedekind discovered from the usual constructive approach, but which Dedekind published in the now common set-theoretical framework. After that, nonconstructive methods which may produce less informative but more easily graspable arguments have become standard (see Metakides and Nerode [1982] for more historical background). Recursion Theory makes the analysis of the *constructive content of classical results* possible, as the following typical case illustrates. Steinitz Theorem shows that a field has an algebraic closure which is unique up to isomorphism. Its original proof does not constructivize: this is an accident for the existence part, but necessary for the uniqueness. The former follows from Rabin [1960] who, using a different existence proof, showed that a recursively presented field (i.e. a field with recursive set of elements and field operations, including equality) always has a recursively presented algebraic closure. The latter comes from Metakides and Nerode [1979], who showed that uniqueness (up to recursive isomorphism) of the recursively presented algebraic closure is equivalent to the existence of a splitting algorithm (to determine whether a polynomial is irreducible or not), a result that uses the priority method introduced in Chapter X. The analysis of the effective content of classical algebra has been thoroughly pursued: see Ershov [1980], Crossley [1981], Nerode and Remmel [1985] for references.

The usefulness of Recursion Theory in the analysis of constructivity in algebra is plausible. But there are unexpected uses too, such as in Higman [1961] who shows that the finitely generated groups embeddable in a finitely presented group are exactly the recursively presented ones (i.e. those for which the set of words equal to 1 is recursively enumerable), thus linking a purely algebraic notion with the notion of recursiveness.

Higman's representation theorem easily implies the undecidability of the *word problem* (to determine whether two words are equal) for finitely presented groups, proposed by Dehn in 1911 and solved by Novikov [1954] and Boone [1959]. The undecidability of the easier word problem for semigroups, proposed by Thue [1914] and solved by Post [1944] and Markov [1947], is historically important, being the first undecidability result of a problem from classical mathematics. These results started a whole area of research, devoted to the determination of which properties of algebraic structures are (un)decidable. See Tarski, Mostowski and Robinson [1953], Ershov, Lavrov, Taimanov and Taislin [1965] and Ershov [1980] for detailed treatments and references.

Analysis

Borel [1912] introduced the notion of computable real number, using the intuitive notion of computability. The very paper in which Turing introduced his influential approach to computability was motivated by the search for a formal definition of computable reals, and was thus the beginning of recursive analysis. Turing isolated a class of *recursive reals* that is independent of the proposed constructivization (in the sense that all classically equivalent definitions of real number remain equivalent when appropriately constructivized), contains all commonly used reals, and is algebraically closed. Subsequent work extending the notion of recursive functional (Section II.4) defined the notion of a recursive function of a real variable as a function defined on all reals, not only on the recursive ones.

This provided the needed tools to analyze the *effective content of analysis*: a result is constructive if whenever it has recursive data it provides us with recursive solutions. As a typical example, Weierstrass proof of the existence of a maximum for a continuous real function on a closed interval is constructive; but an argument at which the maximum is attained cannot be constructively found

(Lacombe [1957], Specker [1959]). Another example is provided by the ordinary differential equation $y' = f(x, y)$: the original proof of Picard that if f satisfies a Lipschitz condition the solution exists and is unique is constructive, but Aberth [1971] and Pour El and Richards [1979] showed that even the existence alone is not constructive if f is only uniformly continuous. See p. 213 for more on the subject.

As for algebra, one can look for undecidability results as well, some of which have been obtained by Richardson [1968], Adler [1969] and Wang [1974]. As an example, the latter proves that there is no recursive procedure to decide whether a real elementary function has a zero.

Set Theory

Recursion Theory and Set Theory have a large overlap in the study of sets of integers of high complexity: the material dealt with in Volume III could hardly be classified as solely belonging to one of them; it is rather a new field sprung from their marriage. But Recursion Theory does have successful applications to pure Set Theory in areas were the latter seems to be classically impotent. The best-developed applications have been two theories about cardinals: recursive equivalence types, and admissible ordinals.

The former deals with sets that, in a constructive sense, are *infinite but Dedekind-finite*, i.e. can be one-one mapped neither to a proper initial segment of ω, nor to a proper subset of themselves. Classically such sets do not exist in the presence of the Axiom of Choice, but their recursive versions have generated a rich theory that provides new insights into the notion of finiteness (see p. 328).

Another branch of Set Theory which is classically unmanageable is the theory of *large cardinals*: even the inaccessible ones, the smallest proposed type, cannot be proved to exist in classical Set Theory. The lack of examples different from ω forces one to resort to trivial cases, such as considering 1 as weakly but not strongly inaccessible because $0^0 = 1$ (Gödel [1964]). Recursion Theory provides a well-developed analogue of the theory of large cardinals, in which the role of the first regular cardinal is taken by the first ordinal which is not the order type of a recursive well ordering of ω (see p. 385). The notion of admissible ordinal (p. 443) takes care of the analogue of regular cardinal in general as an ordinal closed under recursive operations on ordinals, and analogues of a great variety of large cardinals can already be seen to exist among the countable ordinals. The existence of analogues of Ramsey cardinals can be disproved which might prompt some reflection on the role of very large cardinals in Set Theory (see Volume III for details).

Descriptive Set Theory

Cantor's Set Theory, and in particular the unlimited use of the power set, provoked various reactions at the turn of the century, one of which produced Descriptive Set Theory as a study of larger and larger classes of sets of reals which were explicitly defined (see p. 392). This approach, in which hierarchies are one of the main tools, is obviously a forerunner and an analogue of various recursion theoretical hierarchies (see Chapter IV), the main difference being one level of complexity: sets of reals are considered in the first case, sets of integers in the second. But Addison [1954], [1959] discovered that not only are there analogies: the full classical theory can be obtained by relativization of the recursive hierarchy theory by substituting continuous functions and open sets for recursive functions and recursively enumerable sets (see p. 392). This implies that all classical theorems have recursive versions of which they are consequences (but not conversely). This allows a unified approach, with recursion theoretical methods applicable to the classical case, and the theory has been resurrected from the state of lethargy in which it had fallen in the Forties.

Constructive Mathematics

The use of constructivism in classical mathematical theories is conservative: nonconstructive methods are accepted, and the issue is only whether given proofs are constructive as they stand, or can be replaced by constructive ones, a negative answer being interesting and acceptable. But constructivism can be taken more seriously as a philosophy of mathematics that would simply banish nonconstructive notions and proofs from practice. One possible approach to constructive mathematics consists of using the notion of recursiveness as a substitute for the notion of constructivity. This can be taken literally, as in *Markov's school* (see p. 214), which considers only those mathematical objects and operations on them that can be effectively described by recursive procedures as existing. But it can also be taken as a tool of analysis to compare different approaches.

For example, in Kolmogorov [1932] *intuitionism* is seen as a logic of problems: $\alpha \vee \beta$ means to solve one of α and β, $\alpha \to \beta$ to reduce the problem of solving β to that of solving α, $\exists x \alpha(x)$ to solve $\alpha(x)$ for some x, and so on. Kleene [1945] then introduced the notion of recursive realizability for Intuitionistic Number Theory: numbers realize formulas if they code, inductively, recursive procedures that prove the formula according to the constructive meaning of the logical operations. Realizability has been extended to Intuitionistic Set Theory by Kreisel and Troelstra [1970] and, even if not accepted as the only possible way of interpreting intuitionistic provability, it has become a common tool of analysis since it provides for constructive models of theories. See Troelstra

[1973] and Beeson [1985] for detailed treatments of the subject.

Logic

After the first fifty years in which Recursion Theory was mainly motivated by mathematical problems about Arithmetic, the logicians took over. Their main interest was still in Arithmetic, but their point of view was metamathematical. In their hands the theory obtained its most astonishing and revolutionary results which are also the best known applications of the subject and one of the main impulses to its growth. By a balanced use of two of the most fundamental methods of proof of Recursion Theory, arithmetization and diagonalization, a complete characterization of the expressiveness of formal systems was obtained, the result being that (as in the case of diophantine equations) exactly the recursively enumerable sets are (weakly) representable in any consistent formal system having a minimal arithmetical strength. The existence of a recursively enumerable, nonrecursive set then implies the *undecidability and incompleteness* of any such system (see Section II.2), thus showing the inadequacy of the concept of formal system. These are the highlights of the extensional analysis of formal systems provided by recursion theoretical methods, but by no means the only ones (see p. 349). A result of Myhill [1955] (III.7.13) points out the limits of this analysis and shows that, from an extensional point of view, all formal systems of common use look alike in the sense of being all recursively isomorphic.

How to Use the Book

This book has been written with two opposite, and somewhat irreconcilable, goals: to provide for both an adequate textbook, and a reference manual. Supposedly, the audiences in the two cases are different, consisting mainly of students in the former, and researchers in the latter. This has resulted in different styles of exposition, reflecting different primary goals: self-containment and detailed explanations for textbooks, and completeness of treatment for manuals. We have tried to solve the dilemma by giving a detailed treatment of the main topics in the text, and sketches of the remaining arguments in the exercises and in the starred parts.

The **exercises** usually cover material directly connected to the subject just treated and provide hints of proofs in the majority of cases, in various degrees of detail. In a few cases, for completeness of treatment and easiness of reference, some of the exercises use notions or methods of proof introduced later in the book.

The **starred chapters and sections** treat topics that can be omitted on a first reading. The **starred subsections** deal with side material, usually giving broad overviews of subjects that are more or less related to the main flow of thought, but which we believe provide interesting connections of Recursion Theory with other branches of Logic or Mathematics. The style is mostly suggestive: we try to convey the spirit of the subject by quoting the main results and, sometimes, the general ideas of their proofs. Detailed references are usually given, both for the original sources and for appropriate updated treatments.

The general prerequisite for this book is a working knowledge of first year undergraduate mathematics. When dealing with applications, knowledge of the subject will be assumed but, since the treatment is kept separate from the main text, there will be no loss in skipping the relative parts.

The chapters have been kept self-contained as far as possible. We have done our best to keep the style informal and devoid of technicalities, and we have resorted to technical details only when we have not been able to avoid them, no doubt because of our inadequacy.

Instead of the usual complicated diagrams of dependencies, we give suggestions on how the first two volumes of the book can be used as a textbook for classes in which Recursion Theory is the main ingredient.

Elementary Recursion Theory

Chapters I and II provide a number of alternative approaches to recursiveness and the basic development of the theory. Sections 2 to 6 of Chapter I are independent and can be chosen according to the audience in the class. More precisely, **mathematicians** can concentrate on Sections 2 and 3 and cover also the Incompleteness and Undecidability Results, treated in Section II.2. On the other hand, **computer scientists** will find more interest in Sections 4 to 6 of Chapter I and Section II.1, where the foundations of a number of programming languages are laid, and can also cover self-reproducing machines, touched upon in Section II.2, and the tools needed to build models of λ-calculus and combinatory logic, covered in Sections II.3 and II.5. Section I.8 treats Church's Thesis in a less simple-minded way than usual (i.e. facing the problems, instead of sweeping them under the rug), and it is perhaps more appropriate for **philosophers**.

Recursively Enumerable Sets

The elementary theory of r.e. sets and degrees is contained in Chapter III which requires only some background in elementary Recursion Theory. The chapter goes up to the solution to Post's problem (Sections 1 to 5) and the basic classes of r.e. sets. It can be used either as a final section of a course on elementary Recursion Theory (not dealing with alternative definitions of recursiveness), or as the initial segment of an advanced course on r.e. sets. In the latter case, it should be followed by Chapter IX, dealing with the lattice of r.e. sets, and a choice of material from Chapter X, in which priority arguments are introduced. Some of the material here, e.g. the theory of r.e. m-degrees, is not standard, but is useful in various respects: intrinsically, this structure is much better behaved than the schizoid one of r.e. T-degrees, and it reflects the global structure of degrees, which the latter does not; moreover, arguments on T-degrees (such as the coding method) are better understood in their simpler versions for m-degrees.

Degree Theory

Elementary degree theory is treated in Chapter V which, with some background in elementary Recursion Theory, can be read autonomously. We develop the theory up to a point where it is possible to prove a number of global results. This forms the nucleus of a course, and it can be followed by a number of advanced topics including a choice of results from Chapters XI and XII, on degrees of Δ_2^0 and arithmetical sets. Chapter VI, on m-degrees, is often unjustly neglected, but it does provide for the only existing example of global characterization of a structure of degrees. It can be read independently of Chapter V.

Complexity Theory

Chapters VII and VIII deal with abstract complexity theory and complexity classes, and do not require any background, except for a working knowledge of recursiveness and Turing machines (like Sections 1 and 4 of Chapter I). The treatment is fairly complete but, going beyond the usual unbalanced confinement to polynomial time and space computable functions, it also covers unjustly neglected classes of recursive functions, such as elementary, primitive recursive, and ϵ_0-recursive ones which are of interest to the computer scientist.

Notations and Conventions

$\omega = \{0, 1, \dots\}$ is the set of natural numbers, with the usual operations of plus ($+$) and times ($\times$ or $\cdot$), and the order relation $\leq$. $\mathcal{P}(\omega)$ is the power set of ω, i.e. the set of all subsets of ω. ω^ω and $\mathcal{P}$ are, respectively, the sets of total and partial functions from ω to itself.

We reserve certain lower or upper case letters to denote special objects:

- $a, b, c, \dots, x, y, z, \dots$ for natural numbers

- $f, g, h, \ldots$ for total functions of any number of variables

- $\alpha, \beta, \gamma, \ldots, \varphi, \psi, \chi, \ldots$ for partial functions of any number of variables

- $F, G, H, \ldots$ for functionals, i.e. functions with some variables ranging over numbers, and some over functions

- $A, B, C, \ldots, X, Y, Z, \ldots$ for sets of natural numbers

- $P, Q, R, \ldots$ for predicates of any number of variables

- $\sigma, \tau, \ldots$ for strings, i.e. partial functions with finite domain and values in $\{0, 1\}$.

Regarding **sets**:

- $x \in A$ means that x is an element of A

- $|A|$ is the cardinality of A, i.e. the number of its elements

- $A \subseteq B$ and $A \subset B$ are the relations of inclusion and strict inclusion

- $\overline{A}$ is the complement of A, and the prefix 'co-' in front of a property of a set means that the complement has this property (i.e. a set is co-immune if its complement is immune)

- $A \cup B$ is the union of A and B, i.e. the set of elements belonging to at least one of A and B

- $A \oplus B$ is the disjoint union of A and B, i.e. the set of elements of the form $2x$ if $x \in A$, and $2x + 1$ if $x \in B$

- $A \cap B$ is the intersection of A and B, i.e. the set of elements belonging to both A and B

- $A \times B$ is the cartesian product of A and B, i.e. the set of pairs (x, y) whose first and second components are, respectively, in A and B

- $A \cdot B$ is the recursive product of A and B, i.e. the set of codes $\langle x, y \rangle$ of pairs $(x, y) \in A \times B$ (see p. 26 for codings)

- c_A is the characteristic function of A, with value 1 if the given argument is in the set, and 0 otherwise.

Regarding **predicates**:

- $\neg P$, $P \wedge Q$, $P \vee Q$, $P \to Q$, $P \leftrightarrow Q$, $\forall x P$, $\exists x P$ are the usual logical operations of negation, conjunction, disjunction, implication, equivalence, universal and existential quantification.

  The symbols $\to$ and $\leftrightarrow$ will be used in a formal way, to build new properties from given ones. The symbols $\Rightarrow$ and $\Leftrightarrow$ will be used informally, as abbreviations for 'if ... then', and 'if and only if'.

  We use bounded quantifiers as abbreviations:

  $$(\exists x \leq y)P(x) \quad \text{for} \quad (\exists x)[x \leq y \wedge P(x)]$$
  $$(\forall x \leq y)P(x) \quad \text{for} \quad (\forall x)[x \leq y \to P(x)].$$

- c_P is the characteristic function of P, with value 1 if P holds for the given argument and 0 otherwise.

Regarding **binary relations** on a set A, R is:

- reflexive if xRx for every $x \in A$

- antireflexive if $\neg(xRx)$, for every $x \in A$

- symmetric if $xRy \Rightarrow yRx$ for every $x, y \in A$

- transitive if $xRy \wedge yRz \Rightarrow xRz$ for every $x, y, z \in A$

- a (weak) partial ordering if it is reflexive and transitive (weak partial orderings are indicated by $\leq$, $\preceq$, or $\sqsubseteq$)

- a (strict) partial ordering if it is antireflexive and transitive (strong partial orderings are indicated by $<$, $\prec$, or $\sqsubset$)

- a total ordering if it is a partial ordering, and $xRy \vee yRx \vee (x = y)$ for every $x, y \in A$

- an equivalence relation if it is reflexive, transitive, and symmetric; in this case the set A is partitioned into equivalence classes (each consisting of the elements that are in the relation R with each other)

- an uppersemilattice if any pair of elements of A has a l.u.b., and a lattice if any pair of elements of A has both l.u.b. and g.l.b. (given two elements x and y, their least upper bound (l.u.b.) and greatest lower bound (g.l.b.) are, respectively, the smallest element of A greater than both x and y, and the greatest element of A smaller than both x and y).

Regarding **functions**:

- $f \circ g$ or fg denote the composition of f and g

- $f^{(n)}$ denotes the result of n iterations of f, i.e. n successive applications of f (by convention, $f^{(0)}(x) = x$)

- $\varphi(x)\downarrow$ means that φ is defined on x

- $\varphi(x)\uparrow$ means that φ is undefined on x

- the set of elements on which φ is defined is called its domain, and the set of elements which are values of φ for some argument is called its range

- $\varphi \simeq \psi$ means that φ and ψ are equal as partial functions, i.e. on each argument they are either both undefined, or both defined and equal

- the set of pairs (x, y) such that $\varphi(x) \simeq y$ is called the graph of φ

- $\alpha \subseteq \beta$ means that as partial functions β extends α, i.e. if α is defined on an argument, then β is too and has the same value.

Each chapter is divided into numbered sections, and each section is divided into unnumbered subsections. There is a unique progressive numbering inside sections, including definitions, results, and exercises. Internal references in a given chapter may omit the chapter number.

The bibliography only includes papers quoted in the book. We have done our best to attribute results and quote the original sources. In case of unpublished results, when an attribution has been possible through personal communication or other sources we have attached names without references, and the mistakes that may have occurred are unintentional. We are, of course, well aware of the fact that simply quoting original sources is only a ghost of history, and it barely hints at the growth and interaction of ideas. But at least it provides the bare facts.

It is now time to plunge into the real work. We hope you will find the book readable, despite the difficulties imposed partly by the subject, but mostly by our limitations. Try to be patient,

> and remember patience is the great thing, and above all things else we must avoid anything like being or becoming out of patience.
>
> (Joyce, Finnegans Wake)

Chapter I

Recursiveness and Computability

This chapter attacks the problem of characterizing the notion of **effective computability**, by isolating various different proposals. The methods introduced in Section 7 show them all to be equivalent, thus demonstrating that we have certainly found a natural and fundamental class of functions. In Section 8 we discuss whether we have reached a satisfactory solution, and to which extent it is possible to believe that the class of functions so isolated coincides with the class of effectively computable functions.

The various approaches we introduce can be roughly classified into two groups:

Mathematical. We start in Section 1 with a class of functions defined by mimicking the basic arithmetical notions, the principle of induction among them. We note that the functions of this class are naturally defined by means of equations, and thus undertake in Section 2 a general study of systems of equations. We then discover that by adopting special formal rules we can derive the values of a function from a system of equations defining it. In Section 3 we thus investigate the functions whose values can be derived by any logical means in current formal systems suitable for arithmetic.

Computational. By analyzing the human process of routine calculation, we set up in Section 4 a machine-like model of computation and programs for it. In Section 5 we then consider the purely algorithmical skeleton of programs, by abstracting from the specific implementation of the machine. In a final generalization we then set up, in Section 6, a theory of functions as abstract programs.

Of course different classifications are possible. A particularly relevant one, from a computational point of view, would make a distinction between **deterministic** and **nondeterministic** notions. So, e.g., Herbrand-Gödel computability, representability in formal arithmetical systems, λ-definability and, more generally, all notions of derivability in suitable formal systems (the most comprehensive formulation in this direction being Post canonical systems, introduced in Section II.1) are nondeterministic, since they provide rules which can be applied in certain situations, but do not establish the order of application when multiple choices are available. It is however possible to introduce restrictions in nondeterministic approaches to turn them into deterministic ones, usually without affecting their power (and this is actually done when these approaches are taken as basis for programming languages).

It is important to stress that **for much of the later development of Recursion Theory, alternative characterizations of recursiveness, as well as its relation with effective computability, are not needed.** The various sections have been kept mostly independent from each other, so that they can be read separately. The reader not interested in foundational aspects of Recursion Theory can even skip the whole chapter, except for Sections 1 and the first part of Section 7, in which the recursive functions and the fundamental method of **arithmetization** are respectively introduced. Arithmetization is a basic technical tool, which is here applied to produce a normal form for the recursive functions, and to show the equivalence of the various approaches to recursiveness.

As a whole, this introductory chapter (and the first two sections of the next one) may be thought of as a technical version of what Webb [1980] does philosophically and Hofstadter [1979] pyrotechnically. These books may offer various (and sometimes unexpected) complements to the matters here discussed (especially so for those we just hint at). They are recommended reading.

I.1   Induction

The subject of this book is a close look at functions from natural numbers to natural numbers. The interest of our study is evident: the natural numbers are one of the most natural type of mathematical objects, and thus our functions are among the most natural mathematical functions. But to even understand what such functions are, we must first of all have a good grasp of the objects they relate. We then start by analyzing the intuitive picture of the natural numbers, trying to characterize their structure. Something is clear: the natural numbers are all in a single discrete row, with a first but no last element. Since what matters to us is just their mutual relationship and not their ultimate individual nature, we may imagine them as obtained from a first element (the number 0), by iteration of a generation procedure (the successor operation S). Thus the numbers are

$$0 \quad S(0) \quad S(S(0)) \quad \cdots$$

or (by using now the natural numbers metalinguistically, to indicate the number of iterations of S)

$$0 \quad S^{(1)}(0) \quad S^{(2)}(0) \quad \cdots$$

We simply write n for $S^{(n)}(0)$.

Three axioms that we take for granted from our intuitive picture above are the following (in a first-order logic with equality):

Axioms I.1.1 (Dedekind [1888])

A1 $S(x) = S(y) \rightarrow x = y$

A2 $0 \neq S(y)$

A3 $x \neq 0 \rightarrow (\exists y)(x = S(y))$.

They say that the successor induces an isomorphism between ω (the set of natural numbers) and $\omega - \{0\}$. Also, they rule out some unwanted pictures of the natural numbers, like ones with cycles, or with two infinite sequences of elements like

$$a_0 \quad a_1 \quad a_2 \quad \cdots \quad b_0 \quad b_1 \quad b_2 \quad \cdots$$

Unfortunately, they leave space for structures like

$$a_0 \quad a_1 \quad a_2 \quad \cdots \quad \cdots \quad b_{-2} \quad b_{-1} \quad b_0 \quad b_1 \quad b_2 \quad \cdots$$

and to be able to isolate just the initial part of these structures we need to say that every element can be reached from 0 by a *finite number* of applications of S. This seems to involve the very notion of integer that we are trying to characterize, and might seem to be circular (It is actually impossible to do this in a first-order way. See the related remarks on p. 24).

We then take an operational stand, and begin to study how we can deal with functions and properties of natural numbers. There are basically two ways: we may want to define something new, or to check properties of something we already have. Necessarily (according to our intuition) we have to proceed in both cases by **induction**, i.e. starting from 0 and going on by means of the successor operation.

Definitions by inductions

A typical example is given in the following definitions of sum and product, that reduce each of these two binary functions to an infinite family of unary functions (obtained by fixing the first argument).

Definition I.1.2 (Grassmann [1861])

A4 $x + 0 = x$

A5 $x + S(y) = S(x + y)$

A6 $x \cdot 0 = 0$

A7 $x \cdot S(y) = x \cdot y + x$.

In both cases a new function is defined, first for 0 and then for a generic $S(y)$, using the work already done for y. A general formulation of this process (with parameters) is the following, where we write $y + 1$ for $S(y)$, as usual.

Definition I.1.3 (Dedekind [1888]) *A function f is defined from g and h by* **primitive recursion** *if*

$$
\begin{aligned}
f(\vec{x}, 0) &= g(\vec{x}) \\
f(\vec{x}, y + 1) &= h(\vec{x}, y, f(\vec{x}, y)).
\end{aligned}
$$

Proofs by induction

Suppose we have a property φ of natural numbers and we wish to check that it holds for every number. From the way the numbers are generated, this follows if the property holds for 0 and it propagates through the successor operation, since every number is obtained from 0 by a finite iteration of S. This is expressed by the **Axiom of Induction**:

Axiom I.1.4 (Dedekind [1888]) *If φ is a formula with one free variable then*

A8 $\varphi(0) \wedge (\forall x)[\varphi(x) \rightarrow \varphi(S(x))] \rightarrow (\forall y)\varphi(y)$.

In terms of sets this means that any set containing 0 and closed under successor contains ω, or that the numerals $S^{(n)}(0)$ exhaust the natural numbers. This is thus a tentative to restrict the possible models of the axioms A1–A3.

For our purposes it is better to express this principle in the equivalent form of **Complete Induction**, which refers to the natural ordering of the natural numbers, that can be introduced for example as:

$$
\begin{aligned}
x \leq y &\iff (\exists z)(x + z = y) \\
x < y &\iff x \leq y \wedge x \neq y.
\end{aligned}
$$

We then have the following equivalent form of A8:

$$(\forall z)[(\forall x < z)\varphi(x) \rightarrow \varphi(z)] \rightarrow (\forall y)\varphi(y).$$

By writing ψ in place of $\neg\varphi$ and taking the contrapositive, Complete Induction is equivalent to the following **Least Number Principle**:

$$(\exists y)\psi(y) \rightarrow (\exists z)[\psi(z) \wedge (\forall x < z)\neg\psi(x)].$$

Its content is simply that if we know that a number with a certain property exists, then we also know that there is the least number satisfying that property. A general formulation of this principle (with parameters) in terms of functions is:

Definition I.1.5 (Kleene [1936]) *A function f is defined from a relation R by μ-recursion[1] if*

1. *R is a regular predicate, i.e. $(\forall \vec{x})(\exists y) R(\vec{x}, y)$.*

2. *$f(\vec{x}) = \mu y R(\vec{x}, y)$, where $\mu y R(\vec{x}, y)$ is the least number y such that $R(\vec{x}, y)$ holds.*

Similarly, f is defined from g by μ-recursion if

1. *$(\forall \vec{x})(\exists y)(g(\vec{x}, y) = 0)$*

2. *$f(\vec{x}) = \mu y(g(\vec{x}, y) = 0)$.*

Note that the Least Number Principle can be simply written in μ-notation as $(\exists y)\psi(y) \rightarrow (\exists z)(z = \mu y\psi(y))$.

The name *recursion* for both the processes above (primitive recursion and μ-recursion) is justified by the fact that they are both defined by recurrence on the natural numbers.

Recursiveness

We are now ready for our first attack on the notion of effective computability. The idea is simple: the two processes just introduced certainly produce effectively computable functions when applied to effectively computable functions and predicates. We just have to take an inductive approach, by starting from the effectively computable functions corresponding to 0 and S and by successively building up new functions using primitive recursion and μ-recursion. We will also permit a rudimentary logical intuition to contribute to the class, both in initial functions (identities or projections) and in building rules (composition of known functions), to allow for useful manipulations. We are thus led to the following notion.

---

[1] μ is the Greek equivalent of the first letter of 'minimum'.

Definition I.1.6 (Dedekind [1888], Skolem [1923], Gödel [1931])
*The class of **primitive recursive functions** is the smallest class of functions*

1. *containing the initial functions*

$$\mathcal{O}(x) = 0$$
$$\mathcal{S}(x) = x+1$$
$$\mathcal{I}_i^n(x_1,\ldots,x_n) = x_i \qquad (1 \le i \le n)$$

2. *closed under composition, i.e. the schema that given $g_1,\ldots,g_m,h$ produces*

$$f(\vec{x}) = h(g_1(\vec{x}),\ldots,g_m(\vec{x}))$$

3. *closed under primitive recursion.*

A predicate is primitive recursive if its characteristic function is.

Definition I.1.7 (Kleene [1936]) *The class of **recursive functions** is the smallest class of functions*

1. *containing the initial functions*

2. *closed under composition, primitive recursion and μ-recursion.*

A predicate is recursive if its characteristic function is.

Historical roots of Recursion Theory ⋆

Dedekind [1888], improving the work of Grassmann [1861], was the first to succeed in the analysis of the concept of natural number. He was able to isolate the axioms for 0 and $\mathcal{S}$, and the principle of second-order induction. He was immediately faced with the problem of justifying his formal theory as adequately describing the informal notion of number. For this purpose he offered a characterization theorem: the theory described, up to isomorphism, only one structure (in modern terms, it was categorical). The basic idea for the proof was to see that an isomorphism of any two structures satisfying the axioms can be immediately defined by (primitive) recursion. The main task for Dedekind was, thus, the justification of the existence of functions defined by such a principle. By doing this he pulled the trigger of Recursion Theory.

A great impetus for the early work on the field was set by the suggestion of considering only effectively defined functions: this underlay various constructive approaches to mathematics, among which two have been particularly relevant to our subject. **Semi-intuitionists** (like Kronecker [1887], Poincaré [1903], [1913], and Lebesgue [1905]) were interested, on the positive side, in effective solutions to mathematical (especially algebraic) problems. They were also reacting, on the negative side, to Cantor's Set Theory and his use of the power set (which, they thought, should be taken as consisting of just those sets of natural numbers which are somehow explicitly definable). **Finitists** (Hilbert [1904] and his school), stimulated by the discovery of paradoxes, were trying to constructivize Dedekind's second-order result on Arithmetic by bringing it into the realm of first-order logic: one of their main interests was a consistency proof for Arithmetic done by finitary means. This soon led to general problems of characterizing finitistic arithmetical methods and their relationships with primitive recursion.

Formal Arithmetic ⋆

The simplest partial formalization of arithmetic that can be extracted from our treatment is the **Robinson Arithmetic** Q (R.M. Robinson [1950]). It consists, in the language of first-order logic with equality, of a constant 0, functional symbols S, $+$ and $\cdot$, and the axioms A1–A7. Local variations are possible, e.g. both the constants 0 and S can be defined (and thus eliminated) in the following way:

$$x = 0 \quad \Leftrightarrow \quad x + x = x$$
$$x = 1 \quad \Leftrightarrow \quad x \cdot x = x \land x \neq 0$$
$$S(x) = y \quad \Leftrightarrow \quad y = x + 1.$$

Also, axiom A3 can be replaced by the following:

$$x = 0 \lor 0 < x,$$

where $<$ is the predicate so defined:

$$x < y \quad \Leftrightarrow \quad (\exists z)(x + S(z) = y).$$

This system is quite weak, but nevertheless sufficient to represent every recursive function (in a precise sense introduced in definition I.3.1).

Robinson Arithmetic adds the defining equations of plus and times to the axioms for successor. **Primitive Recursive Arithmetic** (Skolem [1923]) actually adds axioms corresponding to the definitions of all the primitive recursive functions. Equivalently, one could add a general schema of primitive recursion:

$$R(f, x, 0) \quad = \quad x$$
$$R(f, x, S(y)) \quad = \quad f(R(f, x, y), y).$$

The next step would be to add the equation for μ-recursion (or equivalently, as we know, the induction principle), and this would result in **Peano**

Arithmetic $\mathcal{PA}$ (a first-order version of Peano [1889]), which can be defined as the extension of $\mathcal{Q}$ obtained by adding to it the axiom A8. Note that A8 is actually a schema of axioms, one for each formula of the language with one free variable. Axiom A3 now becomes derivable from the others, and it is usually dropped. But dramatic simplifications are not possible: $\mathcal{PA}$, *although independently axiomatizable, is not finitely axiomatizable* (Ryll-Nardzewski [1952], Montague and Tarski [1957]).

Leaving aside formal systems, we can consider the structure $\langle \omega, +, \cdot \rangle$ underlying our intuition of natural numbers and take a semantic approach to arithmetic. This pertains to the following chapters of our study and we quote here two extremal cases: **First-Order Arithmetic**, that is, the set of formulas of the first-order language with equality true in the structure; and **Second-Order Arithmetic**, obtained similarly by considering the second-order language. Note that *First-Order Arithmetic, though completely determining the first-order sentences true in the standard model (i.e. determining the structure up to elementary equivalence), admits countable models not isomorphic to it* (Skolem [1934]). This shows that no purely first-order version of Dedekind's isomorphism theorem exists. But, quite appropriately, the notion of recursiveness does provide the missing ingredient: *any recursive model of Peano Arithmetic is isomorphic to the standard one* (Tennenbaum [1959]).

Some primitive recursive functions and predicates

We begin our treatment by noting (after Skolem [1923]) that many interesting arithmetical functions and predicates in common use are primitive recursive. To illustrate the use of identities and composition, we show that addition is primitive recursive. Recall its defining equations

$$
\begin{aligned}
x + 0 &= x \\
x + S(y) &= S(x + y).
\end{aligned}
$$

They define a function $f(x, y) = x + y$ such that

$$
\begin{aligned}
f(x, 0) &= x \\
f(x, y + 1) &= S(f(x, y)).
\end{aligned}
$$

This can be put in the form allowed by Definition 1.3 by letting (by composition of initial functions)

$$
\begin{aligned}
h(x, y, z) &= S(\mathcal{I}_3^3(x, y, z)) \\
f(x, 0) &= \mathcal{I}_1^1(x) \\
f(x, y + 1) &= h(x, y, f(x, y)).
\end{aligned}
$$

In the following we do not attempt to put definitions in the allowed forms, which can be easily done as an exercise by the reader. It should be clear however that, due to the existence of the identity functions $\mathcal{I}_i^n$ and the possibility of substitutions (compositions), *the schemata of primitive recursion and μ-recursion may be applied quite freely by interchanging, identifying and introducing variables when needed.*

The following are primitive recursive:

- Predecessor

$$
\begin{aligned}
pd(0) &= 0 \\
pd(x+1) &= x.
\end{aligned}
$$

- Integer difference[2]

$$
\begin{aligned}
x - 0 &= x \\
x - (y+1) &= pd(x - y).
\end{aligned}
$$

- Bounded sums

$$
\begin{aligned}
\sum_{y \leq 0} f(\vec{x}, y) &= f(\vec{x}, 0) \\
\sum_{y \leq z+1} f(\vec{x}, y) &= \left(\sum_{y \leq z} f(\vec{x}, y) \right) + f(\vec{x}, z+1). \\
\sum_{y < z} f(\vec{x}, y) &= \sum_{y \leq z-1} f(\vec{x}, y).
\end{aligned}
$$

- Bounded products, as above.

This shows that *the primitive recursive predicates are closed under logical connectives and bounded quantifiers,* since e.g. the characteristic functions of $\neg P$, $P \wedge Q$ and $(\forall y \leq z) R(\vec{x}, y)$ are respectively $1 - c_P$, $c_P \cdot c_Q$ and $\prod_{y \leq z} c_R(\vec{x}, y)$. This allows us to translate usual arithmetical definitions into logical terms, and to often find out that they are primitive recursive. For example:

- x divides y

$$
x \mid y \Leftrightarrow (\exists z \leq y)(x \cdot z = y).
$$

- x is a prime

$$
Pr(x) \Leftrightarrow x \geq 2 \wedge (\forall y \leq x)(y \mid x \rightarrow y = 1 \vee y = x).
$$

---

[2]Since we do not consider negative integers, the difference of two numbers is set equal to 0 when it would give a negative value.

To define *the sequence of prime numbers* p_x, the natural guess would be:

$$p_0 = 2$$
$$p_{x+1} = \text{the smallest prime number greater than } p_x$$
$$= \mu y(Pr(y) \wedge y > p_x).$$

This is a permissible application of the μ-operator since the predicate is regular (there are infinitely many prime numbers), but we are not allowed to use the μ-operator in primitive recursion. However $p_{x+1} \leq p_x! + 1$ (by Euclid's proof), i.e. we have a primitive recursive bound (the *factorial* is primitive recursive because $z! = \prod_{y<z}(y+1)$). And *the primitive recursive functions are closed under bounded μ-recursion*

$$\mu y_{\leq z} R(\vec{x}, y) = \begin{cases} \mu y R(\vec{x}, y) & \text{if } (\exists y \leq z) R(\vec{x}, y) \\ 0 & \text{otherwise.} \end{cases}$$

Indeed, if

$$g(\vec{x}, y) = \text{characteristic function of } R(\vec{x}, y) \wedge (\forall z < y) \neg R(\vec{x}, z)$$

then

$$\mu y_{\leq z} R(\vec{x}, y) = \sum_{y \leq z} (y \cdot g(\vec{x}, y)).$$

We can also decompose a number in a primitive recursive way:

- *the exponent of k in the decomposition of y*

$$exp(y, k) = \mu x_{\leq y}[k^x \mid y \wedge \neg(k^{x+1} \mid y)].$$

Here we stop our first taste of primitive recursion: the unsatiated reader can have a bellyful by turning to Chapter VIII.

Codings of the plane

It is possible (after Cantor [1874]) to put the plane (i.e. the set $\omega \times \omega$ of ordered pairs of natural numbers) into a one-one, onto correspondence with the line (i.e. the set ω of natural numbers) in a primitive recursive way. This can be achieved by **dovetailing**: enumerating the pairs of the first row in the picture, and inserting after each of them all the pairs connected to it by arrows.

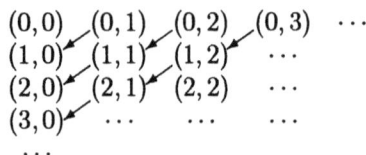

$$
\begin{array}{ccccc}
(0,0) & (0,1) & (0,2) & (0,3) & \cdots \\
(1,0) & (1,1) & (1,2) & \cdots \\
(2,0) & (2,1) & (2,2) & \cdots \\
(3,0) & \cdots & \cdots & \cdots \\
\cdots
\end{array}
$$

Note that all pairs on a same arrow-connection have the same sum. We thus order the pairs by their sum, and the pairs with the same sum in a lexicographical way (i.e. by first component):

$$(x, y) < (x', y') \iff (x + y < x' + y') \vee (x + y = x' + y' \wedge x < x').$$

Since there are $z + 1$ pairs with sum z, the position of the pair (x, y) in the order will be

$$\mathcal{J}(x, y) = \left(\sum_{i < x+y} (i + 1) \right) + x = \frac{(x + y)^2 + 3x + y}{2}$$

and $\mathcal{J}$ is primitive recursive. It also has primitive recursive inverses:

$$\mathcal{R}(z) = \mu y_{\leq z} (\exists x \leq z)(z = \mathcal{J}(x, y))$$
$$\mathcal{L}(z) = \mu x_{\leq z} (\exists y \leq z)(z = \mathcal{J}(x, y)).$$

It is easy to check that the following properties hold:

- $x \leq \mathcal{J}(x, y)$ and $y \leq \mathcal{J}(x, y)$
- $x + y > 1 \Rightarrow x < \mathcal{J}(x, y) \wedge y < \mathcal{J}(x, y)$
- $x < x' \Rightarrow \mathcal{J}(x, y) < \mathcal{J}(x', y)$
- $y < y' \Rightarrow \mathcal{J}(x, y) < \mathcal{J}(x, y')$
- $\mathcal{R}(\mathcal{J}(x, y)) = y$, $\mathcal{L}(\mathcal{J}(x, y)) = x$ and $\mathcal{J}(\mathcal{L}(z), \mathcal{R}(z)) = z$.

Slightly different codings are:

1. By noting that each natural number different from 0 can be uniquely written as the product of an even (2^x) and an odd $(2y + 1)$ number, the map
$$(x, y) \mapsto 2^x (2y + 1) - 1.$$

2. By considering the 1's in the binary expansion of natural numbers as markers, the map
$$(x, y) \mapsto 1 \underbrace{0 \cdots 0}_{x \text{ times}} 1 \underbrace{0 \cdots 0}_{y \text{ times}} = 2^y + 2^{x+y+1}.$$

This is not onto as a coding of pairs. But since every natural number can be thought of as coding a finite sequence of natural numbers this way, this actually provides an **onto coding of all finite sequences** (Minsky [1967]).

Other codings will be considered in the next theorem (Gödel's function β) and in Section 7.

Elimination of primitive recursion

We have introduced two definitional schemata based on induction: primitive recursion and μ-recursion. The next result shows that the former is subsumed under the latter, modulo appropriate initial functions. The intuition comes from Peano Arithmetic, which is built only on plus, times and the induction principle.

Theorem I.1.8 (Gödel [1931], Kleene [1936a]) *The class of recursive functions is the smallest class*

1. *containing sum, product, identities $\mathcal{I}_i^n$ and the characteristic function δ of equality*

2. *closed under composition*

3. *closed under μ-recursion.*

Proof. Let $\mathcal{C}$ be the smallest class satisfying the conditions of the theorem: clearly every function in $\mathcal{C}$ is recursive, and for the converse we only have to show that the constant function $\mathcal{O}$ and the successor $\mathcal{S}$ are in $\mathcal{C}$, and that $\mathcal{C}$ is closed under primitive recursion.

Since 0 is the least number equal to 0,

$$\mathcal{O}_n(\vec{x}) = \mu y(\mathcal{I}_{n+1}^{n+1}(\vec{x}, y) = 0).$$

Also, since

$$\delta(x, y) = \begin{cases} 0 & \text{if } x \neq y \\ 1 & \text{otherwise} \end{cases}$$

and 1 is the least number different from 0,

$$\begin{aligned} 1(x) &= \mu y(\delta(\mathcal{O}_2(x, y), \mathcal{I}_2^2(x, y)) = 0) \\ &= \mu y(\delta(0, y) = 0) = 1 \\ \mathcal{S}(x) &= \mathcal{I}_1^1(x) + 1(x). \end{aligned}$$

The whole problem is thus to show the closure of $\mathcal{C}$ under primitive recursion. The idea is the following: we will show that it is possible to define in $\mathcal{C}$ a function β such that for every finite sequence $a_0, \ldots, a_n$ of natural numbers there is one natural number a coding the sequence via β, i.e. such that

$$(\forall i \leq n)(\beta(a, i) = a_i).$$

Then, if f is defined from g and h in $\mathcal{C}$ by primitive recursion as

$$\begin{aligned} f(\vec{x}, 0) &= g(\vec{x}) \\ f(\vec{x}, y + 1) &= h(\vec{x}, y, f(\vec{x}, y)) \end{aligned}$$

we can code the sequence of values of f from 0 to y by using β:

$$t(\vec{x}, y) = \mu z [\beta(z, 0) = g(\vec{x}) \wedge (\forall i < y)(\beta(z, i+1) = h(\vec{x}, i, \beta(z, i)))].$$

Since then

$$f(\vec{x}, y) = \beta(t(\vec{x}, y), y),$$

f will be in $\mathcal{C}$ if t is, and this is the case if $\mathcal{C}$ is closed under logical operations (including universal bounded quantifier). We thus have three steps to perform:

1. *Existence of β*

   This is a purely number-theoretical argument. We want β s.t. for any sequence $a_0, \ldots, a_n$ there is a s.t. for $i \leq n$, $\beta(a, i) = a_i$. We will try to get the a_i's as remainders of the division of a number c by given numbers $d_0, \ldots, d_n$. That is, we want $a_i = rm(c, d_i)$. Then $c = q_i d_i + a_i$, $a_i < d_i$. Suppose this holds for another c': $c' = q'_i d_i + a_i$. By subtracting we get $c - c' = (q_i - q'_i)d_i$, i.e. the difference $c - c'$ is also divisible by d_i. If the d_i are relatively prime, then $c - c'$ is also divisible by their product $p = d_0 \cdots d_n$. This means that for different $c, c' < p$ we get different sequences of remainders when dividing by d_i (otherwise $c - c' \geq p$). Moreover, any sequence of $n + 1$ numbers less than the d_i's is obtained (since the number of possible sequences is p, and different $c < p$ give different sequences, i.e. we get them all). Then one of these sequences is the given one $a_0, \ldots, a_n$.[3]

   We now have to obtain the d_i's, with the stated conditions that they be relatively prime and $a_i < d_i$.

   The easiest way to get them relatively prime is to consider the sequence

   $$1 + d \quad 1 + 2d \quad \cdots \quad 1 + (n+1)d$$

   for any $d = s!$, $s \geq n$. ($1 + rd$, $1 + r'd$ are relatively prime since if q divides both of them, then it also divides their difference $(r - r')d$, and hence it divides d because $r - r' \leq n$ is a factor of $d = s!$, since $s \geq n$. But then q divides d and $1 + rd$, i.e. it divides 1.) We also want these numbers to be greater than the respective a_i's: for this is enough to have $d \geq a_i$, hence $s \geq a_i$.

   We can then let $d_i = 1 + (i+1)d$ with $d = s!$, for any s greater than n and the a_i's. By coding c and d into $a = \mathcal{J}(c, d)$ we can actually define

   $$\begin{aligned} \beta(a, i) &= rm(c, d_i) = rm(c, 1 + (i+1)d) \\ &= rm(\mathcal{L}(a), 1 + (i+1)\mathcal{R}(a)) \end{aligned}$$

---

[3] This is the Chinese Remainder Theorem, so called because it was known to the Chinese already in the 6th Century B.C.

2. *Closure of C under logical operations*
   For the propositional connectives:

   - negation
     We need a function interchanging 0 and 1. Note that

     $$\delta(x,0) = \begin{cases} 0 & \text{if } x \neq 0 \\ 1 & \text{otherwise} \end{cases}$$

     and thus $c_{\neg R}(x) = \delta(c_R(x),0)$.

   - disjunction
     We need a function which is 0 exactly when both the arguments are 0, i.e. when $x + y = 0$. Then $d(x,y) = \delta(\delta(x+y,0),0)$ satisfies the need, and $c_{R \vee S}(x) = d(c_R(x), c_S(x))$.

   Closure under negation implies that in C we can apply the μ-operator directly to predicates:

   $$\mu y\, R(\vec{x},y) = \mu y\,(c_R(\vec{x},y) = 1) = \mu y\,(c_{\neg R}(\vec{x},y) = 0).$$

   - bounded μ-operator
     We use, in this proof only, the form

     $$\mu y_{<z} R(\vec{x},y) = \begin{cases} \mu y R(\vec{x},y) & \text{if } (\exists y < z) R(\vec{x},y) \\ z & \text{otherwise} \end{cases}$$

     which is expressible in C as

     $$\mu y_{<z} R(\vec{x},y) = \mu y (R(\vec{x},y) \vee y = z).$$

   - bounded existential quantifier
     By the particular form of bounded μ-operator defined above we have:

     $$(\exists y < z) R(\vec{x},y) \Leftrightarrow (\mu y_{<z} R(\vec{x},y)) \neq z.$$

   - The other operations are obtained by composition as usual, e.g.

     $$\begin{aligned} R \wedge S &\Leftrightarrow \neg(\neg R \vee \neg S) \\ (\forall y < z) R(\vec{x},y) &\Leftrightarrow \neg(\exists y < z) \neg R(\vec{x},y) \end{aligned}$$

3. *Definition of β in C*
   We simply need to show that $\mathcal{J}, \mathcal{R}, \mathcal{L}$ and rm are in C.

- $\mathcal{J}$ is defined as

$$\mathcal{J}(x,y) = \frac{(x+y)^2 + 3x + y}{2}$$

  Since in $\mathcal{C}$ we have sum, product and composition, we only need the function $f(z) = z/2$:

$$f(z) = \mu y(2y = z \lor 2y + 1 = z)$$

- $\mathcal{R}$ and $\mathcal{L}$ are in $\mathcal{C}$ by their very definitions, e.g.

$$\mathcal{R}(z) = \mu x(\exists y < z + 1)(z = \mathcal{J}(x,y))$$

- the remainder of the division of x by y is defined as

$$rm(x,y) = \mu z_{<y}(\exists q < x + 1)(x = qy + z). \quad \Box$$

The use of Gödel's function β in the proof above can be avoided by using different codings. Of course the usual coding by prime numbers (see Section 7) would not work here, since the relevant functions and predicates are defined by primitive recursion, which is exactly what we have to avoid. For very elementary coding techniques, see Quine [1946] and Smullyan [1961].

Complementary results have been obtained by J. Robinson [1950], [1968], one of them being that *the recursive functions of any number of variables can be obtained, by composition alone, from the recursive functions of just one variable, plus sum and identities. And the recursive functions of just one variable can be defined independently, without using functions of more variables*, by the operations of composition and inversion of onto functions (the latter clearly corresponding to μ-recursion applied to regular predicates), from two (but not from only one) suitable initial functions.

I.2 Systems of Equations

From a mathematical point of view a function is usually defined from a set of equations, and we have used this fact informally in the previous section. We now undertake an analysis of the expressive power of systems of equations, bearing in mind that our concern is effective computability, and that we therefore require not only a definition of a function, but also some way to compute it.

The formalism of equations

Definitions are linguistic objects, and we then need, first of all, an appropriate language to express them. Also, since we are dealing with numerical functions, we need linguistic analogues of the numbers, which we can generate from 0 by the successor operation. Our formalism thus consists of:

1. **symbols**

   - equality '$=$'
   - constants $\mathbf{0}$ (for the number 0) and S (for the successor operation)
   - parentheses '(' and ')'
   - comma ',' (to separate variables)
   - variables $x_0, x_1, \ldots$ for numbers
   - constants $f_0^n, f_1^n, \ldots$ for n-ary functions (for each n).

2. **numerals**

   - $\mathbf{0}$ is a numeral
   - if a is a numeral, so is $S(a)$
   - nothing else is a numeral.

   We will write $\bar{n}$ for the numeral $S^{(n)}(\mathbf{0})$ representing the number n.

3. **terms**

   - $\mathbf{0}$ is a term
   - variables and 0-ary functional letters are terms
   - if t is a terms then so is $S(t)$
   - if $t_1, \ldots, t_n$ are terms then so is $f_i^n(t_1, \ldots, t_n)$
   - nothing else is a term.

   Note, in particular, that the numerals are terms.

4. **equations**
   If t and s are terms, and t is of the form $f_i^n(t_1, \ldots, t_n)$ for some n, i and $t_1, \ldots, t_n$ not containing any functional letter except possibly S, then $t = s$ is an equation.

   The idea is that equations define functions by their right-hand sides, while left-hand sides just tell which functions are defined.

5. **systems of equations**
   A system of equations is simply a finite set of equations.

   We will write $\mathcal{E}(f_1, \ldots, f_n; \vec{z})$ for a system of equations whose functional letters and numerical variables are all, respectively, among $f_1, \ldots, f_n$, and $\vec{z}$.

Definability by systems of equations

The first thought that comes to mind is to consider systems of equations that define functions uniquely w.r.t. a given letter (which we may always suppose to be f_1):

Definition I.2.1 (Herbrand [1931]) *A function f is* **definable by a system of equations** *if there is a system $\mathcal{E}$ such that:*

1. *there is a solution to the system:*

$$(\exists f_1)\ldots(\exists f_n)(\forall \vec{z})\mathcal{E}(f_1,\ldots,f_n;\vec{z})$$

   (i.e. there are functions satisfying every equation of $\mathcal{E}$)

2. *any solution determines uniquely f_1 as f: for every $f_1,\ldots,f_n$*

$$(\forall \vec{z})\mathcal{E}(f_1,\ldots,f_n;\vec{z}) \Rightarrow (\forall \vec{x})(f_1(\vec{x}) = f(\vec{x}))$$

   (i.e. if $f_1,\ldots,f_n$ are solutions of $\mathcal{E}$, then $f_1 = f$).

Note that the existence condition 1) is necessary, otherwise any system of equations $\mathcal{E}$ without solutions would define any function f, since then the clause $(\forall \vec{z})\mathcal{E}(f_1,\ldots,f_n;\vec{z})$ would be false, and the implication of condition 2) would be vacuously true.

For our purposes of identifying computable functions this notion of definability is however unsatisfactory, because each value $f(\vec{x})$ is determined (independently of $\vec{x}$) by a global infinitary condition involving every $\vec{z}$. Herbrand probably intended the uniqueness condition 2) to be proved constructively, and such a proof would probably give an explicit computation procedure for f. But, we do not have a precise notion of constructiveness (after all, we are precisely trying to characterize the related notion of effectiveness). And from a classical standpoint the class of functions definable by systems of equations is too comprehensive (although mathematically very interesting), and it very much transcends the class of recursive functions (Kalmar [1955]): it coincides with the class of **hyperarithmetical functions** (Grzegorczyck, Mostowski and Ryll-Nardzewski [1958]), which will be studied in Volume III (see also p. 391).

Computations, whatever they might be, are certainly finite objects and thus each value $f(\vec{x})$ should be uniquely determined by just a finite amount of information. We are thus led to the following modification of the notion introduced above.

Definition I.2.2 (Kreisel and Tait [1961]) *A function f is* **finitely definable by a system of equations** *if there is a system $\mathcal{E}$ such that:*

1. $(\exists f_1)\ldots(\exists f_n)(\forall \vec{z})\mathcal{E}(f_1,\ldots,f_n;\vec{z})$

2. for every $\vec{x}$ there is a finite set $\vec{z}_1,\ldots,\vec{z}_p$ (p depending on $\vec{x}$) such that the substitution instances of $\mathcal{E}$ by them determines the value of $f(\vec{x})$, i.e. such that for every $f_1,\ldots,f_n$

$$\mathcal{E}(f_1,\ldots,f_n;\vec{z_1}) \wedge \cdots \wedge \mathcal{E}(f_1,\ldots,f_n;\vec{z_p}) \Rightarrow f_1(\vec{x}) = \overline{f(\vec{x})}.$$

For example, the system of equations (written informally)

$$f(0) = 0 \quad f(x+1) = f(x)+2 \quad g(x) = f(g(x+1))$$

defines (for every x) $f(x) = 2x$ and $g(x) = 0$. Here f is finitely defined (by the first two equations), while g is not: indeed g is uniquely determined by the infinite set of equations

$$g(0) = 2 \cdot g(1) = 4 \cdot g(2) = \ldots$$

but there are infinitely many solutions to any finite subset of these equations.

Theorem I.2.3 (Herbrand [1931], Gödel [1934], Kleene [1935]) *Every recursive function is finitely definable.*

Proof.

1. *initial functions*
   $\mathcal{O}, \mathcal{S}$ and $\mathcal{I}_i^n$ are, respectively, finitely defined by:

$$
\begin{aligned}
f_0^1(x_0) &= \overline{0} \\
f_0^1(x_0) &= \mathcal{S}(x_0) \\
f_0^n(x_1,\ldots,x_n) &= x_i.
\end{aligned}
$$

2. *composition*
   Suppose $g_i(x_1,\ldots,x_n)$ $(i = 1,\ldots,m)$ and h are finitely defined, respectively, by $\mathcal{E}_i$ and $\mathcal{E}$ w.r.t. f_i^n and f_{m+1}^m (by changing the letters if needed, we can always reduce to this case). Then

$$f(x_1,\ldots,x_n) = h(g_1(x_1,\ldots,x_n),\ldots,g_m(x_1,\ldots,x_n))$$

is finitely defined by $\mathcal{E}_1 \cup \cdots \cup \mathcal{E}_m \cup \mathcal{E}$ together with

$$f_0^n(x_1,\ldots,x_n) = f_{m+1}^m(f_1^n(x_1,\ldots,x_n),\ldots,f_m^n(x_1,\ldots,x_n)).$$

Of course we must suppose (here and in the following) that there are no conflicts of functional letters in $\mathcal{E}_1,\ldots,\mathcal{E}_m,\mathcal{E}$ (which can always be achieved by possible changes of letters).

3. *primitive recursion*
   If g and h are finitely defined by $\mathcal{E}_1$ and $\mathcal{E}_2$ w.r.t. f_0^n and f_0^{n+2} then

$$
\begin{aligned}
f(x_1, \ldots, x_n, 0) &= g(x_1, \ldots, x_n) \\
f(x_1, \ldots, x_n, y+1) &= h(x_1, \ldots, x_n, y, f(x_1, \ldots, x_n, y)).
\end{aligned}
$$

   is finitely defined by $\mathcal{E}_1 \cup \mathcal{E}_2$ together with

$$
f_0^{n+1}(x_1, \ldots, x_n, \overline{0}) = f_0^n(x_1, \ldots, x_n)
$$

   and

$$
\begin{aligned}
f_0^{n+1}(x_1, \ldots, x_n, S(x_{n+1})) = \\
f_0^{n+2}(x_1, \ldots, x_n, x_{n+1}, f_0^{n+1}(x_1, \ldots, x_n, x_{n+1})).
\end{aligned}
$$

4. μ-*recursion*
   If g is finitely defined by $\mathcal{E}_1$ w.r.t. f_0^{n+1} and

$$
f(x_1, \ldots, x_n) = \mu y(g(x_1, \ldots, x_n, y) = 0)
$$

   then let $\mathcal{E}$ be $\mathcal{E}_1$ plus the equations defining sum and product (which are primitive recursive, and for which we then already have finite definability), together with

$$
\begin{aligned}
f_0^1(\overline{0}) = \overline{0} \quad & f_0^1(S(x_0)) = \overline{1} \\
f_1^1(\overline{0}) = \overline{1} \quad & f_1^1(S(x_0)) = \overline{0}
\end{aligned}
$$

   and the formal translations of the following:

$$
\begin{aligned}
f_1^{n+1}(x_1, \ldots, x_n, x_{n+1}) = \\
f_0^1(f_0^{n+1}(x_1, \ldots, x_n, x_{n+1})) \cdot f_1^{n+1}(x_1, \ldots, x_n, S(x_{n+1})) \\
+ f_1^1(f_0^{n+1}(x_1, \ldots, x_n, x_{n+1})) \cdot x_{n+1} \\
f_0^n(x_1, \ldots, x_n) = f_1^{n+1}(x_1, \ldots, x_n, \overline{0}).
\end{aligned}
$$

   Then $\mathcal{E}$ finitely defines f w.r.t. f_0^n, because the system just translates the following facts: it defines

$$
h(\vec{x}, y) = \begin{cases} y & \text{if } g(\vec{x}, y) = 0 \\ h(\vec{x}, y+1) & \text{otherwise} \end{cases}
$$

   w.r.t. f_1^{n+1} (by using two auxiliary functions f_0^1 and f_1^1 to allow for case distinction), and then it defines

$$
f(\vec{x}) = h(\vec{x}, 0)
$$

w.r.t. f_0^n. Indeed

$$h(\vec{x},0) = \mu y(g(\vec{x},y) = 0)$$

because if $\mu y[g(\vec{x},y) = 0] = z$ then

$$h(\vec{x},0) = h(\vec{x},1) = \cdots = h(\vec{x},z) = z.$$

Note that $\mathcal{E}$ has solutions: the only trouble might come from f_1^{n+1}, but here the natural interpretation applies: either for a given $\vec{x}$ there are infinitely many y's such that $g(\vec{x},y) = 0$, and then the definition of f_1^{n+1} is the natural step function, or there are only finitely many such y's, and then any function constant from the last one of such y's would work. $\square$

Derivability from systems of equations

Having shown that every recursive function is finitely definable by systems of equations (in logical terms), we would also like to concoct explicit rules to obtain the values of a function from a system of equations finitely defining it, thus determining a syntactical counterpart to the semantical notion of finite definability. Since this amounts to finding an appropriate formal system (capturing a notion of validity for a given interpretation) we have as natural candidates the analogues of well-known rules used for logical formal systems: substitution and cut.

Definition I.2.4 (Gödel [1934]) *An n-ary function f is* **Herbrand-Gödel computable** *if there is a finite system of equations $\mathcal{E}$ such that if f_i^n is the leftmost letter in the last equation of $\mathcal{E}$, then*

$$f(a_1,\ldots,a_n) = b$$

holds if and only if the equation

$$f_i^n(\bar{a}_1,\ldots,\bar{a}_n) = \bar{b}$$

can be derived from $\mathcal{E}$ by means of the following two rules:

R1 Substitution *of a numeral for every occurrence of a particular variable in an equation.*

R2 Replacement *in the right-hand side of an equation of a term of the form $f_j^m(\bar{c}_1,\ldots,\bar{c}_m)$ with a numeral $\bar{d}$, provided $f_j^m(\bar{c}_1,\ldots,\bar{c}_m) = \bar{d}$ has already been derived.*

We say that $\mathcal{E}$ defines f w.r.t. the letter f_i^n.

Theorem I.2.5 (Herbrand [1931], Gödel [1934], Church [1936], Kleene [1936]) *Every recursive function is Herbrand-Gödel computable.*

Proof. We consider here the systems of equations defined in the proof of Theorem I.2.3. There are two things to prove:

1. *Completeness property* (the values can be deduced from the appropriate systems of equations).

   This is quite evident from the informal discussion of I.2.3.

2. *Consistency property* (no other value can be deduced, i.e. the values are uniquely determined).

   This follows by induction on the construction of the systems. As an example, we show the case of primitive recursion (with notations as in I.2.3). Suppose

   $$f_0^{n+1}(\overline{x}_1, \ldots, \overline{x}_n, \overline{y}) = \overline{b}$$

   has been derived. There are two cases:

   - $\overline{y} = \overline{0}$

     Then, since R2 only allows for replacement on the right-hand side, and since there is only one equation with $f_0^{n+1}(\overline{x}_1, \ldots, \overline{x}_n, \overline{0})$ on the left-hand side, it follows that

     $$f_0^{n+1}(\overline{x}_1, \ldots, \overline{x}_n, \overline{0}) = f_0^n(\overline{x}_1, \ldots, \overline{x}_n)$$
     $$f_0^n(\overline{x}_1, \ldots, \overline{x}_n) = \overline{b}$$

     have been previously derived. But the consistency property holds for f_0^n by induction hypothesis.

   - $\overline{y} = S(\overline{z})$

     Then, as in the above case, an equation

     $$f_0^{n+1}(\overline{x}_1, \ldots, \overline{x}_n, S(\overline{z})) = $$
     $$f_0^{n+2}(\overline{x}_1, \ldots, \overline{x}_n, \overline{z}, f_0^{n+1}(\overline{x}_1, \ldots, \overline{x}_n, \overline{z}))$$

     must have been derived. Note that by definition there can be no equation with the left-hand side like the right-hand side of the above equation. Also, R2 only allows for replacement of terms of the form: functional letter followed by numerals. Then equations like

     $$f_0^{n+1}(\overline{x}_1, \ldots, \overline{x}_n, \overline{z}) = \overline{a} \quad f_0^{n+2}(\overline{x}_1, \ldots, \overline{x}_n, \overline{z}, \overline{a}) = \overline{b}$$

     must have been derived. But then again the induction hypothesis applies. □

In Section 7 we will prove that the syntactical notion of Herbrand-Gödel computability and the semantical notion of finite definability are globally equivalent, by showing them equivalent to recursiveness. Also, by the proof just given, *every recursive function is Herbrand-Gödel computable from a system of equations finitely defining it*. But the two notions are not locally equivalent, in the sense that given a system $\mathcal{E}$ the following may happen:

1. *g can be finitely defined by $\mathcal{E}$ without being Herbrand-Gödel computable from it*, as the system

$$f(x) = 0 \quad f(x) = h(x) \quad g(x) = h(x).$$

   and the function $g(x) = 0$ show. This simply results from the fact that the rules R1 and R2 are of a very specific form, and do not even allow for full logical substitution of equal entities.

2. *g can be Herbrand-Gödel computable from $\mathcal{E}$ without being finitely defined by it*, as the system

$$f(0) = 0 \quad f(S(x)) = S(f(S(x))) \quad g(x) = f(0)$$

   and the function $g(x) = 0$ show. Here Herbrand-Gödel computability follows because only $f(0)$ is used among the values of f, but finite definability fails because there is no total function f satisfying every equation (since $f(z) = f(z) + 1$ for $z > 0$).

Kreisel and Tait [1961] isolate a notion of derivability from systems of equations, which is locally equivalent to finite definability. Basically, the rules correspond to the logical axioms for equality and successor. See Statman [1977] for a proof-theoretical analysis.

Herbrand-Gödel computability has the advantage of using simple rules, and the disadvantage of not being complete, in the sense of not allowing the derivation of everything which is logically derivable.

A logical programming language $\star$

Note that an equation can be put into a normal form of the kind

$$R_1 \wedge \cdots \wedge R_n \;\rightarrow\; Q$$

with R_i, Q atomic equations of the form $f_j^n(x_1, \ldots, x_n) = y$ and x_i, y variables or constants (the interpretation of variables being that they are all universally quantified). For example,

$$f(x) = g(h(x))$$

can be written as

$$h(x) = y \land g(y) = z \; \rightarrow \; f(x) = z.$$

Then the previous results show that the values of recursive functions can be logically deduced from axioms of the described kind.

The programming language **PROLOG**(Programming in Logic, Colmerauer, Kanoui, Pasero and Roussel [1972], Kowalski and Van Emden [1976]) is based on logical deductions from clauses of the form above, with R_i, Q atomic relations holding of terms. These are called **Horn clauses**, and are especially interesting because proof procedures for them are particularly manageable. They can be thought of as conditions breaking up a goal Q into a series of subgoals R_i. The results of this section show that *PROLOG, although concocted to handle deductive more than computational problems, has nevertheless the power of computing all the recursive functions*.

I.3 Arithmetical Formal Systems

The general trend of this century's mathematics has been to work in formal systems which are supposed to capture, more or less accurately, some aspects of the objects in which we are interested. From a formalistic point of view we can thus consider a function as computable when we have a consistent formal system representing it, i.e. allowing us to prove for the appropriate numbers (and for nothing else) that they are the function's values for given arguments.

Certainly the approach of Herbrand-Gödel computability falls in this trend, but the formal system involved there, concocted for different goals, is somewhat unnatural from a purely arithmetical point of view. The same will be true of the approach of λ-definability, see Section 6. Since we are considering arithmetical functions, it is natural to investigate which functions are representable in the usual logical systems for arithmetic, for example in Peano Arithmetic (see p. 24).

We attack the problem in a general way, by isolating minimal conditions (which will turn out to be very weak) sufficient to represent every recursive function. *In this section, 'formal system' will always mean 'formal system extending first-order logic with equality, and having constants terms $\bar{n}$, called numerals, for each n'.* For more details on formal systems, see p. 350.

Notions of representability

Definition I.3.1 (Tarski [1931], Gödel [1931], [1934], Tarski, Mostowski and Robinson [1953]) *Given a formal system $\mathcal{F}$ and a function f, we say that:*

1. *f is **weakly representable** in $\mathcal{F}$ if, for some formula φ of the language of $\mathcal{F}$,*

$$f(x_1,\ldots,x_n) = y \quad \Leftrightarrow \quad \vdash_{\mathcal{F}} \varphi(\overline{x}_1,\ldots,\overline{x}_n,\overline{y})$$

2. *f is **representable** in $\mathcal{F}$ if, for some formula φ,*

$$f(x_1,\ldots,x_n) = y \quad \Rightarrow \quad \vdash_{\mathcal{F}} \varphi(\overline{x}_1,\ldots,\overline{x}_n,\overline{y})$$
$$f(x_1,\ldots,x_n) \neq y \quad \Rightarrow \quad \vdash_{\mathcal{F}} \neg\varphi(\overline{x}_1,\ldots,\overline{x}_n,\overline{y})$$

3. *f is **strongly representable** in $\mathcal{F}$ if for some formula φ, f is representable by φ, and moreover the following uniqueness condition holds:*

$$\vdash_{\mathcal{F}} (\forall y)(\forall z)[\varphi(\overline{x}_1,\ldots,\overline{x}_n,y) \wedge \varphi(\overline{x}_1,\ldots,\overline{x}_n,z) \;\rightarrow\; y = z].$$

The relationships among the various notions are: if f is strongly representable then it is representable, and if f is representable in a consistent formal system then it is weakly representable (because if $\mathcal{F}$ is consistent and $\vdash_{\mathcal{F}} \neg\varphi$ then $\nvdash_{\mathcal{F}} \varphi$).

Exercises I.3.2 a) *The two conditions*

$$f(x_1,\ldots,x_n) = y \Rightarrow \vdash_{\mathcal{F}} \varphi(\overline{x}_1,\ldots,\overline{x}_n,\overline{y})$$
$$\vdash_{\mathcal{F}} (\forall y)(\forall z)[\varphi(\overline{x}_1,\ldots,\overline{x}_n,y) \wedge \varphi(\overline{x}_1,\ldots,\overline{x}_n,z) \;\rightarrow\; y = z]$$

are equivalent to the unique condition

$$\vdash_{\mathcal{F}} (\forall y)[\varphi(\overline{x}_1,\ldots,\overline{x}_n,y) \;\leftrightarrow\; y = \overline{f(x_1,\ldots,x_n)}].$$

b) *If $\mathcal{F}$ is such that*

$$x \neq y \Rightarrow \vdash_{\mathcal{F}} \neg(\overline{x} = \overline{y})$$

then strong representability of f in $\mathcal{F}$ is equivalent to the unique condition

$$\vdash_{\mathcal{F}} (\forall y)[\varphi(\overline{x}_1,\ldots,\overline{x}_n,y) \;\leftrightarrow\; y = \overline{f(x_1,\ldots,x_n)}].$$

Proposition I.3.3 (Tarski, Mostowski and Robinson [1953]) *If $\mathcal{F}$ is a consistent formal system with a predicate $<$ satisfying the axiom schemata*

1. $\neg(x < \overline{0})$

2. $x < \overline{n+1} \;\leftrightarrow\; x = \overline{0} \vee \cdots \vee x = \overline{n}$

3. $x < \overline{n} \;\vee\; x = \overline{n} \;\vee\; \overline{n} < x$

then any function representable in $\mathcal{F}$ is strongly representable in it.

Proof. Suppose ψ represents f in $\mathcal{F}$. Then the formula

$$\varphi(x_1,\ldots,x_n,y) \Leftrightarrow \psi(x_1,\ldots,x_n,y) \wedge (\forall z < y)\neg\psi(x_1,\ldots,x_n,z)$$

strongly represents f. Indeed:

- If $f(x_1,\ldots,x_n) = y$ then $f(x_1,\ldots,x_n) \neq z$, for every $z < y$. By representability of f via ψ

$$\vdash_{\mathcal{F}} \neg\psi(\overline{x}_1,\ldots,\overline{x}_n,\overline{0}) \wedge \cdots \wedge \neg\psi(\overline{x}_1,\ldots,\overline{x}_n,\overline{y-1}) \wedge \psi(\overline{x}_1,\ldots,\overline{x}_n,\overline{y}).$$

Axioms 1 and 2 (depending on whether $y = 0$ or $y > 0$) take care of all the $z < \overline{y}$ in the first part of the formula. Thus

$$\vdash_{\mathcal{F}} \psi(\overline{x}_1,\ldots,\overline{x}_n,\overline{y}) \wedge (\forall z < \overline{y})\neg\psi(\overline{x}_1,\ldots,\overline{x}_n,z)$$

(if $y = 0$ the second part is vacuously true, since there is no $z < \overline{y}$), and $\vdash_{\mathcal{F}} \varphi(\overline{x}_1,\ldots,\overline{x}_n,\overline{y})$.

- If $f(x_1,\ldots,x_n) \neq y$ then, by representability, $\vdash_{\mathcal{F}} \neg\psi(\overline{x}_1,\ldots,\overline{x}_n,\overline{y})$ and so $\vdash_{\mathcal{F}} \neg\varphi(\overline{x}_1,\ldots,\overline{x}_n,\overline{y})$.

- To show the uniqueness condition we prove (see I.3.2.a) that

$$\vdash_{\mathcal{F}} (\forall y)[\varphi(\overline{x}_1,\ldots,\overline{x}_n,y) \leftrightarrow y = \overline{f(x_1,\ldots,x_n)}].$$

We have $\vdash_{\mathcal{F}} \varphi(\overline{x}_1,\ldots,\overline{x}_n,\overline{f(x_1,\ldots,x_n)})$ from the first part of the proof. Suppose now $\vdash_{\mathcal{F}} \varphi(\overline{x}_1,\ldots,\overline{x}_n,y)$. By axiom 3 the only possibilities are

$$y < \overline{f(x_1,\ldots,x_n)} \vee y = \overline{f(x_1,\ldots,x_n)} \vee \overline{f(x_1,\ldots,x_n)} < y.$$

The first one is ruled out, since from $\vdash_{\mathcal{F}} \varphi(\overline{x}_1,\ldots,\overline{x}_n,\overline{f(x_1,\ldots,x_n)})$ we have $\vdash_{\mathcal{F}} \neg\psi(\overline{x}_1,\ldots,\overline{x}_n,y)$, while from $\vdash_{\mathcal{F}} \varphi(\overline{x}_1,\ldots,\overline{x}_n,y)$ (assumed by hypothesis) we have $\vdash_{\mathcal{F}} \psi(\overline{x}_1,\ldots,\overline{x}_n,y)$, and $\mathcal{F}$ is consistent. Similarly we can rule out $\overline{f(x_1,\ldots,x_n)} < y$. Then $y = \overline{f(x_1,\ldots,x_n)}$.   □

The notion of representability makes sense for predicates as well:

Definition I.3.4 *Given a formal system $\mathcal{F}$ and a relation R, we say that:*

1. R is **weakly representable** *if, for some φ,*

$$R(x_1,\ldots,x_n) \quad \Leftrightarrow \quad \vdash_{\mathcal{F}} \varphi(\overline{x}_1,\ldots,\overline{x}_n)$$

2. R is **representable** *if, for some φ,*

$$R(x_1,\ldots,x_n) \quad \Rightarrow \quad \vdash_{\mathcal{F}} \varphi(\overline{x}_1,\ldots,\overline{x}_n)$$
$$\neg R(x_1,\ldots,x_n) \quad \Rightarrow \quad \vdash_{\mathcal{F}} \neg\varphi(\overline{x}_1,\ldots,\overline{x}_n).$$

Note that if the characteristic function c_R of R is (weakly) represented by $\varphi(x_1, \ldots, x_n, z)$, then R is (weakly) representable by $\varphi(x_1, \ldots, x_n, \overline{1})$. Also, if $\mathcal{F}$ is such that

$$x \neq y \; \Rightarrow \vdash_{\mathcal{F}} \neg(\overline{x} = \overline{y})$$

and R is represented by $\varphi(x_1, \ldots, x_n)$, then c_R is (strongly) representable by

$$(\varphi(x_1, \ldots, x_n) \wedge z = \overline{1}) \; \vee \; (\neg\varphi(x_1, \ldots, x_n) \wedge z = \overline{0})$$

(this follows from I.3.2.b). Note that the axioms are needed even for simple representability of c_R, because when $c_R(x_1, \ldots, x_n) \neq z$ and $z \neq 0, 1$ we need to know $\overline{z} \neq \overline{0}, \overline{1}$ to be able to infer that the formula intended to represent c_R is not provable.

Formal systems representing the recursive functions

Note that if $f(x_1, \ldots, x_n) = \mu y R(x_1, \ldots, x_n, y)$ then

$$f(x_1, \ldots, x_n) = y \; \Leftrightarrow \; R(x_1, \ldots, x_n, y) \wedge (\forall z < y)\neg R(x_1, \ldots, x_n, z).$$

The axioms of proposition I.3.3 then imply, by means of the same proof, that the strongly representable functions are closed under μ-operator. We turn now to the other conditions.

Proposition I.3.5 (Tarski, Mostowski and Robinson [1953]) *If $\mathcal{F}$ is a formal system such that*

$$x \neq y \; \Rightarrow \vdash_{\mathcal{F}} \neg(\overline{x} = \overline{y})$$

then the functions strongly representable in $\mathcal{F}$ are closed under composition.

Proof. Suppose

$$f(\vec{x}) = g(h_1(\vec{x}), \ldots, h_m(\vec{x}))$$

and g, h_i are strongly represented by, respectively, χ and ψ_i. Then f is strongly represented by

$$\varphi(\vec{x}, y) \; \Leftrightarrow \; (\exists y_1) \ldots (\exists y_m)[\psi_1(\vec{x}, y_1) \wedge \cdots \wedge \psi_m(\vec{x}, y_m) \wedge \chi(y_1, \ldots, y_m, y)].$$

To show this we use (since we have the appropriate axioms for $\mathcal{F}$) the form of strong representability given in I.3.2.b. Then:

- if $f(\vec{x}) = y$ let $h_i(\vec{x}) = y_i$ and $g(y_1, \ldots, y_m) = y$, so that $\vdash_{\mathcal{F}} \psi_i(\vec{\overline{x}}, \overline{y}_i)$ and $\vdash_{\mathcal{F}} \chi(\overline{y}_1, \ldots, \overline{y}_m, \overline{y})$. Then $\vdash_{\mathcal{F}} \varphi(\vec{\overline{x}}, \overline{y})$ and $\vdash_{\mathcal{F}} \varphi(\vec{\overline{x}}, \overline{f(\vec{x})})$.

- if $\vdash_{\mathcal{F}} \varphi(\vec{x}, y)$ let $y_1, \ldots, y_m$ be such that

$$\vdash_{\mathcal{F}} \psi_1(\vec{x}, y_1) \wedge \cdots \wedge \psi_m(\vec{x}, y_m) \wedge \chi(y_1, \ldots, y_m, y).$$

By strong representability it must be $y_i = \overline{h_i(\vec{x})}$ and thus

$$y = \overline{g(h_1(\vec{x}), \ldots, h_m(\vec{x}))} = \overline{f(\vec{x})}. \quad \square$$

We are now ready to conclude our search for axioms which allow representability of every recursive function.

Theorem I.3.6 (Gödel [1936], Mostowski [1947], Tarski, Mostowski and Robinson [1953]) *In any formal system $\mathcal{F}$ with a predicate $<$ and functions $+$ and $\cdot$ satisfying the following axiom schemata, any recursive function is (strongly) representable (and thus, if the system is consistent, also weakly representable):*

B1 $\neg(\overline{x} = \overline{y})$, *for* $x \neq y$

B2 $x < \overline{n} \ \vee \ x = \overline{n} \ \vee \ \overline{n} < x$

B3 $\neg(x < \overline{0})$

B4 $x < \overline{n+1} \ \leftrightarrow \ x = \overline{0} \vee \cdots \vee x = \overline{n}$

B5 $\overline{x} + \overline{y} = \overline{x+y}$

B6 $\overline{x} \cdot \overline{y} = \overline{x \cdot y}$

Proof. We refer to the characterization of recursive functions given in Theorem I.1.8. We have just proved that closure under composition is implied by B1, and closure under μ-recursion follows from B2–B4, as in I.3.3. It then suffices to note that:

- Equality is representable because if $x = y$ then obviously $\vdash_{\mathcal{F}} \overline{x} = \overline{y}$, and if $x \neq y$ then $\vdash_{\mathcal{F}} \neg(\overline{x} = \overline{y})$ by B1. Then its characteristic function is representable too.

- Sum is representable by

$$\varphi(x, y, z) \ \Leftrightarrow \ x + y = z.$$

Indeed, if $x + y = z$ then $\vdash_{\mathcal{F}} \overline{x+y} = \overline{z}$ and (by B5) $\vdash_{\mathcal{F}} \overline{x} + \overline{y} = \overline{z}$, i.e. $\vdash_{\mathcal{F}} \varphi(\overline{x}, \overline{y}, \overline{z})$. And if $x + y \neq z$ then $\vdash_{\mathcal{F}} \neg(\overline{x+y} = \overline{z})$ by B1 and $\vdash_{\mathcal{F}} \neg(\overline{x} + \overline{y} = \overline{z})$ by B5, i.e. $\vdash_{\mathcal{F}} \neg\varphi(\overline{x}, \overline{y}, \overline{z})$.

- Product is similarly represented by

$$\varphi(x, y, z) \ \Leftrightarrow \ x \cdot y = z.$$

- Identities $\mathcal{I}_i^n$ are obviously represented by

$$\varphi(x_1, \ldots, x_n, z) \Leftrightarrow z = x_i. \quad \square$$

The axioms B1–B6 define a theory $\mathcal{R}$ with infinitely many axioms. Tarski, Mostowski and Robinson [1953] show that $\mathcal{R}$ *is not finitely axiomatizable*, since any finite subset of the axioms (and thus, by compactness, any finite set of theorems) admits a natural finite model consisting (for n sufficiently big) of the numbers $\{0, 1, \ldots, n\}$ naturally ordered, and having the operations restricted to value equal to n when the original value would exceed it. Of course $\mathcal{R}$ cannot have a finite model (by B1) and thus it is not reducible to a finite set of theorems. This also proves that any closed theorem of $\mathcal{R}$ is true in some finite model, and thus the power of $\mathcal{R}$ is very limited. Theorem I.3.6 is, thus, very general.

Local improvements of Theorem I.3.6 are possible however, as Cobham (see Vaught [1960]) and Jones and Shepherdson [1983] have shown. The conclusion still holds if B2 and B5 are dropped. Moreover, by changing the language and introducing a predicate $\leq$ in place of $<$, B2–B4 may be reduced to two axiom schemata

$$x \leq \overline{n} \vee \overline{n} \leq x$$
$$x \leq \overline{n} \leftrightarrow x = \overline{0} \vee \cdots \vee x = \overline{n}.$$

Robinson Arithmetic $\mathcal{Q}$ (see p. 23) is an extension of $\mathcal{R}$, and thus provides an example of a finitely axiomatized theory in which every recursive function is representable. Robinson has proved that if any one of the axioms of $\mathcal{Q}$ is removed, then some recursive function is not strongly representable. Thus $\mathcal{Q}$ *is a minimal finitely axiomatizable theory in which every recursive function is strongly representable.* For details and more information, see Tarski, Mostowski and Robinson [1953].

Invariant definability $\star$

Every model $\mathcal{A}$ of $\mathcal{R}$ has a submodel isomorphic to ω, which is called the **standard part** of the model and is still denoted by ω. Given a formula φ and a subset A of the universe of $\mathcal{A}$, we say that A is **defined** by φ in $\mathcal{A}$ if

$$A = \{x : \mathcal{A} \models \varphi(x)\}$$

(where $\models$ is the usual notion of satisfaction in a structure, see IV.1.1). A is **defined on** ω by φ in $\mathcal{A}$ if

$$A = \omega \cap \{x : \mathcal{A} \models \varphi(x)\}.$$

We can also introduce uniform versions of these notions, by saying that A is **invariantly defined (on ω)** by φ if A is defined (on ω) by the same φ in every model $\mathcal{A}$ of $\mathcal{R}$.

The Compactness Theorem implies that *a set is invariantly definable if and only if it is a finite subset of ω*, while the proof of I.3.6 (together with the results of Section 7) shows that *invariant definability on ω and recursiveness coincide* (Kreisel [1965a]). Thus we have a purely model-theoretical reformulation of recursiveness, and by changing the class of models one gets natural generalizations of it (see Mostowski [1962] and Kreisel [1965a]), some of which will be considered in Volume III.

Note that the reformulation of finiteness explains the need of considering the relative notion of invariant definability: ω *is not invariantly definable in models of $\mathcal{R}$*. For models of $\mathcal{PA}$ a stronger fact is true: ω *is not even definable in a single nonstandard model of $\mathcal{PA}$*. Indeed, given a formula φ defining a set A, consider $\neg\varphi$. If there is no element satisfying it then A is the whole universe, which is nonstandard. If there is an element satisfying $\neg\varphi$ then, by the Axiom of Induction, there is a minimal one x, which is not in A because $\neg\varphi(x)$ holds. If x is 0 or the successor of a standard element then A does not contain all standard elements. If x is the successor of a nonstandard element y then, by minimality of x, $\varphi(y)$ holds and A contains a nonstandard element. In any case, A is not ω.

Definability of functions ⋆

We say that f is **definable** in $\mathcal{F}$ if for some term $t(\vec{x})$, $\vdash_{\mathcal{F}} t(\vec{\overline{x}}) = \overline{f(\vec{x})}$. If $\mathcal{F}$ satisfies B1 then definability implies strong representability (by the formula $t(\vec{x}) = z$, see I.3.2). But definability is not an absolute notion, since the fact that a function is definable is quite accidental and simply depends on the functional constants of the theory language. For example, in Robinson Arithmetic only polynomials are definable, because the language has only the function symbols $\mathcal{S}, +$ and $\cdot$.

By extending a theory with axioms for a formal analogue of the μ-operator, namely

$$(\exists z)A(z) \;\rightarrow\; A(\mu y A(y))$$
$$(\forall z)\neg A(z) \;\rightarrow\; (\mu y A(y) = 0)$$
$$(\forall z)[z < \mu y A(y) \;\rightarrow\; \neg A(z)],$$

every function which is strongly representable by $\varphi(\vec{x}, y)$ becomes definable by $\mu y \varphi(\vec{x}, y)$. In particular, *there are formal systems in which every recursive function is definable*.

In any theory with a predicate $<$, a functional symbol f can be introduced in a conservative way (i.e. not producing new theorems in the old language) for any function which is strongly represented by $\varphi(\vec{x}, y)$, together with the axioms

$$\varphi(\vec{x}, f(\vec{x}))$$
$$(\forall z)[z < f(\vec{x}) \rightarrow \neg\varphi(\vec{x}, z)]$$

In any theory with the Axiom of Induction (e.g. Peano Arithmetic and its extensions) the same can be done even more generally, in the sense that only $(\exists y)\varphi(\vec{x}, y)$ is required, instead of strong representability. The provable existence of $\mu y \varphi(\vec{x}, y)$ then follows by the Least Number Principle.

From I.7.7 it follows that *functions definable in consistent formal systems satisfying* B1 *are recursive*, since in this case definability implies representability, as noted above.

I.4 Turing Machines

We have been analyzing the notion of effectiveness by taking our inspiration from mathematical activity. We now switch our point of view and begin an analysis based on computational activity, by providing a model of certain aspects of human behavior during routine computations. The final goal is to understand what we are doing when we compute, in such a way as to be able to simulate this activity by a machine. In this section we present the analysis of Turing [1936] and Post [1936].

In doing a computation we are given an input and then produce an output, usually after an appropriate amount of written calculations. Fortunately, for the pure **writing activity** we already have a mechanical device: the typewriter. On it we base our model. We may imagine our abstract machine as being capable of **printing** symbols on paper. In real life writing is usually done on planar sheets of paper, but since we write in consecutive (horizontal or vertical) lines we may simplify, and just think of our machines as writing on a **linear tape**. Since we do not put *a priori* bounds on the amount of scratch work needed for a computation, we think of our tape as potentially infinite in both directions (i.e. we suppose that we are always able to glue more tape at either end, when needed). Also, since symbols are actually (or can be) written one at a time, we think of our tape as consisting of **cells**, each of which can contain a single symbol.

Although we might, in general, need symbols for infinitely many entities (e.g. one for each number), we can certainly use complex symbols (like words) built up from a finite stock of atomic ones (like letters). Thus our machine will simply be capable of printing symbols from a **finite alphabet**, as in real life.

And as in concrete typewriters, we allow for the possibility of **two directions of movement** along the tape, one cell at a time. Movement is necessary to print symbols in different cells, and the two directions allow for a recall of the work already done. It does not matter whether we picture the situation as an actual movement of the tape, or of the machine, or even just of an extendible telescopic arm examining the tape and transmitting the information to the machine. Since we can come back to cells containing a printed symbol, we allow for the possibility of **erasing** it (as modern typewriters can do).

An abstract human being can carry on his computation by using just such an abstract machine for his explicit writing activity, but his assistance is at this point obviously still necessary (as it would be for writing something by a typewriter). We now take advantage of our restriction to routine work: no step must require ingenuity, and thus each move must be automatically carried on on the basis of the previous work. In particular, the machine needs a **memory**, i.e. a way to recall the crucial features of what it has already done.

One possible way to implement this requirement is to think of the machine as being at each moment in some particular physical state, determined by the previous action up to this point, and determining the successive action. As a simple example, consider once again the typewriter: one possible routine decision that is done when we write something in most western languages is to begin a word coming after a dot with a capital letter. We might thus build a typewriter that automatically steps, after printing a dot, from its current writing state to the state for printing capital letters (in actual machines this change of state is obtained by pushing an appropriate key). If we allow for sufficiently (possibly infinitely) many states we can certainly record any action: it is enough to consider the tree of all possibilities, associate states to its nodes and, by representing possible actions as paths of the tree, to change states by following a given path. But infinitely many states would contradict the intuition both of routine work (as the implementation of an effectively describable task) and of machine (as a device of limited complexity). We thus impose the limitation of having only **finitely many states**.

To be able to work automatically, the machine must perform its elementary operations according to given **instructions**, telling it what to do on the basis of given information. Some of this information reaches the machine through its internal memory (codified by the current state), but other (in particular the input) might be fed from outside. The connections with the outer world (the tape) consists not only in getting, but also in providing information (the output among others) and it may be pictured (following the parallel with the typewriter) as happening through a **head**, which is both able to read and write. Since the tape consists of cells, the head's range of action will just be one single cell, both statically (when reading and printing) and dynamically (when moving).

It is clear at this point that only **finitely many instructions** are needed, specifying what to do in a given situation. Indeed, there are only finitely many possible local situations (determined by state and read symbol) and actions (consisting in printing or erasing a symbol, possibly moving the head one cell left or right, and possibly changing state).

We need not analyze the actual physical implementation of the machine work. Just imagine the existence of a **black box**, which somehow knows the instructions and the way to carry them out (for more on this see the remarks on pp. 52 and 116). Thus the machine **hardware** consists of a tape, a head and a black box. It can be pictured as follows:

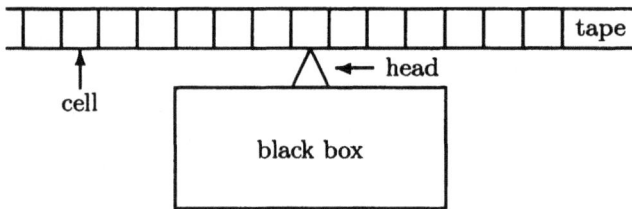

The physical intuition of a machine as a generalized automatic typewriter is certainly useful, but for our abstract purposes nothing more than its **software** is needed:

Definition I.4.1 (Turing [1936], Post [1936]) *A* **Turing machine** M *consists of:*

1. *a finite alphabet* $\{s_0, \ldots, s_n\}$, *with two distinguished symbols* $s_0 = *$ *(blank) and* $s_1 = 1$ *(tally)*

2. *a finite set of states* $\{q_0, \ldots, q_m\}$, *among which distinguished states* q_0 *(initial state) and* q_f *(final state)*

3. *a finite set of consistent instructions* $\{I_1, \ldots, I_p\}$, *each one of the following three basic types:*

   - $q_a\, s_b\, s_c\, q_d$: *the machine in state* q_a *and reading the symbol* s_b *erases it and prints* s_c *in its place, then changes its state to* q_d

   - $q_a\, s_b\, R\, q_d$: *the machine in state* q_a *and reading the symbol* s_b *moves one cell to the right and changes its state to* q_d

   - $q_a\, s_b\, L\, q_d$: *similar, moving one cell to the left.*

   Consistency means that no pair of instructions is contradictory, i.e. with the same premise $q_a\, s_b$ *but with different conclusions.*

Note that instructions are local: their behavior is completely determined by the part of the tape immediately adjacent to the head. Except for this crucial feature, the form of instructions could be different. We could further analyze the first type of instructions into pure erasing and pure writing actions. Note that pure erasing is a special kind of printing, with $s_c = *$. Or we could slightly complicate the basic actions into a unique type $q_a\, s_b\, s_c\, q_d\, j$, telling the machine in state q_a and reading s_b to print s_c in place of s_b, change its state to q_d and (depending on whether $j = 0, 1, 2$) stay still, move one cell to the right or move one cell to the left.

An **instantaneous configuration** of the machine is a complete recording of all the relevant data in a given instant. It can be represented by a sequence

$$\sigma_1\ \ldots\ \sigma_j\ q_a\ \sigma_{j+1}\ \ldots\ \sigma_z$$

in which all the consecutive symbols σ_i (including the blanks) written in the relevant portion of the tape (at least the one between the two outermost non-blank symbols) are indicated, together with the state and the head position (shown by the position of the state symbol among the alphabet symbols). Thus the sequence above records the following situation:

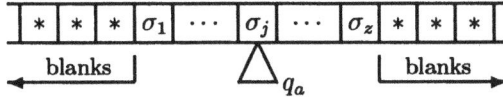

The machine behavior can be represented by a sequence of instantaneous configurations, each obtained from the previous one by application of a necessarily unique instruction, and starting from an **initial configuration** (coding the input and the initial state). A **final configuration** is reached when the machine finds itself in the final state, and we say figuratively that the machine stops. An alternative formulation, not using final states, simply states: the machine stops when no instruction is applicable.

We will show in I.4.2 how a Turing machine can compute a function, which is our real concern. Before that we discuss Turing machines in more detail, but the interested reader can skip this, and go directly to p. 53.

Variations of the Turing machine model

The particular model of a Turing machine introduced in definition I.4.1 is absolutely unimportant, as long as we are only concerned with computational power and not with efficiency. Since in this book we do not deal with machine theory (although we will prove some scattered results in Chapter VIII), we will only quote some facts, and refer to Minsky [1967], Arbib [1969] and Hopcroft and Ullman [1979] for their proofs and for further information.

States. *Although only one state is not in general enough to compute every recursive function* (Shannon [1956], Wang [1957]: basically, a one-state Turing machine must behave in the same way on every cell outside the input), *two states are* (Shannon [1956]). Thus it is not relevant whether we restrict our model to machines with only a fixed number $n \geq 2$ of states, or we allow any number of states.

Symbols. Clearly we must have at least two symbols, since we consider blank as a symbol. *Two symbols are enough to compute any recursive function* (Shannon [1956]), since it is possible to economize on the number of symbols by increasing the number of states. See Priese [1979] for results concerning the simultaneous economy of states and symbols.

Erasing. This is dispensable (Wang [1957]), in the sense that *all the recursive functions can be appropriately computed by machines that never erase.* Basically, given a Turing machine it is possible to simulate it by a new nonerasing machine, that writes on its tape codes of successive configurations of the original machine. The result shows that in principle we do not need erasable material, like magnetic tapes or disks, for the external memory of computers. See also p. 51.

Tapes and heads. Here the freedom is practically absolute. We synthesize it in the following result. *A Turing machine with finitely many tapes, each with its own (finite, or even countable) dimension and its own finite number of heads simultaneously scanning it, can be simulated by a Turing machine with only one linear tape, infinite in just one direction, and scanned by a single head* (Hartmanis and Stearns [1965]). However, we do need the two directions of movement, since to restrict it to one would be compatible only with finite or periodical behavior on cells outside the inputs (Wang [1957]).

Determinism. Our model of a Turing machine is deterministic, in the sense that the instructions are required to be consistent (at most one of them is applicable in any given situation). Randomizing elements in computing devices were introduced early on by Shannon [1948] and De Leeuw, Moore, Shannon and Shapiro [1956]. There are basically two models. *Nondeterministic Turing machines* behave, in an ambiguous situation where conflicting instructions might be applicable, by randomly choosing one of them: their computational power, at least for $0, 1$-valued functions (sets), does not exceed the power of deterministic ones. *Probabilistic* *machines* differ from nondeterministic ones in that the next state has a probability, and thus conflicting instructions do not have the same chance of being chosen by the machine.

Physical Turing machines ⋆

Turing machines are theoretical devices, but have been designed with an eye to physical limitations. In particular, we have incorporated in our model restrictions coming from: (a) **atomism**, by ensuring that the amount of information that can be coded in any configuration of the machine (as a finite system) is bounded; and (b) **relativity**, by excluding actions at a distance, and making causal effect propagate through local interactions. Gandy [1980] has shown that the notion of a Turing machine is sufficiently general to subsume, in a precise sense, any computing device satisfying similar limitations.

However, there is at least one aspect which has been neglected in our discussion, namely **energy consumption**. This is a central problem especially for actual modern computers, whose sizes are getting increasingly bigger. The point is that in devices with packed components, energy dissipation is proportional to volume, while heat removal is proportional only to surface. To provide adequate cooling, it would thus seem necessary to expand machines only in two dimensions, and keep them flat and thin. But this would mean a spread of components in space and an increase in the time needed to transmit information, thus a decrease in speed. It is then crucial to limit energy consumption in computations, not only for costs limitation, but also for physical realization.

At a macroscopical level, energy seems to be needed in at least two different ways. First of all, physical computations involve data handling (like storing and transmitting), and thus measurements: this implies energy consumption. Secondly, Turing machines and real computers are **irreversible devices** (many sets of inputs may produce the same output, and it is usually impossible to invert computations): this implies energy dissipation. Theoretical bounds on energy requirements have been computed by various authors (e.g. from the uncertainty principle $\Delta E \cdot \Delta t \approx h$, by arguing that if Δt represents a switching time, then ΔE must represent energy dissipation). See Landauer [1961], Bremermann [1962], [1982], and Mundici [1981].

At a microscopical level, dynamical physical laws are reversible: it is thus natural to look for reversible models of Turing machines. **Logical reversibility** (of rules) is easy to achieve (Bennett [1973]). First note that a computation can be trivially simulated in a reversible way, by having an extra tape that successively records the quadruples of the Turing machine used, so that the present configuration, and the last record on the history tape, allow for a recovery of the previous configuration. Then note that erasing a computation done by reversible rules is also reversible, and that it can be done after the output has been copied (to be saved) on a separate blank tape: no record is needed at this stage, since copying onto a blank tape is a one-one operation. Then the whole computation, including the erasure of scratch work, can be done by reversible rules. The next step is to look at **energetic reversibility**. Bennet [1973] and Landauer [1976] provide Brownian models for any finite computation on Turing machines, dissipating arbitrarily little energy when proceeding slowly enough. The final step is to look for plainly **conservative models** of Turing machines, not dissipating any energy at all, while computing at finite speed. For classical mechanics, Fredkin and Toffoli [1982] give a ballistic model consisting of hard spheres (whose presence or absence in a constant flow code digits 1 or 0) that collide elastically with each other and with fixed barriers placed inside the computer (see also Landauer [1981] and Toffoli [1981]). For quantum mechanics, models have been given by Benioff [1980], [1981], [1982].

It should be noted that all these models are mathematical. They only show that dissipationless computations are not contrary to current physical laws, not that they are physically realizable. Moreover, energy dissipation *does* occur in these systems, in interactions with external observers (to receive inputs and give outputs): it is not needed for the internal system evolution, i.e. for the computation itself, but it is needed in the end to capture and communicate the information that is wandering inside the process, devoid of meaning.

For more on physics of computations, see the issue devoted to the subject by the *International Journal of Theoretical Physics* (vol. 21 (1982) nos. 3,4), as well as Bennet [1982], Landauer [1985] and Feynman [1996].

Finite automata ⋆

The control box of a Turing machine exemplifies the notion of **finite automaton** (McCulloch and Pitts [1943]), as a machine with finite sets of states and of inputs and outputs, together with functions for next state and for output behavior (both depending on input and current state): this is a mathematically rendition of finite state system. Having only a finite number of possible inputs and outputs, the possible behavior of a finite automaton is a logical function $\{0,1\}^n \to \{0,1\}^m$, and it is thus representable by a **switching circuit** (Shannon [1938]) or a **neuronic net** (McCulloch and Pitts [1943], see also p. 116), by writing a truth-table representation for it in disjunctive normal form, and by using chips for the logical connectives.

A finite automaton is basically a passive device, producing only an output from an input by a series of internal changes. Moreover, it has only a finite **short-term memory**, since the only way it is able to record is by changing its state. A Turing machine is a more complex device, and it may be seen as a finite automaton supplied with an external, potentially infinite **long-term**

memory, and with the ability of interacting with the outer world (the tape) in a dynamical, active way (through the head). Actual modern computers lie somewhere between finite automata and Turing machines.

The theory of finite automata is highly developed, and we will only touch on it in Chapter VIII. For more information see Hartmanis and Stearns [1966], Minsky [1967], Arbib [1969], Trakhtenbrot and Bardzin [1973], Eilenberg [1974], and Hopcroft and Ullman [1979].

Turing machine computability

We have introduced Turing machines as computing devices, but we have not yet spelled out how they compute functions.

Definition I.4.2 (Turing [1936], Post [1936]) *A function $f(x_1, \ldots, x_n)$ is* **Turing machine computable** *if there is a Turing machine that, when starting in the initial configuration*

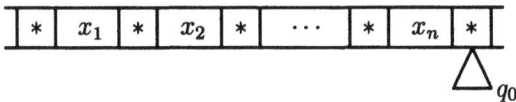

$$\boxed{*\ \big|\ x_1\ \big|\ *\ \big|\ x_2\ \big|\ *\ \big|\ \cdots\ \big|\ *\ \big|\ x_n\ \big|\ *}$$
$$\triangle_{q_0}$$

(with the integers represented in unary notations by tally sequences, and nothing else on the tape except the input representation) and when following its instructions, reaches a final configuration of the kind

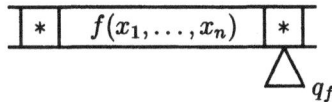

$$\boxed{*\ \big|\ f(x_1, \ldots, x_n)\ \big|\ *}$$
$$\triangle_{q_f}$$

(with $f(x_1, \ldots, x_n)$ represented in unary notation by a tally sequence, and possibly something else on the tape).

It should be noted that the details of the definition are arbitrary. The important thing is that the machine carries configurations coding in some way the inputs, to configurations coding in some (not necessarily the same) way the output. The point is that, as we have already noted, the class of functions computable by Turing machines is widely independent of the details of the definition.

The next result shows the computational power of Turing machines. The proof is not difficult in outline but cumbersome in details, and we are going to give a sketch only: for more details, see e.g. Davis [1958] or Hermes [1965].

Theorem I.4.3 (Turing [1936]) *Every recursive function is Turing machine computable.*

Proof. We proceed by induction on the definition of recursive function. To have a sufficiently strong inductive hypothesis, we prove that every recursive function is computable by Turing machines that: may have initial tape nonempty at the left of the inputs, work only on the half of the tape containing the input and at the right of it, halt in a halting state, and print the value of the function immediately to the right of the original inputs.

1. *constant zero*
   We start in the initial configuration

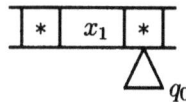

   and want to reach the following final one:

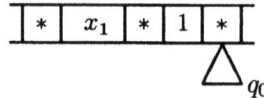

   The following instructions will do (note that q_2 is the final state):

$q_0\ s_0\ R\ q_1$		go right one cell
$q_1\ s_0\ s_1\ q_1$		print a tally
$q_1\ s_1\ R\ q_2$		go right one cell

2. *identities*
   $\mathcal{I}_1^1$ is computed by any machine that copies the input. The flowchart of Figure 1 indicates the sequence of actions to perform.

   Figure 2 exhibits a program written along these lines (the instructions are written in boxes only to indicate explicitly to which parts of the flowchart they correspond). Note that q_{10} is the final state.

   A program for $\mathcal{I}_i^n$ can be written in the same way, by considering x_i as the input, and by moving across $x_{i+1} * \cdots * x_n$ back and forth (which can be easily done by using more changes of states, with the only function of making the head pass through this portion of the tape).

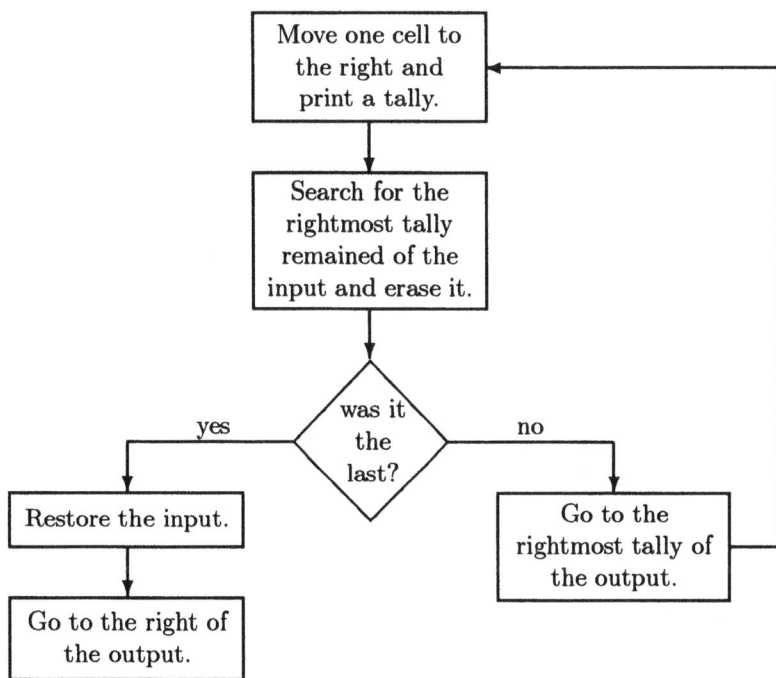

Figure I.1: Flowchart for $\mathcal{T}_1^1$

$$q_0\ s_0\ R\ q_1$$
$$q_1\ s_0\ s_1\ q_2$$

$$q_2\ s_1\ L\ q_2$$
$$q_2\ s_0\ L\ q_3$$
$$q_3\ s_0\ L\ q_3$$
$$q_3\ s_1\ s_0\ q_4$$

$$q_4\ s_0\ L\ q_5$$

$$q_5\ s_0\ R\ q_6$$
$$q_6\ s_0\ s_1\ q_7$$
$$q_7\ s_1\ R\ q_6$$
$$q_6\ s_1\ L\ q_8$$
$$q_8\ s_1\ s_0\ q_9$$

$$q_5\ s_1\ R\ q_{11}$$
$$q_{11}\ s_0\ R\ q_{11}$$
$$q_{11}\ s_1\ R\ q_{12}$$
$$q_{12}\ s_1\ R\ q_{12}$$
$$q_{12}\ s_0\ s_0\ q_1$$

$$q_9\ s_0\ R\ q_{10}$$
$$q_{10}\ s_1\ R\ q_{10}$$

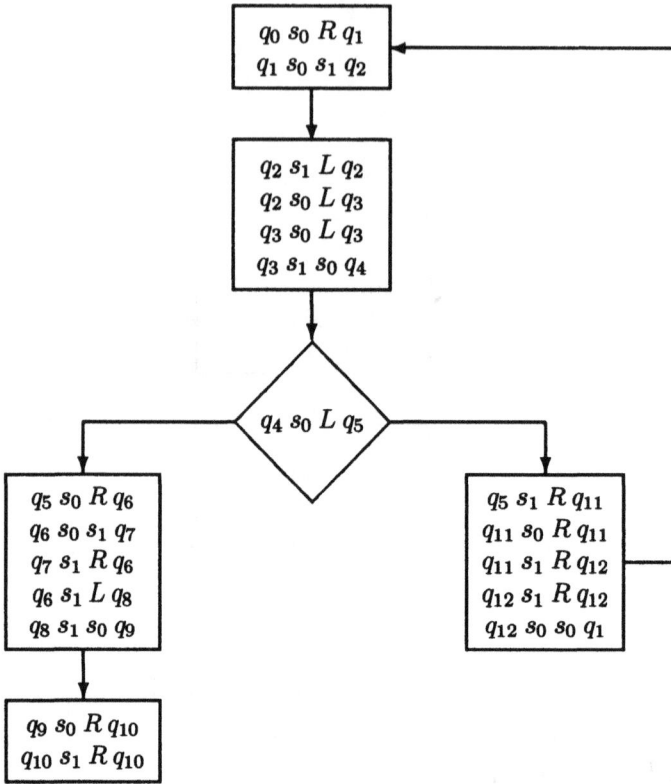

Figure I.2: Program for $\mathcal{I}_1^1$

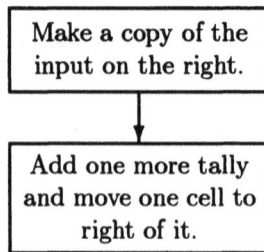

Make a copy of the
input on the right.

Add one more tally
and move one cell to
right of it.

Figure I.3: Flowchart for $\mathcal{S}$

Recopy the inputs on the right.
Simulate M_1 and get output z_1.
Erase the copy of the inputs and
move z_1 to the right of the inputs.

Recopy the inputs to the right of z_1.
Simulate M_2 and get output z_2.
Erase the copy of the inputs and
move z_2 to the right of z_1.

Recopy the inputs to the right of
z_{m-1}. Simulate M_m and get output
z_m. Erase the copy of the inputs
and move z_m to the right of z_{m-1}.

Simulate M_{m+1} and get output z.
Erase $z_1, \ldots, z_m$ and move z
near the inputs

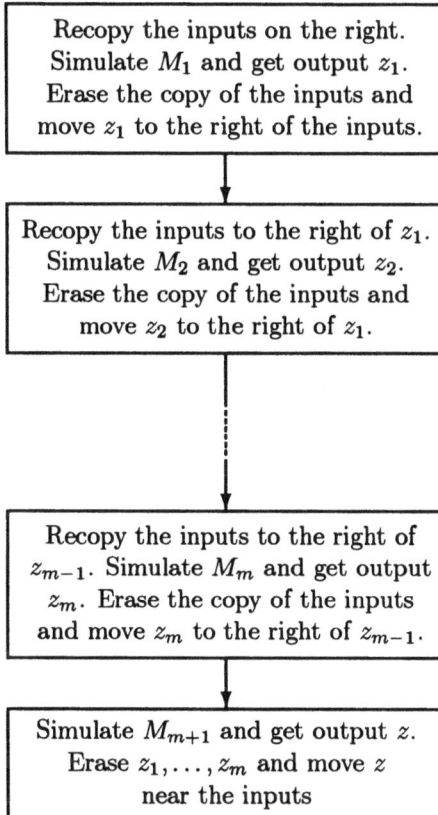

Figure I.4: Flowchart for composition

3. *successor*

The flowchart is given in Figure 3. It is then enough to add, to the program for $\mathcal{I}_1^1$ given above, the following two instructions (corresponding to the last box of the flowchart):

$$q_{10}\ s_0\ s_1\ q_{13}$$
$$q_{13}\ s_1\ R\ q_{13}$$

As we can see from these examples, to write programs is quite routine, once we have the appropriate flowcharts. In the following we will then restrict ourselves to the latter, and leave as an exercise their translation into programs.

4. *composition*

Let

$$f(\vec{x}) = h(g_1(\vec{x}), \ldots, g_m(\vec{x})),$$

and suppose $g_1, \ldots, g_m, h$ are computed, in the sense explained at the beginning of the proof, by machines $M_1, \ldots, M_m, M_{m+1}$. Then f is computed by a machine implementing the flowchart of Figure 4, where simulation of a given machine means changing the name of its states in the appropriate way: by renaming its initial state by the name of the state in which the new machine begins its simulation, renaming the remaining states by new names not yet used, and continuing the work of the simulating machine from the halting state of the simulated one.

5. *primitive recursion*

Let

$$
\begin{aligned}
f(\vec{x}, 0) &= g(\vec{x}) \\
f(\vec{x}, y+1) &= h(\vec{x}, y, f(\vec{x}, y)),
\end{aligned}
$$

and suppose g and h are computed, in the sense explained at the beginning of the proof, by machines M_1 and M_2. Then f is computed by a machine implementing the flowchart of Figure 5. The idea is to compute successive values of f, using the previous one each time, until we reach the one we are interested in. To keep track of the number of iterations still to be done we introduce a counter s, initially set up to the value y, and decreased by one at each step. We do different simulations, depending on whether $y = 0$ or not. In the computation we also need an additional input, which is not present at the first step (computation of g), and increases by one at each step afterwards: we then introduce a second counter t, which is initially empty. The tape then codes a situation of the kind:

$$x_1 * \cdots * x_n * y * s * x_1 * \cdots * x_n * t * f(x_1, \ldots, x_n, t)$$

where $s + t = y$.

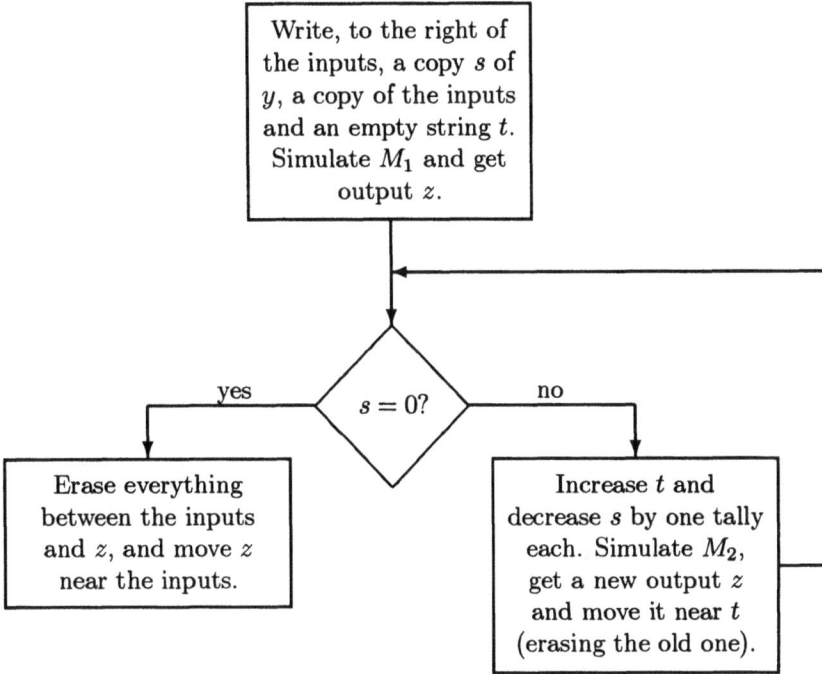

Figure I.5: Flowchart for primitive recursion

6. μ-recursion

Let

$$f(\vec{x}) = \mu y[g(\vec{x}, y) = 0],$$

and suppose g is computed, in the sense explained at the beginning of the proof, by a machine M. Then f is computed by a machine implementing the flowchart of Figure 6. The idea is to compute successive values of $g(\vec{x}, y)$, starting with $y = 0$, until one with output 0 is reached. Then y is the value of f. □

Exercise I.4.4 *Fill in the details of the proof of Theorem I.4.3, by specifying Turing machine programs implementing the flowcharts given there.*

Machine-dependent programming languages ⋆

To show that a certain function is Turing machine computable we have to write a complete program, using only the elementary instructions of defini-

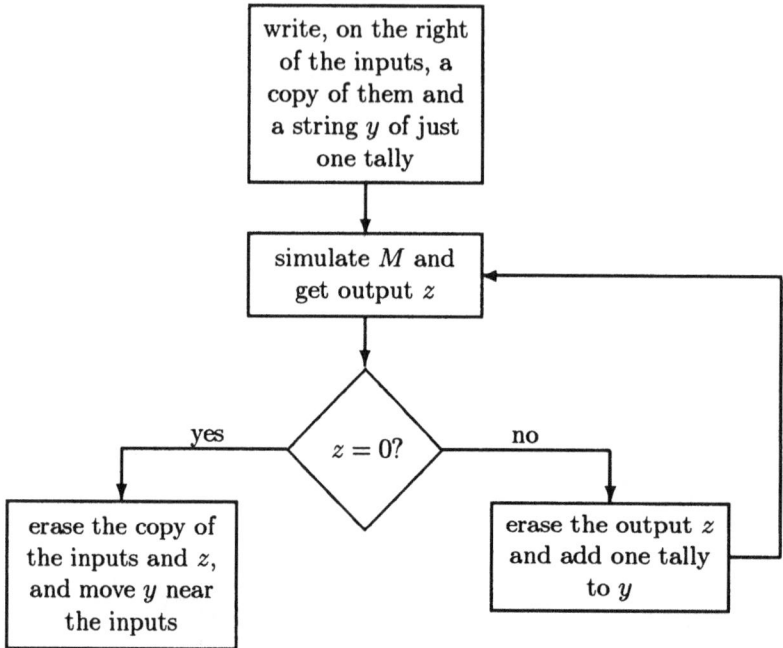

Figure I.6: Flowchart for μ-recursion

tion I.4.1. Since they spell out explicitly the elementary physical operations
that the machine has to perform, such programs are said to be written in **ma-
chine language**. Examples have been given for $\mathcal{O}, \mathcal{I}_1^1$ and $\mathcal{S}$ in the proof of
I.4.3.

The rest of the proof of I.4.3 has been given in a different format, at a
slightly more abstract level: we still described the way to combine some basic
physical operations that the machine has to perform, but left out the task
of specifying how to translate these operations into sets of instructions (to
reduce them to elementary operations that the machine can directly execute).
This is a typical situation of **assembly programs**, whose instructions code
blocks of instructions in machine language, but still refer directly to machine
implementation.

By pursuing this trend toward greater abstraction, we can break the programming task into different assignments: first the algorithm is translated into
a **machine-independent** (or **high-level**) **language**, whose instructions refer to basic abstract operations, and then these operations are programmed in
machine language. This has various advantages:

- It takes into account the difference between man and machine, and the fact that a language suitable for one might not be suitable for the other: machine languages are based on physical commands, high-level languages are closer to ordinary language.

- The translation of the basic abstract operations into machine language can be done once and for all, by means of fixed programs called **compilers** or **interpreters**. The distinction between the two is, at least in origin, that the compiler translates the given high-level program into a program in machine language, which is then executed; the interpreter instead proceeds directly to the execution of the given high-level program, by translating an instruction only if and when it needs it in the computation.

- The abstract analysis of the algorithm is more easily carried out in terms of basic abstract operations, without paying attention to concrete problems of implementation. The advantage is somewhat reflected by the compiler's complexity, since whatever is done automatically does not have to be taken care of directly in the program.

- The same abstract analysis can be implemented by many different types of machines, each one using its own compiler or interpreter.

The general skeleton of **programming** thus takes the following form. On one side the given function is analyzed into an algorithm, which is then translated into a machine-independent program. On the other side the behavior of the given machine is structured into a set of basic instructions in machine (or assembly) language. A compiler or interpreter then relates the two parts and produces a machine-dependent program that can be executed by the machine.

I.5   Flowcharts

Every program written in a machine-dependent language has an abstract core that can be written in a machine-independent language. It is not clear, at this stage of our development, whether the opposite also holds, i.e. whether the most general programs can be implemented on abstract machines (if not step by step at least globally, in the sense of having a machine computing the function defined by the program). Since, however, programs do provide formalizations of algorithms, and our concern is effective computability, we feel compelled to analyze the notion of program as a natural approximation to it. The analysis will make the notion precise and, as a by-product, it will be possible to clarify the relationships between programs and machines.

To define the most general notion of arithmetical program we rely on the intuition given by the experience with Turing machines. It seems that the constituents of abstract programs are reducible to actions of two kinds: perform some basic operations, and ask some basic questions. Both are clearly needed, since without operations we cannot compute, and without conditional actions we cannot exercise judgments or make choices, and then we could only have fixed behaviors.

To picture programs we thus need to distinguish the two types of action, which we do figuratively by using (as in the proof of Theorem I.4.3) *boxes for basic operations, and diamonds for basic questions*. We are then faced with the usual problem of choosing the basic constituents of our programs. For reasons already discussed for the other approaches, we stick to the most elementary arithmetical functions and predicates.

As usual in Computer Science, *we adopt in this section the convention of writing variables in capital letters*.

Definition I.5.1 (Goldstine and Von Neumann [1947]) *A flowchart program is a diagram having exactly one entry and a finite number (possibly zero) of exit points, and built up by connecting (through the outward edges) parts of the following kinds:*

    1. **assignment statements** *of one of the following forms:*

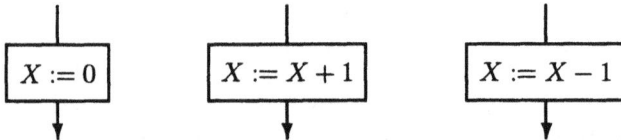

    The sign ':=' means: set the left-hand side equal to the right-hand one. Thus the three assignments correspond, respectively, to instructions of the form: set X equal to 0, increase X by 1, and decrease X by 1.

    2. **conditional statements** *of the form:*

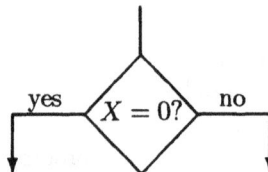

In a flowchart, edges can split only when passing through conditional statements, but they are allowed to merge in a point (thus, even if formally many exit points are permissible, in practice they can be connected and reduced to at most one). Also, the outward edges of conditional statements must exit from the diamond but can go anywhere. In particular, they can re-enter the diagram and thus produce closed paths.

Exercises I.5.2 Choices of primitives. a) *From the assignments $X := X + 1$ and $X := X - 1$ and the conditional '$X = 0$?' one can derive the assignments $X := 0$ and $X := Y$, and the conditional '$X = Y$?'*. (Hint: decrease X and Y simultaneously by 1, until one or both vanish.)

b) *From the assignments $X := X + 1$ and $X := 0$ and the conditional '$X = Y$?' one can derive the assignment $X := X - 1$.* (Hint: use two variables, set initially to 0 and 1, and increased simultaneously by 1 until the greatest reaches X.)

c) *Both choices of primitives above are minimal.* (Hint: without conditionals the arguments cannot influence the form of the computation; without predecessor in a), the only property that can influence a computation is whether an argument is zero or not; without successor there would be no way to write down numbers greater than the arguments.)

Unstructured programming languages ⋆

The motivation that led to the introduction of the planar structure of flowcharts is best explained by its own inventor:

> There is reason to suspect that our predilection for linear codes, which have a simple, almost temporal sequence, is chiefly a literary habit, corresponding to our not particularly high level of combinatorial cleverness, and that a very efficient language would probably depart from linearity. (Von Neumann [1966])

The description of algorithms by means of flowcharts has led to the programming language **GPSS** (General Purpose Simulation System, Gordon [1961]; see Sammett [1969] and Wexelblat [1981] for history and references), which uses block diagram notation directly. This is useful for the simulation of discrete systems, since it allows for a direct representation of the system structure (with blocks standing for operations performed in the system, and edges corresponding to possible sequences of events).

However, in practice, most programming languages represent algorithms as sequences of instructions. We thus have to devise a way to unwind flowcharts, and lay them out (which does not mean eliminating their planar structure, but only representing them in a different way). A simple method is to label statements (e.g. by natural numbers). Then a flowchart can be represented as a juxtaposition of labeled statements (the order in the sequence, not the order of labels, indicating the progressive order of execution), by translating conditional statements via **conditional jumps** (Post[1936]) of the kind

$$\text{if } X = 0 \text{ go to } n$$

(with the meaning: if $X = 0$ then jump to the statement with label n, otherwise continue with the next statement in the list). Note that it is also possible to define **unconditional jumps**

$$\text{go to } n$$

as: if $0 = 0$ then go to n. This permits the treatment of merging edges of a flowchart.

A **'go to' program** is a finite sequence of labeled statements of the kind

$$X := 0 \qquad X := X + 1 \qquad X := X - 1$$
$$\text{if } X = 0 \text{ go to } n,$$

with the conditions that different statements have different labels, and every number mentioned after a 'go to' is the label of a statement. *'Go to' programs and flowcharts programs are mutually translatable, and hence equivalent.*

'Go to' statements are the natural solution to flowchart unwinding: they are typical of unstructured programming languages, like **FORTRAN** (Formula Translator, Backus et al. [1957]; see Sammett [1969] and Wexelblat [1981] for history and references). However, 'go to' statements are currently out of fashion (see p. 68).

Unlimited register, random access machines ⋆

Turing-like machines implementing 'go to' programs have been proposed by Post [1936] (independently of Turing [1936]) and Wang [1957]. A different model is the **unlimited register, random access machine** (Melzak [1961], Lambek [1961], Shepherdson and Sturgis [1963], Peter [1963], Elgot and Robinson [1964]). It consists of a finite number of registers of unbounded capacity, each able to contain an integer, and labeled (by a number). The machine is able to perform the following operations: clear a register (set its content equal to zero), and increase or decrease by one the content of a register. The machine performs the operations by following the instructions of a 'go to' program, with the convention that variables represent the content of associated registers.

This model is somewhat closer to real computers than Turing machines, since the memory consists of labeled registers whose content can be made available by a direct call to the label. The advantage of this kind of **random access memory**, versus a **sequential access memory** like the tape of a Turing machine, is subdivision (and consequent greater accessibility) of information. In real computers both kinds of memory are present: the random access (chips) is internal and limited, while the sequential access (e.g. magnetic tapes or disks) is external and potentially (in theory, at least) unlimited. Of course, unlike random access machines, real computers' registers have only a finite capacity.

Flowchart computability

Flowcharts have a natural semantics:

Definition I.5.3 *A function $f(x_1, \ldots, x_n)$ is* **flowchart computable** *if there is a flowchart program with an output variable Z, and possibly some of the input variables $X_1, \ldots, X_n$, such that whenever at the entry point $X_1, \ldots, X_n$ are set equal to $x_1, \ldots, x_n$, and all the other variables are set equal to 0, and their values are then modified according to the instructions of the program, then the exit point is reached with the variable Z having value $f(x_1, \ldots, x_n)$.*

Theorem I.5.4 (Wang [1957], Peter [1958], Ershov [1960]) *Every recursive function is flowchart computable.*

Proof. We proceed by induction on the definition of recursive function.

1. *initial functions*
   $\mathcal{O}$ is computed by:

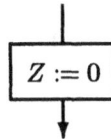

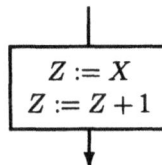

$$\boxed{Z := 0}$$

From I.5.2.a we have that the assignment $X := Y$ is flowchart computable. Then $\mathcal{I}_i^n$ is computed by:

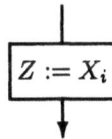

$$\boxed{Z := X_i}$$

S is computed by:

$$\boxed{\begin{array}{l} Z := X \\ Z := Z + 1 \end{array}}$$

Here and in the following, we use a single box with many lines to indicate the concatenation of many boxes with single lines, with instructions read in the order provided by the arrow.

2. *composition*
Suppose $g_1, \ldots, g_m, h$ are computed by flowcharts with appropriately distinct variables. We write statements of the kind

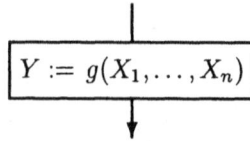

$$\boxed{Y := g(X_1, \ldots, X_n)}$$

as shorthand for the program computing g with respect to input variables $X_1, \ldots, X_n$ and output variable Y, and such that the input variables have, in exit, the same values they had in entry (which can be done by storing their original values by using new variables, and restoring them at the end).

Then if

$$f(x_1, \ldots, x_n) = h(g_1(x_1, \ldots, x_n), \ldots, g_m(x_1, \ldots, x_n)),$$

f is computed by the program of Figure 7.

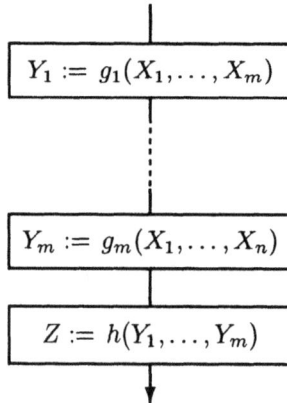

$$\boxed{Y_1 := g_1(X_1, \ldots, X_m)}$$

$$\vdots$$

$$\boxed{Y_m := g_m(X_1, \ldots, X_n)}$$

$$\boxed{Z := h(Y_1, \ldots, Y_m)}$$

Figure I.7: Flowchart for composition

3. *primitive recursion*
Let

$$
\begin{aligned}
f(x_1, \ldots, x_n, 0) &= g(x_1, \ldots, x_n) \\
f(x_1, \ldots, x_n, y+1) &= h(x_1, \ldots, x_n, y, f(x_1, \ldots, x_n, y)),
\end{aligned}
$$

and suppose g and h are computed by programs with appropriately distinct variables. Then f is computed by the program shown in Figure 8.

$$T := 0$$
$$Z := g(X_1, \ldots, X_n)$$

$$T := T + 1$$
$$Z := h(X_1, \ldots, X_n, T, Z)$$

$T = Y?$ — no

yes

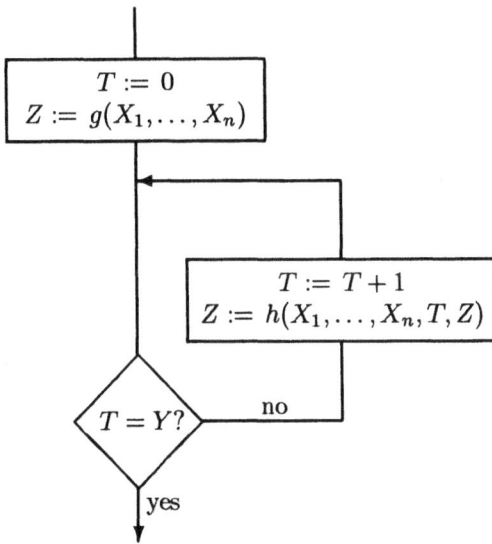

Figure I.8: Flowchart for primitive recursion

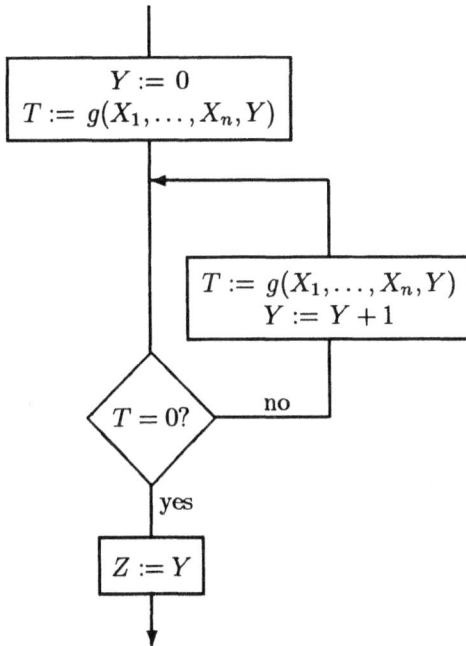

$$Y := 0$$
$$T := g(X_1, \ldots, X_n, Y)$$

$$T := g(X_1, \ldots, X_n, Y)$$
$$Y := Y + 1$$

$T = 0?$ — no

yes

$$Z := Y$$

Figure I.9: Flowchart for μ-recursion

4. *μ-recursion*

Let

$$f(x_1, \ldots, x_n) = \mu y (g(x_1, \ldots, x_n, y) = 0),$$

and suppose g is computed by a program. Then so is f, by the program shown in Figure 9.   □

Structured programming languages ⋆

It has been argued by many (e.g. Dijkstra [1968]) that unstructured flowcharts present definite disadvantages: they may be excessively complicated and difficult to visualize, even for their authors, and errors are consequently easy to make and difficult to debug. The proof of I.5.4 clearly shows that there is no need of complicated flowcharts, with intricate interlacements of edges, at least for the computation of all the recursive functions.

The current trend - with a departure from planarity (see p. 63) and a return to linearity) - is, thus, to stick to **structured flowcharts** (Dahl, Dijkstra and Hoare [1972]), inductively made up from the assignments and conditional statements of blocks with one entry and one exit line, connected only in the ways shown by Figure 10.

One crucial advantage of structured flowcharts is the possibility of a top-down approach, in which the algorithm is successively approximated and expanded, from general to specific diagrams. Related to this is the fact that, while unstructured programs cannot, in general, be thought of as more than sets of isolated instructions, structured programs are really built up from subprograms. This brings in a consideration of **operations on programs** and the need (stressed in Backus [1981]) of an algebraic study of their properties.

Definition I.5.5 *Consider the programming language whose statements are the following:*

1. *assignment statements* $(X := 0, X := X + 1$ and $X := X - 1)$

2. *'while' statements (while $X \neq Y$ do S, with S arbitrary statement)*

3. *compound statements (begin $S_1, \ldots, S_n$ end, with S_i arbitrary statements).*

A **'while'** **program** *is any compound statement.*

'While' programs generate all the recursive functions, as shown by the proof of I.5.4. However, although 'while' programs are sufficient for recursion theory, to simplify the overall picture of programs it is useful in practice to have at hand more instructions than just those which are really needed. The common use instructions of structured programming languages considered in Figure 10

Sequencing (**begin** $S_1, \ldots, S_n$ **end**)

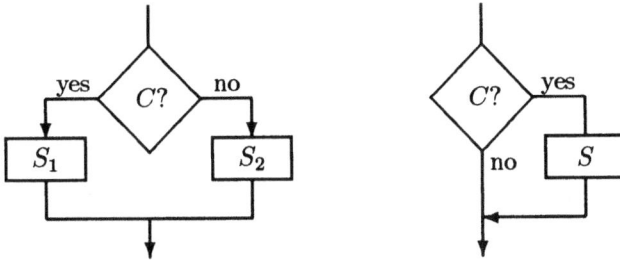

Forward conditionals (**if** C **then** S_1 **else** S_2, **if** C **then** S)

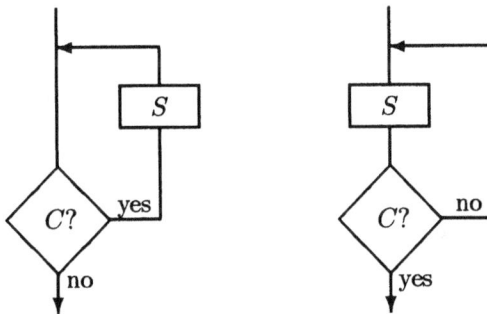

Backward conditionals (**while** C **do** S, **repeat** S **until** C)

Figure I.10: Structured flowcharts

can easily be defined as shorthand expressions for appropriate 'while' programs (for any Boolean combination C of atomic expressions of the kind $a = b$ and $a < b$, with a, b variables or numbers). The language obtained by enriching 'while' programs by these definitions is essentially **PASCAL** (Wirth [1971]) without declarations (variables being directly available) and the special features necessary to handle non-numerical data. PASCAL is a modern derivative of the **ALGOL** family (see Sammett [1969] and Wexelblat [1981] for the saga of its collective gestation, birth and development).

Structured and unstructured flowcharts are equivalent (Böhm and Jacopini [1966]) in the sense that they compute the same functions. This follows indirectly from the fact, to be proved in Section 7, that unstructured flowcharts only compute recursive functions, but this can easily be shown directly by an analysis of unstructured flowcharts. Equivalence is the best result that can be achieved, since we cannot expect, in general, to be able to decompose unstructured flowcharts into parts with only a fixed number of patterns (because we can build unstructured flowcharts with any number of conditional nestings).

Note that our treatment of flowchart and 'while' programs is independent of Turing machines. A different proof, based on I.5.4, of Theorem I.4.3 would then consist of showing how to compute the assignment statements and how to compose Turing machine programs by 'sequencing' and 'while'. In other words, we would just need a compiler (see p. 61). This can be done as a useful exercise (see Davis [1974] for details).

Exercises I.5.6 a) *'If then else' and 'repeat' can be defined as 'while' programs.*

b) *'While' can be defined using sequencing, 'if then else' and 'repeat'.*

Programs for primitive recursion ⋆

An examination of the proof of Theorem I.5.4 shows that primitive recursion and μ-recursion are defined by similar flowcharts programs, with one basic difference: the number of iterations is fixed in advance in the first case, and unknown in the second. We thus isolate, in the class of 'while' programs defined in I.5.5, the following class of structured programs:

Definition I.5.7 (Meyer and Ritchie [1967]) *Consider the programming language whose statements are the following:*

1. *assignment statements*

2. *'for' statements ('for Y do S', with S arbitrary statement)*

3. *compound statements ('begin $S_1, \ldots, S_n$ end', with S_i arbitrary statements).*

A **'for' program** *is any compound statement.*

The statement 'for Y do S' means 'iterate S for Y times', and it is obviously definable as a 'while' program as follows:

$$(\text{begin } T := 0, (\text{while } T \neq Y \text{ do } (\text{begin } T := T + 1, S \text{ end}))\text{end})$$

(with T variable not appearing in S).

Proposition I.5.8 (Meyer and Ritchie [1967]) *A function is 'for' computable if and only if it is primitive recursive.*

Proof. Every primitive recursive function is 'for' computable, by the proof of Theorem I.5.4. The opposite is shown by induction on the depth of nesting of 'for' statements in a program. It is convenient, to have a stronger inductive hypothesis, to show that each variable of the program has, at the end of the program execution, a value which is a primitive recursive function of all the variables (we would need this only for the output variable, as a function of the input variables).

When there is no 'for' (i.e. when the depth of nesting is 0) the program consists of the sequencing of a finite number of assignment statements, and each variable has a primitive recursive value (obtained by composition of $\mathcal{O}$, $\mathcal{S}$ and the predecessor operation).

Consider now 'for Y do S', and suppose S satisfies the inductive hypothesis: if the variables of the program S are among $X_1, \ldots, X_n$, there are primitive recursive functions $g_1, \ldots, g_n$ such that $X_i = g_i(X_1, \ldots, X_n)$. For $1 \leq i \leq n$, let $f_i(X_1, \ldots, X_n, Z)$ be the value of X_i after Z executions of S. Then:

$$\begin{cases} f_1(X_1, \ldots, X_n, 0) = X_1 \\ \cdots \\ f_n(X_1, \ldots, X_n, 0) = X_n \end{cases}$$

$$\begin{cases} f_1(X_1, \ldots, X_n, Z + 1) = g_1(f_1(X_1, \ldots, X_n, Z), \ldots, f_n(X_1, \ldots, X_n, Z)) \\ \cdots \\ f_n(X_1, \ldots, X_n, Z + 1) = g_n(f_1(X_1, \ldots, X_n, Z), \ldots, f_n(X_1, \ldots, X_n, Z)). \end{cases}$$

This is a simultaneous recursion on primitive recursive functions, which is easily seen (by coding, see I.7.2) to define primitive recursive functions. After the execution of 'for Y do S', the variable X_i will have value $f_i(X_1, \ldots, X_n, Y)$. In particular, the output variable will be a primitive recursive function of the input variables. $\square$

The result suggests a natural way of classifying the primitive recursive functions, by measuring the smallest depth of nesting of 'for' statements in programs computing a given function. This is essentially equivalent to the **Grzegor-czyck hierarchy** (Grzegorczyck [1953]), which will be studied in detail in Chapter VIII.

A consequence of I.5.8 is that primitive recursion in the form of definition I.1.3 is not necessary: a simpler operation of **simultaneous iteration** is sufficient. This result can be further refined, and we now show that, at the cost of a slight increase of the stock of initial functions, even simple iteration alone is sufficient.

Definition I.5.9 *A function f is defined by* **iteration** *from a function t if*

$$f(x, n) = t^{(n)}(x),$$

where $t^{(n)}(x)$ denotes the result of n successive applications of t (by convention, $t^{(0)}(x) = x$).

Proposition I.5.10 (Robinson [1947], Bernays) *The class of primitive recursive functions is the smallest class of functions*

1. *containing the initial functions, together with coding and decoding functions for pairs*

2. *closed under composition*

3. *closed under iteration.*

Proof. Let C be the smallest class of functions satisfying the conditions just stated: every function in C is primitive recursive, because the coding and decoding functions for pairs are primitive recursive, and the class of primitive recursive functions is closed under composition and iteration, the former by the definition, and the latter because iteration is a special case of primitive recursion:

$$
\begin{aligned}
f(x, 0) &= x \\
f(x, n+1) &= t(f(x, n)).
\end{aligned}
$$

For the converse, we only have to show that primitive recursion can be reduced to iteration. First of all note that, since C has coding and decoding functions for pairs and is closed under composition, it also has coding and decoding functions for n-tuples, for any fixed n. We continue to use for them the standard notations for coding and decoding functions. Let g and h be functions in C, and

$$
\begin{aligned}
f(\vec{x}, 0) &= g(\vec{x}) \\
f(\vec{x}, y+1) &= h(\vec{x}, y, f(\vec{x}, y)).
\end{aligned}
$$

We will show that the function

$$s(\vec{x}, n) = \langle \vec{x}, n, f(\vec{x}, n) \rangle$$

is in C. Then so is f, by composition, because

$$f(\vec{x}, n) = (s(\vec{x}, n))_{m+2}$$

(where m is the number of elements in the vector $\vec{x}$).
Note that

$$
\begin{aligned}
s(\vec{x}, 0) &= \langle \vec{x}, 0, f(\vec{x}, 0) \rangle \\
&= \langle \vec{x}, 0, g(\vec{x}) \rangle
\end{aligned}
$$

and thus $s(\vec{x}, 0)$ is in C, as a function of $\vec{x}$. Moreover,

$$
\begin{aligned}
s(\vec{x}, n+1) &= \langle \vec{x}, n+1, f(\vec{x}, n+1) \rangle \\
&= \langle \vec{x}, n+1, h(\vec{x}, n, f(\vec{x}, n)) \rangle.
\end{aligned}
$$

Since

$$s(\vec{x}, n) = \langle \vec{x}, n, f(\vec{x}, n) \rangle,$$

$s(\vec{x}, n+1)$ can be obtained from $s(\vec{x}, n)$ by one application of the function

$$t(\langle \vec{x}, n, z \rangle) = \langle \vec{x}, n+1, h(\vec{x}, n, z) \rangle,$$

which is in C by composition, since g, h, the successor, and the coding functions are in C. Finally,

$$s(\vec{x}, n) = t^{(n)}(s(\vec{x}, 0)),$$

and $s(\vec{x}, n)$ is thus the composition of $s(\vec{x}, 0)$ and the iteration of t. Since these are both in C, so is s. $\square$

The result just proved can be variously improved. First of all, Gladstone [1967], [1971] shows that *the introduction of new initial functions can be avoided*: the class of primitive recursive functions is the smallest class containing the initial functions, and closed under composition and iteration.

Second, the iteration schema can be further weakened in the following schema of **pure iteration**:

$$f(n) = t^{(n)}(0)$$

(Robinson [1947]). Some new initial functions are needed here, since by composition and pure iteration we never get, from the initial functions, any function depending on two variables, like $x+y$. Choices of initial functions that generate the primitive recursive functions by pure iteration, and a study of the algebraic structure of the class of primitive recursive functions, are in Robinson [1947], Poliakov [1964], [1964a], Lavrov [1967], Kozmnikh [1968], and Gladstone [1971].

R. Robinson [1947], [1955], and J. Robinson [1955], also show that the pure iteration schema is enough to generate the **unary primitive recursive functions** by composition, starting from two (but not from only one) appropriate unary functions. Thus we do not need to use nonunary functions to get any unary primitive recursive function. See also Peter [1951] for a treatment of this topic, and Georgieva [1976] for further simplifications.

Petri nets ⋆

Systems of separate, interacting components (like finite automata or computer hardware, flowcharts or computer software, physical systems, and so on) can be modeled by **Petri nets** (Petri [1962]), which consist of bipartite, directed multigraphs whose vertices can be either **places** or **transitions** (represented, respectively, by circles and bars), and whose arcs connect places to transitions, or transitions to places. **Tokens** can be assigned to (and can be thought to reside in) the places of a Petri net (which becomes **marked**), and their number in a place may change during the execution.

The **execution** is controlled by the number and distribution of tokens in the net, and consists of transition firings, which remove tokens from their input places, and deposit new ones in their output places. A transition fires when **enabled**, i.e. when each of its input places has at least as many tokens in it as there are arcs from the place to the transition (that is, arcs are seen as conductors of capacity one). Firing continues as long as there exist enabled transitions, then it halts.

Thus, nets model systems, and executions model the flow of information in them. Some of the places may be singled out as **inputs** or **outputs**, to model interactive behavior of a system with the outer world, and thus allowing a computational interpretation of the behavior of a net. Petri nets are quite general devices, apt to model **concurrence**, due to their inherent **parallelism, asynchronicity and nondeterminism**. Events which are both enabled and do not interact may occur independently, there is no inherent measure of time flow in the execution, and no order is placed on transition firings when many transitions are enabled.

An extension of Petri nets (Agerwala [1974], Hack [1975]) allows for **inhibitor arcs** from places to transitions (which are pictured differently, as arrows with the arrowhead substituted by a small circle). The new firing rule is that a transition is enabled if it is in the previous sense, and moreover the input places corresponding to inhibitor arcs are empty. Thus inhibitor arcs permit zero testing. The interest of this extension is that *Petri nets with inhibitor arcs compute all the recursive functions*. This is easily seen by modeling 'go to' programs (p. 64) as Petri nets, with different places corresponding to variables and statements, and transitions corresponding to instructions.

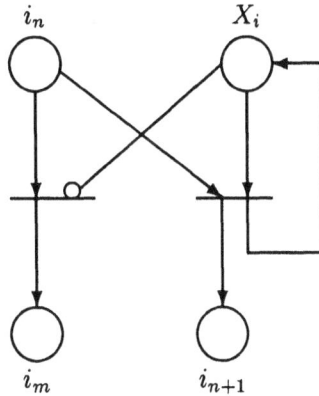

Figure I.11: Petri net for a 'go to' instruction

For example, given a 'go to' program with a list of labeled instructions i_n, the part of the corresponding Petri net relative to a 'go to' instruction

$$i_n : \text{ if } X_i = 0 \text{ go to } m$$

is the one shown in Figure 11, whose places are labeled in the natural way. The program works in the following way: if $X_i = 0$ then the inhibitor arc works, and the instruction i_m is activated, while if $X_i \neq 0$ then the normal arc works, decreases X_i by one, thus allowing the activation of the next instruction i_{n+1}, and then a new arc reintegrates the original value of X_i, which is not supposed to change. The parts corresponding to the other instructions (increasing or decreasing by one the value of a variable) are similar, but easier.

For an introduction to both theory and applications of Petri nets, and an annotated bibliography on the subject, see Peterson [1981].

I.6 Functions as Rules

The post-Dirichelet practice has been to identify functions and their graphs, thus giving a prominent importance to the values, independently of the way they are obtained. The rise of Computer Science has forced people to look at functions in the same way they were originally considered, as intensional objects. We thus set up in this section a theory of functions as explicit rules.

λ-calculus

Rules must be expressed somehow, and they can thus be identified with terms in an appropriate language. A term $t(x)$ can be seen as the value of a function for the argument x. We introduce an abstraction symbol 'λ' to indicate the step from the value $t(x)$ to the function $x \mapsto t(x)$, which we indicate by $\lambda x.\, t(x)$.

To begin with, we work in full generality and consider a language having merely the essential ingredients.

1. **Symbols.**

- variables $x_0, x_1, \ldots$
- the abstraction symbol 'λ'
- parentheses '(' and ')'
- dot '.' .

Note that there is only one kind of variable, and *no theoretical distinction is made between functions and arguments*. The common practice in mathematics, following Russell [1908], is to consider a function as being an object of a type higher than its arguments and values. This was introduced as one possible way out of Russell's paradox (see p. 81), which arose from the consideration of sets belonging to themselves. Since self-membership corresponds, in functional terms, to application of a function to itself, the practice of λ-calculus (of dealing with only one kind of object, that can be function or argument at will) sounds at least suspicious, and will have to be justified. Appropriately, Recursion Theory provides us with various ways of doing this, see pp. 194 and 223.

2. **Terms.**

- a variable x alone is a term, and x occurs free in it
- if M, N are terms then the **application** (MN) of M to N is a term, and free or bound occurrences of variables in M or N remain so in (MN)
- if M is a term and x is a variable then the **λ-abstraction** $(\lambda x.\, M)$ of M w.r.t. x is a term, x is bound in it, and the other variables are free or bound in it according to what they were in M.

By convention,

$$\lambda x_1 \cdots x_n.\, M \quad \text{means} \quad \lambda x_1 (\cdots (\lambda x_n.\, M) \cdots)$$
$$M_1 M_2 \cdots M_n \quad \text{means} \quad (\cdots (M_1 M_2) \cdots M_n).$$

The latter just avoids the need of many parentheses, but the former is theoretically important, since it defines *functions of several arguments as composition
of functions of a single argument*.

As we see, notations in λ-calculus are quite different from the common
practice, and they might be confusing at first sight, both for the treatment of
application (which is just written as juxtaposition) and of multiple arguments
(which are not treated simultaneously, but one at a time). E.g., what is usually
written $f(t_1, t_2)$ is here rendered as $(ft_1)t_2$ or simply, by the conventions just
stated, ft_1t_2.

Terms are names for the objects of λ-calculus (which, as already noted,
may be thought of as both functions and arguments). We now set up a theory
of transformations for them, and introduce rules that allow us to manipulate
terms. The idea is to compute values of functions by purely syntactical transformations of the terms representing the functions and their arguments.

3. **Reduction rules.**

α-**rule.** We can rename bound variables:

$$\lambda x.\, M \xrightarrow{\alpha} \lambda y.\, M[x/y]$$

provided y does not occur in M, where $M[x/y]$ indicates the result
of the substitution of x by y everywhere.

β-**rule.** We can apply a function $\lambda x.\, M$ to an argument N:

$$(\lambda x.\, M)N \xrightarrow{\beta} M[x/N]$$

provided whatever was free in N before the substitution remains
free afterwards. This proviso can always be fulfilled, by a possible
application of the α-rule. For example,

$$(\lambda xy.\, xy)y \to \lambda y.\, yy$$

is an illegal application of the β-rule, and indeed would not preserve
the intended meaning , but it can be replaced by

$$(\lambda xy.\, xy)y \to \lambda z.\, yz,$$

which can be derived by first changing the bound variable y into z
by the α-rule, and then applying the β-rule.

We write $t_1 \stackrel{\beta}{=} t_2$, and say that t_1 and t_2 are **equal** (modulo α or β reductions), if t_1 and t_2 can be reduced to the same term by a finite number of

applications of the α or β rules. Also, a term is said to be in **normal form** if it cannot be β-reduced (i.e. the β-rule is not applicable to any of its subterms), and it is **reducible** otherwise.

A term is **normalizable** if it has a normal form, in the sense of being equal to a term in normal form. Normal forms of terms can be considered as their 'values', and it is natural to investigate when they exist and how to compute them. Unfortunately, *a normal form does not necessarily exist* (thus, in some sense, the functions of λ-calculus are partial ones, see p. 127). To see this, we can simply produce a circular example of a term reducing to itself via the β-rule, say $MN \overset{\beta}{\to} MN$: M must then have a λ-abstraction (to allow for an application of the β-rule), and produce an application. For example,

$$\Delta \;=\; \lambda x.\, xx$$

applies a term to itself:

$$\Delta t \overset{\beta}{=} tt,$$

and hence it self-reproduces:

$$\Delta\Delta \overset{\beta}{=} \Delta\Delta.$$

A second negative point is that *even if a normal form exists, not all sequences of reductions can produce it*, e.g. a term might be reduced to normal form by one reduction and enter a loop by another:

$$(\lambda xy.\, y)\,(\Delta\Delta)\, a \;\overset{\beta}{=}\; a \qquad\qquad \text{by using the outer } \lambda$$
$$(\lambda xy.\, y)\,(\Delta\Delta)\, a \;\overset{\beta}{=}\; (\lambda xy.\, y)\,(\Delta\Delta)\, a \qquad \text{by using the } \lambda \text{ in } \Delta.$$

Terms in which every subterm has a normal form are called **strongly normalizable**, and the example just given shows that a term can be normalizable without being strongly so.

On the positive side, Church and Rosser [1936] have shown (by a difficult proof, outside the scope of this book) that terminating reductions of a same term always give the same result, up to renaming of bound variables. In particular, *the normal form is unique when it exists*. Moreover, *there are reduction strategies that produce the normal form, whenever it exists*. One such strategy is to always reduce the leftmost reducible λ. Then subterms are evaluated exactly as many times as needed, which means that they may be evaluated more than once in some cases, but also that they are not evaluated if not useful.

A different strategy would be to evaluate M and N before doing MN. This does not work in general (as the example above shows), but each term is evaluated only once, which means that the strategy is fast when it works, although some unnecessary terms may happen to be evaluated (and this may be fatal if one of these is a nonterminating one).

We have already noted a peculiar aspect of λ-calculus: the absence of distinction between functions and arguments, which may be seen as a collapse of the concept of type. In Recursion Theory a cumbersome technique (the method of arithmetization of Section 7) is used for the related purpose of assimilating functions with numbers. This is partly achieved by identifying the recursive functions with their finite descriptions, e.g. programs, and by coding these as numbers. Being indirectly obtained, this assimilation does not avoid the need of continuously going back and forth between functions and numbers representing them, and some arguments may at times become darkened. This is the case of the Fixed-Point Theorem II.2.10 and its proof, which are sometimes considered quite mysterious. When recast in terms of λ-calculus, both the statement and the proof of this result (based on methods fully explored in Section II.2) become more transparent.

Theorem I.6.1 (Kleene [1936b], Turing [1937a], Curry [1942], Rosenbloom [1950]) *There is a* **fixed-point operator** $\mathcal{Y}$ *which produces, for every term* M, *a fixed-point for it:*

$$\mathcal{Y}M \overset{\beta}{=} M(\mathcal{Y}M).$$

Proof. We give two different proofs.

1. We start with an informal argument. Recall that we have defined a term

$$\Delta = \lambda x.\, xx$$

such that, for any term t,

$$\Delta t \overset{\beta}{=} tt.$$

By applying Δ to itself, we then have

$$\Delta\Delta \overset{\beta}{=} \Delta\Delta.$$

Given a term M, we want a term z such that $z \overset{\beta}{=} Mz$: then z reproduces not itself, but the application of M to itself. It is then enough to generalize the definition of Δ, which gives what we want when M is the identity operator. Let

$$\Delta_M = \lambda x.\, M(xx).$$

Then, for any term t,

$$\Delta_M t \overset{\beta}{=} M(tt).$$

By applying Δ_M to itself we then have

$$\Delta_M \Delta_M \overset{\beta}{=} M(\Delta_M \Delta_M),$$

and a fixed-point of M is given by

$$\Delta_M \Delta_M = (\lambda x. M(xx))(\lambda x. M(xx)).$$

Since this definition is uniform in M, we can actually abstract M from it, and obtain

$$\mathcal{Y} = \lambda y. (\lambda x. y(xx))(\lambda x. y(xx)).$$

Then

$$\mathcal{Y}M \overset{\beta}{=} \Delta_M \Delta_M \overset{\beta}{=} M(\Delta_M \Delta_M) \overset{\beta}{=} M(\mathcal{Y}M).$$

2. We now give a proof based on the (contrapositive of the) diagonal method of Section II.2. Consider M: a fixed-point z for it is such that $z \overset{\beta}{=} Mz$, i.e. it must be of the form $t_i t_j$ for some terms t_i, t_j. Also, M must be thought of as the result of a transformation of terms of this kind. Consider a possible enumeration of all pairs of λ-terms:

$$
\begin{array}{cccc}
t_0 t_0 & t_0 t_1 & t_0 t_2 & t_0 t_3 \quad \cdots \\
t_1 t_0 & t_1 t_1 & t_1 t_2 & \cdots \\
t_2 t_0 & t_2 t_1 & t_2 t_2 & \cdots \\
t_3 t_0 & \cdots & \cdots & \cdots \\
\cdots
\end{array}
$$

and the effect of M on the diagonal:

$$M(t_0 t_0) \quad M(t_1 t_1) \quad M(t_2 t_2) \quad \cdots$$

This is a sequence of λ-terms, of the form

$$t_n t_i \overset{\beta}{=} M(t_i t_i)$$

for some n. But then it is just the n-th row of the matrix, in particular

$$t_n t_n \overset{\beta}{=} M(t_n t_n)$$

is a fixed-point of M. Explicitly,

$$t_n = \lambda x. M(xx)$$

$$t_n t_n \overset{\beta}{=} (\lambda x. M(xx))(\lambda x. M(xx)),$$

and this is uniform in M, i.e. it produces a fixed-point operator

$$\mathcal{Y} = \lambda y.\,(\lambda x.\,y(xx))(\lambda x.\,y(xx)).$$

Since all this might look quite mysterious, we can check that $\mathcal{Y}$ really gives us a fixed-point for M:

$$
\begin{aligned}
\mathcal{Y}M &= (\lambda y.\,(\lambda x.\,y(xx))(\lambda x.\,y(xx)))M &\text{(I.1)}\\
&\overset{\beta}{=} (\lambda x.\,M(xx))(\lambda x.\,M(xx)) &\text{(I.2)}\\
&\overset{\beta}{=} M(\underbrace{(\lambda x.\,M(xx)(\lambda x.\,M(xx))}_{\mathcal{Y}M}) &\text{(I.3)}\\
&\overset{\beta}{=} M(\mathcal{Y}M) &\text{(I.4)}
\end{aligned}
$$

where (1) is obtained by definition of $\mathcal{Y}$, (2) by β-reducing the λy, and (3) by β-reducing the outer λx. Note that (3) is still reducible, using the outer λx. This is a typical property of the terms of the form $\mathcal{Y}M$, and somewhat illustrates the circular aspect of fixed-point definitions. □

Exercise I.6.2 Y *is a fixed-point operator if and only if it is itself a fixed-point of* $G = \lambda ym.\,m(ym)$. (Böhm, Van Der Mey) (Hint: identify $\lambda m.\,Ym$ and Y.)

$\mathcal{Y}$ is sometimes called the **paradoxical combinator**, because it embodies the argument used in **Russell's paradox** (Russell [1903]). The connection between Set Theory and λ-calculus can be established by the following correspondences:

element	argument
set	function
membership	application
set formation	λ-abstraction
set equality	term equality.

Russell's paradox is obtained by considering the set

$$A = \{x : x \notin x\}.$$

Then

$$x \in A \Leftrightarrow x \notin x,$$

and thus

$$A \in A \Leftrightarrow A \notin A,$$

contradiction.

In terms of λ-calculus, the negation operator can be considered as a term N that is never the identity. Since membership corresponds to application, self-membership corresponds to self-application, and then the set A corresponds to the term

$$\lambda x.\, N(xx).$$

By β-reduction,

$$(\lambda x.\, N(xx))x \overset{\beta}{=} N(xx),$$

and thus

$$(\lambda x.\, N(xx))(\lambda x.\, N(xx)) \overset{\beta}{=} N((\lambda x.\, N(xx))(\lambda x.\, N(xx))),$$

i.e.

$$\mathcal{Y}N \overset{\beta}{=} N(\mathcal{Y}N).$$

Note however that here there is no paradox: from the last assertion, which is true, we just deduce that there is no term N that is never the identity.

Other formulations of the λ-calculus $\star$

Our approach (Church [1933]) has been to *define* (by means of λ-abstractions) terms, called combinators, reflecting the way functional letters can be effectively combined to produce names for intensionally presented functions. Our term formation rules allow $\lambda x.\, M$ to be a term whenever M is. This version is called the **λK-calculus**. If a restriction were imposed on M, requiring that x occur free in it, we would obtain a version called the **λI-calculus**, in which functions with fictitious arguments are excluded. This restriction has been introduced to avoid some pathologies, like the existence of normalizable, not strongly normalizable terms, as seen above.

The α and β rules define a system called the calculus of **β-conversion**. Various modifications of it are possible, e.g. by adding an extensionality law for λ-terms:

η-rule. We can identify every term with a function:

$$\lambda x.\, Mx \overset{\eta}{\to} M$$

provided x is not free in M. Note that, by the rules of term formation, for any term M the expression $\lambda x.\, Mx$ is also a term, and thus it represents a function: the rule ensures that this is exactly the function that is represented by the term M itself.

The reason we call this an extensionality law is that it implies that if M and N behave extensionally in the same way, i.e. $\lambda x.\, Mx$ and $\lambda x.\, Nx$ are equal, then so are M and N.

An equivalent but different approach to the λ-calculus (called the **theory of combinators**, Schönfinkel [1924], Curry [1930]) consists in *postulating* some, actually very few, of the combinators as primitive, and to *deduce* all the others from these. This produces a kind of synthetical (bottom-up) analysis of the global concept of λ-definability. In particular, it can be shown that only two combinators are needed:

$$\boldsymbol{S} = \lambda xyz.\, xz(yz) \qquad \boldsymbol{K} = \lambda xy.\, x.$$

$\boldsymbol{S}$ can be seen as a kind of interpreted application, where x and y are first interpreted in the environment z, and then applied one to the other. Since the identity $\boldsymbol{I}$ can be defined as $\boldsymbol{SKK}$:

$$\boldsymbol{I}x = \boldsymbol{SKK}x = \boldsymbol{K}x(\boldsymbol{K}x) = x,$$

the λ operator can be defined by induction on the terms obtained from the variables by application, as follows:

$$
\begin{aligned}
\lambda x.\, x &= \boldsymbol{I} \\
\lambda x.\, c &= \boldsymbol{K}c \quad \text{if } c \text{ is } \boldsymbol{K},\, \boldsymbol{S} \text{ or a variable } y \neq x \\
\lambda x.\, MN &= \boldsymbol{S}(\lambda x.\, M)(\lambda x.\, N).
\end{aligned}
$$

We can then *prove* the β-reduction rule, by induction.

Standard references on the subject are Church [1941], Curry and Feys [1958], Curry, Hindley and Seldin [1972], Barendreght [1981], Hindley and Seldin [1986]. Historical accounts on the origins of λ-calculus and its interaction with Recursion Theory are in Kleene [1981] and Rosser [1984].

λ-definability

We are interested in numerical functions, but it would seem that until now we have just set up a logical basis, and that we still need to add to the language of λ-calculus numerical terms and some basic numerical functions. But then we would face the problem of not knowing exactly what to add, and we would have to turn back to different approaches, with the λ-calculus relegated to a mere role of convenient notation. Instead, and this is a most interesting aspect of this approach, it turns out that there is no need of additional notions.

The natural numbers appear obliquely in this general setting, when we consider the number of iterations of a function application. We can thus define λ-terms $\bar{n}$ that, applied to a function f and an argument x, give the result

$f^{(n)}(x)$ of n iterations of the function f on x, and take them as representing the integers. Since we want

$$\bar{n}fx = f^{(n)}(x)$$

we can just let:

Definition I.6.3 (Peano [1891], Wittgenstein [1921], Church [1933])
The numeral $\bar{n}$ is the λ-term $\lambda fx.\, f^{(n)}(x)$.

We write f and x to help the intuitive understanding, but note that we just have one type of variable (with no distinction between functions and arguments), and thus we should just write

$$\bar{n} = \lambda xy.\, x^{(n)}y$$

Note that the terms $\bar{m}$ and $\bar{n}$ are different if $m \neq n$, by the theorem of Church and Rosser quoted on p. 78, because they are in normal form and distinct. Inductively we have, by definition of iteration,

$$\begin{aligned}
\bar{0} &= \lambda fx.\, x \\
\overline{n+1} &= \lambda fx.\, f(\bar{n}fx)
\end{aligned}$$

because $f^{(0)}(x) = x$ and $f^{(n+1)}(x) = f(f^{(n)}(x))$. Since these terms represent in some way the constant function $\mathcal{O}$ and the successor operation $\mathcal{S}$, we get from this the idea of representing numerical functions:

Definition I.6.4 (Church [1933], Kleene [1935]) *An n-ary function f is* **λ-definable** *if there is a λ-term F such that*

$$f(a_1, \ldots, a_n) = b$$

holds if and only if

$$F\bar{a}_1 \ldots \bar{a}_n \overset{\beta}{=} \bar{b}.$$

Theorem I.6.5 (Church [1933], Rosser [1935], Kleene [1935], [1936b])
Every recursive function is λ-definable.

Proof. We proceed by induction on the definition of recursive function. To simplify the technical details, we rely on the alternative characterization of the class of primitive recursive functions given in I.5.10.

1. *initial functions*
   $\mathcal{O}$, $\mathcal{S}$ and $\mathcal{I}_i^n$ are, respectively, λ-defined by:

$$\lambda x.\bar{0}$$
$$\lambda zfx.\, f(zfx)$$
$$\lambda x_1 \cdots x_n.\, x_i$$

2. *coding and decoding functions for pairs*
   This is obtained, in analogy with the representation of natural numbers, as:

   $$\overline{(n,m)} = \lambda fgx.\, f^{(n)}(g^{(m)}(x)).$$

   Thus the pairing function is

   $$\lambda yzfgx.\, yf(zgx).$$

   Decoding is then immediate: if I is the representation of $\mathcal{I}_1^1$, i.e. of the identity function, then

   $$\begin{aligned}
   \lambda fx.\,\overline{(n,m)}fIx &= \lambda fx.\, f^{(n)}x &= \overline{n} \\
   \lambda gx.\,\overline{(n,m)}Igx &= \lambda gx.\, g^{(m)}x &= \overline{m}.
   \end{aligned}$$

   Thus the decoding functions are

   $$\lambda yfx.\, yfIx \quad \text{and} \quad \lambda ygx.\, yIgx.$$

3. *composition*
   Suppose h, g_i are λ-defined by H, G_i. Then

   $$f(x_1,\ldots,x_n) = h(g_1(x_1,\ldots,x_n),\ldots,g_m(x_1,\ldots,x_n))$$

   is λ-defined by

   $$\lambda x_1,\ldots,x_n.\, H(G_1 x_1 \cdots x_n)\cdots(G_m x_1 \cdots x_n).$$

4. *iteration*
   This is an immediate consequence of the representation of natural numbers: if T is a term representing the function t, then the iteration

   $$f(x,n) = t^{(n)}(x)$$

   is represented by

   $$\lambda xy.\, yTx.$$

5. *μ-recursion*
   This is obtained as in the proof of I.2.3. Recall that if

   $$f(\vec{x}) = \mu y(g(\vec{x},y) = 0)$$

   then

   $$f(\vec{x}) = h(\vec{x},0),$$

where

$$h(\vec{x}, y) = \begin{cases} y & \text{if } g(\vec{x}, y) = 0 \\ h(\vec{x}, y + 1) & \text{otherwise} \end{cases}$$

By the first part of the proof case definition, successor, and composition are all λ-representable, because primitive recursive. By induction hypothesis, so is g. Thus there is a term M representing $\lambda h \vec{x} y. F$, where

$$F(\vec{x}, y, h) = \begin{cases} y & \text{if } g(\vec{x}, y) = 0 \\ h(\vec{x}, y + 1) & \text{otherwise.} \end{cases}$$

Then the function h defined above, being a fixed-point of F, is represented by $\mathcal{Y}M$. And f is finally represented by $\lambda \vec{x}. (\mathcal{Y}M)\vec{x}\overline{0}$.

To conclude the proof we note that:

- The *completeness property*, stating that the values can be deduced from the appropriate λ-terms, is quite evident from the informal discussion just given.

- The *consistency property*, stating that no other value can be deduced, follows from the theorem of Church and Rosser quoted on p. 78, which ensures that if the process of β-reduction of a term produces a term in normal form (like $\overline{n}$), then this term is uniquely determined (up to renaming of the bound variables). $\square$

Exercises I.6.6 a) *Alternative coding and decoding functions are, respectively,* $\lambda xyu. uxy$, $\lambda z. zT$ *and* $\lambda z. zF$, *where* $T = \lambda xy. x$ *and* $F = \lambda xy. y$. *Being distinct* λ-*terms*, T *and* F *can be taken as representation of the truth values 'true' and 'false'.*

b) $\delta = \lambda z. z(\lambda u. F)T$ *represents a function that, when its argument is a numeral, is* T *if* z *is* $\overline{0}$, *and* F *otherwise. Then* $\lambda xyz. (\delta z)xy$ *represents definition by cases, i.e. a function that returns the first or the second argument, according to whether the third is 0 or not. (Hint: a numeral applied to two terms produces the second if it is* $\overline{0}$, *and applies the first at least once otherwise.)*

c) *The predecessor function can be directly represented, using only representations of successor, coding, and decoding functions.* (Kleene [1935]) *(Hint: the predecessor of* n *is the second component of the* n-*th iteration of the function on pairs defined as* $t((x, y)) = (x + 1, x)$, *started on* $(0, 0)$.)

By II.2.15 this provides an alternative proof of I.6.5, not using I.5.10.

Functional programming languages $\star$

The programming languages discussed in Sections 4 and 5 are **imperative** in the sense that they specify a sequence of instructions and an order of execution to be followed to produce an output $f(\vec{x})$. The **functional** approach, suggested by λ-calculus, defines f directly, and computes the required values by

β-reductions. The name 'functional' reflects the emphasis put on the functions themselves as intensional objects, as opposed to extensional emphasis on the values.

Relying on the characterization of recursive functions given in II.2.15, McCarthy [1960] notes that a function is recursive if and only if:

- there is a definition of it from identities, successor, and predecessor by means of λ-abstraction, composition and the conditional operator

$$\text{if } t(\vec{x}) = 0 \text{ then } g(\vec{x}) \text{ else } h(\vec{x}).$$

- the function is computable from this definition by a *call by value procedure*, i.e. a computation which evaluates first the innermost (and leftmost, if there is more than one) occurrence of the letter defining the function.

Clearly the conditional operator replaces the definition by cases, and the fixed-point operator is eliminated, in favor of a computational approach that just produces it.

This formulation of recursive functions can be extended from numerical functions to functions on words of a given alphabet and, as such, it has furnished the computational basis for the programming language **LISP** (List Processing, McCarthy [1960]; see Sammett [1969] and Wexelblat [1981] for history and references).

Since there is no reason that functional programming languages should be forced to run on machines designed for imperative ones, work has been done to design hardware directly inspired by the functional approach, based on λ-calculus (the **SECD machine**, Landin [1963], implementing β-reductions) or on the theory of combinators (the **SK machine**, Turner [1979], implementing reductions of graphs that represent the definition of a function, in terms of the combinators S and K, see p. 83).

I.7   Arithmetization

Arithmetization simply means translation into the language of arithmetic. We will give one detailed example of the method, and show how to code the machinery of computation of recursive functions in a primitive recursive way. The details are quite cumbersome, and in the rest of our work we will content ourselves to sketch similar arguments, leaving the details to the reader.

Historical remarks ⋆

The first attempt to find number-like connections between propositions of various sorts probably goes back to Lullus' *Ars Magna*, but it was Leibniz [1666] who dreamt of arithmetization as a general method to replace reasoning in natural language by arithmetical propositions, with the goal of substituting arguments with computations. Leibniz went further than just this oneiric activity, and devised (see [1903]) a precise coding method, by first assigning numbers to primitive notions, and then showing how to associate composite numbers (e.g. by multiplication) to composite notions. However, his tentative work was left unpublished until 1903, and did not influence modern developments.

Hilbert [1904] again envisaged arithmetization in his idea of formalizing consistency proofs into Arithmetic, but it was Gödel [1931] who first used it explicitly and formally to translate the concepts relative to formal systems into an arithmetical language. Tarski [1936] independently arrived at the method in his investigations of the concept of truth.

Contrary to Leibniz's dreams, the effect of arithmetization was, ironically, not to shield the language against its oddities, but rather to spring a leak in Arithmetic, through which the linguistic paradoxes poured only to reveal the inadequacies of formalism (see II.2.17).

Numerical tools for arithmetization

Primitive recursive functions and predicates are more than enough to carry out arithmetizations. We will use prime numbers and factorizations (recall, see p. 26, that the sequence $\{p_x\}_{x \in \omega}$ of prime numbers is primitive recursive), because this coding is particularly simple. It should however be noted that we could use functions from much smaller classes: this will be done in Chapter VIII, when the necessity for more efficient codings will arise.

To code the sequence $\langle x_0, \ldots, x_n \rangle$, the simplest way would be to use the number $p_0^{x_0} \cdots p_n^{x_n}$. But then we could not uniquely decode a number, since we would not know whether a prime in the decomposition has exponent 0 accidentally or meaningfully (in the sense that it is coding the number 0). Thus, either we rule out 0 as a meaningful exponent, and let

$$\langle x_0, \ldots, x_n \rangle = p_0^{x_0+1} \cdots p_n^{x_n+1},$$

or we tell in advance how many numbers we are coding, and let

$$\langle x_1, \ldots, x_n \rangle = p_0^n \cdot p_1^{x_1} \cdots p_n^{x_n}.$$

Since we have to make a choice for the following, we decide to use the second proposal. The decoding system is given by the following functions and predicates, all primitive recursive (see p. 26):

$$
\begin{aligned}
(x)_n &= exp(x, p_n) \\
ln(x) &= (x)_0 \\
Seq(x) &\Leftrightarrow (\forall n \le x)[n > 0 \wedge (x)_n \ne 0 \ \rightarrow \ n \le ln(x)].
\end{aligned}
$$

We call $ln(x)$ the **length** of x, and $(x)_n$ the n-th **component** of x. If $Seq(x)$ holds then we say that x is a **sequence number**. In this case,

$$x = \langle (x)_1, \ldots, (x)_{ln(x)} \rangle.$$

We will also need a **concatenation** operation $*$, such that

$$\langle x_1, \ldots, x_n \rangle * \langle y_1, \ldots, y_m \rangle = \langle x_1, \ldots, x_n, y_1, \ldots, y_m \rangle.$$

This is formally defined as:

$$
\begin{aligned}
x * y &= p_0^{ln(x)+ln(y)} \cdot p_1^{(x)_1} \cdots p_{ln(x)}^{(x)_{ln(x)}} \cdot p_{ln(x)+1}^{(y)_1} \cdots p_{ln(x)+ln(y)}^{(y)_{ln(y)}} \\
&= p_0^{ln(x)+ln(y)} \cdot \prod_{i<ln(x)} p_{i+1}^{(x)_{i+1}} \cdot \prod_{i<ln(y)} p_{ln(x)+i+1}^{(y)_{i+1}}
\end{aligned}
$$

if $Seq(x) \wedge Seq(y)$ holds, and 0 otherwise.

Finally, we will need the notion of **initial subsequence**, defined as

$$
\begin{aligned}
x \sqsubseteq y &\Leftrightarrow Seq(x) \wedge Seq(y) \wedge (\exists u \leq y)(Seq(u) \wedge x * u = y) \\
x \sqsubset y &\Leftrightarrow x \sqsubseteq y \wedge x \neq y.
\end{aligned}
$$

As a first use of sequence numbers we get the following useful result.

Proposition I.7.1 Course-of-values recursion (Skolem [1923], Peter [1934]) *The class of primitive recursive functions is closed under recursions in which the definition of $f(\vec{x}, y+1)$ may involve not only the last value $f(\vec{x}, y)$, but any number of (and possibly all) the values $\{f(\vec{x}, z)\}_{z \leq y}$ already obtained.*

Formally, let $\hat{f}$ be the **history function** *of f, defined as:*

$$\hat{f}(\vec{x}, y) = \langle f(\vec{x}, 0), \ldots, f(\vec{x}, y) \rangle.$$

Then, if f is defined as

$$
\begin{aligned}
f(\vec{x}, 0) &= g(\vec{x}) \\
f(\vec{x}, y+1) &= h(\vec{x}, y, \hat{f}(\vec{x}, y))
\end{aligned}
$$

and g, h are primitive recursive, so is f.

Proof. It is enough to show that $\hat{f}$ is primitive recursive, since then also f is:

$$f(\vec{x}, y) = \left(\hat{f}(\vec{x}, y) \right)_{y+1}.$$

But this is immediate, since

$$\hat{f}(\vec{x},0) = \langle f(\vec{x},0) \rangle$$
$$= \langle g(\vec{x}) \rangle$$
$$\hat{f}(\vec{x},y+1) = \langle f(\vec{x},0), \ldots, f(\vec{x},y), f(\vec{x},y+1) \rangle$$
$$= \hat{f}(\vec{x},y) * \langle f(\vec{x},y+1) \rangle$$
$$= \hat{f}(\vec{x},y) * \langle h(\vec{x},y,\hat{f}(\vec{x},y)) \rangle.$$

Thus $\hat{f}(\vec{x},y+1)$ only uses the last previous value $\hat{f}(\vec{x},y)$, and $\hat{f}$ is primitive recursive because so are coding and concatenation. □

Exercise I.7.2 Simultaneous primitive recursion. *The class of primitive recursive functions is closed under simultaneous recursion on more than one function.* (Hilbert and Bernays [1934]) (Hint: reduce simultaneous primitive recursion of, say, $f_1(\vec{x})$ and $f_2(\vec{x})$, to primitive recursion of the single function $\langle f_1(\vec{x}), f_2(\vec{x}) \rangle$, by coding.)

The Normal Form Theorem

We are now in position to give our first and only complete example of arithmetization, by reducing the recursive functions to a normal form. Any approach to recursiveness would produce similar results, and we will sketch the versions relative to the approaches of Sections 2–6 later in this section, but here we give a self-contained treatment based on recursiveness alone.

Theorem I.7.3 Normal Form Theorem (Kleene [1936]) *There is a primitive recursive function $\mathcal{U}$ and (for each $n \geq 1$) primitive recursive predicates T_n, such that for every recursive function f of n variables there is a number e (called **index** of f) for which the following hold:*

1. $\forall x_1 \ldots \forall x_n \exists y T_n(e, x_1, \ldots, x_n, y)$

2. $f(x_1, \ldots, x_n) = \mathcal{U}(\mu y T_n(e, x_1, \ldots, x_n, y)).$

Proof. The idea of the proof is to associate numbers to functions and computations in such a way that the predicate $T_n(e, x_1, \ldots, x_n, y)$ translates the assertion: y is the number of a computation of the value of the function with associated number e, on inputs $x_1, \ldots, x_n$. Having this, $\mu y T_n(e, x_1, \ldots, x_n, y)$ will give the number of one such computation, and the function $\mathcal{U}$ will extract the value of the output from it. This is more easily said than done, and to achieve it we need to carry out a number of steps.

1. *associate numbers to recursive functions*
   This is done according to the inductive procedure that generates the recursive functions. The details are obviously irrelevant, and we just give one possible assignment:

   - $\langle 0 \rangle$ to $\mathcal{O}$
   - $\langle 1 \rangle$ to $\mathcal{S}$
   - $\langle 2, n, i \rangle$ to $\mathcal{I}_i^n$, for $1 \le i \le n$
   - $\langle 3, b_1, \ldots, b_m, a \rangle$ to $f(\vec{x}) = g(h_1(\vec{x}), \ldots, h_m(\vec{x}))$, where $b_1, \ldots, b_m$ and a are numbers respectively associated to $h_1, \ldots, h_m$, and g
   - $\langle 4, a, b \rangle$ to $f(\vec{x}, y)$ defined by primitive recursion from g and h, where a and b are respectively associated to g and h
   - $\langle 5, a \rangle$ to $f(\vec{x}) = \mu y(g(\vec{x}, y) = 0)$, if $\forall \vec{x} \exists y (g(\vec{x}, y) = 0)$ and a is associated to g.

   Any number associated to a recursive function is called an **index** of this function. Since there are many ways to define the same function, there will be many indices for each recursive function (see II.1.6). Also, many numbers are not indices of recursive functions, either because they are not sequence numbers of the right form, or because some of their relevant components are not indices of recursive functions. We will see (p. 146) that in general there is no effective way to tell whether a number is indeed the index of a recursive function, because basically there is no way to tell whether $\forall \vec{x} \exists y (g(\vec{x}, y) = 0)$.

2. *put computations in a canonical form*
   A natural way to organize a computation for the values of a given recursive function, is by way of **computation trees**. Each node of such a tree will tell how a value needed in the computation can be inductively obtained. Of course, the only possibilities are those given by the permissible schemata of definition I.1.7, namely:

   - nodes without predecessors:

   $$\begin{aligned} f(x) &= 0 & &\text{if } f = \mathcal{O} \\ f(x) &= x + 1 & &\text{if } f = \mathcal{S} \\ f(x_1, \ldots, x_n) &= x_i & &\text{if } f = \mathcal{I}_i^n \end{aligned}$$

   - composition
     If $f(\vec{x}) = g(h_1(\vec{x}), \ldots, h_m(\vec{x}))$ then the node $f(\vec{x}) = z$ has $m + 1$ predecessors:

$$f(\vec{x}) = z$$

$$h_1(\vec{x}) = z_1 \quad \cdots \quad h_m(\vec{x}) = z_m \quad g(z_1, \ldots, z_m) = z$$

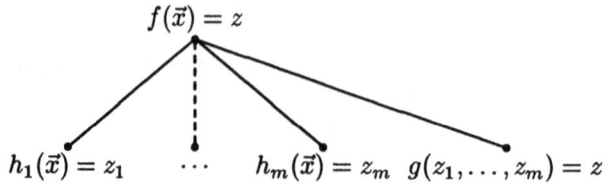

- primitive recursion
  If $f(\vec{x}, y)$ is defined by primitive recursion from g and h there are two cases, respectively with one or two predecessors:

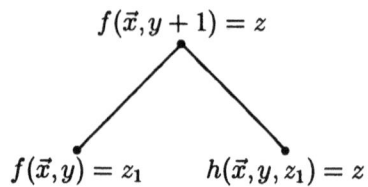

$$f(\vec{x}, 0) = z \qquad\qquad\qquad f(\vec{x}, y+1) = z$$

$$g(\vec{x}) = z \qquad\qquad f(\vec{x}, y) = z_1 \qquad h(\vec{x}, y, z_1) = z$$

- μ-recursion
  If $f(\vec{x}) = \mu y (g(\vec{x}, y) = 0)$ then there is no fixed pattern to the predecessors of $f(\vec{x}) = y$. The general situation is:

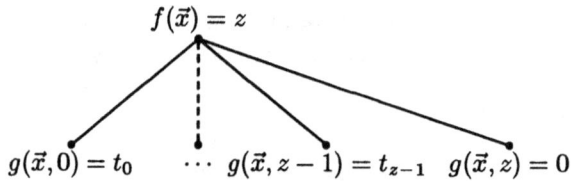

$$f(\vec{x}) = z$$

$$g(\vec{x}, 0) = t_0 \quad \cdots \quad g(\vec{x}, z-1) = t_{z-1} \quad g(\vec{x}, z) = 0$$

where $t_0, \ldots, t_{z-1}$ are all different from 0.

3. *associate numbers to computations*
   This is done by induction on the construction of the computation tree. First of all, we assign numbers to nodes: since they are expressions of the kind $f(x_1, \ldots, x_n) = z$, we give them numbers

$$\langle e, \langle x_1, \ldots, x_n \rangle, z \rangle,$$

where e is a given index of f. Thus a node is represented by three numbers, corresponding respectively to the function, the inputs and the output.

We then assign numbers to trees: each tree T consists of a vertex v with associated number v, and of a certain number (finite, and possibly equal to zero) of ordered predecessors, each one being a subtree T_i. By

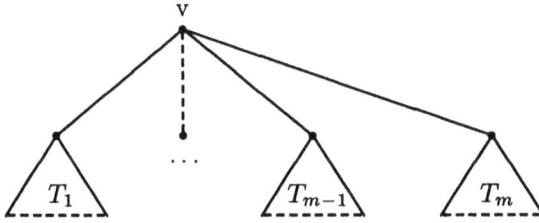

Figure I.12: A tree T with vertex v and subtrees T_i's

induction, we assign to this tree the number

$$\widehat{T} = \langle v, \widehat{T_1}, \ldots, \widehat{T_m} \rangle,$$

where $\widehat{T_i}$ is the number assigned to the subtree T_i. In particular, if the vertex with number v does not have predecessors, then it has number $\langle v \rangle$ as a tree.

4. *translate in a primitive recursive predicate $T(y)$ the property that y is a number coding a computation tree*
   To increase readability we use commas instead of nested parentheses, and write e.g. $(a)_{i,j,k}$ in place of $(((a)_i)_j)_k$. To keep track of what we are doing, check Figure 13 and recall that

$$y = \langle v, \widehat{T_1}, \ldots, \widehat{T_m} \rangle,$$

and hence:

$$
\begin{array}{rcl}
(y)_1 & = & \langle e, \langle x_1, \ldots, x_n \rangle, z \rangle \\
(y)_{1,1} & = & \text{various types, depending on } e \\
(y)_{1,2} & = & \langle x_1, \ldots, x_n \rangle \\
(y)_{1,3} & = & z \\
(y)_{i+1} & = & \widehat{T_i} \\
(y)_{i+1,1} & = & \text{number of the vertex of } T_i.
\end{array}
$$

First we let:

$$
\begin{aligned}
A(y) \quad \Leftrightarrow \quad & Seq(y) \wedge Seq((y)_1) \wedge ln((y)_1) = 3 \wedge \\
& Seq((y)_{1,1}) \wedge Seq((y)_{1,2}).
\end{aligned}
$$

This expresses the most trivial properties of y. We then have four cases, corresponding to the possible situations spelled out in Part 2 above.

$$\langle\langle 3, b_1, \ldots, b_m, a\rangle, \langle x_1, \ldots, x_n\rangle, z\rangle$$

$$\langle b_1, \langle x_1, \ldots, x_n\rangle, z_1\rangle \qquad \langle b_m, \langle x_1, \ldots, x_n\rangle, z_m\rangle \qquad \langle a, \langle z_1, \ldots, z_m\rangle, z\rangle$$
$$= \qquad\qquad = \qquad\qquad =$$
$$(y)_{2,1} \qquad\qquad (y)_{m+1,1} \qquad\qquad (y)_{m+2,1}$$

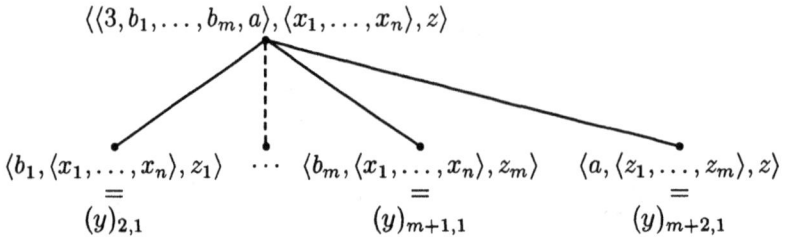

a) composition

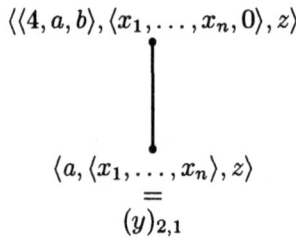

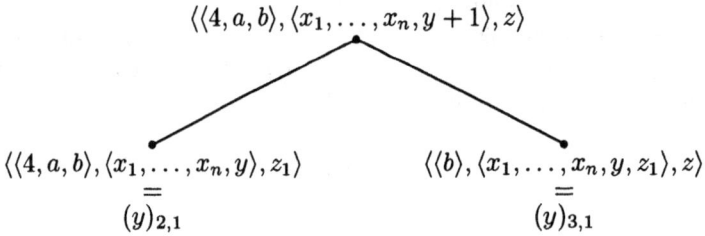

$$\langle\langle 4, a, b\rangle, \langle x_1, \ldots, x_n, 0\rangle, z\rangle$$

$$\langle a, \langle x_1, \ldots, x_n\rangle, z\rangle$$
$$=$$
$$(y)_{2,1}$$

$$\langle\langle 4, a, b\rangle, \langle x_1, \ldots, x_n, y+1\rangle, z\rangle$$

$$\langle\langle 4, a, b\rangle, \langle x_1, \ldots, x_n, y\rangle, z_1\rangle \qquad\qquad \langle\langle b\rangle, \langle x_1, \ldots, x_n, y, z_1\rangle, z\rangle$$
$$= \qquad\qquad\qquad\qquad\qquad =$$
$$(y)_{2,1} \qquad\qquad\qquad\qquad\qquad (y)_{3,1}$$

b) primitive recursion

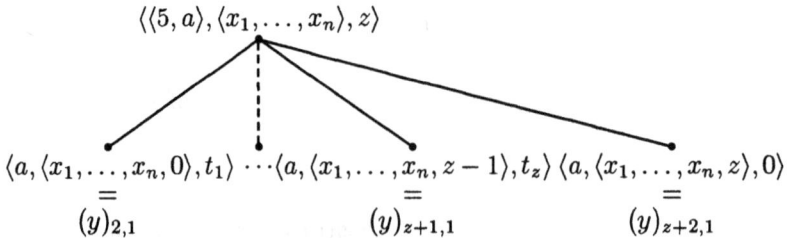

$$\langle\langle 5, a\rangle, \langle x_1, \ldots, x_n\rangle, z\rangle$$

$$\langle a, \langle x_1, \ldots, x_n, 0\rangle, t_1\rangle \quad \cdots \langle a, \langle x_1, \ldots, x_n, z-1\rangle, t_z\rangle \ \langle a, \langle x_1, \ldots, x_n, z\rangle, 0\rangle$$
$$= \qquad\qquad\qquad\qquad = \qquad\qquad\qquad =$$
$$(y)_{2,1} \qquad\qquad\qquad\qquad (y)_{z+1,1} \qquad\qquad (y)_{z+2,1}$$

c) μ-recursion

Figure I.13: Cases for the definition of $\mathcal{T}$

For the initial functions, there are three possibilities for $v = (y)_1$:

$$\langle\langle 0\rangle, \langle x\rangle, 0\rangle$$
$$\langle\langle 1\rangle, \langle x\rangle, x + 1\rangle$$
$$\langle\langle 2, n, i\rangle, \langle x_1, \ldots, x_n\rangle, x_i\rangle.$$

We then let:

$B(y) \quad \Leftrightarrow \quad ln(y) = 1 \wedge$
$\qquad \{[(y)_{1,1} = \langle 0\rangle \wedge ln((y)_{1,2}) = 1 \wedge (y)_{1,3} = 0] \vee$
$\qquad [(y)_{1,1} = \langle 1\rangle \wedge ln((y)_{1,2}) = 1 \wedge (y)_{1,3} = (y)_{1,2,1} + 1] \vee$
$\qquad [ln((y)_{1,1}) = 3 \wedge (y)_{1,1,1} = 2 \wedge (y)_{1,1,2} = ln((y)_{1,2}) \wedge$
$\qquad 1 \leq (y)_{1,1,3} \leq (y)_{1,1,2} \wedge (y)_{1,3} = ((y)_{1,2})_{(y)_{1,1,3}}]\}.$

For composition we let:

$C(y) \quad \Leftrightarrow \quad ln((y)_{1,1}) \geq 3 \wedge (y)_{1,1,1} = 3 \wedge ln(y) = ln((y)_{1,1}) \wedge$
$\qquad (\forall i)_{2 \leq i < ln(y)}[(y)_{i,1,1} = (y)_{1,1,i} \wedge (y)_{i,1,2} = (y)_{1,2}] \wedge$
$\qquad (y)_{ln(y),1,1} = (y)_{1,1,ln(y)} \wedge (y)_{ln(y),1,3} = (y)_{1,3} \wedge$
$\qquad (y)_{ln(y),1,2} = \langle (y)_{2,1,3}, \ldots, (y)_{ln(y)-1,1,3}\rangle.$

For primitive recursion, recall that there are two possible cases:

$D(y) \quad \Leftrightarrow \quad ln((y)_{1,1}) = 3 \wedge (y)_{1,1,1} = 4 \wedge$
$\qquad \{[(y)_{1,2,ln((y)_{1,2})} = 0 \wedge ln(y) = 2 \wedge (y)_{2,1,1} = (y)_{1,1,2} \wedge$
$\qquad (y)_{2,1,2} * \langle 0\rangle = (y)_{1,2} \wedge (y)_{2,1,3} = (y)_{1,3}] \vee$
$\qquad [(y)_{1,2,ln((y)_{1,2})} > 0 \wedge ln(y) = 3 \wedge$
$\qquad (y)_{2,1,1} = (y)_{1,1} \wedge ln((y)_{2,1,2}) = ln((y)_{1,2}) \wedge$
$\qquad (\forall i)_{1 \leq i < ln((y)_{1,2})}((y)_{2,1,2,i} = (y)_{1,2,i}) \wedge$
$\qquad (y)_{2,1,2,ln((y)_{1,2})} + 1 = (y)_{1,2,ln((y)_{1,2})} \wedge$
$\qquad (y)_{3,1,1} = \langle (y)_{1,1,3}\rangle \wedge (y)_{3,1,3} = (y)_{1,3} \wedge$
$\qquad (y)_{3,1,2} = (y)_{2,1,2} * \langle (y)_{2,1,3}\rangle]\}.$

For μ-recursion we let:

$E(y) \quad \Leftrightarrow \quad ln((y)_{1,1}) = 2 \wedge (y)_{1,1,1} = 5 \wedge$
$\qquad ln(y) \geq 2 \wedge (y)_{1,3} = ln(y) - 2 \wedge$
$\qquad (\forall i)_{2 \leq i \leq ln(y)}[(y)_{i,1,1} = (y)_{1,1,2} \wedge$
$\qquad (y)_{i,1,2} = (y)_{1,2} * \langle i - 2\rangle] \wedge$
$\qquad (\forall i)_{2 \leq i < ln(y)}[(y)_{i,1,3} \neq 0] \wedge (y)_{ln(y),1,3} = 0.$

These conditions take care of all possible cases, and thus we may define inductively:

$$T(y) \quad \Leftrightarrow \quad A(y) \wedge [B(y) \vee C(y) \vee D(y) \vee E(y)] \wedge$$
$$[ln(y) > 1 \rightarrow (\forall i)_{2 \leq i \leq ln(y)} T((y)_i)].$$

Then T is primitive recursive because it is defined by using only primitive recursive clauses and values (of its characteristic function) for previous arguments because, by definition of coding, $(y)_i < y$. That is, T is defined by course of value recursion, which is a primitive recursive operation by I.7.1.

5. define T_n and $\mathcal{U}$

We are now ready to conclude our work. Namely, for each $n \geq 1$ we let:

$$T_n(e, x_1, \ldots, x_n, y) \Leftrightarrow T(y) \wedge (y)_{1,1} = e \wedge (y)_{1,2} = \langle x_1, \ldots, x_n \rangle$$

and

$$\mathcal{U}(y) = (y)_{1,3}$$

These are obviously primitive recursive.

Let now f be a recursive n-ary function with index e. Since f is total, for every $x_1, \ldots, x_n$ there is a computation tree for $f(x_1, \ldots, x_n)$ relative to the computation procedure coded by e. This is formally expressed by:

$$\forall x_1 \ldots \forall x_n \exists y T_n(e, x_1, \ldots, x_n, y).$$

Moreover, from any computation tree (in particular from the one with the smallest code number) we can extract the value of the function by looking at the third component of its vertex. This is formally expressed by:

$$f(x_1, \ldots, x_n) = \mathcal{U}(\mu y T_n(e, x_1, \ldots, x_n, y)). \quad \square$$

Exercise I.7.4 *There is a recursive function enumerating the unary primitive recursive functions*, i.e. a recursive function $f(e, x)$ such that: for every e the function $\lambda x. f(e, x)$ is primitive recursive, and every primitive recursive function is equal to $\lambda x. f(e, x)$ for some e. (Peter [1935]) (Hint: let

$$f(e, x) = \begin{cases} \mathcal{U}(\mu y T_1(e, x, y)) & \text{if } e \text{ is a primitive recursive index} \\ 0 & \text{otherwise}, \end{cases}$$

where being a primitive recursive index means to define a recursive function from the initial functions by composition and primitive recursion alone, without using the μ-operator.)

Equivalence of the various approaches to recursiveness

By using the method of arithmetization we can get a cascade of results relative to the various approaches introduced in Sections 2–6. We will not go beyond sketches because we believe that, once the method is understood, the translation of these into formal proofs should not present theoretical difficulties. It is, however, a very useful exercise to try to fill in the cumbersome details of some of these sketches.

We begin by dealing with the notions of Section 2.

Proposition I.7.5 (Kreisel and Tait [1961]) *Every finitely definable function is recursive.*

Proof. Suppose f is finitely definable by $\mathcal{E}$ w.r.t. f_1. We know that if $f(\vec{x}) = z$ then $f_1(\vec{x}) = \bar{z}$ is a *logical consequence* of

$$\mathcal{E}(f_1, \ldots, f_m; \vec{\bar{z}_1}) \wedge \cdots \wedge \mathcal{E}(f_1, \ldots, f_m; \vec{\bar{z}_p}),$$

for some $\vec{z}_1, \ldots, \vec{z}_p$. By the completeness of the predicate calculus, and the fact that whenever we have a model we also have an ω-model (i.e. one with domain the integers, and with 0 and S interpreted as zero and successor), this is equivalent of saying that if $f(\vec{x}) = z$ then $f_1(\vec{x}) = \bar{z}$ is *derivable* in any complete formalization of the predicate calculus with equality, from the premises

$$\mathcal{E}(f_1, \ldots, f_m; \vec{\bar{z}_1}) \wedge \cdots \wedge \mathcal{E}(f_1, \ldots, f_m; \vec{\bar{z}_p})$$

and the axioms for the successor operation:

$$(\forall x)(S(x) \neq 0)$$
$$(\forall x)(S^{(n)}(x) \neq x) \qquad \text{(for } n > 0)$$
$$(\forall x)(\forall y)(S(x) = S(y) \rightarrow x = y)$$

(with the $f_1, \ldots, f_m$ held fixed in the derivation). By arithmetization we can define a primitive recursive predicate $\mathcal{T}_n(e, \vec{x}, y)$ (where n is the number of components of the vector $\vec{x}$) meaning:

> y codes a derivation of an equation of the form $f_1(\vec{x}) = \bar{z}$, in the predicate calculus with equality, from the axioms for successor and a finite conjunction of substitution instances of the system of equations coded by e.

Let $\mathcal{U}$ be a primitive recursive function such that

> whenever y codes a derivation, then $\mathcal{U}(y)$ gives the value of the numeral on the right-hand side of the last equation coded by y.

Then we have that f is recursive, because

$$f(\vec{x}) = \mathcal{U}(\mu y T_n(e, \vec{x}, y)). \quad \Box$$

Proposition I.7.6 (Kleene [1936]) *Every Herbrand-Gödel computable function is recursive.*

Proof. By arithmetization we can define $T_n(e, x_1, \ldots, x_n, y)$ primitive recursive, meaning:

> y codes a derivation, by means of the rules R1 and R2 and from the system of equations coded by e, of an equation of the form $f_i^n(\overline{x}_1, \ldots, \overline{x}_n) = \overline{z}$, where f_i^n is the leftmost letter in the last equation of the system coded by e.

Let $\mathcal{U}$ be a primitive recursive function such that

> if y codes a derivation then $\mathcal{U}(y)$ is the value of the numeral in the right-hand side of the last equation coded by y.

If f is Herbrand-Gödel computable from the system of equations coded by e then f is recursive, because

$$f(x_1, \ldots, x_n) = \mathcal{U}(\mu y T_n(e, x_1, \ldots, x_n, y)). \quad \Box$$

We turn now to the representability approach of Section 3.

Proposition I.7.7 (Gödel [1936], Church[1936]) *Every function weakly representable in a consistent formal system (with recursive sets of axioms and of recursive rules) is recursive.*

Proof. By arithmetization we can define $T_n(e, x_1, \ldots, x_n, y)$ primitive recursive, meaning:

> y codes a derivation of a sentence of the form $\phi(\overline{x}_1, \ldots, \overline{x}_n, \overline{z})$, where ϕ is the formula coded by e, from the axioms of the given system and by means of its rules.

Let $\mathcal{U}$ be a primitive recursive function such that

> if y codes a derivation then $\mathcal{U}(y)$ is the value of the numeral which instantiates the last variable of the last formula of the derivation coded by y.

If f is weakly representable by the formula coded by e then f is recursive, because

$$f(x_1, \ldots, x_n) = \mathcal{U}(\mu y T_n(e, x_1, \ldots, x_n, y)). \quad \Box$$

Corollary I.7.8 *For any consistent formal system extending* $\mathcal{R}$, *the following are equivalent:*

1. *f is weakly representable*

2. *f is representable*

3. *f is strongly representable.*

Proof. Indeed,

$$
\begin{aligned}
\text{recursive} &\Rightarrow \text{strongly representable} & &\text{(by I.3.6)} \\
&\Rightarrow \text{representable} & &\text{(by definition)} \\
&\Rightarrow \text{weakly representable} & &\text{(by consistency)} \\
&\Rightarrow \text{recursive} & &\text{(by the proposition).} \quad \square
\end{aligned}
$$

We now turn to the computational approaches of Sections 4 and 5.

Proposition I.7.9 (Turing [1936], [1937]) *Every Turing machine computable function is recursive.*

Proof. By arithmetization we can define $T_n(e, x_1, \ldots, x_n, y)$ primitive recursive, meaning:

> y codes a computation carried out by the Turing machine coded by e, on inputs $x_1, \ldots, x_n$.

Let $\mathcal{U}$ be a primitive recursive function such that

> if y codes a computation then $\mathcal{U}(y)$ is the value of the number written on the tape to the left of the head, in the last configuration of the computation coded by y.

If f is computed by the Turing machine coded by e then f is recursive, because

$$f(x_1, \ldots, x_n) = \mathcal{U}(\mu y T_n(e, x_1, \ldots, x_n, y)). \quad \square$$

Proposition I.7.10 (Wang [1957], Peter [1959]) *Every flowchart computable function is recursive.*

Proof. To simplify the details we can make the following conventions: any program has only variables named $X_0, X_1, \ldots$; the inputs are indicated by $X_1, \ldots, X_n$, and the output by X_0. By arithmetization we can define a primitive recursive predicate $T_n(e, x_1, \ldots, x_n, y)$, meaning:

> y codes a computation of the program coded by e when, at the beginning, the input variables are set equal to $x_1, \ldots, x_n$, and all the remaining variables are set equal to 0.

Let $\mathcal{U}$ be a primitive recursive function such that

> if y codes a computation, then $\mathcal{U}(y)$ is the value of the output variable, in the last step of the computation coded by y.

If f is computed by the program coded by e then f is recursive, because

$$f(x_1, \ldots, x_n) = \mathcal{U}(\mu y T_n(e, x_1, \ldots, x_n, y)). \quad \square$$

We finally turn to the λ-definability approach of Section 6.

Proposition I.7.11 (Church [1936], Kleene [1936b]) *Every λ-definable function is recursive.*

Proof. By arithmetization we can define $T_n(e, x_1, \ldots, x_n, y)$ primitive recursive, meaning:

> y codes a reduction to a numeral, via α or β rules, of the term coded by e, applied to $\bar{x}_1, \ldots, \bar{x}_n$.

Let $\mathcal{U}$ be a primitive recursive function such that

> if y codes a reduction then $\mathcal{U}(y)$ is the value of the numeral obtained in its last step.

If f is λ-definable by the term coded by e then f is recursive, because

$$f(x_1, \ldots, x_n) = \mathcal{U}(\mu y T_n(e, x_1, \ldots, x_n, y)). \quad \square$$

The basic result of the foundations of Recursion Theory

The results proved in this first part of the book imply that all the approaches to effective computability introduced so far are equivalent, and thus show that *the notion of recursiveness is absolute and very stable*. This is a striking fact, and to stress its importance we isolate it in a theorem of its own, which captures the essence of this chapter:

> ancor dirò, perché tu veggi pura
> la verità che là giú si confonde,
> equivocando in sí fatta lettura.[4]
> (Dante, *Paradiso*, XXIX)

Theorem I.7.12 Basic result. *The following are equivalent:*

---

[4]I shall say more, so that you may see clearly
the truth that, there below, has been confused
by teaching that may be ambiguous.

1. *f is recursive*

2. *f is finitely definable*

3. *f is Herbrand-Gödel computable*

4. *f is representable in a consistent formal system extending* $\mathcal{R}$

5. *f is Turing computable*

6. *f is flowchart (or 'while') computable*

7. *f is λ-definable.*

Proof. The direction showing that all these notions are not more extensive than recursiveness is given by the results just proved in this section.

The opposite direction, showing that all these notions are at least as extensive as recursiveness, is given by Theorems I.2.3, I.2.5, I.3.6, I.4.3, I.5.4 and I.6.5.   □

It should be noted that *the equivalence proofs among different notions of computability are effective and efficient.* Effectiveness means that for any pair of notions there is a recursive function that, given the code of a recursive function relative to one notion, produces a code of the same recursive function relative to the other notion. This is the basis of the definition of acceptable system of indices, see II.5.2. A precise statement of efficiency requires concepts introduced in Chapter VII, and will be given there. The intuitive idea is that the code of a function not only defines the function, but also shows a method to compute it, and the translation roughly preserves the computational efficiency of the methods.

I.8   Church's Thesis ⋆

In this section we discuss the assertion that every effectively computable function is recursive, by considering physical and biological computers. For the former we rely on physical theory, and try first to determine how far the particular model of determinism provided by recursiveness accounts for the general model of determinism, and then to establish the extent of determinism itself. For the latter, due to lack of theory, we pursue the synthetical and analytical approaches, by analyzing the brain structure and formulations of constructive reasoning.

Due to the generality of the discussion, we will freely quote results which either will be proved later in the book, or will not be proved at all, being outside the scope of our work. However, appropriate references will be given, whenever needed.

Introduction to Church's Thesis

The work done so far shows that the class of recursive functions is certainly a very **basic** one, since it arises in fields as varied as mathematics, logic, computer science and linguistics, with quite independent approaches (and each one interesting in its own right), that turn out to be equivalent *a posteriori* (by the Basic Result I.7.12). The generality of the method of *arithmetization*, that allows for these equivalence results, also leads us to believe that other possible approaches to the notion of computability are likely to produce notions not more extensive than recursiveness, if not outright equivalent.

The fact that many variations in the details of the various approaches do not produce changes in the defined class (see e.g. the discussion on p. 49), shows that the notion of recursiveness is very **stable**.

By Theorem I.7.12, the class of recursive functions is not sensitive to changes in the formal systems considered to represent its functions: that is, the same functions are representable in any consistent formal system having a least minimal power, independently of the system strength. And even more is true: Kreisel [1972] shows that not only in formal systems, but even in vast classes of recursive transfinite progressions of formal systems, only recursive functions are representable. Thus the notion is **absolute** in a certainly astonishing way, with few (if any) analogues among other logical notions. To quote Gödel [1946]:

> With this concept one has for the first time succeeded in giving an absolute definition of an interesting epistemological notion, i.e. one not depending on the formalism chosen. In all other cases treated previously, such as demonstrability or definability, one has been able to define them only relative to a given language, and for each individual language it is clear that the one thus obtained is not the one looked for. For the concept of computability however, although it is merely a special kind of demonstrability or definability, the situation is different.

Gödel referred to the situation as 'a kind of miracle'.

These facts point out the exceptional importance of the class of recursive functions, and have led (see the next subsection for historical notes) to propose the following as a working hypothesis:

Church's Thesis (Church [1936], Turing [1936]) *Every effectively computable function is recursive.*

The Thesis, if true, would have a great relevance as a piece of applied philosophy, since it imposes a precise, mathematical upper bound to the vague, intuitive but basic notion of **algorithm** that underlies the concept of effective computability, and that has permeated technique and mathematical experience for thousands of years. Post [1944] emphasizes that

> if general recursive functions is the formal equivalent of effective calculability, its formalization may play a role in the history of combinatorial mathematics second only to that of the formulation of the concept of natural number.

In applications, the Thesis has an *essential use in metamathematics*. Limiting the extension of the concept of algorithm allows for proofs of **absolute unsolvability**: if we prove that a function is not recursive then, by the Thesis, it is not computable by any effective means. Thus, to see that a problem is effectively unsolvable, is enough to faithfully translate it into a function, and prove that this is not recursive.

> Like the classical unsolvability proofs, these proofs are of unsolvability by means of given instruments. What is new is that in the present case these instruments, in effect, seem to be the only instruments at man's disposal. (Post [1944])

Thus *undecidability proofs rest on two conceptually different bases: a mathematical proof of recursive unsolvability (independent of the Thesis), and an appeal to the Thesis, to deduce from it absolute unsolvability.*

There is another *avoidable use of the Thesis, in Recursion Theory*. Giving an algorithm for a function amounts, by the Thesis, to showing that this function is recursive. Although theoretically not important, and in principle always avoidable (if the Thesis is true), this use is often quite convenient, since it avoids the need for producing a precise recursive definition of a function (which might be cumbersome in details). Strictly speaking, however, this use does not even require a Thesis: it is just an expression of a general preference, widespread in mathematics, for informal (more intelligible) arguments, whenever their formalization appears to be straightforward, and not particularly informative. We will do this (and have already done it) throughout the book.

The meaning of the above formulation of Church's Thesis is ambiguous in at least two respects. First of all, the statement can be taken as saying that each effectively computable *function* is extensionally equivalent to a recursive one or, more strongly, that every effective *rule* is intensionally equivalent to, say, some program for an idealized computer. Following Kreisel [1971] we will distinguish the two meanings, and refer to them, respectively, as **Thesis** and **Superthesis**. Second, and this is the crux of the matter, there are various possible meanings for the word 'effective', partly depending on one's philosophy of mathematics.

Extremal attitudes are possible. One (Church [1936]) is to take recursiveness as a precise *definition* of the otherwise vague notion of effective computability: this makes the Thesis empty. An opposite one (Kalmar [1959]) is to consider effective computability as an open concept, that can only be successively approximated: this only allows for partial verifications of the Thesis, relative to given approximations (although it would allow for a disproval of it). The popular attitude is, however, to consider the facts that various attempts to characterize the notion of effectiveness have all led to the same class of functions, and that no counterexample to the Thesis has ever been found, as conclusive arguments in favor of it, and to regard the matter as (positively) settled. Kreisel [1972] has stressed the fact that equivalence of attempts is not particularly significant (there might be systematic errors), and that only the intrinsic values of each model can be relevant (one good reason is better than many bad ones).

Church's Thesis can be analyzed from various points of view, some of which dealt with in the special issue of the *Notre Dame Journal of Formal Logic* (vol. 28, no. 4, 1987) on the subject. The task we set for ourselves in this section is to analyze some meanings of the word 'effective', and to discuss the (lack of) evidence of the Thesis for these meanings. Precisely, we consider physical and biological computers.

A **physical computer**, as described here, is a discrete physical system together with a theory for its behavior (according to which the values are under experimental control). We restrict our attention to discrete systems because we are considering discrete functions (from natural numbers to natural numbers), although continuous systems can be treated via approximations (see below). The fact that we have a theory (physical laws) to work with, is what makes the Thesis in this case less pretentious, therefore less simple-minded, than in the original intended meaning (considered afterwards): it allows us to compare an abstract model of computability with descriptions of classes of physical devices. Obviously we do not question here the validity of the world description in terms of (present day) physical laws: the relevance of our discussion will be proportional to the degree of confidence we have in it. Since Turing machines are locally deterministic devices, to ask whether any physical computer computes only recursive functions actually splits into two questions: it means first to determine how far a particular model of determinism accounts for all of it, then to establish the extent of determinism itself. Clearly the former is less problematic, and it therefore produces a more satisfactory analysis.

For the **biological computer**, we do not have yet a theory, and discussions of human computability are mostly rambling talk. We pursue both the synthetical (bottom-up) and the analytical (top-down) approaches, by analyzing the brain structure and theories of constructive reasoning, but we reach a dead end soon in both cases.

Before we plunge into our work, we would like to warn about what effective computability does *not* mean: it is not **practical (feasible) computability**. The relationship between these two notions is the distinction between Recursion

Theory and Computer Science, i.e. between ideal computers and real ones. The issue in Computer Science is not Church's Thesis (whether the class of recursive functions is broad enough), but its dual (which recursive functions are practically computable): from this the attempt to define restricted versions of recursiveness, like polynomial-time computability (considered in Chapter VIII). Actually, from a strict point of view, practical computability is not even interested in asymptotical behavior, and it will never use infinitely many values. In this respect even the attempt to restrict the class of recursive functions to other abstract models may be irrelevant.

Historical remarks

Post was working, at the beginning of the Twenties, toward a general formulation of the undecidability results he had obtained. He defined the notion of canonical systems (p. 143) as an abstraction of the notion of formal systems, and proposed (on the basis of reductions he had of known formal systems to canonical ones) the identification of the notions of a set of strings effectively generable on one side, and generable by canonical systems on the other. This is equivalent to saying, in modern terminology, that the effectively generable sets are the recursively enumerable ones, and it is thus indirectly equivalent to Church's Thesis (for partial functions). However, Post had a platonist philosophical view, and saw his proposal as something that had to be proved somehow, by a kind of

> psychological analysis of the mental processes involved in combinatorial mathematical processes.

In particular, he believed that the analysis he had at the moment was 'fundamentally weak', and thus that the proposal was not completely convincing. All this work (Post [1922]) was left unpublished, and so did not influence later developments.

At the beginning of the thirties Church formulated the λ-calculus, in a foundational attempt to develop a system of logic from the primitive notion of function (Section 6). It gradually turned out (in 1932–33) that there was a natural way to represent integers in λ-notation, and that a great number of functions were λ-definable (ultimately that all the recursive ones were, Theorem I.6.5). In 1934 Church proposed his Thesis (Church [1936]), as a mathematical *definition* of the informal concept of computability.

Meanwhile Gödel, dissatisfied with Church's approach, believed (somewhat following Hilbert [1926]) that the computable functions could all be defined by some general kind of recursion. This again turns out to be equivalent to Church's Thesis, through the Fixed-Point Theorem. Gödel [1934] even ventured to formulate the notion of Herbrand-Gödel computability (Section 2) as a test, but was not at all convinced that this concept really comprises all possible recursion. His proposal was similar to Post's: to analyze the notion of computability, aiming at the isolation of its essential features.

This was done (in a way satisfactory to Gödel) by Turing [1936], who - unaware of the work referred to above - proposed his model of an abstract computer (Section 4), and an equivalent version of Church's Thesis. Simultaneously and independently, Post [1936] attained a very similar analysis and stated that

> a fundamental discovery in the limitations of the mathematicizing power of Homo Sapiens has been made.

Gödel thought otherwise (see Davis [1965], p. 73):

> The results mentioned ... do not establish any bounds for the power of human reason, but rather for the potentiality of pure formalism in mathematics.

For more information on the history of Church's Thesis see Kleene [1981], [1981a], [1987], Davis [1982], Shanker [1987], Webb [1980] and the original papers in Davis [1965].

Computers and physics

The notion of a deterministic reality that evolves according to mathematically explicit laws is typical of **classical mechanics**. Galilei [1623], [1638] introduces the modern scientific methodology of experimenting in order to verify the results of theoretical reasoning, and stresses the importance of mathematics ('the language the book of nature is written in'). Newton [1687] achieves an informal axiomatization of mechanics, for the first time unifying large tracts of experience into a coherent picture. With him the mechanization of the world picture (see Dijksterhuis [1961] for an historical account) is accomplished: a system with k degrees of freedom needs only $2k$ parameters (positions and moments) to completely specify every value of physical quantity for the system at a given time and the evolution in time of the system state. If some of the parameters are unknown, by averaging over them in some way it is still possible to obtain statistical prevision (like in **thermodynamics**, where the hidden $2k$ parameters needed to describe a system of k molecules produce a statistical description in terms of pressure and temperature alone).

Plank's discovery in 1900 of energy packets ignited a new physics (**quantum mechanics**, see Feynman, Leighton and Sands [1963] for background), with philosophical foundations as distant from those of classical mechanics as they can be. The concept itself of reality is at stake: matter has a double appearance, as waves and as particles (Einstein, De Broglie), and its physical quantities cannot in general be simultaneously measured with absolute precision (Heisenberg). A system with k degrees of freedom is now described by a wave function $\Psi(q_1, \ldots, q_k)$, which still evolves deterministically in time (Schrödinger), but allows only statistical prevision on the values of physical quantities for the system at a given time (Born). This last fact can be variously interpreted as necessary (change at subatomic level is casual, and can only be accounted for probabilistically), accidental (as in the case of thermodynamics, hidden variables might deterministically account for change, and give quantum-mechanics states as average), or contingent (the wave function represents not only possibilities, but realities in simultaneously coexisting worlds: a measurement, forcing a possibility into an actuality, corresponds to choosing a path in the tree of all possible universes, see DeWitt and Graham [1973]).

Classical mechanics

The first aspect that we examine of Church's Thesis can be phrased as follows: the notion of recursiveness (a technical isolation of a restricted class of mechanical processes) captures the essence of mechanism. We can formulate, more precisely:

> **Thesis M (for 'mechanical') (Kreisel [1965])** *The behavior of any discrete physical system evolving according to local mechanical laws is recursive.*

This clearly implies our real interest: that any function computable by such a device is recursive as well (each output being obtained by a finite iteration of a recursive procedure applied to the input). The Thesis formulation in terms of behavior of physical systems, in addition to being more general, has the advantage of being directly suitable for analysis (since we do not need to know details on how the device computes a function).

The Turing-Post analysis of Section 4 is certainly not sufficient to prove Thesis M since, being explicitly patterned on human behavior, it sees computations as well-ordered sequences of atomic steps, and thus (at least) it does not account for parallel computations. Arguments in favor of Thesis M fall into three distinct categories, which we analyze separately.

a) A general theory of discrete, deterministic devices

The analysis (Church [1957], Kolmogorov and Uspenskii [1958], Gandy [1980]) starts from the assumptions of atomism and relativity. The former reduces the structure of matter to a finite set of basic particles of bounded dimensions, and thus justifies the theoretical possibility of dismantling a machine down to a set of basic constituents. The latter imposes an upper bound (the speed of light) on the propagation speed of causal changes, and thus justifies the theoretical possibility of reducing the causal effect produced in an instant t on a bounded region of space V, to actions produced by the region whose points are within distance $c \cdot t$ from some point of V. Of course, the assumptions do not take into account systems which are continuous, or which allow unbounded action-at-a-distance (like Newtonian gravitational systems).

Gandy's analysis shows that *the behavior is recursive, for any device with a fixed bound on the complexity of its possible configurations* (in the sense that both the levels of conceptual build-up from constituents, and the number of constituents in any structured part of any configuration, are bounded), *and fixed finite, deterministic sets of instructions for local and global action* (the former telling how to determine the effect of action on structured parts, the latter how to assemble the local effects). Moreover, the analysis is optimal in the sense that, when made precise, any relaxing of conditions becomes compatible with any behavior, and it thus provides a sufficient and necessary description of recursive behavior.

b) Numerical approximations of the local differentiable equations of classical mechanics

The work in classical mechanics, from Newton to Hamilton, has led to a description of the evolution of mechanical systems by local differentiable equations. More precisely, a conservative Hamiltonian system is defined, in local coordinates, by Hamilton's equations:

$$\dot{q}_i = \frac{\partial H}{\partial p_i} \qquad \dot{p}_i = -\frac{\partial H}{\partial q_i}$$

where $q = (q_1, \ldots, q_k)$ and $p = (p_1, \ldots, p_k)$ are the vectors of, respectively, positions and momenta of the system, k being the number of degrees of freedom of the system. Then the evolution in continuous time of the system state s (completely describing the relevant variables of the system) can be expressed by a vector differential equation of the form $\dot{s} = f(s)$. By assuming sufficient smoothness conditions on the derivative involved, and stepping from continuous to discrete time (in which the evolution of the system is sampled at regular, sufficiently small, discrete time intervals) we can linearly approximate the rate of change given by the previous equation, as

$$s(t + \Delta t) \approx s(t) + f(s(t)) \cdot \Delta t.$$

By taking Δt as the unit interval of sampling, we get

$$s(t + 1) \approx s(t) + f(s(t))$$

and this gives, *for mechanical (not necessarily discrete) systems, a recursively described system evolution* (in the form of a simultaneous recursive definition of all the relevant variables explicit in s) (Kreisel [1965]).

Note that the other, equivalent, way to describe the evolution of systems in classical mechanics, namely by global variational principles (like Maupertuis' principle of least action), does not seem to be useful for a similar analysis, because of its teleological approach (see Von Neumann [1954]).

c) Discrete models for classical mechanics

In classical mechanics discrete data (coming from experiments) are used to build continuous models, from which discrete data have to be deduced by numerical approximation methods. The step from discrete to discrete through a continuous model seems logically unbalanced: as Feynman [1982] puts it,

> It is really true, somehow, that the physical world is representable in a discretized way, and ... we are going to have to change the laws of physics.

Discrete models, studying the dynamical behavior of systems entirely in terms of (high speed) arithmetic, have been obtained for classical mechanics, including special relativity and conservative Hamiltonian theory (Greenspan [1973], [1980], [1982], La Budde [1980]). Their dynamical equations are difference (opposed to differential) equations, whose solutions are discrete functions. This approach still yields various conservation and symmetry laws of continuous mechanics, and it also has direct applications to non-linear physical behavior. Related to this, *cellular automata have been investigated as a basis for the representation of partial differential equation models in a direct computer simulation*, again avoiding indirect numerical approximation (Vichniac [1984], Toffoli [1984]). See also Ord-Smith and Stephenson [1975] for a general treatment of computer simulation of continuous systems.

To sum up the discussion above, *it is plausible that the behavior of a discrete physical system, evolving according to the local and causal laws of classical mechanics, can be simulated by a computer, and it is thus, in particular, recursive.*

Probabilistic physics

We try now to formulate Church's Thesis for abstract machines, in the most general way. We will have to account for **analog computers**, that is any physical system computing some function, by representing numerical data 'by analogy' (based on any physical, and possibly continuous, quantity, like intensity of an electrical current, or rotation angles of a watch hand). More precisely, an analogue computation is a combination of physical processes, behaving (mathematically) in the same way as some other process, which is the real object of study, but which for some reason is more manageable or better observable than it (e.g., because of difference in scale). In the extreme case, *any physical process is an analog calculation of its own behavior.*[5] We thus formulate:

> **Thesis P (for 'probabilistic') (Kreisel [1965])** *Any possible behavior of a discrete physical system (according to present day physical theory) is recursive.*

Possible behavior means a sequence of states with non-zero probability: we cannot simply talk of behavior according to present day physical theory, because this (as opposed to classical mechanics) is formulated also in terms of probability.

We collect our observations under categories parallel to those used for classical mechanics.

a) A general theory of analog machines

Nothing similar in spirit to the theory of Gandy [1980] for discrete deterministic devices has yet been developed. A first step has been undertaken by Shannon [1941], who generalized the notion of finite automaton into that of **general purpose analog computer**. This is a device consisting of electronic circuits, and a series of black-boxes (hooked up with lots of instantaneous feedback), of four elementary kinds: constant (producing any desired constant voltage), adder and multiplier (producing sum and product of the inputs), and integrator (producing, given inputs u and v, the output $\int_0^t u(s)dv(s) + C$, where C is the 'initial setting' of the integrator). Once the connections and the initial settings are made, the device is permitted to run in real time, and any voltage that can be read in the circuit (as a function of time) is an output.

A characterization of the behavior of these devices has been obtained (Shannon [1941], Pour El [1974], Lipschitz and Rubel [1987]): *a function $f(x)$ is the output of a general purpose analog computer if and only if it is differentially algebraic*, i.e. the solution of an algebraic differential equation

$$P(x, f(x), f'(x), \ldots, f^{(n)}(x)) = 0$$

where P is a polynomial (over the complex field) in all its variables. Such functions provide an extremely rich class, including almost all the special functions

---

[5] In this case, Church's Thesis amounts to saying that the universe is, or at least can be simulated by, a computer. This is reminiscent of similar tentatives to assimilate nature to the most sophisticated available machine, like the mechanical clock in the 17th Century, and the heat engine in the 19th Century, and it might soon appear to be as simplistic.

in common use (algebraic, trigonometric, Bessel functions), and with strong closure properties (see Rubel [1982], Rubel and Singer [1985]), including the existence of analogues of the universal function (Rubel [1981]). Also, despite the fact that some transcendental functions (notably, Euler's Γ and Riemann's ζ) are not exact solutions of algebraic differential equations, any continuous function can be approximated, with arbitrary preassigned accuracy, by differential algebraic functions.

This approach thus describes a wide variety of physical phenomena, but it is still only a first step toward a general theory of analog computers. Some extensions have been recently proposed, e.g. allowing black-boxes for convolution (which would add memory to the device, since convolutions involve the whole past history of their inputs and introduce time delays).

Digital simulations with arbitrary precision (which are our real concern, since we talk about functions of integral, not real or complex, variables) should be possible, by replacing the black boxes by digital approximations to them (e.g., integration can be performed by some appropriate numerical integration, say via Simpson's rule), but details have not yet been worked out.

b) Analysis of the formulation of probabilistic physical laws

The dynamics of classical physical systems with probabilistic behavior may be described by **Markov chains**, which consist of a finite set of states $\{q_1, \ldots, q_n\}$, together with a $n \times n$ stochastic matrix $P = (p_{ij})_{1 \leq i,j \leq n}$, whose interpretation is: if the system is in a given state q_i at a certain instant of time, the probability that it be in state q_j at the next instant (in a discrete time scale) is p_{ij}. A system described by a Markov chain satisfies Thesis P, and actually something more general holds: *any sequence of states with non-zero probability, in a stochastic process with infinitely many discrete states and recursive matrix of transition probabilities, is recursive* (Kreisel [1970a]). The reason is simply that such a sequence is an isolated branch of a finitely branching recursive tree (the tree of possible sequences of states), since there are only finitely many possible sequences of a given non-zero probability.

The remark just made teaches a more general lesson. Suppose we consider a **structurally stable system**, i.e. such that slight changes of the parameters in the equations describing the system behavior produce only slight changes in the behavior itself. The stability of the solutions tends to require that they be isolated in the relevant spaces, and if these spaces are recursively described, the solutions are recursive as well. Thus *it is likely that if Thesis P fails, counterexamples have to be looked for in unstable systems.*

It can be noted that it is known that some differential equations in recursive analysis (of the kind arising in the description of physical phenomena) have recursive data but no recursive solutions (see e.g. Pour El and Richards [1979],

[1981], [1983]). These results are, however, not directly relevant to Thesis P, since their data are mathematically concocted, and do not apparently arise from the description of physical phenomena (see Kreisel [1982] for a review).

c) Deterministic models of quantum mechanics

We have already noted above that quantum mechanics is not deterministic, as it stands. **Hidden-variables theories** were postulated to leave open the possibility of a deterministic description of subatomic phenomena: their existence would prove quantum mechanics observably inadequate, but at the same time quantum theory - although incomplete - could be complemented to obtain a full description of individual systems.

Impossibility proofs of the existence of hidden-variables theories had been proposed, from Von Neumann [1932] on, but with unsatisfactory features analyzed in Bell [1966]. A breakthrough was Bell [1964]: he proved that *realism and hidden variables are not only philosophically, but also theoretically incompatible with quantum theory*. He devised (in the style of Einstein, Podolski and Rosen [1935]) a simple experiment, and computed probabilistic lower bounds to the outcome predictions, assuming that well-defined states really exist, prior to their observation. This bound is greater than the one obtained by quantum theory considerations.

Recently (1981–82) the experiment has been actually carried out, and seemingly conclusive evidence provided that *the quantistic predictions are correct* (see Mermin [1985] for an elementary description, and references). This moves the incompatibility of realism with quantum theory from philosophical and theoretical ground to the experimental one, and seems to settle the matter.

At first sight it might seem impossible to simulate Bell's experiment deterministically, since the theoretical outcome predictions would clash with experimental evidence, but we should not forget that these predictions are obtained by using a particular kind of inductive inference based, in particular, on classical probability theory. Now the same inference theory is used in quantum mechanics, and this flatly produces its incompatibility with determinism. But, as Feynman [1982] points out, *there could be a problem with probability theory itself, at quantum level*: we assume that we can always do and repeat any experiment that we want, without taking into account the constrictions (stressed by quantum theory!) imposed by the fact that we are all part of the same universe, and that the universe does not remain the same. On the other hand, we do not know of any version of Bell's experiment that avoids probabilistic computations.

Besides probability, logic is the other tool used in the inductive inference of Bell's theorem, and *classical logic itself seems to be inadequate to describe phenomena at quantum level*. See e.g. Birkoff and Von Neumann [1936], where it is argued that the experimental propositions concerning a system in classical mechanics form a Boolean algebra, while (due to the fact that only compatible observations commute, and incompatible observations cannot be independently performed) they are a complemented but nondistributive lattice in quantum mechanics.

To sum up the discussion, *we have only scarce evidence in favor of Thesis P and, despite the fact that no outright refutation exists, there is plenty of room to doubt its validity.*

Computers and thought

The reduction of soul to (atomistic) physics has an old pre-Socratic tradition, centering around Leucippus and Democritus. The Socratic revolution, and the standing success of Plato and Aristotle, has led to a tradition of organismic physics that has left little space for pure mechanism until more recent times. A notable exception was Lucretius, who devoted two books of his *De Rerum Natura* to an atomistic account of the mind and its functions.

Descartes [1637] laid the foundation of the modern mechanistic world view, by trying to devitalize the human organism as much as was logically possible, in particular by including into physics (as he envisaged it) a great deal of what later came to be called psychology, but not the mind itself. He saw self-consciousness and language sophistication (in particular, the ability to see the meaning of signs and events) as the privilege and exclusive ability of an immaterial, unextended mind.

Hobbes [1655] provided the dissenting note: by relying on the apparently effective manipulation, in reasoning, of names as symbols for thoughts, and on Pascal's construction of the first calculating machine in 1645, he defended a global mechanism, and did not hesitate to obliterate the difference between mind and matter. In his extreme dedication to mechanism, Hobbes was a rather lonely figure in his day, but with the advent and the success of machines, it was inevitable that mechanism would attract more advocates. La Mettrie [1748] provided a most notorious attempt, aimed at wholesale reductionism.

Once the terms of the debate had been set up, endless arguments developed, and new life to the dispute has been provided by modern advances in the areas that supported Hobbes and Descartes, respectively. On the one hand, much current mathematical practice has been formalized (from Boole to Bourbaki), and thus indeed mechanized (and Gödel Completeness Theorem [1930] shows that, for what concerns first-order logic, the formalization is complete). Moreover, the quality of computers (from Babbage to the Fifth Generation) has improved enormously, and machines are now capable of quite sophisticated behavior themselves. On the other hand, the undecidability and incompleteness results (Section II.2) expose the limitations of the formalization and mechanization programs. Moreover, these results have sometimes been used to infer the superiority of men over machines, basically with the following arguments:

Undecidability. Church's Theorem II.2.18 shows that first-order logical reasoning is not mechanically decidable. It might thus appear that man, the bearer of this reasoning, is capable of nonmechanical behavior, and thus not a machine. It is easy to see the weak point of this argument: a mechanism is such because of local mechanical behavior, and the Completeness Theorem for Predicate Calculus (Gödel [1930]) does indeed show that classical logical reasoning can be formalized, and thus simulated by locally mechanical steps. But a mechanism does not need to have a global, mechanically predictable behavior, as discussed on p. 151.

Undefinability. Tarski's Theorem (see p. 166) shows that, for a classical formal system, truth is not representable in it. Again, it would seem that man has a notion of truth, and thus that thought has nonmechanical elements. The difficulty of this argument is two-fold: on the one hand, it is only global truth that it is not representable, while for each fixed bound of logical complexity there is a representable notion of truth (at the next level of complexity); on the other hand, not only does man not appear to have a global notion of truth: he even seems unable to handle (by direct intuition) more than four or five alternations of quantifiers, and hence local truth itself, beyond very small levels of complexity.

Incompleteness. Gödel's Theorem II.2.17 tells us that any consistent and sufficiently strong formal system is incomplete, in the sense that it does not prove some numerical sentence which we know is true. It would seem that man, being able to produce, for any machine (formal system), a task that can be solved by him but not by the machine, is not himself a machine.

The first objection to this is that the proof of the result is effective, i.e. there is a machine that, given the number of a formal system, produces the undecidable sentence (this effectiveness is actually one of the crucial features of Gödel's result, since - by showing the incompleteness of *every* sufficiently strong formal system - it points to inadequacies in the concept of formal system itself). Post [1922] has noted that such an effectiveness does not show up by chance: given an argument intended to prove that man can fool any machine, if this argument can be made sufficiently precise, then it becomes itself mechanizable, and it backfires.

Another important feature of Gödel's proof is that the undecidable sentence is shown to be true only under the hypothesis of the consistency of the system. Certainly the problem of deciding the consistency of given formal systems is not recursively solvable, but there is no hint that man has a decision procedure for such a problem (and hence for the truth of the undecidable sentence relative to a given consistent system). If it were so, *this* would be a direct refutation of mechanism.

Finally, although Gödel's Theorem does not by itself refute mechanism, it does so when combined with the assumptions that there are no number-theoretical questions undecidable for the human mind, and that the mind is somehow consistent. But this does not solve the problem, it just moves it to a different level. Also, although the second assumption seems quite reasonable, the first one is certainly more problematic and controversial, even if understood (as it should be) as saying that the human mind has no limitation regarding the problems it poses itself in the limit (given enough time and resources), and taking the words 'deciding a question' as meaning 'settling the problem', possibly by showing it to be unsolvable.

See the papers in Anderson [1964], and Hofstadter and Dennett [1981] for some discussions on the relevance of these topics for the mind-brain debate, and Popper and Eccles [1977] for a modern philosophical and neurological introduction to the latter.

The brain

Our first global approach to Church's Thesis for human thought is to look at the brain, the physical basis of intelligence. We begin by discussing:

Superthesis B (for 'brain') (Descartes [1637]) *The brain is a machine.*

It should be stressed that the statement is not to be taken as reducing the brain complexity to the roughness of present-day machines: a proof of Superthesis B would probably revolutionize the contemporary idea of machine, and precisely in this lies its interest. In particular, Turing (see p. 164) has stressed the significance of the limitation results (Section II.2) in showing how a purely deterministic model of machine cannot fully account for intelligence.

It is, however, instructive to compare the brain and the most sophisticated available machine, the computer[6] (for information on the brain see Von Neu-

---

[6]History teaches us that this should not be taken too literally: Descartes [1664] saw the brain as a complex hydraulic system, permitting the periodic flow of vital spirits from the central reservoir into the muscles; Pearson [1892] described it as a telephone exchange, consisting of fixed wires and mobile switches (a model that proved useful for an understanding of spinal reflex response); Ashby [1952] provided a cybernetic model as a collection of self-controlling systems. The computer model might look as simplistic in the not too distant future.

mann [1958], Arbib [1964], [1972], Eccles [1973], Young [1978], Kandel and Schwartz [1981]). Computers are electromagnetic devices with fixed wiring between more or less linearly connected elements, operating mostly sequentially, and at high speed. Brains are dynamical electrochemical organs with extensively branched connections, operating with massive parallel action at a slow speed and low energetic cost, continuously capable of generating new elements, and perhaps making new connections. The architectural differences are great: see e.g. Haddon and Lamola [1985] for a survey of the technical foreseeable advances regarding chip dimensions. The holistic logic employed by the brain is simply out of reach: we do not know how it concentrates on essential information and experiences it as structured.

We are thus forced to *suspend judgement on the validity of Superthesis B*, until enough might be known on these problems. If ever, since it is certainly conceivable (La Mettrie [1748], Von Neumann [1951]) that, due to the extreme complexity involved, a linguistic (mechanical) description of the cerebral functions might be simply unfeasible or uninforming: that is, the system itself could be its own most intelligible description (in the terminology of p. 151, *the brain could be a random object*).

On the positive side there are some results worth mentioning which show the *mechanical behavior of some simplified neuron nets*. As a first approximation to the great complexity of natural neuron systems, McCulloch and Pitts [1943] introduce regularity assumptions for artificial neurons: they are infallible all-or-nothing devices with fixed synaptic threshold, firing synchronously at discrete intervals (when the algebraic sum of the adjacent neurons effects reaches the threshold). The behavior of an isolated system of artificial neurons is completely characterized by the input conditions, and the system then works as an abstract machine (actually, this is simply an equivalent description, and the original one, of finite automata, see p. 52). Since the control box of a Turing machine can be regarded as a finite automaton, it can thus be seen as an **abstract brain of the Turing machine**. *We thus have two complementary analyses: Turing's analytical (top-down) approach describes the functioning of the computing device, without further analyzing the way it is actually built, while McCulloch and Pitts's synthetical (bottom-up) analysis shows how to obtain the same functioning by organizing, in a possibly very complex way, simple parts of described structure.*

Much work has been done toward a relaxation of the restrictive assumptions on artificial neuron nets. Von Neumann [1956], and Winograd and Cowan [1963] consider systems in which the **unreliability of the components** does not affect the reliability of the whole net, by transmitting the same information in a highly redundant way, along multiple parallel lines or, respectively, blocks of components. Hebb [1949] and Eccles [1953] permit variable synaptic thresholds: their systems have feedback information, by which they can somehow realize whether the outputs are conforming to the expectations. By trials and errors it is possible to gather sufficient information and determine thresholds that give the expected output, that is the systems have the **ability to learn**. This represents short-term memory by the variable states of the system, and **long-term memory** by the level of the synaptic thresholds. Hopfield [1982], [1984] has considered very general neural networks, with backward coupling (where neurons can act indirectly, through other neurons, on themselves), asynchronous firing, graded continuous response (in the form of a sigmoid input-output relation, as opposed to a 0,1-valued step function) and integrative time delays (due to capacitance). He has shown that (if the connections between neurons are symmetric) these general networks are still computational devices, since any set of inputs leads to stable states. This provides a model of **content-addressable memory**, where a stable state represents memorized information, that can be retrieved by setting the right input that would lead to it. Various other models have been proposed, see e.g. Arbib [1973], Bienestock, Fogelman Soulié and Weisbuch [1985], Selverston [1985], McClelland and Rumelhart [1986], and recent issues of *Biological Cybernetics*.

The models just discussed are all digital (based on neuron nets), and have been obviously inspired by the structure of digital computers (finite automata, in particular). But it is known that parts of the central nervous system function analogically, e.g. many neurons never fire, and are engaged in different activities (see Rakic [1975], Shepherd [1979], Roberts and Bush [1981], Crick and Asanuma [1986]). To be closer to reality, the digital model of the brain should thus be supplemented, and substituted by a hybrid one, partly digital and partly analog. A natural approach seems to be the use of the **general purpose analogue computer** of p. 110 (Rubel [1985]): neurons or neuron-circuits that perform the functions of the black-boxes have been already identified, and thus at least the basic components of the general purpose analog computer are present in the central nervous system.

It is only fair to note that (some of) the results quoted might be more relevant to the problem of whether machines can think (in the operative sense, introduced by Turing [1950], of being able to simulate aspects that we believe to be characteristic of thought) than to our discussion of Superthesis B. It is not clear whether the models proposed above really describe the brain's own solutions to problems of unreliability, learning, memorization and organization of information. However, they are certainly relevant to the following 'Promethean irreverence':

Thesis AI (for 'Artificial Intelligence') (Wiener [1948], Turing [1950]) *The mental functions can be simulated by machines.*

All work in Artificial Intelligence (pattern recognition, language reproduction, problem solving, theorem proving, game playing, learning and understanding) produces inductive evidence for Thesis AI.

Note that Thesis AI is not simply the extensional version of Superthesis B. The step from the latter to the former is not automatic: it requires **psychological materialism**, in the form (La Mettrie [1748]) of human thought being completely determined by the brain, with no intervention of an extraphysical mind. But, stepping down to the simulation level introduced by Thesis AI, we are not interested - in our discussion of Church's Thesis - in its full version, since our present concern is just mathematical thought.

Constructivism

We thus isolate our real interest in:

> **Thesis C (for 'constructive') (Kleene [1943], Beth [1947])**
> *Any constructive function is recursive.*

In view of the discussion in the last part of the previous subsection, even establishing Superthesis B would not automatically establish Thesis C. Conversely, failure of Thesis C would not disprove Superthesis B. Also, a failure of Thesis C does not disprove materialism, unless (the extensional version of) Superthesis B holds simultaneously.

Actually, establishing both Superthesis B and Thesis AI is probably as far as we can go toward a possible justification of (the above form of) materialism. It is obviously impossible to disprove the existence of mind without using Ockham razor, i.e. beyond showing its unusefulness in explaining thought activities. On the other hand, as Gödel has suggested (Wang [1974], p. 326), it might be possible to disprove mechanism by showing that there is not sufficient structure (at nerve level) to perform all tasks actually performed by man.

We turn now to a discussion of Thesis C. A first step has already been carried out by Turing and Post, with an analysis of routine computations (Section 4). This provides, at the same time, more and less of what we need: it gives an intensional argument, but it concerns only a portion of the intended meaning of 'constructive'. We can however say that in the limited context to which it applies, this analysis is conclusive. We isolate what is proved in the following:

> **Superthesis R (for 'routine') (Turing [1936], Post [1936])**
> *Any computation performed by an abstract human being working in a routine way, is isomorphic to a computation performed by a Turing machine.*

But this is still a far cry from Thesis C. We cannot rely on the analysis of mechanical reasoning given by Turing machines: *constructive and mechanical*

are apparently independent concepts. It seems that we understand (by grasp-
ing abstract objects) nonmechanical rules, and do not understand rules which
(although mechanical) are too long or too detailed.

We thus have to lean on analyses of the notion of mathematical constructive
reasoning. There are many of them, based on different approaches. Following
the detailed treatments of (and, for more information, referring to) Gödel [1958]
and Kreisel [1965], [1966], we just hint at the basic features of the most popular,
increasingly more comprehensive ones:

formalism (Frege, Russell) considers as constructive what can be done on
    physical objects (symbols), by purely combinatorial (hence mechanical)
    means (formal rules of derivation for symbols sequences).

finitism (Hilbert) accepts also what can be seen by pure intuition, provided
    only concrete (spatio-temporal) objects are used, and claims that any
    thinking process of this kind must be finite (though not necessarily me-
    chanical).

intuitionism (Brouwer) allows for whatever is mentally understandable, pos-
    sibly using also abstract objects (like higher-type objects, or generalized
    inductive definitions).

Nonconstructive reasoning enters only into **platonism**, which regards mathematical objects not as thoughts but as real objects, that the mental process
does not create, but only discovers. Their properties are thus perfectly defined
as those of physical objects, and this justifies use of, for example, tertium non
datur and actual infinity.

Formalism is directly related to Thesis C. On the one hand, the very formalistic program (of compressing mathematical knowledge into formal systems)
rests on the belief that some form of the Thesis holds (that this knowledge can
be mechanically reproduced); on the other hand, since everything computable
in formal systems is recursive (by arithmetization, see I.7.7), each success of
formalism is partial proof of the Thesis validity.

Thesis C is true when constructive is taken in the finitistic meaning as well:
indeed, a finitistically defined arithmetical function is certainly (as shown by
an analysis of computations) finitely defined by a system of equations (Section
2), and hence recursive (by I.7.12).

Thus *the whole problem of Thesis C lies in the intuitionistic meaning of constructive (law-like) function.* Since allowing for abstract objects (which are not
necessarily finitely representable) might make arithmetization of the involved
mental processes troublesome, we will concentrate our discussion on formal
systems capturing aspects of the intuitionistic (constructive) reasoning. The
fact that formal systems usually capture semantical notions of reasoning only
extensionally is not important here, since the Thesis is precisely extensional.

The technical advantage of formal systems is not only a matter of convenience (of having a syntactical, concrete version for semantical, abstract notions): it could be instrumental in outright *proving* Thesis C. For this, since we know that only recursive functions are representable in formal systems (by arithmetization), it would be enough to *find a formal system equivalent to constructive arithmetical intuition* (by Kreisel [1972], even a recursive transfinite progression of formal systems would suffice). This is certainly a delicate point: such an equivalence proof might require unfamiliar principles of evidence, and would certainly provide a better insight into the notion of constructive validity. On the other hand, failure of Thesis C would show the unfeasibility of this reductionist program. Of course, no consistent formal system can be arithmetically complete in the classical sense, by II.2.17 (and the same holds for recursive progressions of formal systems, by Feferman and Spector [1962]). But this is not relevant to the reductionist program, since we do not expect constructive intuition to be itself classically complete: the problem would be to succeed in deriving what is constructively valid, not to decide (let alone constructively) everything.[7]

The realization of this reductionist program does not appear easy, also in light of a result of Kreisel [1962], [1965], by which Thesis C implies that the set of constructively valid formulas of first-order logic is not recursively enumerable. In particular, *if Thesis C holds then there is no formal system capturing constructive logical validity*: thus, if a formal system capturing constructive arithmetic validity exists, it cannot be obtained by just extending (by means of arithmetical axioms) a logical system that can be detached from it by recursive means. We thus have the amusing situation that, in the process of searching for a complete formalization of constructive arithmetical reasoning, we might begin by a formulation of the purely logical constructive reasoning, and discover that we lose the war by overwinning a battle: if we are completely successful with the logical formalization, then Thesis C does not hold, and we are bound to fail in the arithmetical formalization. Also, and this is a situation with no analogue in classical mathematics, *constructive validity for first-order logical formulas somehow depends on what the constructive arithmetical functions are* (in particular, on their being or not all recursive). Otherwise said, first-order constructive validity is actually a second-order notion.

Short of proving Thesis C by the reductionist program, we may consider related questions, technically more manageable but, as we will see, more moderately interesting. We isolate two of them.

Given an intuitionistic formal system $\mathcal{F}$, we might see whether in $\mathcal{F}$ the recursive functions provide uniformization, in the sense of II.1.13. This is

---

[7]Note that, as Gödel himself has admitted (see Wang [1974], p. 324), it might even be possible to find, or have already found, a formal system equivalent to full, not only constructive, mathematical intuition, although of course, in this case, not provably so.

expressed in a weak form by the following rule:

> **Church's Rule CR.** *If* $\vdash_{\mathcal{F}} \forall x \exists y R(x,y)$ *then, for some recursive function* f *and all* x, $\vdash_{\mathcal{F}} R(\overline{x}, \overline{f(x)})$.

By Kreisel [1972], CR is actually equivalent to the following:

> **Constructive ∃-Rule.** $\vdash_{\mathcal{F}} \exists y \varphi(y) \Rightarrow \vdash_{\mathcal{F}} \varphi(\overline{y})$, *for some* y.

To be sure (Kreisel [1972]), there are *ad hoc* intuitionistic systems for which CR fails. But this does not automatically disprove Thesis C: it could merely be a symptom of incompleteness, since $\forall x \varphi(x, f(x))$ might hold for some recursive f, but we might not be able to prove even its numerical instances in $\mathcal{F}$. On the other hand, *no intuitionistic system is known to be inconsistent with CR*, something that *would* disprove Thesis C. Moreover, *for all current intuitionistic systems CR has actually been established*, even in the stronger form:

$$\vdash_{\mathcal{F}} \forall x \exists y R(x,y) \Rightarrow \vdash_{\mathcal{F}} \exists e \forall x \exists z [T_1(e,x,z) \wedge R(x, \mathcal{U}(z))]$$

(see e.g. Kleene [1945], Kreisel and Troelstra [1970]). This is however only a very indirect evidence in favor of Thesis C: it merely excludes the inconsistency of CR with these systems, and it thus shows that they cannot be used to *disprove* the Thesis.

We might be tempted (on the acceptable argument that constructive validity of an existential statement should exhibit explicit witnesses) to consider only those systems for which the Constructive ∃-Rule holds. But, since we know that there must be incompleteness (for any sufficiently strong arithmetical system, see II.2.17), there is no reason to expect it to show up necessarily somewhere else than in numerical instantiations of existential theorems. Only a formal system complete for constructive reasoning would automatically satisfy the Constructive ∃-Rule (but then not only CR would hold: Thesis C would indeed be true).

Another property at least formally related to Thesis C, is its formal version:

CT1 $\forall f \exists e \forall x \exists z [T_1(e,x,z) \wedge f(x) = \mathcal{U}(z)]$

CT2 $\forall x \exists y R(x,y) \rightarrow \exists e \forall x \exists z [T_1(e,x,z) \wedge R(x, \mathcal{U}(z))]$.

The former tells, via the Normal Form Theorem, that every function is recursive and is suitable for second-order systems with functional variables. The latter is the axiom of choice (extracting a function from a $\forall \exists$ form), plus the fact that every function is recursive and is also suitable for first-order systems. Of course, both forms are false in usual classical systems, and thus *CT1 and CT2 are not provable in usual intuitionistic systems* (in which the corresponding classical systems are interpretable).

The relevance of the two principles to our discussion is quite feeble. *Even if we can disprove one of them in a formal system for constructive mathematics, this would not disprove Thesis C*: it would simply mean that it is absurd that all functions can be proved recursive in the system (not that some functions are not recursive). Kripke has given a formalization of Brouwer's theory of the creative subject, and has shown that it implies the negation of CT. However, *for all current intuitionistic systems (not involving the concept of choice sequence) the consistency with CT has actually been established* (see e.g. Kleene [1945], Kreisel and Troelstra [1970]). Once again this is not evidence in favor of Thesis C, not even indirect (as it was for CR): indeed, *even a proof of CT would just show that every function we can talk about in the system is recursive* and, once again, this would be interesting only for a system complete for constructive reasoning (since these functions would then be all the constructive functions).

The reader will find more information on the philosophical analysis and (proofs of) the technical results of this subsection in Kreisel [1970], [1972], Troelstra [1973] and McCarty [1987].

To sum up, the arguments for Thesis C point out how it could be proved by a formal analysis of constructive reasoning (reductionist program), and disproved by showing - for any acceptable constructive arithmetical formal system - the inconsistency of closure under Church's Rule. Both validity and failure of Thesis C have interesting consequences for constructive mathematics. Except for these methodological remarks, we have collected only *very weak, and certainly inconclusive, evidence in favor of Thesis C, whose validity must be retained as unproved* (which is after all not surprising, since we do not even fully understand it: we still have only a partial grasp of what 'constructive' means).

Conclusion

To recapitulate our discussion, recursiveness seems to be a model of discrete, deterministic processes general enough to account for mechanical phenomena, according to classical physics. The notion certainly reaches beyond this, e.g. it takes care of probabilistic phenomena described by Markov's chains, and of a wide variety of structurally stable systems. But we have no positive results, and actually some positive doubts, for what concerns subatomic phenomena governed by quantum mechanics.

Turning to biological computers, only very rough simplifications allow us to look at the brain as a kind of machine, and we are still far from a complete theory. The analysis of human computations and reasoning produces a recursive description only under assumptions of routinnes and formal (at most finitistic) manipulation of symbols, respectively. It is an open problem whether constructive reasoning in intuitionistic sense is recursive.

Despite the weak evidence for some of them, the various theses have been proposed not out of empire-builder rashness (with the tacit ambition of convincing, short of proving, that recursiveness is somehow a universally permeating concept), but rather out of experimenters circumspection (with the manifest hope of understanding the exact limits of the notion). The validity of Church's Thesis (presently proved to some, but certainly not full, extent) is not what would *give* importance to Recursive Function Theory, although undoubtedly it *adds* to it (to the extent it holds). The notion of recursiveness has more than sufficient motivations (reviewed in the introduction to this section) to deserve a thorough mathematical study, disregarding its - certainly fascinating - connections with mechanism, neurophysiology, and constructivism. But, independently of its practical relevance, work along the line of this section has an abstract importance. To quote Kreisel [1970]:

> The principal interest is philosophical: not to confine oneself to
> what is necessary for (current) practice, but to see what is possible
> by way of theoretical analysis.

Chapter II

Basic Recursion Theory

This chapter contains the core of Recursion Theory, and introduces its basic notions, methods and results. We start, in Section 1, with an extension of the notion of recursiveness, by dropping a weak point in the various definitions of Chapter I (the request, not effectively verifiable, of totality for an algorithm). This leads to the class of **partial recursive functions** and their set-theoretical counterparts, the **recursively enumerable sets**. The elementary properties of these functions and sets are explored throughout the chapter, while a deeper structural analysis will begin in Chapter III, and continue in Volume II.

Two fundamental tools for nontrivial results are the method of **diagonalization** and the notion of **degree**. The former, one of the innovative inventions of Cantor, is an extremely helpful technique which has become, in various disguises, a pervasive element of Recursion Theory. In Section 2 we introduce the fundamentals of the method, including a codified version of it called the **Fixed-Point Theorem**. This is a powerful and somewhat mysterious result which underlies the famous undecidability results of the Thirties, also treated and discussed in Section 2.

The notion of degree is introduced in Section 3, which is devoted to **relative computability** as opposed to the absolute computability dealt with so far. We generalize computations that can be performed solely by machines, and allow the machine to stop, from time to time, and ask questions. The model still describes real computations, but the machine is not autonomous anymore, and may rely on interactions with the external world (that is, also during the computation and not only, as previously, in the input-output activity). The distinction between absolute and relative computations is the one between fully automatic and interactive (man-machine) behavior of computers, the latter being the common practice in sophisticated (not purely computational) projects, e.g. in automatic theorem provers, or in Artificial Intelligence tasks. The decision to relax the autonomy of the machine still leaves various possibilities open in terms of the amount and the structure of the interaction with the outer world. In Section 3 we deal with Turing computability, the most general and fundamental case, imposing no limitation on the help given to the machine, except for an obvious finiteness requirement. Other, more restrictive, notions of relative computability will be introduced in Chapter III.

A fundamental property of partial recursive functions is the possibility of enumerating their programs in an effective way, and thus of assigning **indices** to them, according to their place in the enumeration. Indices thus code descriptions of partial recursive functions, and can be used to refer to a function in an oblique (intensional) way. Section 4 deals with the **effective operations** that can be defined intensionally on the (partial) recursive functions (by working on their indices), and their relations with the **partial recursive functionals**, which are their extensional analogues (working directly on functions). Section 5 considers various topics connected with indices.

The results of this chapter collectively show that *the class of partial recursive functions is very comprehensive* as a result of its striking closure properties. First of all, the *universal partial function* (Theorem II.1.8) provides a descriptional closure. Second, the recursive functions are closed under *recursive diagonalization*, with a two-fold escape from contradiction: for total recursive functions there is no universal function (Theorem II.2.1), hence any recursive class of total recursive functions is not exhaustive, and diagonalization just produces another recursive function, which is not in the given class; for partial recursive functions, diagonalization simply produces particular undefined values (see p. 152). Finally, and this accounts for the name of the class, the *Recursion Theorems* II.2.10 and II.3.15 ensure closure under recursion of any kind (where 'recursion' can be taken to mean, in its greatest generality, the definition of a function in terms of itself and of known functions).

II.1 Partial Recursive Functions

We have introduced in Chapter I various independent approaches to the notion of effective computability, and the methods of Section I.7 showed them to be all equivalent. We might thus be quite satisfied, but there is still a point that seems a bit out of tune: we have been longing for a precise notion of effective computable function, and all our definitions have a strongly noneffective element in them, namely the infinitary restriction that we consider devices computing only total functions. Having a device potentially computing a function, we did not accept it as an algorithm until we had somehow recognized that it produces answers for any input: since it is possible to prove (see p. 146) that this cannot be done in general by any recursive means, the class of recursive functions seems to depend on something external to it, and it is even conceivable that it depends on the methods of proof allowed for the recognition of the totality of an algorithm.

All this might sound quite discouraging, but the final solution to the problem of characterizing effective procedures is at hand: we only have to set a missing brick, and the construction will be completed. Since it is the verification of totality that troubles us, we simply decide to drop it.

The notion of partial function

A **partial function** is simply a function that may be undefined for some (and possibly all) arguments. The set of arguments for which it is defined is called its **domain**. Of course a partial function is total on its domain, but here we give a privileged status to the set of natural numbers, and consider a function whose domain is properly included in ω as only partially defined.

The step from total to partial functions should be appreciated: it was a longstanding philosophical position that there cannot be precise logical laws for propositions about incompletely defined objects, from Aristotle (*Metaphysica*, Γ 7, 1012a, 21–24) to this century. It was probably Brouwer [1919] (see also [1927]) who first corrected this position with his work on choice sequences.

We use Greek letters to indicate partial functions, and an extended equality relation '$\simeq$', meaning that both sides are equal as partial functions (i.e. their respective values are either both undefined, or both defined and with the same value). Also, $\varphi(\vec{x}) \downarrow$ means that φ is defined (also said: it **converges**) for the arguments $\vec{x}$, while $\varphi(\vec{x}) \uparrow$ means the opposite (also said: it **diverges**). Finally, partial functions can be partially ordered by the **inclusion relation** $\subseteq$, naturally defined as:

$$\alpha \subseteq \beta \Leftrightarrow \forall x [\alpha(x) \downarrow \Rightarrow \beta(x) \simeq \alpha(x)].$$

Thus whenever α is defined so is β, and with the same value.

Partial recursive functions

We adapt definition I.1.7 to partial functions:

Definition II.1.1 (Kleene [1938]) *The class of* **partial recursive functions** *is the smallest class of functions*

1. *containing the initial functions $\mathcal{O}$, $\mathcal{S}$ and $\mathcal{I}_i^n$*

2. *closed under composition, i.e. the schema that given $\gamma_1, \ldots, \gamma_m, \psi$ produces*
$$\varphi(\vec{x}) \simeq \psi(\gamma_1(\vec{x}), \ldots, \gamma_m(\vec{x})),$$

where the left-hand side is undefined when at least one of the values of $\gamma_1, \ldots, \gamma_m, \psi$ for the given arguments is undefined

3. *closed under primitive recursion, i.e. the schema that given ψ, γ produces*

$$\varphi(\vec{x}, 0) \simeq \psi(\vec{x})$$
$$\varphi(\vec{x}, y+1) \simeq \gamma(\vec{x}, y, \varphi(\vec{x}, y))$$

4. *closed under unrestricted μ-recursion, i.e. the schema that given ψ produces*

$$\varphi(\vec{x}) \simeq \mu y[(\forall z \leq y)(\psi(\vec{x}, z) \downarrow) \wedge \psi(\vec{x}, y) \simeq 0],$$

where $\varphi(\vec{x})$ is undefined if there is no such y.

At first sight, we may think to define the μ-recursion schema as:

$$\varphi(\vec{x}) \simeq \mu y(\psi(\vec{x}, y) \simeq 0).$$

This would mean to look for the least y such that $\psi(\vec{x}, y) \simeq 0$ and, for every $z < y$,

$$\psi(\vec{x}, z) \downarrow \Rightarrow \psi(\vec{x}, z) \not\simeq 0$$

(thus allowing for $\psi(\vec{x}, z) \uparrow$), but it is unacceptable for various reasons. First, on a computational ground: to discover whether a given z has the stated property we can only compute $\psi(\vec{x}, z)$, and since the computation gives an answer only if it converges, this brings us back to the original proposal. Second, there is no recursive method to decide whether a partial recursive function converges (see II.2.7), and thus the schema just proposed would have the same flaw of the regularity condition for the (restricted) μ-recursion. In particular it would again give rise to a notion which is not self-contained. Finally, even after the facts the proposal does not work: *the partial recursive functions are not closed under the schema*

$$\varphi(\vec{x}) \simeq \mu y(\psi(\vec{x}, y) \simeq 0)$$

(Kleene [1952]). This is easy to see, using later results. Let A be an r.e. nonrecursive set (II.2.3), and define:

$$\psi(x, y) \simeq 0 \Leftrightarrow (y = 0 \wedge x \in A) \vee y = 1.$$

Then ψ is partial recursive (II.1.11), but if

$$f(x) = \mu y(\psi(x, y) \simeq 0)$$

then f (total) is not partial recursive, otherwise it would be recursive (by II.1.3), and since

$$f(x) = 0 \Leftrightarrow x \in A$$

so would be A.

The next theorem shows that *we get an equivalent definition of partial recursive function if we consider the schema*

$$\varphi(\vec{x}) \simeq \mu y R(\vec{x}, y)$$

with R recursive relation (where $\varphi(\vec{x})$ is undefined if there is no y such that $R(\vec{x}, y)$ holds). In terms of later definitions and results, the counterexample just given above shows that *the partial recursive functions are not closed under the same schema, with R recursively enumerable.*

Theorem II.1.2 Normal Form Theorem for partial recursive functions (Kleene [1938]) *There is a primitive recursive function $\mathcal{U}$ and (for each $n \geq 1$) primitive recursive predicates T_n, such that for every partial recursive function φ of n variables there is a number e (called **index** of φ) for which the following hold:*

1. $\varphi(x_1, \ldots, x_n) \downarrow \Leftrightarrow \exists y T_n(e, x_1, \ldots, x_n, y)$

2. $\varphi(x_1, \ldots, x_n) \simeq \mathcal{U}(\mu y T_n(e, x_1, \ldots, x_n, y)).$

Proof. The proof of Theorem I.7.3 shows exactly this, by using $\langle 5, a \rangle$ as the index associated to

$$\varphi(\vec{x}) \simeq \mu y[(\forall z \leq y)(\psi(\vec{x}, z) \downarrow) \wedge \psi(\vec{x}, y) \simeq 0],$$

when a is associated to ψ. Note that the computation tree in the case of μ-recursion uses indeed only the values $\psi(\vec{x}, z)$ for $z \leq y$. $\square$

Corollary II.1.3 *The recursive functions are exactly the partial recursive functions which happen to be total.*

Proof. Obviously, a recursive function is partial recursive and total. Conversely, if φ is partial recursive then, for some e,

$$\varphi(x_1, \ldots, x_n) \simeq \mathcal{U}(\mu y T_n(e, x_1, \ldots, x_n, y)).$$

If φ is total then $\forall x_1 \ldots \forall x_n (\varphi(x_1, \ldots, x_n) \downarrow)$, hence by the theorem

$$\forall x_1 \ldots \forall x_n \exists y T_n(e, x_1, \ldots, x_n, y)$$

and $T_n(e, x_1, \ldots, x_n, y)$ is regular. Then φ is recursive by I.1.7. $\square$

By referring to the corollary there can be no confusion when talking of **total recursive functions**, meaning recursive functions as in definition I.1.7, or partial recursive functions which are total.

The Normal Form Theorem says that every partial recursive function has an index, but this was true for the recursive functions as well. The advantage given by the introduction of partial functions is that we can now invert the theorem, and consider every number e as the index of the partial recursive function

$$\mathcal{U}(\mu y(\mathcal{T}_n(e, x_1, \ldots, x_n, y))),$$

since the condition of regularity for the application of the μ-operator has been dropped. This will be applied throughout the book, and we can then set up a special notation:

Definition II.1.4

1. φ_e^n (or $\{e\}^n$) is the e-th partial recursive function of n variables:

$$\varphi_e^n(\vec{x}) \simeq \{e\}^n(\vec{x}) \simeq \mathcal{U}(\mu y \mathcal{T}_n(e, \vec{x}, y))$$

2. $\varphi_{e,s}^n$ (or $\{e\}_s^n$) is the finite approximation of φ_e^n of level s:

$$\varphi_{e,s}^n(\vec{x}) \simeq \{e\}_s^n(\vec{x}) \simeq \left\{ \begin{array}{ll} \varphi_e^n(\vec{x}) & \textit{if } (\exists y < s)\mathcal{T}_n(e, \vec{x}, y)) \\ \textit{undefined} & \textit{otherwise} \end{array} \right.$$

Intuitively, $\varphi_{e,s}^n$ may be thought of as the approximation to φ_e^n obtained by considering the computation of φ_e^n, and cutting it at step s. Note that if $\varphi_{e,s}^n(\vec{x}) \downarrow$ then, by the properties of the coding functions and the fact that a computation codes everything relevant, inputs and output must be less than s.

For simplicity of notations *we will drop the indication of the number of variables when this is either not important or understood*, and just write φ_e and $\varphi_{e,s}$ in that case.

We can now state the symmetric version of the Normal Form Theorem:

Theorem II.1.5 Enumeration Theorem (Post [1922], Turing [1936], Kleene [1938]) *The sequence $\{\varphi_e^n\}_{e \in \omega}$ is a partial recursive enumeration of the n-ary partial recursive functions, in the sense that:*

1. *for each e, φ_e^n is a partial recursive function of n variables*

2. *if ψ is a partial recursive function of n variables, then there is e such that $\psi \simeq \varphi_e^n$*

3. *there is a partial recursive function φ of $n + 1$ variables such that*

$$\varphi(e, \vec{x}) \simeq \varphi_e^n(\vec{x}).$$

Proof. Everything follows from the Normal Form Theorem for partial recursive functions, and the definition of φ_e^n. It is enough to let

$$\varphi(e, \vec{x}) \simeq \mathcal{U}(\mu y T_n(e, \vec{x}, y)). \quad \square$$

The Enumeration Theorem exposes the *basic double role of numbers in Recursion Theory*: apart from its intended and natural meaning (as a number), a number also has a hidden, second-level meaning as a code of a function. This is the basis of the self-referential phenomena underlying the results of Section 2 (see, in particular, p. 165), and it also produces a natural interpretation of λ-calculus (see p. 223).

We give now two basic properties related to indices, and refer the reader to Section 5 for more results on the subject.

Proposition II.1.6 Padding Lemma. *Given one index of a partial recursive function, we can effectively generate infinitely many other indices of the same function.*

Proof. Given an index e for φ as a partial recursive function, we get infinitely many others by attaching to the description coded by e any finite number of redundant equations. $\quad \square$

If we fix a certain number of variables in a partial recursive function ψ, we still get a partial recursive function γ of the remaining variables. Moreover, given a program for ψ, we can effectively get a program for γ. The next theorem says that this can be done uniformly in the fixed variables.

Proposition II.1.7 S_n^m-Theorem (Kleene [1938]) *Given m, n there is a primitive recursive, one-one function $S_n^m(e, x_1, \ldots, x_n)$ such that*

$$\varphi_{S_n^m(e, x_1, \ldots, x_n)}(y_1, \ldots, y_m) \simeq \varphi_e(x_1, \ldots, x_n, y_1, \ldots, y_m).$$

Proof. Suppose we have a description (coded by e) of a partial recursive function $\psi(x_1, \ldots, x_n, y_1, \ldots, y_m)$. We want from it a description of the function defined as

$$\gamma(y_1, \ldots, y_m) \simeq \psi(x_1, \ldots, x_n, y_1, \ldots, y_m).$$

We might think to use the description coded by e followed by the above equation, but then γ would be ambiguously defined (depending on the values of $x_1, \ldots, x_n$ which appear in its definition). What we want instead is to define a function for each fixed value of $x_1, \ldots, x_n$. But then, instead, we must use the constant 0-ary functions $C_{x_1}, \ldots, C_{x_n}$, corresponding to these values. Thus we have to find an index of the function whose description is the one coded by e, followed by the equation

$$\gamma(y_1, \ldots, y_m) \simeq \psi(C_{x_1}, \ldots, C_{x_n}, y_1, \ldots, y_m).$$

This is now a good description of γ, depending uniformly on e and the values given to $x_1, \ldots, x_n$. Its index is thus a function $S_n^m(e, x_1, \ldots, x_m)$, and can be made primitive recursive by the method of arithmetization used in Section I.7. It is one-one by its definition and the properties of the coding functions. $\square$

The S_n^m-Theorem (also called the **Parametrization Theorem**, or the **Iteration Theorem**) looks, at first sight, innocuous, and simply appears to be stating that *data can be effectively incorporated into a program*. But we should not forget the fundamental double role of numbers in Recursion Theory: data can code programs themselves, and thus incorporating them into a program may have the effect that the program interprets them as subprograms. Thus *the S_n^m-Theorem actually embodies a notion of subcomputation and an effective version of composition.*

In a precise sense, enumeration and parametrization are inverse translations, and provide the technical tools needed to handle the basic duality of numbers: by enumeration an index can be considered as an argument, and by parametrization an argument can be considered as an index. This explains their fundamental role, analyzed in Section 5.

It should be noted that all the other approaches of Chapter I could be similarly adapted to the treatment of partial functions by dropping the totality requirements. Thus we could consider partial functions computed by Turing machines, and say that for given inputs a Turing machine computes a value if it halts in the prescribed way, and it does not otherwise (i.e. if it does not halt, or it does, but not in the prescribed way). Flowchart programs, Herbrand-Gödel computability, λ-definability, and so on are treated similarly. As it was the case for total functions, the various approaches remain equivalent for partial functions as well, with similar proofs. Thus *the class of partial recursive functions retains the absoluteness and stability of the class of recursive functions, and it has the extra quality of admitting an intrinsic definition, without reference to nonconstructive notions.*

Universal Turing machines and computers $\star$

The Enumeration Theorem admits a stronger formulation, due to the uniformities of the definition of T_n w.r.t. n.

Theorem II.1.8 Universal Partial Function (Post [1922], Turing [1936], Kleene [1938]) *There is a partial recursive function $\varphi(e, x)$, called* **universal partial function,** *which generates all the partial recursive functions of any number of variables, in the sense that for every partial recursive* *function ψ of n variables there is e such that*

$$\psi(x_1, \ldots, x_n) \simeq \varphi(e, \langle x_1, \ldots, x_n \rangle).$$

Proof. With notations as in the proof of I.7.3, let

$$\varphi(e, x) \simeq \mathcal{U}(\mu y (\mathcal{T}(y) \wedge (y)_{1,1} = e \wedge (y)_{1,2} = x))$$

and recall that, by definition,

$$\mathcal{T}_n(e, x_1, \ldots, x_n, y) \Leftrightarrow \mathcal{T}(y) \wedge (y)_{1,1} = e \wedge (y)_{1,2} = \langle x_1, \ldots, x_n \rangle.$$

Then, for every n,

$$\varphi(e, \langle x_1, \ldots, x_n \rangle) \simeq \varphi_e^n(x_1, \ldots, x_n). \quad \Box$$

Any Turing machine computing a universal partial function φ is called a **universal Turing machine**. Since φ is partial recursive, universal Turing machines exist by I.4.3. This is an indirect proof, and explicit constructions of universal Turing machines are in Turing [1936], Wang [1957a] and in many textbooks, e.g. Hermes [1965], Minsky [1967], Arbib [1969], Hopcroft and Ullman [1979]. More information on the topic is in Davis [1956], [1957], Shannon [1956], Rogers [1967] and Priese [1979].

The interest of the notion is that *a universal Turing machine is a computer in the modern sense of the word, and it works as an interpreter*, decoding the program e given to it as data (in the same form as the other inputs) and simulating it. In other words, a universal Turing machine is not a special-purpose machine: it is instead programmable in essence, and thus all-purpose. In particular, all universal Turing machines are equivalent in power, and they differ only in speed and efficiency.

Conversely, *any of the present-day automatic electronic computers (if abstracted from physical malfunctioning) is equivalent to a universal Turing machine, if it is given the possibility of having a potentially infinite memory* (that is, of always being able to add more memory units, and have access to the units already used). In addition to unlimited memory, the only necessary properties of a universal Turing machine are the abilities of performing coding and decoding operations (which enable the machine to read the instructions of a given Turing machine out of its index), and simulation. Once this level of complexity is reached, the machine can perform tasks more complicated than those for which it was directly built (actually, any possible task performable by Turing machines): the needed complication may be turned over to the software, and does not need to be built-in. Thus, *modulo a universal Turing machine, hardware and software are interchangeable.*

The realization that a machine could be universal-purpose, by being able to simulate other machines through their programs, antedates both Recursion Theory and Computer Science: it goes back to Babbage's [1837] conception of the **analytical engine**. This crucial notion appears natural nowadays, but it did not always look so. This was the case not only in Babbage's time, but even after Turing's abstract development, and well into the process of building real computers: see e.g. Hodges [1983] for an account of the resistance Turing had to face in his own computer project, against (in his words)

> the tradition of solving one's difficulties by means of much equip-
> ment rather than by thought

which meant a privilege of hardware and special-purpose machines over software.

Recursively enumerable sets

Having looked at the notion of partial recursive function, we turn now to its analogue in terms of sets and relations. For total recursive functions we had no doubts: the analogues were just those sets and relations whose characteristic functions were recursive. But since a characteristic function is always total, partial recursive characteristic functions would again give the recursive sets, by II.1.3.

Natural sets associated to partial functions are their domains.

Definition II.1.9 (Post[1922], Kleene [1936]) *An n-ary relation is* **recursively enumerable** *(abbreviated* **r.e.**) *if it is the domain of an n-ary partial recursive function.*

We indicate by $\mathcal{W}_e^n$ and $\mathcal{W}_{e,s}^n$, respectively, the domains of φ_e^n and $\varphi_{e,s}^n$.

As we have already done for functions, *we will drop the mention of the number of arguments for relations as well, when no confusion arises. Also, we will identify sets and unary relations, and thus write $x \in \mathcal{W}_e$ for $\mathcal{W}_e^1(x)$.*

From the definition we have immediately:

Theorem II.1.10 Normal Form Theorem for r.e. relations (Kleene [1936], Rosser [1936], Mostowski [1947]) *An n-ary relation P is r.e. if and only if there is an $n+1$-ary recursive relation R such that*

$$P(\vec{x}) \quad \Leftrightarrow \quad \exists y R(\vec{x}, y),$$

i.e. if and only if there is a number e (called **index** *of P) such that*

$$P(\vec{x}) \quad \Leftrightarrow \quad \mathcal{W}_e^n(\vec{x}) \quad \Leftrightarrow \quad \exists y T_n(e, \vec{x}, y).$$

Proof. If P is r.e. then P is the domain of a recursive function φ_e, i.e. P is equal to $\mathcal{W}_e$. Then, by II.1.2,

$$\mathcal{W}_e(\vec{x}) \quad \Leftrightarrow \quad \varphi_e(\vec{x})\downarrow$$
$$\Leftrightarrow \quad \exists y T_n(e, \vec{x}, y).$$

Conversely, if

$$P(\vec{x}) \quad \Leftrightarrow \quad \exists y R(\vec{x}, y)$$

with R recursive, then P is the domain of the partial recursive function

$$\varphi(\vec{x}) \quad \simeq \quad \mu y R(\vec{x}, y). \quad \Box$$

R.e. relations appear naturally and abundantly in mathematics. Consider e.g. a **diophantine equation** $p(\vec{x}, y) = q(\vec{x}, y)$, where p and q are polynomials in $\vec{x}, y$ with coefficients in the natural numbers. The set of non-negative, integral solutions of the equation, defined as:

$$y \in D \Leftrightarrow \exists \vec{x} \, [p(\vec{x}, y) = q(\vec{x}, y)]$$

is r.e., by the theorem just proved. Matiyasevitch [1970] has shown that the converse also holds, and thus *the r.e. sets are exactly the sets of non-negative[1] integral solutions of diophantine equations*. See Matiyasevitch [1972] or Davis [1973] for an exposition of this remarkable result, which improves the Normal Form Theorem, and also solves **Hilbert's Tenth Problem** (Hilbert [1900]). A discussion of its significance is in Davis, Matiyasevitch and Robinson [1976], and an easy proof of a slightly weaker version of it is in Jones and Matiyasevitch [1984].

The notion of r.e. set has been defined from that of partial recursive function, but the next result shows that the opposite approach is also possible.

Proposition II.1.11 Graph Theorem. *Let φ and f be, respectively, a partial and a total function. Then:*

1. *φ is partial recursive if and only if its graph is r.e.*

2. *f is recursive if and only if its graph is recursive.*

Proof. Recall that the graph G_φ of φ is the set so defined:

$$G_\varphi(\vec{x}, z) \Leftrightarrow \varphi(\vec{x}) \simeq z.$$

---

[1]This condition cannot be eliminated: it is known e.g. that there is no diophantine equation whose integral solutions are exactly the prime numbers.

If φ is partial recursive then $\varphi \simeq \varphi_e$, for some e. It follows that

$$G_\varphi(\vec{x}, z) \Leftrightarrow \varphi_e(\vec{x}) \simeq z$$
$$\Leftrightarrow \mathcal{U}(\mu y T_n(e, \vec{x}, y)) \simeq z$$
$$\Leftrightarrow \exists y[T_n(e, \vec{x}, y) \wedge (\forall t < y)\neg T_n(e, \vec{x}, t) \wedge \mathcal{U}(y) = z]$$

and thus (by II.1.10) G_φ is r.e. Conversely, if G_φ is r.e. then

$$G_\varphi(\vec{x}, z) \Leftrightarrow \exists y R(\vec{x}, z, y)$$

for some recursive R, again by II.1.10. Thus

$$\varphi(\vec{x}) \downarrow \Leftrightarrow \exists z \exists y R(\vec{x}, z, y).$$

By coding z and y into a single number $t = \langle z, y \rangle$, we have

$$\varphi(\vec{x}) \simeq (\mu t R(\vec{x}, (t)_1, (t)_2))_1 .$$

Intuitively, this is nothing more than a dovetailed verification of $G_\varphi(\vec{x}, z)$ for every z, until one of these verifications succeeds. The reason we cannot simply verify the z's one by one, is that we could get stuck with the first one which is not a value of $\varphi(\vec{x})$, and never get to consider the remaining possible values.

If f is recursive, so is G_f:

$$c_{G_f}(\vec{x}, z) = \begin{cases} 1 & \text{if } f(\vec{x}) = z \\ 0 & \text{otherwise.} \end{cases}$$

Conversely, let G_f be recursive. Since

$$f(\vec{x}) = \mu z G_f(\vec{x}, z),$$

and the hypothesis that f is total can be written as

$$\forall \vec{x} \exists z \, G_f(\vec{x}, z),$$

we have that f is defined by μ-recursion over a regular predicate (I.1.5), and it is then recursive. $\square$

Exercises II.1.12 Partial functions with recursive graph. a) *If φ is a partial function, then φ has a recursive graph if and only if there is a recursive R such that $\varphi(\vec{x}) \simeq \mu y R(\vec{x}, y)$.*

b) *There are partial recursive, nontotal functions with recursive graph.* (Hint: let W_e be an r.e. set different from ω, and let $\varphi(x) \simeq \mu s(x \in W_{e,s})$.)

c) *There are partial recursive functions with nonrecursive graph.* (Hint: let A be an r.e. nonrecursive set, see II.2.3, and $\varphi(x) \simeq 0 \Leftrightarrow x \in A$.)

See VIII.2.11 for a refinement of this topic.

The close relationship between partial recursive functions and r.e. relations is also indicated by the following property, defined by Lusin [1930] in a descriptive set-theoretical context.

Proposition II.1.13 Uniformization Property (Kleene [1936])

1. *If P is r.e., there is a partial recursive function φ such that*

$$\exists y P(\vec{x}, y) \;\Rightarrow\; \varphi(\vec{x}){\downarrow} \wedge P(\vec{x}, \varphi(\vec{x}))$$

2. *If P is r.e. and regular, there is a recursive function f such that*

$$\forall \vec{x} P(\vec{x}, f(\vec{x})).$$

Proof. This is similar to part 2 of the proof of II.1.11. Since P is r.e., there is R recursive such that

$$P(\vec{x}, y) \Leftrightarrow \exists z R(\vec{x}, y, z).$$

Then it is enough to let

$$\varphi(\vec{x}) \simeq (\mu t R(\vec{x}, (t)_1, (t)_2))_1\,.$$

Intuitively, φ chooses the first element y such that $P(\vec{x}, y)$ has been verified (which, as the next exercise shows, is not necessarily the first one for which P holds). If P is regular (I.1.5), then such a y always exists, and φ is then recursive. $\quad\square$

Exercise II.1.14 *If P is recursive, a uniformizing function is simply $\mu y P(\vec{x}, y)$. This does not hold in general, for P r.e.* (Uspenskii [1957]) (Hint: see the remarks after II.1.1.)

Exercises II.1.15 Choice functions for r.e. sets. a) *There is a partial recursive choice function for the r.e. sets*, i.e. a partial recursive function φ such that

$$W_e \neq \emptyset \;\Rightarrow\; \varphi(e){\downarrow} \wedge \varphi(e) \in W_e.$$

(Hint: uniformize $P(e, y) \Leftrightarrow y \in W_e$.)
    b) *There is no invariant, partial recursive choice function for the r.e. sets*, i.e. a choice function φ such that

$$W_i = W_e \neq \emptyset \;\Rightarrow\; \varphi(i){\downarrow} \wedge \varphi(e){\downarrow} \wedge \varphi(i) = \varphi(e).$$

Thus no partial recursive analogue of Hilbert's ε-operator exists for r.e. sets. (Kleene [1952]) (Hint: suppose such a function exists. Assume there exists a nonrecursive r.e. set A, see II.2.3, and let:

$$y \in W_{g(e)} \quad\Leftrightarrow\quad y = 0 \vee (y = 1 \wedge e \in A)$$
$$y \in W_{h(e)} \quad\Leftrightarrow\quad y = 1 \vee (y = 0 \wedge e \in A)$$

Then both $\varphi g(e)$ and $\varphi h(e)$ converge and

$$e \in A \Leftrightarrow \varphi g(e) = \varphi h(e),$$

i.e. A would be recursive.)

c) *There is no recursive choice function for the r.e. sets.* (Hint: let f be one such, set $\mathcal{W}_{h(e)} = \{f(e) + 1\}$ and apply the Fixed-Point Theorem II.2.10.)

Having analyzed the analogies of r.e. sets versus partial recursive functions, we turn now to the differences between r.e. and recursive sets. We first characterize both notions in terms of enumeration properties, and in so doing we account for the name 'recursively enumerable' (which suggests ranges, more than domains). The next result is very useful and will be applied repeatedly.

Theorem II.1.16 Characterization of the r.e. sets (Kleene [1936]) *The following are equivalent:*

1. *A is r.e.*

2. *A is the range of a partial recursive function φ*

3. *$A = \emptyset$ or A is the range of a recursive function f.*

Proof. We prove the result in a round robin style.

- $1 \Rightarrow 2$

  If A is r.e. then $A = \mathcal{W}_e$, for some e. Let

$$\varphi(x) \simeq x \Leftrightarrow \varphi_e(x)\downarrow.$$

  Then the domain of φ_e is equal to the range of φ, and φ is partial recursive, e.g. because $\varphi(x) \simeq x + 0 \cdot \varphi_e(x)$.

- $2 \Rightarrow 3$

  Let A be nonempty, and the range of a partial recursive function φ_e. Choose $a \in A$: we would like to set

$$f(x) = \begin{cases} z & \text{if } \varphi_e(x)\downarrow \wedge \varphi_e(x) \simeq z \\ a & \text{otherwise.} \end{cases}$$

  As it stands f is not however recursive, because we cannot decide recursively whether $\varphi(x)\downarrow$ (see II.2.7). To avoid being stuck while waiting for some undefined value to converge, we dovetail the computation of all possible values, and put them in the range of f as soon as they appear, the value a being used for the stages in which no new value appears (to keep f total). Thus the following modified version of f is recursive:

$$f(\mathcal{J}(x,s)) = \begin{cases} z & \text{if } \varphi_{e,s}(x)\downarrow \wedge \varphi_{e,s}(x) \simeq z \\ a & \text{otherwise.} \end{cases}$$

Here $\mathcal{J}$ is a recursive, onto pairing function (see e.g. p. 27), and ontoness is required to have f total.

- $3 \Rightarrow 1$

If $A = \emptyset$ then A is the domain of the completely undefined function, which is obviously partial recursive. If A is the range of a recursive f, we want a partial recursive φ with domain A, i.e.

$$\varphi(x){\downarrow} \Leftrightarrow \exists z(f(z) = x).$$

Then we can just let $\varphi(x) \simeq \mu z(f(z) = x)$. $\quad \square$

Also the recursive sets can be characterized in terms of enumerating functions.

Proposition II.1.17 *The following are equivalent:*

1. *A is recursive*

2. *$A = \emptyset$ or A is the range of a nondecreasing, recursive function f.*

Proof. If A is recursive and nonempty, let a be its smallest element, and

$$
\begin{aligned}
f(0) &= a \\
f(n+1) &= \begin{cases} n+1 & \text{if } n+1 \in A \\ f(n) & \text{otherwise.} \end{cases}
\end{aligned}
$$

Conversely, if A is finite then it is recursive. If A is infinite and the range of a nondecreasing recursive function f, to know whether $z \in A$, search for the smallest x such that $f(x) > z$ (which exists because A is infinite). Since f is nondecreasing, then

$$z \in A \Leftrightarrow z \in \{f(0), \ldots, f(x)\}. \quad \square$$

Exercises II.1.18 One-one enumerating functions. a) *An infinite r.e. set is the range of a one-one recursive function.* (Kleene [1936], Rosser [1936]) (Hint: define f by primitive recursion, looking at each stage for the next element generated in the set, which has not been previously generated.)

b) *An infinite recursive set is the range of an increasing recursive function.* (Kleene [1936]) (Hint: define f by primitive recursion, looking at each stage for the next element in the set.)

The description of a set consists of the infinitely many facts that tell, for any given element, if it is in the set or not. In general, although they always answer yes or no to a question, these facts are just a sequence of accidents, with no common pattern. The recursive sets are those for which a pattern exists, and a general procedure to give effective answers can be finitely described. *The recursively enumerable sets present a basic asymmetry between membership, that can be effectively determined by a finite amount of information, and non-membership, whose determination may instead require an infinite amount of it* (the partial test for membership of a given element being simply to recursively generate the set, and wait for that element to appear). The r.e. sets are thus somehow only 'half-recursive'.[2]

The distinction between recursive and recursively enumerable can then be traced back to the informal distinction between a decision procedure and a generating procedure, envisaged once again by Leibniz [1666], when he talked of **ars iudicandi** (checking the correctness of a proof) and **ars inveniendi** (finding a proof).

Another reason to see the r.e. sets as half-recursive is given by the next result, which also characterizes recursiveness in terms of recursive enumerability. It is sometimes called **Post's Theorem**, and it will be used repeatedly.

Theorem II.1.19 (Post [1943], Kleene [1943], Mostowski [1947]) *A set is recursive if and only if both the set and its complement are recursively enumerable.*

Proof. If A is recursive then both A and $\overline{A}$ are r.e., since e.g. functions with domain A and $\overline{A}$ are

$$\varphi(x) \simeq \begin{cases} 1 & \text{if } c_A(x) = 1 \\ \uparrow & \text{otherwise} \end{cases}$$

$$\psi(x) \simeq \begin{cases} 0 & \text{if } c_A(x) = 0 \\ \uparrow & \text{otherwise.} \end{cases}$$

Let now both A and $\overline{A}$ be r.e., and suppose they are nonempty (if one is empty the other is ω, and they are already both recursive as wanted). Then there are recursive functions f and g generating them:

$$A = \{f(0), f(1), \ldots\}$$
$$\overline{A} = \{g(0), g(1), \ldots\}.$$

The two lists are disjoint and exhaustive, and to know whether a given x is in A or in $\overline{A}$ is enough to generate them simultaneously, until x appears in one of the two lists.

---

[2]For this reason the r.e. sets are sometimes called semirecursive. We will use this term in a different context, see p. 294.

More formally, if A and $\overline{A}$ are r.e., there are recursive relations R and Q such that

$$x \in A \;\Leftrightarrow\; \exists y R(x,y)$$
$$x \in \overline{A} \;\Leftrightarrow\; \exists y Q(x,y).$$

Since

$$\forall x \exists y (R(x,y) \vee Q(x,y))$$

holds, the function

$$f(x) = \mu y (R(x,y) \vee Q(x,y))$$

is recursive, and exactly one of $R(x, f(x))$ and $Q(x, f(x))$ holds. Then A is recursive, since

$$c_A(x) = \begin{cases} 1 & \text{if } R(x, f(x)) \\ 0 & \text{otherwise} \end{cases} \qquad \square$$

Proposition II.1.20 (Post [1944]) *Every infinite r.e. set has an infinite recursive subset.*

Proof. Let A be infinite, and the range of a recursive function f. Define g recursive and increasing as:

$$
\begin{aligned}
g(0) &= f(0) \\
g(n+1) &= \text{the first element generated in } A \text{ and greater than } g(n) \\
&= f(\mu y (f(y) > g(n))).
\end{aligned}
$$

Then the range of g is recursive by II.1.17, and it is an infinite subset of A by definition. $\square$

Infinite sets which do not have infinite r.e. (equivalently, by the previous proposition, infinite recursive) subsets are called **immune**, and will be studied in Sections 6 and III.2.

Proposition II.1.21 Set-theoretical properties of r.e. sets (Post [1943], Mostowski [1947])

1. *With respect to set-theoretical inclusion, the r.e. sets form a distributive lattice with smallest and greatest element, and with the recursive sets as the only complemented elements.*

2. *The property of being r.e. is preserved under images and inverse images via partial recursive functions.*

Proof. Smallest and greatest elements are clearly $\emptyset$ and ω. The part relative to complementation follows from II.1.19. Finally, if A and B are r.e. then so are $A \cap B$ and $A \cup B$: generate A and B simultaneously, put in $A \cap B$ the elements appearing in both lists, and in $A \cup B$ those appearing in at least one list.

Given A r.e. and φ partial recursive, $\varphi(A)$ consists of all elements $\varphi(x)$, for $x \in A$. To generate $\varphi(A)$ is thus enough to generate A and, simultaneously, to dovetail the computations of $\varphi(x)$, for the various x's which are found to be in A. Similarly, $\varphi^{-1}(A)$ consists of all x such that $\varphi(x) \downarrow$ and $\varphi(x) \in A$. $\square$

Corollary II.1.22 Set-theoretical properties of recursive sets.

1. *With respect to set-theoretical inclusion, the recursive sets form a Boolean algebra.*

2. *The property of being recursive is preserved under inverse images via recursive functions.*

Proof. If A is recursive both A and $\overline{A}$ are r.e., and then so are $f^{-1}(A)$ and $f^{-1}(\overline{A})$. But $f^{-1}(\overline{A}) = \overline{f^{-1}(A)}$, and so both $f^{-1}(A)$ and its complement are r.e., and $f^{-1}(A)$ is recursive by II.1.19. $\square$

Note that *if A and f are recursive, then $f(A)$ is r.e. by the proposition above, but is not necessarily recursive.* Indeed, any nonempty r.e. set A is the range of a recursive function f, and thus the image of ω (which is recursive) via f, but not every r.e. set is recursive (see II.2.3).

A detailed study of the set-theoretical structure of both recursive and r.e. sets will be made in Chapter IX. For now we just prove an additional property, defined by Kuratowski [1936] in a descriptive set-theoretical context.

Proposition II.1.23 Reduction Property (Rosser [1936], Kleene [1950]) *The union of two r.e. sets can be reduced to the union of two disjoint r.e. sets. Precisely, given two r.e. sets A and B there are two r.e. sets $A' \subseteq A$ and $B' \subseteq B$ such that*

$$A' \cap B' = \emptyset \quad and \quad A' \cup B' = A \cup B.$$

Proof. Given A and B r.e., the only problem is to decide where to put the elements of $A \cap B$. We let speed of generation decide: if $z \in A \cap B$ and z is generated by A faster than by B, then z goes into A', otherwise it goes into B'. More formally, if

$$x \in A \quad \Leftrightarrow \quad \exists y\, R(x,y)$$
$$x \in B \quad \Leftrightarrow \quad \exists y\, Q(x,y)$$

with R and Q recursive, let

$$x \in A' \quad \Leftrightarrow \quad \exists y[R(x,y) \wedge (\forall z \leq y)\neg Q(x,z)]$$
$$x \in B' \quad \Leftrightarrow \quad \exists y[Q(x,y) \wedge (\forall z < y)\neg R(x,z)]. \quad \square$$

Exercise II.1.24 *The reduction property follows from the uniformization property.*

R.e. sets as foundation of Recursion Theory $\star$

We have derived the notion of recursive enumerability from that of partial recursive function, but we have already noted that, by II.1.11, the opposite is also possible: the partial recursive functions are those with an r.e. graph. What is needed to avoid circularities, is an independent characterization of recursive enumerability.

This has been provided by Post [1922], [1943] (incidentally, quite before the notion of recursiveness had been isolated). His formulation comes from an analysis of derivations in formal systems, and it is thus the natural conclusion of our journey of Chapter I (see p. 18). The underlying idea is that effective mathematical and, more generally, linguistical activity can be seen as a way of generating words from words (of a given language), according to rules. With considerations similar to those of Section I.4, one is quickly led to restrict attention to **canonical systems** consisting of finite alphabets (possibly with distinguished symbols), finite sets of axioms and finite sets of finitary rules (called **productions**), telling how to decompose a word and rearrange its parts (by possibly dropping some and adding others). Formally, a production has the form

$$x_0 \boxed{1} x_1 \cdots x_{n-1} \boxed{n} x_n \longrightarrow y_0 \boxed{i_1} y_1 \cdots y_{m-1} \boxed{i_m} y_m$$

where $i_j \in \{1, \ldots, n\}$, with the meaning: if a word can be decomposed in the way written on the left (by somehow filling up the boxes), then it can be transformed into the word written on the right (where the boxes on the right with a given label are supposed to contain exactly what the boxes on the left with same label do). Since decompositions of words are usually not unique, productions are not deterministic rules.

It is possible to show that *the sets of words produced by canonical systems are exactly (under suitable coding) the r.e. ones* (one direction comes by arithmetization; the other can be seen by noting that Turing machines transitions between successive configurations can be written as productions, operating on words expressing the configurations).

Post also showed that systems with **multiple-premise productions** are reducible to canonical systems, and that these are in turn reducible to **normal**

systems, with just one axiom, and with productions of the very specific form

$$x\boxed{1}\ \rightarrow\ \boxed{1}y.$$

For a proof of this and a treatment of the whole subject, see Minsky [1967].

The approach sketched above is relevant not only to computability but also (and even more naturally) to **linguistics**, and it provides a framework for studying structural descriptions of sentences, in formal approximations to natural languages. A system with productions only of the form

$$\boxed{1}x\boxed{2}\ \rightarrow\ \boxed{1}y\boxed{2}$$

(that can be written simply as $x \rightarrow y$), is called a **grammar** (Thue [1914]). The notion is sufficiently general, since the simulation of Turing machines referred to above can be naturally carried out by means of grammar productions (instructions act only locally on the tape of a Turing machine). Chomsky [1956], [1959] has introduced a hierarchy on the types of grammar productions, which has turned out to be strongly connected with machine models. Since we only prove some scattered results about this in Chapter VIII, we refer to Hopcroft and Ullman [1979] for a detailed study of the subject, and to Greibach [1981] for a broad overview and an historical account.

A programming language based on r.e. sets ⋆

Although Post's productions are intended to provide a basis for recursive enumerability, they can be used directly to define partial algorithms for partial functions on strings, by introducing restrictions that make the production process deterministic.

A finite sequence of grammar productions (some of which are singled out as final) computes a partial function on strings φ if, given any string w, the following partial algorithm produces the string $\varphi(w)$: at each step (starting from w), search for the first production in the sequence which can be applied (i.e. such that the premise of the production matches a substring of the given string), and apply the production to the leftmost possible substring to which it can be applied; then stop the process if the production is a final one, and repeat the process otherwise. This approach has been introduced by Markov [1951], [1954] and it is easily seen (Detlovs [1953], [1958]) to be equivalent to partial recursiveness (one direction follows as sketched above for Turing machines, the other by arithmetization). It is usually referred to as **Markov algorithms**.

This suggests the possibility of *using the production approach as a basis for a programming language for string operations*, which has been done with the introduction of **SNOBOL** (String Oriented Symbolic Language) by Farber,

Griswold and Polonsky [1964] (see Sammett [1969] and Wexelblat [1981] for history and references). The instructions of this language are labelled, and are either assignment statements (giving values to the variables) or replacement statements (telling to substitute the leftmost occurrence of a substring in a string by another string), the latter together with conditional jump instructions (sending to other instructions, depending on whether the given substitutions was successfully applied, or could not be applied). The language is thus unstructured (see p. 63), and it does not need any primitive operation (pattern matching being sufficient for all purposes: the needed operations can be specified each time, as production rules).

II.2 Diagonalization

In this section we introduce one of the basic methods of proof in Recursion Theory: diagonalization. We will give a number of applications, but the method will be used throughout the book either directly or in some codified way (like the unsolvability of the Halting Problem, or the Fixed-Point Theorem, both proved below).

The essence of diagonalization

Given a set S, a function $d : S \to S$ which is never the identity on S (i.e. $d(a) \neq a$ for every element a of S) and an infinite matrix of elements of S

$$
\begin{array}{cccc}
a_{0,0} & a_{0,1} & a_{0,2} & \cdots \\
a_{1,0} & a_{1,1} & a_{1,2} & \cdots \\
a_{2,0} & a_{2,1} & a_{2,2} & \cdots \\
\cdots & \cdots & \cdots & \cdots
\end{array}
$$

we get a transformed diagonal sequence of elements of S

$$d(a_{0,0}) \ \ d(a_{1,1}) \ \ d(a_{2,2}) \ \ \cdots$$

which is not equal to any row of the matrix, because it differs from the n-th row on the n-th element (by the hypothesis on d). That's all.

The ingredients of the method are two:

1. *the use of the diagonal $\{a_{n,n}\}_{n \in \omega}$*
   This was systematically done by Du Bois Reymond, in his study of orders of infinity (see Hardy [1910] for a neat exposition).

2. *the use of the switching function d*
   This crucial part was introduced by Cantor [1874] to prove his celebrated theorem that the set of subsets of ω is not denumerable.

In many applications, like the results proved or quoted from p. 165 on, the two ingredients take the form of self-reference and negation.

Recursive undecidability results

As a sample application of the method we prove a constructive analogue of Cantor's Theorem, whose original form can be stated by saying that there is no function on ω which enumerates all subsets of ω (or, equivalently, the set of characteristic functions, i.e. the set of total functions from ω to $\{0,1\}$).

Proposition II.2.1 Recursive version of Cantor's Theorem (Kleene [1936], Turing [1936]) *There is no recursive function which enumerates (at least one index of) each recursive (0,1-valued) function.*

Proof. Let f be a recursive function such that $\varphi_{f(x)}$ is total for every x, and define

$$g(x) \simeq 1 - \varphi_{f(x)}(x).$$

Then g is a 0,1-valued function, which is partial recursive by the Enumeration Theorem II.1.5, and total by the hypothesis on f. Moreover, g is different from $\varphi_{f(x)}$ (on the element x) for every x, and thus no index of g is in the range of f. $\square$

The proof just given falls under the general framework of diagonalization, by letting $a_{i,j} = \varphi_{f(i)}(j)$, and $d(a) = 1 - a$. It also implies that *the set*

$$\mathbf{Tot} = \{x : \varphi_x \text{ is total}\}$$

is not r.e. In particular it is not recursive, and thus *there is no recursive way to detect whether a number codes a total recursive function or not.*

Note that we expressed our results in terms of 0,1-valued functions, and not of sets. This is because there are many different ways to associate numbers to recursive sets, and different results hold for them (see p. 226).

Exercises II.2.2 a) A function is called **potentially recursive** (Church [1936]) if it has a total recursive extension. *There is a partial recursive function which is not potentially recursive.* (Kleene [1938]) (Hint: let $\varphi(x) \simeq 1 - \varphi_x(x)$.)

b) *There is a recursive, not primitive recursive set.* (Sudan [1927], Ackermann [1928]) (Hint: consider 0,1-valued functions, and use the function of I.7.4.)

We now prove one of the crucial results of Recursion Theory.

Theorem II.2.3 Combinatorial core of the undecidability results (Post [1922], Gödel [1931], Kleene [1936]) *There is an r.e. nonrecursive set. Explicitly, the set defined by*

$$x \in \mathcal{K} \Leftrightarrow x \in \mathcal{W}_x \Leftrightarrow \varphi_x(x){\downarrow}$$

is r.e. and nonrecursive.

Proof. By the Enumeration Theorem II.1.5, there is a partial recursive function φ such that

$$\varphi(x) \simeq \varphi_x(x).$$

Then $\mathcal{K}$ is r.e., because

$$x \in \mathcal{K} \Leftrightarrow \varphi(x){\downarrow} \, .$$

To show that $\mathcal{K}$ is not recursive we give two different proofs, based on the two equivalent definitions of $\mathcal{K}$ given above (in terms of partial recursive functions and of r.e. sets).

- If $\mathcal{K}$ were recursive, the function

$$\varphi(x) = \begin{cases} 0 & \text{if } x \in \overline{\mathcal{K}} \\ \text{undefined} & \text{otherwise} \end{cases}$$

would be partial recursive. Then, for some e, $\varphi \simeq \varphi_e$ and

$$\varphi_e(e){\downarrow} \Leftrightarrow e \in \overline{\mathcal{K}},$$

contradicting the definition of $\mathcal{K}$.

- If $\mathcal{K}$ were recursive, then $\overline{\mathcal{K}}$ would be r.e. But

$$x \in \overline{\mathcal{K}} \Leftrightarrow x \notin \mathcal{W}_x,$$

and so $\overline{\mathcal{K}}$ differs on the element x from the x-th r.e. set, and cannot itself be r.e. □

The argument to show that $\mathcal{K}$ is not recursive falls under the general framework of diagonalization, by letting

$$a_{i,j} = \begin{cases} 1 & \text{if } j \in \mathcal{W}_i \\ 0 & \text{otherwise,} \end{cases}$$

and $d(a) = 1 - a$.

$\overline{\mathcal{K}}$ obviously resembles the set used in **Russell's paradox** (Russell [1903]), namely the set of sets not belonging to themselves (see p. 81). Here $\overline{\mathcal{K}}$ is the set of numbers not belonging to the r.e. sets they code. There is no paradox here because Russell's argument simply shows that such a set is not r.e. itself.

We have used the word 'undecidability' in the theorem head, and we will use it over and over again throughout the book, interchangeably with the word 'unsolvability'. In both cases we really mean undecidability and unsolvability *by recursive means*. The reason we do not write this down explicitly is that there are good reasons to suspect that in fact something much stronger is involved here, namely *absolute* undecidability and unsolvability. The step from recursive to absolute unsolvability requires an appeal to Church's Thesis (see Section I.8, and p. 103 in particular).

We now strengthen the theorem just proved. The starting point is the fact that A is recursive if and only if both A and $\overline{A}$ are r.e. (II.1.19). This suggests the possibility of extending the theory of r.e. sets to pairs of disjoint r.e. sets. The next property is a recursive version of one defined by Lusin, in a descriptive set-theoretical context.

Definition II.2.4 (Kleene [1950], Trakhtenbrot [1953]) *Two disjoint sets A and B are called:*

1. **recursively separable** *if there is a recursive set C such that $A \subseteq C$ and $B \subseteq \overline{C}$*

2. **recursively inseparable** *if they are not recursively separable.*

Clearly, A is recursive if and only if A and $\overline{A}$ are recursively separable. The existence of a disjoint pair of recursively inseparable r.e. sets is thus a stronger result than the simple existence of r.e. nonrecursive sets.

Theorem II.2.5 (Rosser [1936], Kleene [1950], Novikov, Trakhten-brot [1953]) *There are two disjoint, recursively inseparable r.e. sets.*

Proof. We give two different proofs, which generalize the two of Theorem II.2.3.

- Define two disjoint r.e. sets as

$$x \in A \quad \Leftrightarrow \quad \varphi_x(x) \simeq 0$$
$$x \in B \quad \Leftrightarrow \quad \varphi_x(x) \simeq 1.$$

Suppose there is a recursive set C such that $A \subseteq C$ and $B \subseteq \overline{C}$. Then the function

$$f(x) = \begin{cases} 1 & \text{if } x \in C \\ 0 & \text{otherwise} \end{cases}$$

is recursive. If e is an index for it, we get a contradiction:

$$\varphi_e(e) \simeq 0 \Rightarrow e \in A \Rightarrow e \in C \Rightarrow \varphi_e(e) \simeq f(e) \simeq 1$$
$$\varphi_e(e) \simeq 1 \Rightarrow e \in B \Rightarrow e \in \overline{C} \Rightarrow \varphi_e(e) \simeq f(e) \simeq 0.$$

• Define two r.e. sets as

$$x \in A \iff x \in \mathcal{W}_{(x)_1}$$
$$x \in B \iff x \in \mathcal{W}_{(x)_2}.$$

They are not necessarily disjoint, but $A - B$ and $B - A$ are recursively inseparable. Indeed, suppose that $A - B \subseteq C$ and $B - A \subseteq \overline{C}$, for some recursive set C. Let $C = \mathcal{W}_a$ and $\overline{C} = \mathcal{W}_b$ (since both C and $\overline{C}$ are r.e.), and set $x = \langle b, a \rangle$. Then $x \notin A \cap B$, because

$$x \in A \iff x \in \mathcal{W}_b \iff x \in \overline{C}$$
$$x \in B \iff x \in \mathcal{W}_a \iff x \in C.$$

Moreover,

$$x \in A \Rightarrow x \in \overline{C} \quad \text{(contradicting } A - B \subseteq C)$$
$$x \in B \Rightarrow x \in C \quad \text{(contradicting } B - A \subseteq \overline{C}),$$

and so C cannot exist.

The only trouble is that $A - B$ and $B - A$ are not necessarily r.e., but clearly any two disjoint r.e. supersets of them will still be recursively inseparable. Then it is enough to reduce (II.1.23) A and B, to get a pair of disjoint r.e. sets which extend $A - B$ and $B - A$, and which are thus recursively inseparable. □

Exercises II.2.6 a) *Another proof of the existence of recursively inseparable r.e. sets* can be obtained by first getting an enumeration $\{(A_n, B_n)\}_{n \in \omega}$ of the disjoint pairs of r.e. sets, and then letting

$$x \in A \iff x \in A_x$$
$$x \in B \iff x \in B_x.$$

(Hint: for the first part, consider a double enumeration of the r.e. sets, and reduce each pair uniformly. Then prove that there cannot be a pair (A_a, B_a) such that $A \subseteq B_a$ and $B \subseteq A_a$.)

b) *Any two disjoint co-r.e. sets A and B are recursively separable* (Sierpinski [1924], Laventrieff [1925]). (Hint: reduce $\overline{A}$ and $\overline{B}$, and show that the reduced sets are complementary, and hence recursive, r.e. sets.)

Limitations of mechanisms ⋆

A slight reformulation of Theorem II.2.3 is the following, which rules out the existence of a recursive procedure to decide whether a partial recursive function converges for given arguments. The name comes from its original formulation, which was in terms of Turing machines, and in that setting it shows that there is no Turing machine that decides whether a universal Turing machine halts or not on given arguments.

Theorem II.2.7 Unsolvability of the Halting Problem (Turing [1936])
The set defined by

$$\langle x, e \rangle \in \mathcal{K}_0 \Leftrightarrow x \in \mathcal{W}_e \Leftrightarrow \varphi_e(x){\downarrow}$$

is r.e. and nonrecursive.

Proof. $\mathcal{K}_0$ is shown to be r.e. as in II.2.3, by the Enumeration Theorem. And if $\mathcal{K}_0$ were recursive so would be $\mathcal{K}$, because

$$x \in \mathcal{K} \Leftrightarrow \langle x, x \rangle \in \mathcal{K}_0. \quad \square$$

Actually, the unsolvability of the Halting Problem is just the tip of the iceberg. To measure the complexity of a problem about recursive functions, we introduce the following notion.

Definition II.2.8 *A set A is the* **index set** *of a class $\mathcal{A}$ of partial recursive functions if*

$$A = \{x : \varphi_x \in \mathcal{A}\}.$$

If A is the index set of $\mathcal{A}$, we write $A = \theta\mathcal{A}$.

Index sets contain all possible programs that compute functions belonging to a given class, and are useful in classifying the complexity of such classes. In particular, a class of partial recursive functions is called **completely recursive** if its index set is recursive (Dekker [1953a], Rice [1953]). Completely recursive classes of partial recursive functions correspond to (recursively) solvable problems about them, and are characterized by the next result.

Theorem II.2.9 Rice's Theorem (Rice [1953]) *A class of partial recursive functions $\mathcal{A}$ is completely recursive if and only if it is trivial, i.e. either empty or containing all partial recursive functions.*

Proof. If $\mathcal{A} = \emptyset$ then its index set is $\emptyset$, and if $\mathcal{A}$ contains all partial recursive functions then its index set is ω. Suppose then that $\mathcal{A}$ is nontrivial: there are a, b such that $\varphi_a \in \mathcal{A}$ and $\varphi_b \notin \mathcal{A}$. If the completely undefined function is not in $\mathcal{A}$, let f be a recursive function such that

$$\varphi_{f(x)} = \begin{cases} \varphi_a & \text{if } x \in \mathcal{K} \\ \text{undefined} & \text{otherwise.} \end{cases}$$

Then

$$x \in \mathcal{K} \Leftrightarrow \varphi_{f(x)} \in \mathcal{A} \Leftrightarrow f(x) \in A,$$

where A is the index set of $\mathcal{A}$. Then A is not recursive, otherwise so would be $\mathcal{K}$. The case of the completely undefined function being in $\mathcal{A}$ is treated similarly, this time using φ_b, and showing that $\overline{A}$ is not recursive.   □

Rice's Theorem is quite powerful, since it incorporates a number of undecidability results. Its content is that *any nontrivial property of partial recursive functions is undecidable*. Thus, undecidability proofs of given properties are reduced to proofs of nontriviality, which are usually immediate. E.g., the following problems for partial recursive functions are undecidable: being the completely undefined function, being defined for a given fixed argument, being total, having a given number as value, being onto, being equal to a given partial recursive function, and so on.

A **mechanism** is a device with a predictable *local* behavior, in the sense that each move is governed by a mechanical rule. The unsolvability results just proved show that *the behavior of a mechanism does not need to be globally predictable*, and that a purely mechanical analysis of mechanisms is bound to fail.

This has consequences for the possible **description of mechanisms** (Von Neumann [1951], [1966]). These are of two kinds: purely descriptive (telling how the device is made, out of its constituent parts), and operational (telling how it behaves in given situations). The two descriptions do not need to be at the same logical level: the former is always a *number* (the index of the machine), but the latter is a number only if the mechanism has a sufficiently simple behavior (describable by a recursive function, and thus again by a number). This is however not necessary: in general the behavior is not recursive, and then a *function* is needed to describe it for every possible input. Thus *for sufficiently complicated mechanisms, the device itself is its own best (logically simplest) description*, and it might be impossible to effectively say substantially more about it than how it is made.

To measure the complexity of (codes of) finite objects, we can introduce the so called **Kolmogorov complexity** (Kolmogorov [1963], [1965], Solomonov [1964], Chaitin [1966]):

$$K(x) \overset{\text{def}}{=} \mu e(\varphi_e(0) \simeq x),$$

whose intuitive meaning is to pick up the smallest description of the number x in terms of programs printing it (on a fixed input, like 0), and thus to measure the quantity of information carried by x. A number x and the object coded by it are called **random** (Church [1940]) if x is its shortest description, i.e. if $x \leq K(x)$. Randomness of an object can be determined by reasons antithetical in nature, but indistinguishable: extreme structural complexity and chaos. Thus *a sufficiently complicated mechanism is a random object*, and II.2.7 can be taken to mean that a universal Turing machine is a random object. See p. 261 for a discussion of random numbers and their properties.

The confusion between the two meanings of predictability is somewhat widespread and harmful. For example, scientific theories describing local mechanical behavior of biological (Darwin [1859]), historical (Marx and Engels [1848]) and psychological (Freud [1917]) evolution, are often rejected or opposed on social grounds, on the false belief that the local mechanisms explained by them might imply global predictability (i.e. a forecast of the final outcome of evolution), something which might be felt as antihumanistic.

Fixed-Point Theorem

We try now to generalize the proof of II.2.1 and, given a recursive function f, we try to get a partial recursive function ψ which is not in the set $\{\varphi_{f(x)}\}_{x \in \omega}$. Of course we do not suppose anymore that the $\varphi_{f(x)}$ be always total, since we have already disposed of this case. The natural idea is to diagonalize as in II.2.1, and let

$$\psi(x) \simeq 1 - \varphi_{f(x)}(x).$$

The trouble here is that ψ, although partial recursive, is not necessarily different from $\varphi_{f(x)}$, since this might be undefined on x, and then so would be ψ too. Thus *the notion of partial recursive function seems to have a built-in defense against diagonalization*. That it is indeed so is the content of the next theorem, one of the most tricky and useful applications of the diagonal method.

Io sentiva osannar di coro in coro
al punto fisso che li tiene alli ubi[3]
(Dante, *Paradiso*, XXVIII)

Theorem II.2.10 Fixed-Point Theorem (Kleene [1938]) *Given a recursive function f, there is an e such that e and $f(e)$ compute the same function, i.e.*

$$\varphi_e \simeq \varphi_{f(e)} \quad \text{and hence} \quad \mathcal{W}_e = \mathcal{W}_{f(e)}.$$

Proof. We give two different proofs.

1. We start with an informal argument. Since f can be thought of as a program transformation, it would be enough to take a program as follows:

---

[3]I heard 'Hosanna' sung, from choir to choir,
to that fixed point that holds each to his place.

transform this very program according to f, and then apply the result to x.

If e codes this program then, by definition, $\varphi_e \simeq \varphi_{f(e)}$. But this program is self-referential, and thus not well-formed. We have to unravel the self-reference, which we will achieve by two successive diagonalizations.

Since we are searching for a code e of the program written above, we know that e must also code the program:

transform the program with number e according to f, and then apply the result to x.

First of all we compute a number coding the program just written. This number will depend on e, and it will thus be $h(e)$, for some recursive function h. If $h \simeq \varphi_a$, this number will be $\varphi_a(e)$. We now know that if such a program exists, it must have a number of the form $\varphi_a(e)$, for some a and e, and this number must code the program:

transform the program with number $\varphi_a(e)$ according to f, and then apply the result to x.

Now this program depends on two parameters a and e, and if we compute a number coding it we are going to add a new one, and so on. If we want to avoid an infinite regress, we have to stop adding parameters. We thus try to find a program with code number of the form $\varphi_e(e)$ (**first diagonalization**), since this depends on just one parameter, and it has the right form. In other words, we consider the program:

transform the program with number $\varphi_e(e)$ according to f, and then apply the result to x.

Now there is a recursive function φ_b giving the code of this program depending on e, and the program has thus number $\varphi_b(e)$.

We now just have to let $e = b$ (**second diagonalization**), since then $\varphi_b(b)$ codes the program:

transform the program with number $\varphi_b(b)$ according to f, and then apply it to x.

Apart from the motivation, we can extract from the argument just given a formal proof. Let b be an index of the recursive function defined as:

$$\varphi_{\varphi_b(e)} \simeq \varphi_{f(\varphi_e(e))}$$

(i.e. of a function giving a code of the program coded by $f(\varphi_e(e))$, as a function of e). Then, for $e = b$, we have

$$\varphi_{\varphi_b(b)} \simeq \varphi_{f(\varphi_b(b))}.$$

Thus $\varphi_b(b)$ is a fixed-point of f.

2. (Owings [1973]) Referring to the general framework for diagonalization given at the beginning of the section, note that the diagonal method can also be expressed in a contrapositive form: given a function $d : S \to S$ and a matrix $\{a_{i,j}\}_{i,j \in \omega}$ of elements of S, if the transformed diagonal sequence is a row of the matrix, then d has a fixed-point, i.e. there is $a \in S$ such that $d(a) = a$.

Let $d(\varphi_e) \simeq \varphi_{f(e)}$: we want a fixed-point for d, that is an e such that $d(\varphi_e) \simeq \varphi_e$. Let then S be the set of partial recursive functions. Since $f \simeq \varphi_a$ for some a, the values of d are of the form $\varphi_{\varphi_a(e)}$. Then consider the matrix

$$a_{i,j} = \varphi_{\varphi_i(j)}$$

where, if $\varphi_i(j)$ diverges, $a_{i,j}$ is the completely undefined function. The transformed diagonal sequence is

$$\varphi_{f(\varphi_0(0))} \quad \varphi_{f(\varphi_1(1))} \quad \cdots$$

and this is a recursive sequence of partial recursive functions, and thus a row of the matrix. Then a fixed-point for d exists.

If we also want to know exactly what the fixed-point is, just note that it must be the element of the transformed sequence that lies on the diagonal. So let g be a recursive function (which exists by the Enumeration Theorem and the S_n^m-Theorem) such that

$$\varphi_{g(e)} \simeq \varphi_{f(\varphi_e(e))}.$$

If b is any index of g, then $g(b) = \varphi_b(b)$ is a fixed-point for f, since

$$\varphi_{g(b)} \simeq \varphi_{f(\varphi_b(b))} \simeq \varphi_{f(g(b))}. \quad \square$$

Some remarks might be worthwhile. First, if $\varphi_{g(e)} \simeq \varphi_{f(\varphi_e(e))}$ then $g(e)$ is not $f(\varphi_e(e))$: indeed $g(e)$ is always defined, as an index of $\varphi_{f(\varphi_e(e))}$, while $f(\varphi_e(e))$ is undefined when $\varphi_e(e)$ is, in which case $g(e)$ is an index of the completely undefined function.

Second, f does not need to be extensional, i.e. it does not need to induce a map of functions such that

$$\varphi_e \simeq \varphi_i \implies \varphi_{f(e)} \simeq \varphi_{f(i)}.$$

In particular, in the second proof of the theorem the diagonal function d need not be a function on the set $\{\varphi_{\varphi_i(j)}\}_{i,j \in \omega}$ as a set of partial recursive functions, but this does not affect the proof.

Third, *the proofs of the Fixed-Point Theorem given above are uniform and constructive*: they not only tell that a fixed-point exist, but they also explicitly produce it, in a way depending only on (an index of) the given function f. Thus there actually exists a recursive function h such that

$$\varphi_a \text{ total} \Rightarrow \varphi_{h(a)} \simeq \varphi_{\varphi_a(h(a))}.$$

Finally, *the constructions of fixed-points in the proofs given above are nothing else than a version of the fixed-point operator $\mathcal{Y}$ in λ-calculus* (see I.6.1), with the complications produced by the fact that we have to distinguish between numbers as arguments, and numbers as codes of functions. Specifically,

$$
\begin{array}{lll}
\varphi_e(e) & \text{corresponds to} & xx \\
\varphi_{f(\varphi_e(e))} & \text{corresponds to} & y(xx) \\
\varphi_b(e) & \text{corresponds to} & \lambda x.\, y(xx) \\
\varphi_b(b) & \text{corresponds to} & (\lambda x.\, y(xx))(\lambda x.\, y(xx)).
\end{array}
$$

Here the two diagonalizations are explicit, in the form of terms applied to themselves.

Exercises II.2.11 a) **Fixed-Point Theorem with parameters.** *If f is a recursive $n+1$-ary function, there is a recursive n-ary function h such that*

$$\varphi_{f(x_1,\ldots,x_n,h(x_1,\ldots,x_n))} \simeq \varphi_{h(x_1,\ldots,x_n)}.$$

Moreover, h may be taken to be one-one. (Hint: for the last part recall, from II.1.7, that the S_n^m functions can be taken to be one-one.)

b) **Double Fixed-Point Theorem.** *If f, g are recursive functions of two variables, there are a and b such that*

$$\varphi_a \simeq \varphi_{f(a,b)} \quad \text{and} \quad \varphi_b \simeq \varphi_{g(a,b)}.$$

(Muchnik [1958a], Smullyan [1961]). (Hint: first get h such that $\varphi_{h(x)} \simeq \varphi_{f(h(x),x)}$, by part a). Then get b such that $\varphi_b \simeq \varphi_{g(h(b),b)}$, and let $a = h(b)$.) Note that it is in general impossible to find a such that $\varphi_a \simeq \varphi_{f(a)} \simeq \varphi_{g(a)}$, since f and g could be constant functions giving indices of two different functions.

Exercises II.2.12 Sets of fixed-points. a) *No recursive function has only finitely many fixed-points.* (Rogers [1967]) (Hint: if f has only a finite set A of fixed-points, let ψ be a partial recursive function different from all those whose index is in A. Then the recursive function g such that

$$
\varphi_{g(x)} \simeq
\begin{cases}
\psi & \text{if } x \in A \\
\varphi_{f(x)} & \text{otherwise}
\end{cases}
$$

would have no fixed-point.)

b) *There are nontrivial recursive sets which are sets of fixed-points of some recursive function.* (Hint: let

$$\varphi_{f(x)} \simeq \begin{cases} \varphi_x & \text{if } x \in R \\ \text{undefined} & \text{otherwise,} \end{cases}$$

with $\overline{R}$ a recursive set not containing any index for the completely undefined function.)

c) *There are nontrivial recursive sets which are not sets of fixed-points of any recursive function.* Actually, a recursive set is not a set of fixed-points if its complement is. (Shore) (Hint: if $R, \overline{R}$ are sets of fixed-points for f, g, and

$$\varphi_{h(x)} \simeq \begin{cases} \varphi_{g(x)} & \text{if } x \in R \\ \varphi_{f(x)} & \text{otherwise,} \end{cases}$$

then h has no fixed-point.)

d) *There are nonrecursive sets of fixed-points of recursive functions*, e.g. the set of indices of the constant function 0. In a sense this is a worst-case example, since (with terminology to be introduced later, see IV.1.6) it is Π_2^0-complete, and a set of fixed-points must be Π_2^0.

We now give an equivalent form of the Fixed-Point Theorem, with Kleene's original proof. The reason for the name is purely contingent, namely the order of presentation in Kleene's classical book [1952]. The First Recursion Theorem will be given in II.3.15.

Theorem II.2.13 Second Recursion Theorem (Kleene [1938]) *If ψ is a partial recursive function, there is an index e such that*

$$\varphi_e(x) \simeq \psi(e, x).$$

Proof. Fix any recursive function h. Since the function $\psi(h(z), x)$ is partial recursive, it has an index a (depending on h). By the S_n^m-Theorem,

$$\psi(h(z), x) \simeq \varphi_{S_1^1(a,z)}(x).$$

In particular, this holds for $h(z) = S_1^1(z, z)$, for the appropriate a:

$$\psi(S_1^1(z, z), x) \simeq \varphi_{S_1^1(a,z)}(x).$$

By letting $e = S_1^1(a, a)$ we have

$$\psi(e, x) \simeq \psi(S_1^1(a, a), x) \simeq \varphi_{S_1^1(a,a)}(x) \simeq \varphi_e(x). \quad \square$$

Exercise II.2.14 *The Fixed-Point Theorem and the Second Recursion Theorem are equivalent.* (Hint: in one direction use the S_n^m-Theorem, in the other the Enumeration Theorem.)

Although the results are equivalent, *the proofs of the Fixed-Point and the Second Recursion Theorems require different tools*: the S_n^m-Theorem is used in both, but the first also uses the Enumeration Theorem. Thus versions of the Second Recursion Theorem and the Fixed-Point Theorem may respectively hold and fail for classes of functions having the S_n^m, but not the enumeration property (such as the classes studied in Chapter VIII).

The Fixed-Point Theorem, in any of its forms, and the extensions considered in the exercises, assure that *it is possible to define (any finite number of) partial recursive functions in a (simultaneous) self-referential way*, by using the indices of the functions in their own recursive definitions. Otherwise said, *the partial recursive functions are closed under fixed-point definitions*. But more is true, as the following result shows.

Theorem II.2.15 (Kleene [1952]) *The class of partial recursive functions is the smallest class of functions:*

1. *containing initial functions and predecessor*

2. *closed under composition*

3. *closed under definition by cases*

4. *closed under fixed-point definitions.*

Proof. By definition by cases we mean the schema that produces

$$f(\vec{x}) = \begin{cases} g(\vec{x}) & \text{if } t(\vec{x}) = 0 \\ h(\vec{x}) & \text{otherwise} \end{cases}$$

from g, h and t. Note that, having composition, closure under definition by cases follows by adding the following to the initial functions:

$$f(x, y, z) = \begin{cases} x & \text{if } z = 0 \\ y & \text{otherwise.} \end{cases}$$

Let $\mathcal{C}$ be the smallest class of functions satisfying the stated conditions. Clearly $\mathcal{C}$ is contained in the class of the partial recursive functions, by the Fixed-Point Theorem, and because case definition is a primitive recursive operation. To prove the converse, we proceed by induction on the definition of partial recursive function. Since identities, successor, and composition are given, we only have to consider:

1. *primitive recursion*
   If

$$
\begin{aligned}
f(\vec{x},0) &= g(\vec{x}) \\
f(\vec{x},y+1) &= h(\vec{x},y,f(\vec{x},y))
\end{aligned}
$$

then f can be defined as the fixed-point of the following equation:

$$
f(\vec{x},y) = \begin{cases} g(\vec{x}) & \text{if } y = 0 \\ h(\vec{x},y-1,f(\vec{x},y-1)) & \text{otherwise.} \end{cases}
$$

This uses only predecessor, definition by cases, and composition.

2. μ-*recursion*
   If

$$
f(\vec{x}) = \mu y[g(\vec{x},y) = 0],
$$

let h be the unique fixed-point of the following equations:

$$
h(\vec{x},y) = \begin{cases} 0 & \text{if } g(\vec{x},y) = 0 \\ h(\vec{x},y+1)+1 & \text{otherwise.} \end{cases}
$$

Then, as in the proof of Theorem I.2.3,

$$
f(\vec{x}) = h(\vec{x},0).
$$

This uses only $\mathcal{O}$, $\mathcal{S}$, definition by cases, and composition. □

It follows that *the Fixed-Point Theorem, together with composition and case definition, generates the partial recursive functions*, in an approach with indices. Usually this is done indirectly, by postulating the S^m_n-Theorem and the Enumeration Theorem (see Section 6 for a discussion of the central role of these two properties in Recursion Theory), since from these the Fixed-Point Theorem follows immediately, as in II.2.13. This approach has been taken by Kleene [1959], and it has proved useful in contexts like recursion on higher types (see p. 199) or on abstract domains (see p. 203), where no analogue of the μ-operator is available.

For further comments on the role of the Fixed-Point Theorem, see pp. 182 and 184.

Limitations of formalism ⋆

We are now ready to prove the celebrated results of the Thirties on the limitations of formalism. They rest on two main foundations:

1. *a combinatorial argument*
   This is embodied in Theorem II.2.3, whose proof, as we have already noted, is a positive recasting of the diagonalization used by Russell in his paradox, but actually goes back to much older paradoxes (see the next subsection).

2. *an analysis of formal systems expressiveness*
   This is the content of the next theorem, which determines exactly what is representable and what is not in sufficiently powerful formal systems. This provides the link with the combinatorial part, by allowing the representation of r.e. nonrecursive sets in formal systems.

We now prove the missing result, which has interest on its own: it points out another difference between recursiveness and recursive enumerability, this time in terms of representability notions. The reader interested only in Theorem II.2.17 should note that, for its proof, only part 1 of the next result is needed (and it has already been established, see I.7.12). Part 2 can thus be skipped, if wanted, but it will provide a different proof of II.2.17.

Theorem II.2.16 Expressiveness of formal systems (Gödel [1931], [1936], Mostowski [1947], Tarski, Mostowski and Robinson [1953], Ehrenfeucht and Feferman [1960], Shepherdson [1960]). *In any consistent formal system extending $\mathcal{R}$:*

1. *a relation is representable if and only if it is recursive*

2. *a relation is weakly representable if and only if it is recursively enumerable.*

Proof. The part relative to recursiveness is immediate, from previous results: the remarks following definition I.3.4 show that, by the axioms of $\mathcal{R}$, a relation is representable if and only if its characteristic function is representable, and this last condition is equivalent to recursiveness (of the characteristic function, and hence of the relation itself), by I.7.12.

Again immediate is the fact that if P is weakly representable in a formal system $\mathcal{F}$, then P is r.e. Suppose indeed that

$$P(x_1, \ldots, x_n) \iff \vdash_{\mathcal{F}} \varphi(\bar{x}_1, \ldots, \bar{x}_n),$$

for some φ. By arithmetization, the predicate

$$T(x_1, \ldots, x_n, y) \iff y \text{ codes a proof of } \varphi(\bar{x}_1, \ldots, \bar{x}_n) \text{ in } \mathcal{F}$$

is recursive, and so P is r.e. by the Normal Form Theorem, since

$$P(x_1, \ldots, x_n) \iff \exists y T(x_1, \ldots, x_n, y).$$

The only real thing to prove is thus the weak representability of P r.e., in any consistent formal system $\mathcal{F}$ extending $\mathcal{R}$. For simplicity of notations, we restrict ourselves to the case of sets. There is a recursive R such that $P(x) \Leftrightarrow \exists y R(x,y)$ and, by the part of the theorem already proved, there must be φ which represents R:

$$R(x,y) \;\Rightarrow\; \vdash_{\mathcal{F}} \;\; \varphi(\overline{x},\overline{y})$$
$$\neg R(x,y) \;\Rightarrow\; \vdash_{\mathcal{F}} \neg\varphi(\overline{x},\overline{y}).$$

- The idea would be to represent P by the formula

$$\psi(x) \;\Leftrightarrow\; \exists y \varphi(x,y).$$

One direction follows immediately: if $P(x)$ holds, then so does $R(x,y)$, for some y, and thus $\vdash_{\mathcal{F}} \varphi(\overline{x},\overline{y})$. In particular, $\vdash_{\mathcal{F}} \psi(\overline{x})$.

But suppose now that $\vdash_{\mathcal{F}} \exists y \varphi(\overline{x},y)$: if we could deduce that, for some y, $\vdash_{\mathcal{F}} \varphi(\overline{x},\overline{y})$, then we would have $R(x,y)$, and hence $P(x)$. But

$$\vdash_{\mathcal{F}} \exists y \varphi(x,y) \;\Rightarrow\; \text{for some } y, \vdash_{\mathcal{F}} \varphi(x,\overline{y})$$

is a strong assumption: it would follow from an infinitary axiom of the kind

$$y = \overline{0} \vee y = \overline{1} \vee y = \overline{2} \vee \cdots$$

But in $\mathcal{R}$ we only have the finitary axioms

$$y < \overline{n+1} \;\leftrightarrow\; y = \overline{0} \vee \cdots \vee y = \overline{n},$$

from which it only follows

$$\vdash_{\mathcal{F}} (\exists y \leq \overline{n})\varphi(x,y) \;\Rightarrow\; \text{for some } y, \vdash_{\mathcal{F}} \varphi(x,\overline{y}).$$

Actually, we would just need the weaker assumption

$$\vdash_{\mathcal{F}} \exists y \varphi(x,y) \;\Rightarrow\; \text{for some } y, \text{ not } \vdash_{\mathcal{F}} \neg\varphi(x,\overline{y}),$$

because then, from $\vdash_{\mathcal{F}} \exists y \varphi(\overline{x},y)$, we would know that, for some y, not $\vdash_{\mathcal{F}} \neg\varphi(\overline{x},\overline{y})$. Since φ strongly represents R, for that y it could not be $\neg R(x,y)$, otherwise $\vdash_{\mathcal{F}} \neg\varphi(\overline{x},\overline{y})$ would follow, and hence $R(x,y)$ would hold, which is what we wanted.

But even this weaker assumption is a strong one, called **ω-consistency** (Gödel [1931], Tarski [1933]), and it does not follow from simple consistency. Thus we have only proved the theorem for ω-consistent systems.

- We now modify this naive approach, and define a new formula $\psi(x)$ that still says $\exists y \varphi(x, y)$, but which moreover safely bounds y below a numeral, whenever it is provable (recall that this is exactly where we ran into troubles above, and that bounded quantifiers can be handled in $\mathcal{R}$). But what exactly do we know when $\psi(x)$ is provable? Precisely this, and then we play a trick (the **Rosser trick**) and use a number coding a proof for it, to bound y: to be sure we stay below the number of any proof, we pick up only those y's which do not bound any number coding a proof of $\psi(x)$. This is the intuition behind the definition that follows:

$$\psi(x) \Leftrightarrow \exists y \, [\varphi(x, y) \wedge (\forall z \le y)(z \text{ does not code a proof of } \psi(x))].$$

If we succeed in defining such a formula (which, as it stays, is self-referential), then we pay a price by having a more difficult proof for the direction that was trivial before, but we win in the troublesome direction:

1. If $P(x)$ holds, then so does $R(x, y)$ for some y, and $\vdash_{\mathcal{F}} \varphi(\overline{x}, \overline{y})$ by strong representability of R. Suppose $\psi(\overline{x})$ is not provable: then no number z codes a proof for it, and in particular this is true for $z \le \overline{y}$, and it is provable in $\mathcal{F}$ because the predicate 'coding a proof' is recursive, and thus strongly representable. By the axioms on $\mathcal{R}$, we can also show (see e.g. the proof of I.3.3)

$$(\forall z \le \overline{y})(z \text{ does not code a proof of } \psi(\overline{x})).$$

Since we know already that we can prove $\varphi(\overline{x}, \overline{y})$, we can then prove $\psi(\overline{x})$, contradiction. Then $\psi(\overline{x})$ is provable.

2. If $\psi(\overline{x})$ is provable, there is a number z coding a proof of it and (by strong representability of the recursive predicate 'coding a proof') we can prove this in $\mathcal{F}$. But then the definition of ψ implies that $\vdash_{\mathcal{F}} (\exists y < \overline{z})\varphi(\overline{x}, y)$. This time we can indeed apply the axioms of $\mathcal{R}$, and get $\vdash_{\mathcal{F}} \varphi(\overline{x}, \overline{y})$, for some y. Hence $R(x, y)$ and $P(x)$ hold.

It only remains to show how to find such a formula ψ. We use diagonalization as in the Fixed-Point Theorem: let $\{\psi_n\}_{n \in \omega}$ be an effective enumeration of the formulas of $\mathcal{F}$ with two free variables, and $\varphi_1(z, x, n)$ be a formula strongly representing the recursive predicate 'z codes a proof of $\psi_n(x, n)$'. Let e be such that

$$\psi_e(x, n) \Leftrightarrow \exists y \, [\varphi(x, y) \wedge (\forall z \le y) \neg \varphi_1(z, x, n)]$$

(where, recall, φ strongly represents R). Then

$$\psi(\overline{x}) \Leftrightarrow \psi_e(x, e)$$

satisfies the requirements (since then $\varphi_1(z, x, e)$ represents the predicate 'z codes a proof of $\psi_e(x, e)$', i.e. 'z codes a proof of $\psi(x)$').

Following the informal sketch given above, it should now be easy to formalize the argument and show that ψ weakly represents P. $\square$

The proof of this theorem uses a number of the arguments we have introduced until now: arithmetization (in showing that the predicate 'coding a proof' is recursive), diagonalization in self-referential form, like in the Fixed-Point Theorem (in the construction of ψ), and the so called Rosser trick (used, in a simpler context, in the proof of the Reduction Property II.1.23). For similar proofs, see the next subsection.

The difficulty that lies behind the theorem just proved is that *weak representability, unlike representability, is not preserved in consistent extensions*: if $x \in A \Leftrightarrow \vdash_{\mathcal{F}} \varphi(\overline{x})$, and A is not recursive, then A is not strongly representable. Thus there must be $x \in \overline{A}$ for which $\vdash_{\mathcal{F}} \neg\varphi(\overline{x})$ fails. If we take the consistent system $\mathcal{F} \cup \{\varphi(\overline{x})\}$, then φ no longer weakly represents A in it.

We now have all the necessary results to approach **Gödel's First Theorem** in a modern version. This is one of the jewels of logic in this century, and it should be contemplated with due reverence.

> Leva dunque, lettore, all'alte ruote
> meco la vista[4]...
>
> (Dante, *Paradiso*, X)

Theorem II.2.17 Limitations of logical systems (Post [1922], Gödel [1931], [1934], Rosser [1936], Church [1936], Tarski, Mostowski and Robinson [1953])

1. *Every consistent extension $\mathcal{F}$ of $\mathcal{R}$ (i.e. any consistent set of formulas closed under logical consequence, and containing all axioms of $\mathcal{R}$) is undecidable.*

2. *If, moreover, $\mathcal{F}$ is a formal system (i.e. the set of its theorems is r.e.), then $\mathcal{F}$ is incomplete.*

Proof. We give two proofs, exploiting different properties of $\mathcal{F}$.

- Let $\{\psi_n\}_{n \in \omega}$ be an effective enumeration of the formulas in the language of $\mathcal{F}$, with one free variable. If $\mathcal{F}$ is decidable, the diagonal set

$$n \in F \Leftrightarrow \vdash_{\mathcal{F}} \psi_n(\overline{n})$$

---

[4] Then, reader, lift your eyes with me to see the lofty wheels ...

is recursive, and then so is $\overline{F}$. Every recursive set is representable in $\mathcal{F}$, by the proof of II.2.16 (which does not use, in this direction, the fact that $\mathcal{F}$ is a formal system) and hence, by consistency of $\mathcal{F}$, weakly representable. Then there is an a such that

$$n \in \overline{F} \Leftrightarrow \vdash_{\mathcal{F}} \psi_a(\overline{n}).$$

For $n = a$ we get a contradiction.

If, moreover, $\mathcal{F}$ is a formal system then, by arithmetization, the set of its theorems is an r.e. set. If $\mathcal{F}$ were complete then we would know that either a sentence is a theorem, or its negation is. But then $\mathcal{F}$ would be decidable: to know whether a sentence is a theorem, generate the theorems until either the sentence or its negation appear.

- This second proof works only for formal systems. Since $\mathcal{K}$ is r.e., by the proof of II.2.16 there is a formula φ that weakly represents it:

$$x \in \mathcal{K} \Leftrightarrow \vdash_{\mathcal{F}} \varphi(\overline{x}).$$

Then $\mathcal{F}$ cannot be recursively decidable, as otherwise $\mathcal{K}$ would be recursive, contradicting II.2.3.

Moreover, $\mathcal{K}$ cannot be represented by φ, otherwise it would be recursive. Then there is at least one x such that

$$x \in \overline{\mathcal{K}} \wedge \text{ not } \vdash_{\mathcal{F}} \neg\varphi(\overline{x}).$$

By weak representability, from $x \in \overline{\mathcal{K}}$ we also have

$$\text{not } \vdash_{\mathcal{F}} \varphi(\overline{x}).$$

Then $\mathcal{F}$ is incomplete, since $\varphi(\overline{x})$ and $\neg\varphi(\overline{x})$ are not provable. $\quad\square$

The two proofs of the theorem are quite different. The first requires only the weak representability of all recursive sets, which is given by I.7.12, as well as a simple diagonal argument, given directly in the proof (showing that *the set of non-theorems of any consistent system is not weakly representable in it*). The second requires the weak representability of some nonrecursive set, and thus the full version of II.2.16, whose proof uses the Fixed-Point Theorem techniques. For a discussion of the two methods of undecidability proofs, see p. 352.

As far as incompleteness is concerned, the proofs given in II.2.17 are indirect, and do not explicitly exhibit undecidable sentences, which are neither provable nor disprovable. To obtain this, even under the hypothesis of ω-consistency, a full use of the self-referential diagonalization must be made (see the next subsection), but even the examples thus obtained have been regarded as somewhat artificial, from a mathematical point of view. Great efforts have been made to obtain natural undecidable sentences, for various systems in common use: two extreme and classical examples are the **Continuum Hypothesis** for Set Theory (Gödel [1938], Cohen [1963]), and a finite version of **Ramsey's Theorem** for Peano Arithmetic (Paris and Harrington [1977]). See Harrington, Morley, Sčedrov and Simpson [1985] for an account of recent results in this area, due mainly to H. Friedman.

The fatal consequences for formalism embodied in Theorem II.2.17 can be expressed as follows: *any classical formal system is inadequate, being either inconsistent, or undecidable (and hence also incomplete), or not sufficiently strong* (to prove at least the elementary arithmetical facts expressed by the axioms of $\mathcal{R}$). Turing (see Hodges [1983], p. 361) interprets these facts as

> saying almost exactly that if a machine is expected to be infallible,
> it cannot also be intelligent

thus isolating in the rigid, purely deterministic approach to knowledge the source of formal systems limitations.

A limitation of a different kind (which holds, by Gödel's First Theorem, for every sufficiently strong, consistent formal system) comes from the following observation: *in any undecidable formal system there are infinitely many theorems with arbitrarily long proofs, with respect to the length of their statements.* Indeed, take any recursive function f. If $f(n)$ were a bound for the length of at least one proof of any theorem of length n, then the system would be decidable: to find out, for a formula of length n, whether it is a theorem or not, produce all the proofs of length at most $f(n)$, and see if one proves it. Thus there must be (infinitely many) n for which a theorem of length n has its shortest proof longer than $f(n)$.

The combinatorial part of Gödel's Theorem (II.2.3), together with the Matiyasevich result quoted on p. 135, shows that arithmetic is already complicated at low levels of complexity: *there is no decision method for one-quantifier formulas in the language of plus and times.* Consider indeed a diophantine representation of $\mathcal{K}$:

$$y \in \mathcal{K} \iff \exists \vec{x}\,[p(\vec{x}, y) = q(\vec{x}, y)].$$

If we could decide one-quantifier formulas, then $\mathcal{K}$ would be recursive.

We conclude our presentation of limitation results by proving another famous one: **Church's Theorem**. It concerns first-order Predicate Calculus, and shows that the dream of Leibniz [1666], of having a **calculus ratiocinator** that would decide the logical truths, is an unfulfillable dream.

Theorem II.2.18 Unsolvability of the Entscheidungsproblem (Church [1936a], Turing [1936]) *The Predicate Calculus is undecidable.*

Proof. Let Q be Robinson Arithmetic (p. 23): since Q is a consistent extension of R, it is undecidable. Let ψ be the conjunction of its axioms (recall, and this is the crucial point, that Q is finitely axiomatized). By the Deduction Theorem of Predicate Calculus, $\vdash_R \varphi$ holds if and only if $\psi \to \varphi$ is provable in Predicate Calculus, and thus any decision procedure for this would give one for Q, contradicting II.2.17. □

The extent of undecidability and decidability of subsystems of the Predicate Calculus has been thoroughly analyzed. See Ackermann [1954], Dreben and Goldfarb [1979] for the former, and Lewis [1979] for the latter.

Self-reference ⋆

> And God said unto Moses:
> 'I am that I am'.
> (*Exodus*, III, 14)

The Fixed-Point Theorem can be used directly to find programs exhibiting self-referential features. E.g., by considering f recursive such that $\varphi_{f(e)}(x) = e$, we get an index e such that $\varphi_e(x) = e$, which can be interpreted as the code of *a program printing itself* (Lee [1963]). Thatcher [1963] has explicitly written down such a program. A more sophisticated self-referential program, able not only to print itself, but also to simulate any given recursive function g, is coded by e such that $\varphi_e(x) = \langle e, g(x) \rangle$, and can be obtained in a similar way.

The technique underlying the Fixed-Point Theorem (or its equivalent form, the Second Recursion Theorem) is the tool allowing the unraveling of self-referential statements, and their replacement by incontrovertible versions. In particular (being obviously impossible to have a finite phrase with itself as a proper part) *self-reference is never direct: it comes from a controlled confusion of two levels of meaning for integers, which are seen both as numbers and as names for formulas*. Then a formula telling some arithmetical fact about an integer may be seen as the translation - by arithmetization - of a metamathematical property.

It is easy to adapt the methods used in II.2.10 to build *a sentence that, for a given property P weakly representable in an extension of R, says of itself that it has the property P* (Carnap [1934], Gödel [1934]). It is enough to consider an enumeration $\{\psi_n\}_{n \in \omega}$ of the formulas with one free variable, and let (with the notations of p. 145)

$$d(\psi) \quad = \quad \text{the sentence '}\psi\text{ has the property } P\text{'}$$

$$a_{i,j} \quad = \quad \text{the sentence '}\psi_j \text{ has the property expressed by } \psi_i\text{'.}$$

The transformed diagonal sequence is still a row of the matrix (up to provable equivalence), and thus there is a ψ such that $d(\psi)$ is provably equivalent to ψ, i.e. ψ says of itself that it has the property P.

An old example of paradoxical self-reference goes back to Epimenides (6th century B.C.), a Cretan who said that Cretans are always liars. This is known as **the liar paradox,** and it was quoted (not very sympathetically) by Paul (*Epistle to Titus*, 1.12), as an example of teaching by

> unruly and vain talkers and deceivers,[5] ... whose mouths must be stopped, who subvert whole houses, teaching things which they ought not, for filthy lucre's sake.

For more elaborated discussions of the liar paradox, see Martin [1978], [1984]. The modern version of the liar paradox is the positive result (usually referred to as **Tarski's Theorem**), that *truth cannot be weakly representable in any consistent extension of* $\mathcal{R}$ (Gödel [1934], Tarski [1936]): otherwise its negation would be weakly representable too, and by the general result obtained above we would get a contradictory sentence asserting its own falsehood.

Provability is instead weakly representable in consistent extensions of $\mathcal{R}$, and in this case the general result obtained above gives *a sentence asserting its own unprovability* (Gödel [1931]), an explicit example of a sentence not provable in a system which proves only truths, and hence true. From this we immediately get the limitation results already proved in the previous subsection. Specifically:

1. *undecidability of any consistent extension* $\mathcal{F}$ *of* $\mathcal{R}$ (Church [1936], [1936a])
   Suppose $\mathcal{F}$ is decidable: this means that 'being a theorem of $\mathcal{F}$' is recursive, hence representable (since $\mathcal{F}$ extends $\mathcal{R}$) by some ψ:

$$\vdash_{\mathcal{F}} \psi_x \quad \Rightarrow \quad \vdash_{\mathcal{F}} \quad \psi(\overline{x}) \tag{II.1}$$

$$\text{not } \vdash_{\mathcal{F}} \psi_x \quad \Rightarrow \quad \vdash_{\mathcal{F}} \neg\psi(\overline{x}). \tag{II.2}$$

Let now ψ_x assert its own unprovability, i.e.

$$\vdash_{\mathcal{F}} \psi_x \leftrightarrow \neg\psi(\overline{x}). \tag{II.3}$$

Then

$$\vdash_{\mathcal{F}} \psi_x \Rightarrow \vdash_{\mathcal{F}} \psi(\overline{x}) \Rightarrow \vdash_{\mathcal{F}} \neg\psi_x$$

by II.1 and II.3, contradicting consistency, and

$$\text{not } \vdash_{\mathcal{F}} \psi_x \Rightarrow \vdash_{\mathcal{F}} \neg\psi(\overline{x}) \Rightarrow \vdash_{\mathcal{F}} \psi_x$$

by II.2 and II.3, contradiction. Thus $\mathcal{F}$ cannot be decidable.

---

[5] Presumably including, nowadays, the logicians.

2. *an explicit example of incompleteness, for any ω-consistent formal system*
   $\mathcal{F}$ extending $\mathcal{R}$ (Gödel [1931])
   Since $\mathcal{F}$ is a formal system, 'y codes a proof of ψ_x' is recursive, and so
   representable (since $\mathcal{F}$ extends $\mathcal{R}$) by some φ:

   $$y \text{ codes a proof of } \psi_x \;\Rightarrow\; \vdash_{\mathcal{F}} \varphi(\overline{x},\overline{y})$$
   $$y \text{ does not code a proof of } \psi_x \;\Rightarrow\; \vdash_{\mathcal{F}} \neg\varphi(\overline{x},\overline{y}).$$

Then the formula

$$\psi(x) \Leftrightarrow \exists y \varphi(x,y)$$

weakly represents provability in $\mathcal{F}$ (see the proof of II.2.16, where the
hypothesis of ω-consistency is used to show this).

Let now ψ_x assert its own unprovability, i.e.

$$\vdash_{\mathcal{F}} \psi_x \leftrightarrow \neg\psi(\overline{x}) \tag{II.4}$$

Then:

- ψ_x is not provable in $\mathcal{F}$: if it were, both $\psi(\overline{x})$ and $\neg\psi(\overline{x})$ would be
  provable (contradicting consistency), the former because ψ weakly
  represents provability, the latter because of II.4.

- $\neg\psi_x$ is not provable in $\mathcal{F}$: if it were, II.4 would imply that $\psi(\overline{x})$
  is provable, and ψ_x would then be provable (contradicting consis-
  tency), because ψ weakly represents provability.

Note that we can decide, from the outside, that ψ_x is true: it is not
provable, and it asserts its own unprovability.

3. *an explicit example of incompleteness, for any consistent formal system*
   $\mathcal{F}$ extending $\mathcal{R}$ (Rosser [1936])
   Let φ be as above, $\psi_{neg\,x} \Leftrightarrow \neg\psi_x$, and

   $$\psi(x) \;\Leftrightarrow\; \psi_x \text{ is provable before } \neg\psi_x \text{ is}$$
   $$\Leftrightarrow\; \exists y \, [\varphi(x,y) \wedge (\forall z \leq y)\neg\varphi(neg\,x,z)].$$

We cannot assert that ψ weakly represents provability in $\mathcal{F}$ (see II.2.16
for a formula that does this). But at least we do have

$$\vdash_{\mathcal{F}} \psi_x \;\Rightarrow\; \vdash_{\mathcal{F}} \psi(\overline{x}). \tag{II.5}$$

Suppose indeed that ψ_x is provable. Then $\vdash_{\mathcal{F}} \varphi(\overline{x},\overline{y})$, for some y. By
consistency of $\mathcal{F}$, $\neg\psi_x$ is not provable, and hence $\vdash_{\mathcal{F}} \neg\varphi(\overline{neg\,x},\overline{z})$, for

every $z \leq y$. By the axioms of $\mathcal{R}$, $\vdash_{\mathcal{F}} (\forall z \leq \overline{y}) \neg \varphi(\overline{neg\,x}, z)$. Thus $\vdash_{\mathcal{F}} \psi(\overline{x})$.

Let now ψ_x assert that it is not provable before its own negation, i.e.

$$\vdash_{\mathcal{F}} \psi_x \leftrightarrow \neg \psi(\overline{x}). \tag{II.6}$$

Then:

- ψ_x is not provable in $\mathcal{F}$, otherwise so would be $\psi(\overline{x})$ and $\neg \psi(\overline{x})$, respectively by II.5 and II.6, contradicting consistency.

- $\neg \psi_x$ is not provable in $\mathcal{F}$, otherwise so would be $\psi(\overline{x})$ and $\neg \psi(\overline{x})$ (contradicting consistency), the former by II.6, the latter by the following reasoning. Note that

$$\neg \psi(x) \Leftrightarrow \forall y[\varphi(x, y) \rightarrow (\exists z \leq y) \varphi(neg\,x, z)].$$

If $\neg \psi_x$ is provable, let z code a proof of it. Then $\vdash_{\mathcal{F}} \varphi(\overline{neg\,x}, \overline{z})$. By the axioms of $\mathcal{R}$, $y < \overline{z} \vee \overline{z} \leq y$, for any y. In the first case y is a numeral less than $\overline{z}$, by the axioms of $\mathcal{R}$, and thus $\vdash_{\mathcal{F}} \neg \varphi(\overline{x}, y)$, because if $\varphi(\overline{x}, y)$ holds then ψ_x would be provable, contradicting consistency. In the second case $\vdash_{\mathcal{F}} (\exists z \leq y) \varphi(\overline{neg\,x}, z)$, because $\vdash_{\mathcal{F}} \varphi(\overline{neg\,x}, \overline{z})$. Then

$$\vdash_{\mathcal{F}} (\forall y)[\neg \varphi(\overline{x}, y) \vee (\exists z \leq y) \varphi(\overline{neg\,x}, z),$$

and hence $\vdash_{\mathcal{F}} \neg \psi(\overline{x})$.

As before, ψ_x is true because it is not provable, in particular not provable before its negation.

It is interesting to note that *the sentence asserting its own unprovability is equivalent to the assertion of consistency of the system*: if the system is consistent then ψ_x is not provable (otherwise both $\psi(\overline{x})$ and $\neg \psi(\overline{x})$ would follow, the first by the properties of ψ, the second by definition of ψ_x), and hence ψ_x holds; and if ψ_x holds then it is not provable, and the system is consistent (otherwise everything would be provable). This equivalence proof is informal, and the simple assumption of representability of provability in $\mathcal{F}$ is not sufficient to reproduce the proof inside the system. But under some stronger assumptions (see below), it is possible to prove it inside $\mathcal{F}$, i.e.

$$\vdash_{\mathcal{F}} Con_{\mathcal{F}} \leftrightarrow \psi_x,$$

where $Con_{\mathcal{F}}$ is a formal translation of consistency, for example the assertion that '$0 = 1$' is not provable. It then follows, from the unprovability of ψ_x, that *the consistency of a consistent formal system is not provable inside the system itself* (Gödel [1931]), a result known as **Gödel's Second Theorem**, and which destroyed Hilbert's program of justifying abstract mathematics by proving the consistency of formal systems of common use by elementary (finitary) means, e.g. in $\mathcal{PA}$, or in any other sufficiently weak formal system in which only finitistically acceptable reasoning could be formalized.

The assumptions on the provability predicate ψ needed for the proof of Gödel's Second Theorem (Gödel [1933], Hilbert and Bernays [1939], Löb [1955], Jeroslow [1973]) are worth examining also for another reason, directly related to the subject of self-reference. Since they are outside the scope of Recursion Theory, because not purely extensional, we just quote them:

1. $\vdash_{\mathcal{F}} \psi_x \Rightarrow \vdash_{\mathcal{F}} \psi(\overline{x})$
   This says that if a formula is a theorem of $\mathcal{F}$, we can prove inside $\mathcal{F}$ that it is. It expresses half of the condition for weak representability of the provability predicate.

2. $\vdash_{\mathcal{F}} \psi(\overline{x}) \to \psi(\overline{pr(x)})$
   where $\psi_{pr(x)} = \psi(\overline{x})$. The first condition was external to $\mathcal{F}$, saying that any single provable formula can be recognized to be provable by $\mathcal{F}$. This second condition is internal to $\mathcal{F}$, and says that $\mathcal{F}$ is aware of the first condition: inside $\mathcal{F}$ we know that if a formula is provable, then we can prove this fact.

3. $\vdash_{\mathcal{F}} \psi(\overline{x}) \wedge \psi(\overline{impl(x,y)}) \to \psi(\overline{y})$
   where $\psi_{impl(x,y)} = \psi_x \to \psi_y$. This says that $\mathcal{F}$ is aware of the fact that the provability relation is closed under modus ponens.

These conditions are satisfied by usual first-order formal systems extending $\mathcal{PA}$ (see Bezboruah and Shepherdson [1976] for a version of Gödel's Second Theorem for $\mathcal{Q}$), and can be loosely stated as: *in $\mathcal{F}$ we can prove that a provable formula is provable, and we are aware of this fact and of modus ponens.*

Going back to self-reference, we have seen that the sentence asserting its own unprovability is true and not provable. We now want to show that, under the conditions stated above, *the sentence asserting its own provability* is true and provable. This provides an example of true self-referential statement which makes no use of negation. The general result is that, under the conditions stated above, *a sentence $\psi(\overline{x}) \to \psi_x$* (asserting that if ψ_x is provable then it is true, and thus expressing a form of soundness) *is provable if and only if ψ_x is* (**Löb's Theorem**, Löb [1955]). The only nontrivial direction is to prove that if $\psi(\overline{x}) \to \psi_x$ is provable then so is ψ_x. Suppose ψ_x is not provable. Then $\mathcal{F} \cup \{\neg\psi_x\}$ is consistent, and its consistency cannot be proved in it, by Gödel's Second Theorem. I.e. we cannot prove in it that ψ_x cannot be proved in $\mathcal{F}$, which is expressed by the formula $\neg\psi(\overline{x})$. But if $\mathcal{F}\cup\{\neg\psi_x\}$ cannot prove $\neg\psi(\overline{x})$ then, by the Deduction Theorem and contrapositive, $\mathcal{F}$ cannot prove $\psi(\overline{x})\to\psi_x$.

Since the formula that asserts its own provability has the property that $\vdash_{\mathcal{F}}\psi(\overline{x})\leftrightarrow\psi_x$, we have in particular that $\psi(\overline{x})\to\psi_x$ is provable, and then so is ψ_x itself, by Löb's Theorem. A similar example, relying on the same conditions on the provability predicate but not using Gödel's Second Theorem, is *the sentence asserting of itself that if it is provable then it is true*, i.e.

$$\vdash_{\mathcal{F}}\ \psi_x\leftrightarrow(\psi(\overline{x})\to\psi_x).$$

Suppose ψ_x is provable. Then $\vdash_{\mathcal{F}}\psi(\overline{x})$ by representability of provability, and $\vdash_{\mathcal{F}}\psi(\overline{x})\to\psi_x$ by definition of ψ_x. By modus ponens, $\vdash_{\mathcal{F}}\psi_x$. Thus we have proved that if ψ_x is provable, i.e. if $\psi(\overline{x})$ holds, then so does ψ_x. This establishes that if ψ_x is provable then it is true. By ψ_x asserts exactly this, and thus it is true and provable.

For more information on the topics of this subsection, see Smorynski [1977].

Self-reproduction and cellular automata ⋆

> Vergine madre, figlia del tuo figlio[6]
> (Dante, *Paradiso*, XXXIII)

On a linguistic level, to build self-reproducing objects is not difficult. A nontrivial example (quoted by Hofstadter [1985]) is:

> Alphabetize and append, copied in quote, these words: 'these append, in Alphabetize and words: quote, copied'

which both lists its parts at the word level, and tells how to put them together to reconstitute itself. By acting according to the rules stated in it on the words quoted in it we get the same sentence, but this is somewhat unsatisfactory as an example of self-reproduction, since it obviously requires an external agent that understands the rules and performs them.

On a mechanical level, at first sight it might appear that a machine can only reproduce less complicated machines, since the building machine must somehow contain a complete description of the built one, together with some additional device to do the actual building. This was a stumbling block for Descartes [1637], who thought that animals and human bodies are machines, but had to resort to miracles to explain biological reproduction. The crucial point missed by the above discussion is that descriptions of machines are at a lower logical level than machines themselves, and can be brought down to

---

[6]Virgin mother, daughter of your son

the same logical level of their inputs. Thus *a machine can reproduce itself, whenever it is able to build machines by following their descriptions, and is fed its own description.* When this minimal complexity is attained, obstacles are removed not only for self-reproduction, but even for reproduction of more complicated machines. In this subsection we flesh out these observations.

The ideas of the Fixed-Point Theorem can be used to build a **self-reproducing mechanical automaton** (Von Neumann [1951]). The first machine we need is a universal constructor A, with the property that, when it is fed the description d_X of a machine X, it searches in the surrounding environment for the needed mechanical parts, and builds X. In symbols:

$$(A, d_X) \longrightarrow X.$$

Of course universality refers to a fixed class of machines, all particular specimen from a common species, that can be described in a uniform way. Thus A has the ability to understand all plans of a given type, and realize them. A alone is not yet self-reproducing because, when it is given its own description, it does build a copy of itself, but the result is lacking the description that was fed at the beginning:

$$(A, d_A) \longrightarrow A.$$

We thus introduce a second machine B, with the simple task of reproducing any description given to it:

$$(B, d_X) \longrightarrow d_X.$$

To coordinate the two machines A and B, a third one C is introduced, and the compound machine $A + B + C$ will now work in the following way. Given a description d_X, A builds a copy of X, while B reproduces a copy of d_X; then the copy of X is fed the copy of d_X:

$$(A + B + C, d_X) \longrightarrow (X, d_X).$$

Then, if we call D the resulting machine $A + B + C$, we get

$$(D, d_D) \longrightarrow (D, d_D).$$

Thus $S = (D, d_D)$ is self-reproducing.

The same ideas allow not only for self-reproduction, but even for production of more complicated machines (a sort of **evolutionary process**). Indeed, it is enough to insert in D a description d_{D+F} of a machine composed of D and some other machine F. Then S will produce itself, together with F.

Some observations on the ideas used in the construction just sketched are in order. First, note that there is no circularity involved, since we first obtained

D, and then we fed it its own description; also, none of the three parts A, B, C is self-reproducing by itself, but their combination is. Second, the crucial fact that circumvents the difficulties pointed out at the beginning is that B, while being of fixed size, can copy descriptions of any length: thus the description of the constructed machine need not be incorporated in the parent machine, but only coupled to it. Third, descriptions (i.e. indirect reproductions) are necessary in actual realizations: if we wish to reproduce a machine X directly, piece by piece, we need to interfere with it (to know how it is built) in some possibly disrupting way, while the reproduction process presupposes an unchanging original. For similar reasons, descriptions must be directly given, and cannot in general be deduced by machine observation. Note that descriptions are used at two different levels: one time they are purely duplicated (and so used as raw material), and another time they are followed as projects (and so interpreted as instructions). Finally, alterations of some parts of S might produce different effects: a change in D itself might inhibit or perturb the whole production process; a change in d_{D+F} might affect the constructed machine, directly in the copy of D (thus producing a machine that might not be self-reproducing anymore), or possibly only in the by-product F.

Note that *real life reproduces exactly this way*: living cells contain universal constructors, basically the same for plants and animals, and only the genetical material (the program) is different. More precisely, a suggestive **biological reinterpretation** of the construction of S is the following: d_X works like a gene (a segment of DNA) that codifies the reproduction information; B (a special enzyme - RNA polymerase) has the function of duplicating the genetic material into a segment of RNA; A (a set of ribosomes) builds proteins by following (a segment of RNA containing) the reproduction information; S is a self-reproducing cell. This is thus an abstract, simplified representation of **genetical reproduction**. One of the simplifications occurs in the fact that the gene has only a partial codification of the reproduction information, and this allows for a partial modification of the reproduced object. The additional parts possibly produced are the analogues of enzymes that the gene produces, or whose production it stimulates. The effect of alterations on S are reminiscent of **mutations**: they may be lethal or sterilizing (killing the organism or inhibiting reproduction), produce modified (and possibly sterile) successors, or produce fertile successors with changed hereditary strain (generating different by-products). See Watson [1970] for general information, Arbib [1969a] for a discussion of the relevance of self-reproducing automata in biological contexts, and Burks and Farmer [1984] for a model of DNA sequences as automata.

The troublesome part in the discussion above lies in the assumption of the existence of a **universal constructor**. We might reason by analogy, and think that an argument like the one used for universal Turing machines (p. 132) might suffice. However, for Turing machines the actions needed for universality are of quite limited complexity (writing, reading, moving), and at the same level as those of the simulated machines. For actual constructors much more is certainly needed (recognizing, moving, manipulating and assembling components). Thus the possibility again arises of an infinite regress, in the sense of needing an additional level of complexity, to build machines at a certain level.

Von Neumann [1966] was, however, able to solve the problem by abstracting some of the properties of the mechanical model seen above. He noted that constructions can be seen as series of events taking place in a space, and he thus considered - by a sort of cartesian representation - a space of cells that can be in a certain number of states (intended to represent presence or absence of certain parts of the mechanical model, so that motion of parts is represented by change of state in cells). Von Neumann thus used particular **cellular automata**, which are simply potentially infinite, directed graphs (spaces), whose nodes (cells) are finite state machines (see p. 52): the global behavior consists of the simultaneous and coordinated behavior (change of state) of the single cells. Von Neumann's particular automaton consisted of a planar space with 29-state cells of a single type, each connected to the four orthogonally adjacent neighbors: what he did was to find a finite quiescent configuration (of around 200,000 cells) that, given any other finite quiescent configuration, reproduces it in a different part of the space, without erasing itself: he thus found a universal constructor for the class of quiescent configurations. See some of the essays in Burks [1970] for more explanations on this model, Codd [1968] and Arbib [1969] for improved treatments (with opposite emphasis: the former on local simplicity of cells, the latter on global simplicity of construction), and Langton [1984] for a simple, nontrivial (although not construction universal) cellular model of self-reproduction.

The simplest automaton admitting self-reproducing configurations we know of is Conway's **Life automaton**, so called because modelled on life-like behavior. It consists of a planar space, with each cell connected to the eight adjacent ones. The cells have just two states: 0 (death) and 1 (life), and the Life rules are the following: a dead cell gets born when exactly 3 neighbors are alive; a live cell survives if 2 or 3 neighbors are alive, and it dies otherwise (by overpopulation or starvation). Conway (see Berlekamp, Conway and Guy [1982]) has shown that *there are self-reproducing Life configurations*, and that *the Life automaton is universal*, in the sense that any cellular automaton can be constructed by taking sufficiently large squares of Life cells as its basic cells. These results are also interesting because, unlike other cellular automata admitting self-reproductive behaviors, Life's rules were not introduced with the purpose of making self-reproduction possible. Also, Life is about as simple as it can be (it is known that 2-state cells with a Von Neumann neighborhood do not admit nontrivial self-reproduction), and it shows that *self-reproduction does not need a complicated universe* (since it is logically possible from simple physical models). For more information on Life, see Poundstone [1985].

A notion somewhat opposite to self-reproduction is the impossibility of being reproduced. A finite pattern in a cellular automaton is called a **Garden-of-Eden configuration** if it cannot be obtained from any previous configuration: it corresponds to a machine that cannot be built, in the sense that it cannot arise as a result of any past state of its universe. Obviously, a self-reproducing configuration (being obtainable from itself) cannot contain any Garden-of-Eden configuration: this places limits on the possibilities of universal constructors (since it exhibits nonconstructible configurations), as well as on the possible patterns of self-reproducing machines (this being relevant for the determination of the simplest possible conditions for self-reproduction, and hence for the computation of the probability of getting living organisms by chance interaction of nonliving ones). *Garden-of-Eden configurations exist on a cellular automata (satisfying some general conditions) if and only if the automata is not backwards-deterministic*, i.e. if there are configurations that can be obtained in more than one way (Moore [1962], Myhill [1963]). This condition obviously applies when erasing is possible, erasing being an irreversible process that loses information.

Note that cellular automata are quite general devices: given any Turing machine, it is possible to build a cellular automaton that simulates the computation of the machine on any given input (the automaton is a linear tape; the state of each cell will reflect the situation of the corresponding cell in the tape by telling which symbol is written there, and whether the head is scanning such a cell or not and, if so, in which state the machine is). In particular, *a cellular automata can simulate a universal Turing machine* (Smith [1972]). The universal constructor we have been quoting above can be made complex, in the sense of being simultaneously able to simulate a universal Turing machine (and thus being both computation and construction universal).

For more on cellular automata see Toffoli and Margolus [1987], and the papers in Burks [1970], Farmer, Toffoli and Wolfram [1984], Demongeot, Golès and Tchuente [1985], and Wolfram [1986]. Actual computer machines modelled on cellular automata are studied in Preston and Duff [1984], and Toffoli and Margolus [1987]. Problems of reversibility for cellular automata, analogous to those considered on p. 51 for Turing machines, are considered in Toffoli [1977], [1981] (where the existence of computation and construction universal, reversible cellular automata is shown).

II.3   Partial Recursive Functionals

In Chapter I we introduced the notion of recursive functions, in various equivalent formulations. In Section II.1 we then extended this notion to encompass the case of partial functions. Here we present a further, substantial generalization of recursiveness by considering effective procedures that act not only on numbers, but on functions as well. That is, we extend the notion of recursiveness from functions to functionals.

Oracle computations and Turing degrees

Let us revisit the definition of recursiveness: we had a set of initial functions, and a set of operations, transforming given functions into new functions. The idea was that the initial functions were effectively computable, and the operations transformed effectively computable functions into effectively computable ones. If a function g is added to the initial functions, then the class obtained is the same if g is recursive, but it is otherwise more comprehensive.

Definition II.3.1 (Turing [1939]) *If g is a total function, the class of* **functions recursive in g** *is the smallest class of functions*

1. *containing the initial functions and g*

2. *closed under composition, primitive recursion and restricted μ-recursion.*

If A is a set, the class of **functions recursive in A** *is the class of functions recursive in c_A.*
A predicate is recursive in g or A if its characteristic function is.

The extension of recursiveness relative to a given function corresponds to the algebraic procedure of transcendental extension. The functions recursive in g are not all outright computable, unless g itself is, but they are still 'computable modulo g'. A pictorial way to express this state of affairs is to say that they are computable with the help of an **oracle**, a term introduced by Turing [1939] and now standard.

> χαὶ δή ποτε χαὶ εις Δελφοὺς ελθὼν
> ετόλμησε τοῦτο μαντεύσασθαι·
> τί ποτε λέγει ο θεός;
> ου γὰρ δήπου ψεύδεταί γε.[7]
> (Plato, *Socrates' Apology*)

The oracle, as the word emphasizes, is an extrarecursive entity, helping the computation of any function recursive in g in its troublesome spots, when a call to g (which could be effectively answered only if g were recursive) is made. The oracle supplies the answer to any such call for free.

---

[7] And thus once, gone even to Delphi, he dared to consult the oracle on this: what does the god say? He certainly does not lie.

An oracle f may also help the computation of f itself, in some nontrivial way. For example, some of the information coded by f may be redundant, and recoverable from the rest of it in various ways (see III.5.9 and V.5.15 for some precise formulations).

Definition II.3.2 *Given two functions f and g, we say that:*

1. *f is* **Turing reducible** *to g ($f \leq_T g$) if f is recursive in g*

2. *f is* **Turing equivalent** *to g ($f \equiv_T g$) if $f \leq_T g$ and $g \leq_T f$.*

Note that $\leq_T$ is a reflexive and transitive relation, and thus $\equiv_T$ is an equivalence relation. Although trivial, this is a crucial fact for later development, since then $\equiv_T$ partitions the class of total functions (and in particular the class of characteristic functions, i.e. the class of sets) into equivalence classes, called **degrees of unsolvability**.

Definition II.3.3 (Post [1948]) *The equivalence classes of total functions w.r.t. Turing equivalence are called* **Turing degrees** *(or T-degrees).* $(\mathcal{D}, \leq)$ *is the structure of Turing degrees, with the partial ordering $\leq$ induced on them by $\leq_T$.*

Two functions are Turing equivalent (in the same T-degree) when they are recursive in each other: thus, they help each other's computation, and they are in the same relationship as the recursive functions are among themselves. The continuum is thus classified from a recursion-theoretical point of view, and the study of such a classification (and of related ones) is one of the main subjects of our book.

Exercises II.3.4 a) *Every T-degree contains a set.* (Hint: consider the graph of a given function, and code it into a set.)
  b) *There is a smallest T-degree.* (Hint: the degree of recursive sets.)

The notion of relative recursiveness is easily generalized to partial functions:

Definition II.3.5 *If β is a partial function, the class of* **functions partial recursive in β** *is the smallest class of functions*

1. *containing the initial functions and β*

2. *closed under composition, primitive recursion and unrestricted μ-recursion.*

Relative partial recursiveness is still called Turing reducibility, and we still write $\alpha \leq_T \beta$ when α is partial recursive in β.

By substituting the concept of recursiveness with its version relativized to a given (total) function or set, notions and results dealt with so far generalize. For example, we can define a set as r.e. in A if it is the domain of a function partial recursive in A, and prove that

$$A \leq_T B \Leftrightarrow \text{both } A \text{ and } \overline{A} \text{ are r.e. in } B.$$

In the following we will refer to theorems in relativized form when needed, but it must be clear that *the fact that a result relativizes should not be taken for granted*: claiming a relativized form of a result which holds unrelativized requires a proof, even if only a check that everything relativizes in the original proof. In Chapter V we will actually see that some results do *not* relativize (V.7.13).

The notion of functional

A **functional** is simply a function whose variables range over numbers or over functions of numbers, and whose values are numbers. For simplicity of notations, we will mostly consider the function variables of a functional to range over unary functions, and will leave to the reader the care of extending notions and results to the general case. Our convention is not a restrictive one, since any function with many variables can be recursively reduced to a unary function by coding its arguments into a single one.

We may suppose that the function variables of a functional range only over total functions, or we may allow them to range over partial functions as well. We use the word functional to refer to the latter case, and talk of **restricted functional** in the former.

As for functions, a functional can be undefined for some of its arguments. We call **total functional** a functional which is always defined whenever its function arguments are total. Thus a total functional can be undefined for some of its partial arguments, but a total restricted functional is always defined. Examples of total and nontotal functionals are, respectively, the application functional

$$Ap(\alpha, x) \simeq \alpha(x),$$

and the (unrestricted) μ-operation schema

$$Mu(\alpha, x) \simeq \mu y(\alpha(\langle x, y \rangle) \simeq 0).$$

It is precisely the existence of nontotal, restricted functionals that makes the consideration of nonrestricted functionals natural.

We can also talk of **(restricted) relations of numbers and functions**, by just referring to the characteristic (restricted) functional.

Partial recursive functionals

We now use the notion of relative partial recursiveness to introduce the idea of recursive functional.

Definition II.3.6 (Kleene [1952], [1969], Sasso [1971]) *The functional* $F(\alpha_1, \ldots, \alpha_n, \vec{x})$ *is a* **partial recursive functional** *if it can be obtained from* $\alpha_1, \ldots, \alpha_n$ *and the initial functions by composition, primitive recursion and unrestricted μ-recursion.*

A **relation** *of function and number variables is recursive if its characteristic functional is.*

The definition just given contains more than the mere fact that the function $\lambda\vec{x}.\, F(\alpha_1, \ldots, \alpha_n, \vec{x})$ is partial recursive in the α_i's for each choice of its functions variables. This is certainly implied, but more is true: actually $\lambda\vec{x}.\, F(\alpha_1, \ldots, \alpha_n, \vec{x})$ is **uniformly partial recursive** in the α_i's, in the sense that there is a master way of showing the partial recursiveness of it in the α_i's, in which these appear as parameters.

Exercises II.3.7 a) *β is partial recursive in α if and only if there is a partial recursive functional F such that $\beta(x) \simeq F(\alpha, x)$.*

b) *There are functions f and g such that f is recursive in g, but there is no partial recursive, total functional F such that $f(x) = F(g, x)$.* (Post [1944]) (Hint: from II.3.10 we will get a notion of index for partial recursive functionals. Let F_e be the functional with index e, and A be the set of indices of partial recursive, total functionals of one function variable and one number variable. If

$$x \in B \Leftrightarrow x \in A \wedge F_x(c_A, x) = 0$$

then $B \leq_T A$, but the existence of a partial recursive, total functional F such that $c_B(x) = F(c_A, x)$ leads to a contradiction.)

The partial recursive functionals are closed under composition in a strong form, which allows for substitution not only in the number arguments, but also in the function ones.

Proposition II.3.8 Substitution Property (Kleene [1952]) *If $F(\alpha, z)$ and $G(\beta, x)$ are partial recursive functionals, then so is*

$$H(\alpha, x) \simeq G(\lambda z.\, F(\alpha, z), x).$$

Proof. By induction on the definition of G. From a computational point of view, this is quite clear: when we get to a call of β in the computation of G, we substitute it by the corresponding call of $\lambda z.\, F(\alpha, z)$, and continue the computation. Thus the only calls to the oracle which are not discharged are those relative to α. $\square$

Exercises II.3.9 a) *Substitution holds for partial recursive, restricted total functionals.* (Kleene [1955])

b) *Substitution fails for partial recursive, restricted functionals.* (Kleene [1963], Kreisel) (Hint: let $G(f, x) \simeq 0$, and $F(x, z) \simeq 0 \Leftrightarrow x \in \overline{\mathcal{K}_z}$ Then the function $H(x) \simeq G(\lambda z. F(x, z), x)$ would converge if and only if $\lambda z. F(x, z)$ is total, i.e. when $x \notin \mathcal{K}$, and $\mathcal{K}$ would be recursive.)

Theorem II.3.10 Normal Form Theorem for partial recursive functionals (Kleene [1952], [1969], Davis [1958], Sasso [1971]) *There is a primitive recursive function $\mathcal{U}$ and (for each $m, n \geq 1$) recursive predicates $\mathcal{T}_{m,n}$ such that, for every partial recursive functional F of m function variables and n number variables, there is a number e (called* **index** *of F) for which the following hold:*

1. $F(\alpha_1, \ldots, \alpha_m, x_1, \ldots, x_n) \downarrow \, \Leftrightarrow \, \exists y \mathcal{T}_{m,n}(e, x_1, \ldots, x_n, \alpha_1, \ldots, \alpha_m, y)$

2. $F(\alpha_1, \ldots, \alpha_m, x_1, \ldots, x_n) \simeq \mathcal{U}(\mu y \mathcal{T}_{m,n}(e, x_1, \ldots, x_n, \alpha_1, \ldots, \alpha_m, y))$.

Proof. We cannot use the functions α_i in computations, since they might not be recursive, and we then use finite approximations to them. We refer to the proofs of I.7.3 and II.1.2, and just indicate the appropriate changes to be made.

1. The index of a partial recursive functional can be defined by just adding clauses for the function arguments α_i's, as if they were initial functions.

2. Computations are put in canonical form as usual. Now some of the nodes can be expressions of the kind $\alpha_i(x) = z$.

3. Numbers to nodes of the computation trees are assigned in the form

$$\langle e, \langle x_1, \ldots, x_n \rangle, \langle a_1, \ldots, a_m \rangle, z \rangle$$

where e is the index of the functional, and a_i is the code of a fixed finite approximation to α_i, e.g. of the following function:

$$\tilde{a}_i(x) \simeq \begin{cases} exp(a_i, p_x) - 1 & \text{if } exp(a_i, p_x) > 0 \\ \text{undefined} & \text{otherwise.} \end{cases}$$

Numbers to computations are assigned in the usual way.

4. The definition of $\mathcal{T}$ remains the same, except for a local modification of clause B, to take care of the new case corresponding to nodes of the kind $\alpha_i(x) = z$, where the evaluation is made by means of the finite function $\tilde{a}_i$, in place of α_i.

5. $\mathcal{T}_{m,n}$ is the predicate defined as:

$$\mathcal{T}_{m,n}(e, x_1, \ldots, x_n, \alpha_1, \ldots, \alpha_m, y) \Leftrightarrow \mathcal{T}(y) \wedge$$
$$(y)_{1,1} = e \wedge (y)_{1,2} = \langle x_1, \ldots, x_n \rangle \wedge$$
$$ln((y)_{1,3}) = m \wedge (\forall i)_{1 \le i \le m}(\alpha_i | (y)_{1,3,i}).$$

Here $\alpha | a$ means that α extends the finite function coded by a, i.e.

$$(\forall x \le a)(exp(a, p_x) \neq 0 \Rightarrow exp(a, p_x) = \alpha(x) + 1).$$

6. Finally, since now nodes are quadruples, and the value is the last component,

$$\mathcal{U}(y) = (y)_{1,4}. \quad \square$$

In the case of restricted functionals, we get a smoother version (where, recall, $\hat{g}_i$ is the course-of-value of g_i, see I.7.1):

Theorem II.3.11 Normal Form Theorem for partial recursive restricted functionals (Kleene [1952], Davis [1958]) *There is a primitive recursive function $\mathcal{U}$ and (for each $m, n \ge 1$) primitive recursive predicates $\mathcal{T}_{m,n}$ of only numerical variables such that, for every partial recursive restricted functional F of m function variables and n number variables, there is a number e (called* **index** *of F) for which the following hold:*

1. $F(g_1, \ldots, g_m, \vec{x}) \downarrow \Leftrightarrow \exists y \mathcal{T}_{m,n}(e, \vec{x}, \hat{g}_1(y), \ldots, \hat{g}_m(y), y)$

2. $F(g_1, \ldots, g_m, \vec{x}) \simeq \mathcal{U}(\mu y \mathcal{T}_{m,n}(e, \vec{x}, \hat{g}_1(y), \ldots, \hat{g}_m(y), y)).$

Proof. It is enough to modify, in the previous theorem, the definition of $\mathcal{T}_{m,n}$ as:

$$\mathcal{T}_{m,n}(e, x_1, \ldots, x_n, z_1, \ldots, z_m, y) \Leftrightarrow \mathcal{T}(y) \wedge$$
$$(y)_{1,1} = e \wedge (y)_{1,2} = \langle x_1, \ldots, x_n \rangle \wedge ln((y)_{1,3}) = m \wedge$$
$$(\forall i)_{1 \le i \le m}(Seq(z_i) \wedge ln(z_i) = y + 1 \wedge z_i | (y)_{1,3,i}).$$

Here $z | a$ now means

$$(\forall x \le a)(exp(a, p_x) \neq 0 \Rightarrow exp(a, p_x) = (z)_{x+1} + 1).$$

The idea is that we have to take a computation coded by y, consider the finite functions used in it (coded by $(y)_{1,3,i} < y$), and note that only values up to y of the function arguments g_i can be needed in the computation. But these values are all coded in the sequence numbers $\hat{g}_i(y)$. The definition of $\mathcal{T}_{m,n}$ is slightly complicated by the fact that the g_i's do not appear directly in it, but only through the numbers $\hat{g}_i(y)$, whose role is taken by z_i. □

Of course the improved normal form does not work for nonrestricted functionals, since $\hat{\alpha}(y)$ could be undefined for partial α (when $\alpha(z) \uparrow$ for some $z < y$), even if all the converging values of α needed for the computation are relative to arguments up to y.

There is a special case which is going to be particularly useful, and we introduce special notations for it, in analogy with the notations used for partial recursive functions (II.1.4). It is the case of oracle computations w.r.t. a set A.

Definition II.3.12

1. φ_e^A (or $\{e\}^A$) is the e-th function of n variables, partial recursive in A:

$$\varphi_e^A(\vec{x}) \simeq \{e\}^A(\vec{x}) \simeq \mathcal{U}(\mu y T_{n,1}(e, \vec{x}, \hat{c}_A(y), y))$$

2. $\varphi_{e,s}^A$ (or $\{e\}_s^A$) is the finite approximation of φ_e^A of level s:

$$\varphi_{e,s}^A(\vec{x}) \simeq \{e\}_s^A(\vec{x}) \simeq \begin{cases} \varphi_e^A(\vec{x}) & \text{if } (\exists y < s)T_{n,1}(e, \vec{x}, \hat{c}_A(y), y)) \\ \text{undefined} & \text{otherwise} \end{cases}$$

First Recursion Theorem

The Normal Form Theorem for partial recursive functionals implies that a computation tree of $F(\alpha, x)$ is finite, and thus uses only a finite number of values of α. We state explicitly the basic properties of oracle computations that can be deduced from it. These are key to the proof of the First Recursion Theorem.

Corollary II.3.13 (Kleene [1952], Davis [1958]) *If $F(\alpha, x)$ is a partial recursive functional, and $F(\alpha, x) \simeq y$, then:*

1. **compactness:** *for some finite function $u \subseteq \alpha$, $F(u, x) \simeq y$*

2. **monotonicity:** *if $\alpha \subseteq \beta$, then $F(\beta, x) \simeq y$.*

Proof. By definition of $T_{m,n}$, a value of F is computed by using a finite approximation of α. Thus compactness is immediate, and monotonicity follows from the fact that a finite approximation of α is also a finite approximation of any extension of it. □

Exercise II.3.14 *Compactness and monotonicity are equivalent to the unique condition*

$$F(\alpha, x) \simeq y \Leftrightarrow (\exists u \text{ finite})(u \subseteq \alpha \wedge F(u, x) \simeq y).$$

We can define a function α by writing down conditions that its values must satisfy. If these conditions involve α itself, then the definition takes the general form

$$\alpha(x) \simeq F(\alpha, x),$$

for some partial functional F (not necessarily recursive). Any function satisfying the equation just written is called a **fixed-point** of F, and could be considered as defined by the given conditions. But the usual intent of a definition is not only to specify all the necessary information, but also to rule out all additional, not explicitly stated information. Thus we may consider a function as defined by the given conditions if it is the **least fixed-point** of F: in this case the function does not contain arbitrary information, on top of what is directly implied by the definition.

Let us consider some examples:

1. $F(\alpha, x) \simeq \alpha(x)$.
   Every partial function is a fixed-point of F, and the least fixed-point is thus the completely undefined function. In particular, F is a total functional with total fixed-points, but the least one is not total.

2. $F(\alpha, x) \simeq \alpha(x) + 1$,
   Now there is exactly one fixed-point, the completely undefined function. In particular, F is a total functional with no total fixed-point.

3. Given $R(x, y)$ recursive, let

$$F(\alpha, x, y) \simeq \begin{cases} y & \text{if } R(x, y) \\ \alpha(x, y+1) & \text{otherwise.} \end{cases}$$

   Then F is a partial recursive functional, and if α is the smallest fixed-point, then α is partial recursive and

$$\alpha(x, y) \simeq \mu z(z \geq y \wedge R(x, z)).$$

   In particular, $\alpha(x, 0) \simeq \mu y R(x, y)$ (see also p. 158).

Theorem II.3.15 First Recursion Theorem (Kleene [1952]) *Every partial recursive functional $F(\alpha, x)$ admits a least fixed-point, and this is recursive. In other words, there is a partial recursive function α such that:*

1. $\forall x(\alpha(x) \simeq F(\alpha, x))$

2. $\forall x(\beta(x) \simeq F(\beta, x)) \Rightarrow \alpha \subseteq \beta$.

Proof. Define by induction :

α_0 everywhere undefined
$$\alpha_{n+1}(x) \simeq F(\alpha_n, x) \ .$$

Intuitively, we first take all the values of F that can be computed without any call to α (i.e. by using the completely undefined function as oracle). Then, at any given stage, we compute those values that use calls to α that can be answered because they have already been computed at previous stages.

By induction on n, we show that $\alpha_n \subseteq \alpha_{n+1}$.

- For $n = 0$ this follows because α_0 is completely undefined.

- If $\alpha_n \subseteq \alpha_{n+1}$, let $\alpha_{n+1}(x) \simeq F(\alpha_n, x) \simeq y$. By monotonicity we have $F(\alpha_{n+1}, x) \simeq y$. Then $\alpha_{n+2}(x) \simeq y$, and $\alpha_{n+1} \subseteq \alpha_{n+2}$.

It thus makes sense to consider the limit α of the α_n's:

$$\alpha(x) \simeq y \Leftrightarrow \exists n(\alpha_n(x) \simeq y).$$

α is partial recursive, because the α_n's are partial recursive, uniformly in n. Moreover:

1. α is a fixed-point of F.
   If $\alpha(x) \downarrow$ then, for some n,

   $$\alpha(x) \simeq \alpha_{n+1}(x) \simeq F(\alpha_n, x).$$

   Since $\alpha_n \subseteq \alpha$, by monotonicity $\alpha(x) \simeq F(\alpha, x)$.
   If $F(\alpha, x) \downarrow$ then, by compactness, only finitely many values of α are used in the computation, and thus α_n, for a big enough n, will suffice. Then

   $$F(\alpha, x) \simeq F(\alpha_n, x) \simeq \alpha_{n+1}(x) \simeq \alpha(x).$$

2. α is the smallest fixed-point of F.
   Suppose $F(\beta, x) \simeq \beta(x)$, for all x. By induction on n we have $\alpha_n \subseteq \beta$, and thus $\alpha \subseteq \beta$:

   - $\alpha_0 \subseteq \beta$ because α_0 is completely undefined.

   - If $\alpha_n \subseteq \beta$ then, by monotonicity,

   $$\alpha_{n+1}(x) \simeq F(\alpha_n, x) \simeq F(\beta, x) \simeq \beta(x)$$

   and thus $\alpha \subseteq \beta$. $\quad\Box$

The proof of the First Recursion Theorem gives *a computation procedure for the least fixed-point* of $F(\alpha, x)$: take the usual computation tree for F, and any time a node calling for a computation of a value $\alpha(z)$ is reached, continue the computation by substituting $F(\alpha, z)$ to $\alpha(z)$. If every branch of the tree comes to an end, then α has been completely discharged, and the value is obtained.

This shows that there is nothing mysterious about the definition of a function in terms of itself. What actually happens is that the values are defined in terms of previously obtained ones (in a precise sense, although not always a trivial one, like in the case of primitive recursion). Of course, in general, there will be values which can be computed without any call to previous values. Otherwise the function is really defined circularly, and it is then the completely undefined function.

As for the Fixed-Point Theorem, *the First Recursion Theorem, together with composition and case definition, generates the partial recursive functions.* Kleene [1978], [1981a], [1985] reverses our approach, and develops Recursion Theory by taking the least fixed-point as a primitive schema. This approach is quite natural, and it accounts for the name 'recursive' for functions whose values are somehow defined (by recurrence) using other values of the same function. The disadvantage of this approach is that it needs partial recursive functionals for the definition of partial recursive functions, and thus it does not provide an intrinsic characterization of the latter (which is defined simultaneously with the former). The approach with indices and the Fixed-Point Theorem (II.2.15) is similar, but without this drawback.

This brings us to the obvious analogies and differences between the First Recursion Theorem and the Fixed-Point Theorem II.2.10 (or, equivalently, the Second Recursion Theorem II.2.13). The former produces not only a fixed-point, but the least one. The latter has a wider range of application, since it does not require any extensionality (in the sense that $\varphi_{f(e)}$ is not necessarily a functional on partial recursive functions, see p. 154). Actually, a case can be made that the Second Recursion Theorem is more general than the First Recursion Theorem:

Proposition II.3.16 (Rogers [1967]) *Let $F(\alpha, x)$ be a partial recursive functional, and ψ be the partial recursive function*

$$\psi(e, x) \simeq F(\varphi_e, x).$$

The fixed-point of ψ produced by the proof of the Second Recursion Theorem is the least fixed-point of F.

Proof. First note that ψ is indeed partial recursive, by the Substitution Property II.3.8. Let e be the fixed-point of ψ produced by the proof of II.2.13: then

$$\varphi_e(x) \simeq F(\varphi_e, x),$$

and φ_e is a fixed-point of F.

Recall that $\varphi_e(x) \simeq \varphi_{S^1_1(a,a)}(x)$ for an appropriate a and, by definition of S^m_n (see II.1.7), to compute $\varphi_{S^1_1(a,a)}(x)$ we must first compute $\psi(S^1_1(a,a), x)$, i.e. $\psi(e, x)$. In other words, whenever $\varphi_e(x)$ converges, its computation is longer than that of $\psi(e, x)$.

To see that φ_e is the least fixed-point of F, let β be any fixed-point of F. We show that $\varphi_e \subseteq \beta$, by induction on the length of computations. Suppose $\varphi_e(x) \downarrow$: then $F(\varphi_e, x) \downarrow$ and, by compactness, only a finite subfunction u of φ_e is used in the computation. Choose u minimal: obviously, all the values of u have to be computed before we get $\psi(e, x) \simeq F(\varphi_e, x)$. Thus all the computations of values of u have length smaller than the length of computation of $\varphi_e(x)$, and by induction hypothesis we have $u \subseteq \beta$. Then

$$\varphi_e(x) \simeq F(\varphi_e, x) \simeq F(u, x) \simeq F(\beta, x) \simeq \beta(x),$$

by monotonicity. $\quad\square$

The last result shows that *the fixed-point operator defined by the proof of II.2.10 (which is the analogue of the fixed-point operator $\mathcal{Y}$ in λ-calculus) is actually a least fixed-point operator.*

Recursive programs $\star$

Let us look at extended 'while' programs (I.5.5), obtained by adding to the assignment statements the name P (for an unspecified program). A **recursive 'while' program** is an extended 'while' program, preceded by an instruction of the kind:

procedure P.

The program is interpreted as follows: the extended 'while' program defines a program depending on the unspecified program P, and the recursive clause added at the beginning tells that P must be the program defined by the extended 'while' program itself. Thus P is a program with a self-referential flavor.

As 'while' programs define partial recursive functions, *extended 'while' programs define partial recursive functionals*. The First Recursion Theorem tells that *there is a 'while' program defining a recursive function which is the least fixed-point of the extended 'while' program*. Thus recursive programs are in principle avoidable, in the sense of being equivalent to usual 'while' programs (or flowcharts) with no recursive calls. A direct method to translate recursive programs into flowcharts equivalent to them is given by Strong [1971].

However, in practice, using recursive programs is a very useful tool, since it often allows us to write easy and concise programs (with self-referential calls), in situations which might require elaborate nonrecursive programs. Thus recursive programs are widely used (after McCarthy [1963]) in programming languages allowing definitions (like those of the ALGOL family, and LISP), as well in languages naturally suited for recursive calls (like PROLOG).

A computational approach to recursive programs can be deduced from the remarks after the First Recursion Theorem: it corresponds to *substituting every occurrence of P in the computation, with the whole program* (complete replacement). A detailed study of computation procedures more efficient than this (i.e. requiring less substitutions), but still sufficient to compute the least fixed-point of a recursive program, can be found in Manna [1974].

Topological digression

An instructive way to look at the results on partial recursive functionals is by taking a more general stand, and considering the set $\mathcal{P}$ of partial (unary) functions as a topological space, following Uspenskii [1955] and Nerode [1957] (see Kelley [1955] for a reference on topology). This is easily done, by noting that $\mathcal{P}$ can be viewed as a product space S^ω, with $S = \omega \cup \{\uparrow\}$, $\uparrow$ being a distinguished element (for the undefined value). A natural topology on S is defined by taking as open sets all subsets of ω, and the space S (thus no nontrivial open set contains $\uparrow$, and S is not a Hausdorff space). Then $\mathcal{P}$ can be given the product topology. This is called the **positive information topology**, since a countable basis for it consists of the finite functions (together with $\mathcal{P}$ itself), which contain a finite amount of positive information (specifying the values for a finite set of arguments). More precisely, the basic open sets are

$$\hat{u} = \{\alpha : u \subseteq \alpha\},$$

where u is a finite function. These sets form a basis because they are closed under intersection:

$$\hat{u} \cap \hat{v} = \{\alpha : u \cup v \subseteq \alpha\},$$

where $u \cup v$ is undefined on x if both $u(x)$ and $v(x)$ are undefined, or both are defined and different.

Proposition II.3.17 $X \subseteq \mathcal{P}$ *is open if and only if both of the following hold:*

1. $\alpha \in X \Rightarrow$ for some finite $u, u \in X \wedge u \subseteq \alpha$.

2. $\alpha \in X \wedge \alpha \subseteq \beta \Rightarrow \beta \in X$.

Proof. Let X be open. Then X is a union of $\hat{u}$'s. If $\alpha \in X$, then $\alpha \in \hat{u}$ for some $\hat{u} \subseteq X$. But then $u \subseteq \alpha$, and $u \in X$ because $u \in \hat{u}$. If moreover $\alpha \subseteq \beta$, then $u \subseteq \beta$ and $\beta \in \hat{u}$, so $\beta \in X$.

Suppose now that the two conditions hold for X. Then X is the union of the $\hat{u}$'s such that $u \in X$ (and hence it is open), because: if $\alpha \in X$ then, by 1, there is $u \in X$ such that $u \subseteq \alpha$, so $\alpha \in \hat{u}$; and if $\alpha \in \hat{u}$ for some $u \in X$, then $u \subseteq \alpha$ and $\alpha \in X$ by 2. $\square$

An equivalent way of restating the characterization of open sets is: X is open if and only if

$$\alpha \in X \Leftrightarrow (\exists u \text{ finite})(u \in X \wedge u \subseteq \alpha).$$

We call X **effectively open** if the set of finite functions belonging to it is r.e. This requires the identification between finite functions and numbers in any effective way, e.g. let u be the number coding the finite function $\tilde{u}$ in the following way:

$$(\forall x \leq u)(exp(u, p_x) \neq 0 \Leftrightarrow \tilde{u}(x)\downarrow \wedge exp(u, p_x) = \tilde{u}(x) + 1).$$

In the following, for simplicity of notations, we will not distinguish between a finite function $\tilde{u}$ and the number u coding it.

Proposition II.3.18 (Uspenskii [1955], Nerode [1957]) *A function* $F : \mathcal{P} \to \mathcal{P}$ *is continuous if and only if it is compact and monotone.*

Proof. Since the $\hat{u}$'s are a basis for the topology, F is continuous if and only if $F^{-1}(\hat{u})$ is open, for every finite u.

- Let F be continuous, and $F(\alpha)(x) \simeq y$ (note that $F(\alpha)$ is a function from ω to ω). The finite function $u(x) \simeq y$ (undefined otherwise) defines the open set $\hat{u}$, so $F^{-1}(\hat{u})$ is open, and α is in it. By the characterization of open sets, $\alpha \in \hat{v}$ for some $\hat{v} \subseteq F^{-1}(\hat{u})$: since $v \in F^{-1}(\hat{u})$, we have $F(v)(x) \simeq y$ and $v \subseteq \alpha$, hence F is compact. If moreover $\alpha \subseteq \beta$, then $v \subseteq \beta$ and $\beta \in \hat{v}$, so $F(\beta)(x) \simeq y$, and F is monotone.

- Let now F be compact and monotone. We want to show that $F^{-1}(\hat{u})$ is open, using the characterization of open sets given above.

  If $\alpha \in F^{-1}(\hat{u})$ then $u \subseteq F(\alpha)$: u is finite, and for each pair (x_i, y_i) such that $u(x_i) \simeq y_i$ is $F(\alpha)(x_i) \simeq y_i$. By compactness of F, $F(u_i)(x_i) \simeq y_i$ for some finite $u_i \subseteq \alpha$. Let w be the union of the u_i's: then $w \subseteq \alpha$ and $u \subseteq F(w)$, so $w \in F^{-1}(\hat{u})$ and 1 of II.3.17 is verified.

  If $\alpha \in F^{-1}(\hat{u})$ and $\alpha \subseteq \beta$, $F(\alpha) \in \hat{u}$ and so $u \subseteq F(\alpha)$. By monotonicity of F, $u \subseteq F(\beta)$. So $\beta \in F^{-1}(\hat{u})$ and 2 of II.3.17 is verified. Then $F^{-1}(\hat{u})$ is open. $\square$

The behavior of a continuous function on $\mathcal{P}$ is then completely determined by its behavior on finite functions:

$$F(\alpha) = \bigcup\{F(u) : u \text{ finite} \wedge u \subseteq \alpha\}$$

(notice that $F(u)$ is not a finite function, in general). The function $\beta(u) \simeq F(u)$ for finite u's is called the **modulus of continuity** of F, and under suitable coding it can be thought of as a partial function from ω to ω. We call a function **effectively continuous** if its modulus of continuity is partial recursive.

We have considered functionals, i.e. functions from $\mathcal{P} \times \omega$ to ω, but there is an obvious correspondence between functionals and functions from $\mathcal{P}$ to $\mathcal{P}$: if $F(\alpha, x)$ is a functional then $\lambda x.\, F(\alpha, x)$ is such a function, and if F is such a function then $F(\alpha)(x)$ is a functional. Thus the result just proved shows that:

Proposition II.3.19 (Uspenskii [1955], Nerode [1957]) *The partial recursive functionals are effectively continuous.*

On the other hand, the converse fails.

Proposition II.3.20 (Sasso [1971], [1975]) *There are effectively continuous functionals which are not partial recursive.*

Proof. The idea is simple: define $F(\alpha)$ on x by using two possible values of α, e.g.

$$F(\alpha)(x) \simeq \begin{cases} 0 & \text{if } \alpha(2x) \simeq 0 \vee \alpha(2x+1) \simeq 0 \\ \text{undefined} & \text{otherwise.} \end{cases}$$

Then F is effectively continuous, but it should not be partial recursive, because a computation tree of $F(\alpha)(x)$ might get stuck on one undefined value of α, even when the other is defined. To turn this into a proof, we build α in such a way that $F(\alpha)$ is not the value of any partial recursive functional. Let

$$F_e(\alpha) \simeq \lambda x.\mathcal{U}(\mu y T_{1,1}(e, x, \alpha, y))$$

be an enumeration of the partial recursive functionals of one function variable. We want α such that $F(\alpha) \not\simeq F_e(\alpha)$ for every e, and we define it by stages. Start with α_0 being the completely undefined function. Having α_n, choose x such that both $2x$ and $2x + 1$ are greater than all the values for which α_n is defined, as well as those for which α_n is and must remain undefined because so required by the construction at previous stages. Consider $F_n(\alpha_n)(x)$. There are two cases:

- $F_n(\alpha_n)(x) \downarrow$.
  Then only values of α_n which have already been defined have been used in the computation. Since $\alpha_n \subseteq \alpha$, $F_n(\alpha)(x) \downarrow$ by monotonicity. It is then enough to have $F(\alpha)(x) \uparrow$, and this only requires leaving α undefined on both $2x, 2x + 1$ at this and later stages.

- $F_n(\alpha_n)(x)\uparrow$.

  Then either it diverges by using only convergent values of α_n in the computation, and then it will remain divergent, or some value of α_n used in the computation is undefined. In the latter case, pick up one: it is enough that also α be undefined on it, to force $F_n(\alpha)(x)\uparrow$ as well, since the computation is always going to be stuck on this value. We then want $F(\alpha)(x)\downarrow$, and this is ensured but letting one of $\alpha(2x)$ or $\alpha(2x+1)$ be 0. Since one of $2x, 2x+1$ is free (because both were at the beginning of this stage, and at most one of them is required to remain so by the previous work), define α_{n+1} as the extension of α_n that gives it value 0.

Then $\alpha = \bigcup_{n \in \omega} \alpha_n$ satisfies the requirements. $\square$

An equivalent way of looking at $\mathcal{P}$ is by considering it as a partially ordered set, under $\subseteq$. In this case $\mathcal{P}$ is a **chain-complete partial ordering**, in the sense that every linearly ordered chain has a least upper bound (which is just the union of the chain). We can state the First Recursion Theorem in full generality, by considering any chain-complete partial ordering $(D, \sqsubseteq)$.[8] Call $\sqcup$ the l.u.b. operation, and $\bot$ the least element (which exists, being the l.u.b. of the empty chain). A function $f : D \to D$ is:

monotone if it preserves the partial ordering

continuous if it preserves l.u.b.'s of chains.

A continuous function on an uppersemilattice D is monotone, since

$$
\begin{aligned}
x \sqsubseteq y &\Rightarrow y = x \sqcup y \\
&\Rightarrow f(y) = f(x) \sqcup f(y) \\
&\Rightarrow f(x) \sqsubseteq f(y).
\end{aligned}
$$

The general existence theorem for least fixed-points is the following:

Theorem II.3.21 (Knaster [1928], Tarski [1955], Abian and Brown [1961]) *If $(D, \sqsubseteq)$ is a chain-complete partial ordering, and f is a monotone function on it, then f has a least fixed-point.*

Proof. Iterate f transfinitely, starting with $\bot$ and taking l.u.b.'s at limit stages (which always exist, since D is chain-complete):

$$
\begin{aligned}
x_0 &= \bot \\
x_{\alpha+1} &= f(x_\alpha) \\
x_\beta &= \bigsqcup_{\alpha < \beta} f(x_\alpha), \text{ if } \beta \text{ limit.}
\end{aligned}
$$

---

[8]The partial ordering $\sqsubseteq$ should not be confused with the partial ordering of sequence numbers introduced on p. 89.

This defines a nondecreasing chain in D (because f is monotone), whose length cannot exceed the maximal length of chains of D. Let x_{α_0} be its last element, which must exist by the definition at limit stages (otherwise the l.u.b. of the chain would produce a bigger member). Then $f(x_{\alpha_0}) = x_{\alpha_0}$, otherwise x_{α_0} would not be the last element of the chain.

And x_{α_0} is not only a fixed-point of f, but the least one: if y is any fixed-point of f, by induction we have $x_\alpha \sqsubseteq y$ for each α (in particular $x_{\alpha_0} \sqsubseteq y$):

- $x_0 \sqsubseteq y$, because $\bot$ is the least element of D

- if $x_\alpha \sqsubseteq y$ then, by monotonicity and the fact that y is a fixed-point of f,

$$x_{\alpha+1} = f(x_\alpha) \sqsubseteq f(y) = y$$

- if $x_\alpha \sqsubseteq y$ for all $\alpha < \beta$, then $x_\beta \sqsubseteq y$ by definition of l.u.b.   □

In particular, continuous functions on a chain-complete partial ordering have least fixed-points. Continuity being stronger than monotonicity, we might however expect a stronger result. Indeed, *if f is continuous then only ω iterations are necessary to reach the least fixed- point*:

$$f(x_\omega) = f(\bigsqcup_{n \in \omega} x_n) = \bigsqcup_{n \in \omega} f(x_n) = \bigsqcup_{n \in \omega} x_{n+1} = x_\omega.$$

Exercises II.3.22 More on Fixed-Points. a) Call f expansionary if $x \sqsubseteq f(x)$. *Every expansionary function on a chain-complete partial ordering has a fixed-point.*

b) A complete lattice is a partially ordered set in which every subset has l.u.b. and g.l.b. There is *a purely algebraic proof of the existence of least fixed-points for monotone functions on a complete lattice.* (Knaster [1928], Tarski [1955]) (Hint: consider the set $\{x : f(x) \sqsubseteq x\}$, which contains every fixed-point of f. Its g.l.b. is the least fixed-point of f.)

c) *A nonmonotone function on a chain-complete partial ordering need not have a least fixed-point.* (Hint: on $\mathcal{P}$, consider

$$F(\alpha, x) \simeq \begin{cases} 1 & \text{if } \alpha(x) \simeq 0 \\ 0 & \text{otherwise.} \end{cases}$$

Note that $F(\alpha, x)$ is defined also when $\alpha(x)$ is not.)

Before we can apply II.3.21 to $\mathcal{P}$, we have to show that the two notions of continuity obtained by viewing it as a topological space and as a chain-complete partial ordering coincide.

Proposition II.3.23 *For a functional F on $\mathcal{P}$ the following are equivalent:*

1. F is compact and monotone

2. F preserves l.u.b.'s of (countable) chains.

Proof. Let F be compact and monotone, and $\{\alpha_\beta\}_{\beta<\beta_0}$ be a chain of partial functions with l.u.b. α. We want

$$F(\alpha) = \bigcup_{\beta<\beta_0} F(\alpha_\beta).$$

By monotonicity, from $\alpha_\beta \subseteq \alpha$ we have $F(\alpha_\beta) \subseteq F(\alpha)$, and thus

$$\bigcup_{\beta<\beta_0} F(\alpha_\beta) \subseteq F(\alpha).$$

Now suppose $F(\alpha)(x) \simeq y$. By compactness, $F(u)(x) \simeq y$ for some finite function $u \subseteq \alpha$, and there must be β such that $u \subseteq \alpha_\beta$. By monotonicity we then have $F(\alpha_\beta)(x) \simeq y$, and hence

$$F(\alpha) \subseteq \bigcup_{\beta<\beta_0} F(\alpha_\beta).$$

Thus F preserves l.u.b.'s of arbitrary chains.

Let F now preserve l.u.b.'s of countable chains. It is automatically monotone (see p. 189). For compactness, suppose $F(\alpha)(x) \simeq y$. Since α is the l.u.b. of a chain $\{\alpha_n\}_{n\in\omega}$ of finite functions (e.g. α_n can be taken to be the restriction of α to the arguments from 0 to n), we have

$$F(\alpha) = \bigcup_{n\in\omega} F(\alpha_n),$$

and thus $F(\alpha_n)(x) \simeq y$ for some finite $\alpha_n \subseteq \alpha$. $\quad\square$

We then have the general fixed-point existence result:

Theorem II.3.24 (Uspenskii [1955], Nerode [1957]) *Let F be a functional on $\mathcal{P}$:*

1. *if F is continuous, F has a least fixed-point*

2. *if F is effectively continuous, its least fixed-point is partial recursive.*

We have proved the results of this subsection by using only a few particular properties of $\mathcal{P}$, namely:

1. Two partial functions are compatible if they have a common extension.

2. A chain of partial functions is a set of compatible elements, containing the l.u.b. of every pair of functions in it.

3. Every chain has a l.u.b.

4. Every function is the least upper bound of a chain of finite subfunctions.

We can then guess that the results just proved would extend to partial orderings with similar properties: we only have to turn them into definitions. Consider a partially ordered set $(D, \sqsubseteq)$. Two elements are **compatible** if they have a common extension. A non-empty set of compatible elements is **directed** if it contains, together with each pair of elements, a common extension of them. D is a **complete partial ordering (c.p.o.)** if every directed set has a l.u.b. in D. A function on D is **continuous** if it preserves l.u.b.'s of directed sets.

To be able to extend property 4 above we need a notion of finite element in D. The idea comes from the observation that in $\mathcal{P}$ the finite functions are exactly those functions which, whenever covered by the l.u.b. of a chain of functions, are already covered by some element of the chain. We then call an element of D **compact** if, whenever it is bounded by the l.u.b. of a directed set A, it is already bounded by an element of A. The compact elements play the role of finite elements, and we call D **algebraic** if, for every element x, the compact elements below it form a directed set with l.u.b. x. Different names used for algebraic c.p.o.'s are **complete f_0-spaces** (Ershov [1972]) and **domains** (Scott [1982]).

Having defined a notion of finite element we can now impose on D the **Scott topology** (Scott [1972]), generated by the basic open sets

$$\hat{u} = \{x : \ u \text{ compact } \wedge u \sqsubseteq x\},$$

and consider the associated notion of continuity. *On algebraic c.p.o.'s the two notions of continuity coincide, and the behavior of a continuous function is completely determined by its behavior on compact elements.*

To be able to extend the notion of effective continuity, we need the possibility of effectively manipulating the compact elements. We call D an **effective algebraic c.p.o.** if there is an enumeration (see p. 237) of the set of compact elements, which makes the relations of partial ordering and of compatibility, as well as the operation of l.u.b. for compatible elements, recursive. A continuous function on D is then **effectively continuous** if its modulus of continuity is partial recursive, w.r.t. the given enumeration.

Iteration and fixed-points ⋆

The method of proof used for the First Recursion Theorem and its refinements is an old and fruitful one. It was first applied by Newton [1669] to obtain zeros of differentiable real functions f, by starting from any point x_0 and iterating

$$F(x_n) = x_n - \frac{f(x_n)}{f'(x_n)}.$$

By the geometrical interpretation of derivative this procedure converges to a fixed-point x of F such that $f(x) = 0$, whenever the starting point x_0 is sufficiently close to x and the derivative of f is not zero (under these hypotheses, each step doubles the number of the correct decimal digits of the approximation).

It is natural to ask whether the method would always produce a zero, starting from any point x_0. Cayley [1879] showed that for quadratic functions on the complex plane every point not on the line bisecting the segment connecting the two zeros converges to one of them. Barna [1956] showed that in general Newton's method works for real polynomials with real zeros, except on a set of measure 0. The result fails for complex coefficients (Curry, Garnett, and Sullivan [1983]).

This prompts the more general question of when the iteration of a given function with fixed-points would lead to one of them (in our setting that is always the case if f is expansionary, see II.3.22.a). A modern study case is the iteration of the quadratic function $x^2 + c$ on the complex numbers, which has at most two fixed-points. For a given c, the set of points whose set of iterations is bounded (containing in particular the points converging to a fixed-point, and those with a periodic behavior) has an interesting boundary, called the **Julia set** J_c (Julia [1918]), which is either connected or pulverized. The shapes of these sets describe, for different c's, an incredible variety of forms, whose behavior is related to the position of c in the **Mandelbrot set**, defined as the set of c such that J_c is connected (Mandelbrot [1980]). This amazing set contains, on a smaller scale, a reproduction of all Julia sets J_c. It also has a sort of universality, since it appears in the study of the iteration of a great number of functions (precisely, of any function one of whose iterations behaves like x^2 in some portion of the plane). One aspect of Julia (and, less stringently, of Mandelbrot) sets is their self-resemblance: they contain copies of themselves and present, at any scale of observation, the same global character. Sets with this property are called **fractals**, and are useful to describe natural phenomena involving chaotic behavior. See Mandelbrot [1982] and Peitgen and Richter [1986] for more on this.

Another example of a situation in which the iteration of a function always produces a fixed-point is the one described by **Banach Fixed-Point Theorem** (Banach [1922]): *a contraction function on a complete metric space has a unique fixed-point, given by the limit of the iterations of f starting on any point, where f is a contraction if $|f(x) - f(y)| \leq c \cdot |x - y|$ for some fixed constant c such that $0 \leq c < 1$ ($|x - y|$ being the distance between x and y).* Indeed, given any x then, by induction,

$$|f^{(n+1)}(x) - f^{(n)}(x)| \leq c^n \cdot |f(x) - x|.$$

By the triangular inequality,

$$|f^{(n+m)}(x) - f^{(n)}(x)| \leq \sum_{i<m} |f^{(n+i+1)}(x) - f^{(n+i)}(x)|$$

$$\leq \left(\sum_{i<m} c^{n+i}\right) \cdot |f(x) - x|.$$

Thus $\{f^{(n)}(x)\}_{n \in \omega}$, and hence $\{f^{(n+1)}(x)\}_{n \in \omega}$, converge to a point x_0. Since f is continuous, the latter also converges to $f(x_0)$, and thus $f(x_0) = x_0$. Moreover, if x_1 is another fixed-point of f then

$$|x_0 - x_1| = |f(x_0) - f(x_1)| \leq c \cdot |x_0 - x_1|$$

and, being $c < 1$, it must be $x_0 = x_1$.

Models of λ-calculus (part I) $\star$

The only objects of λ-calculus are terms, and thus a model will be a set D over which terms are interpreted, in such a way that provably equal terms are interpreted by the same object. The presence of the λ-operator forces some terms to be interpreted as functions acting on terms, and thus some elements of D will have to be interpreted as functions on D, i.e. an appropriate set $[D \to D]$ of functions from D to D will have to be identified with a subspace of D. This rules out, by cardinality considerations, the naive approach of taking as $[D \to D]$ the set of *all* functions from D to D. Finally, both λ-abstraction and functional application will have to be interpreted over D, the first being a term formation operator, the second because of the β-conversion rule. See p. 223 and Meyer [1982] for a discussion of the notion of model of λ-calculus.

Methods and results of the last subsection are all relevant to the present subject. To start with, $\mathcal{P}$ *is a model of λ-calculus, when $[\mathcal{P} \to \mathcal{P}]$ is taken to be the set of continuous functionals* (this model is connected to the **graph model** of Plotkin [1972] and Scott [1975], see below). The reason is that continuous functionals are completely determined by their behavior on finite functions, and thus they can be coded by a single function (the modulus of continuity). The only additional piece of information needed to turn this into a model of λ-calculus is to note that functional application of continuous functionals is continuous, and thus again representable in $\mathcal{P}$. This also applies to λ-abstraction, as well as to the operator Fix, that produces the least fixed-point of continuous functionals. By II.3.16, *the interpretation of Fix coincides with the interpretation of the combinator $\mathcal{Y}$* (Park [1970]). What $\mathcal{P}$ does not automatically provide for, is a model of extensionality (η-rule, p. 82, according to which every term is identified with a function): the embedding of $[\mathcal{P} \to \mathcal{P}]$

into $\mathcal{P}$ does not cover $\mathcal{P}$ itself (two different moduli of continuity can code the same continuous function).

As it might be expected, there is nothing special about $\mathcal{P}$: its place can be taken by any other **reflexive c.p.o.**, i.e. any (algebraic) c.p.o. in which the set $[D \rightarrow D]$ of continuous functions is embeddable (meaning that there are continuous functions

$$i : D \rightarrow [D \rightarrow D] \quad \text{and} \quad j : [D \rightarrow D] \rightarrow D$$

such that $i \circ j$ is the identity on $[D \rightarrow D]$). E.g., in the literature $\mathcal{P}$ is usually replaced by $\mathcal{P}(\omega)$, the set of subsets of ω, with the positive information topology generated by the finite sets. Here functions are represented by their graphs, hence the name 'graph model' quoted above.

To get *models of extensionality* as well, we need not only to embed $[D \rightarrow D]$ into D, but to identify the two. This obviously sounds like a fixed-point, and we do have a general existence theorem (II.3.21). To be able to apply it, we need to consider the function $F(D) = [D \rightarrow D]$ as monotone over a giant chain-complete partial ordering, whose members are (algebraic) c.p.o.'s. We thus need an appropriate partial ordering on them, and it turns out that the following natural one will work:

$$D \trianglelefteq D' \quad \Leftrightarrow \quad \text{there are continuous functions}$$
$$i : D \rightarrow D' \text{ and } j : D' \rightarrow D$$
$$\text{such that } j(i(x)) = x \text{ and } i(j(y)) \sqsubseteq y.$$

The intuition behind this definition is that D' is supposed to contain more information than D, and D has to be embedded into it. So $i(x)$ is the element in D' that corresponds to x in D and, going back, we simply recover it (since x and $i(x)$ have the same information content). But an element y in D' might not correspond to any element in D, and $j(y)$ is only the closest approximation to it in D: going back, we might lose some information, and thus only $i(j(y)) \sqsubseteq y$ holds.

With this notion of partial ordering for c.p.o.'s, it is easy to show that the l.u.b.'s of a chain $\{D_n\}_{n \in \omega}$ exists: we just have to take the c.p.o. D_∞ that sums up exactly the information contained in the chain. Note that if $j_n : D_{n+1} \rightarrow D_n$, then $j_n(x)$ is an element in D_n that approximates x. Thus we only need to consider chains $\langle x_n \rangle_{n \in \omega}$ such that $x_n \in D_n$ and $j_n(x_{n+1}) = x_n$, as elements of D_∞. Then x_n may be thought of as the approximation of the chain at stage n (think of x as a real number given by its decimal expansion: then x_n is the expansion truncated at the n-th digit, and the projection j_n simply cuts out the last digit of x_{n+1}).

First, F is a function on c.p.o.'s (since $[D \rightarrow D]$ can be turned into a c.p.o. by pointwise ordering its elements). Second, F commutes with l.u.b.'s of chains (and it is thus continuous). Moreover, F is expansionary, in the sense that $D \trianglelefteq [D \to D]$. Thus not only F admits least fixed-points, and they can be reached in at most ω iterations of F: they exist above any given D. Thus *any (algebraic) c.p.o. D can be embedded in an extensional model D_∞ of λ-calculus* (Scott [1972]).

All this machinery can be appropriately formalized in the cartesian closed category of (algebraic) c.p.o.'s (where continuous functions, l.u.b.'s and monotonicity of F become, respectively, morphisms, inverse limits and functorial covariance, and D is a reflexive object if $D^D \trianglelefteq D$). Again, there is nothing special about this category: *any reflexive object with enough points in a cartesian closed category gives rise to a model of λ-calculus and, conversely, any model of λ-calculus is a reflexive object with enough points in a cartesian closed category* (the condition of having enough points basically meaning that different functions must behave differently on some point) (Scott [1980a], Koymans [1982]). This completely characterizes the models of λ-calculus.

For history, exposition, and philosophy see Scott [1973], [1975], [1975a], [1976], [1977], [1980], [1980a]. For technical development and other models see Stoy [1977], Barendregt [1981], Beeson [1985], Hindley and Seldin [1986].

Different notions of recursive functionals $\star$

We have defined partial recursive functionals, obtained from a uniformization of the notion of relative recursiveness $\leq_T$ for partial functions. The topological approach developed above suggests the consideration of the broader class of effectively continuous functionals as well. These are expressible in the form

$$F(\alpha, x) \simeq z \Leftrightarrow (\exists u)(u \subseteq \alpha \wedge \varphi(x, u) \simeq z),$$

with φ partial recursive and u, here and in the following, (code of a) finite function. By II.1.11, an equivalent formulation is

$$F(\alpha, x) \simeq z \Leftrightarrow (\exists u)(u \subseteq \alpha \wedge R(x, u, z)),$$

with R r.e. Of course, such an expression defines a functional only if the relation R (that embodies the graph of the continuity modulus) gives consistent answers. We thus have two ways to turn this general form into a definition of functional:

1. A **partial recursive operator** (Myhill [1961a], Rogers [1967]) is a functional F that can be defined as

$$
\begin{aligned}
F(\alpha, x) \simeq z \quad \Leftrightarrow \quad & (\exists u)(u \subseteq \alpha \wedge R(x, u, z)) \wedge \\
& (\forall u')(\forall z')(u' \subseteq \alpha \wedge R(x, u', z') \Rightarrow z = z'),
\end{aligned}
$$

with R r.e. relation.

2. A **recursive operator** (Davis [1958], Rogers [1967]) is a functional F that can be defined as

$$F(\alpha, x) \simeq z \Leftrightarrow (\exists u)(u \subseteq \alpha \wedge R(x, u, z))$$

with R consistent r.e. relation, i.e. such that

$$R(x, u, z) \wedge R(x, u', z') \wedge u, u' \text{ compatible} \Rightarrow z = z'.$$

Thus the difference between partial recursive operators and recursive operators is in the consistency condition required on R: only relative to the input function for partial recursive operators, and global for recursive operators.

It is clear that, from a computational point of view, the first notion is not satisfactory, since it requires a consistency check relative to a given partial function. It is not surprising that *there is no Enumeration Theorem for partial recursive operators*. A better way of thinking of partial recursive operators is in terms of functionals on multivalued functions or, better still, sets, rather than on partial functions, and the associated reducibility notion for sets is called **enumeration reducibility**:

$$A \leq_e B \quad \Leftrightarrow \quad \text{for some r.e. relation } R,$$
$$x \in A \Leftrightarrow (\exists u)(D_u \subseteq B \wedge R(x, u)).$$

The structure of degrees associated with this reducibility (called **partial degrees**) will be studied in Chapter XIV.

An *Enumeration Theorem for recursive operators* (Rogers [1967]) can easily be obtained by stepping from an enumeration $\{W_e^3\}_{e \in \omega}$ of all the r.e. ternary relations, to an enumeration $\{W_{f(e)}^3\}_{e \in \omega}$ of the consistent ones, where $W_{f(e)}^3$ is obtained from an enumeration of W_e^3, by dropping the triple (x, u', z') whenever a triple (x, u, z) with u, u' consistent has already been generated.

The basic difference between recursive operators on the one hand, and partial recursive functionals on the other, is computational: the latter are **serial**, and a computation gets stuck if it tries to query the oracle for an undefined value; the former are **parallel**, and can get around undefined values by dovetailing computations (for general discussions of parallelism, see Elgot, Robinson and Rutledge [1967], Shepherdson [1975], Cook [1982]). We thus arrive at the notion of partial recursive functional by relativizing deterministic approaches to computability, like recursiveness (by adding a function to the initial ones, as we did in this section), Turing machine computability (by adding an additional state that calls for the oracle, Turing [1939]) or flowchart computability (by adding assignment instructions of the kind

$$\boxed{X := g(Y_1, \ldots, Y_n)}$$

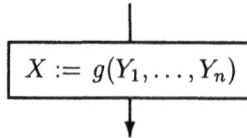

as in Ianov [1958]). On the other hand, we arrive at the notion of (partial) recursive operator by relativizing nondeterministic approaches, like Herbrand-Gödel computability, or representability in formal systems (by adding a functional letter to the constants of the language, Kleene [1943]).

We then have three notions of relative computability for partial functions: if $\alpha \simeq F(\beta)$, we say that

$$\alpha \leq_T \beta \quad \text{if } F \text{ is a partial recursive functional}$$
$$\alpha \leq_{wT} \beta \quad \text{if } F \text{ is a recursive operator}$$
$$\alpha \leq_e \beta \quad \text{if } F \text{ is a partial recursive operator.}$$

They are respectively called **Turing, weak Turing** and **enumeration reducibilities**.

Proposition II.3.25 (Myhill [1961a], Sasso [1971])

1. $\alpha \leq_T \beta \Rightarrow \alpha \leq_{wT} \beta$, but not conversely

2. $\alpha \leq_{wT} \beta \Rightarrow \alpha \leq_e \beta$, but not conversely

3. $\leq_T, \leq_{wT}$ and $\leq_e$ coincide on total functions.

Proof. The first two implications are obvious, and II.3.20 gives a counterexample to the converse of the first. A counterexample to the converse of the second is given by

$$F(\alpha, x) \simeq z \quad \Leftrightarrow \quad (z = 1 \wedge \alpha(2x) \simeq 1) \vee$$
$$(z = 0 \wedge \alpha(2x+1) \simeq 1).$$

Clearly, F is not consistent in general, but it is consistent for those α's coding a set A in the following way:

$$\alpha(2x) \quad \simeq \quad \begin{cases} 1 & \text{if } x \in A \\ \text{undefined} & \text{otherwise} \end{cases}$$

$$\alpha(2x+1) \quad \simeq \quad \begin{cases} 1 & \text{if } x \notin A \\ \text{undefined} & \text{otherwise.} \end{cases}$$

In this case $F(\alpha)$ is the characteristic function of the set coded by α, and $F(\alpha) \leq_e \alpha$. But if $F(\alpha) \leq_{wT} \alpha$, then $F(\alpha) \leq_{wT} \beta$ for any β extending α, (this uses the consistency property, and it fails in general for $\leq_e$). In particular, this holds e.g. when β is the constant function with value 1, and then α codes a recursive set (because $F(\alpha)$ is the characteristic function of this set, and it is recursive since β is). Thus, if α codes a nonrecursive set, $F(\alpha) \leq_e \alpha$ but $F(\alpha) \not\leq_{wT} \alpha$.

To show that the three reducibilities agree on total functions, suppose that $\alpha \leq_e g$, i.e.

$$\alpha(x) \simeq z \Leftrightarrow (\exists u)(u \subseteq g \wedge R(x, u, z))$$

with R r.e. Then there is Q recursive such that

$$\alpha(x) \simeq z \Leftrightarrow (\exists u)(\exists y)(u \subseteq g \wedge Q(x, u, z, y)),$$

and

$$\alpha(x) \simeq (\mu t[(t)_1 \subseteq g \wedge Q(x, (t)_1, (t)_2, (t)_3)])_2 .$$

This is a correct application of the μ-operator, since $u \subseteq g$ is recursive in g (because we know that g is total, and we only have to check the values). Thus $\alpha \leq_T g$. Note that, when β is partial, $u \subseteq \beta$ would only be r.e. in β, and thus the μ-operator could not be directly applied. □

Since we are only dealing with degrees of sets, we will not study the degrees of partial functions under $\leq_T$ and $\leq_{wT}$. We refer to Sasso [1975] for a survey of results (see also Casalegno [1985] for later advances). For degrees under $\leq_e$, see Chapter XIV.

Higher Types Recursion Theory $\star$

Let T_n, the set of **objects of type n**, be defined as:

$$
\begin{aligned}
T_0 &= \omega \\
T_{n+1} &= \text{total functions from } \bigcup_{i \leq n} T_i \text{ to } \omega.
\end{aligned}
$$

We have so far introduced notions of recursiveness for objects of type 1 (functions) and 2 (functionals). Recursiveness can be successively extended to objects of higher types by iterating the process that has led to the theory of this section, namely by relativizing in a uniform way the notion at lower types. This has led (at least for restricted functionals, defined only on total objects) to a highly developed theory, begun by Kleene in two epochal papers ([1959], [1963]).

A basic fact which is a source of divergence with the classical theory, and of technical complications, is the *failure of compactness*: a computation tree still has finite branches when a value is defined, but it does not need to be finitely branching (and hence finite), since a higher type object might need to know infinitely many values of its arguments, already at level two, as for:

$$E(f, x) \simeq \begin{cases} 0 & \text{if } (\forall z)(f(z) = 0) \\ 1 & \text{otherwise.} \end{cases}$$

This rules out a Normal Form Theorem (since a computation can no longer be coded by a number), and *the μ-operator loses its central role* (as it is restricted to a search on ω, and as no natural analogue is available at higher levels). The normal form is replaced by the *Enumeration Theorem*, which states that a functional may see one of its number arguments as an index and simulate the functional coded by it. This principle is taken as primitive, something which requires an index approach from the very beginning, see p. 158. The μ-operator is partly replaced by *Selection Theorems* (Gandy [1967]), which are analogues of the Uniformization Theorem II.1.13, and provide for choice operators at the number level (the choice being now made according to the stage of generation in the inductive definition of the predicate to be uniformized).

Another source of trouble is *equality* for objects of a given type, which is recursive only at the integer (type 0) level. The theory has shown a dichotomy between objects strong enough to compute equality at the level of their arguments (**normal objects**, for which strong regularity results - like the Selection Theorems quoted above - hold, and which are, from a topological point of view, effectively discontinuous, see Grilliott [1971], Hinman [1973]), and those which are not (see below). A treatment of various parts of Recursion Theory for normal, higher types objects may be found in Hinman [1978], and Fenstad [1980]. For an elegant introduction, see Kechris and Moschovakis [1977].

A completely different way to extend Recursion Theory to higher types is by taking compactness and monotonicity, and hence continuity, as a basis. This has been proposed by Davis [1959], Kleene [1959a], and Kreisel [1959], and has led to a theory of **countable functionals**, quite similar to the one of this section. The basic aspect is that *computation trees are still finite*, and hence at the same level of numbers. A functional $F(g)$ of type 2 is continuous if and only if its values are determined by finite pieces of its arguments: this can be formally coded by an associate $f : \omega \to \omega$ which, on increasingly long segments of g, takes value 0 for a certain time, and a constant greater than 0 from a certain point on. Thus

$$F(g) = f(\hat{g}(n)) - 1, \text{ for any sufficiently large } n.$$

This notion can easily be generalized to higher types. If F is of type $n+1$, and G of type n, an **associate** for F is any function f as above such that, for any associate g of G,

$$F(G) = f(\hat{g}(n)) - 1, \text{ for any sufficiently large } n.$$

A (recursive) countable functional is any functional with a (recursive) associate. Note that countable functionals act on finite pieces not of their arguments (at type > 2), but of associates for them. Thus associates are algorithms to compute functionals, that act on algorithms for their arguments.

It is clear that a notion of continuity is underlying that of countable functionals. One way to make this precise (Hyland [1979]) is to consider **filter spaces** (see Kelley [1955]), i.e. sets with collections of abstract approximations for each element (each approximation being a filter of sets, and at least the following being approximations to an element: the principal ultrafilter generated by that element, as well as any filter including an approximation to it). A continuous function between filter spaces is a function that respects approximations. Products and sets of continuous maps on filter spaces are still filter spaces, and they nicely commute. Thus a hierarchy, coinciding exactly with that of countable functionals, can be obtained inductively, by starting with ω (organized as a filter space by taking, as only approximation to n, the principal ultrafilter containing n).

Another way to look topologically at the countable functionals (Ershov [1972]) is via **effective algebraic c.p.o.'s** (see p. 192). Products of and sets of continuous maps on effective algebraic c.p.o.'s are still such, and they nicely commute. Then a class $\mathcal{C}$ of **partial continuous functionals of higher type** can be defined inductively, by letting:

$$
\begin{aligned}
C_0 &= \text{finite sets with the canonical enumeration} \\
C_{(\sigma,\tau)} &= C_\sigma \times C_\tau \\
C_{\sigma \to \tau} &= [C_\sigma \to C_\tau].
\end{aligned}
$$

The objects in these spaces are not, in general, extensional or total. *The (effective) extensional, hereditarily total functionals are exactly the (recursive) countable functionals* (Ershov [1974]). The condition on totality is required because the countable functionals are, by definition, restricted functionals.

Other approaches to countable functionals can be found in Feferman [1977], Normann [1981], Longo and Moggi [1984]. For an exposition of the theory, see Ershov [1977], and Normann [1980], [1981].

There are relationships between the two extensions of recursiveness to higher types quoted above, surveyed in Gandy and Hyland [1977]. In one direction, *a recursive higher type object restricted to countable arguments is countable* (Kleene [1959]). In the other direction, *there are recursive countable functionals which are not recursive as higher type objects*, e.g. the functional that gives a modulus of continuity for any continuous functional of type 2 on compact sets of functions (Tait). However, *the countable functionals can be generated by an extension of the schemata generating the higher type recursive objects* (Normann [1981]).

Computability on abstract structures ⋆

Recursion Theory is a notion of computability on the structure of the natural numbers. Recursiveness relative to given oracles suggests one possible way of extending the notion of computability to abstract structures

$$\langle A, f_1, \ldots, f_n, R_1, \ldots, R_m \rangle$$

with a given domain A, and functions and relations on it. Of course, the various approaches to relative computability, which are equivalent in the case of the structure of natural numbers (see p. 197), generalize in ways which are not necessarily equivalent on abstract structures. We briefly describe the most popular ones, and leave to Kreisel [1971], Ershov [1981] and Shepherdson [1985] the discussion of other possible approaches, including the case of partial structures.

Formalized algorithmic procedures (fap) (Ianov [1958], Ershov [1960], Luckham and Park [1964], Paterson [1968], Friedman [1971a], Kfoury [1974]) are finite lists of labelled instructions (called *program schemata*), each one of the following kind:

$$y := f_i(\vec{x})$$
$$\text{if } R_j(\vec{x}) \text{ go to } p, \text{ else go to } q$$
$$\text{stop.}$$

This is an extension of (unstructured) flowchart computability, using the functions and relations of the structure as primitives, and with the variables ranging over the elements of the domain. See Manna [1974] for an extensive treatment.

Fap with counting (fapc) (Friedman [1971a], Kfoury [1974]) are extensions of fap using also natural numbers, over which range special variables $X, Y, \ldots$ The following counting instructions are also available:

$$X := 0$$
$$X := X + 1$$
$$\text{if } X = Y \text{ go to } p, \text{ else go to } q.$$

This is equivalent to expanding the given structure to a structure

$$\langle A \cup \omega, f_1, \ldots, f_n, R_1, \ldots, R_m, 0, pd, S \rangle$$

(with pd and S the predecessor and successor functions for natural numbers), and considering fap over it.

Recursive schemata (McCarthy [1963], Platek [1966]) are based on fixed-points, and allow for functions explicitly definable over the structure by case definitions (on the relations of the structure), and fixed-point operators. Alternatively, one could simply use recursive schemata

$$g_1(\vec{x_1}) = \text{if } P_1(\vec{x_1}) \text{ then } h_{1,1}(\vec{x_1}) \text{ else } h_{1,2}(\vec{x_1})$$
$$\cdots$$
$$g_p(\vec{x_p}) = \text{if } P_p(\vec{x_p}) \text{ then } h_{p,1}(\vec{x_p}) \text{ else } h_{p,2}(\vec{x_p}),$$

where the P_i's are predicates in their variables over the structure, and the h's are terms in their variables over the structure, possibly using the g_i's.

Generalized Turing algorithms (Friedman [1971a]) are programs for extended Turing machines which, on top of performing the usual operations and working with their own alphabet, can also handle elements of the domain of the structure on the tape, in the following way. First, the inputs are placed on the tape at the beginning, in adjacent cells to the left of the head, the rest of the tape being empty. Elements on the tape can be moved around, and possibly copied, one cell at a time. A new element can be introduced in the scanned cell, but only if it is the value of one of the functions of the structure for the arguments, placed in consecutive cells to the right of the head. Finally, the machine can test the validity of a predicate of the structure for the arguments, placed in consecutive cells to the right of the head.

Effective definitional schemata (Friedman [1971a], Gordon [1974]) are r.e. sequences of specifications, each giving a value (in the form of a term over the structure) under mutually exclusive conditions, expressed as quantifier free formulas over the structure. They correspond to effective infinitary case definitions, and generalize the approach to recursiveness by systems of equations (see p. 39).

Prime computability (Moschovakis [1969]) extends the original definition of (relative) recursiveness reformulated, as on p. 158, with indices and enumeration. The approach with indices is used because the μ-operator obviously works only for well-ordered structures. Indices are here elements of the domain of the structure, and to make this approach work we also need a coding mechanism. This is obtained by considering the schemata for recursion on the expanded structure

$$\langle A \cup I, f_1, \ldots, f_n, R_1, \ldots, R_m, 0, pd, \mathcal{S}, \mathcal{J}, \mathcal{R}, \mathcal{L} \rangle,$$

where I is a set of indices containing ω, $\mathcal{J}$ is a one-one pairing function on I, and $\mathcal{R}$ and $\mathcal{L}$ are its decoding functions.

Search computability (Moschovakis [1969]) adds to prime computability an unordered search operator, performed by an oracle that searches through arbitrarily large parts of the domain. This requires an approach with multiple-valued functions because the effect of the search operator is to take all elements satisfying a given condition, instead of selecting one (since there is, in general, no canonical choice). Prime and search computability correspond, respectively, to deterministic and nondeterministic computations on the given structure.

Despite the great variety of approaches, some of the equivalences which hold for the classical notions retain their validity. E.g., generalized Turing algorithms, effective definitional schemata, and prime computability are equivalent (Friedman [1971a], Gordon [1974]). We thus have only five notions available, namely fap, fapc, recursive schemata, prime and search computability. It should be clear that *both fapc and recursive schemata extend fap, in incomparable ways, and prime computability extends both.* Obviously, search computability is stronger than prime computability.

The obvious problem at this point is to choose the 'right' notion of computability on an abstract domain, among the nonequivalent ones. By Friedman [1971a] and Kfoury [1974], *fap's are equivalent to fapc's if the structure has analogues of the natural numbers, and the counting instructions can be defined.* By Moldestad, Stoltenberg-Hansen, and Tucker [1980a], *fapc's are equivalent to prime computability if term evaluation is fapc-computable (uniformly in the number of variables).* Thus these four approaches are equivalent for sufficiently rich structures (like all the infinite algebraic structures of common use, e.g. rings and fields of characteristic zero). It seems a reasonable assumption to allow the use of the natural numbers in computations as an auxiliary tool (for example, as the order of an element in a group, or as the characteristic of an element in a field), and to have term evaluation as a computable function. Under these assumptions, the four notions discussed above coincide with prime computability.

Search computability is also equivalent to a great number of other notions, see Moschovakis [1969a], Gordon [1970], [1974], and Grilliot [1971a]. In particular, it coincides with extended effective definitional schemata, using existential formulas in place of quantifier free ones. Also, for countable domains (with computable equality) search computability coincides with the very natural notion of ∀-**recursiveness** (Lacombe [1964]), so defined. Since the structure is countable, there are one-one onto functions between the domain and ω, each of which allows us to translate functions and relations from the domain to ω. A function or relation over the domain is ∀-recursive if its translation is, for all one-one onto translating functions.

It thus seems that *prime and search computability are the natural notions of deterministic and nondeterministic computability on an abstract domain.*

II.4 Effective Operations

The previous section introduced various notions of functionals, with the common characteristic that the function arguments are treated extensionally. This is not avoidable in general, but for the particular case of partial recursive functions we could also treat the arguments intensionally, through (codes for) their programs. In this section we study the intensional version of extensional recursive functionals, and their mutual relationships.

Effective operations on partial recursive functions

Having studied (effective) continuity on $\mathcal{P}$, it is natural to look at the same notion on its effective part, namely the set $\mathcal{PR}$ of partial recursive functions. We will consider *functionals on $\mathcal{PR}$* or, equivalently, *extensional functions on indices*, where f is extensional if

$$\varphi_e \simeq \varphi_{e'} \;\Rightarrow\; \varphi_{f(e)} \simeq \varphi_{f(e')}.$$

One approach consists in seeing $\mathcal{PR}$ *as a subspace of $\mathcal{P}$*. This automatically induces a topology, whose (effectively) open sets are the intersections of (effectively) open sets of $\mathcal{P}$, and whose (effectively) continuous functionals are the restrictions of (effectively) continuous functionals on $\mathcal{P}$.

Another approach exploits the fact that the members of $\mathcal{PR}$ have indices. We call a class of partial recursive functions a **completely r.e. class** if its index set (see p. 150) is r.e. (Dekker [1953a], Rice [1953]). The **Ershov topology** on $\mathcal{PR}$ (Ershov [1972]) is the topology generated by taking the completely r.e. classes of partial recursive functions as basic open sets. Thus open sets are unions of completely r.e. classes, and effectively open sets can be defined as r.e. unions of basic open sets (since r.e. unions of r.e. sets are still r.e., the effectively open sets are just the completely r.e. classes). Unraveling the definition of continuity shows that a function f induces a continuous functional if, for every e and every r.e. index set A such that $f(e) \in A$, there is an r.e. index set containing e and included in $f^{-1}(A)$. This suggests the following notion of effective continuity, obtained by considering extensional recursive functions (since then $f^{-1}(A)$ is itself r.e.).

Definition II.4.1 *An **effective operation** on $\mathcal{PR}$ is a functional F on $\mathcal{PR}$ induced by a recursive function, i.e. $F(\varphi_e) \simeq \varphi_{f(e)}$ for some extensional, recursive function f.*

We now wish to compare the two approaches just introduced. To start with, the next result implies that *the two topologies on $\mathcal{PR}$ (and hence the two notions of continuity) coincide.* Recall (see p. 187) that we identify finite functions and numbers by explicitly coding the graphs of finite functions in some effective way.

Theorem II.4.2 (Myhill and Shepherdson [1955], McNaughton, Shapiro) *A class $\mathcal{A}$ of partial recursive functions is completely r.e. if and only if there is an r.e. set A such that*

$$\varphi_e \in \mathcal{A} \Leftrightarrow (\exists u \text{ finite})(u \in A \wedge u \subseteq \varphi_e).$$

Proof. If there is such an r.e. set A, then $\mathcal{A}$ is completely r.e. To see if $\varphi_e \in \mathcal{A}$, generate simultaneously A and the graph of φ_e, and wait until, for some $u \in A$, the graph of the finite function coded by u (which can be completely decoded from u in finitely many steps) is contained in the part already generated of the graph of φ_e.

Conversely, let $\mathcal{A}$ be nonempty and completely r.e., and $\theta\mathcal{A}$ be its index set. The obvious guess for an r.e. set of finite functions generating $\mathcal{A}$ is the set of its finite functions: they are indeed an r.e set (let $g(u)$ be an index of the finite function (coded by) u: to see if $u \in \mathcal{A}$, it is enough to see if $g(u) \in \theta\mathcal{A}$, because $\theta\mathcal{A}$ contains *all* indices of any function in $\mathcal{A}$). We then have to prove the following.

1. $\mathcal{A}$ *is monotone on partial recursive functions*, i.e. if $\alpha \in \mathcal{A}$, β is partial recursive, and $\alpha \subseteq \beta$, then $\beta \in \mathcal{A}$.

   Suppose that α and β are partial recursive functions such that $\alpha \in \mathcal{A}$, $\alpha \subseteq \beta$, but $\beta \notin \mathcal{A}$. Let f be a recursive function such that

$$\varphi_{f(e)}(x) \simeq y \Leftrightarrow \alpha(x) \simeq y \vee [e \in \mathcal{K} \wedge \beta(x) \simeq y].$$

   The function $\varphi_{f(e)}$ is well-defined, because $\alpha \subseteq \beta$. Then

$$e \in \overline{\mathcal{K}} \Leftrightarrow \varphi_{f(e)} \in \mathcal{A}.$$

   Indeed, if $e \in \mathcal{K}$ then $\varphi_{f(e)} \simeq \beta$, while if $e \notin \mathcal{K}$ then $\varphi_{f(e)} \simeq \alpha$.

   Since $\mathcal{A}$ is completely r.e., $\overline{\mathcal{K}}$ is r.e., contradiction.

2. $\mathcal{A}$ *is compact*, i.e. if $\alpha \in \mathcal{A}$ then, for some finite function $u \subseteq \alpha$, $u \in \mathcal{A}$.

   Suppose that α is a partial recursive function such that α is in $\mathcal{A}$, but no finite subfunction of it is. Let f be a recursive function such that

$$\varphi_{f(e)}(x) \simeq y \Leftrightarrow e \notin \mathcal{K}_x \wedge \alpha(x) \simeq y.$$

Then
$$e \in \overline{\mathcal{K}} \Leftrightarrow \varphi_{f(e)} \in \mathcal{A}.$$

Indeed, if $e \in \mathcal{K}$ then $e \in \mathcal{K}_x$ for all sufficiently big stages x, and hence $\varphi_{f(e)}$ is a finite subfunction of α, and cannot be in $\mathcal{A}$. Conversely, if $e \notin \mathcal{K}$ then $e \notin \mathcal{K}_x$, for any x, and $\varphi_{f(e)} \simeq \alpha$.

Since $\mathcal{A}$ is completely r.e., $\overline{\mathcal{K}}$ is r.e., contradiction.   □

By the definition of an effectively open set of functions (p. 187), the result shows that a completely r.e. class is the intersection of $\mathcal{PR}$ with an (effectively) open set of $\mathcal{P}$. Conversely, it is immediate that the intersection of $\mathcal{PR}$ with a basic open set of $\mathcal{P}$ is completely r.e. This shows the coincidence of the two topologies (and hence of the two notions of continuity) on $\mathcal{PR}$ introduced above. Moreover, *the effectively open sets (in both topologies) are exactly the completely r.e. classes*. It only remains to show that *the two notions of effective continuity coincide* as well.

Theorem II.4.3 (Myhill and Shepherdson [1955], Uspenskii [1955])
The effective operations on $\mathcal{PR}$ are exactly the restrictions to $\mathcal{PR}$ of effectively continuous functionals on $\mathcal{P}$.

Proof. Let F be an effectively continuous functional: then the function $\varphi(e, x) \simeq F(\varphi_e, x)$ is partial recursive because, for some r.e. relation R,

$$F(\varphi_e, x) \simeq z \Leftrightarrow (\exists u \text{ finite})(u \subseteq \varphi_e \wedge R(x, u, z)),$$

and thus the graph of φ is r.e. (see II.1.11). By the S^m_n-Theorem, there is f recursive such that

$$\varphi_{f(e)}(x) \simeq \varphi(e, x) \simeq F(\varphi_e, x),$$

and F is an effective operation.

Let now f define an effective operation F on $\mathcal{PR}$. It is enough to show that F is compact and monotone on $\mathcal{PR}$, i.e. that for some r.e. relation R,

$$F(\varphi_e, x) \simeq z \Leftrightarrow (\exists u \text{ finite})(u \subseteq \varphi_e \wedge R(x, u, z)).$$

Then we can extend F to all of $\mathcal{P}$, by just letting

$$F(\alpha, x) \simeq z \Leftrightarrow (\exists u \text{ finite})(u \subseteq \alpha \wedge R(x, u, z)).$$

But since

$$F(\varphi_e, x) \simeq z \Leftrightarrow \varphi_{f(e)}(x) \simeq z,$$

for fixed x and z the class

$$\{\varphi_e : F(\varphi_e, x) \simeq z\}$$

is completely r.e. Then all the work has already been done in II.4.2: there is an r.e. set A such that

$$F(\varphi_e, x) \simeq z \Leftrightarrow (\exists u \text{ finite})(u \subseteq \varphi_e \wedge u \in A).$$

This holds, as said, for fixed x, z. But they appear as parameters in the argument, and then A depends uniformly on them. In other words, A is really an r.e. relation R depending on u, x, z. Then, this time for every x and z,

$$F(\varphi_e, x) \simeq z \Leftrightarrow (\exists u \text{ finite})(u \subseteq \varphi_e \wedge R(x, u, z))$$

as wanted.   □

Note that the proof shows that every effective operation is induced by a *unique* continuous functional (which turns out to be recursive), since its values are completely determined by the behavior on finite functions (which are all partial recursive). This, in turn, uniquely determines a continuous functional.

Exercise II.4.4 *There is a continuous functional whose restriction to $\mathcal{PR}$ is not an effective operation.* (Dekker and Myhill [1958]) (Hint: there are only countably many effective operations on $\mathcal{PR}$, so it is enough to build uncountably many continuous functionals, with distinct restrictions to $\mathcal{PR}$. Given a partial function β, let

$$F_\beta(\alpha) \simeq \begin{cases} \text{constant function 1} & \text{if } dom\,\alpha \cap dom\,\beta \neq \emptyset \\ \text{completely undefined} & \text{otherwise.} \end{cases}$$

If $dom\,\beta_0 \neq dom\,\beta_1$, then F_{β_0} and F_{β_1} differ on $\mathcal{PR}$.)

Effective operations on total recursive functions

The notion of effective operation can be generalized from $\mathcal{PR}$ to any class $\mathcal{A}$ of partial recursive functions:

Definition II.4.5 *An* **effective operation on a set** $\mathcal{A}$ *of partial recursive functions is a functional F mapping $\mathcal{A}$ to $\mathcal{A}$ and such that, for some partial recursive function ψ,*

$$\varphi_e \in \mathcal{A} \Rightarrow \psi(e){\downarrow} \wedge F(\varphi_e) \simeq \varphi_{\psi(e)}.$$

Under which conditions for $\mathcal{A}$ does an analogue of Theorem II.4.3 hold? We first prove an important special case, and then discuss the general situation. In the following, $\mathcal{R}$ is going to be the class of total recursive functions.

Theorem II.4.6 (Kreisel, Lacombe and Shoenfield [1957], Čeĭtin [1959]) *The effective operations on $\mathcal{R}$ are exactly the restrictions of the effectively continuous functionals mapping $\mathcal{R}$ to $\mathcal{R}$.*

Proof. As in the previous theorem, an effectively continuous functional mapping $\mathcal{R}$ to $\mathcal{R}$ induces an effective operation on $\mathcal{R}$. Let now F be an effective operation on $\mathcal{R}$. We approximate the result by a series of steps.

1. If f is recursive and $F(f, x) = z$, for any preassigned length k there is a function u of finite support (i.e. 0 almost everywhere) which agrees with f up to k, and such that $F(u, x) = z$.
   This provides a sort of *compactness*, with functions of finite support taking place of finite functions (which cannot be considered here because they are not total). Suppose the claim fails: then a recursive function t can be defined, such that $\varphi_{t(e)}$ agrees with f up to the maximum of k and (if $e \in \mathcal{K}$) the least stage in which e is generated in $\mathcal{K}$, and afterwards agrees with a function u of finite support and such that $F(u, x) \neq z$ (which exists by the hypothesis that the claim fails). Then, by definition,

$$e \in \mathcal{K} \Leftrightarrow F(\varphi_{t(e)}, x) = z,$$

   and $\mathcal{K}$ is recursive, contradiction.

2. If f is recursive and $F(f, x) = z$, there is a fixed length k_f (which can be found effectively) such that, for any function u of finite support which agrees with f up to k_f, $F(u, x) = z$.
   This provides a sort of *modulus of continuity*: the value of $F(f, x)$ is determined by the initial segment of f up to k_f. The existence proof is the same as given above (dropping k), but we want to obtain k_f effectively. Then define $\varphi_{t(e)}$ as above, but as a partial function (i.e. look for the appropriate u, and if this is found, proceed as above). Notice that the set

$$e \in C \Leftrightarrow F(\varphi_{t(e)}, x) \simeq z$$

   is r.e. If a is an index for it, it cannot be $a \in \overline{\mathcal{K}}$, otherwise by construction $\varphi_{t(a)} = f$: then $a \in C = W_a$, and hence $a \in \mathcal{K}$. Thus it must be $a \in \mathcal{K}$: let k_f be the least stage in which a is generated in $\mathcal{K}$. There cannot exist a function u of finite support which agrees with f up to k_f, and such that $F(u, x) \neq z$, otherwise $\varphi_{t(a)}$ would be one such, forcing $a \notin C$, against the fact that $a \in \mathcal{K}$. But F is defined for any recursive function, and then if u agrees with f up to k_f, it must be $F(u, x) = z$.

3. The two parts just shown prove that, for f recursive,

$$F(f, x) = z \Leftrightarrow (\exists u \text{ of finite support})$$
$$(u \text{ agrees with } f \text{ up to } k_f \wedge F(u, x) = z).$$

   One direction follows from part 1, since if $F(f, x) = z$ there are functions u of finite support, agreeing with f up to any preassigned length, and such that $F(u, x) = z$.

Conversely, let u agree with f up to k_u, and $F(u, x) = z$. By part 1 there is v of finite support, agreeing with f up to the maximum of k_f, k_u. By part 2 then $F(v, x) = F(f, x)$, because v agrees with f up to k_f. And $F(v, x) = F(u, x)$ because v agrees with u up to k_u (since v agrees with f up to k_u, and so does u).

4. F can then be extended to an effectively continuous functional on all of $\mathcal{P}$, by letting

$$F(\alpha, x) \simeq z \Leftrightarrow (\exists u \text{ of finite support})$$
$$(u \text{ agrees with } \alpha \text{ up to } k_u \wedge F(u, x) = z)$$

for any partial function α. □

Note that the condition that an effectively continuous functional map $\mathcal{R}$ to $\mathcal{R}$ is weaker than totality (which requires mapping any total function to a total function, not only recursive ones).

Exercise II.4.7 *There is an effective operation on $\mathcal{R}$, which is not the restriction to $\mathcal{R}$ of any total effectively continuous functional.* (Rogers [1967]) (Hint: let $F(\alpha, x)$ be the first z generated in $\mathcal{K}$ such that $\alpha(z) = \varphi_z(z)$, if it exists. F is an effective operation on $\mathcal{R}$, but if G is a total effectively continuous functional agreeing with F on $\mathcal{R}$, let

$$\alpha(z) = \begin{cases} \varphi_z(z) + 1 & \text{if } z \in \mathcal{K} \\ 0 & \text{otherwise.} \end{cases}$$

Consider any value $G(\alpha, x)$, the finite part of α used in the computation, and a function β agreeing with α on that part and on all elements $\leq G(\alpha, x)$, and equal to $\varphi_a(a)$ otherwise, where a is the first element generated in $\mathcal{K}$ and $> G(\alpha, x)$. Then $F(\beta, x)$ is both equal and different to a, contradiction.)

Effective operations in general ⋆

An analysis of the proofs of Theorems II.4.3 and II.4.6 from a constructive point of view has been provided by Beeson [1975], and Beeson and Ščedrov [1984]. In particular, both results are derivable in Intuitionistic Arithmetic from Markov's Principle, but without this principle they are not derivable even in Intuitionistic Set Theory. Both results describe phenomena which hold in settings much more general than those here considered.

For the first, given an **effective algebraic c.p.o.** X (p. 192) we can consider its effective part X_e, consisting of the elements for which the set of compact approximations is r.e. (w.r.t. the given enumeration). The enumeration of the compact elements of X generates an enumeration of X_e, and we can then impose on it the **Ershov topology**, with the completely r.e. sets (w.r.t.

the enumeration of X_e) as basic sets. Then II.4.2 and II.4.3 are respectively generalized as: *Ershov's topology on X_e is induced by Scott's topology on X,* and *the morphisms on X_e as an enumerated set (i.e. functions commuting with the enumeration, through a recursive function) are exactly the restrictions to X_e of effectively continuous functions on X* (Ershov [1973]).

For the second, Čeitin [1959], [1962] and Moschovakis [1964] show that it holds on every **effective complete separable metric space**, i.e. a space with a recursive metric function on the recursive reals (see p. 213), with a dense r.e. subset, and in which one can effectively compute limits of recursive, recursively convergent Cauchy sequences. A more general framework for the theorem is the notion of **effective topological space**, see Nogina [1966], [1978].

Spreen and Young [1984] give a uniform generalization of both results above by considering **countable T_0-spaces** satisfying certain effectivity requirements, which hold for both $\mathcal{PR}$ and $\mathcal{R}$.

We return now to the question formulated above, and look for conditions on $\mathcal{A}$ ensuring that the effective operations on $\mathcal{A}$ are exactly the restrictions of effectively continuous functionals mapping $\mathcal{A}$ to $\mathcal{A}$. The positive results proved above apply to more general situations, but some conditions on $\mathcal{A}$ are necessary.

Exercises II.4.8 *The effective operations on $\mathcal{A}$ are exactly the restrictions of effectively continuous functionals mapping $\mathcal{A}$ on $\mathcal{A}$, in the following cases:*

a) *$\mathcal{A}$ is completely r.e.* (Myhill and Shepherdson [1955]) (Hint: see II.4.3.)

b) *$\mathcal{A}$ is the intersection of a completely r.e. class with $\mathcal{R}$. Such classes are called* **totally r.e.** (Kreisel, Lacombe and Shoenfield [1957], Čeitin [1959]) (Hint: see II.4.6 and, when defining $\varphi_{t(e)}$, use functions of finite support which are in $\mathcal{A}$.)

Proposition II.4.9 (Pour El [1960], Myhill) *There is an effective operation on a class $\mathcal{A}$, which is not compact (and thus not the restriction of a continuous functional).*

Proof. Let $\mathcal{A}$ consist of the constant functions 0 and 1, and of the recursive functions which take a value different from 0 for an argument smaller than their minimal index. Let f be such that

$$\varphi_{f(e)}(x) \simeq \begin{cases} 0 & \text{if } (\forall z \leq e)(\varphi_e(z) \simeq 0) \\ 1 & \text{if } (\exists z \leq e)(\varphi_e(z)\downarrow \wedge \varphi_e(z) \not\simeq 0). \end{cases}$$

Then f defines an effective operation F on $\mathcal{A}$ which sends the constant function 0 to itself, and the other functions of $\mathcal{A}$ to the constant function 1. Suppose that only finitely many values determine the fact that $F(\alpha, x) \simeq 0$, and let k be bigger than all of them. There is a recursive function which is 0 up to k, and which takes a value different from 0 below its smaller index, e.g.

$$\alpha(x) \simeq \begin{cases} 0 & \text{if } x \neq k+1 \\ a & \text{if } x = k+1 \end{cases}$$

where a is different from $\varphi_e(k+1)$, for all $e \leq k$. Then $F(\alpha)$ is the constant function 1, despite the fact that α is 0 up to k. □

Proposition II.4.10 (Yates, Young [1968], Helm [1971]) *There is an effective operation on a class $\mathcal{A}$ which is effectively continuous on $\mathcal{A}$, and is the restriction to $\mathcal{A}$ of a continuous functional, but not of any effectively continuous functional.*

Proof. Let $\mathcal{A}$ be the class of (partial) recursive functions α such that $\alpha(0)$ is defined, and either is not in $\mathcal{K}$, or is promptly generated in it, at a stage smaller than the minimal index of α. Let f be such that

$$\varphi_{f(e)}(x) \simeq \begin{cases} 0 & \text{if } \varphi_e(0)\!\downarrow \wedge \varphi_e(0) \in \mathcal{K}_e \\ 1 & \text{if } \varphi_e(0)\!\downarrow \wedge \varphi_e(0) \notin \mathcal{K}_e. \end{cases}$$

Then f defines an effective operation F on $\mathcal{A}$, which sends α to the constant function 0 if $\alpha(0)$ is in $\mathcal{K}$, and to the constant function 1 otherwise. And the functional F' which does the same on all of $\mathcal{P}$ is an extension of F. Both F and F' are clearly compact (only $\alpha(0)$ is needed to determine the value) and monotone on their domains, hence continuous.

Suppose F has an extension G which is effectively continuous on all of $\mathcal{P}$. Then $\overline{\mathcal{K}}$ is r.e., because

$$x \in \overline{\mathcal{K}} \Leftrightarrow (\exists u \text{ finite})(G(u,0) \simeq 1 \wedge u(0) = x).$$

Indeed, if there is u as stated and $x \in \mathcal{K}$, then x is generated at some stage s. It is enough to take a partial recursive function α extending u, and with minimal index greater than s. Then $G(\alpha, 0) \simeq 1$ by monotonicity, but $F(\alpha, 0) \simeq 0$, contrary to the fact that G extends F: thus $x \in \overline{\mathcal{K}}$. Conversely, if $x \in \overline{\mathcal{K}}$, let $u(0) = x$. Then $F(u, 0) \simeq 1$ and, since G extends F, $G(u, 0) \simeq 1$. □

The two examples succeed because they both use not only values of the arguments, but also algorithms for them. The proofs are similar but quite different, because they fail to extend to all of $\mathcal{P}$ for distinct reasons. They provide examples as far apart as they can be: the first does not extend for purely topological reasons (not being compact), the second for purely computational ones (being extendable to a continuous functional, but not to an effectively continuous one).

Exercise II.4.11 Weak effective operations. A weak effective operation on $\mathcal{A}$ is defined as in II.4.5, without the condition that F map $\mathcal{A}$ to itself.

a) *If $\mathcal{A}$ is completely r.e., the weak effective operations on $\mathcal{A}$ are restrictions of effectively continuous functionals.* (Myhill and Shepherdson [1955]) (Hint: see II.4.3.)

b) *There is a weak effective operation on $\mathcal{R}$ which is not compact.* (Friedberg [1958b], Muchnik) (Hint: let $\varphi_{f(e)}(x)$ be 0 if φ_e is either 0 for all arguments up to e, or it coincides, up to the first argument in which is not 0, with a function φ_i that has index i smaller than that argument. Then f induces a weak effective operation F on $\mathcal{R}$ because it is extensional, and F is not compact, as in II.4.9.)

We have only touched on the subject of effective operations. For more information see Grzegorczyk [1955], Friedberg [1958a], Kreisel, Lacombe and Shoenfield [1959], Pour El [1960], Čeitin [1962], Lachlan [1964], Young [1968], [1969], Löb [1970], Helm [1971], and Freivald [1978].

Recursive analysis $\star$

The real numbers can be classically defined in many equivalent ways, either as particular sequences or sets of rational numbers, or axiomatically (up to isomorphism) as archimedean complete ordered fields. Since rational numbers can be effectively coded by integers, we have notions of recursive sequences and recursive sets of rational numbers, and different recursive analogues of the real numbers (Borel [1912], Turing [1936], Specker [1949]), using e.g. recursive decimal expansions, recursive Cauchy sequences with a recursive modulus of convergence, recursive Dedekind cuts, recursive sequences of nested intervals with length approaching zero. All of these notions turn out to be equivalent (Robinson [1951], Myhill [1953]), and thus a stable notion of a **recursive real number** is available. *The recursive real numbers form a countable subfield $\mathcal{R}$ of the reals, which contains all the algebraic reals, as well as the real zeros of the Bessel functions, and becomes algebraically closed by the adjunction of the imaginary unit* (Rice [1954]).

Recursive real numbers are coded by integers (the indices of their recursive presentations), and thus a notion of a **recursive function of recursive real variable** is available, as an effective operation on the codes. Under this definition, the field operations on $\mathcal{R}$ are recursive (on the codes obtained from the Cauchy sequences definition), and Moschovakis [1965] characterizes $\mathcal{R}$ up to recursive isomorphism, by a recursive analogue of the notion of an archimedean complete ordered field. Thus $\mathcal{R}$ appears to be a good recursive counterpart to the reals.

There are two different approaches to what may be called **recursive analysis**, which are respectively parts of constructive and classical mathematics. The first has been followed by the 'Markov school', which advocates a philosophy (Markov [1954]) in which every mathematical object is described by a word of a given alphabet, manipulated by Markov algorithms (p. 144) and discussed by means of constructive logic (intuitionistic logic plus Markov's principle). Under these philosophical assumptions, analysis *is* the study of $\mathcal{R}$ and of the recursive functions on it.

The second approach remains in the realm of classical logic and mathematics, and it can be seen as a study of the extent of constructivity in classical analysis, by an examination of which results remain valid when constructivized (either by the same proofs or by new ones), and which results instead fail. Once we live in a classical world, there is no particular reason to stick to recursive functions on $\mathcal{R}$, and a notion of **recursive functions of real variable** has been introduced (Grzegorczyck [1955], [1957], Lacombe [1955], [1957]) in analogy to the effectively continuous functionals, as opposed to effective operations. This provides a new level of analysis, since it isolates computable functions (defined on all the reals) among the classical ones.

Typical theorems of recursive analysis are that *every recursive function on $\mathcal{R}$ is continuous, although not necessarily uniformly so* (the positive part rephrases II.4.6, the negative one follows from the failure of the recursive analogue of König lemma, see V.5.25), and that *the least upper bound principle fails* (Specker [1949]) in the sense that there is a bounded, strictly increasing, recursive sequence of rationals which does not converge to any recursive real number (simply, the sequence of the $r_m = \sum_{n=0}^{m} 1/2^{f(n)}$, with f a recursive one-one function with a nonrecursive range).

For a treatment of recursive analysis, see Grzegorczyck [1959], Goodstein [1961], Kušner [1973], Aberth [1980], Pour El and Richards [1983a], Kreitz and Weihrauch [1984], and Beeson [1985]. Although not especially on recursive analysis, Bishop [1967] is also relevant.

A broader perspective can be obtained by looking at recursive interpretations of (intuitionistic) set theory, in which analysis can then be formalized. The **effective topos** (Hyland [1982]) and the **recursive topos** (Mulry [1982]) generalize, respectively, the constructive and classical approach to analysis (the first using recursive realizability, and the second using forcing). Analogies and differences are investigated in Sčedrov [1984], [1987], and Rosolini [1986].

II.5 Indices and Enumerations $\star$

In II.1.4 and II.1.9 we introduced a particular system of indices for partial recursive functions and r.e. sets, and later proved some of its properties. In this section we investigate systems of indices in general, both for all partial recursive functions and r.e. sets (giving new examples, some equivalent to the one we already know, and some radically different) and for subclasses of them.

Acceptable systems of indices

A number of results have been proved so far for recursive functions, and some of them mention indices in their statements, or use them in their proofs. The question naturally arises: do these results depend on the particular formalism chosen to work with (through which the indices were defined), or are they instead somehow model-independent?

To study the problem, we first note that the basic results mentioning indices are:

- Enumeration Theorem II.1.5

- S_n^m-Theorem II.1.7

- Padding Lemma II.1.6.

Indeed, the remaining results mentioning indices, or using them in their proofs, merely refer to these (e.g. the Fixed-Point Theorem II.2.10 follows from Enumeration and S_n^m-Theorem alone).

We now characterize the systems of indices for which these theorems hold.

Definition II.5.1 (Uspenskii [1956], Rogers [1967]) *We call a* **system of indices** *any family ψ of maps ψ^n from ω onto the set of n-ary partial recursive functions (by ψ_e^n we will indicate the partial recursive function corresponding to the index e). We say that:*

1. ψ satisfies **enumeration** *if for every n there is a number a such that*

$$\psi_a^{n+1}(e, x_1, \ldots, x_n) \simeq \psi_e^n(x_1, \ldots, x_n)$$

2. ψ satisfies **parametrization** *if for every m, n there is a total recursive function s such that*

$$\psi_{s(e,x_1,\ldots,x_n)}^m(y_1, \ldots, y_m) \simeq \psi_e^{m+n}(x_1, \ldots, x_n, y_1, \ldots, y_m).$$

As usual we will drop the mention of the number of variables, when this is understood.

An example of a system of indices is obviously given by the usual φ_e^n, which also satisfies both enumeration and parametrization: we call it the **standard system**. A system may resemble the standard one, in the sense that it is possible to go effectively from the former to the latter, and conversely:

Definition II.5.2 (Rogers [1958]) *A system of indices ψ is* **acceptable** *if, for every n, there are total recursive functions f and g such that*

$$\psi_e^n \simeq \varphi_{f(e)}^n \qquad \varphi_e^n \simeq \psi_{g(e)}^n.$$

Note that in practice the equivalence among different definitions of the class of partial recursive functions is usually proved precisely by showing that the systems of indices induced by them resemble one another, in the sense just given (see the proof of I.7.12). The next result shows that they satisfy the same basic properties, and also gives an intrinsic characterization of the notion of an acceptable system of indices.

Proposition II.5.3 (Rogers [1967]) *A system of indices is acceptable if and only if it satisfies both enumeration and parametrization.*

Proof. Suppose ψ satisfies both enumeration and parametrization. Then ψ is acceptable:

- Given ψ_e^n, by the enumeration property this is a recursive function of $n+1$ variables. Since systems of indices must enumerate all the partial recursive function, there is an index a for ψ_e^n with respect to φ. Then

$$\psi_e^n(\vec{x}) \simeq \varphi_a^{n+1}(e, \vec{x}) \simeq \varphi_{S_1^1(a,e)}^n(\vec{x}).$$

  Thus we can let $f(e) = S_1^1(a, e)$, and have $\psi_e^n \simeq \varphi_{f(e)}^n$.

- g is obtained, symmetrically, by using the Enumeration Theorem for φ and the parametrization property for ψ.

Conversely, suppose that ψ is acceptable.

- To show enumeration, note that

$$\psi_e^n(\vec{x}) \simeq \varphi_{f(e)}^n(\vec{x}).$$

  But $\varphi_{f(e)}^n(\vec{x})$ is partial recursive as a function of $e, \vec{x}$ (by the Enumeration Theorem), and thus it must admit an index a in the system ψ.

- To show parametrization, consider $\psi_e^{n+m}(\vec{x}, \vec{y})$. By the enumeration property just proved, this is a partial recursive function of $m+n+1$ variables (including e), and thus there is an index a for it with respect to φ:

$$\psi_e^{n+m}(\vec{x}, \vec{y}) \simeq \varphi_a^{m+n+1}(e, \vec{x}, \vec{y}) \simeq \varphi_{S_{n+1}^m(a,e,\vec{x})}^m(\vec{y}).$$

  Since g allows us to go back to the system ψ, if we let

$$s(e, \vec{x}) = g(S_{n+1}^m(a, e, \vec{x}))$$

  we get

$$\psi_e^{n+m}(\vec{x}, \vec{y}) \simeq \psi_{s(e,\vec{x})}^m(\vec{y}). \quad \square$$

Corollary II.5.4 *Every acceptable system of indices satisfies the Fixed-Point Theorem. In other words, given a recursive function f there is an index e such that* $\psi_e \simeq \psi_{f(e)}$.

Proof. The only properties used in the proof of the Fixed-Point Theorem are enumeration and parametrization. □

Exercises II.5.5 a) *A system satisfying enumeration alone is not necessarily acceptable.* (Friedberg [1958]) (Hint: a recursive enumeration without repetitions of the partial recursive functions, see II.5.23, does not satisfy II.5.6.)

b) *A system satisfying enumeration and fixed-point, i.e. the existence of a recursive function f such that, for* ψ_i *total,* $\psi_{\psi_i(f(i))} \simeq \psi_{f(i)}$, *is not necessarily acceptable.* (Machtey, Winklmann and Young [1978]) (Hint: let $\varphi_{h(e)}$ be a recursive enumeration without repetitions of the partial recursive functions, see II.5.23. Then $\psi_e \simeq \varphi_{h(\varphi_e(0))}$ satisfies enumeration and fixed-point, the latter because ψ_e depends extensionally on φ_e, and thus fixed-points transfer from φ to ψ. Note that any partial recursive function is equal to $\varphi_{h(i)}$ for a unique i, and then can be equal to ψ_e only when $\varphi_e(0) \simeq i$: thus it has an r.e. set of indices w.r.t. ψ. But e.g. the completely undefined function cannot have an r.e. set of indices in any acceptable system, and thus ψ is not acceptable.)

c) *A system satisfying enumeration and composition, i.e. the existence of a recursive function c such that* $\psi_e \circ \psi_i \simeq \psi_{c(e,i)}$, *is acceptable.* (Machtey, Winklmann and Young [1978]) (Hint: by coding the arguments, parametrization can be reduced to composition and its special case for sequence numbers, e.g. if $\psi_{f(x)}(y) = \langle x, y \rangle$ then

$$\psi_e(\langle x, y \rangle) \simeq \psi_e(\psi_{f(x)}(y)) \simeq \psi_{c(e,f(x))}(y).$$

And f can be defined by successive compositions of the two functions $g(y) = \langle 0, y \rangle$ and $h(\langle x, y \rangle) = \langle x + 1, y \rangle$.)

The results just proved show that *the notion of a universal machine, embodied in the enumeration property, is not sufficient for elementary Recursion Theory: it has to be supplemented with the notion of subcomputation*, through the possibility of either effectively incorporating data into a program (expressed by the parametrization property), or effectively concatenating programs (expressed by composition). On the other hand, the Fixed-Point Theorem, and thus the possibility of having self-referential programs (so-called recursive programs), does not appear to be as fundamental.

Notice that enumeration and parametrization imply, among other things, a back and forth translation between the function spaces

$$\omega^n \times \omega^m \to \omega \quad \text{and} \quad \omega^n \to (\omega^m \to \omega).$$

Indeed, a partial recursive function of $n + m$ variables becomes, by parametrization, a partial recursive function of n variables, whose values are (indices of) partial recursive functions of m variables. Conversely, any partial recursive function of n variables whose values are indices of partial recursive functions of m variables becomes, by enumeration and composition, a partial recursive function of $n + m$ variables.

To the reader acquainted with Category Theory (Mac Lane [1971]), this is reminiscent of the typical property of **cartesian closed categories**, i.e. those categories in which the sets of morphisms between objects are at the same level of the objects themselves. More precisely, in a cartesian closed category products $A \times B$ and exponents B^A (which generalize the notions, respectively, of cartesian product of A and B, and of sets of functions from A to B) nicely commute, in the sense that the following sets of morphisms are isomorphic, in a natural way:

$$Hom(A \times B, C) \quad \text{and} \quad Hom(A, C^B).$$

For a development of Recursion Theory from a categorical point of view, see Eilenberg and Elgot [1970], and Di Paola and Heller [1987].

We still have to analyze the role of the Padding Lemma. For explicitly given systems of indices (like the ones induced by the approaches to partial recursiveness of Chapter I and Section II.1) this property holds trivially, and it simply amounts to adding redundant information in the given description of the function. But having infinitely many indices in one acceptable system does not help in another one, since the translations provided by the definition of acceptability are not necessarily one-one. Nevertheless, the result holds in general.

Proposition II.5.6 Padding Lemma (Rogers [1958]) *In any acceptable system, given one index of a partial recursive function, we can effectively generate infinitely many indices of the same function.*

Proof. It is enough to show that, given α partial recursive and D finite set of indices for it in the acceptable system ψ, we can effectively find an index for α which is not in D. Define, by parametrization, a recursive function f such that

$$\psi_{f(e)} \simeq \begin{cases} \alpha & \text{if } e \notin D \\ \text{undefined} & \text{otherwise.} \end{cases}$$

By the Fixed-Point Theorem for ψ, there is e such that $\psi_e \simeq \psi_{f(e)}$. Now two cases may occur:

- $e \notin D$

  Then we must be in the first case of the definition of f, and hence $\psi_e \simeq \psi_{f(e)} \simeq \alpha$. Thus e is an index of α which is not in D.

- $e \in D$

  Now we must be in the second case of the definition, and ψ_e is the completely undefined function. But e, being in D, is an index of α, and thus α is also completely undefined. Then we can play a symmetric game, this time letting g be such that

  $$\psi_{g(a)} \simeq \begin{cases} \text{constant } 0 & \text{if } a \in D \\ \text{undefined} & \text{otherwise.} \end{cases}$$

  By the Fixed-Point Theorem, there is a such that $\psi_a \simeq \psi_{g(a)}$. Now it cannot be the case that $a \in D$, otherwise (as above) α would be the constant function 0, while we know it is completely undefined. Then we must be in the second case, i.e. $a \notin D$ and ψ_a is completely undefined. Thus a is an index of α which is not in D. $\quad\Box$

Now we have, for acceptable systems of indices, all the results proved for the standard one, but this does not guarantee that the same will remain true for future results. We will now prove two general theorems, showing how much any acceptable system must resemble the standard one. They will ensure that acceptable systems are really very much alike.

The first theorem shows that *any acceptable system, viewed as a sequence of partial recursive functions*

$$\psi_0 \ \psi_1 \ \psi_2 \ \cdots$$

is nothing else than a recursive permutation of the standard sequence

$$\varphi_0 \ \varphi_1 \ \varphi_2 \ \cdots$$

Theorem II.5.7 (Rogers [1958]) $\{\psi_e\}_{e \in \omega}$ *is an acceptable system of indices if and only if there is a recursive permutation h such that*

$$\psi_e(x) \simeq \varphi_{h(e)}(x).$$

Proof. One direction obviously holds: if h exists, its inverse h^{-1} is also a recursive permutation, and h and h^{-1} provide the needed translations of the systems ψ and φ into each other.

Suppose now that ψ is acceptable: we define a recursive permutation h that interchanges indices in ψ and φ. Since h has to be total, we ensure at even stages that the least element not yet in the domain gets into it (by defining h on it). Similarly, h has to be onto, and at odd stages we ensure that the least element not yet in the range gets into it (by letting it be the value of h for some argument). We then have to show how to ensure that h is both one-one and a function. But h is a function when h^{-1} is one-one, and thus we will have a symmetric construction that alternates steps to make h total and one-one, to steps to make it onto and a function. We just show the steps that have to be alternately taken. Suppose h is defined on $x_0, \ldots, x_n$.

If we want to add one element to the domain of h (even stages), let e be the least number not in $\{x_0, \ldots, x_n\}$: we have to define $h(e)$ in such a way that $h(e) \notin \{h(x_0), \ldots, h(x_n)\}$ (to have h one-one), and $\psi_e \simeq \varphi_{h(e)}$. Since ψ is acceptable, given ψ_e we can effectively find an index of the same function w.r.t. φ. The Padding Lemma II.1.6 ensures that we can effectively generate infinitely many others, and thus one can be found that is not in $\{h(x_0), \ldots, h(x_n)\}$. The first such one is the needed value of $h(e)$.

If we want to add one element to the range of h (odd stages), let y be the least number not in $\{h(x_0), \ldots, h(x_n)\}$: we have to define e in such a way that $e \notin \{x_0, \ldots, x_n\}$ (since h has to be a function), and $\psi_e \simeq \varphi_y$. Then we will let $h(e) = y$. Since ψ is acceptable, given φ_y we can effectively find an index of the same function w.r.t. ψ. The Padding Lemma II.5.6 for ψ ensures that we can effectively generate infinitely many others, and thus one can be found that is not in $\{x_0, \ldots, x_n\}$. The first such one is the needed value for e.   □.

The previous result is not completely satisfactory, because it misses the basic duality of Recursion Theory (see p. 131). It translates a number when it is a code of a function, but it leaves the same number untouched when this behaves as a number (i.e. as an argument or a value). The next result shows that *any acceptable system is really the standard one, and merely works with an appropriated reinterpretation of the numbers* (with no distinction made on whether they code programs or not). Thus the isomorphism provided here is not an isomorphism of the sequence of function, but rather an isomorphism of the underlying structure of the natural numbers.

Theorem II.5.8 (Blum) *If $\{\psi_e\}_{e \in \omega}$ is an acceptable system of indices, then there is a recursive permutation h such that*

$$h(\psi_e(x)) \simeq \varphi_{h(e)}(h(x)).$$

Proof. There are actually two different things to ensure, namely two reinterpretations of the numbers, one taking care of the case when they are seen as arguments and values, and the other when they are seen as codes for programs. Moreover, the two have to coincide.

If we only had to define a recursive permutation h that produces the given isomorphism on the numbers as codes for a given reinterpretation α of the numbers as values and arguments, then we could simply proceed as in the previous theorem. Namely, let α^{-1} be any partial recursive function that inverts α, i.e. any function that, given z, dovetails computations of α until it finds, and outputs, a number x such that $\alpha(x) \simeq z$. We have to ensure that

$$\alpha(\psi_e(x)) \simeq \varphi_{h(e)}(\alpha(x)),$$

and this can be restated as both:

$$\psi_e(x) \simeq \alpha^{-1}(\varphi_{h(e)}(\alpha(x)))$$

and

$$\varphi_{h(e)}(z) \simeq \alpha(\psi_e(\alpha^{-1}(z))).$$

If e has to be added to the domain of h, we use the fact that $\alpha\psi_e\alpha^{-1}$ is partial recursive, and an infinite number of indices w.r.t. φ for it can then effectively be found.

If y has to be added to the range of h, we use the fact that $\alpha^{-1}\varphi_y\alpha$ is partial recursive, and an infinite number of indices w.r.t. ψ for it can then effectively be found.

Thus, given any partial recursive function α, we obtain, uniformly in it, a recursive permutation h_α that satisfies the given conditions. If we could start with a function α equal to the function h_α we obtain from it, then we would have what we want. That this can be done is ensured by the Fixed-Point Theorem. Formally, given $\alpha \simeq \varphi_z$, let $\varphi_{f(z)} \simeq h_\alpha$ be the recursive permutation obtained from it:

$$\varphi_z(\psi_e(x)) \simeq \varphi_{\varphi_{f(z)}(e)}(\varphi_z(x)).$$

Let a be such that $\varphi_a \simeq \varphi_{f(a)}$. Then φ_a is a recursive permutation such that

$$\varphi_a(\psi_e(x)) \simeq \varphi_{\varphi_a(e)}(\varphi_a(x)),$$

as wanted.   □

We thus see that *acceptable systems of indices provide the same structure theory for recursive functions as the standard one* we have been using so far, and from now on we will just suppose that φ_e^n is any acceptable system. In particular, any of the approaches to partial recursiveness discussed in Chapter I and Section II.1 provides a perfectly adequate basis for Recursion Theory.

For more on acceptable systems of indices see Rogers [1958], Lachlan [1964], Schnorr [1975], Hartmanis and Baker [1975], Machtey, Winklmann and Young [1978], and Hartmanis [1982].

Axiomatic Recursion Theory ⋆

The central role of enumeration and parametrization, and the fact that they are purely algebraic properties seemingly having nothing to do with computations, suggest their use in an axiomatic treatment of the part of Recursion Theory developed so far for abstract domains and collections of partial functions over them.

Wagner [1969] and Strong [1968] introduce the notion of **Basic Recursive Function Theory BRFT** as a structure $\langle D, \mathcal{F}, \{\varphi^n\}_{n \in \omega}\rangle$, with D an infinite set, $\mathcal{F}$ a set of partial functions on D, and φ^n an $n+1$-ary function of $\mathcal{F}$ enumerating (over D) the n-ary functions of $\mathcal{F}$. Moreover, $\mathcal{F}$ contains the identities, the constant functions on D, a function

$$f(x, a, b, c) = \begin{cases} b & \text{if } x = a \\ c & \text{otherwise} \end{cases}$$

(for case definition), all the φ^n, and parametrization functions. The notion of $BRFT$ can be variously polished (see e.g. Moschovakis [1971], for an alternative equivalent notion of **precomputation theory**).

BRFT basically captures the essence of elementary Recursion Theory, that is, the part that does not explicitly involve the notion of length of computation (like the proof of the Reduction Property II.1.23). It thus appears that this part of Recursion Theory does not use any particular property of the set of natural numbers (like being countable, well-ordered, etc.).

A particularly interesting special case is the one of $\boldsymbol{\omega}$-**BRFT**, where D is ω, and the successor function is in $\mathcal{F}$. It is not surprising (see p. 158) that *every ω-BRFT contains all partial recursive functions*: this means that the partial recursive functions, together with an acceptable system of indices, are a minimal ω-BRFT. Thus the axiomatization extends the properties of the class of partial recursive functions to bigger classes. The result holds in general, and provides a connection between the axiomatic approach and computability on abstract structures (see p. 202): *the minimal BRFT on an abstract domain is the set of prime computable functions over it* (Moschovakis [1971]).

A strengthening of II.5.3 shows that *any two ω-BRFT with the same set $\mathcal{F}$ of partial functions are mutually translatable by functions in $\mathcal{F}$,* in the sense that, given $\langle \omega, \mathcal{F}, \{\varphi^n\}_{n \in \omega}\rangle$ and $\langle \omega, \mathcal{F}, \{\psi^n\}_{n \in \omega}\rangle$, there are, for every n, functions f and g in $\mathcal{F}$ such that

$$\varphi^n(e, x_1, \ldots, x_n) \simeq \psi^n(f(e), x_1, \ldots, x_n)$$
$$\psi^n(e, x_1, \ldots, x_n) \simeq \varphi^n(g(e), x_1, \ldots, x_n).$$

A strengthening of II.5.8 shows that *any two such ω-BRFT are isomorphic as structures* (Friedman [1971]), in the sense that there is a one-one, onto function i (in $\mathcal{F}$) such that, for every n,

$$i(\varphi^n(e, x_1, \ldots, x_n)) \simeq \psi^n(i(e), i(x_1), \ldots, i(x_n)).$$

In particular, *there is only one ω-BRFT with the set of all partial recursive functions as the set of partial functions, up to recursive isomorphism*, see II.5.8. Note that these ω-*BRFT*'s are exactly the ones corresponding to acceptable systems of indices. On the other hand, Friedman [1971] shows that *there are uncountably many ω-BRFT's with the set of total recursive functions as the set of total functions*.

As we have noted, *BRFT* captures only elementary Recursion Theory. As a first step toward an extension that also covers arguments like II.3.16, Moschovakis [1971] introduces the notion of **computation theory**, basically by adding the primitive notion of *length of computation*, and postulating the fact that the computation of $\varphi_{S_n^m(e, \vec{x})}(\vec{y})$ takes longer than the computation of $\varphi_e(\vec{x}, \vec{y})$. Then the First Recursion Theorem becomes provable in a computation theory, as in II.3.16. An equivalent formulation of computation theory, based on the primitive notion of *immediate subcomputation*, is given by Fenstad [1974].

For more information on *BRFT* and computation theories, see Barendregt [1975], Fenstad [1980], Beeson [1985], Byerly [1985].

To consider abstract domains and collections of functions over them is taking the point of view of Category Theory. Not surprisingly, Recursion Theory can be formulated and developed in this setting, see Eilenberg and Elgot [1970] for the basics, and Di Paola and Heller [1987] for the consequences of enumeration and parametrization.

Models of λ-calculus (part II) ⋆

We have seen (p. 83) that the two combinators

$$\boldsymbol{S} = \lambda xyz.\, xz(yz) \quad \text{and} \quad \boldsymbol{K} = \lambda xy.\, x$$

allow us to define a version of λ-abstraction, and thus to interpret λ-terms as combinators built up from $\boldsymbol{S}$ and $\boldsymbol{K}$. Conversely, combinators built up from $\boldsymbol{S}$ and $\boldsymbol{K}$ can be naturally translated into λ-terms. This suggests the introduction of a first-order theory with equality, called **combinatory logic**, with two primitive symbols $\boldsymbol{S}$ and $\boldsymbol{K}$, an application operation between terms (written as left-associative juxtaposition), and with axioms reflecting the behavior of $\boldsymbol{S}$ and $\boldsymbol{K}$:

$$\boldsymbol{K}xy = x \quad \text{and} \quad \boldsymbol{S}xyz = xz(yz).$$

Then combinatory logic and λ-calculus can be interpreted one in the other, although (without additional assumptions, such as extensionality) the interpretations are not inverses.

Models of combinatory logic can be easily defined: a **combinatory algebra** is a structure $(A, \cdot, k, s)$, with $\cdot$ a binary operation, and k and s distinct elements of A satisfying the axioms above. An **extensional combinatory algebra** is a combinatory algebra such that, whenever $a \cdot c = b \cdot c$ for every c in A, a and b are equal.

The notions of combinatory logic and combinatory algebra can be easily extended to allow for partial application, with strong equality $=$ substituted by a partial one $\simeq$ (Klop [1982], see Beeson [1985] for details). Then *the structure $(\omega, \cdot)$, with partial application defined as*

$$e \cdot x \simeq \varphi_e(x),$$

is a partial combinatory algebra, because there are numbers k and s such that

$$\varphi_k(x,y) \simeq x \quad \text{and} \quad \varphi_s(x,y,z) \simeq \varphi_{\varphi_x(z)}(\varphi_y(z)).$$

By II.5.8, this structure is independent (up to recursive isomorphism) of the acceptable system $\{\varphi_e\}_{e \in \omega}$. Since only enumeration and S_n^m-Theorem are used to determine s and k, *any ω-BRFT is naturally interpreted as a partial combinatory algebra*, and thus provides a model of partial combinatory logic.

Models of partial combinatory logic are not automatically models of λ-calculus. A first problem is that application is total in the latter, and thus only total combinatory algebras are eligible (but this could be solved by defining a version of partial λ-calculus). A more serious problem is due to the fact that the translations of λ-calculus and combinatory logic are not inverses. The solution here consists of additional requirements that force the interpretation of λ-terms defined in a total combinatory algebra to more fully reflect the structure of λ-terms. There are various possibilities.

1. Clearly, a model of λ-calculus should identify terms that are equal (modulo α or β reductions). A combinatory algebra that preserves equality of λ-terms is called a **λ-algebra**.

2. Unfortunately, a λ-algebra does not necessarily satisfy the requirement that terms behaving the same way (i.e. such that $Mx = Nx$ for every x) are equal as functions (i.e. $\lambda x.\, Mx = \lambda x.\, Nx$), a kind of weak extensionality that is obviously valid in λ-calculus. A λ-algebra which is also weakly extensional in this sense is called a **λ-model**, and this notion is considered to be the correct notion of model (Scott [1980a], Meyer [1982]).

3. A stronger notion of extensionality requires terms that behave the same way to be equal not only as functions, but as terms. An **extensional combinatory algebra** satisfies this, and can be interpreted as a λ-model (and actually as a model of extensional λ-calculus, see p. 82) in a unique way. The existence of an extensional *partial* combinatory algebra follows from II.5.23.a (Kreisel [1971]).

Not every combinatory algebra can be extended to a λ-model and, when it can, the extension is not necessarily unique. Typical examples of combinatory algebras which are λ-models in a unique way are D_∞ and $\mathcal{P}(\omega)$ (p. 194). The former, but not the latter, is also an extensional combinatory algebra.

For a treatment of this topic, see Barendregt [1981], Beeson [1985], Hindley and Seldin [1986]. See also Byerly [1982], and Freyd and Ščedrov [1987].

Indices for recursive and finite sets

Recursive and finite sets, being r.e., have r.e. indices which code ways to generate them. But they also have special properties that allow for different representations.

Definition II.5.9 *A recursive set A may be given three different types of indices:*

1. **characteristic indices,** *i.e. the indices of its characteristic function. We write $A = C_e$ if $\varphi_e \simeq c_A$.*

2. **complementary indices,** *i.e. the pairs of indices of the set and its complement, as r.e. sets. We write $A = R_e$, with $e = \langle a, b \rangle$, if $A = W_a$ and $\overline{A} = W_b$.*

3. **r.e. indices,** *i.e. the indices of the set as an r.e. set.*

Exercise II.5.10 *Characteristic and complementary indices are equivalent,* in the sense that it is possible to go effectively from one to the other, and conversely. (Hint: see II.1.19.)

Proposition II.5.11 (Suzuki [1959]) *It is possible to go effectively from complementary indices to r.e. indices of recursive sets, but not conversely.*

Proof. The positive assertion is obvious. As a counterexample to the converse, suppose there is φ partial recursive such that

$$W_e \text{ recursive } \Rightarrow \varphi(e) \!\downarrow \wedge W_e = R_{\varphi(e)}.$$

Then there is also ψ partial recursive such that

$$W_e \text{ recursive } \Rightarrow \psi(e) \!\downarrow \wedge W_e = \overline{W_{\psi(e)}}.$$

Let f be a recursive function such that

$$W_{f(e)} = \begin{cases} \omega & \text{if } e \in \mathcal{K} \\ \emptyset & \text{otherwise.} \end{cases}$$

Then $\mathcal{W}_{f(e)}$ is always recursive, so $\psi f(e)$ is total and

$$
\mathcal{W}_{\psi f(e)} = \begin{cases} \emptyset & \text{if } e \in \mathcal{K} \\ \omega & \text{otherwise.} \end{cases}
$$

Hence

$$
e \in \overline{\mathcal{K}} \;\Leftrightarrow\; \mathcal{W}_{\psi f(e)} \text{ is nonempty,}
$$

and $\overline{\mathcal{K}}$ would be r.e., contradicting II.2.3. $\quad\square$

Exercise II.5.12 *There is no partial recursive function φ such that*

$$
\mathcal{W}_e \text{ recursive } \;\Rightarrow\; \varphi(e){\downarrow} \wedge (\exists i \le \varphi(e))(\mathcal{W}_i = \overline{\mathcal{W}_e}).
$$

Thus not only can an r.e. index for the complement of a recursive set not be found effectively, it cannot even be bounded effectively. (Gold [1967]) (Hint: define $\mathcal{W}_{h(e)}$ to contain an element from each nonempty $\mathcal{W}_i$, for $i \le \varphi(e)$, so that its complement cannot be any of the $\mathcal{W}_i$, and apply the Fixed-Point Theorem.)

The content of the results just proved is that the characteristic function of a recursive sets and a pair of enumerations of it and its complement have the same information, while a simple enumeration of the set is less informative than both.

Note that proposition II.2.1 shows that there is no recursive function which enumerates at least one characteristic index for each recursive set, and that *the set*

$$
\textbf{Char} = \{x : \varphi_x \text{ is total and } 0,1\text{-valued}\,\}
$$

is not r.e. Different results hold for r.e. indices. By II.5.19 it is still true that *the set*

$$
\textbf{Rec} = \{x : \mathcal{W}_x \text{ is recursive}\,\}
$$

is not r.e. (a computation of its complexity will be given in X.9.12). But a recursive function does exist that enumerates at least one r.e. index for each recursive set (II.5.26).

Definition II.5.13 *A finite set A may be given three different types of indices:*

1. **canonical index**, *i.e. a number explicitly coding all the elements of the set. We write $A = D_e$ if $e = 2^{x_0} + \cdots + 2^{x_n}$, and A consists of the distinct elements $x_0, \ldots, x_n$. By convention, $D_0 = \emptyset$.*

2. **characteristic indices**, *i.e. the indices of its characteristic function. We write $A = C_e$ if $\varphi_e \simeq c_A$.*

3. **r.e. indices**, *i.e. the indices of the set as an r.e. set.*

The requirement that the x_i be distinct ensures that every canonical index e defines a unique set D_e. Moreover, the relevant initial segment of the characteristic function of D_e is coded by the digits of the binary expansion of e, read from right to left (for example, $e = 1001101$ codes the set $D_{77} = \{0, 2, 3, 6\}$).

As it might be imagined, canonical, characteristic and r.e. indices are increasingly less effective, and give less and less information on the sets they code:

Proposition II.5.14 (Rogers [1967]) *It is possible to go effectively from canonical indices to characteristic indices, and from characteristic indices to r.e. indices, but not conversely.*

Proof. The positive directions are obvious. Suppose it is possible to go effectively from characteristic indices to canonical ones, i.e. that for some partial recursive function ψ:

$$\varphi_e \text{ characteristic function of a finite set } A \;\Rightarrow\; \psi(e){\downarrow} \wedge A = D_{\psi(e)}.$$

Define a recursive function g such that $\varphi_{g(e)}$ is the characteristic function of a nonempty finite set if $e \in \mathcal{K}$, and of $\emptyset$ otherwise, e.g.

$$\varphi_{g(e)}(x) \simeq \begin{cases} 1 & \text{if } e = f(x) \\ 0 & \text{otherwise} \end{cases}$$

where f is a one-one enumeration of $\mathcal{K}$. Then $\psi g(e)$ is total, and

$$e \in \overline{\mathcal{K}} \Leftrightarrow D_{\psi g(e)} = \emptyset \Leftrightarrow \psi g(e) = 0,$$

and $\mathcal{K}$ would be recursive.

Similarly, suppose that, for some partial recursive function ψ,

$$\mathcal{W}_e \text{ finite } \;\Rightarrow\; \psi(e){\downarrow} \wedge \varphi_{\psi(e)} \text{ characteristic function of } \mathcal{W}_e.$$

If

$$W_{g(e)} = \begin{cases} \{e\} & \text{if } e \in \mathcal{K} \\ \emptyset & \text{otherwise} \end{cases}$$

then $\psi g(e)$ is total, and again $\mathcal{K}$ would be recursive, since

$$e \in \overline{\mathcal{K}} \Leftrightarrow \varphi_{\psi g(e)}(e) \simeq 0. \quad \Box$$

Exercise II.5.15 *The r.e. indices may be arbitrarily smaller than the canonical ones: if h is recursive, there are x, y such that $W_x = D_y$ and $y > h(x)$. Thus less information is more easily coded. (Hint: given h, let*

$$W_{g(x)} = \{1 + max(\bigcup_{z \leq h(x)} D_z)\},$$

and apply the Fixed-Point Theorem.)

As already for the r.e. indices of recursive sets, by II.5.19 *the set*

$$\mathbf{Fin} = \{x : \mathcal{W}_x \text{ is finite}\}$$

is not r.e. A computation of its complexity will be given in X.9.6.

The next definition introduces useful terminology, that we will use repeatedly.

Definition II.5.16 *An* **array** *is an r.e. set whose elements code finite sets. It is called:*

1. **weak** *if its elements are viewed as r.e. indices*

2. **strong** *if its elements are viewed as canonical indices*

3. **disjoint** *if its elements code pairwise disjoint, finite sets.*

A strong array is nothing more than a weak array together with a recursive function giving the cardinality of the members of the array. Since, if we know how to enumerate a set and how many elements there are in it, then we can obtain all its elements: enumerate the set, until the right number of elements has been generated.

We will still call a (weak or strong) array the collection of finite sets coded by the elements of a given (weak or strong) array.

Exercise II.5.17 *There are weak arrays which are not strong.* (Hint: let A be r.e. and nonrecursive, and $A_n = \{z : z \in A \wedge z \leq n\}$.)

Enumerations of classes of r.e. sets

In this section we consider classes of r.e. sets, and possible ways of enumerating them. Some of the results we prove could also be stated for classes of partial recursive functions, but the consideration of r.e. sets sometimes allows for smoother formulations and proofs.

Definition II.5.18 (Dekker [1953a], Rice [1953]) *A class $\mathcal{A}$ of r.e. sets is called:*

1. **completely r.e.** *if its index set (i.e. the set containing all indices of each member of $\mathcal{A}$) is r.e.*

2. **r.e.** *if there is a recursive function f which enumerates at least one index of each member of $\mathcal{A}$, i.e. $\mathcal{A} = \{\mathcal{W}_{f(x)}\}_{x \in \omega}$.*

3. **r.e. without repetitions** *if there is a recursive function f which enumerates exactly one index of each member of $\mathcal{A}$, i.e.*

$$\mathcal{A} = \{\mathcal{W}_{f(x)}\}_{x \in \omega} \quad and \quad (\mathcal{W}_{f(x)} = \mathcal{W}_{f(y)} \Rightarrow x = y).$$

The characterization of completely r.e. classes of r.e. sets is the same as that of completely r.e. classes of partial recursive functions (see II.4.2), and we leave the routine modification of the proof to the reader.

Proposition II.5.19 (Myhill and Shepherdson [1955], McNaughton, Shapiro) *A class of r.e. sets $\mathcal{A}$ is completely r.e. if and only if it consists of the r.e. supersets of the elements of a strong array, i.e. for some r.e. set A:*

$$\mathcal{W}_x \in \mathcal{A} \iff (\exists y)(y \in A \wedge D_y \subseteq \mathcal{W}_x).$$

We can now show that a number of classes of r.e. sets are not completely r.e. For example: any finite class, any class containing only finite (or only infinite) sets, the class of recursive sets (since any finite set admits an r.e. nonrecursive extension), and so on. On the other hand, the class of the r.e. supersets of finitely many finite sets is completely r.e.

Exercises II.5.20 a) *Not every completely r.e. class is the class of the r.e. supersets of the elements of a recursive strong array.* (Rice [1956]) (Hint: let $\mathcal{A}$ be the class of r.e. supersets of the strong array $\{\{x\} : x \in \mathcal{K}\}$.)

b) *There is a completely r.e. class such that any strong array that generates it must contain superfluous information.* (Rice [1956]) (Hint: the only array not containing superfluous information is the core, i.e. the set of minimal finite sets belonging to the class. Let $\mathcal{A}$ be the class of r.e. supersets of the finite sets $\{f(x), f(x)+1, \ldots, f(y)-1\}$ for $f(x) < f(y)$, with f recursive one-one function enumerating an r.e. nonrecursive set B containing 0. Then the core of $\mathcal{A}$ is not r.e., otherwise B would be recursive.)

c) *Not every class consisting of the r.e. supersets of the elements of a weak array is completely r.e.* (Myhill and Shepherdson [1955]) (Hint: there is an r.e. set A which is hypersimple but not hyperhypersimple, see III.4.12. Then there is a disjoint weak array B, but no disjoint strong array, whose members intersect $\overline{A}$. Let $\mathcal{A}$ be the class of r.e. supersets of the elements of B.)

We turn now to r.e. classes. Obviously, the class of r.e. sets is r.e. (any number codes an r.e. set), but there seems to be no natural way to extract an enumeration without repetitions.

Exercise II.5.21 *There is no invariant recursive choice function for indices of r.e. sets*, i.e. a recursive function f such that $f(e)$ is an index of $\mathcal{W}_e$, and

$$\mathcal{W}_i = \mathcal{W}_e \Rightarrow f(i) = f(e).$$

(Hint: suppose f exists, and let

$$W_{g(e)} = \begin{cases} \omega & \text{if } e \in \mathcal{K} \\ \emptyset & \text{otherwise.} \end{cases}$$

For any $a \in \mathcal{K}$, $e \in \mathcal{K} \Leftrightarrow fg(e) = fg(a)$, and $\mathcal{K}$ would be recursive.)

Nevertheless, we have the following result.

Theorem II.5.22 (Friedberg [1958]) *The class of r.e. sets is r.e. without repetitions.*

Proof. A natural idea would be to pick up, for each r.e. set, its minimal index, but we know from the exercise that this cannot be done recursively. The idea of the proof is to try anyway, with an indirect approach: we simulate each r.e. set, until we discover that it looks too much like some other r.e. set with a smaller index, in which case we drop the finite approximation of the former. By doing so, we actually introduce a number of additional finite sets to the original enumeration of the r.e. sets, but this does not interfere with our goal as finite sets are r.e. anyway.

Consider any acceptable enumeration $\{W_x\}_{x \in \omega}$ of the r.e. sets, such that $W_0 = \omega$. We are going to define an r.e. sequence $\{S_x\}_{x \in \omega}$ of r.e. sets, in which every r.e. set appears exactly once (the existence of f recursive such that $S_x = W_{f(x)}$ will follow by the S_n^m-Theorem, since the sequence S_x is r.e.). Since we enumerate the sets S_x's, we let $S_{x,s}$ be the part of S_x enumerated by the end of stage s. We call x a *follower* of e if we try to make, in the end, $S_x = W_e$. If and when we decide, for whatever reason, that we no longer want to pursue $S_x = W_e$, we *release* x, and x will never return to be a follower. If, instead, x is never released, it is a *permanent follower*. At any stage each e has at most one follower, and x is called *unused* at a certain stage if at that stage it is not, and has never been, a follower.

The construction starts by letting 0 be a follower of 0 (i.e. we try to make $S_0 = W_0$), and $S_{x,0} = \emptyset$.

At step $n+1$, suppose x is a follower of e (hence, by construction, we have $S_{x,n} = W_{e,n}$). We release it in the following situations:

1. for some $i < e$, $W_{i,n}$ and $W_{e,n}$ look the same up to x (i.e. for each $z \leq x$), the reason being that e does not look as a minimal index for W_e up to x.

2. $x > 0$ and, for some y already released, $S_{x,n} = S_{y,n}$, the reason being that W_e might just be the finite set $S_{y,n} = S_y$, and e might be a minimal index for it, so if we let x continue to be a follower of e we would get, in the end, $S_x = S_y$.

Note that 0 is never released. If x has just been released, S_x will not change anymore after this stage. Since we want to have all the S_i's distinct, let b be an element larger than $\bigcup_{e \in \omega} S_{e,n}$ (which is a finite set), and let

$$S_x = S_{x,n+1} = \{z : z < x + b\}.$$

Certainly S_x is different from all the other sets at this stage, and $S_{x,n} \subseteq S_x$.

To keep things going we have to add followers, at least from time to time, to elements that do not have them: e.g. we assign the smallest unused element as a follower to e, if e does not have one (either because it never did, or because it lost it) and $n = \langle e, t \rangle$ (so that e gets infinitely many chances to receive a follower).

Finally, if x is a follower of e at this stage, we let

$$S_{x,n+1} = \mathcal{W}_{e,n+1}.$$

We now prove that the construction works.

1. *every r.e. set appears at least once:* $\forall e \exists x (\mathcal{W}_e = S_x)$.
   We may suppose that e is a minimal index. Then there is a stage n_0 such that

   $$(\forall n \geq n_0)(\forall z \geq n_0)(\forall i < e)(\mathcal{W}_{i,n} \text{ and } \mathcal{W}_{e,n} \text{ look different up to } z).$$

   If e has a permanent follower x, then $S_x = \mathcal{W}_e$. And this must be the case, otherwise e keeps on getting new followers that sooner or later become released. After stage n_0 the only possibility for release is the second one, and this means that for each follower x there is a stage n and a released element y such that

   $$\mathcal{W}_{e,n} = S_{x,n} = S_{y,n} = \{z : z < c_y\}$$

   for increasingly big c_y's. Hence $\mathcal{W}_e = \omega$ and, by minimality of e, $e = 0$. But 0 has a permanent follower, namely 0.

2. *every r.e. set appears at most once:* $x \neq y \Rightarrow S_x \neq S_y$.
   By construction no element is always unused (there are infinitely many r.e. sets, and we always choose as followers the smallest unused elements). So four cases may happen, for different x and y:

   - both are permanent followers, say $S_x = \mathcal{W}_e$ and $S_y = \mathcal{W}_i$
     Since an index has at most one permanent follower, $i \neq e$. Suppose e.g. that $i < e$. If $S_x = S_y$ then $\mathcal{W}_i = \mathcal{W}_e$, so there is n_0 such that

     $$(\forall n \geq n_0)(\mathcal{W}_{i,n} \text{ and } \mathcal{W}_{e,n} \text{ look the same up to } x).$$

     Then x is released at stage $n + 1$, contradiction.

- both x and y are released
  If they are released at the same stage then $S_x \neq S_y$, because (for the appropriate b) $x + b \neq y + b$. If e.g. x is released after y is, then (for the appropriate b) $b \in S_x$ but $b \notin S_y$.

- one is released and the other is permanent
  Say x is released and y is permanent. If $S_x = S_y$ then S_y is finite, so $y \neq 0$ (since $S_0 = \omega$). Hence, at some stage n, $S_{x,n} = S_{y,n}$ and y is released at stage $n + 1$, contradiction.   □

The method of proof just used (where, as II.5.21 shows, release of followers is not avoidable) is a weak version of the **priority method**, in which there are requirements, positive (trying to put elements in a set) or negative (trying to leave them out), that have to be satisfied, and actions to satisfy one type might interfere with the satisfaction of the other type. Here the positive requirements try to put $\mathcal{W}_e$ in the enumeration, while the negative requirements tend to make the S_x's distinct. The crucial fact is that the action to satisfy a given positive requirement can be interfered with (by releasing a follower) only finitely often, and thus the construction succeeds. The priority method will be introduced and fully exploited in Chapter X.

Exercises II.5.23 a) *The class of partial recursive functions is r.e. without repetitions.* (Friedberg [1958]) (Hint: start with an enumeration of the partial recursive functions, given by their graphs, and make the set S_x different from all the other S_i's as a partial function, instead of simply as a set, when x is released.)

b) *An r.e. class $\mathcal{A} = \{A_x\}_{x \in \omega}$ of disjoint, nonempty r.e. sets is r.e. without repetitions if and only if it satisfies the effective choice principle*, i.e. there is a recursive function f such that $f(x) \in A_x$, and $A_x = A_y \Rightarrow f(x) = f(y)$. (Pour El and Howard [1964])

Corollary II.5.24 (Pour El and Putnam [1965]) *Every r.e. class containing all finite sets is r.e. without repetitions.*

Proof. The proof above did not use any assumption about the r.e. sets, except the fact that they are an r.e. class containing all finite sets and ω. Given any r.e. class containing the finite sets, augment an enumeration of it by putting ω in the first place: the proof above will give an enumeration without repetitions of its elements. If ω was not in the given class, it is enough to drop the first element from the newly obtained class.   □

Some more criteria for enumeration without repetitions are in Pour El and Howard [1964], Lachlan [1965a], [1967], Khutorezkii [1969], and Marchenkov [1971]. The one just given by the corollary provides a number of examples, in view of the fact that the r.e. classes containing all finite sets are completely characterized by the following result (expressed in terms of a classification of sets introduced in Chapter IV, see IV.1.6).

Proposition II.5.25 (Yates [1969]) *A class of r.e. sets containing all finite sets is r.e. if and only if its index set is Σ_3^0.*

Proof. If $\mathcal{A} = \{W_{f(x)}\}_{x \in \omega}$ then

$$
\begin{aligned}
W_e \in \mathcal{A} \ &\Leftrightarrow\ \exists x (W_e = W_{f(x)}) \\
&\Leftrightarrow\ \exists x \forall y (y \in W_e \leftrightarrow y \in W_{f(x)}) \\
&\Leftrightarrow\ \exists x \forall y [(y \in W_e \to y \in W_{f(x)}) \wedge (y \in W_{f(x)} \to y \in W_e)]
\end{aligned}
$$

and $\theta \mathcal{A}$ is Σ_3^0.

Conversely, suppose $\mathcal{A}$ contains all finite sets, and $\theta \mathcal{A}$ is Σ_3^0, i.e.

$$
W_e \in \mathcal{A} \ \Leftrightarrow\ \exists x \forall y \exists z R(e, x, y, z),
$$

with R recursive. Define

$$
t \in A_{\langle e,x \rangle} \ \Leftrightarrow\ t \in W_e \wedge (\forall y \leq t)(\exists z) R(e, x, y, z).
$$

If $A_{\langle e,x \rangle}$ is finite, then it is in $\mathcal{A}$ by hypothesis; if it is infinite, then it is in $\mathcal{A}$ because $A_{\langle e,x \rangle} = W_e \in \mathcal{A}$. Finally, if $W_e \in \mathcal{A}$ then $\forall y \exists z R(e, x, y, z)$, for some x, and so $W_e = A_{\langle e,x \rangle}$. Thus $\mathcal{A} = \{A_{\langle e,x \rangle}\}_{e,x \in \omega}$, and $\mathcal{A}$ is r.e. $\square$

A particularly interesting example of r.e. class is the one of recursive sets. This follows from the criterion just given, but it can be easily proved directly:

Proposition II.5.26 (Muchnik [1958a], Suzuki [1959]) *The class of recursive sets is r.e. (without repetitions).*

Proof. Let $W_{f(e)}$ be the following r.e. set, obtained uniformly from e: generate W_e, and put a new element appearing in it in $W_{f(e)}$ only if it is greater than all the elements which are already in $W_{f(e)}$. Each $W_{f(e)}$ is either finite or enumerated in increasing order, hence recursive by II.1.17. And each recursive set has an index coding the instructions to enumerate it in increasing order, and so it appears among the $W_{f(e)}$'s. Since every finite set is recursive, from simple enumerability we get enumerability without repetitions, by II.5.24. $\square$

What the proposition tells us is that *the recursive sets are uniformly r.e.*, in the sense that there is an r.e. relation $R(e, x)$ such that, while e ranges over ω, the set $\{x : R(e, x)\}$ ranges over the recursive sets, and each recursive set is $\{x : R(e, x)\}$, for some e. Obviously, by diagonalization and closure with respect to complementation, *the recursive sets are not uniformly recursive.*

Also, the result just proved shows that there is an r.e. subset of *Rec* (defined on p. 226), which contains at least one index of each recursive set, although the set *Rec* is not itself r.e. (by II.5.19). On the other hand, no such property holds for characteristic indices of recursive sets in place of r.e. indices, by II.2.1.

Exercises II.5.27 a) *The classes of the infinite r.e. sets and of the infinite recursive sets are not r.e.* (Uspenskii [1955], [1957], Dekker and Myhill [1958a]) (Hint: if they were, an infinite recursive set intersecting each element of the given class could be built by enumerating it in increasing order.)

b) *The class of the coinfinite r.e. sets is not r.e.* (Hint: criterion II.5.25 can be applied, but it is difficult to show that the index set of this class is not Σ_3^0 (see X.9.11). A direct diagonalization is easier, although not trivial, and requires the priority method introduced in Chapter X. Elements from complements of r.e. sets, instead of from the sets themselves, are needed. Positive requirements ask for a nonempty intersection with the complement of each set in the class, and negative requirements ask for the infinity of the complement of the set thus built.)

We briefly discuss now classes of finite sets, which are at the opposite extreme of those taken care of by the criterion II.5.25.

Proposition II.5.28 (Lachlan [1965a], Pour El and Putnam [1965])
There is a class of finite sets which is r.e. (and it actually admits an enumeration in which every element has at most two indices), but is not r.e. without repetitions.

Proof. Let R_{2x}, R_{2x+1} be generated as follows:

- R_{2x}: put $2x$ in it, then generate $\mathcal{K}$ and, if $x \in \mathcal{K}$, add $2x + 1$.

- R_{2x+1}: put $2x + 1$ in it, then generate $\mathcal{K}$ and, if $x \in \mathcal{K}$, add $2x$.

Then:

- if $x \in \mathcal{K}$, $R_{2x} = R_{2x+1} = \{2x, 2x + 1\}$

- if $x \notin \mathcal{K}$, $R_{2x} = \{2x\}$ and $R_{2x+1} = \{2x + 1\}$.

If there were an enumeration without repetitions of $\mathcal{A} = \{R_e\}_{e \in \omega}$ then $\overline{\mathcal{K}}$ would be r.e., since $x \in \overline{\mathcal{K}}$ if and only if there are two sets in the enumeration, one with $2x$ in it, the other with $2x + 1$. $\square$

Exercises II.5.29 (Pour El and Putnam [1965]) a) *For every $n \geq 1$, there is an r.e. class of finite sets which is r.e with at most $n + 1$ repetitions, but not with at most n repetitions.* (Hint: use sets with at most $n + 1$ elements.)

b) *There is an r.e. class of finite sets which is r.e. with at most finitely many repetitions, but not with at most n repetitions, for any fixed n.* (Hint: use sets with unbounded cardinality.)

c) *Every r.e. class of disjoint finite sets is r.e. with finitely many, possibly unbounded, repetitions.*

d) *There is an r.e. class of finite sets which is not r.e. with at most finitely many repetitions.* (Hint: instead of considering, as above, an r.e. nonrecursive set, consider a set $\Sigma_2^0 - \Pi_2^0$, see IV.1.13.)

e) *There is no r.e. class of finite sets such that every enumeration of it repeats each element infinitely often.* (Hint: if $\mathcal{A}$ is r.e. and $A \in \mathcal{A}$ is finite, then $\mathcal{A} - \{A\}$ is still an r.e. class, since we can simply enumerate the elements of $\mathcal{A}$ which contain some element which is not A.)

For more on this topic see Pour El and Putnam [1965], Young [1966], and Florence [1967], [1969], [1975].

Exercises II.5.30 Standard classes of r.e. sets. (Lachlan [1964a]) An r.e. class $\mathcal{A} = \{S_x\}_{x \in \omega}$ of r.e. sets is standard if, whenever $\mathcal{W}_e \in \mathcal{A}$, $S_e = W_e$. Thus standard classes are r.e. classes indexed in the same way as the class of the r.e. sets.

a) *The Fixed-Point Theorem holds for standard classes.* (Hint: let f be a recursive function such that $S_x = W_{f(x)}$. Given $S_{g(x)}$ with g recursive, by the Fixed-Point Theorem there is e such that $S_{g(e)} = W_{fg(e)} = W_e$. Then $W_e \in \mathcal{A}$ and $W_e = S_e$, i.e. $S_{g(e)} = S_e$.)

b) *The class $\{\emptyset, A\}$, with A nonempty r.e. set, is standard.* (Hint: define $W_{f(x)}$ as $\emptyset$ or A, depending on whether W_x is empty or not.)

c) *If a standard class contains a finite set, it contains a least member.* (Hint: let A be a finite set in $\mathcal{A}$, and consider the supersets of A, and the sets intersecting $\overline{A}$: since they are complementary classes, by an analogue of Rice's Theorem for $\mathcal{A}$ their union can cover $\mathcal{A}$ only if one of them does already. Since A is in $\mathcal{A}$, either A is the least member of $\mathcal{A}$, or a finite subset of it is in $\mathcal{A}$.)

d) *A finite class is standard if and only if it has a least member.* (Hint: suppose $\mathcal{A} = \{A_1, \ldots, A_n\}$, and choose finite subsets B_i of A_i such that

$$B_i \subseteq A_j \quad \Leftrightarrow \quad A_i \subseteq A_j$$
$$B_i \subseteq B_j \quad \Leftrightarrow \quad A_i \subseteq A_j.$$

If $\mathcal{A}$ is standard, as in c) we can show that it has a least member, by considering the supersets of the B's. And if $\mathcal{A}$ has a least member A_{i_0}, then we may suppose $B_{i_0} = \emptyset$, and construct a standard enumeration by letting S_e be the union of the A's corresponding to a maximal strictly increasing sequence of B's included in W_e.)

e) *A strong array is a standard class if and only if there is no infinite r.e. set which is the union of an increasing sequence of members of the array.* (Hint: one direction is like part d) above, by letting $\mathcal{A} = \{A_x\}_{x \in \omega}$, $A_0 = \emptyset$ and $B_i = A_i$. We may suppose that $\emptyset$ is in $\mathcal{A}$, since by b) above there is a least member, and the class obtained by subtracting it to every element of $\mathcal{A}$ is still standard. For the other direction, let C be an infinite r.e. set which is the union of an increasing sequence of members of $\mathcal{A}$. We may suppose $C = \bigcup_{x \in \omega} C_x$, for a strong array $\{C_x\}_{x \in \omega}$ of members of $\mathcal{A}$. Consider

$$W_{f(e)} = C_0 \cup \bigcup \{C_{n+1} : C_n \subseteq A_e\},$$

and choose e such that $\mathcal{W}_{f(e)} = \mathcal{W}_e$. Then $\mathcal{W}_e$ is finite by definition of f, but at the same time it should also, inductively, include every C_n, and hence C.)

We conclude by giving the relationships among the various concepts of enumerability for classes of r.e. sets:

Proposition II.5.31 (Pour El and Howard [1964])

1. *Any completely r.e. class is r.e. without repetitions, but not conversely.*

2. *Any class r.e. without repetitions is r.e., but not conversely.*

Proof. Let $\mathcal{A}$ be completely r.e.: then $\mathcal{A}$ is the class of r.e. supersets of a strong array. To get an enumeration of $\mathcal{A}$ without repetitions, start with such an enumeration $\{S_x\}_{x \in \omega}$ for the r.e. sets, and put in the new enumeration of $\mathcal{A}$ only the S_x's for which it is discovered that they extend one element of the strong array, in a dovetailed generation of each S_x and of the elements of the strong array. Thus a completely r.e. class is r.e. without repetitions. The converse does not hold, as the example of recursive sets shows (see II.5.26).

A class r.e. without repetitions is obviously r.e., but the converse does not hold, as II.5.28 shows. $\Box$

The Theory of Enumerations ⋆

The results of the last subsection suggest the possibility (proposed by Kolmogorov) of a systematic study (begun by Uspenskii [1955], [1955a], [1956]) of the recursive **enumerations** of an r.e. class $\mathcal{A}$ of r.e. sets, defined as functions $\nu : \omega \xrightarrow{\text{onto}} \mathcal{A}$ such that the sets $\mathcal{W}_{\nu(x)}$ are uniformly r.e. Given two such enumerations ν_0 and ν_1, we let
  $\nu_0 \leq \nu_1$ if there is a total recursive function f such that $\nu_0 = \nu_1 \circ f$
  $\nu_0 \equiv \nu_1$ if $\nu_0 \leq \nu_1 \wedge \nu_1 \leq \nu_0$.
Then $\equiv$ is an equivalence relation (already used in II.5.3). $\mathcal{L}^\circ(\mathcal{A})$ is the structure of equivalence classes of the r.e. enumerations of $\mathcal{A}$, partially ordered by the order induced by $\leq$. The Theory of Enumerations studies the algebraic structure of $\mathcal{L}^\circ(\mathcal{A})$, for any class $\mathcal{A}$. Its emphasis is thus somewhat opposite to that of BRFT (p. 221), being on subclasses of the class of r.e. sets, instead of on superclasses of it.

The scattered results of this section can be seen in a new light in this framework. Call ν **principal** if it is in the greatest element of $\mathcal{L}^\circ(\mathcal{A})$ (so that $\nu_0 \leq \nu$, for every ν_0), and **minimal** if it is in a minimal one (so that, for every ν_0, if $\nu_0 \leq \nu$ then $\nu_0 \equiv \nu$). Then, *for the class of all r.e. sets, an enumeration is principal if and only if it is acceptable, and is minimal if it is an enumeration* *without repetitions.* The enumeration without repetitions given by II.5.22 is not the only possible one, up to equivalence: there are countably many other, pairwise inequivalent ones (Pour El [1964], Khutorezkii [1969]). On the other hand, there are also countably many, pairwise inequivalent minimal enumerations, which are not equivalent to enumerations without repetitions (Ershov [1968b]). Finally, there are countably many, pairwise inequivalent enumerations, which do not bound minimal enumerations (Khutorezkii [1969a]). Note also that *the fact that the r.e. sets are not uniformly recursive implies that there is no least enumeration* (i.e. such that $\nu \leq \nu_0$, for every ν_0).

The case of *finite classes* $\mathcal{A}$ has been thoroughly examined, and it is known that the structure of $\mathcal{L}^\circ(\mathcal{A})$ depends only on the set-theoretical structure of $\mathcal{A}$ under inclusion. Moreover, $\mathcal{L}^\circ(\mathcal{A})$ is a distributive uppersemilattice with least and greatest element, and it is either trivial (a single element), or very rich (having an ideal isomorphic to the structure of r.e. m-degrees, studied in Chapter X). Since, for $\mathcal{L}^\circ(\mathcal{A})$ and $\mathcal{L}^\circ(\mathcal{A}')$ to be isomorphic, the number of minimal elements (under inclusion) of $\mathcal{A}$ and $\mathcal{A}'$ must be equal, *there are countably many isomorphism types*, but a complete classification of them has not yet been obtained.

The case of *infinite classes* $\mathcal{A}$ presents an even greater variety: not even the greatest or the least element exist necessarily in $\mathcal{L}^\circ(\mathcal{A})$ (as shown, respectively, by the classes of recursive sets, and of all the r.e. sets). But, even in this case, $\mathcal{L}^\circ(\mathcal{A})$ either has only one element (e.g. when $\mathcal{A}$ is a disjoint strong array), or it is infinite and complicated (not linear and not a lattice).

A particularly useful notion is that of **complete enumeration** (Malc'ev [1961], [1963], Lacombe [1965]). This is an enumeration of $\mathcal{A}$ for which there exists an element e such that, whenever φ is a partial recursive function, there is a total recursive function f such that

$$\nu f(x) = \begin{cases} \nu\varphi(x) & \text{if } \varphi(x)\downarrow \\ e & \text{otherwise.} \end{cases}$$

Clearly, for the class of all r.e. sets, the standard enumeration is complete (with $e = \emptyset$). *The classes $\mathcal{A}$ possessing a complete enumeration are exactly those containing a smallest set* (Malc'ev [1964]). The interest of the notion comes from the fact that it implies versions of (and thus it generalizes) the Fixed-Point Theorem (for every recursive function f there is e such that $\nu f(e) = \nu e$), of Rice's Theorem (a subset $\mathcal{A}'$ of $\mathcal{A}$ has a recursive index set $\nu^{-1}(\mathcal{A}')$ if and only if it is trivial) and of the Padding Lemma (for every element $a \in \mathcal{A}$, its index set $\nu^{-1}(\{a\})$ is infinite).

Lacombe [1960] and Malc'ev [1961] have suggested (with motivations from constructive algebra) the *extension of the notion of enumeration, from classes of r.e. sets to any countable, nonempty set S.* They drop effectiveness requirements, and consider any function $\nu : \omega \xrightarrow{\text{onto}} S$ as an enumeration of S. The equivalence relation $\equiv$ still makes sense, and $\mathcal{L}(S)$ is the set of equivalence classes of enumerations of S: it is now an uppersemilattice, either trivial or uncountable.

The theory has been here very successful: *there are only three possible isomorphism types for $\mathcal{L}(S)$*, corresponding to cardinalities of S equal to 1, finite and greater than 1, and infinite. In the first case there is only one element. In the second, the structure has been characterized (Ershov [1975]) up to isomorphism, as a strongly universal uppersemilattice with a least element (it is actually isomorphic to the structure of m-degrees, which is the special case $\mathcal{L}(\{0,1\})$, see VI.4.1). The last case differs from the other two, e.g. because there is no least element.

For detailed treatments of the Theory of Enumerations see Lacombe [1965], Malc'ev [1965], and Ershov [1977].

II.6   Retraceable and Regressive Sets $\star$

Chapter II has been characterized by the search for possible extensions of the notion of recursiveness. Our last attempt is inspired by the property of recursive sets of being effectively enumerable in increasing or decreasing order.

Consider a recursive set $A = \{a_0 < a_1 < a_2 < \dots\}$. There are two recursive functions f and g, such that:

1. $f(a_n) = a_{n+1}$, unless a_n is the maximum element of A, in which case $f(a_n) = a_n$

2. $g(a_{n+1}) = a_n$ and $g(a_0) = a_0$.

Exercise II.6.1 *Property 1 is characteristic of the recursive sets*, i.e. if there is a recursive function f as above, then A is recursive. (Hint: use II.1.17)

It is natural to see if property 2 is also characteristic of recursive sets, and if not, to study the sets which share this property with them.

Definition II.6.2 (Tennenbaum, Dekker [1962])

1. *If $\{a_0, a_1, a_2, \dots\}$ is an enumeration without repetitions of A, and φ is a partial recursive function such that*

$$\varphi(a_{n+1}) \simeq a_n \quad \text{and} \quad \varphi(a_0) \simeq a_0,$$

*then A is called **regressive** via φ, and with respect to the given enumeration.*

2. *Under the same conditions, and if the enumeration is in order of magnitude (principal enumeration), then A is called* **retraceable**.

We stress the fact that the enumeration of A does not have to be recursive.

Exercises II.6.3 a) *If A is retraced by φ, we can always suppose $\varphi(x) \le x$, whenever $\varphi(x)$ is defined.*

b) *If A is regressed by φ, we may always suppose range $\varphi \subseteq$ domain φ and, whenever $\varphi(x)$ is defined, $\varphi^n(x) \simeq a_0$, for some n.*

There is a surface analogy

$$\frac{\text{r.e.}}{\text{recursive}} = \frac{\text{regressive}}{\text{retraceable}}$$

and we now explore the extent to which it holds.

Retraceable versus recursive

The following notion will be helpful in our study:

Definition II.6.4 *A set A is* **immune** *if it is infinite, but it does not contain infinite r.e. subsets.*

Note that, by II.1.20, an infinite set is immune if and only if it does not contain infinite *recursive* subsets.

Proposition II.6.5 (Dekker and Myhill [1958])
Recursive $\Rightarrow$ retraceable $\Rightarrow$ recursive or immune.

Proof. If A is recursive, then there is a recursive function f which enumerates it in increasing order, i.e.

$$A = \{f(0) < f(1) < \cdots\}.$$

It is then enough to define

$$\varphi(f(0)) \simeq f(0) \qquad \varphi(f(n+1)) \simeq f(n)$$

to have a partial recursive function retracing A. Actually, since f is increasing, we can get φ total recursive, e.g. by letting it be 0 in the intervals between any two successive values of f.

Let now A be retraceable (via φ) and infinite (if A is finite, it is certainly recursive), and suppose A is not immune: then A has an infinite recursive subset B. But then A is recursive: given x, find an element $g(x)$ of B greater than it (which exists and can be found recursively, because B is infinite and recursive). Now $g(x)$ is certainly in A (because $B \subseteq A$), and we can then repeatedly apply φ to it, to generate the elements of A smaller than $g(x)$ in decreasing order, until one is repeated (which means we hit the smallest one). And x, being smaller than $g(x)$, is in A if and only if it is generated in this process. □

We will see, in the subsection of existence theorems, that it is far from being true that every retraceable set is recursive, but there is a nontrivial special case in which this holds:

Proposition II.6.6 (Mansfield) *If A and $\overline{A}$ are both retraceable, then A is recursive.*

Proof. The proof is in two steps:

1. *If A and $\overline{A}$ are retraceable, so is the set*

$$A \oplus \overline{A} = \{2x : x \in A\} \cup \{2x + 1 : x \in \overline{A}\}.$$

   The basic fact here is that exactly one element of each pair $(2x, 2x + 1)$ is in $A \oplus \overline{A}$. Let φ and φ' retrace A and $\overline{A}$, respectively. Then $A \oplus \overline{A}$ is retraced by ψ so defined:

$$\psi(2x) \quad \simeq \quad \begin{cases} 2x - 2 & \text{if } \varphi(x) \simeq x - 1 \\ 2x - 1 & \text{if } \varphi(x) \simeq x \vee \varphi(x) < x - 1 \end{cases}$$

$$\psi(2x + 1) \quad \simeq \quad \begin{cases} 2x - 1 & \text{if } \varphi'(x) \simeq x - 1 \\ 2x - 2 & \text{if } \varphi'(x) \simeq x \vee \varphi'(x) < x - 1. \end{cases}$$

   Indeed, if $x \in A$ then $2x \in A \oplus \overline{A}$, and we need to define $\psi(2x)$ as $2x - 1$ or $2x - 2$, depending on whether $x - 1$ is in A or in $\overline{A}$. But the first case happens when $\varphi(x) \simeq x - 1$, and the second when either x is the smallest element of A (and $\varphi(x) \simeq x$) or the next element of A in descending order is smaller than $x - 1$ (and $\varphi(x) < x - 1$). The case $x \in \overline{A}$, i.e. $2x + 1 \in A \oplus \overline{A}$, is symmetric, using φ'.

2. *If $A \oplus \overline{A}$ is retraceable, A is recursive.*
   We prove that any set B retraceable and with exactly one element in each pair $(2x, 2x + 1)$, is r.e. (and hence recursive: to know which one of $2x, 2x + 1$ is in B, generate B until one of them appears). If ψ retraces B, two cases are possible:

   - If $2x, 2x + 1$ are both sent by ψ onto the same element c (meaning that some iteration of ψ on $2x$ is c, and similarly for some iteration, not necessarily the same number of times, of ψ on $2x + 1$), then

$c \in B$ (because one of $2x, 2x+1$ is in B). And then ψ automatically generates every element of B smaller than c. Then, if there are infinitely many pairs $2x, 2x + 1$ as such it is enough to look for them, and each of them will generate an initial segment of B, via ψ.

- If there are only finitely many such pairs, let $b \in B$ be greater than all of their elements (if it does not exist, B is finite and hence recursive). To generate B is now enough to generate all the x's such that $x > b$ and x is sent by ψ onto b. Indeed, there cannot be one such element that is not in B, since otherwise the other element of the pair to which x belongs would be sent to b by ψ as well, against the choice of b. $\square$

Note that we proved that A is recursive, but did not produce an algorithm to compute it, since the proof is by nonconstructive cases (we only produced two algorithms, one of which will work, but we do not know which one).

Exercises II.6.7 Introreducible sets. The proof of II.6.5 suggests the following notion: A is **introreducible** if it is recursive in each of its infinite subsets (Tennenbaum). Not much is known on the results which generalize from retraceable to introreducible sets. In particular it is open whether A is recursive, whenever A and $\overline{A}$ are introreducible. A stronger and more tractable notion is: A is **uniformly introreducible** if there is an index e such that, whenever B is an infinite subset of A, then $A = \varphi_e^B$ (Jockusch [1968]).

a) *Every retraceable set is uniformly introreducible.* (Hint: see the proof of II.6.5.)

b) *Not every uniformly introreducible set is retraceable.* (Dekker and Myhill [1958]) (Hint: use the intervals $(2^x, 2^{x+1})$, and put in A exactly one element of each interval, extremes excluded, by letting $2^x + 1 + f(x)$ be in A if x is, where f is any nonrecursive 0,1-valued function. Then A is retraceable, via the function that sends the interval $(2^x, 2^{x+1})$ to x, and nonrecursive, hence immune. Let B code A on the extremes of the intervals, i.e. $2^x \in B$ if and only if $x \in A$. Then B is retraceable, since A is. But $A \cup B$ is not, otherwise an infinite recursive subset of A could be generated, since if x is in A then 2^x is in B, and a retracing function would give an element of A greater than x. And $A \cup B$ is uniformly introreducible, because if C is an infinite subset of $A \cup B$ then one of $C \cap A$ and $C \cap B$ is infinite, and then one of A and B is recoverable, and so is $A \cup B$.)

c) *Not every introreducible set is uniformly introreducible.* (Lachlan) (Hint: this uses the priority method, introduced in Chapter X. Build A r.e. such that $\overline{A}$ is as wanted. Given B r.e. and co-retraceable, and such that $K \leq_T B$, see II.6.16, encode $\overline{B}$ into $\overline{A}$, allowing finitely many errors to ensure that $\overline{A}$ is not uniformly introreducible. Since A is r.e., $A \leq_T B$. The coding ensures $B \leq_T A$. Since then $A \equiv_T B$, if A is uniformly introreducible there is e such that, whenever C is an infinite subset of A, $B = \varphi_e^C$. Then spoil uniform introreducibility, by looking at B and using Sacks' agreement method, which succeeds because B is not recursive.)

d) *If A and $\overline{A}$ are uniformly introreducible, then A is recursive.* (Jockusch [1968]) (Hint: prove that if B is uniformly introreducible and immune, then any infinite r.e.

set of disjoint finite sets intersecting B has members of unbounded cardinality. Apply this to $A \oplus \overline{A}$, which is uniformly introreducible, and to the set of pairs $\{2x, 2x+1\}$.)

Regressive versus r.e.

We follow the path set up by the previous subsection.

Proposition II.6.8 (Dekker and Myhill [1958])
R.e. $\Rightarrow$ *regressive* $\Rightarrow$ *r.e. or immune.*

Proof. If A is r.e. and infinite, there is a recursive function f which enumerates it without repetitions:

$$A = \{f(0), f(1), \dots\}.$$

It is then enough to define

$$\varphi(f(0)) \simeq f(0) \qquad \varphi(f(n+1)) \simeq f(n)$$

to have a partial recursive function regressing A. Note that (unlike in the proof of II.6.5) φ is not immediately extendable to a total recursive function (see II.6.11 for a reason).

Let now A be regressive (via φ) and infinite, and suppose A is not immune: then A has an infinite recursive subset B. But then A is r.e.: to generate it, repeatedly apply φ to any element of B, until the smallest element of A is hit (this can be recognized, since it is left fixed by φ). □

It is not true that if both A and $\overline{A}$ are regressive, then A is recursive: any nonrecursive, co-regressive r.e. set (see II.6.16) is a counterexample. By symmetry, not even the conclusion that A is r.e. holds. The correct generalization of II.6.6 is the following:

Proposition II.6.9 (Appel and McLaughlin [1965]) *If A and $\overline{A}$ are both regressive, then one of A and $\overline{A}$ is r.e.*

Proof. We try to extend the proof of II.6.6. Part 2 of it generalizes (using now any enumeration of B), and shows that if B is regressive and there is an infinite r.e. set of disjoint pairs intersecting B, then B is r.e. (note that the facts that exactly one element of each pair was in B, and that the pairs covered ω, were used only to deduce the stronger conclusion that B was recursive).

It is not however true that if both A and $\overline{A}$ are regressive, then so is $A \oplus \overline{A}$ (see the exercises below). We then proceed directly. Let A be regressed by φ, and define

$$x \in S_n \Leftrightarrow \varphi^{(n+1)}(x) \simeq \varphi^{(n)}(x) \not\simeq \varphi^{(n-1)}(x).$$

The S_n's are disjoint r.e. sets, and each contains exactly one element of A (the n-th in the given enumeration).

If there are only finitely many S_n's with at least two elements, then A has an infinite r.e. subset (each S_n with exactly one element contributes to it), and is not immune. Being regressive, A is then r.e.

If there are infinitely many S_n's with at least two elements, we can get an infinite r.e. set of disjoint pairs intersecting $\overline{A}$ (each S_n with at least two elements contributes two elements, and at most one of them can be in A). Then $\overline{A}$ is r.e. by the first part of the proof, being regressive.   □

Note that again, as in II.6.6, we proved that one of A and $\overline{A}$ is r.e., but we do not know which one, since the proof is by nonconstructive cases.

Exercises II.6.10 (Appel and McLaughlin [1965]) a) *If $A \oplus \overline{A}$ is regressive, then A is recursive.* (Hint: use the first part of the proof above.)

b) *If B is regressive and there is an infinite r.e. set of disjoint finite sets of bounded cardinality intersecting B, then B is r.e.* (Hint: consider the greatest number n such that there are infinitely many n-tuples from the same set, all regressing on the same element.)

We have shown that an r.e. set is regressive, but the proof of II.6.8 does not give a total regressive function. The reason is that such a function does not always exist.

Proposition II.6.11 (Appel and McLaughlin [1965]) *There are r.e. sets which are not regressed by any total recursive function.*

Proof. We use the fact that there is an r.e. set A with immune complement (a simple set, see III.2.11), and we prove that any function regressing such a set cannot have a total recursive extension. Suppose f is such an extension, and let

$$x \in S_n \Leftrightarrow f^{(n+1)}(x) = f^{(n)}(x) \neq f^{(n-1)}(x).$$

S_n contains exactly one element of A (the n-th in the given enumeration). It is enough to prove that there are infinitely many S_n's with at least two elements, since then we can get an infinite r.e. set of disjoint pairs intersecting $\overline{A}$, and from it an infinite r.e. subset of $\overline{A}$ (against simplicity): either from a certain point on all pairs are contained in $\overline{A}$, or there are infinitely many pairs intersecting A, and then it is enough to generate A to discriminate which ones do, and choose the element of $\overline{A}$.

The claim is easy to prove:

- The set

$$\{x, f(x), f^{(2)}(x), \dots\}$$

(called the **splinter** of f at x) is finite for any x, because either it contains an element of A, and then it stops after the first element in the enumeration of A is hit, or it is an r.e. subset of $\overline{A}$, and it is finite by simplicity ($\overline{A}$ cannot contain infinite r.e. subsets).

- The set $S = \bigcup_{n \in \omega} S_n$ is then recursive: to check if x is in it, it is enough to generate the splinter of f at x, and see whether the conditions for membership in S_n are satisfied for some n (f being total, and the splinter being finite, f must cycle over the elements of the splinter, and thus the conditions can be checked recursively).

- Since $A \subseteq S$, $\overline{S}$ is finite (being an r.e. subset of $\overline{A}$). But $\overline{A}$ is infinite, and then so is $S \cap \overline{A}$: thus the S_n's have to cover an infinite part of $\overline{A}$. But each S_n is finite (it contains only one element of A, and the rest of it is an r.e. subset of $\overline{A}$), and each contains one element of A: then infinitely many S_n's must contain more than one element.   □

Exercises II.6.12 Splinters. A set A is a splinter if, for some recursive function g and some x,

$$A = \{x, g(x), g^{(2)}(x), \dots\}.$$

A splinter is obviously r.e., but the converse fails by III.7.10.a.

a) *There are nonrecursive splinters.* (Ullian [1960]) (Hint: let A be an r.e. nonrecursive set enumerated by a recursive function g, and define f recursive as follows. On a sequence number of the form

$$\langle g(x), g(0), g(1), \dots, g(x), g(x+1) \rangle$$

f gives $\langle g(x+1) \rangle$. On all other sequence numbers x, f gives $x * \langle g(ln(x) - 1) \rangle$. Then the splinter of f at $\langle g(0) \rangle$ contains $\langle x \rangle$ if and only if $x \in A$, and it is nonrecursive.) A different proof will be given in III.7.10.c, but the present one actually shows that *every r.e. degree contains a splinter.*

b) *An r.e. set is regressed by a total recursive function if and only if it is a splinter.* (Degtev [1970]) (Hint: if A is regressed by g w.r.t. $\{a_0, a_1, \dots\}$, the enumeration can be recovered because A is r.e. Extract an infinite recursive subset $B = \{b_0 < b_1 < \dots\}$, with $b_0 = a_0$ and $a_1 < b_1$. Define $f(x) = g(x)$ if $x \in \overline{B} \wedge x \neq a_1$, so that f sweeps the intervals between b's, and

$$f(x) = \begin{cases} g(b_1) & \text{if } x = a_0 = b_0 \\ g(b_2) & \text{if } x = a_1 \\ g(b_{n+2}) & \text{if } x = b_n \wedge n > 0 \end{cases}$$

so that f visits successive intervals. The converse is similar.)

For more information on splinters see III.7.10 and Myhill [1959], Ullian [1960], Young [1965], [1966], [1967].

Results proved later (see III.4.9) will show the existence of:

1. *a retraceable set which is not regressed by total recursive functions*

2. *a set retraced by a total recursive function, which is not regressed by total many-one recursive functions.*

Existence theorems and nondeficiency sets

We state our results in strong form, using the notion of degree introduced in II.3.3.

Proposition II.6.13 (Dekker and Myhill [1958]) *Every T-degree contains a retraceable set.*

Proof. If A is finite then it is recursive and retraceable. Let A be infinite, and f be the enumeration of A in order of magnitude (f is not recursive, in general). Let B be enumerated (in order of magnitude) by the function

$$g(n) = \langle f(0), f(1), \ldots, f(n) \rangle.$$

Then B is retraceable via the recursive function that chops off the last component of a sequence number of length greater than 1, and leaves unchanged the remaining numbers. Clearly $B \leq_T A$ by definition, and $A \leq_T B$ because

$$x \in A \Leftrightarrow x \text{ is a component of } g(x),$$

since $x \leq g(x)$.   □

Corollary II.6.14 *There are $2^{\aleph_0}$ retraceable sets.*

The recursive sets are the simplest retraceable sets. R.e. nonrecursive sets cannot be retraceable (by II.6.5 a nonrecursive, retraceable set must be immune), and thus the next level of complexity, and the first nontrivial one, for retraceable sets is being co-r.e.

Exercise II.6.15 *If A is retraceable and co-r.e., it is retraced by a total recursive function.* (Hint: given x, see if $\varphi(x)\downarrow$ or $x \in \overline{A}$.)

We now prove that such sets not only exist, but are as abundant as they can be.

Theorem II.6.16 (Dekker and Myhill [1958]) *Every r.e. T-degree contains a retraceable, co-r.e. set.*

Proof. If A is recursive then it is itself co-retraceable and r.e. Let then A be r.e. nonrecursive, and let f be a recursive, one-one enumeration of it. Let

$$x \in B \quad \Leftrightarrow \quad (\exists y > x)(f(y) < f(x))$$
$$x \in \overline{B} \quad \Leftrightarrow \quad (\forall y > x)(f(y) > f(x)).$$

The elements of $\overline{B}$ are called **stages of nondeficiency**, or **true stages**, in the enumeration of A given by f, because no new element of A smaller than $f(x)$ is generated by f in the future. Hence, for $x \in \overline{B}$,

$$\{f(0), \ldots, f(x)\} \cap \{0, \ldots, f(x)\} = A \cap \{0, \ldots, f(x)\}.$$

B is clearly r.e. Moreover:

- $A \leq_T B$
  To see if $z \in A$ it is enough to find $x \in \overline{B}$ such that $f(x) > z$, and see if

  $$z \in \{f(0), \ldots, f(x)\}.$$

  And x exists because f is one-one and $\overline{B}$ is infinite (given an element $b \in \overline{B}$, a greater one can be obtained by taking first the smallest element $a \in A$ which is not in $\{f(0), \ldots, f(b)\}$, and then the stage in which a is generated by f).

- $B \leq_T A$
  $x \in B$ if and only if there is some element in

  $$(A \cap \{0, \ldots, f(x)\}) - \{f(0), \ldots, f(x)\}.$$

- B is co-retraceable
  Given $x \in \overline{B}$, we want to give an effective procedure to find the greatest element of $\overline{B}$ smaller than x. Since for $y > x$ is $f(y) > f(x)$, it is enough to check the values of f for arguments below x. In other words, for $z < x$,

  $$z \in \overline{B} \quad \Leftrightarrow \quad (\forall y > z)(f(y) > f(z))$$
  $$\Leftrightarrow \quad (\forall y)(z < y \leq x \Rightarrow f(z) < f(y)).$$

  Then it is enough to define $g(x)$ as the biggest $z < x$ such that

  $$(\forall y)(z < y \leq x \Rightarrow f(z) < f(y))$$

  if there is one, and x otherwise (so that the first element of B is left fixed). $\square$

The idea of using nondeficiency stages is an ingenious one, invented by Dekker [1954]. It will show its usefulness time and again, in many different contexts (including infinite injury priority arguments, see Chapter X). As far as retraceable co-r.e. sets are concerned, this idea completely captured the heart of the matter:

Proposition II.6.17 (Yates [1962]) *A co-r.e. set A is retraceable if and only if, for some recursive function f,*

$$x \in A \quad \Leftrightarrow \quad (\forall y > x)(f(y) > f(x))$$
$$x \in \overline{A} \quad \Leftrightarrow \quad (\exists y > x)(f(y) \leq f(x)).$$

Proof. If a function f as stated exists, the proof of II.6.16 shows that A is retraceable. If A is finite, it can easily be seen that a function f as stated exists. Let then A be an infinite, retraceable and co-r.e. set: we want to find f. We have g recursive retracing A, and we may suppose that g is total (II.6.15) and $g(x) \leq x$ (II.6.3). Consider the recursive height function

$$h(x) = \mu z \, [g^{(z+1)}(x) = g^{(z)}(x)],$$

and the recursive height sets

$$x \in H_n \Leftrightarrow h(x) = n.$$

Clearly the height sets are disjoint, cover ω (since g is total and descending) and have exactly one element of A each (since A is infinite).

The idea is to define f on H_n (ordered by magnitude), by letting it be n until the first element of A is hit, and greater than n afterwards. Thus, if $x \in H_n \cap A$ and $y < x \wedge y \in H_n$, then y is a deficiency stage, while x becomes a nondeficiency one. Given $x \in H_n$, consider the set

$$\hat{x} = \{y : y < x \wedge y \in H_n\}.$$

If $\hat{x}$ is empty, x is the first element of H_n, so let $f(x) = n$. If $\hat{x}$ is not empty, enumerate $\overline{A}$ until exactly one element z of $\hat{x} \cup \{x\}$ has not yet been generated in it. This is possible because $\overline{A}$ is r.e., and H_n has exactly one element in A.

If $z = x$ then all smaller elements in H_n are in $\overline{A}$, and we can still let $f(x) = n$. If $z \neq x$ then x has been enumerated in $\overline{A}$, and we now know that x has to be a deficiency stage. Then we want $y > x$ such that $f(y) \leq f(x)$. Note that $f(x) = n + 1$ is not enough, since it might be that H_{n+1} contains no element $y > x$. And $m > n$ such that H_m has an element $y > x$ is not enough either, since the unique element $t \in H_m \cap A$ might be smaller than x, and setting $f(x) = m$ would make $f(t) = f(x)$, while $t < x$, and t has to be a nondeficiency stage (since $t \in A$). But then let $f(x) = m$ for $m > n$ such that H_m contains no element smaller than x. □

Exercises II.6.18 Nondeficiency sets. A set A is a nondeficiency set if, for some recursive function f,

$$x \in A \Leftrightarrow (\forall y > x)(f(y) > f(x)).$$

a) *If f is not finite-one, the nondeficiency set of f is finite.*

b) *A co-r.e. set is retraceable if and only if it is the nondeficiency set of a finite-one function.* (Yates [1962]) (Hint: see the proof above.)

c) *Not every co-r.e. retraceable set is the nondeficiency set of a one-one function.* (Degtev [1970]) (Hint: let $A(g)$ be the deficiency set of g, and $A_n(g)$ be its approximation up to n, i.e.

$$x \in A_n(g) \Leftrightarrow x < n \wedge (\exists y)(x < y \le n \wedge g(y) \le g(x)).$$

Define f as follows. Given $f(0), \ldots, f(n)$, let $n = \langle e, x \rangle$ be the first stage in which $\varphi_{e,n}$ is total and one-one on $\{0, \ldots, x\}$, and $A_{n_0}(f) \subseteq A_x(\varphi_e)$, where

$$n_0 = max\{x \le n : f(x) = e\}.$$

Then let $f(n+1) = e$. Otherwise, let $f(x) = n$. If φ_e is one-one and total, then $\overline{A(\varphi_e)}$ is infinite, and it cannot be $A(\varphi_e) = A(f)$, otherwise by construction f takes the value e infinitely often, and then $\overline{A(f)}$ is finite.)

Proposition II.6.19 (Marchenkov [1976a]) *The class of r.e. co-retraceable sets is r.e. without repetitions.*

Proof. Since every finite set is co-retraceable, it is enough (by II.5.24) to show that the class of r.e. co-retraceable sets is r.e. Let

$$x \in A_e \Leftrightarrow (\exists y)[(\forall z \le y)(\varphi_e(z){\downarrow}) \wedge y > x \wedge \varphi_e(y) \le \varphi_e(x)].$$

Then $\{A_e\}_{e \in \omega}$ is an r.e. class, and each A_e is either finite (if φ_e is not total) or co-retraceable (by II.6.17). $\square$

The class of r.e. co-retraceable sets is not completely r.e. (by II.4.2), since it is not closed under supersets (see II.6.21.b).

Regressive versus retraceable

We now briefly come back to the original question of the extent of the analogy with recursive and r.e. sets, and we show that it fails quite strongly. First we give a positive result, which is the analogue of II.1.20.

Proposition II.6.20 (Dekker [1962]) *Every infinite regressive set has an infinite retraceable subset.*

Proof. Let A be regressed by φ w.r.t. $\{a_0, a_1, \dots\}$, and

$$
\begin{aligned}
b_0 &= a_0 \\
b_{n+1} &= \text{the first element in the list of } A \text{ greater than } b_n.
\end{aligned}
$$

Then $B = \{b_0, b_1, \dots\}$ is infinite, because A is. To define the retracing function for B on a given x, we first iterate φ until we stop, which we must if $x \in A$. Then we recreate the initial segment of B from b_0 to x, by dropping the elements that break the monotone growing, and take the biggest element obtained which is smaller than x. If $x \in B$ then we do recreate the initial segment of B, and we do choose the right element. $\square$

There are two properties that we consider essential to claim a nontrivial analogy with recursive and r.e. sets, namely:

1. The recursive sets are closed under complementation.

2. A set which is r.e., together with its complement, is recursive (II.1.19).

They both fail here:

1. *There is a retraceable set with a nonretraceable complement.*
   Take any retraceable, nonrecursive set: its complement is not retraceable, by II.6.6.

2. *There is a set regressive together with its complement, but not retraceable.*
   Take any set A r.e. and nonrecursive, with a retraceable complement. Then both A (being r.e.) and $\overline{A}$ are regressive, but if A were retraceable then it would also be recursive, again by II.6.6.

Exercises II.6.21 (Dekker and Myhill [1958]) a) *If A and B are retraceable, so is $A \cap B$.* (Hint: given x, consider the greatest element smaller than it on which x is sent by both functions retracing A and B.) Appel [1967] has shown that this fails for regressive sets.
   b) *There are retraceable sets A and B such that $A \cup B$ is not regressive.* (Hint: let A be an infinite recursive set, and B a nonrecursive, retraceable and co-r.e. set. If $A \cup B$ were regressive, it would be r.e. because A is an infinite recursive subset of it, and B would be recursive.)

Despite the failure of the analogy with r.e. and recursive sets, retraceable and regressive sets are interesting on their own, and are useful in some parts of Recursion Theory. See McLaughlin [1982] for a detailed study of them.

Chapter III

Post's Problem and Strong Reducibilities

One theme of this chapter is **relative computability**. In Chapter II we introduced the most general and fundamental case: Turing reducibility. It will be recalled that no limitation was imposed there on the help given to the machine by the oracle, except for an obvious finiteness requirement. Here we take an opposite stand, and look at various possible limitations. Section 2 deals with the most restrictive case of m-reducibility, in which only one question is allowed to the machine during a computation, and only at the very end of it. Section 3 treats the case of Boolean combinations of atomic questions, called tt-reducibility, while a number of other, less fundamental, reducibilities are dealt with in Sections 4, 7 and 8.

Relative computations induce equivalence classes, by identifying functions and sets which have the same degree of difficulty of computation. A second theme in the chapter is **Post's problem**, introduced in Section 1, which asks whether there are only two such classes of r.e. sets. The solution, obtained in Section 5, will tell whether the r.e. sets can be distinguished, from a computational point of view, only between recursive and nonrecursive, or whether instead this rough dichotomy can somehow be essentially refined. The strategy for a solution to the problem is to analyze the possible structure of r.e. sets (as opposed to giving direct constructions, an alternative strategy pursued in Chapter X), and the tactic is to solve the problem first for m-reducibility, in Section 2, and then gradually improve the solution for weaker and weaker reducibilities, in Sections 3 and 4, until we reach the one we are really interested in.

The original motivation for the study of r.e. sets was that they code (by arithmetization) the sets of theorems of formal systems. A third theme of this chapter is the **analysis of formal systems**, from this abstract point of view. We make the relationship between formal systems and r.e. sets precise in Section 10, where we also revisit some of the notions and results obtained in the chapter, and discuss their bearing on the subject of formal systems.

III.1   Post's Problem

We have so far encountered only two different kinds of r.e. sets, namely the recursive sets and $\mathcal{K}$, and they generate different degrees.

Definition III.1.1 *An* **r.e. T-degree** *is a degree containing at least one r.e. set. Two r.e. T-degrees are:*

1. *the T-degree* **0** *of the recursive sets*

2. *the T-degree* **0′** *of* $\mathcal{K}$.

Note that, because of Post's Theorem and the fact that a set and its complement are in the same degree (being obviously computable one from the other), a degree contains only r.e. sets if and only if it contains only recursive sets. This explains why we only require the existence of an r.e. set in an r.e. degree.

Recall that there is a partial order on the degrees, induced by the relation $\leq_T$. It is obvious that **0** is the least degree with respect to it, and the next result shows that **0′** is the greatest r.e. degree.

Proposition III.1.2 (Post [1944]) *If A is any r.e. set then $A \leq_T \mathcal{K}$.*

Proof. We prove that there is a recursive function f such that

$$x \in A \Leftrightarrow f(x) \in \mathcal{K} \Leftrightarrow f(x) \in \mathcal{W}_{f(x)},$$

where the last equivalence holds by definition of $\mathcal{K}$. By the S_n^m-Theorem, let f be a recursive function such that

$$\mathcal{W}_{f(x)} = \begin{cases} \omega & \text{if } x \in A \\ \emptyset & \text{otherwise.} \end{cases}$$

Then:

- $x \in A \Rightarrow \mathcal{W}_{f(x)} = \omega \Rightarrow f(x) \in \mathcal{W}_{f(x)} \Rightarrow f(x) \in \mathcal{K}$

- $f(x) \in \mathcal{K} \Rightarrow f(x) \in \mathcal{W}_{f(x)} \Rightarrow \mathcal{W}_{f(x)} \neq \emptyset \Rightarrow x \in A.$ □

Thus, since $A \leq_T \mathcal{K}$ automatically holds for r.e. sets, an r.e. set A is in the greatest r.e. degree **0′** if and only if $\mathcal{K} \leq_T A$.

Definition III.1.3 *A set A is* **Turing complete** *(or T-complete) if it is r.e. and its degree is* $0'$, *i.e.* $\mathcal{K} \leq_T A$.

In general, given a reducibility $\leq_r$, we will call an r.e. set A **r-complete** if $\mathcal{K} \leq_r A$, and **r-incomplete** otherwise.

As noted above, the r.e. sets we know at this point are all recursive or T-complete. It is natural to ask whether there are others.

> **Post's Problem (Post [1944])** Are there r.e. T-degrees different from 0 and $0'$? Equivalently, are there r.e. sets which are neither recursive nor T-complete?

The reasons to isolate this natural problem and give it a name are many. First, despite its technical formulation, the problem was motivated by deep methodological questions, related to the undecidability results, and reviewed in the next subsection. Second, the solution to the problem escaped the researchers for many years and provided, as a by-product, new techniques and results, some of them treated in this chapter. Finally, versions of the problem arise in different areas of Generalized Recursion Theory, and their solution is usually regarded as a proof of maturity for the new areas.

Origins of Post's Problem ⋆

Post arrived at the formulation of his problem after an exciting intellectual development, which is worth reviewing. In his dissertation, completed in 1920, he started by analyzing the system of *Principia Mathematica*, and attacking the problem of its *decidability*. He was able to solve a particular case, namely the decision problem for propositional calculus, by proving a completeness theorem that showed that the theorems were exactly the tautologies. He published this in [1921].

In the academic year 1920–21, as a postgraduate, Post set down to generalize this decidability result, by attacking the general case. Trying to capture the essence of formal systems he considered, by successive abstractions, a sequence of notions, finally obtaining the canonical systems (see p. 143). By showing that the system of *Principia Mathematica* could be translated in a canonical system, he convinced himself that he had a sufficiently general notion.

Post then turned to the decidability problem for canonical systems, hoping that the generality of the notion would make the proof simpler, because independent of details related to particular systems. He soon concentrated on a special problem, called the *tag*, of which he was able to handle some particular cases, but that turned out to be intractably complicated in general (with good reasons: it was undecidable, see Minsky [1961]).

At this point, the unsuccessful attempts prompted a revision of the plan, and Post turned to *undecidability*. He defined a universal canonical system, which is just a version of the set $\mathcal{K}_0$ of p. 150, and showed its diagonal set, i.e. $\mathcal{K}$, to be, in modern terms, r.e. but nonrecursive. To be able to deduce from this a general unsolvability result Post needed a version of Church's Thesis, which he stated in the form: every effectively generable set can be generated by a canonical system. From this the existence of incomplete formal systems followed easily.

All this work, concluded in 1921, and anticipating a number of results by Gödel, Church, and Turing that would follow much later, was left unpublished (see [1922]), because Post was not convinced of (his version of) Church's Thesis. He took it as a working hypothesis that needed verification, in the form of psychological analysis of the computational process. A step toward such analysis was [1936], in which Post proposed a version of Turing machines, independently of Turing. The canonical systems were published only in [1943], and the form of Gödel's theorem based on canonical systems only in [1944].

The mutual reductions among various notions of canonical systems, as well as particular formal systems like *Principia Mathematica*, led Post to the concept of m-reducibility between sets. And the fact that known undecidability proofs, by Post, Church, and others, were all obtained by appropriately reducing $\mathcal{K}$ to the problem in question, prompted the problem of whether the only undecidable systems were the universal ones, in which $\mathcal{K}$ could be interpreted. Post [1944] was able to disprove this for m-reducibility, and then he asked the same question for the general notion of T-reducibility, introduced by Turing [1936]. Despite a good deal of intermediate work, he could not reach a solution. Sections 2 to 4 are a report of Post's work and of modern improvements, and Section 5 provides the missing brick.

Turing reducibility on r.e. sets

Since the solution to Post's Problem will require a detailed study of the structure of the r.e. sets, we prove some technical results that will facilitate the task.

Proposition III.1.4 T-reducibility on r.e. sets (Rogers [1967]) *If A and B are r.e. sets, then $A \leq_T B$ if and only if, for some r.e. relation R,*

$$x \in \overline{A} \Leftrightarrow (\exists u)(D_u \subseteq \overline{B} \wedge R(x, u)).$$

Proof. By compactness and monotonicity of oracle computations (II.3.13) and the Normal Form Theorem for restricted functionals (II.3.12) we have, for a given e and some r.e. relation Q,

$$\varphi_e^B(x) \simeq z \Leftrightarrow (\exists v)(\exists u)(D_v \subseteq B \wedge D_u \subseteq \overline{B} \wedge Q(x, u, v, z)),$$

because D_v and D_u together specify a finite subfunction of c_B. If $A \leq_T B$ then $c_A \simeq \varphi_e^B$, for some e. In particular

$$x \in \overline{A} \Leftrightarrow (\exists v)(\exists u)(D_v \subseteq B \wedge D_u \subseteq \overline{B} \wedge Q(x, u, v, 0)).$$

If moreover B is r.e., the expression $D_v \subseteq B$ is r.e. itself, and thus there is R r.e. such that

$$x \in \overline{A} \Leftrightarrow (\exists u)(D_u \subseteq \overline{B} \wedge R(x, u)).$$

Conversely, if this expression for $\overline{A}$ holds for some r.e. relation R, then $\overline{A}$ is r.e. in B. If moreover A is r.e., then it is r.e. in B and, by the relativization of Post's Theorem to B, A is recursive in B. □

The next result is a strong generalization of the Fixed-Point Theorem, to any function which is recursive in an incomplete r.e. set.

Theorem III.1.5 T-completeness of r.e. sets (Martin [1966], Lachlan [1968], Arslanov [1981]) *An r.e. set A is T-complete if and only if there is a function $f \leq_T A$ without fixed-points, i.e. such that $\forall x(W_x \neq W_{f(x)})$.*

Proof. If A is T-complete, the set $\{x : 0 \notin W_x\}$ is recursive in A, being co-r.e. By the relativized S_n^m-Theorem, there is $f \leq_T A$ such that

$$W_{f(x)} = \begin{cases} \omega & \text{if } 0 \notin W_x \\ \emptyset & \text{otherwise.} \end{cases}$$

Then $W_{f(x)} \neq W_x$, because the two sets differ on 0.

Suppose now that A is r.e., and $f \leq_T A$ has no fixed-points. The idea to get $\mathcal{K} \leq_T A$, thus showing that A is T-complete, is the following. Consider the function

$$s_x = \begin{cases} \mu s(x \in \mathcal{K}_s) & \text{if } x \in \mathcal{K} \\ 0 & \text{otherwise.} \end{cases}$$

Then $x \in \mathcal{K} \Leftrightarrow x \in \mathcal{K}_{s_x}$. We want to majorize s_x recursively in A, i.e. to find ψ recursive in A, such that $\psi(x) \geq s_x$. Then $x \in \mathcal{K} \Leftrightarrow x \in \mathcal{K}_{\psi(x)}$, and this implies $\mathcal{K} \leq_T A$.

Since $f \leq_T A$, there is e such that $f \simeq \varphi_e^A$. Moreover A is r.e., and it can be recursively approximated by an enumeration $\{A_s\}_{s \in \omega}$. Although φ_e^A is only recursive in A, it can be approximated by the recursive function $\varphi_{e,s}^{A_s}$. Fix now $x \in \mathcal{K}$: by the Fixed-Point Theorem, there is z such that

$$W_z = W_{\varphi_{e,s_x}^{A_{s_x}}(z)}.$$

By the assumption on f, it cannot be $W_z = W_{f(z)}$. This means that, while $\varphi_{e,s_x}^{A_{s_x}}(z)$ is an approximation of $f(z)$, it must be a wrong one. Recursively in

A we can first compute $\varphi_e^A(z)$, and find the length of the least initial segment $\hat{c}_A(y)$ that gives the right value. Then we can recursively enumerate A to find a stage $s \geq y$ in which all the elements of A used in the computation have been generated, i.e. $\hat{c}_{A_s}(y) = \hat{c}_A(y)$. From this point on the approximations will give the right value, i.e. $\varphi_{e,t}^{A_t}(z) = f(z)$, for any $t \geq s$. Then it must be $s \geq s_x$.

We have only to do this in general now. By the Fixed-Point Theorem with parameters (II.2.11), there is g recursive such that

$$ W_{g(x)} = \begin{cases} W_{\varphi_{e,s_x}^{A_{s_x}}(g(x))} & \text{if } x \in \mathcal{K} \\ \emptyset & \text{otherwise.} \end{cases} $$

Let $\psi(x)$ be a stage in which A has generated all the elements needed to compute the right value of $f(g(x))$ from then on, as above. Then ψ is recursive in A, and $\psi(x) \geq s_x$ (if $x \in \mathcal{K}$ as above, and if $x \notin \mathcal{K}$ because then $s_x = 0$). □

Note that the condition that A be r.e. is essential, since without it the result fails in general.

Exercise III.1.6 *There is a set $A \leq_T \mathcal{K}$, and a function $f \leq_T A$, such that f has no fixed-points, but $\mathcal{K} \not\leq_T A$.* (Arslanov [1981]) (Hint: by the proof of III.2.18, it is enough to find A effectively immune, recursive in $\mathcal{K}$, and such that $\mathcal{K} \not\leq_T A$. And such a set exists by the Low Basis Theorem V.5.32, applied to the Π_1^0 class

$$ \{A : A \subseteq \overline{S} \wedge (\forall x)(D_{g(x)} \cap A \neq \emptyset)\}, $$

where S is Post's simple set, see III.2.11, and g enumerates a strong array intersecting $\overline{S}$.)

For more results on fixed-points, see Arslanov [1981], Jockusch, Lerman, Soare, and Solovay [1989], Kučera [1986], [1989], and Jockusch [1989]. In particular, Kučera [1986] (see X.4.7) has solved Post's problem by showing that any degree below $\mathbf{0}'$ and containing no function without fixed-points bounds an r.e. nonrecursive degree.

III.2   Simple Sets and Many-One Degrees

The path we shall follow to attack Post's Problem is the one suggested by Post himself, which is also a natural mathematical practice: confronted with a difficult problem, try first with simpler versions of it and, once a solution is found for them, proceed to more difficult cases, in the hope of finally reaching the solution of the original problem. This will be a long path, but it will finally pay off in III.5.20, leaving us with a deep knowledge of the structure of the r.e. sets.

Our present approach is **structural**, since it tries to solve the problem by isolating nonempty properties of r.e. sets that imply nonrecursiveness and T-incompleteness. A different approach consists of trying to build ad hoc solutions by **brute force**, and will be considered in Chapter X.

Many-one degrees

The simplest special case of Turing reducibility, which we have been using already in many of the previous proofs, is the following:

Definition III.2.1 (Post [1944]) *A is **m-reducible** to B $(A \leq_m B)$ if, for some recursive function f, the following equivalent conditions are satisfied:*

1. $\forall x (x \in A \Leftrightarrow f(x) \in B)$

2. $A = f^{-1}(B)$

3. $f(A) \subseteq B \wedge f(\overline{A}) \subseteq \overline{B}.$

*A is **m-equivalent** to B $(A \equiv_m B)$ if $A \leq_m B$ and $B \leq_m A$.*

Exercises III.2.2 a) *If A is recursive, then $A \leq_m B$ for any set $B \neq \emptyset, \omega$.*
  b) *If $A \leq_m B$ and B is recursive, so is A.*
  c) *If $A \leq_m B$ and B is r.e. then so is A.*
  d) *If A is r.e. then $A \leq_m \overline{A}$ if and only if A is recursive and $A \neq \emptyset, \omega$.*
  e) *There is a nonrecursive set A such that $A \leq_m \overline{A}$. (Hint: consider $\mathcal{K} \oplus \overline{\mathcal{K}}$.)*
  f) *If A and B differ finitely, then $A \equiv_m B$.*

Note that $\leq_m$ is a reflexive and transitive relation, and thus $\equiv_m$ is an equivalence relation.

Definition III.2.3 *The equivalence classes of sets w.r.t. m-equivalence are called **m-degrees**, and $(\mathcal{D}_m, \leq)$ is the structure of m-degrees, with the partial ordering $\leq$ induced on them by $\leq_m$.*
  *The m-degrees containing r.e. sets are called **r.e. m-degrees**, and two of them are:*

1. $\mathbf{0}_m$, *the m-degree of the recursive sets different from $\emptyset$ and ω*

2. $\mathbf{0}'_m$, *the m-degree of $\mathcal{K}$.*

Note that an r.e. m-degree contains only r.e. sets. The m-degrees containing recursive sets are three: $\mathbf{0}_m$, $\{\emptyset\}$, $\{\omega\}$. The last two are incomparable and smaller than $\mathbf{0}_m$, but every other m-degree is greater than or equal to $\mathbf{0}_m$. In the following we will always consider nontrivial sets, and thus $\mathbf{0}_m$ may be considered as the least m-degree. We now show that $\mathbf{0}'_m$ is the greatest r.e. m-degree.

Proposition III.2.4 (Post [1944]) *A set A is r.e. if and only if* $A \leq_m \mathcal{K}$.

Proof. If A is r.e. then $A \leq_m \mathcal{K}$ by the proof of III.1.2. Conversely, if $x \in A \Leftrightarrow f(x) \in \mathcal{K}$ then the elements of A are exactly the counterimages of those in the intersection of $\mathcal{K}$ and the range of f, and hence A is r.e. $\square$

Definition III.2.5 *A set A is* **m-complete** *if it is r.e. and its m-degree is* $0'_m$, *i.e.* $\mathcal{K} \leq_m A$.

Exercises III.2.6 Both $\mathcal{K}$ and the notion of m-completeness are defined w.r.t. the class of the r.e. sets. Let a recursive enumeration $\{\mathcal{W}_{h(x)}\}_{x \in \omega}$ of the recursive sets be given (see II.5.26).

a) *The set* $x \in \mathcal{K}^* \Leftrightarrow x \in \mathcal{W}_{h(x)}$, *obtained by diagonalization over the recursive sets, is m-complete.* (Muchnik [1958a]) (Hint: let

$$\mathcal{W}_{h(g(x))} = \begin{cases} \omega & \text{if } x \in \mathcal{K} \\ \emptyset & \text{otherwise.} \end{cases}$$

Then $x \in \mathcal{K} \Leftrightarrow g(x) \in \mathcal{K}^*$.)

b) *If the recursive sets are uniformly m-reducible to an r.e. set A, then A is m-complete.* (Smullyan [1961]) (Hint: if $z \in \mathcal{W}_{h(x)} \Leftrightarrow f(z, x) \in A$, then $\mathcal{K}^* \leq_m A$.)

The analogue of Post's Problem for m-reducibility is: are there r.e. sets which are neither recursive, nor m-complete? The beginning of our story is the following observation.

Proposition III.2.7 (Post [1944]) *If A is m-complete, then* $\overline{A}$ *contains an infinite r.e. subset.*

Proof. First consider the special case of the m-complete set $\mathcal{K}$. Note that

$$\mathcal{W}_x \subseteq \overline{K} \Rightarrow x \in \overline{K} - \mathcal{W}_x.$$

Suppose indeed that $x \in \mathcal{W}_x$: then x is in $\mathcal{K}$ by definition, and in $\overline{\mathcal{K}}$ because $\mathcal{W}_x \subseteq \overline{K}$, contradiction. Then it must be $x \notin \mathcal{W}_x$, and thus also $x \in \overline{\mathcal{K}}$.

Then the index of an r.e. subset of $\overline{\mathcal{K}}$ is an element of $\overline{\mathcal{K}}$ but not of the given subset, and this permits us to generate an infinite r.e. subset of $\overline{\mathcal{K}}$, by starting with the emptyset, and getting new elements at each stage. Formally, if f is a recursive function such that:

$$\begin{aligned} f(0) &= \text{ an index of } \emptyset \\ f(n+1) &= \text{ an index of } \{f(0), \ldots, f(n)\}, \end{aligned}$$

then the range of f is an infinite r.e. subset of $\overline{\mathcal{K}}$.

To extend this to any m-complete set A, let

$$x \in \mathcal{K} \Leftrightarrow g(x) \in A.$$

Given an r.e. subset B of $\overline{A}$, first pull it back (through g) to an r.e. subset of $\overline{\mathcal{K}}$. Use the property of $\overline{\mathcal{K}}$ proved above, to get an element in $\overline{\mathcal{K}}$ but not in the pull back of B, then project it via g to an element in $\overline{A}$ but not in B. Formally, let f be a recursive function such that:

$$\begin{aligned} f(0) &= g(a_0) & \text{with } a_0 \text{ index of } \emptyset \\ f(n+1) &= g(a_{n+1}) & \text{with } a_{n+1} \text{ index of the r.e. set} \\ & & g^{-1}(\{f(0), \dots, f(n)\}). \end{aligned}$$

Then the range of f is an infinite r.e. subset of $\overline{A}$. $\square$

Exercise III.2.8 *A different proof of the result above* consists of noting that any infinite r.e. set of indices of $\emptyset$ is an infinite r.e. subset of $\overline{\mathcal{K}}$, and can be projected to $\overline{A}$ because the function that reduces $\mathcal{K}$ to A can be taken to be one-one.

Simple sets

Since coinfinite recursive sets and m-complete sets have complement with an infinite r.e. subset, a solution to Post's Problem for m-degrees would be given by sets without this property. Recall (p. 141, and II.6.4) that an infinite set is immune if it does not contain infinite r.e. (or recursive) subsets.

Definition III.2.9 (Post [1944]) *A set is **simple** if it is r.e. and coimmune, i.e. its complement is infinite and does not contain infinite r.e. subsets.*

Exercises III.2.10 a) *A coinfinite r.e. set is simple if and only if it has no coinfinite, recursive superset.* (Hint: use II.1.20.)
   b) *If A and B are simple then $A \cap B$ is simple, and $A \cup B$ is simple or cofinite.* Thus simple or cofinite sets are a filter in the lattice of the r.e. sets under inclusion. (Dekker [1953])
   c) *If A is simple and $\mathcal{W}_x$ is infinite, then $A \cap \mathcal{W}_x$ is infinite.*

It only remains to show that the notion of simplicity is not empty.

Theorem III.2.11 (Post [1944]) *There exists a simple set.*

Proof. We give two different constructions.

1. *Post's simple set S.*
   The idea is to build S intersecting each infinite r.e. set, so that $\overline{S}$ does not contain any infinite r.e. subset. We cannot recursively enumerate the

infinite r.e. sets (II.5.27.a), so we will intersect S with each r.e. set with enough elements. To prevent the collapse of the complement of S, we do not want to put too many elements in S, so we make sure that each time a new element goes into S, another one will stay out. Precisely, we put at most half of $\{0, 1, \ldots, 2x\}$ into S, for each x. The set S is defined as follows: dovetail an enumeration of all the r.e. sets and, for each e, put into S the first element greater than $2e$ enumerated into $\mathcal{W}_e$.

S is r.e. by construction. $\overline{S}$ is infinite since an element of $\{0, \ldots, 2x\}$ can enter S only if it comes from some $\mathcal{W}_e$ such that $2e < 2x$: but there are at most x such sets, and each contributes at most one element. Thus $\overline{S}$ has at least $x + 1$ elements, for each x, and it is then infinite. And S is simple, because if $\mathcal{W}_e$ is infinite then it has elements greater than $2e$, and one of them will be in S.

2. *A direct construction.*
We build a simple set A by stages. At stage s we will have A_s, and we will let $\overline{A_s} = \{a_0^s < a_1^s < \cdots\}$. In the end $\overline{A} = \{a_0 < a_1 < \cdots\}$, where $a_n = \lim_{s \to \infty} a_n^s$. We want to satisfy the following requirements:

$$P_e : \mathcal{W}_e \text{ infinite} \Rightarrow \mathcal{W}_e \cap A \neq \emptyset$$
$$N_e : \overline{A} \text{ has at least } e \text{ elements, or } \lim_{s \to \infty} a_e^s < \infty.$$

P stands for positive, because to satisfy the P_e's we have to do something positive on A (namely, to put some element in it). N stands for negative, because to satisfy the N_e's we have to do something negative on A (namely, to leave some element out of A).

The construction is as follows. We start with $A_0 = \emptyset$ (hence $a_n^0 = n$). At stage $s + 1$ we search for the smallest $e \leq s$ such that:

- $\mathcal{W}_{e,s} \cap A_s = \emptyset$
- for some $n \geq e$, $a_n^s \in \mathcal{W}_{e,s}$.

Note that both A_s and $\mathcal{W}_{e,s}$ are finite, and we only look at $e \leq s$, so the search is effective. If e does not exist, we go to the next stage. Otherwise, P_e is the condition with smallest index which looks unsatisfied, and with a chance to be satisfied. Then we put a_n^s into A, where n is the smallest one such that $n \geq e$ and $a_n^s \in \mathcal{W}_{e,s}$. This makes all the a_m^s with $m \geq n$ move to the next one (since they enumerate the complement). In other words,

$$a_m^{s+1} = a_m^s \quad \text{if } m < n$$
$$a_m^{s+1} = a_{m+1}^s \quad \text{if } m \geq n.$$

Since the construction is effective, $A = \bigcup_{s \in \omega} A_s$ is r.e. Moreover:

- $\overline{A}$ is infinite.

  It is enough to prove that $\lim_{s \to \infty} a_m^s$ exists. Indeed, a_m^s may move at a certain stage $s+1$ only if it happens that $a_n^s \in W_e$, for some $e \leq n \leq m$. Since each W_e contributes at most one element for each e, and there are only finitely many $e \leq m$, a_m^s moves only finitely many times. So all negative requirements are satisfied.

- A is simple.

  By induction, suppose that s_0 is such that $s_0 \geq e$, and all P_i's with $i < e$ have been satisfied at stage s_0 (i.e. $W_{e,s_0} \cap A_{s_0} \neq \emptyset$ if W_e is infinite). If W_e is an infinite subset of $\overline{A}$, there are $s \geq s_0$ and $n \geq e$ such that $a_n^s = a_n \in W_{e,s}$. Then one such a_n goes into A at stage $s+1$ (because e is the smallest index for which P_e looks unsatisfied, and with a chance to be satisfied), contradiction. $\square$

The second proof given above is a typical **priority argument with no injury**, with

$$P_0 > N_0 > P_1 > N_1 > \cdots$$

as order of priority for the satisfaction of the requirements. Indeed, we allow a_n^s to move only to satisfy P_e for some $e \leq n$, so e.g. a_0^s can move only to satisfy P_0, a_1^s only to satisfy P_0 or P_1, and so on. Moreover, we choose the smallest possible positive requirement for satisfaction, and this says that the positive requirements are ordered by their indices. There is no injury, because once a positive requirement is satisfied it remains so forever (since we never take elements out of A). The priority method will be discussed in full generality in Chapter X.

The two examples of simple sets given above are direct constructions, and thus somehow unnatural. There are however sets which can be naturally defined, and that turn out to be simple. The first uses the notion (p. 151) of **random number**, as a number that is its own shorter description. In terms of the Kolmogorov complexity K, defined (on p. 151) as

$$K(x) = \mu e(\varphi_e(0) \simeq x),$$

a number is random if $x \leq K(x)$.

Proposition III.2.12 (Kolmogorov [1963], [1965]) *The set of nonrandom numbers is simple.*

Proof. Let $A = \{x : K(x) < x\}$. Then A is r.e., because

$$x \in A \Leftrightarrow (\exists e < x)(\varphi_e(0) \simeq x).$$

To show that $\overline{A}$ is infinite consider, given any n, the converging values among

$$\varphi_0(0), \varphi_1(0), \ldots, \varphi_n(0).$$

If x is any number different from them all, then $n + 1 \leq K(x)$ by definition. And if x is the least number not among them, $x \leq n + 1$ (since we considered only $n + 1$ possible values, and in the worst case they are the numbers from 0 to n). Then $x \leq K(x)$, and x is random. Thus there is a random number with complexity at least $n+1$. Since this holds for every n, there are infinitely many random numbers, and $\overline{A}$ is infinite.

To show that there is no infinite r.e. set of random numbers, first note that the index e of an r.e. set W_e provides a uniform description of the elements of the set: if $W_e = \{x_0, x_1, \ldots\}$, then $x_n =$ the n-th element enumerated in W_e. By the S_n^m-Theorem (II.1.7), there is a one-one recursive function h such that $\varphi_{h(e,n)}(0) \simeq x_n$. If we could prove that for some n we have $h(e, n) < x_n$, we would know that x_n is not random. This is certainly the case if n is big enough and x_n is greater than

$$t(n) = \max_{e \leq n} h(e, n),$$

since then $h(e, n)$ is bounded by $t(n)$ almost everywhere.

It is now enough to note that, given W_e, we can uniformly obtain an r.e. subset $W_{g(e)}$ of it whose n-th element is bigger than $t(n)$, by waiting until such an element is generated in W_e. If W_e is infinite, then $W_{g(e)}$ is infinite and it contains a nonrandom element. Thus there is no infinite r.e. set of random numbers and A is simple. $\square$

The proof just given is a positive use of **Berry's paradox** (Russell [1906]), a version of which is: given n, consider 'the least number that cannot be defined in less than n characters'. This defines, in $c + |n|$ characters (where $|n|$ is the length of n and c is a constant), a number whose definition needs more than n characters, and it is paradoxical for all n such that $c + |n| \leq n$, i.e. for almost every n.

This proof contains additional information, exploited on p. 265. A quick, less informative proof is provided by the Fixed-Point Theorem. Suppose A is an infinite r.e. set of random numbers, and let h be a recursive function such that

$$\varphi_{h(e)}(0) \simeq \text{ the smallest } x > e \text{ generated in } A.$$

By the Fixed-Point Theorem, there is e such that $\varphi_e \simeq \varphi_{h(e)}$. By definition $\varphi_e(0)$ is a random number (being in A), hence it cannot be bigger than e. But this is exactly how it was defined, contradiction.

The immunity of the set of random numbers is a strong result, implying:

1. *a version of the incompleteness results*: in any consistent formal system sound for arithmetic, we can prove that a random number is so only in finitely many cases, since the set of provably random numbers is r.e. (Chaitin [1974]).

2. *the undecidability of the halting problem*: if we could decide whether $\varphi_e(0) \downarrow$, then we could compute $K(x)$ and decide whether a number is random or not. Actually, *the halting problem and the Kolmogorov complexity function have the same T-degree*: K is obviously recursive in $\mathcal{K}$, by definition, and the converse holds because the set of nonrandom numbers is T-complete, see p. 265.

Exercise III.2.13 *The existence of infinitely many random numbers implies the existence of infinitely many prime numbers.* (Chaitin [1979]) (Hint: if there are only $n + 1$ primes, say $p_0, \ldots, p_n$, then for every x, $x = p_0^{x_0} \cdots p_n^{x_n}$. But $x_i \leq \log x$, and so x can be described in $\approx n \cdot \log x$ bits of information. For big enough x, then x is not random.)

The next example of simple sets provides with such sets in every nonzero T-degree, and thus shows that a simple set is not necessarily T-incomplete.

Proposition III.2.14 (Dekker [1954]) *Every nonrecursive r.e. T-degree contains a simple set.*

Proof. Given A r.e. and nonrecursive, let B be its deficiency set (II.6.16). Then $A \equiv_T B$, and B is a coinfinite and coretraceable r.e. set. So $\overline{B}$ is (by II.6.5) recursive or immune. Since A is nonrecursive, $\overline{B}$ is immune, and then B is simple. $\square$

Exercise III.2.15 *Give a direct proof that the deficiency set of f recursive and one-one is recursive or immune.* (Hint: if $\overline{B}$ has an infinite recursive subset C, given x find $y \in C$ such that $f(y) > x$. Then $x \in A \Leftrightarrow x \in \{f(0), \ldots, f(y)\}$. Thus A, and hence B, are recursive.)

Effectively simple sets $\star$

We have just seen that a simple set can be T-complete, and thus simple sets do not automatically solve Post's Problem. But before we go on with different trials, we want to make sure that none of the simple sets built above is already T-incomplete. The idea is to effectivize the notion of simplicity.

Definition III.2.16 (Smullyan [1964]) *A is **effectively simple** if it is a coinfinite r.e. set, and there is a recursive function g such that*

$$\mathcal{W}_e \subseteq \overline{A} \;\Rightarrow\; |\mathcal{W}_e| \leq g(e).$$

Exercise III.2.17 We show that other natural attempts to define effective simplicity are either equivalent to the one proposed, or impossible.

a) *A coinfinite r.e. set A is effectively simple if and only if there is a partial recursive function φ such that $W_e \subseteq \overline{A} \Rightarrow \varphi(e) \downarrow \wedge |W_e| \leq \varphi(e)$.* (Sacks [1964])

b) *For every simple set A there is a partial recursive function φ such that if W_e is infinite then $\varphi(e) \downarrow \wedge \varphi(e) \in W_e \cap A$.*

c) *For no simple set there is a similar total recursive function.* (Hint: consider g recursive such that $W_{g(x)} = \overline{\{f(x)\}}$, and apply the Fixed-Point Theorem.)

Thus for simple sets we only know that an r.e. subset of the complement is finite, while for effectively simple sets we also know a recursive bound on its cardinality. The interest of the notion is that it implies T-completeness.

Proposition III.2.18 (Martin [1966]) *Every effectively simple set is T-complete.*

Proof. Let

$$W_e \subseteq \overline{A} \Rightarrow |W_e| \leq g(e).$$

Define $f \leq_T A$ such that

$$W_{f(e)} = \{\text{the first } g(e) + 1 \text{ elements of } \overline{A}\}.$$

Then f has no fixed-points: if $W_e = W_{f(e)}$ then $W_e \subseteq \overline{A}$, but $|W_e| = g(e) + 1$. By the criterion for T-completeness III.1.5, A is then T-complete. $\square$

Exercises III.2.19 a) *There are simple sets which are not effectively simple.* (Sacks [1964]) (Hint: a direct proof is difficult, requiring priority and Fixed-Point Theorem. To build A as wanted, we want to destroy

$$W_x \subseteq \overline{A} \Rightarrow |W_e| \leq \varphi_e(x).$$

Given x and e, we build an r.e. set $W_{f(e,x)}$ with the property that

$$W_{f(e,x)} \subseteq \overline{A} \wedge |W_{f(e,x)}| > \varphi_e(x).$$

Then we need the Fixed-Point Theorem to step from the sets $W_{f(e,x)}$ to sets of the form W_x. What III.2.18 accomplishes is to separate the use of the Fixed-Point Theorem, to give the T-completeness of effectively simple sets, and of priority, to build a T-incomplete simple set or just, due to III.2.14, a nonrecursive, T-incomplete r.e. set.)

b) *There are simple, not effectively simple, and T-complete sets.* (Hint: take a simple T-complete set A, a simple not effectively simple set B, and consider $A \oplus B$.)

c) *A simple set A is T-complete if and only if there is $g \leq_T A$ such that*

$$W_e \subseteq \overline{A} \Rightarrow |W_e| \leq g(e).$$

(Lachlan [1968]) (Hint: T-completeness is proved as in III.2.18. If A is T-complete, it is recursive in A to ask if $W_e \subseteq \overline{A}$, i.e. if $\exists x(x \in W_e \cap A)$. If $W_e \subseteq \overline{A}$ then W_e is finite, by simplicity, and recursively in A we can search for an x such that $(\exists y > x)(y \in W_e)$ fails.)

We can now show that *the simple sets constructed above are all effectively simple, and thus T-complete.* We refer to the proofs of III.2.11 and III.2.12, and use the same notations as there.

1. *Post's simple set.*
   If $W_e \subseteq \overline{S}$ then $|W_e| \le 2e + 1$, since otherwise W_e has more than $2e + 1$ elements, so one of them is greater than $2e$, and goes into S.

2. *The set A built in III.2.11.*
   If $W_e \subseteq \overline{A}$ then $|W_e| \le e$, since otherwise there is $n \ge e$ such that $a_n \in W_e$. By going to a stage s after which the W_i's with $i < e$ do not contribute anymore elements to A, the elements of $\overline{A}$ up to a_n have settled, and $a_n \in W_{e,s}$, we would then put one element of W_e into A.

3. *The set of nonrandom numbers.*
   If W_e contains at least $t(g(e))$ elements, then $W_{g(e)}$ contains at least $g(e)$ elements by its definition. And if $W_{g(e)}$ contains at least $g(e)$ elements, then it contains a nonrandom number (because $h(g(e), n) \le t(n)$ for any $n \ge g(e)$). This means that if W_e contains only nonrandom elements, then it must have at most $t(g(e))$ elements, i.e. $|W_e| \le t(g(e))$.

We consider now the simple sets provided by III.2.14.

Proposition III.2.20 (Smullyan [1964]) *The deficiency set of $\mathcal{K}$ is effectively simple.*

Proof. Let f be a one-one recursive function with range $\mathcal{K}$, and

$$x \in B \Leftrightarrow (\exists y > x)(f(y) < f(x)).$$

We want to find g recursive such that

$$W_e \subseteq \overline{B} \Rightarrow |W_e| \le g(e).$$

Suppose then $W_e \subseteq \overline{B}$. Since we already know that B is simple, W_e is finite, say $\{x_1 < \cdots < x_n\}$. Consider the behavior of f on these elements.

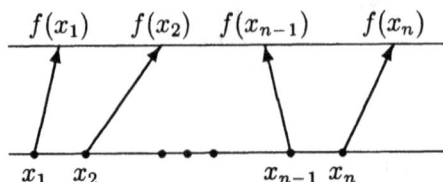

Since each x_i is a nondeficiency stage, there is no crossover. It is then enough to find effectively $a > f(x_n)$: since f is one-one, $a \geq |\mathcal{W}_e|$.

To find a, we use the fact that

$$\mathcal{W}_a \subseteq \overline{\mathcal{K}} \Rightarrow a \in \overline{\mathcal{K}} - \mathcal{W}_a.$$

Since $\mathcal{K}$ is the range of f, it is enough to choose $\mathcal{W}_a$ containing all the elements of $\overline{\mathcal{K}}$ below $f(x_n)$. This is possible because $x_n \in B$, so

$$\overline{\mathcal{K}} \cap \{0, \ldots, f(x_n)\} = \{0, \ldots, f(x_n)\} - \{f(0), \ldots, f(x_n)\}.$$

We then do this in general. Given $\mathcal{W}_e$, let

$$\mathcal{W}_{g(e)} = \{z : (\exists y \in \mathcal{W}_e)(z < f(y) \wedge z \notin \{f(0), \ldots, f(y)\})\}.$$

Then

$$\mathcal{W}_e \subseteq \overline{B} \Rightarrow \mathcal{W}_{g(e)} \subseteq \overline{\mathcal{K}} \Rightarrow g(e) \in \overline{\mathcal{K}} - \mathcal{W}_{g(e)} \Rightarrow |\mathcal{W}_{g(e)}| \leq g(e). \quad \square$$

Exercises III.2.21 Strongly effectively simple sets. (McLaughlin [1965]) A is **strongly effectively simple** if it is a coinfinite r.e. set, and there is a partial recursive function ψ such that

$$\mathcal{W}_e \subseteq \overline{A} \Rightarrow \psi(e) {\downarrow} \wedge (\max \mathcal{W}_e) < \psi(e).$$

An example is given by Post's simple set.

a) ψ *may always be supposed to be total.*

b) *There are effectively simple sets which are not strongly effectively simple.* (Martin [1966]) (Hint: this follows from results in III.4.24, since a strongly effectively simple set is not maximal, and there are maximal effectively simple sets. A direct construction, e.g. in the style of III.2.19.a, is also possible.)

McLaughlin [1973], and Cohen and Jockusch [1975], prove in different ways that *the deficiency set of $\mathcal{K}$ is not strongly effectively simple.*

Exercises III.2.22 Immune sets. a) *There are $2^{\aleph_0}$ sets immune and coimmune, called **bi-immune**.* (Hint: enumerate a set of pairs (x_n, y_n) such that x_n and y_n are different elements of the n-th infinite r.e. set, not yet enumerated in the previous pairs. Any set with exactly one element of each pair is bi-immune.)

b) A set A is **effectively immune** if it is infinite and, for some partial recursive function ψ,

$$W_e \subseteq A \;\Rightarrow\; \psi(e)\!\downarrow \wedge |W_e| \le \psi(e).$$

ψ *may always be supposed to be total.* Similarly for **strongly effectively immune** sets, where instead

$$W_e \subseteq A \;\Rightarrow\; \psi(e)\!\downarrow \wedge (\max W_e) \le \psi(e).$$

(Hint: let g be a recursive function such that

$$W_{g(e)} = \begin{cases} W_e & \text{if } \psi(e)\!\downarrow \\ \emptyset & \text{otherwise.} \end{cases}$$

Then one of $\psi(e)$ and $\psi(g(e))$ converges.)

c) *There are effectively bi-immune sets*, i.e. sets effectively immune and effectively coimmune. (Ullian) (Hint: a natural construction, in which each W_e contributes at most two elements, one for A and one for $\overline{A}$, gives a bi-immune set A such that if $W_e \subseteq A$ or $W_e \subseteq \overline{A}$, then $|W_e| \le 2e + 1$.)

d) *If A is strongly effectively immune, then $\overline{A}$ cannot be immune.* (McLaughlin [1965]) (Hint: suppose

$$W_e \subseteq A \;\Rightarrow\; \max(W_e) < f(e).$$

Let s_x be the smallest stage in which x appears in $\mathcal{K}$, if it ever does, and 0 otherwise. Define

$$W_{t(x)} = \begin{cases} \{s_x\} & \text{if } x \in \mathcal{K} \\ \emptyset & \text{otherwise.} \end{cases}$$

Then $x \in \mathcal{K} \wedge s_x \ge f(t(x)) \Rightarrow s_x \in \overline{A}$. There are infinitely many such s_x's, otherwise for almost every x is $x \in \mathcal{K} \Leftrightarrow x \in \mathcal{K}_{f(t(x))}$, and $\mathcal{K}$ is recursive. But the condition $x \in \mathcal{K} \wedge s_x \ge f(t(x))$ is r.e., so $\overline{A}$ contains an infinite r.e. subset.)

e) A set A is **constructively immune** if, for some partial recursive φ,

$$W_e \text{ infinite } \Rightarrow \varphi(e)\!\downarrow \wedge \varphi(e) \in W_e \cap \overline{A}.$$

If A is constructively immune, then $\overline{A}$ cannot be immune. (Li Xiang [1983]) (Hint: an infinite r.e. subset of $\overline{A}$ can be built, by starting with an r.e. index of ω.)

f) *The notions of effective and constructive immunity are independent.* (Li Xiang [1983]) (Hint: every simple set is constructively coimmune; for the other directions, see c) and e) above.)

III.3 Hypersimple Sets and Truth-Table Degrees

We have solved Post's Problem for m-reducibility by constructing a simple set, but have noticed that this does not automatically imply a solution to Post's Problem for T-reducibility, because there are simple sets which are T-complete (actually, all the simple sets we built were such). The next step is to relax m-reducibility somewhat, and solve Post's Problem for the weaker notion.

In m-reducibility we only allowed one positive query to the oracle. There are many possible extensions, depending on the number of queries allowed (a fixed bounded number, or unboundedly many), their nature (only positive, i.e. asking whether some elements are *in* the oracle, or also negative, asking whether some elements are *not* in it), and the way they are combined (conjunctions, disjunctions, or any possible combination). Some of the possible reducibilities, called respectively **conjunctive**, **disjunctive**, and **positive** (Jockusch [1966]) are the following:

$$A \leq_c B \quad \Leftrightarrow \quad \text{for some recursive function } f,$$
$$x \in A \Leftrightarrow D_{f(x)} \subseteq B.$$
$$A \leq_d B \quad \Leftrightarrow \quad \text{for some recursive function } f,$$
$$x \in A \Leftrightarrow D_{f(x)} \cap B \neq \emptyset.$$
$$A \leq_p B \quad \Leftrightarrow \quad \text{for some recursive function } f,$$
$$x \in A \Leftrightarrow \exists u(u \in D_{f(x)} \wedge D_u \subseteq B).$$

The consideration of these reducibilities makes good sense for the study of r.e. sets, due to the fact that they only use positive information on the oracle, which is exactly what we may obtain from r.e. sets (note that A *is r.e. if and only if* $A \leq_p \mathcal{K}$). We will not be too much concerned with them, since the picture we get from m-reducibility is finer, but we will prove some scattered results here and in Chapter X.

Since our present concern is the solution to Post's Problem, we might as well consider the strongest possible generalization along these lines: to allow for any (finite) number of questions, both positive and negative, to the oracle. To define this notion precisely, let $\{\sigma_n\}_{n \in \omega}$ be an effective enumeration of all the propositional formulas, built from the atomic ones '$m \in X$', for $m \in \omega$. These are also called **truth-table conditions**, since they can be arranged in truth-tables. Given a set B, $B \models \sigma_n$ means that B satisfies σ_n, i.e. that the propositional formula σ_n becomes true when X in the atomic formulas is interpreted as B.

Definition III.3.1 (Post [1944]) *A is* **tt-reducible** *to B* ($A \leq_{tt} B$) *if, for some recursive function* f,

$$x \in A \Leftrightarrow B \models \sigma_{f(x)}.$$

A is **tt-equivalent** *to B* ($A \equiv_{tt} B$) *if* $A \leq_{tt} B$ *and* $B \leq_{tt} A$.

In terms of connectives, the various truth-table reducibilities correspond to truth-table conditions built up from the atomic ones by means of the following connectives:

$$\leq_{tt} = \{\neg, \wedge, \vee\} \quad \leq_p = \{\wedge, \vee\} \quad \leq_c = \{\wedge\} \quad \leq_d = \{\vee\},$$

while $\leq_m$ correspond to using only atomic formulas. We will see in III.8.4 that, on the r.e. sets, $\leq_m = \{\neg\}$. Since $\neg$, together with any one of $\wedge$ and $\vee$, generates all the propositional formulas, this would seem to take care of all the possible types of truth-table-like reducibilities. Bulitko [1980] and Selivanov [1982] have shown that it is almost so, in the sense that only another one such reducibility exists, called **linear reducibility**, corresponding to the logical sum (i.e. addition modulo 2). Its definition can be put as:

$$A \leq_l B \quad \Leftrightarrow \quad \text{for some recursive function } f$$
$$x \in A \Leftrightarrow |D_{f(x)} \cap B| \equiv 1 \pmod 2.$$

Truth-table degrees

Our first concern is to make the difference between truth-table and Turing reducibility clear. Recall that a relative computation consists of two different kinds of actions, one purely computational (performed by a machine), and the other interactive (queries answered by the oracle). In Turing computations the two parts can be strongly interwoven, and impossible to unravel: we may come to know the questions we need to ask the oracle only during the computation itself, and there might even be no recursive bound on their number or size (as a function of the input). On the other hand, truth-table computations clearly separate the two parts of the relative computation, computing ahead of time not only the elements which need to be queried, but also the outcome of the computation for any possible answer the oracle is going to provide for them.

Another way to see the difference is in terms of functionals. Recall that $A \leq_T B$ if and only if $c_A \simeq F(c_B)$, for some partial recursive functional F.

Proposition III.3.2 (Trakhtenbrot [1955], Nerode [1957]) $A \leq_{tt} B$ if and only if $c_A \simeq F(c_B)$ for some partial recursive, total functional F.

Proof. Let

$$x \in A \Leftrightarrow B \models \sigma_{f(x)},$$

and define $F(\alpha, x)$ as follows. First see if $\alpha(z) \downarrow$ for every z such that the atomic formula $z \in X$ occurs in $\sigma_{f(x)}$. If so, consider any set C such that, for any z as just said, $z \in C \Leftrightarrow \alpha(z) \simeq 1$, and let

$$F(\alpha, x) \simeq \begin{cases} 1 & \text{if } C \models \sigma_{f(x)} \\ 0 & \text{otherwise.} \end{cases}$$

F is partial recursive and total by definition, and $c_A \simeq F(c_B)$.

Suppose now that F is a partial recursive, total functional. For each x and X, $F(c_X, x)$ is defined, and for every branch of the space 2^ω of sets, the information on X needed to compute $F(c_X, x)$ is bounded, by compactness. By König's Lemma, there is a bound which works for all branches. Another way to see this is to note that F is continuous and total on 2^ω, which is a compact space in the positive information topology: then F is uniformly continuous, and there is a modulus of continuity that works for every member of 2^ω. In any case, we may thus write down a truth-table that gives $F(c_X, x)$ for any X, because only values of X up to the bound are needed, and there are only finitely many possible combinations. $\square$

By II.3.7.b we then know that T-reducibility and tt-reducibility do not coincide. We now turn to a study of tt-reducibility.

Exercises III.3.3 a) *If A is recursive, then $A \leq_{tt} B$ for any set B.*
b) *If $A \leq_{tt} B$ and B is recursive, so is A.*
c) *$A \leq_{tt} \overline{A}$.*

Note that $\leq_{tt}$ is a reflexive and transitive relation, and thus $\equiv_{tt}$ is an equivalence relation.

Definition III.3.4 *The equivalence classes of sets w.r.t. tt-equivalence are called **tt-degrees**, and $(\mathcal{D}_{tt}, \leq)$ is the structure of tt-degrees, with the partial ordering $\leq$ induced on them by $\leq_{tt}$.*

*The tt-degrees containing r.e. sets are called **r.e. tt-degrees**, and two of them are:*

1. *0_{tt}, the tt-degree of the recursive sets*

2. *$0'_{tt}$, the tt-degree of $\mathcal{K}$.*

*An r.e. set A is **tt-complete** if its tt-degree is $0'_{tt}$, i.e. if $\mathcal{K} \leq_{tt} A$.*

Note that an r.e. tt-degree, being closed under complementation, contains only r.e. sets if and only if it contains only recursive sets. Also, 0_{tt} and $0'_{tt}$ are, respectively, the least and the greatest r.e. tt-degrees.

The analogue of Post's Problem for tt-reducibility is: are there r.e. sets which are neither recursive, nor tt-complete? We already know that simple sets are not m-complete, and we make sure that they are not automatically tt-incomplete.

Proposition III.3.5 (Post [1944]) *There is a simple, tt-complete set.*

Proof. Consider Post's simple set S (III.2.11). Any coinfinite r.e. superset of it will still be simple (since any infinite r.e. subset of its complement would also be a subset of $\overline{S}$). We then look for an r.e. set S^* such that:

1. $\mathcal{K} \leq_{tt} S^*$, i.e. $\mathcal{K}$ is coded into S^* by truth tables

2. $S \subseteq S^*$, and S^* coinfinite.

Let $\{F_x\}_{x \in \omega}$ be a strong array of disjoint finite sets intersecting $\overline{S}$. It exists, because the construction of S ensures that at most z elements of $\{0, \ldots, 2z\}$ go into S, so $\overline{S}$ intersects each subset of $\{0, \ldots, 2z\}$ with at least $z + 1$ elements. It is then enough to let

$$F_x = \{n : 2^x - 1 \leq n < 2^{x+1} - 1\}.$$

We can then use the set F_x to code the fact that $x \in \mathcal{K}$, by putting it into S^* if this holds, and not otherwise. We also want S^* to be a superset of S, and we thus let

$$S^* = S \cup \bigcup_{x \in \mathcal{K}} F_x.$$

Then:

1. $\mathcal{K} \leq_{tt} S^*$, because $x \in \mathcal{K} \Leftrightarrow F_x \subseteq S^*$. Indeed, if $x \in \mathcal{K}$ then $F_x \subseteq S^*$ by construction. And if $x \notin \mathcal{K}$ then F_x contains an element of $\overline{S}$, which never goes into S^* because the F_y's are disjoint: thus $F_x \not\subseteq S^*$.

2. $\overline{S^*}$ is infinite, because there are infinitely many elements x not in $\mathcal{K}$, and thus infinitely many F_x not contributing to S^*. Each one contains an element of $\overline{S}$, which cannot go in S^* because the F_y's are disjoint. $\square$

Note that S^* is effectively simple, being a superset of S. Thus *there are effectively simple, tt-complete sets*. But this does not mean that every effectively simple set is not only T-complete, but actually tt-complete (since hypersimple sets, which are not tt-complete, may be effectively simple, see III.3.15.b).

The tt-complete, simple set obtained above is a modification of Post's simple set. It is natural to wonder whether Post's simple set is already tt-complete itself. The surprising answer is that this depends on the acceptable system of indices we are working with (one half of the result is given below, the other half in III.9.2). Thus *Recursion Theory is not completely independent of the acceptable system of indices chosen to work with*. This was first shown by Jockusch and Soare [1973], who proved that another set constructed by Post [1944] (namely his hypersimple set) could be T-complete or T-incomplete, depending on the acceptable system of indices. This cannot be the case for Post's simple set, which is always T-complete (being effectively simple).

Exercise III.3.6 *There is an acceptable system of indices for which Post's simple set is tt-complete.* (Lachlan [1975]) (Hint: given $\{\mathcal{W}_x\}_{x \in \omega}$ acceptable, define $\{\widehat{\mathcal{W}}_x\}_{x \in \omega}$ acceptable, $\widehat{S}$ relative to it, and a strong array $\{B_x\}_{x \in \omega}$ such that

$$x \in \mathcal{K} \Leftrightarrow B_x \subseteq \widehat{S}.$$

To have all the r.e. sets in the enumeration, let $\widehat{\mathcal{W}}_{2^x} = \mathcal{W}_x$. Note that, by construction, if $z > 2e$ and $\widehat{\mathcal{W}}_e = \{z\}$ then $z \in \widehat{S}$. If $x \in \mathcal{K}$ thus let

$$\widehat{\mathcal{W}}_{2^x + n} = \{2^{x+1} + 2n + 1\}$$

for $0 < n < 2^x$. If $x \in \overline{\mathcal{K}}$, then let $\widehat{\mathcal{W}}_{2^x + n} = \emptyset$. For definiteness, let $\widehat{\mathcal{W}}_0 = \emptyset$. Finally, let

$$B_x = \{2^{x+1} + 2n + 1 : 0 < n < 2^x\}.)$$

Hypersimple sets

Since simple sets do not solve Post's Problem for tt-reducibility, we look for a stronger notion. The idea comes from the fact that tt-reducibility uses finite questions about sets, while m-reducibility uses only one question. We may thus think of replacing, in the definition of simple sets, infinite r.e. sets by disjoint strong arrays. Of course, a disjoint strong array contained in $\overline{A}$ produces an infinite r.e. subset of it (by choosing one number in each element of the array), and to rule out disjoint strong arrays contained in $\overline{A}$ is thus equivalent to simplicity. We thus relax the condition a bit:

Definition III.3.7 (Post [1944]) *A set A is **hyperimmune** if it is infinite, and there is no disjoint strong array (with members all) intersecting it, i.e. there is no recursive function f such that:*

- $x \neq y \Rightarrow D_{f(x)} \cap D_{f(y)} = \emptyset$

- $D_{f(x)} \cap A \neq \emptyset.$

*A set is **hypersimple** if it is r.e. and co-hyperimmune.*

We give some conditions equivalent to hyperimmunity.

Proposition III.3.8 (Medvedev [1955], Rice [1956a], Uspenskii [1957], Kuznekov) *For an infinite set A, the following are equivalent:*

1. *A is hyperimmune*

2. *there is no recursive function f such that*

   - $x \neq y \Rightarrow D_{f(x)} \cap D_{f(y)} = \emptyset$

- $D_{f(x)} \cap A \neq \emptyset$
- $\bigcup_{x \in \omega} D_{f(x)} \supseteq A$.

3. there is no recursive function f such that, for each n, A has at least n elements smaller than $f(n)$

4. there is no recursive function f such that, for each n, $a_n \leq f(n)$, where a_n is the n-th element of A in increasing order.

Proof. The last two conditions are clearly equivalent. Moreover:

- $1 \Rightarrow 2$ by definition

- $2 \Rightarrow 3$ because, if A has at least n elements below $f(n)$, there is an element of A between n and $f(n+2)$, and thus a disjoint strong array intersecting A and covering ω can be obtained by iteration:

$$\{0, \ldots, f(1) - 1\}, \quad \{f(1), \ldots, f(f(1) + 2)\}, \quad \cdots$$

- $3 \Rightarrow 1$ because if $\{D_{h(x)}\}_{x \in \omega}$ is any disjoint strong array intersecting A, then

$$f(n) = \max\left(\bigcup_{i < n} D_{h(i)}\right)$$

has at least n elements of A below it. $\square$

The intuitive content of the various characterizations of hyperimmunity is that *a hyperimmune set has very sparse elements, from a recursion theoretical point of view*: for any recursive function f, no matter how fast growing, there are infinitely many elements a_n in the ordering of A by magnitude, such that $f(n) < a_n$.

Exercises III.3.9 Domination properties. We say that f **dominates** φ if, for almost every argument x, $f(x) \geq \varphi(x)$ whenever $\varphi(x) \downarrow$. Let p_A be the function enumerating the infinite set A by magnitude (principal enumeration), so that $p_A(n)$ is the n-th element of A.

a) *A is hyperimmune if and only if p_A is not dominated by any recursive function.*

b) Call a set A **dense immune** if p_A dominates every total recursive function, and an r.e. set **dense simple** if its complement is dense immune (Martin [1963]). Then *a dense simple set is hypersimple.*

c) *A strongly effectively simple set is not dense simple.* (Cohen and Jockusch [1975]) (Hint: let A be dense, and strongly effectively simple via g:

$$W_e \subseteq \overline{A} \Rightarrow (\max W_e) < g(e).$$

Let a_n and a_n^s be the n-th element of $\overline{A}$ and $\overline{A}_s$. If h is a recursive function going to infinity, there is n_0 such that $(\forall n \geq n_0)(a_n \geq h(n))$. We get that A is cofinite if we show

$$(\forall n \geq n_0)(\forall x)(a_n \geq h(x)),$$

since h goes to infinity. This follows by induction if, for $n \leq x$,

$$a_n \geq h(x) \wedge a_{n+1} \geq h(x+1) \Rightarrow a_n \geq h(x+1).$$

It is enough to define f and g such that $h(x) \geq g(f(n,x))$, and

$$W_{f(n,x)} = \{a_n^s\}, \text{ with } s \text{ minimal such that } a_n^s \geq h(x) \wedge a_{n+1}^s \geq h(x+1).$$

This can be done by the Fixed-Point Theorem. Then $a_n^s \neq a_n$ because A is strongly effectively simple, hence $a_n \geq a_{n+1}^s \geq h(x+1)$.)

d) *If p_A dominates every partial recursive function, then $\mathcal{K} \leq_T A$.* (Tennenbaum [1961]) (Hint: see the proof of III.1.5.)

The following result shows that the intuition that led to the definition of hypersimple sets was correct.

Theorem III.3.10 (Post [1944]) *A hypersimple set is not tt-complete.*

Proof. Let $x \in \mathcal{K} \Leftrightarrow A \models \sigma_{f(x)}$. We prove that A is not hypersimple by finding effectively, given n, a number $m \geq n$ such that

$$\overline{A} \cap \{n, n+1, \ldots, m\} \neq \emptyset.$$

This will automatically produce, by iteration, a disjoint strong array intersecting $\overline{A}$. Let

$$z \in A^* \quad \Leftrightarrow \quad (z \in A \wedge z < n) \vee z \geq n$$
$$x \in C \quad \Leftrightarrow \quad \neg(A^* \models \sigma_{f(x)}).$$

C is r.e. because A^* is recursive (being cofinite). So let $C = W_a$:

- if $a \in \mathcal{K}$ then $A \models \sigma_{f(a)}$ but $\neg(A^* \models \sigma_{f(a)})$, since $a \in W_a = C$

- if $a \in \overline{\mathcal{K}}$ then $\neg(A \models \sigma_{f(a)})$ but $A^* \models \sigma_{f(a)}$, since $a \notin W_a = C$.

So $A \models \sigma_{f(a)} \Leftrightarrow \neg(A^* \models \sigma_{f(a)})$, and since A and A^* agree below n, they must disagree above it. But by definition everything above n is in A^*, so one of the elements used in $\sigma_{f(a)}$ is not in A.

The first thought would be to let m be greater than all the elements used in $\sigma_{f(a)}$, but we cannot get this effectively from n, because the definition of A^* uses A, which is only r.e. But since only elements below n are considered, there are only finitely many subsets B_i ($i < 2^n$) of $\{0, \ldots, n-1\}$, and one of them is really $A \cap \{0, \ldots, n-1\}$. So if we consider them all, and let $A_i^* = B_i \cup \{z : z \geq n\}$, we can proceed as before, getting $C_i = \mathcal{W}_{a_i}$. Now we cannot anymore assert in general that one of the elements used in $\sigma_{f(a_i)}$ is not in A, since A_i^* might not be A^*, but if we let

$$m = 1 + \max_{i < 2^n} \{\text{elements used in } \sigma_{f(a_i)}\}$$

then m is in particular greater than all elements used in $\sigma_{f(a)}$, and we still have $\overline{A} \cap \{n, \ldots, m\} \neq \emptyset$. □

Exercises III.3.11 a) *A coinfinite r.e. set A is hypersimple if and only if it has no c-complete superset.* (Hint: if $A \subseteq B$ and B is hypersimple, then B is hypersimple or cofinite, hence not tt-complete. For the converse, see the proof of III.3.5.)

b) *If A and B are hypersimple sets then $A \cap B$ is hypersimple, and $A \cup B$ is hypersimple or cofinite.* Thus hypersimple or cofinite sets form a filter in the lattice of the r.e. sets under inclusion. (Dekker [1953])

We have now to turn to the existence of hypersimple sets. Post's simple set is obviously not hypersimple, but a modification of the second existence proof of simple sets in III.2.11 will produce a hypersimple one.

Theorem III.3.12 (Post [1944]) *There exists a hypersimple set.*

Proof. We build a hypersimple set A by stages. At stage s we will have A_s, and we will let $\overline{A_s} = \{a_0^s < a_1^s < \cdots\}$. In the end $\overline{A} = \{a_0 < a_1 < \cdots\}$, where $a_n = \lim_{s \to \infty} a_n^s$. We want to satisfy the following requirements:

P_e : $\mathcal{W}_e$ infinite disjoint strong array $\Rightarrow (\exists z \in \mathcal{W}_e)(D_z \subseteq A)$
N_e : $\overline{A}$ has at least e elements, or $\lim_{s \to \infty} a_e^s < \infty$.

The construction is as follows. We start with $A_0 = \emptyset$ (hence $a_n^0 = n$). At stage $s + 1$ we search for the smallest $e \leq s$ such that:

- $z \in \mathcal{W}_{e,s} \Rightarrow D_z \cap \overline{A_s} \neq \emptyset$

- for some $z \in \mathcal{W}_{e,s}$, $D_z \subseteq [a_e^s, \infty)$.

Note that both A_s and $\mathcal{W}_{e,s}$ are finite, and we only look at $e \leq s$, so the search is effective. If e does not exist, we go to the next stage. Otherwise, P_e is the condition with smallest index which looks unsatisfied and with a chance to be satisfied. Then we put all of D_z into A, where z is the smallest one such that $z \in \mathcal{W}_{e,s}$ and $D_z \subseteq [a_e^s, \infty)$.

Since the construction is effective, $A = \bigcup_{s \in \omega} A_s$ is r.e. Moreover:

- $\overline{A}$ is infinite.

  It is enough to prove that $\lim_{s \to \infty} a_n^s$ exists. Indeed, a_n^s may move at a certain stage $s + 1$ only if it happens that $a_n^s \in D_z$, for some $z \in W_e$, $e \le n$. Since each W_e contributes at most one finite set for each e, and there are only finitely many $e \le n$, a_n^s moves only finitely many times. So all N_e's are satisfied.

- A is hypersimple.

  By induction, suppose that s_0 is such that $s_0 \ge e$, all P_i with $i < e$ have been satisfied at stage s_0, and no finite set with index in W_i ($i < e$) goes into A after stage s_0. If W_e is a disjoint strong array intersecting $\overline{A}$, there are $s \ge s_0$ and $z \in W_{e,s}$ such that $D_z \subseteq [a_e^s, \infty)$. Then, for one of these z, D_z goes into A at stage $s+1$ (because e is the smallest index for which P_e looks unsatisfied and with a chance to be satisfied), contradiction. $\square$

Also III.2.14 generalizes, and shows that a hypersimple set is not necessarily T-incomplete.

Proposition III.3.13 (Dekker [1954]) *Every nonrecursive r.e. T-degree contains a hypersimple set.*

Proof. Given A r.e. nonrecursive enumerated by f recursive, let B be its deficiency set (II.6.16):

$$x \in B \Leftrightarrow (\exists y > x)(f(y) < f(x)).$$

We already know that B is r.e. and coinfinite, and $A \equiv_T B$. Suppose there is a recursive function g majorizing $\overline{B}$, i.e. $g(x)$ greater than the x-th element of $\overline{B}$, and in particular of x itself. Then $f(g(x))$ is also greater than x, because f is increasing on nondeficiency stages, and after stage $g(x)$ no element smaller than $f(g(x))$ is going to be enumerated. Then

$$x \in A \Leftrightarrow x \in \{f(0), f(1), \ldots, f(g(x))\},$$

and A would be recursive. $\square$

Note that, although the proof given above produces $B \le_{tt} A$, we only have $A \le_T B$, and this is necessary: not every nonrecursive r.e. tt-degree contains a hypersimple set (because a hypersimple set cannot be tt-complete). Jockusch [1981a] actually shows that *not every nonrecursive r.e. tt-degree contains a simple set.*

Exercises III.3.14 a) *Every r.e. nonrecursive coregressive set is hypersimple.* (Dekker and Myhill [1958], Dekker [1962]) (Hint: see III.5.6 and III.5.3.)

b) *There are nonrecursive retraceable sets which are not hyperimmune.* (Dekker and Myhill [1958]) (Hint: see II.6.7.b.)

Exercises III.3.15 A is **effectively hyperimmune** if it is coinfinite, and there is a recursive function f such that, whenever $\mathcal{W}_e$ is a disjoint strong array intersecting A, $|\mathcal{W}_e| \le f(e)$. A set is **effectively hypersimple** if it is r.e., and its complement is effectively hyperimmune.

a) *An effectively hypersimple set is effectively simple, and hence T-complete.* (Hint: given $\mathcal{W}_e$, let $\mathcal{W}_{h(e)}$ be the set of canonical indices of the singletons consisting of the elements of $\mathcal{W}_e$. Then $\mathcal{W}_e \subseteq \overline{A} \Rightarrow |\mathcal{W}_e| \le f(h(e))$.)

b) *The hypersimple set constructed in III.3.12 is effectively hypersimple.* (Hint: let $f(e) = e$.) Thus there are effectively simple sets which are not tt-complete, and III.2.18 cannot be improved.

c) *There are hypersimple, not effectively hypersimple, T-complete sets.* (Arslanov [1970]) (Hint: see III.2.19.b.)

d) *A hypersimple set A is T-complete if and only if there is $f \le_T A$ such that, whenever $\mathcal{W}_e$ is a disjoint strong array intersecting $\overline{A}$, $|\mathcal{W}_e| \le f(e)$.* (Arslanov [1970]) (Hint: see III.2.19.c.)

e) *A and $\overline{A}$ cannot both be effectively hyperimmune.* Note that this is not true for effective simplicity, see III.2.22.c. (Arslanov [1969]) (Hint: by taking the maximum of the two functions, A and $\overline{A}$ can be supposed effectively immune via the same function. Build a disjoint strong array intersecting both of them. Put in $\mathcal{W}_{g(x)}$ the canonical indices of the singletons $\{y\}$, for $y \le f(x) + 1$, and choose e such that $\mathcal{W}_{g(e)} = \mathcal{W}_e$. Then $\mathcal{W}_e$ is a disjoint strong array with more than $f(e)$ elements, so it intersects both A and $\overline{A}$. Thus $\{0, \ldots, f(e) + 1\}$ is the first set of the wanted array. Continue similarly.)

For different notions of effective hypersimplicity see Arslanov [1969], [1970], [1985], Arslanov and Soloviev [1978], and Kanovich [1975].

The permitting method $\star$

The proof of the existence of hypersimple sets in each nonrecursive r.e. T-degree used deficiency stages, and was thus elegant but a bit artificial. In particular, it does not appear to be useful to prove different results. But we can isolate from it a useful tool that was used there implicitly, and that has instead a vast range of applications.

Proposition III.3.16 Permitting method (Dekker [1954], Muchnik [1956], Friedberg [1957], Yates [1965]) *If A and C are r.e. sets, $\{A_s\}_{s \in \omega}$ and $\{C_s\}_{s \in \omega}$ are recursive enumerations of them, and for every x*

$$x \in A_{s+1} - A_s \Rightarrow (\exists y \le x)(y \in C_{s+1} - C_s),$$

then $A \le_T C$.

Proof. To see if $x \in A$, look at the stages in which some $y \le x$ is generated in C: recursively in C we may determine if $y \in C$, and if so we just have to generate C until y appears in it. Then $x \in A$ if and only if x is generated at one of these stages.   □

The name of the method comes from the fact that an element may go into A only if it is permitted by some element of C (smaller than it). The method is very useful when we have a construction of an r.e. set with certain properties, and we want to build one such set below any given (nonrecursive) r.e. set.

Exercise III.3.17 *The permitting method does not produce $A \leq_{tt} C$.* (Hint: by using the priority method, introduced in Chapter X, we build disjoint r.e. sets A and B, such that $A \not\leq_{tt} A \cup B$. If $C = A \cup B$ and $C_s = A_s \cup B_s$, $x \in A_{s+1} - A_s \Rightarrow x \in C_{s+1} - C_s$. To satisfy $x \notin A \Leftrightarrow A \cup B \models \sigma_{\varphi_e(x)}$, pick up a witness $z_e \notin A_s \cup B_s$, and wait until $\varphi_e(z_e)$ converges. Let $A_{s+1} \cup B_{s+1} = A_s \cup B_s \cup \{z_e\}$. To decide where z_e should go, see if $A_{s+1} \cup B_{s+1} \models \sigma_{\varphi_e(z_e)}$. If so, let $z_e \in B$, otherwise let $z_e \in A$.)

As an example, we show how to apply permitting to the construction of Post's simple set.

Proposition III.3.18 (Yates [1965]) *Every nonrecursive r.e. T-degree contains a simple not hypersimple set.*

Proof. Given C r.e. nonrecursive, we build A simple, nonhypersimple and such that $A \leq_T C$, by adding permitting to the first proof of III.2.11: at stage s, for every $e \leq s$ such that $\mathcal{W}_{e,s} \cap A_s = \emptyset$, put in A the smallest x such that

$$x > 2e \wedge x \in \mathcal{W}_{e,s} \wedge x \text{ permitted by } C \text{ at stage } s,$$

where x is permitted by C at stage s if $(\exists y \leq x)(y \in C_{s+1} - C_s)$.

Since $A \subseteq S$, where S is Post's simple set, it is immediate that A is coinfinite and not hypersimple. By permitting, $A \leq_T C$. It remains to show that A is simple, and this is not automatic, because less elements go into A than in S. Suppose $\mathcal{W}_e$ is infinite, and $\mathcal{W}_e \subseteq \overline{A}$. Then, by construction,

$$x > 2e \wedge x \in \mathcal{W}_{e,s} \wedge e \leq s \Rightarrow x \text{ is not permitted by } C \text{ at stage } s.$$

We prove that then C must be recursive. Given y, to see if $y \in C$ look for x and s such that
$$x > 2e \wedge x \in \mathcal{W}_{e,s} \wedge e \leq s \wedge y \leq x$$

(which exist, because $\mathcal{W}_e$ is infinite). Then $y \in C \Leftrightarrow y \in C_s$, because if $y \in C - C_s$ then, for some $t \geq s$, $y \in C_{t+1} - C_t$, and x is permitted by y at stage t (since $\mathcal{W}_{e,s} \subseteq \mathcal{W}_{e,t}$).

We still have to get $C \leq_T A$. Let $B = A \oplus A^*$, where A^* is a hypersimple set such that $A^* \equiv_T C$ (III.3.13). Then $C \leq_T A^* \leq_T B$. Moreover, both A

and A^* are reducible to C, so $B \leq_T C$. And B is simple and not hypersimple, because so are A and A^* (an infinite subset of $\overline{B}$ induces an infinite subset of $\overline{A}$ or $\overline{A^*}$). $\quad \square$

Exercises III.3.19 a) $C \leq_T A$ *can be ensured directly by a* **coding procedure,** *as follows: at stage s, if y is the element of C enumerated at stage $s + 1$, put a_y^s (the y-th element of $\overline{A}_s$) into A.* (Hint: to see if $y \in C$, search recursively in A a stage s such that $a_y^s = a_y$. Then $y \in C \Leftrightarrow y \in C_s$. Note also that a_n^s moves only finitely many times for the coding, at most once for each $y \leq n$.) *If this method is used, then we have a different proof of III.2.14.*

b) $C \leq_T A$ *is automatic in the proof above. This generalizes the fact that Post's simple set is automatically T-complete.* (Jockusch and Soare [1972a]) (Hint: by the Fixed-Point Theorem, find g recursive and one-one such that

$$W_{g(e)} = \begin{cases} \{a_0^{s_e}, \dots, a_{2 \cdot g(e)}^{s_e}\} & \text{if } e \in C \\ \emptyset & \text{otherwise,} \end{cases}$$

where

$$s_e = \begin{cases} \mu s(e \in C_s) & \text{if } e \in C \\ 0 & \text{otherwise.} \end{cases}$$

Let also $r_e = \mu s(\forall n \leq 2g(e))(a_n^s = a_n)$. Since r_e is recursive in A, if $e \in C \Leftrightarrow e \in C_{r_e}$ then $C \leq_T A$. So consider $\widehat{C} = \{e : e \in C - C_{r_e}\}$. This is r.e. in A, since C is r.e. and r_e is recursive in A. If it is finite, then $C \leq_T A$. So suppose it is infinite. Then there are infinitely many e such that $s_e > r_e$, in particular $W_{g(e)} \subseteq \overline{A}$, and $W_{g(e)}$ has $2g(e) + 1$ elements, so there is $x \in W_{g(e)} \wedge x > 2g(e)$. Given y, to test for $y \in C$ look, recursively in A, for:

> $e \in \widehat{C}$ such that $2g(e) > y$ ($\widehat{C}$ is infinite, and g is one-one)
> $x \in W_{g(e)} \wedge x > 2g(e)$
> a stage s such that $x \in W_{g(e),s} \wedge g(e) \leq s$.

Then, since $x \in W_{g(e),s} \wedge x > 2g(e) \wedge g(e) \leq s$ but $W_{g(e)} \subseteq \overline{A}$, x is never permitted. So no element smaller than it can enter C after that stage. Hence $y \in C \Leftrightarrow y \in C_s$.)

c) *The existence of hypersimple sets in any nonrecursive r.e. T-degree can be proved directly, by permitting and coding.*

To avoid false expectations, we must stress the fact that *none of the following is true:*

1. permitting can be applied to any construction of r.e. sets

2. when permitting can be applied, it can also be combined with a coding procedure to give a set with the highest possible degree.

As we will see in Chapter X, the first fails for maximal sets (since no maximal set can be built below a low degree), the second for contiguous degrees or

η-maximal semirecursive sets (since they exist below any given nonrecursive degree, but not in every degree).

It is however true that there seems to be a sort of **maximum degree principle** (Jockusch and Soare [1972a]), according to which *natural constructions with 'weak negative requirements' usually give sets with the highest possible degree not explicitly ruled out by the construction.*

Another fact, related to the one above and which might be called **effectivity principle**, is that *sets which are constructed in some natural way to satisfy some requirements, tend to satisfy them in some effective way.*

For example, the constructions of simple sets (III.2.11) and of hypersimple sets (III.3.12) automatically produced sets which are both T-complete, and effectively simple or hypersimple. We will see many other manifestations of the two principles in the following. In particular, Section 6 is an elaboration of the fact that $\mathcal{K}$ is effectively nonrecursive.

III.4   Hyperhypersimple Sets and Q-degrees

We have solved Post's problem for tt-reducibility by constructing a hypersimple set, but have noticed that this does not automatically imply a solution to Post's problem for T-reducibility, because there are T-complete hypersimple sets. Once again the next step is to relax tt-reducibility, and solve Post's problem for the weaker notion.

Since there seems to be no easy way to weaken tt-reducibility without falling into T-reducibility, we pursue a complementary tactic, and strengthen T-reducibility instead. Recall (III.1.4) that, for r.e. sets A and B, $A \leq_T B$ is equivalent to the existence of an r.e. relation R such that

$$x \in \overline{A} \Leftrightarrow (\exists u)(D_u \subseteq \overline{B} \wedge R(x, u)).$$

This means, intuitively, that from time to time we ask the oracle questions about containment of a finite set D_u. We thus propose to ask the oracle only questions about singletons. This can be expressed by the existence of an r.e. relation R such that

$$x \in \overline{A} \Leftrightarrow (\exists u)(u \in \overline{B} \wedge R(x, u)).$$

By the S_n^m-Theorem, the existence of a binary r.e. relation R is equivalent to the existence of a recursive function f such that $R(x, u) \Leftrightarrow u \in \mathcal{W}_{f(x)}$. We can then express the previous formulation as

$$x \in \overline{A} \Leftrightarrow (\exists u)(u \in \overline{B} \wedge u \in \mathcal{W}_{f(x)}) \Leftrightarrow \mathcal{W}_{f(x)} \cap \overline{B} \neq \emptyset.$$

Q-reducibility

The discussion above leads us to the following notion.

Definition III.4.1 (Tennenbaum) A *is* **Q-reducible** *to* B *($A \leq_Q B$) if, for some recursive function f,*

$$x \in A \Leftrightarrow W_{f(x)} \subseteq B.$$

Thus Q-reducibility is similar to c-reducibility (p. 268), only using r.e. sets of questions in place of finite sets. On r.e. sets, which are our real concern, the similarity is even greater, since we can actually use only finite r.e. sets of questions.

Proposition III.4.2 (Soloviev [1974]) *If A and B are r.e. sets, then $A \leq_Q B$ if and only if there is g recursive such that, for every x,*

1. $W_{g(x)}$ is finite

2. $x \in A \Leftrightarrow W_{g(x)} \subseteq B$.

Proof. Given f as in the original definition of $\leq_Q$, let $W_{g(x)}$ be the r.e. set generated as follows. Generate simultaneously $W_{f(x)}, A,$ and B. At each stage of the enumeration, put the elements already generated in $W_{f(x)}$ into $W_{g(x)}$, unless either x has already been generated in A (this is a permanent block), or some z has already been generated in $W_{g(x)}$, but not yet in B (this is a temporary block, that could later be removed if z is generated in B).

If $x \in A$ then $W_{f(x)} \subseteq B$, and hence $W_{g(x)} \subseteq B$, because $W_{g(x)}$ is a subset of $W_{f(x)}$. Moreover, $W_{g(x)}$ is finite because its enumeration stops no later than x has been generated in A.

If $x \notin A$ then $W_{f(x)} \cap \overline{B} \neq \emptyset$: if z is any element of the intersection, the enumeration of $W_{g(x)}$ stops permanently no later than z has been generated in $W_{g(x)}$, and thus $W_{g(x)}$ is finite. Moreover, $W_{g(x)}$ contains one such z, because x is not in A, and thus $W_{g(x)} \not\subseteq B$. $\square$

Notice that if $A \leq_Q B$ then A is not necessarily recursive in B. To know if x is in A we must check whether $W_{f(x)}$ is contained in B, and there are two problems: first of all, $W_{f(x)}$ may be infinite, and thus the check could not be done in finite time; second, even if $W_{f(x)}$ were finite, we would not know when its enumeration had been completed. What we can say in general is only that $\overline{A}$ is r.e. in B, since to know if x is in $\overline{A}$ we only have to find an element of $W_{f(x)}$ which is not in B. If A is r.e. (in B), then we also have the missing half, and A is recursive in B by Post's Theorem.

Exercises III.4.3 Q-reducibility. We have just noted that $\leq_Q$ is a nonstandard reducibility, and we introduce it mostly for the sake of solution to Post's problem. Here are some properties of $\leq_Q$ and the associated notion of Q-degree, some of which depend on the definition of the Arithmetical Hierarchy, see IV.1.6.

a) $\leq_Q$ *is reflexive and transitive.*

b) *If $A \leq_Q B$ and $B \in \Pi_n^0$ then $A \in \Pi_n^0$.*

c) *There is a least Q-degree, containing exactly the Π_1^0 sets.* (Hint: if $A \in \Pi_1^0$ and $B \neq \omega$ is any set, let $b \notin B$, and

$$W_{f(x)} = \begin{cases} \{b\} & \text{if } x \in \overline{A} \\ \emptyset & \text{otherwise.} \end{cases}$$

Then $x \in A \Leftrightarrow W_{f(x)} \subseteq B$, and $A \leq_Q B$.)

d) *$A \leq_Q \mathcal{K}$ if and only if $A \in \Pi_2^0$.* (Hint: if $A \in \Pi_2^0$ then $x \in A \Leftrightarrow (\forall y)R(x,y)$, with R r.e. Then there is a recursive function f such that $R(x,y) \Leftrightarrow f(x,y) \in \mathcal{K}$. If $W_{g(x)} = \{f(x,y) : y \in \omega\}$, then $x \in A \Leftrightarrow W_{g(x)} \subseteq \mathcal{K}$.)

e) *If $A \leq_m B$ then $A \leq_Q B$.*

f) *If $A \leq_Q B$ then $A \leq_T B$ for r.e. sets, but not in general.* (Hint: for the counterexample, see e.g. part d) above.)

g) *Neither of $A \leq_{tt} B$ and $A \leq_Q B$ implies the other, even on the r.e. sets.* (Hint: there is a hypersimple Q-complete set, see p. 297, and thus Q-reducibility does not imply tt-reducibility. For the converse we build by the priority method, introduced in Chapter X, two r.e. sets A and B such that $\langle x,y \rangle \in A \Leftrightarrow x \in B \vee y \in B$, so that $A \leq_{tt} B$. To spoil the e-th Q-reduction $x \in A \Leftrightarrow W_{\varphi_e(x)} \subseteq B$, pick up distinct witnesses x_e, y_e and enumerate $W_{\varphi_e(z_e)}$, where $z_e = \langle x_e, y_e \rangle$. If at stage s we find, for the first time, $W_{\varphi_e(z_e),s} \not\subseteq B_s$ then pick up $u_e \in W_{\varphi_e(z_e)} \cap \overline{B_s}$. Since x_e and y_e are distinct, one of them is distinct from u_e: put it into B (so $z_e \in A$), and restrain u_e from entering B. If such a stage is never found, then $W_{\varphi_e(z_e)} \subseteq B$, but we never put either x_e or y_e into B, so $z_e \notin A$. Intuitively, the reason why this works is that A is defined from B in a disjunctive way, whereas Q-reducibility uses a conjunctive request on B.)

As usual we have the notion of completeness:

Definition III.4.4 *A set A is **Q-complete** if it is r.e. and $\mathcal{K} \leq_Q A$.*

The analogue of Post's problem for Q-reducibility is: are there r.e. sets which are neither recursive, nor Q-complete? A solution is not automatically provided by hypersimple sets, since they can be Q-complete (p. 297).

Hyperhypersimple sets

The idea to get r.e. sets which are automatically not Q-complete comes from the fact that the hypersimple sets are not tt-complete, and in particular not c-complete. Since Q-reducibility corresponds to c-reducibility, with finite sets given by canonical indices replaced by finite sets given by r.e. indices, we change strong arrays into weak arrays in the definition of hypersimple sets.

Definition III.4.5 (Post [1944]) *A is* **hyperhypersimple** *if it is a coinfinite r.e. set, and there is no disjoint weak array (with members all) intersecting $\overline{A}$, i.e. if there is no recursive function f such that*

- $W_{f(x)}$ *finite*

- $x \neq y \Rightarrow W_{f(x)} \cap W_{f(y)} = \emptyset$

- $W_{f(x)} \cap \overline{A} \neq \emptyset$.

This definition is the point where Post got stuck in his epochal paper [1944]. He left open the problems of existence and T-completeness of hyperhypersimple sets, both of which will be solved in this section.

First we give some characterizations of hyperhypersimple sets. The first one shows that, in place of r.e. sets of disjoint finite sets given by their r.e. indices, we may only look at r.e. sets of (not necessarily finite) r.e. sets, disjoint on $\overline{A}$.

Proposition III.4.6 (Yates [1962]) *A is hyperhypersimple if and only if it is a coinfinite r.e. set, and there is no recursive function f such that:*

- $x \neq y \Rightarrow W_{f(x)} \cap W_{f(y)} \cap \overline{A} = \emptyset$

- $W_{f(x)} \cap \overline{A} \neq \emptyset$.

Proof. Suppose such an f exists. We first build g recursive such that:

- $x \neq y \Rightarrow W_{g(x)} \cap W_{g(y)} = \emptyset$

- $W_{g(x)} \cap \overline{A} \neq \emptyset$.

For this it is enough to generate the $W_{f(x)}$'s simultaneously, and eliminate repetitions. If at some stage we consider $W_{f(x)}$, and a new element of it comes out, we put it into $W_{g(x)}$, unless it has already been put in some other $W_{g(y)}$ before. Then, if $x \neq y$, $W_{g(x)} \cap W_{g(y)} = \emptyset$. And since $W_{f(x)} \cap \overline{A} \neq \emptyset$, we also have $W_{g(x)} \cap \overline{A} \neq \emptyset$, because on $\overline{A}$ the $W_{f(x)}$'s are disjoint.

Now from g we build h recursive with the additional property that $W_{h(x)}$ is finite. To define $W_{h(x)}$, simultaneously generate $W_{g(x)}$ and A, and at each stage put the elements already generated in $W_{g(x)}$ into $W_{h(x)}$, unless some z has already been generated in $W_{h(x)}$, but not yet in A (this is a temporary block, that could later be removed if z is generated in A). Since $W_{g(x)} \cap \overline{A} \neq \emptyset$, as soon as the first element of the intersection is generated in $W_{h(x)}$ the enumeration stops permanently, and $W_{h(x)}$ is finite. But since such an element has entered $\mathcal{W}_{h(x)}$, its intersection with $\overline{A}$ is nonempty. $\square$

The second characterization of hyperhypersimple sets parallels the definition of simple sets, with r.e. or recursive sets replaced by regressive or retraceable ones.

Proposition III.4.7 (Yates [1962]) *The following are equivalent, for r.e. coinfinite sets:*

1. *A is hyperhypersimple*

2. *$\overline{A}$ does not contain infinite retraceable sets*

3. *$\overline{A}$ does not contain infinite regressive sets.*

Proof. The last two conditions are equivalent by II.6.20, so we can consider any of them.

If $\overline{A}$ contains B retraceable via φ we may suppose that $\varphi(x) \le x$, whenever $\varphi(x)\!\downarrow$ (by II.6.3.a). Let

$$\mathcal{W}_{f(n)} = \{z : n = \mu x(\varphi^{(x)}(z) \simeq \varphi^{(x+1)}(z))\}.$$

Then the $\mathcal{W}_{f(n)}$'s are disjoint. Since the n-th element of B in order of magnitude is in $\mathcal{W}_{f(n)}$, $\mathcal{W}_{f(n)} \cap \overline{A} \ne \emptyset$, and A is not hyperhypersimple.

Let now A be not hyperhypersimple, and f be a recursive function such that:

- $x \ne y \Rightarrow \mathcal{W}_{f(x)} \cap \mathcal{W}_{f(y)} = \emptyset$

- $\mathcal{W}_{f(x)} \cap \overline{A} \ne \emptyset$.

If we could effectively pick up an element $z_n \in \mathcal{W}_{f(n)} \cap \overline{A}$, we could simply let φ send $\mathcal{W}_{f(0)}$ over z_0, and $\mathcal{W}_{f(n+1)}$ over z_n. Then φ would regress the infinite subset $\{z_n\}_{n\in\omega}$ of $\overline{A}$. This we cannot do, but since the effective choice of z_n is needed only to have φ partial recursive (because the enumeration of the regressed set need not be effective), there is an easy way out: we just have to do the same for each element of each $\mathcal{W}_{f(n)}$, not just for z_n. E.g., let φ' send $\mathcal{W}_{f(0)}$ over z_0, and $\mathcal{W}_{f(\langle x,y\rangle+1)}$ over the x-th element generated in $\mathcal{W}_{f(y)}$. Then φ' regresses the set inductively defined as follows: $z'_0 = z_0$, and if z'_n is the x-th element generated in $\mathcal{W}_{f(y)}$, then $z'_{n+1} \in \mathcal{W}_{f(\langle x,y\rangle+1)} \cap \overline{A}$. $\square$

Exercises III.4.8 Hyperhyperimmune sets. Two definitions are possible. A is **strongly hyperhyperimmune** if there is no r.e. set of disjoint r.e. sets intersecting it, and it is **hyperhyperimmune** if there is no disjoint weak array intersecting it. The difference is that in the first case the r.e. sets are not necessarily finite.

a) *An infinite set is strongly hyperhyperimmune if and only if it does not have infinite retraceable or regressive sets.* (Yates [1962]) (Hint: see the proof of III.4.7.)

b) *A strongly hyperhyperimmune set is hyperhyperimmune, but not conversely.* (Hint: build a hyperhyperimmune set in the natural way, with at least one element from each $B_n = \{\langle n, x \rangle : x \in \omega\}$. Then the set has an infinite retraceable subset.) III.4.6 shows that the converse implication holds for A co-r.e. Cooper [1972] shows that it holds for $A \in \Delta_2^0$.

c) *There are $2^{\aleph_0}$ hyperhyperimmune, co-hyperhyperimmune sets.* (Hint: see III.2.22.a.)

Exercises III.4.9 Strongly hypersimple sets. (Young [1966a]) The following definitions are patterned on III.3.8. A coinfinite r.e. set is **(finitely) strongly hypersimple** if there is no r.e. set of r.e. indices of disjoint (finite) r.e. sets intersecting $\overline{A}$, and with union covering it.

a) *Hyperhypersimple $\Rightarrow$ strongly hypersimple $\Rightarrow$ finitely strongly hypersimple $\Rightarrow$ hypersimple, but none of the converse implications holds.* (Young [1966a], Robinson [1967]) (Hint: by the proof of III.5.7, a semirecursive hypersimple set is not finitely strongly hypersimple. In Chapter IX we will introduce r-maximal sets, which are strongly hypersimple, and show the existence of r-maximal sets which are not hyperhypersimple (see IX.2.17 and IX.2.20). Finally, we build a finitely strongly hypersimple set which is not strongly hypersimple. Define

$$\langle x, y \rangle \in A^* \Leftrightarrow x \in A \vee (x < y).$$

If we picture $\omega \times \omega$ as a matrix, then in A^* go full rows if they correspond to elements of A, and only the part beyond the diagonal otherwise. If A is coinfinite, then the columns intersect $\overline{A^*}$, and A^* is not strongly hypersimple. Let A be hyperhypersimple, by III.4.18: then A^* is finitely strongly hypersimple. Indeed, suppose there is a disjoint weak array f intersecting $\overline{A^*}$, and covering it. We can build a disjoint weak array g intersecting $\overline{A}$, as follows. The idea is to try to put the first component of each element of $\mathcal{W}_{f(e)}$ into $\mathcal{W}_{g(e)}$: since $\mathcal{W}_{f(e)}$ intersects $\overline{A^*}$, $\mathcal{W}_{g(e)}$ then intersects $\overline{A}$. The trouble is that different $\mathcal{W}_{f(e)}$'s might have elements with the same first component, and the $\mathcal{W}_{g(e)}$'s would then not be disjoint. So we start as we said, and at stage $x + 1$ we look at the x-th row: generate simultaneously A and the $\mathcal{W}_{f(e)}$s', until we find either $x \in A$, or the $\mathcal{W}_{f(e)}$'s covering $\{\langle x, 0 \rangle, \langle x, 1 \rangle, \ldots, \langle x, x \rangle\}$. One of the two cases must happen, since the second case does if $x \notin A$, and the weak array f covers $\overline{A^*}$. Choose the $\mathcal{W}_{f(e)}$ assigned to $\mathcal{W}_{g(n)}$, with n minimal, and put x in $\mathcal{W}_{g(n)}$. If some of the remaining $\mathcal{W}_{f(e)}$'s was assigned to some other $\mathcal{W}_{g(m)}$'s, define new assignments. Note that, by induction, the assignment to $\mathcal{W}_{g(e)}$ can change only finitely many times. E.g., $\mathcal{W}_{g(0)}$ always has $\mathcal{W}_{f(0)}$ assigned and, since this is finite, it can force a change of assignment to $\mathcal{W}_{g(1)}$ only finitely often.)

b) *A coinfinite r.e. set A is (finitely) strongly hypersimple if and only if there is no r.e. set of characteristic indices of disjoint (finite) recursive sets intersecting $\overline{A}$.* Thus hypersimple, finitely strongly hypersimple, and hyperhypersimple sets are obtained from the same definition, but using respectively canonical, characteristic, and r.e. indices of finite sets. The condition of finiteness may be dropped for r.e.

indices, but not for characteristic ones. (Martin [1966a]) (Hint: given $\mathcal{W}_{f(x)}$ such that $\bigcup_{x \in \omega} \mathcal{W}_{f(x)} \supseteq \overline{A}$, let

$$z \in R_{g(x)} \Leftrightarrow z \text{ shows up in } \mathcal{W}_{f(x)} \text{ before that in } A.$$

Conversely, given $R_{f(x)}$, it is possible to suppose $R_{f(x)} \cap \overline{A}$ infinite, by taking appropriate unions. Let

$$z \in \mathcal{W}_{g(x)} \Leftrightarrow (\forall y < x)(z \notin R_{f(y)}) \wedge [z = x \vee (z > x \wedge z \in R_{f(x)})].)$$

c) *A coinfinite r.e. set is (finitely) strongly hypersimple if and only if $\overline{A}$ does not have infinite subsets retraced by total (and many-one) recursive functions.* (Martin [1966a]) (Hint: use part b) above, and see III.4.7.)

Other parallel characterizations of strongly hypersimple and hyperhypersimple sets, in terms of arrays, are given by Robinson [1967a].

The following result shows that the intuition that led to the definition of hyperhypersimple sets was correct.

Theorem III.4.10 (Soloviev [1974], Gill and Morris [1974]) *A hyperhypersimple set is not Q-complete.*

Proof. Let $x \in \mathcal{K} \Leftrightarrow \mathcal{W}_{g(x)} \subseteq A$, with $\mathcal{W}_{g(x)}$ finite. To prove that A is not hyperhypersimple, suppose we already have $\mathcal{W}_{f(0)}, \ldots, \mathcal{W}_{f(n)}$, all finite. We want $\mathcal{W}_{f(n+1)}$ finite and such that:

- $\mathcal{W}_{f(n+1)} \cap \overline{A} \neq \emptyset$

- $\mathcal{W}_{f(n+1)} \cap (\bigcup_{i \leq n} \mathcal{W}_{f(i)}) = \emptyset$.

We try to have $f(n+1) = g(a)$ for some $a \in \overline{\mathcal{K}}$, so that $\mathcal{W}_{g(a)} \cap \overline{A} \neq \emptyset$. The idea is to define an r.e. set $B \subseteq \overline{\mathcal{K}}$ such that

$$x \in \overline{B} \Rightarrow \mathcal{W}_{g(x)} \cap (\bigcup_{i \leq n} \mathcal{W}_{f(i)}) = \emptyset.$$

Then, for a index of B, we have

$$B = \mathcal{W}_a \subseteq \overline{\mathcal{K}} \Rightarrow a \in \overline{\mathcal{K}} - B,$$

and the two conditions are satisfied. To have $B \subseteq \overline{\mathcal{K}}$ we ask, by the properties of g,

$$x \in B \Rightarrow \mathcal{W}_{g(x)} \cap \overline{A} \neq \emptyset.$$

Putting things together, we try the definition

$$x \in B \Leftrightarrow \mathcal{W}_{g(x)} \cap \overline{A} \cap (\bigcup_{i \leq n} \mathcal{W}_{f(i)}) \neq \emptyset.$$

This gives a sequence of sets which are disjoint only on $\overline{A}$, but we know from III.4.6 that this is enough. Since $\overline{A}$ is not r.e., the true definition will be:

$$x \in B \;\Leftrightarrow\; (\exists s \geq x)[\mathcal{W}_{g(x),s} \cap \overline{A_s} \cap (\bigcup_{i \leq n} \mathcal{W}_{f(i),s}) \neq \emptyset].$$

The reason to have $s \geq x$ is that, since $\bigcup_{i \leq n} \mathcal{W}_{f(i)}$ is finite, for x (and hence s) big enough it will be true that x is in $\overline{\mathcal{K}}$ when x is in B. Indeed, there is a stage s_0 such that after it all elements of $A \cap (\bigcup_{i \leq n} \mathcal{W}_{f(i)})$ (a finite set) have been generated. For $s \geq s_0$, any element of $\overline{A_s} \cap (\bigcup_{i \leq n} \mathcal{W}_{f(i),s})$ is not in A, and hence is in $\overline{A} \cap (\bigcup_{i \leq n} \mathcal{W}_{f(i)})$. Thus, for $x \geq s_0$, if x is in B then it is in $\mathcal{W}_{g(x)} \cap \overline{A} \cap (\bigcup_{i \leq n} \mathcal{W}_{f(i)})$, and hence in $\overline{\mathcal{K}}$, because $\mathcal{W}_{g(x)} \cap \overline{A} \neq \emptyset$.

Thus, even if we do not know whether $B \subseteq \overline{\mathcal{K}}$, at least we know that we miss this only for finitely many numbers. Let now $B = \mathcal{W}_{a_0}$: if $a_0 \in \overline{B}$ then, by definition of $\mathcal{K}$, $a_0 \in \overline{\mathcal{K}}$, as wanted. But if $a_0 \in B$ then $a_0 \in \mathcal{K}$, thus $\mathcal{W}_{g(a_0)} \subseteq A$, and this does not work. In this case, let $\mathcal{W}_{a_1} = B - \{a_0\}$, and see if $a_1 \in \mathcal{W}_{a_1}$, and so on. We thus define a sequence $a_0, a_1, \ldots$ After a finite number of steps we must hit $a_i \in \mathcal{W}_{a_i}$, because all the a_i's are in B, and only finitely many of them are not in $\overline{\mathcal{K}}$. Then we can just let $\mathcal{W}_{f(n+1)} = \bigcup \mathcal{W}_{g(a_i)}$, and note that the a_i's in $\mathcal{K}$ contribute $\mathcal{W}_{g(a_i)} \subseteq A$, and thus do not interfere with what occurs on $\overline{A}$. □

Exercises III.4.11 a) *A coinfinite r.e. set A is hyperhypersimple if and only if it has no Q-complete superset.* (Hint: if $A \subseteq B$ is hyperhypersimple, then B is hyper-hypersimple or cofinite, hence not Q-complete. For the converse, if f is a weak array intersecting $\overline{A}$ then $B = A \cup (\bigcup_{x \in \mathcal{K}} \mathcal{W}_{f(x)})$ is Q-complete.)

b) *If A and B are hyperhypersimple sets then $A \cap B$ is hyperhypersimple, and $A \cup B$ is hyperhypersimple or finite.* Thus hyperhypersimple or cofinite sets are a filter in the lattice of r.e. sets under inclusion. (Martin, McLaughlin) (Hint: if A and B are r.e. and $\overline{A \cap B}$ has an infinite retraceable set, so does one of $\overline{A}$, $\overline{B}$.)

We now want to turn to the existence of hyperhypersimple sets. We have a great number of hypersimple sets, namely the deficiency sets of any recursive function enumerating any nonrecursive r.e. set (III.3.13), but none of them is hyperhypersimple (III.4.7), because their complement is retraceable (II.6.16). This actually gives:

Proposition III.4.12 (Yates [1962]) *Every r.e. T-degree contains a hypersimple set which is not hyperhypersimple.*

The existence of hyperhypersimple sets seems to be a problematic matter: for the first time we are in a situation in which *the obvious attack fails*. Namely, if we try to follow the line of III.3.12 then we face the trouble that, at some given stage s, we would like to put into A the elements of a finite set $\mathcal{W}_z$, but we only have $\mathcal{W}_{z,s}$, and there is no way to know whether $\mathcal{W}_z$ has been generated completely at stage s, or not.

Moreover, and differently from all the cases dealt with so far, the construction of a hyperhypersimple set cannot produce a set that satisfies the definition effectively. To make this precise, let A be **effectively hyperhypersimple** as in III.3.15, i.e. if there is a recursive function f such that $|\mathcal{W}_e| \leq f(e)$, whenever $\mathcal{W}_e$ is a disjoint weak array intersecting $\overline{A}$. Then *there is no effectively hyperhypersimple set*. Indeed, given a coinfinite r.e. set A, and f as just said, we can build $\mathcal{W}_{g(e)}$ as an r.e. set containing the indices of $f(e)+1$ finite disjoint r.e. sets intersecting $\overline{A}$ (start with n in the n-th set, for $n \leq f(e)$, and when the last element put in a set is generated in A, put in the smallest element that is not yet in any of the sets). The Fixed-Point Theorem produces e such that $\mathcal{W}_e = \mathcal{W}_{g(e)}$, and thus $|\mathcal{W}_e| > f(e)$, contradiction.

These observations show that something new is required for the construction of hyperhypersimple sets. As we have said, Post [1944] left the problem open, and it took a few years to solve it. However, for the solution of Post's problem the existence of hyperhypersimple sets is not necessary, and the reader may turn directly to the next section, if (s)he wishes.

Maximal sets ⋆

The following definition provided the starting point for the construction of hyperhypersimple sets.

Definition III.4.13 (Myhill [1956], Rose and Ullian [1963]) *A set A is* **cohesive** *if it is infinite, and cannot be split into two infinite parts by an r.e. set, i.e. if B is r.e. then either $B \cap A$ or $\overline{B} \cap A$ is finite.*

A set is **maximal** *if it is r.e. and its complement is cohesive.*

The name maximal comes from the fact that *maximal sets are the maximal elements in the lattice of r.e. sets under inclusion, modulo finite sets*. Indeed, a coinfinite r.e. set is maximal if and only if it has only trivial supersets, i.e. if B is r.e. and $B \supseteq A$ then either B is cofinite, or $B - A$ is finite.

Proposition III.4.14 (Friedberg [1958]) *A maximal set is hyperhypersimple.*

Proof. Let A be not hyperhypersimple: then there is a disjoint weak array $\{\mathcal{W}_{f(x)}\}_{x \in \omega}$ intersecting $\overline{A}$. We can split $\overline{A}$ by taking 'half' of the array, e.g. by letting $B = \bigcup_{x \in \omega} \mathcal{W}_{f(2x)}$: B is an infinite r.e. set, and $B \cap \overline{A}$ and $\overline{B} \cap \overline{A}$ are both infinite. □

Exercise III.4.15 *A maximal set is dense simple.* (Martin [1963], Tennenbaum)
(Hint: this follows from the fact, proved in IX.2.25.b, that hyperhypersimple sets are
dense simple, but a direct proof is easier. Given an enumeration $\{a_0 < a_1 < \cdots\}$ of $\overline{A}$,
and a recursive function f, suppose there are infinitely many n such that $a_n < f(n)$.
Let $h(n) = \max_{x \le 2n} f(x)$. Then infinitely many of the finite sets

$$W_{g(n)} = \{z : h(n) \le z < h(n+1)\}$$

have at least two elements of $\overline{A}$. Build B r.e. by putting in it, for each n, the elements
of $W_{g(n)}$ up to and including the first not in A. Then B splits $\overline{A}$ into two infinite
parts.)

Despite their being stronger, the requirements to build maximal sets seem to
be easier to deal with than the ones for hyperhypersimple sets, since a natural
approach to satisfy them works. As a warm up, we first dispose of cohesive
sets.

Proposition III.4.16 (Dekker and Myhill) *There exists a cohesive set.*

Proof. For future reference, we view the construction as building a set A with
cohesive complement, by stages. At stage s we will have A_s, and we will let
$\overline{A}_s = \{a_0^s < a_1^s < \cdots\}$. In the end $\overline{A} = \{a_0 < a_1 < \cdots\}$, where $a_n = \lim_{s \in \omega} a_n^s$,
i.e. $A = \bigcup_{s \in \omega} A_s$. We want to satisfy the following requirements:

P_e: $\mathcal{W}_e \cap \overline{A}$ finite or $\overline{\mathcal{W}_e} \cap \overline{A}$ finite
N_e : $\overline{A}$ has at least e elements, i.e. $\lim_{s \to \infty} a_e^s < \infty$.

The construction is as follows. We start with $A_0 = \emptyset$ (hence $a_n^0 = n$). At stage
$e+1$ we satisfy P_e. Let B_e be $\mathcal{W}_e \cap \overline{A}_e$ if it is infinite, and $\overline{\mathcal{W}_e} \cap \overline{A}_e$ otherwise.
Note that B_e is infinite because $\overline{A}_e$ is, by construction, and its elements are
either all in $\mathcal{W}_e$, or all in $\overline{\mathcal{W}_e}$. To satisfy P_e is thus enough to have almost all
the elements of $\overline{A_{e+1}}$ (and hence of $\overline{A}$) in B_e. We then let

$$
\begin{aligned}
a_n^{e+1} &= a_n^e & &\text{if } n \le e \\
a_{n+1}^{e+1} &= \text{the smallest element of } B_e \text{ greater than } a_n^{e+1} & &\text{if } n \ge e.
\end{aligned}
$$

Note that the construction is highly noneffective, for two reasons: we first ask
whether given sets are infinite, and then we ask membership questions about
B_e. $\overline{A}$ is however cohesive: the P_e's are satisfied by construction, and $\overline{A}$ is
infinite because a_n^s may move at a certain stage $s+1$ only if $n > s$, hence only
finitely many times. Thus $\lim_{s \in \omega} a_n^s$ exists, and all N_e's are satisfied.   □

Exercises III.4.17 Cohesive sets. a) *Every infinite set has a cohesive subset.*
(Dekker and Myhill) (Hint: given C infinite, let $A_0 = \overline{C}$ and proceed as above.)

b) *There are $2^{\aleph_0}$ cohesive sets.* (Hint: any infinite subset of a cohesive set is cohesive.)

c) *There is no cohesive and co-cohesive set.* (Hint: if A is such, and B is an infinite coinfinite r.e. set, then $B \cap A$ is finite, by cohesiveness of A. Since B is infinite, $B \cap \overline{A}$ is infinite. By cohesiveness of $\overline{A}$, $\overline{B} \cap \overline{A}$ is finite. So A differs finitely from $\overline{B}$ or from B for each such B, contradiction.)

d) *No finite union of disjoint cohesive sets can cover ω, but an infinite one can.* (Hint: see parts a) and c) above.)

We now constructivize the proof just given, and build a cohesive set with r.e. complement.

Theorem III.4.18 (Friedberg [1958]) *There exists a maximal set.*

Proof. We use notations as in III.4.16. The requirements are the same as there, plus the fact that A has to be r.e. This forces us to have a recursive construction, and we cannot anymore ask whether given sets are infinite. This forbids the simple strategy used before, of satisfying one positive condition at each stage, and forces us to work with approximations.

Let us examine P_0 first: since we cannot ask whether W_0 is infinite, and at any given stage we only have a finite approximation to it, we want to play two different strategies, one of which will win. Suppose we knew that W_0 is infinite: then we could force a_n^0 on it, by waiting long enough. We can push this as far as possible and, at any given stage $s + 1$, force the longest possible initial segment of $\overline{A}_{s+1}$ in W_0. This means that, whenever there are elements of $\overline{A}_s$ in $\overline{W_0}$ which are smaller than elements of $\overline{A}_s$ in W_0, we drop them (by putting them into A). This produces, at any stage s, a picture as follows:

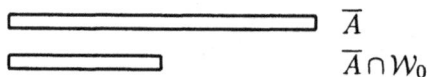

$$\begin{array}{ll} \boxed{} & \overline{A} \\[4pt] \boxed{} & \overline{A} \cap W_0 \end{array}$$

If W_0 is really going to be infinite, then $\overline{A}$ is going to be included in W_0. Otherwise, except for a finite initial segment, $\overline{A}$ is going to be included in $\overline{W_0}$, and P_0 is going to be satisfied either way (this does not take care of the other requirements, see III.4.19).

We turn now to P_1. If we knew the final outcome of the strategy for P_0, we would not worry: we could play the same strategy inside the part of $\overline{A}$ on which the characteristic function of W_0 is eventually constant, betting on the fact that W_1 is going to be infinite on it, and having at each stage an initial segment in W_1, and the rest in $\overline{W_1}$. Even if we do not know the final outcome, nothing forbids us to play the same strategy on both approximations, and so on. We thus try to have $\overline{A}$ look like this:

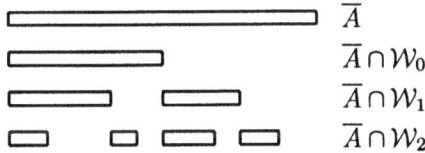

This can easily be formalized, by using the concept of **e-state**, which codes the membership of a given element x into the first e r.e. sets, at a given stage s:

$$\sigma(e, x, s) = \sum \{2^{e-i} : i \leq e \wedge x \in \mathcal{W}_{i,s}\}.$$

Another way to visualize e-states is by viewing them as binary strings of 0's and 1's, ordered lexicographically, and with the i-th digit from the left giving the value of the characteristic function of $\mathcal{W}_i$. The following properties of e-states are immediate:

- there are only finitely many e-states, namely 2^{e+1}

- the e-state of an element at a given stage can only increase at later stages, i.e. if $s \leq t$ then $\sigma(e, x, s) \leq \sigma(e, x, t)$ since, for each i, $\mathcal{W}_{i,s} \subseteq \mathcal{W}_{i,t}$

- e-states give absolute priority to membership in r.e. sets with lower indices, i.e. if $\sigma(e, x, s) < \sigma(e, y, s)$ then $\sigma(i, x, s) < \sigma(i, y, s)$, for any $i \geq e$.

The construction of A can now be easily formulated, by trying to let the e-th element of $\overline{A}$ have maximal e-state. Precisely, we start with $A_0 = \emptyset$ (hence $a_e^0 = e$). At stage $s+1$ we let, by induction, a_e^{s+1} be the smallest $x \in \overline{A}_s$ greater than a_{e-1}^{s+1}, and with maximal e-state. Note that, by induction, $\overline{A}_s$ is infinite and recursive, so the construction is effective, and A is r.e. Moreover:

- $\overline{A}$ is infinite.
  It is enough to prove that $\lim_{s \in \omega} a_e^s$ exists. Suppose, by induction, that $\lim_{s \in \omega} a_i^s$ exists, for each $i < e$. Let s_0 be a stage after which all a_i^s's have reached their final position. Then, for $s > s_0$, a_e^s may move only to reach a higher e-state, and hence only finitely many times.

- A is maximal.
  Suppose, by induction, that P_i is satisfied, for all $i < e$. Then there is n_0 such that

$$(\forall i)[(\forall n \geq n_0)(a_n \in \mathcal{W}_i) \vee (\forall n \geq n_0)(a_n \in \overline{\mathcal{W}_i})].$$

Suppose $\mathcal{W}_e \cap \overline{A}$ and $\overline{\mathcal{W}_e} \cap \overline{A}$ are both infinite. There is $n \geq n_0, e$ such that $a_n \in \overline{\mathcal{W}_e}$ and $a_{n+1} \in \mathcal{W}_e$. Then there is also s_0 such that, for all

$s \geq s_0,$

$$(\forall x \leq n+1)(a_x^s = a_x)$$
$$(\forall i < e)(a_n \in \mathcal{W}_i \Leftrightarrow a_n \in \mathcal{W}_{i,s})$$
$$(\forall i \leq e)(a_{n+1} \in \mathcal{W}_i \Leftrightarrow a_{n+1} \in \mathcal{W}_{i,s}).$$

Then, by definition of e-state,

$$\sigma(e, a_n^s, s) < \sigma(e, a_{n+1}^s, s)$$

and hence, since $n \geq e$,

$$\sigma(n, a_n^s, s) < \sigma(n, a_{n+1}^s, s).$$

By construction we then have that a_n^{s+1} should be the smallest element of $\overline{A_s}$ greater than $a_{n-1}^s = a_{n-1}$ and with maximal n-state, hence it cannot be a_n, contradiction. $\square$

Exercise III.4.19 *There is an acceptable system of indices $\{\hat{\mathcal{W}}_e\}_{e \in \omega}$ of the r.e. sets such that $\hat{\mathcal{W}}_0$ is infinite, but the complement of the maximal set constructed above has no element in $\hat{\mathcal{W}}_0$.* (Hint: define $\hat{\mathcal{W}}_{x+1} = \mathcal{W}_x$, and generate elements in $\hat{\mathcal{W}}_0$ only after they have already been generated in A.)

The **e-state method** used in the proof above is a kind of priority method, with the usual order of priority

$$P_0 > N_0 > P_1 > N_1 > \ldots$$

The basic difference between the constructions of maximal and simple (or hypersimple) sets is that *the positive requirements P_e are infinitary*, and we cannot hope to satisfy them with a finite action. Actually, since they only allow for finitely many exceptions, each requirement has to be considered cofinitely many times, and at any given stage we have to consider many positive requirements all together. But the requirements have different priorities, and *e-states are a device to assign priorities not to single requirements, but to groups of them*. E.g., the order of priorities assigned by 2-states is

$$(P_0, P_1, P_2) \geq (P_0, P_1) \geq (P_0, P_2) \geq (P_0) \geq (P_1, P_2) \geq (P_1) \geq (P_2).$$

We might say that locally this is a **finite injury argument**, since every element of $\overline{A}$ moves at most finitely many times, but globally it is an **infinite injury argument**, since a positive requirement may be injured infinitely often, each time for different elements of $\overline{A}$.

Exercises III.4.20 a) *Maximal sets are not closed under intersection.* (Yates [1962]) (Hint: take A maximal, and let $x \in B \Leftrightarrow x + 1 \in A$.)

b) *There are hyperhypersimple sets which are not maximal.* (Hint: the hyperhypersimple sets are closed under intersection.)

Maximal sets have thinnest possible complement. The existence of maximal T-complete sets would show that a notion of thin complement alone is not sufficient to solve Post's problem. In fact, the maximal set just built is T-complete, as we now prove, as usual, by showing that it is maximal in an effective way.

Definition III.4.21 (Lachlan [1968]) *A is **effectively maximal** if it is a coinfinite r.e. set, and there is a recursive function g such that, for every e, the sequence of 0's and 1's consisting of the values of the characteristic function of W_e on the elements of $\overline{A}$ in increasing order has at most $g(e)$ alternations.*

Note that having finitely many alternations simply means that one of $W_e \cap \overline{A}$ and $\overline{W_e} \cap \overline{A}$ is finite.

Proposition III.4.22 (Lachlan [1968]) *Every effectively maximal set is T-complete.*

Proof. Let g witness that A is effectively maximal, and define $f \leq_T A$ such that

$$W_{f(e)} = \text{ a finite set with } g(e) + 1 \text{ alternations on } \overline{A}.$$

This is possible because $\overline{A}$ is infinite, e.g.

$$W_{f(e)} = \{a_{2i} : i \leq g(e) + 1\}.$$

Then f has no fixed-points and, by the criterion for T-completeness III.1.5, A is T-complete. $\square$

Corollary III.4.23 (Yates [1965]) *There exists a T-complete maximal set.*

Proof. The maximal set built in III.4.18 is effectively maximal, since W_e can have at most $g(e) = 2^{e+1}$ alternations on $\overline{A}$. $\square$

Exercises III.4.24 a) *For no coinfinite r.e. set A there is a recursive function g such that*

$$W_e \cap \overline{A} \text{ infinite } \Rightarrow |\overline{W_e} \cap \overline{A}| \leq g(e).$$

Similarly for

$$\overline{W_e} \cap \overline{A} \text{ infinite } \Rightarrow |W_e \cap \overline{A}| \leq g(e).$$

Thus these natural candidates for effective maximality fail. (Lachlan [1968]) (Hint: for the second property, put in $W_{h(e)}$ the first $g(e)+1$ elements of $\overline{A}$, by starting with $\{0, \ldots, g(e)\}$ and, each time that one element goes into A, adding the first element not yet in $\mathcal{W}_{h(e)}$, and not yet generated in A. By the Fixed-Point Theorem there is e such that $\mathcal{W}_e = \mathcal{W}_{h(e)}$. Thus $\overline{\mathcal{W}_e} \cap \overline{A}$ is infinite, but $|\mathcal{W}_e \cap \overline{A}| = g(e) + 1$.)

b) *There is a maximal, effectively simple set.* (Cohen and Jockusch [1975]) (Hint: modify the construction of a maximal set given above by inserting steps to make A simple, as in the second proof of III.2.11. Note that each a_n^s may move more times than it did for maximality alone, but still only finitely often.)

c) *A maximal set is not strongly effectively simple.* (Cohen and Jockusch [1975]) (Hint: by III.4.15 a maximal set is dense, and by III.3.9 is not strongly effectively simple. See III.6.21 for a different proof.)

d) *A strongly effective simple set is not contained in maximal sets.* Thus Post's simple set is not contained in maximal sets. (Cohen and Jockusch [1975]) (Hint: a coinfinite r.e. superset of a strongly effectively simple set is still strongly effectively simple.)

III.5   A Solution to Post's Problem

Post concluded his paper [1944] by saying:

> we are left completely on the fence as to whether there exists a recursively enumerable set of positive integers of absolutely lower degree of unsolvability than the complete set $\mathcal{K}$, or whether, indeed, all recursively enumerable sets of positive integers with recursively unsolvable decision problems are absolutely of the same degree of unsolvability. On the other hand, if this question can be answered, that answer would seem to be not far off, if not in time, then in the number of special results to be gotten on the way.

This section can be seen as the missing conclusion to Post's paper, and shows that he was indeed right, regarding the number of special results needed to solve his problem.

Semirecursive sets

We know that hyperhypersimple sets are not Q-complete. We are then looking for a notion that, together with T-completeness, would imply Q-completeness. By coupling it with hyperhypersimplicity we would then have a notion implying T-incompleteness, and Post's problem would be solved.

Since the difference between Q-reducibility and T-reducibility is that we query the oracle on single elements in the first case, and on finite sets in the second, we need a notion that would reduce a question of inclusion of a finite set to the question of membership of a single element.

Definition III.5.1 (Jockusch [1968a]) *A set A is* **semirecursive** *if there is a recursive function f of two variables such that*

1. $f(x,y) = x$ *or* $f(x,y) = y$

2. $x \in A \vee y \in A \Rightarrow f(x,y) \in A$.

Clearly *a recursive set is semirecursive*, since we can simply decide whether one of x and y is in the set. Also, *the complement of a semirecursive set is semirecursive*: if f witnesses the semirecursiveness of A, then the function that always chooses, between x and y, the element not chosen by f, witnesses the semirecursiveness of $\overline{A}$.

The next result is not unexpected, and is actually the reason why we introduced the notion of semirecursiveness.

Proposition III.5.2 (Marchenkov [1976]) *If an r.e. set A is semirecursive and T-complete, then it is also Q-complete.*

Proof. We only have to show how to reduce finite questions of the form $D_u \subseteq \overline{A}$ to single questions $g(u) \in \overline{A}$, for some recursive g. Let f be as in the definition III.5.1. Given $D_u = \{x_1, \ldots, x_n\}$, let

$$\begin{aligned}
y_1 &= x_1 \\
y_{i+1} &= f(y_i, x_{i+1}) \quad (i < n) \\
g(u) &= y_n.
\end{aligned}$$

Then we obviously have $D_u \subseteq \overline{A} \Leftrightarrow g(u) \in \overline{A}$, because if one element of D_u is in A, then so is $g(u)$. $\square$

Exercises III.5.3 a) *A semirecursive simple set is hypersimple.* (Jockusch [1968a]) (Hint: if A is semirecursive so is $\overline{A}$, and thus a finite set intersecting $\overline{A}$ effectively produces an element of $\overline{A}$.)

b) *A semirecursive set is not p-complete.* (Jockusch [1968a]) (Hint: a semirecursive p-complete set would be m-complete, and thus every r.e. set would be semirecursive, contradicting the existence of simple, nonhypersimple sets.)

We are obviously interested in knowing which sets are semirecursive. Since the definition of semirecursive sets might appear somewhat ad hoc, we first give an interesting alternative characterization.

Proposition III.5.4 (Appel, McLaughlin) *A set is semirecursive if and only if it is a cut of a recursive linear ordering of ω.*

Proof. Let A be semirecursive via f. We define, by induction, a recursive linear ordering $\prec$ such that

$$x \prec y \wedge y \in A \Rightarrow x \in A.$$

Let $x_0 \prec x_1 \prec \ldots \prec x_n$ be the ordering of the numbers up to n. We want to extend it to $n+1$. Three cases are possible:

1. $f(n+1, x_0) = n+1$
   If $x_0 \in A$ then $n+1 \in A$, by the properties of f, and we can then let $n+1 \prec x_0$.

2. the first case fails, and $f(n+1, x_n) = x_n$
   Similarly, if $n+1 \in A$ then $x_n \in A$, and we let $x_n \prec n+1$.

3. the first two cases fail
   Then $f(n+1, x_0) = x_0$ and $f(n+1, x_n) = n+1$. Then there is i such that

   $$f(n+1, x_i) = x_i \wedge f(n+1, x_{i+1}) = n+1,$$

   and thus

   $$x_{i+1} \in A \Rightarrow n+1 \in A \Rightarrow x_i \in A.$$

   Then we can let $x_i \prec n+1 \prec x_{i+1}$.

Let now A be the lower cut of a recursive linear ordering $\prec$. If

$$f(x,y) = \text{least element of } \{x,y\} \text{ w.r.t. } \prec$$

then $f(x,y) = x$ or $f(x,y) = y$ by definition, and if one of x and y is in A then so is $f(x,y)$, because A is closed downward w.r.t. $\prec$. Thus f satisfies the conditions of III.5.1, and A is semirecursive. $\square$

Exercises III.5.5 a) *Every tt-degree contains a semirecursive set.* (Jockusch [1968a]) (Hint: let A be infinite and coinfinite, and define $r = \sum_{n \in A} 2^{-n}$. For any x, let $r_x = \sum_{n \in D_x} 2^{-n}$. Then $x \prec y \Leftrightarrow r_x < r_y$ is recursive. Let B be the lower cut determined by r. $A \leq_T B$ because if $D_x = (A \cap \{0, \ldots, n-1\}) \cup \{n\}$ then, by induction,

$$n \in A \Leftrightarrow D_x \subseteq A \Leftrightarrow r_x < r \Leftrightarrow x \in B.$$

And $B \leq_T A$ because $x \in B$, i.e. $r_x < r$, if and only if there is a finite set D_y contained in $\{0, \ldots, \max D_x\}$, such that $r_x \leq r_y$ and $D_y \subseteq A$.)
  b) *There are $2^{\aleph_0}$ semirecursive sets.* (Martin, McLaughlin)

We turn now to the investigation of which sets, among the ones introduced so far for the solution of Post's problem, are semirecursive.

Proposition III.5.6 (Jockusch [1968a]) *Every coregressive r.e. set is semi-recursive.*

Proof. Let A be an r.e. set, coregressive via φ. Given x and y, note that either one of x and y is in A, or they are both in $\overline{A}$, and thus one of them follows the other in the given enumeration of $\overline{A}$, and it is then sent over it by φ (meaning that some iteration of φ on the former produces the latter). To define f generate simultaneously A, $\{\varphi^{(n)}(x)\}_{n \in \omega}$ and $\{\varphi^{(n)}(y)\}_{n \in \omega}$, and let

$$f(x,y) = \begin{cases} x & \text{if } x \in A \text{ or } y \in \{\varphi^{(n)}(x)\}_{n \in \omega} \\ y & \text{if } y \in A \text{ or } x \in \{\varphi^{(n)}(y)\}_{n \in \omega}. \end{cases}$$

Suppose $f(x,y) \in \overline{A}$: then e.g. $f(x,y) = y$ and $x \in \{\varphi^{(n)}(y)\}_{n \in \omega}$, so $y \in \overline{A}$ and $x \in \overline{A}$, and A is semirecursive. $\square$

The result implies, by II.6.16 and III.3.13, that *many hypersimple sets are semirecursive.* Actually, every r.e. T-degree contains a semirecursive hypersimple set (in particular *there are hypersimple sets which are Q-complete,* namely any T-complete semirecursive hypersimple set). But the result we were really looking for escapes us.

Proposition III.5.7 (Martin) *No hyperhypersimple set can be semirecursive.*

Proof. We know (III.4.7) that the complement of a hyperhypersimple set cannot contain infinite retraceable sets. Since the complement of a set is semirecursive when the set is, it is enough to show that *an infinite semirecursive set contains an infinite retraceable set.* Let A be infinite and semirecursive. We may suppose A immune, otherwise it has infinite recursive, hence retraceable, subsets. Let $\prec$ be a recursive linear ordering of which A is a lower cut: then A is well-ordered by $\prec$ with order type ω, because for any $x \in A$ the set $\{y : y \prec x\}$ is a recursive subset of A, hence finite by immunity. Let

$$f(x) = \begin{cases} n & \text{if } x \text{ is the } n\text{-th element of } A \text{ w.r.t. } \prec \\ \infty & \text{if } x \notin A. \end{cases}$$

Consider the elements corresponding to nondeficiency stages of f:

$$\begin{aligned} x \in B &\Leftrightarrow (\forall y > x)(f(y) > f(x)) \\ &\Leftrightarrow (\forall y)(y > x \Rightarrow y \succ x). \end{aligned}$$

B is a subset of A because, given $x \in B$, there is $y \in A$ greater than it (since A is infinite): then $y \succ x$, and $x \in A$ because A is closed downward w.r.t. $\prec$. And B is infinite, as usual for nondeficiency stages (note that f is one-one on A).

To show that B is retraceable, we cannot appeal directly to the fact that so are nondeficiency sets, because f is not recursive, but an argument similar to that of II.6.16 can be reproduced directly. Given $x \in B$, we want to get an effective procedure to find the greatest element of B smaller than x. Since for $y > x$ we have $y \succ x$, we only have to check numbers below x. In other words, for $z < x$,

$$z \in B \quad \Leftrightarrow \quad (\forall y)(z < y \Rightarrow z \prec y)$$
$$\Leftrightarrow \quad (\forall y)(z < y \leq x \Rightarrow z \prec y).$$

Thus it is enough to define $g(x)$ as the biggest $z < x$ such that

$$(\forall y)(z < y \leq x \Rightarrow z \prec y)$$

if $b_0 \prec x$ (where b_0 is the least element of B), and x otherwise.   □

Actually, the proof shows that *no finitely strongly hypersimple set is semirecursive*, because the retracing function defined above is total and many-one (see III.4.9).

Exercises III.5.8 Hereditary sets. A set A is **hereditary** if there is a recursive function f such that $x \in A \wedge y \in A \Leftrightarrow f(x,y) \in A$.

a) *A set A is hereditary if and only if there is a recursive function g such that $D_x \subseteq A \Leftrightarrow g(x) \in A$.* (Lavrov [1968])

b) *For any set A, the set $A^\omega = \{x : D_x \subseteq A\}$ is hereditary.* (Degtev [1972])

c) *Semirecursive $\Rightarrow$ hereditary, but not conversely.* (Degtev [1972]) (Hint: $\mathcal{K}$ is not semirecursive, see III.5.3.b, but is hereditary because it is recursively isomorphic to $\mathcal{K}^\omega$, see III.7.14.)

Degtev [1972] has shown that some of the properties of semirecursive sets, e.g. III.5.7, hold for hereditary sets as well.

Exercises III.5.9 Verbose and terse sets. (Beigel, Gasarch, Gill, and Owings [1993]) We say that we can do 'n for m' on a set A if n questions of membership to A can be answered by a recursive procedure that asks at most m question to the oracle A. A set A is **verbose** if, for every n, we can do $2^n - 1$ for n on it, and it is **terse** if we cannot do n for $n - 1$.

a) *If we can do 2^n for n on A, then A is recursive.* Thus the definition of verbose sets is optimal. (Hint: we show the proof for $n = 1$. Suppose we can do 2 for 1 on A: given two elements x and y, we can compute $A(x)$ and $A(y)$ by a recursive procedure that asks only one question to the oracle A. Since the answer to the query can be only 0 or 1, there are two partial recursive functions φ_0 and φ_1, respectively using the answer 0 and 1, one of which gives the correct values for $A(x)$ and $A(y)$.

To compute A recursively, consider two cases. If for every x there is y such that φ_0 and φ_1 compute the same value of $A(x)$, then this must be the right value of $A(x)$, and to compute it it is enough to look for such a y. If there is x such that, for every y, φ_0 and φ_1 give different answers for $A(x)$, then fix x, and consider the right value $A(x)$. For every y, the right value of $A(y)$ is computed by the one, between φ_0 and φ_1, that computes the right value of $A(x)$.) Note that A plays no role as an oracle, and thus A would be recursive even if we could answer 2^n membership questions on A by a recursive procedure asking n questions to an oracle B.

b) $\mathcal{K}$ is verbose. (Hint: we show how to do 3 for 2. To know which of x_1, x_2, x_3 are in $\mathcal{K}$, first ask if at least two are. This is an r.e. question, hence $\mathcal{K}$ can answer it. If the answer is yes, ask if all three are, otherwise ask if at least one is. The whole procedure requires only two questions, and determines how many of the three elements are in $\mathcal{K}$. To know exactly which ones are, generate $\mathcal{K}$ until that many appear.) Since verbose sets are closed are under m-equivalence, *every m-complete set is verbose*.

c) *A semirecursive set is verbose.* (Hint: let A be the upper cut of a recursive linear ordering, by III.5.4. To determine which of $2^n - 1$ elements are in A, first order them according to the linear ordering. To know which are is A and which are not, it is enough to find the first element which is in A, and this can be done by a binary search, thus requiring only n steps and n questions to the oracle A.)

d) *(R.e.) verbose sets exist in every (r.e.) T-degree.* (Hint: by III.5.5.a and III.5.6, because verbose sets are closed under complements.)

e) *Every nonzero T-degree contains terse sets.* (Hint: given a semirecursive, nonrecursive set A, notice that, given elements $x_1, \ldots, x_n$, if we know how many are in A, then we can determine exactly which ones: order them according to the ordering w.r.t. which A is an upper cut, and count from the biggest one. Given $2^n - 1$ elements, less than 2^n can be in A, and thus their number can be written, in binary, with at most n digits. Define a set B, by putting $\langle x_1, \ldots, x_{2^n-1}, i \rangle$ in it if and only if the i-th digit in the binary representation of $|A \cap \{x_1, \ldots, x_{2^n-1}\}|$ is 1. Clearly $A \equiv_T B$. And $2^n - 1$ questions on A can be reduced to n questions on B. If we could do n for $n-1$ on B then we could do $2^n - 1$ for $n-1$ on A, with oracle B, and hence 2^n for n with oracle $B \oplus A$, contradicting part a).)

η-hyperhypersimple sets

Despite the fact that hyperhypersimple sets are not semirecursive, we are not willing to give up and waste all the work done so far, especially so after having been so close to a solution of Post's problem. The idea is to look for a weakening of the notion of hyperhypersimplicity that is still incompatible with Q-completeness, but is compatible with semirecursiveness. We simply generalize the notion of number, and substitute it with the notion of equivalence class w.r.t. an r.e. equivalence relation.

Definition III.5.10 (Malcev [1965], Ershov [1971]) *An equivalence relation η is called* **positive** *if $R(x,y) \Leftrightarrow x\eta y$ is r.e.*

Exercise III.5.11 *The class of positive equivalence relations is a sublattice of the complete lattice of all the equivalence relations on ω under inclusion.* (Ershov [1971]) (Hint: the smallest element is the equality relation, the greatest one the trivial equivalence relation. The g.l.b. is the set-theoretical intersection, the l.u.b. the smallest equivalence relation including the given ones.)

Definition III.5.12 *A set A is η-closed if it consists of equivalence classes w.r.t. η, i.e. $x \in A \wedge x\eta y \Rightarrow y \in A$.*
    The **η-closure** *$[A]_\eta$ of a set A is the smallest η-closed set containing it.*

Exercises III.5.13 (Ershov [1971]) a) *If A is r.e. then*

$$x\eta_A y \Leftrightarrow (x \in A \wedge y \in A) \vee x = y$$

is a positive equivalence relation, with (if A has more than one element) A as the only nontrivial equivalence class. A set is η_A-closed if and only if it either contains A, or is disjoint from it.
    b) *If φ is a partial recursive function, then*

$$x\eta_\varphi y \Leftrightarrow (\varphi(x) \simeq \varphi(y)\downarrow) \vee x = y$$

is positive. A set A is η_φ-closed if and only if $\varphi^{-1}(\varphi(A)) \subseteq A$.
    c) *If φ is a partial recursive function, then*

$$x\eta_\varphi^i y \Leftrightarrow \exists m \exists n (\varphi^{(m)}(x) \simeq \varphi^{(n)}(y)\downarrow)$$

is positive. (Hint: recall that, by definition, $\varphi^{(0)}(x) \simeq x$.)
    d) *η is a positive equivalence relation if and only if, for some partial recursive function φ, $\eta = \eta_\varphi^i$.* (Hint: let η be approximated monotonically by η_s. Approximate φ as follows: given φ_s, extend it by defining, when possible, $\varphi_{s+1}(x) \simeq \mu y(x\eta_{s+1}y)$.)

Definition III.5.14 *An η-closed set A is η-finite or η-infinite, according to whether it consists of a finite or an infinite number of equivalence classes of η.*

By restricting our attention to η-closed sets, and replacing the notion of finiteness by η-finiteness, we can relativize to η the notions introduced so far.

Definition III.5.15 (Ershov [1971]) *Let η be a positive equivalence relation, and A be an η-closed and co-η-infinite r.e. set. Then A is:*

1. *η-simple if any η-closed r.e. set contained in $\overline{A}$ is η-finite*

2. *η-hypersimple if there is no recursive function f such that*

- $x \neq y \Rightarrow [D_{f(x)}]_\eta \cap [D_{f(y)}]_\eta = \emptyset$
- $[D_{f(x)}]_\eta \cap \overline{A} \neq \emptyset$

3. **η-hyperhypersimple** *if there is no recursive function f such that*

- $W_{f(x)}$ *finite*
- $x \neq y \Rightarrow [W_{f(x)}]_\eta \cap [W_{f(y)}]_\eta = \emptyset$
- $[W_{f(x)}]_\eta \cap \overline{A} \neq \emptyset$

4. **η-maximal** *if, for every η-closed r.e. set $B \supseteq A$, one of $B - A$ and $\overline{B}$ is η-finite.*

The implications among the various concepts still hold, but nontriviality conditions are not automatically satisfied.

Proposition III.5.16 *An η-maximal set can be empty.*

Proof. Let A be any maximal set, and

$$x\eta_A y \Leftrightarrow (x \in A \wedge y \in A) \vee x = y.$$

If B is r.e. and η_A-closed, then either it contains A or it is contained in $\overline{A}$. In the second case it is finite by maximality, hence η_A-finite. In the first, either $B - A$ is finite, hence B is η_A-finite because A consists of just one class, or $\overline{B}$ is finite, and hence η_A-finite. But then the empty set is η_A-maximal. $\square$

Exercise III.5.17 *Every nonrecursive η-maximal set A whose equivalence classes on $\overline{A}$ are all finite is simple, although not necessarily hypersimple. (Hint: if B is r.e. and $B \subseteq \overline{A}$, consider the closure of $A \cup B$. It cannot be co-η-finite, otherwise A is recursive. So B is contained in finitely many classes, and thus finite.)*

It is convenient to picture a positive equivalence relation as a series of boxes, representing the equivalence classes. Since the equivalence relations we deal with are positive, there is an effective procedure that builds up the boxes. At some given stage, boxes that were previously separated may be put together to form a bigger box. We may also suppose that, at each stage, all boxes are finite.

We are now ready for the final steps of our long journey.

Proposition III.5.18 (Marchenkov [1976]) *An η-hyperhypersimple set is not Q-complete.*

Proof. The proof is like the one of III.4.10, after the following modifications are made, for A η-closed:

- if $x \in \mathcal{K} \Leftrightarrow W_{g(x)} \subseteq A$, then $W_{g(x)}$ is η-closed. If it is not so, take its η-closure: it still works, because A is η-closed itself.

- A consists of equivalence classes with finitely many elements each. If it is not so, define a new positive equivalence relation η' coinciding with η on $\overline{A}$, hence with no effect on η-hyperhypersimplicity, as follows. At stage $s + 1$, a given box of η is also a box of η' if either it does not intersect A_s, or s is the first stage in which it does intersect A. □

Theorem III.5.19 (Degtev [1973]) *There exists an r.e. set A which is nonrecursive, semirecursive, and η-maximal, for some positive equivalence relation η.*

Proof. We define A, which may thought of as one big box, and infinitely many boxes, which are going to be the equivalence classes of $\overline{A}$. Thus A will automatically be η-closed and η-infinite. Let $\{B_n^s\}_{n \in \omega}$ be an enumeration of the boxes of $\overline{A}_s$ at stage s, such that if $m < n$ then $\max B_m^s < \min B_n^s$.

We start with $A_0 = \emptyset$, and $B_n^0 = \{n\}$. In the construction we alternately take care of the positive requirements (for nonrecursiveness and semirecursiveness) and of the negative ones (for η-maximality). Thus at stage $s + 1$ we take different actions, according to whether s is even or odd.

1. s even

   To make A nonrecursive, we make it simple. Thus we search for the smallest $e \leq s$ such that:

   - $W_{e,s} \cap A_s = \emptyset$
   - for some $n \geq e$, $B_n^s \cap W_{e,s} \neq \emptyset$.

   If there is such an e, we put all of B_n^s into A, where n is the smallest one such that $n \geq e$ and $B_n^s \cap W_{e,s} \neq \emptyset$. If there is no such e, we go to the next stage.

   To ensure semirecursiveness we also put in A, together with B_n^s, all the following boxes up to the one containing s. This ensures, because of the way the boxes are grouped together during the construction, that whenever an element z goes into A at stage $s + 1$, so do all elements y such that $z \leq y \leq s$. To see that this has the desired effect, let

   $$f(x, y) = \begin{cases} x & \text{if } x \in A_{\max\{x,y\}} \\ y & \text{if } x \notin A_{\max\{x,y\}} \wedge y \in A_{\max\{x,y\}} \\ \max\{x, y\} & \text{otherwise.} \end{cases}$$

   We then have $x \in A \vee y \in A \Rightarrow f(x, y) \in A$. This is clear if f is defined in any of the first two cases. In the last, if one of x and y goes into A, it must be at a stage bigger than their maximum (otherwise one of the two other cases would apply). The only case of concern is when the smallest of x, y goes in (since $f(x, y)$ is then the other one). But in this case the construction also puts the other in A, and thus $f(x, y) \in A$.

2. s odd

   We consider the usual e-states

$$\sigma(e, x, s) = \sum \{2^{e-i} : i \leq e \wedge B_x^s \cap W_{i,s} \neq \emptyset\},$$

and proceed by induction, by letting B_e^{s+1} be the union of all boxes between B_e^s and B_x^s, where x is the smallest element greater than $e - 1$ and with maximal e-state.

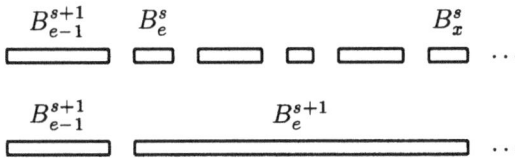

Note that, unlike the case of maximal sets, the construction of η-maximal sets does not force us to put elements into A, as we already know from III.5.16, and thus the two parts of the construction do not interfere.

The proof that the construction works is similar to the ones for simple and maximal sets. The only slightly different part is the fact that $\overline{A}$ consists of infinitely many classes. But to show that $\lim_{s \to \infty} B_e^s$ exists, it is enough to note that B_e^s can change only for two reasons:

- some B_n^s with $n < e$ goes into A. Note that this must happen because, for some index $i < n$, the box B_n^s or one with smaller index is satisfying the simplicity requirement for i. But this can happen only once for each $i < e$, hence at most finitely often.

- B_e^s reaches a higher e-state, and once again this can happen only finitely often. $\square$

We have thus finally reached the end of our journey. We put things together, for completeness and the reader's sake.

Or quel che t'era dietro t'è davanti:
ma perché sappi che di te mi giova,
un corollario voglio che t'ammanti.[1]
(Dante, *Paradiso*, VIII)

Corollary III.5.20 Solution to Post's problem (Muchnik [1956], Friedberg [1957a]) *There exists an r.e. set which is neither recursive nor T-complete.*

Proof. Consider a nonrecursive, semirecursive, η-hyperhypersimple r.e. set: it is not Q-complete by III.5.18, and hence not T-complete by III.5.2. ☐

This shows in particular that *there are more than two r.e. T-degrees*, and immediately raises the question of how complicated the structure of r.e. T-degrees is. A complete description of it as a partial ordering is not known, but Chapter X will provide a good deal of information about it, as well as different methods to solve Post's problem.

III.6   Creative Sets and Completeness

For an r.e. set A, to be nonrecursive means that $\overline{A}$ is not r.e. The fact that $\forall x(\overline{A} \neq W_x)$ may be formalized in different ways:

1. $\exists y(y \in A \Leftrightarrow y \in W_x)$

2. $W_x \subseteq \overline{A} \Rightarrow \exists y(y \in \overline{A} - W_x)$

3. $\overline{A} \subseteq W_x \Rightarrow \exists y(y \in A \cap W_x)$.

We cannot expect in general to find y recursively in x, and we now investigate what happens if we impose this requirement.

Effectively nonrecursive sets

We begin by constructivizing the first notion, which also appears to be the most natural and symmetric.

Definition III.6.1 (Dekker [1955a]) *A set A is **completely productive** if there is a recursive function f such that, for every x,*

$$f(x) \in \overline{A} \Leftrightarrow f(x) \in W_x.$$

---

[1] Now that which stood behind you, stands in front:
but so that you may know the joy you give me,
I now would cloak you with a corollary.

A set is **effectively nonrecursive** *if it is r.e., and its complement is completely productive.*

$\mathcal{K}$ is effectively nonrecursive because, by definition, $x \in \mathcal{K} \Leftrightarrow x \in W_x$.

Proposition III.6.2 (Myhill [1955]) *A set is effectively nonrecursive if and only if it is m-complete.*

Proof. Let f be a recursive function such that $f(x) \in A \Leftrightarrow f(x) \in W_x$. If

$$W_{g(x)} = \begin{cases} \omega & \text{if } x \in \mathcal{K} \\ \emptyset & \text{otherwise} \end{cases}$$

then $x \in \mathcal{K} \Leftrightarrow f(g(x)) \in A$, because:

- $x \in \mathcal{K} \Rightarrow W_{g(x)} = \omega \Rightarrow f(g(x)) \in W_{g(x)} \Rightarrow f(g(x)) \in A$
- $f(g(x)) \in A \Rightarrow f(g(x)) \in W_{g(x)} \Rightarrow W_{g(x)} \neq \emptyset \Rightarrow x \in \mathcal{K}$.

Thus A is m-complete.

Let now $x \in \mathcal{K} \Leftrightarrow f(x) \in A$, and $z \in W_{g(x)} \Leftrightarrow f(z) \in W_x$. Then

$$f(g(x)) \in A \Leftrightarrow g(x) \in \mathcal{K} \Leftrightarrow g(x) \in W_{g(x)} \Leftrightarrow f(g(x)) \in W_x,$$

and A is effectively nonrecursive. $\quad\square$

This shows that any set built by effective diagonalization over the r.e. sets must be m-complete, and thus *Post's problem cannot be solved by effectively satisfying the requirements for nonrecursiveness*.

Exercises III.6.3 (Rogers [1967]) a) B is **effectively nonrecursive in** A if, for some recursive function f, $f(x) \in B \Leftrightarrow f(x) \in W_x^A$. *Post's problem cannot be solved by effectively satisfying the requirements for T-incompleteness,* in the sense of building an r.e. set A such that $\mathcal{K}$ is effectively nonrecursive in it. (Hint: if A and B are r.e. and B is effectively nonrecursive in A then A is recursive, because if

$$W_{g(x)}^A = \begin{cases} \omega & \text{if } x \in \overline{A} \\ \emptyset & \text{otherwise} \end{cases}$$

then $x \in \overline{A} \Leftrightarrow f(g(x)) \in B$.)
    b) B is **effectively not m-reducible** to A if, for some recursive function f,

$$\varphi_e \text{ total } \Rightarrow [f(e) \in B \Leftrightarrow \varphi_e(f(e)) \in \overline{A}],$$

i.e. $f(e)$ witnesses the fact that φ_e is not an m-reduction of B to A. *Post's problem for m-reducibility cannot be solved by effectively satisfying the requirements for m-incompleteness,* in the sense of building an r.e. set A such that $\mathcal{K}$ is effectively not

m-reducible to it. (Hint: as above, by considering as $\varphi_{g(x)}$ the constant function with value x.)

Effective nonrecursiveness gives a nice criterion for m-completeness. With little effort it is possible to generalize it, and get similar criteria for all the reducibilities so far introduced.

Exercises III.6.4 (Friedberg and Rogers [1959]) Let $\leq_r$ be defined as:

$$A \leq_r B \Leftrightarrow \text{ for some recursive function } f, \; x \in \overline{A} \Leftrightarrow Q_r(f(x), B, \overline{B})$$

where Q_r is a predicate with the following properties:

$$Q_r(z, W_x, W_y) \text{ is r.e. as a predicate of } x, y, z$$
$$Q_r(z, X, Y) \text{ is monotone in } Y.$$

Note that monotonicity of Q_r for r.e. arguments follows from the first property, and that if Q_r is also monotone in X then $A \leq_r B \Rightarrow A \leq_T B$.

a) *All the reducibilities introduced so far can be so expressed.* (Hint: $Q_{tt}(z, X, Y)$ holds if σ_z is satisfied when positive and negative atomic formulas are respectively interpreted over Y and X. $Q_p(z, X, Y)$ holds when every finite set with canonical index in D_z intersects Y. For r.e. sets, $Q_T(z, X, Y)$ holds when $\exists u(D_u \subseteq Y \wedge u \in W_z)$.)

b) *An r.e. set A is r-complete if and only if, for some recursive function g,*

$$Q_r(g(x), A, \overline{A}) \Leftrightarrow \neg Q_r(g(x), A, W_x).$$

Explicitly, the interesting criteria are (for some recursive function g):

d-completeness:	$D_{g(x)} \subseteq \overline{A}$	$\Leftrightarrow$	$D_{g(x)} \not\subseteq W_x$	
c-completeness:	$D_{g(x)} \cap \overline{A} \neq \emptyset$	$\Leftrightarrow$	$D_{g(x)} \subseteq \overline{W_x}$	
Q-completeness:	$W_{g(x)} \cap \overline{A} \neq \emptyset$	$\Leftrightarrow$	$W_{g(x)} \subseteq \overline{W_x}.$	

Creative sets

We now constructivize the second notion of nonrecursiveness given at the beginning.

Definition III.6.5 (Post [1944], Dekker [1953]) *A set A is* **productive** *if there is a recursive function f such that, for every x,*

$$W_x \subseteq A \Rightarrow f(x) \in A - W_x.$$

A set is **creative** *if it is r.e. and coproductive.*

As we have already noted, $\mathcal{K}$ *is creative*: if $W_x \subseteq \overline{\mathcal{K}}$ then x cannot be in W_x (otherwise $x \in \mathcal{K}$ by definition, and $x \in \overline{\mathcal{K}}$ from $W_x \subseteq \overline{\mathcal{K}}$, contradiction). Then $x \notin W_x$, and thus also $x \in \overline{\mathcal{K}}$.

The term 'creative' was introduced by Post [1944] because $\mathcal{K}$ (as well as $\mathcal{K}_0$, defined in II.2.7) embodied the essence of the incompleteness theorems of the Thirties (II.2.17), and the property of productiveness captured the main consequence of these results, namely that

> every symbolic logic is incomplete and extendible relative to the class of propositions constituting $\mathcal{K}_0$. The conclusion is inescapable that even for such a fixed, well defined body of mathematical propositions, *mathematical thinking is, and must remain, essentially creative.*

That Post got the right notion is proved by the next result: if $\mathcal{F}$ is any consistent extension of $\mathcal{R}$, then every r.e. set is weakly representable in it (see II.2.16), and hence m-reducible to the set of (codes of) its theorems, which is then m-complete.

Theorem III.6.6 (Myhill [1955]) *A set is creative if and only if it is m-complete.*

Proof. If A is m-complete it is effectively nonrecursive, and hence creative. Directly, if $x \in \mathcal{K} \Leftrightarrow g(x) \in A$, let $z \in W_{h(x)} \Leftrightarrow g(z) \in W_x$. Then

$$W_x \subseteq \overline{A} \implies W_{h(x)} \subseteq \overline{\mathcal{K}} \implies h(x) \in \overline{\mathcal{K}} - W_{h(x)} \implies g(h(x)) \in \overline{A} - W_x,$$

and $f(x) = g(h(x))$ is a productive function for $\overline{A}$.

Suppose now that A is creative, and $W_x \subseteq \overline{A} \Rightarrow f(x) \in \overline{A} - W_x$. We want to find a recursive function h such that $z \in \mathcal{K} \Leftrightarrow h(z) \in A$. Since we want to use f, we define g such that $z \in \mathcal{K} \Leftrightarrow f(g(z)) \in A$.

- If $z \in \overline{\mathcal{K}}$, f gives naturally an element of $\overline{A}$, if we start from $W_{g(z)} \subseteq \overline{A}$. The simplest way to ensure this is to let $W_{g(z)} = \emptyset$ when $z \in \overline{\mathcal{K}}$: then we have $f(g(z)) \in \overline{A}$.

- We try now the converse, i.e. to have $f(g(z)) \in \overline{A} \Rightarrow z \in \overline{\mathcal{K}}$. If $f(g(z)) \in \overline{A}$ then $\{f(g(z))\} \subseteq \overline{A}$, and by productivity it cannot be $W_{g(z)} = \{f(g(z))\}$, otherwise $f(g(z)) \in \overline{A} - W_{g(z)}$. We thus define $W_{g(z)} = \{f(g(z))\}$ when $z \in \mathcal{K}$: then $f(g(z)) \in A$.

Let then g be a recursive function such that

$$W_{g(z)} = \begin{cases} \{f(g(z))\} & \text{if } z \in \mathcal{K} \\ \emptyset & \text{otherwise.} \end{cases}$$

The existence of g is ensured by the Fixed-Point Theorem, e.g. let

$$W_{\varphi_{t(e)}(z)} = \begin{cases} \{f(\varphi_e(z))\} & \text{if } z \in \mathcal{K} \\ \emptyset & \text{otherwise,} \end{cases}$$

and choose e such that $\varphi_e \simeq \varphi_{t(e)}$. $\square$

Exercises III.6.7 a) *An r.e. set A is creative if and only if, for some recursive one-one function f, $f(x) \in A \Leftrightarrow f(x) \in W_x$.* (Rogers [1967], Gill and Morris [1974]) (Hint: if A is creative then $\overline{A}$ is completely productive and, by III.7.6, it has a one-one completely productive function f: then A can be represented as stated. Conversely, if f is one-one then there is a set A satisfying the given conditions, since membership of $f(x)$ in A is determined solely by W_x. Note that, if we had $f(x) = f(y)$ with $x \neq y$, then W_x and W_y could give contradictory answers for membership of $f(x)$ in A.)

  b) *An r.e. set A is creative if and only if, for some acceptable system of indices $\{\hat{W}_x\}_{x \in \omega}$ of the r.e. sets, $A = \hat{\mathcal{K}}$.* (Rogers [1958], Lynch [1974]) (Hint: use III.7.14 to get a recursive isomorphism f between A and $\mathcal{K}$, and let $x \in \hat{W}_e \Leftrightarrow f(x) \in W_{f(e)}$.)

  c) *A creative set is neither recursive nor simple.* (Hint: see III.2.7.)

  d) *Every infinite r.e. set is the disjoint union of a creative and a productive set.* (Dekker [1955a]) (Hint: if A is the range of a one-one recursive function f, consider the images of $\mathcal{K}$ and $\overline{\mathcal{K}}$.)

  e) *Every infinite r.e. set is the union of a creative and an infinite recursive set.* (Myhill [1959]) (Hint: let A be r.e. and infinite, and f be a recursive one-one function whose range is an infinite recursive subset B of A. $f(\mathcal{K})$ is creative, and then so is $f(\mathcal{K}) \cup (A - B)$, and $A = f(\mathcal{K}) \cup (A - B) \cup B$.)

The next result is an analogue of III.1.5, and provides a criterion for m-completeness based on fixed-points. We use the following generalization of m-reducibility to functions: $f \leq_m A$ if and only if there are recursive functions f_1, f_2, and g such that

$$f(x) = \begin{cases} f_1(x) & \text{if } g(x) \in A \\ f_2(x) & \text{otherwise.} \end{cases}$$

Clearly, $B \leq_m A$ if and only if $c_B \leq_m A$.

Proposition III.6.8 (Arslanov) *An r.e. set A is m-complete if and only if there is a function $f \leq_m A$ without fixed-points.*

Proof. If A is m-complete, the set $\{x : 0 \in W_x\}$ is m-reducible to A, being r.e. Let g be a recursive function such that

$$0 \in W_x \Leftrightarrow g(x) \in A,$$

and define $f \leq_m A$ as:

$$f(x) = \begin{cases} a & \text{if } g(x) \in A \\ b & \text{otherwise,} \end{cases}$$

where a and b are indices of, respectively, $\emptyset$ and ω. Then $\mathcal{W}_{f(x)} \neq \mathcal{W}_x$, because the two sets differ on 0.

Suppose now that A is r.e., and $f \leq_m A$ has no fixed-points. We show that A is m-complete by proving that it is creative. Given $\mathcal{W}_x$, let

$$\varphi(z) \simeq \begin{cases} f_1(x) & \text{if } g(z) \text{ shows up in } A \text{ before than in } \mathcal{W}_x \\ f_2(x) & \text{if } g(z) \text{ shows up in } \mathcal{W}_x \text{ before than in } A \\ \text{undefined} & \text{otherwise.} \end{cases}$$

Note that if $g(z) \in A \cup \mathcal{W}_x$ then $\varphi(z) \downarrow$, and if $\mathcal{W}_x \subseteq \overline{A}$ then $\varphi(z) = f(z)$, if convergent. Consider a recursive function h such that $\mathcal{W}_{h(z)} = \mathcal{W}_{\varphi(z)}$, i.e. such that $\mathcal{W}_{h(z)} = \emptyset$ if $\varphi(z) \uparrow$: if $\mathcal{W}_x \subseteq \overline{A}$, which is the only case of interest, then $\mathcal{W}_{h(x)} = \mathcal{W}_{f(x)}$ if $\varphi(x)$ converges. Let e be a fixed-point for h: $\mathcal{W}_{h(e)} = \mathcal{W}_e$. Since f has no fixed-points $\varphi(e)$ must diverge, and then it must be $g(e) \notin A \cup \mathcal{W}_x$. Since the procedure is uniform in x (by using the Fixed-Point Theorem with parameters II.2.11.a), we have a recursive function that produces an element of $\overline{A} - \mathcal{W}_x$ whenever $\mathcal{W}_x \subseteq \overline{A}$, and A is creative. $\square$

Exercises III.6.9 Weakenings of creativeness. a) *An r.e. set A is creative if and only if there is a partial recursive function φ such that*

$$\mathcal{W}_x \text{ empty or singleton} \wedge \mathcal{W}_x \subseteq \overline{A} \Rightarrow \varphi(x) \downarrow \wedge \varphi(x) \in \overline{A} - \mathcal{W}_x.$$

(Dekker [1955a], Myhill [1955]) (Hint: see the proof of III.6.6.)
   b) *For any r.e. nonrecursive set A there is a partial recursive function φ such that*

$$R_x \subseteq \overline{A} \Rightarrow \varphi(x) \downarrow \wedge \varphi(e) \in \overline{A} - R_x.$$

(Mitchell [1966]) This cannot be interpreted as saying that every nonrecursive r.e. set is effectively nonrecursive, because the system of indices $\{R_x\}_{x \in \omega}$ for the recursive sets is not effective itself. (Hint: if $R_x = \mathcal{W}_a = \overline{\mathcal{W}_b}$ let $\varphi(x)$ be the first element generated in $A \cap \mathcal{W}_b$. Since A is nonrecursive, if $R_x = \mathcal{W}_a \subseteq \overline{A}$ then $A \cap \mathcal{W}_b \neq \emptyset$.)
   c) *An r.e. set A is T-complete if and only if $\overline{A}$ has a productive function recursive in A.* (Lachlan [1968]) (Hint: if A is T-complete, is recursive in A to ask if $\mathcal{W}_x \subseteq \overline{A}$, i.e. if $\exists z (z \in \mathcal{W}_x \cap A)$. If $\mathcal{W}_x \subseteq \overline{A}$, $\mathcal{W}_x$ is properly included in $\overline{A}$, since A is not recursive, and an element in $\overline{A} - \mathcal{W}_x$ can be found, recursively in A. Conversely, if $f \leq_T A$ is a productive function for $\overline{A}$, the function $g \leq_T A$ such that

$$\mathcal{W}_{g(x)} = \begin{cases} \{f(x)\} & \text{if } f(x) \in \overline{A} \\ \emptyset & \text{otherwise} \end{cases}$$

has no fixed-points, and A is T-complete by III.1.5.)

Exercises III.6.10 Productive sets. a) *A set A is productive if and only if there is a partial recursive function φ such that*

$$\mathcal{W}_x \subseteq A \Rightarrow \varphi(x) \downarrow \wedge \varphi(x) \in A - \mathcal{W}_x.$$

(Dekker and Myhill [1958]) (Hint: given φ and

$$W_{h(x)} = \begin{cases} W_x & \text{if } \varphi(x)\downarrow \\ \emptyset & \text{otherwise,} \end{cases}$$

then one of $\varphi(x)$ and $\varphi(h(x))$ converges.)

b) *A set A is productive if and only if $\overline{K} \leq_m A$. Thus the T-degrees containing productive sets are exactly those above $\mathbf{0'}$.* (Dekker and Myhill [1958]) (Hint: see the proof of III.6.6.)

c) *A set is productive if and only if it is completely productive.* (Myhill [1955])

d) *There are $2^{\aleph_0}$ productive and co-productive sets, and the T-degrees containing such sets are exactly those above $\mathbf{0'}$.* (Karp) (Hint: if A is productive, $A \oplus \overline{A}$ is productive and co-productive.)

e) *The set $\{x : W_x \subseteq \overline{K}\}$ is properly contained in $\overline{K}$. Thus even transfinite iterations of the process starting from $\emptyset$, and generating new elements of $\overline{K}$ (as in III.2.7), does not exhaust $\overline{K}$.* (Dekker [1955a]) (Hint: the Fixed-Point Theorem produces x such that $W_x = \overline{\{x\}}$: then $x \in \overline{K}$, but $W_x \not\subseteq \overline{K}$.)

f) *There is a set A productive w.r.t. f, such that $A = \{f(x) : W_x \subseteq A\}$.* (Dekker [1955a]) (Hint: let A be the intersection of all sets for which f is a productive function.)

We still have to take care of the last notion considered at the beginning of this section.

Definition III.6.11 (Dekker [1955a]) *A set A is* **contraproductive** *if there is a recursive function f such that, for every x,*

$$A \subseteq W_x \Rightarrow f(x) \in \overline{A} \cap W_x.$$

Proposition III.6.12 (Muchnik [1958a], McLaughlin [1962]) *An r.e. set A is creative if and only if $\overline{A}$ is contraproductive.*

Proof. A creative set is effectively nonrecursive, and then $\overline{A}$ is contraproductive. For the converse, suppose $\overline{A} \subseteq W_x \Rightarrow f(x) \in A \cap W_x$, and let

$$W_{g(x)} = \begin{cases} \omega & \text{if } x \in K \\ \omega - \{f(g(x))\} & \text{otherwise.} \end{cases}$$

Then $x \in K \Leftrightarrow f(g(x)) \in A$, because:

- $x \in K \Rightarrow W_{g(x)} = \omega \supseteq \overline{A} \Rightarrow f(g(x)) \in A$

- $f(g(x)) \in A \Rightarrow W_{g(x)} \supseteq \overline{A} \Rightarrow f(g(x)) \in W_{g(x)} \Rightarrow x \in K$

Thus A is m-complete, and hence creative. $\square$

Exercises III.6.13 a) *A set is contraproductive if and only if it is productive.*
b) *If A is an r.e. nonrecursive set, there is φ partial recursive such that*

$$\overline{A} \subseteq W_x \;\Rightarrow\; \varphi(x)\!\downarrow \wedge \varphi(x) \in A \cap W_x.$$

(Dekker [1955a])

For more information on productiveness and creativeness see Dekker [1955a], Friedberg and Rogers [1959], McLaughlin [1964], Lachlan [1965], Mitchell [1966], Soloviev [1976], and Omanadze [1978].

Quasicreative sets ⋆

Exercise III.6.4 considers reducibilities $\leq_r$ generated by relations Q_r, and generalizes the criterion of m-completeness given by effective nonrecursiveness, by considering r.e. sets A such that, for some recursive function f,

$$Q_r(f(x), A, \overline{A}) \;\Leftrightarrow\; \neg Q_r(f(x), A, W_x).$$

We might then try to extend the criterion of m-completeness given by creativeness, by considering r.e. sets A such that, for some recursive function f,

$$W_x \subseteq \overline{A} \;\Rightarrow\; Q_r(f(x), A, \overline{A}) \wedge \neg Q_r(f(x), A, W_x).$$

Exercise III.6.14 *The criterion fails for c-reducibility.* (Omanadze [1978a]) (Hint: let A be a strongly effectively simple, not hypersimple, tt-incomplete set. It exists because there is an acceptable system of indices for which Post's simple set is tt-incomplete, see III.9.2, and all the acceptable systems are isomorphic, see II.5.7. Such a set is thus c-incomplete. Let h be a strong array intersecting $\overline{A}$, and g be such that $W_x \subseteq \overline{A} \Rightarrow |W_x| < g(x)$. To get f such that

$$W_x \subseteq \overline{A} \Rightarrow D_{f(x)} \not\subseteq A \wedge D_{f(x)} \subseteq \overline{W}_x,$$

let $D_{f(x)} = (\bigcup_{z \leq g(x)+1} D_{h(z)}) - \{0, \ldots, g(x)\}$.)

Thus this criterion does not always work, and we do not know of any other formulation as general as III.6.4 (see Lachlan [1965], Soloviev [1976], Omanadze [1976] for weaker results). There are however two special interesting cases, which we treat in this and the following subsection, in which the proposed formulation succeeds. We begin with $\leq_d$ which, we recall, is defined by $Q_d(z, X, Y) \Leftrightarrow D_z \subseteq Y$. The proposed criterion for d-completeness is thus the following:

Definition III.6.15 (Shoenfield [1957]) *A is **quasicreative** if it is r.e. and, for some recursive function f,*

$$W_x \subseteq \overline{A} \;\Rightarrow\; D_{f(x)} \subseteq \overline{A} \wedge D_{f(x)} \not\subseteq W_x.$$

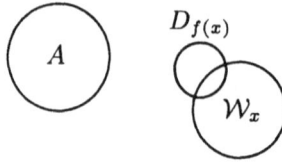

Obviously, *a creative set is quasicreative*: if h is a productive function for $\overline{A}$, and $D_{f(x)} = \{h(x)\}$, then

$$W_x \subseteq \overline{A} \Rightarrow h(x) \in \overline{A} - W_x \Rightarrow D_{f(x)} \subseteq \overline{A} \wedge D_{f(x)} \not\subseteq W_x.$$

Proposition III.6.16 (Shoenfield [1957]) *A set is quasicreative if and only if it is d-complete.*

Proof. That a d-complete set is quasicreative follows from the stronger criterion for d-completeness given in III.6.4. Or directly, as in the case of m-completeness, if $x \in \mathcal{K} \Leftrightarrow D_{g(x)} \cap A \neq \emptyset$ let $z \in W_{h(x)} \Leftrightarrow D_{g(z)} \subseteq W_x$. Then $f(x) = g(h(x))$ witnesses the quasicreativeness of A, since

$$W_x \subseteq \overline{A} \Rightarrow W_{h(x)} \subseteq \overline{\mathcal{K}} \Rightarrow h(x) \in \overline{\mathcal{K}} - W_{h(x)}$$
$$\Rightarrow D_{g(h(x))} \subseteq \overline{A} \wedge D_{g(h(x))} \not\subseteq W_x.$$

Conversely, as in III.6.6, if f witnesses the quasicreativeness of A and g is a recursive function (which exists by the Fixed-Point Theorem) such that

$$W_{g(z)} = \begin{cases} D_{f(g(z))} & \text{if } z \in \mathcal{K} \\ \emptyset & \text{otherwise,} \end{cases}$$

then $z \in \mathcal{K} \Leftrightarrow D_{f(g(z))} \cap A \neq \emptyset$, and A is d-complete. $\square$

Proposition III.6.17 (Shoenfield [1957]) *There exists a quasicreative, not creative set.*

Proof. The idea is to build a quasicreative set A with a simple superset B, such that $B - A$ is r.e. Then A is not creative, otherwise the usual procedure (see the proof of III.2.7) would give an infinite r.e. subset of $\overline{B}$, starting with $B - A$.

Fix a strong array $\{F_x\}_{x \in \omega}$ intersecting Post's simple set (see III.3.5). We proceed as in the construction of a simple tt-complete, with the role of the set $\mathcal{K} = \{x : x \in W_x\}$ taken by $\{x : F_x \subseteq W_x\}$. If

$$B = S \cup \bigcup_{F_x \subseteq W_x} F_x$$

then B is simple, if coinfinite. Moreover

$$F_x \subseteq B \Leftrightarrow F_x \subseteq W_x.$$

To have A quasicreative, we can ensure the stronger condition

$$F_x \subseteq \overline{A} \Leftrightarrow F_x \not\subseteq W_x,$$

equivalent to

$$F_x \cap A \neq \emptyset \Leftrightarrow F_x \subseteq W_x \Leftrightarrow F_x \subseteq B.$$

To build A generate B and, when $F_x \subseteq B$ is found, put into A the element of F_x which has been generated last in B. This makes $A \subseteq B$, and both A and $B - A$ r.e.

Since A is quasicreative $\overline{A}$ is not r.e., and so $\overline{B}$ must be infinite (otherwise $\overline{A} = (B - A) \cup \overline{B}$ would be r.e.). $\square$

A different proof, relying on completeness (i.e. building a d-complete set which is not Q-complete, and thus not m-complete) will be given on p. 342.

Exercises III.6.18 a) *A quasicreative set is not simple.* (Shoenfield [1957]) (Hint: let $x \in \overline{\mathcal{K}} \Leftrightarrow D_{f(x)} \subseteq \overline{A}$, for some recursive function f. Given distinct elements $x_1, \ldots, x_n$ of $\overline{A}$, let $x \in C \Leftrightarrow D_{f(x)} \subseteq \{x_1, \ldots, x_n\}$. If $C = W_a$ then $D_{f(a)} \subseteq \overline{A}$ but $D_{f(a)} \not\subseteq \{x_1, \ldots, x_n\}$. This builds an infinite r.e. subset of $\overline{A}$.)

b) *An r.e. set A is c-complete if and only if there is a recursive function f such that*

$$\overline{A} \subseteq W_x \Rightarrow D_{f(x)} \subseteq A \wedge D_{f(x)} \not\subseteq \overline{W}_x.$$

This provides an analogue of contraproductiveness. (Lachlan [1965]) (Hint: see III.6.12.)

c) An r.e. set A is **semicreative** if, for some recursive function f,

$$W_x \subseteq \overline{A} \Rightarrow W_{f(x)} \subseteq \overline{A} \wedge W_{f(x)} \not\subseteq W_x.$$

(Dekker [1955a]). This provides an analogue of quasicreativeness. *Every nonrecursive r.e. T-degree contains a semicreative set.* (Yates [1965]) (Hint: use permitting, see p. 277. The most natural example of semicreative set is $\mathcal{K}$, but it is not useful for permitting, because each W_x contributes at most one element. As a variation consider $\langle x, e \rangle \in A \Leftrightarrow \langle x, e \rangle \in W_e$, and $W_{f(e)} = \{\langle x, e \rangle : x \in \omega\}$. A is semicreative because, if $W_e \subseteq \overline{A}$, then $\langle x, e \rangle \in \overline{A} - W_e$, and hence $W_{f(e)} \subseteq \overline{A}$ and $W_{f(e)} \not\subseteq W_e$. Given an r.e. nonrecursive set C, modify the construction of A by adding permitting, in such a way to obtain $A \oplus C$ semicreative.)

d) *A semicreative set is not simple.*

Subcreative sets ⋆

We examine now the completeness criterion proposed in the last subsection, in the case of Q-reducibility. Recall that $\leq_Q$ is defined by

$$Q_Q(z, X, Y) \Leftrightarrow \exists u (u \in Y \wedge u \in W_z).$$

The criterion thus takes the following form:

Definition III.6.19 (Blum and Marques [1973]) *A is* **subcreative** *if it is r.e. and, for some recursive function f,*

$$W_x \subseteq \overline{A} \Rightarrow W_{f(x)} \not\subseteq A \wedge W_{f(x)} \subseteq \overline{W_x}.$$

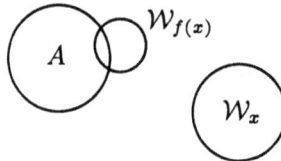

Equivalently, by taking unions with A, we might require the existence of a recursive function g such that $W_x \subseteq \overline{A} \Rightarrow A \subset W_{g(x)} \subseteq \overline{W_x}$.

Obviously, *a creative set is subcreative*: if h is the productive function for A, and $W_{f(x)} = \{h(x)\}$, then

$$W_x \subseteq \overline{A} \Rightarrow h(x) \in \overline{A} - W_x \Rightarrow W_{f(x)} \not\subseteq A \wedge W_{f(x)} \subseteq \overline{W_x}.$$

Proposition III.6.20 (Blum and Marques [1973], Gill and Morris [1974]) *A set is subcreative if and only if it is Q-complete.*

Proof. That a Q-complete set is subcreative follows from the stronger criterion for Q-completeness given in III.6.4. Or directly, as in the case of m-completeness, let $x \in \mathcal{K} \Leftrightarrow W_{g(x)} \subseteq A$, and

$$z \in W_{h(x)} \Leftrightarrow W_{g(z)} \cap W_x \neq \emptyset.$$

Then $f(x) = g(h(x))$ witnesses the quasicreativeness of A, since

$$W_x \subseteq \overline{A} \Rightarrow W_{h(x)} \subseteq \overline{\mathcal{K}} \Rightarrow h(x) \in \overline{\mathcal{K}} - W_{h(x)}$$
$$\Rightarrow W_{g(h(x))} \not\subseteq A \wedge W_{g(h(x))} \subseteq \overline{W_x}.$$

Suppose now that A is subcreative:

$$W_x \subseteq \overline{A} \Rightarrow W_{f(x)} \not\subseteq A \wedge W_{f(x)} \subseteq \overline{W_x}.$$

We want to find a recursive function h such that $z \in \mathcal{K} \Leftrightarrow W_{h(z)} \subseteq A$. Since we want to use f, we define g such that $z \in \mathcal{K} \Leftrightarrow W_{f(g(z))} \subseteq A$.

- If $z \in \overline{\mathcal{K}}$, f gives naturally a set not contained in A, if we only start from $W_{g(z)} \subseteq \overline{A}$. The simplest way to ensure this is to let $W_{g(z)} = \emptyset$ when $z \in \overline{\mathcal{K}}$: then $W_{f(g(z))} \not\subseteq A$.

- We try now the converse, i.e. we want $W_{f(g(z))} \not\subseteq A \Rightarrow z \in \overline{\mathcal{K}}$. If $W_{f(g(z))} \not\subseteq A$, there is an element $a \in W_{f(g(z))} \cap \overline{A}$. By subcreativity it cannot be $W_{g(z)} = \{a\}$, otherwise $W_{g(z)} \subseteq \overline{A}$ and $W_{f(g(z))} \subseteq \overline{W}_{g(z)}$, while $a \in W_{f(g(z))}$. We thus define, when $z \in \mathcal{K}$, $W_{g(z)}$ as a set that chooses an element of $W_{f(g(z))} \cap \overline{A}$ if there is one, and is empty otherwise. Then $W_{f(g(z))} \cap \overline{A} \neq \emptyset \Rightarrow z \in \overline{\mathcal{K}}$.

We now have to define g. The fact that g is self-referential is taken care of by the Fixed-Point Theorem, but the naive approach leads to another problem. If we try the natural procedure to pick up an element from $W_{f(g(z))} \cap \overline{A}$, we simultaneously generate $W_{f(g(z))}$ and A. At each stage of the enumeration, we put the elements already generated in $W_{f(g(z))}$ into $W_{g(z)}$, unless some element has already been generated in $W_{g(z)}$ but not yet in A. This certainly puts an element of $\overline{A}$ in $W_{g(z)}$ if there is one in $W_{f(g(z))}$, but may also put other elements of A (ones which are in A, but are generated in it only after having been generated in $W_{f(g(z))}$). The effect is that $W_{g(z)}$ is not contained in $\overline{A}$, and the discussion above fails.

To override this we would like to make sure that, if $W_{f(g(z))} \cap \overline{A} \neq \emptyset$, one element of the intersection is generated in $W_{f(g(z))}$ as the first element. Think of the process of generating an r.e. set, and to pass from it to a new r.e. set which consists of the elements of the given set generated until the first stage in which it is found that the first element of the enumeration is in A. If the given set is infinite, it can be a fixed-point of this process only if the first element enumerated in it is in $\overline{A}$. Note that $W_{f(x)}$ may always be supposed infinite (otherwise consider $A \cup W_{f(x)}$).

To define $W_{g(z)}$, first wait until it is discovered that $z \in \mathcal{K}$. Then let $W_{f(g(z))}$ be a fixed-point of the process described above, and let $W_{g(z)}$ consist of the first element generated in it (note that the definition of g actually requires a *double use of the Fixed-Point Theorem*). Then:

- $z \in \overline{\mathcal{K}} \Rightarrow W_{g(z)} = \emptyset \subseteq \overline{A} \Rightarrow W_{f(g(z))} \not\subseteq A$.

- $W_{f(g(z))} \not\subseteq A \Rightarrow z \in \overline{\mathcal{K}}$.
  Otherwise $W_{g(z)} = \{a\}$, for some $a \in W_{f(g(z))} \cap \overline{A}$ (actually, for the first element generated in $W_{f(g(z))}$). Then

$$W_{g(z)} \subseteq \overline{A} \Rightarrow W_{f(g(z))} \subseteq \overline{W}_{g(z)},$$

contradiction. $\square$

Exercises III.6.21 Strongly effectively simple sets. a) *Every strongly effectively simple set is Q-complete.* (Gill and Morris [1974]) (Hint: if g is a recursive function such that $W_e \subseteq \overline{A} \Rightarrow (\max W_e) < g(e)$, let $W_{f(e)}$ consist of all the elements greater than $g(e)$. Then f witnesses the subcreativeness of A, because $\overline{A}$ is infinite.)

b) *A strongly effectively simple set is neither hyperhypersimple, nor contained in maximal sets.* (Cohen and Jockusch [1975]) (Hint: by III.4.10, and the fact that coinfinite r.e. supersets of strongly effectively hyperhypersimple sets are such.)

Effectively inseparable pairs of r.e. sets

The fact that a set A is recursive if and only if both A and $\overline{A}$ are r.e. suggests the possibility of extending the theory of r.e. sets to pairs of disjoint r.e. sets. The first step was taken in II.2.4, with the definition of the notion of recursive inseparability as an analogue of nonrecursiveness. The existence of recursively inseparable pairs of r.e. sets was proved in II.2.5, and we now strengthen that result.

Proposition III.6.22 (Shoenfield [1958]) *Every nonrecursive r.e. T-degree contains a recursively inseparable pair of r.e. sets.*

Proof. Let C be a nonrecursive r.e. set. We modify the first proof of II.2.5, and define

$$x \in A \quad \Leftrightarrow \quad (x)_1 \in C \wedge \varphi_{(x)_2}(x) \simeq 0 \text{ before } (x)_1 \in C$$
$$x \in B \quad \Leftrightarrow \quad (x)_1 \in C \wedge \varphi_{(x)_2}(x) \simeq 1 \text{ before } (x)_1 \in C.$$

Then A and B are a disjoint pair of r.e. sets. Moreover:

- $A \leq_T C$ and $B \leq_T C$
  To see if $x \in A$ first see, recursively in C, if $(x)_1 \in C$. If not, then $x \notin A$. If so, simultaneously generate C and compute $\varphi_{(x)_2}(x)$. Then $x \in A$ if and only if the computation of $\varphi_{(x)_2}(x)$ converges to 0 before than $(x)_1$ appears in C. Thus $A \leq_T C$, and $B \leq_T C$ similarly.

- $C \leq_T A$ and $C \leq_T B$
  To see if $z \in C$, let $x = \langle z, a \rangle$, where a is an index of the constant function 0. See, recursively in A, if $x \in A$. If so, then $z \in C$. If not, simultaneously generate C and compute $\varphi_a(x)$. Then $z \in C$ if and only if it has been generated in it by the time $\varphi_a(x)$ converges. Thus $C \leq_T A$, and $C \leq_T B$ similarly.

- A and B are recursively inseparable
  Suppose D is a recursive set such that $A \subseteq D$ and $B \subseteq \overline{D}$, and let

$$\varphi_e(x) = \begin{cases} 1 & \text{if } x \in D \\ 0 & \text{otherwise.} \end{cases}$$

Then

$$z \in C \Leftrightarrow z \text{ is generated in } C \text{ before } \varphi_e(\langle z, e \rangle) \text{ converges.}$$

Indeed, if $z \in C$ and $\varphi_e(\langle z, e \rangle)$ converges before z is generated in C, then

$$\varphi_e(\langle z, e \rangle) \simeq 0 \Rightarrow \langle z, e \rangle \in A \Rightarrow \langle z, e \rangle \in D \Rightarrow \varphi_e(\langle z, e \rangle) \simeq 1$$
$$\varphi_e(\langle z, e \rangle) \simeq 1 \Rightarrow \langle z, e \rangle \in B \Rightarrow \langle z, e \rangle \in \overline{D} \Rightarrow \varphi_e(\langle z, e \rangle) \simeq 0.$$

But then C is recursive, contradiction. $\square$

Exercises III.6.23 a) $\mathcal{K}$ *is part of a recursively inseparable pair of r.e. sets.* (Hint: let A and B be a recursively inseparable pair of r.e. sets. Then $A \leq_m \mathcal{K}$, and there is a one-one recursive function f such that $x \in A \Leftrightarrow f(x) \in \mathcal{K}$, namely the function given by III.1.2, being obtained by the S_n^m-Theorem. Then $\mathcal{K}$ and $f(B)$ are recursively inseparable.)

b) *If $B \leq_m A$ and B is part of a recursively inseparable pair of r.e. sets, then A is not simple.* This generalizes the fact that m-complete sets are not simple. (Hint: if f reduces B to A, and B and C are recursively inseparable, then $D = f(C)$ is infinite, otherwise $f^{-1}(D)$ is a recursive set separating B and C.)

c) *If $B \leq_{tt} A$ and B is part of a recursively inseparable pair of r.e. sets, then B is not hypersimple.* This generalizes the fact that tt-complete sets are not hypersimple. (Denisov [1974]) (Hint: if $x \in B \Leftrightarrow A \models \sigma_{f(x)}$, and B and C are recursively inseparable, then

$$x \in B \wedge y \in C \Rightarrow A \models \sigma_{f(x)} \wedge \neg(A \models \sigma_{f(y)}).$$

With notations as in III.3.10, if

$$x \in B \wedge y \in C \wedge (A^* \models \sigma_{f(x)} \Leftrightarrow A^* \models \sigma_{f(y)})$$

then A and A^* differ on some element used in $\sigma_{f(x)}$ or $\sigma_{f(y)}$. There must be x and y such that

$$x \in B \wedge y \in C \wedge (\forall i < 2^n)(A_i^* \models \sigma_{f(x)} \Leftrightarrow A_i^* \models \sigma_{f(y)}),$$

otherwise we could recursively separate B and C, because

$$x \eta y \Leftrightarrow (\forall i < 2^n)(A_i^* \models \sigma_{f(x)} \Leftrightarrow A_i^* \models \sigma_{f(y)})$$

is a recursive equivalence relation with only finitely many classes, and we could take the union of the equivalence classes containing elements of B.)

d) *Not every nonrecursive r.e. tt-degree contains a recursively inseparable pair of r.e. sets.* (Hint: from c) above.)

Effective nonrecursiveness can be generalized as follows:

Definition III.6.24 (Kleene [1950], Uspenskii [1953]) *A and B are effectively inseparable if they are disjoint r.e. sets and, for some recursive function* f,

$$A \subseteq W_x \wedge B \subseteq W_y \wedge W_x \cap W_y = \emptyset \Rightarrow f(x,y) \in \overline{W_x \cup W_y}.$$

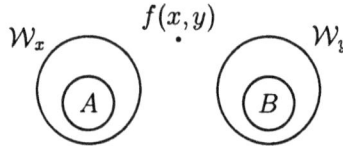

The recursively inseparable pairs constructed in II.2.5 are both effectively inseparable. For example, if

$$x \in A \Leftrightarrow \varphi_x(x) \simeq 0 \text{ and } x \in B \Leftrightarrow \varphi_x(x) \simeq 1,$$

let $f(x,y)$ be an index of the partial recursive function which gives z value 1 if the stage in which z appears in W_x is not greater than the stage in which z appears in W_y, and 0 in the opposite case. Then, if $A \subseteq W_x$ and $B \subseteq W_y$ and W_x, W_y are disjoint, $f(x,y) \in \overline{W_x \cup W_y}$, because e.g.

$$f(x,y) \in W_x \Rightarrow \varphi_{f(x,y)}(f(x,y)) \simeq 1 \Rightarrow f(x,y) \in B \Rightarrow f(x,y) \in W_y,$$

contradiction.

Proposition III.6.25 *If A and B are effectively inseparable, any r.e. superset of one disjoint from the other is creative. In particular, so are A and B.*

Proof. Let $A \subseteq C$ and $B \subseteq \overline{C}$, $W_{g(x)} = W_x \cup B$ and $C = W_a$. If f witnesses the effective inseparability of A and B, then

$$W_x \subseteq \overline{C} \Rightarrow A \subseteq W_a \wedge B \subseteq W_{g(x)} \Rightarrow f(a, g(x)) \in \overline{C} - W_x. \quad \square$$

Exercises III.6.26 a) *There are recursively inseparable, not effectively inseparable pairs of r.e. sets.* (Muchnik [1956a], Shoenfield [1957], Tennenbaum [1961a]) (Hint: from III.6.22 and III.5.20, since effectively inseparable pairs must be T-complete. Or directly, as in the construction of Post's simple set, by building A and B r.e. and disjoint, intersecting each infinite r.e. set, and with $\overline{A \cup B}$ infinite. To achieve the last condition, consider only elements of W_e greater that $3e$.)
b) *The sets A and B − A in the proof of III.6.17 are recursively inseparable, but not effectively inseparable.* (Shoenfield [1957]) (Hint: they cannot be effectively inseparable, since A is not creative. Suppose $A \subseteq C$ and $B - A \subseteq \overline{C}$, with C recursive. Since A is quasicreative we can build, by iteration, an r.e. set $D \supseteq \overline{C}$ such that $D \cap C$ is infinite, and then B is not simple, contradiction.)

c) *There are two disjoint creative sets which are recursively inseparable, but not effectively inseparable.* (McLaughlin [1962a]) (Hint: use the fact, proved in IX.2.4, that every infinite nonrecursive r.e. set is the disjoint union of two infinite nonrecursive r.e. sets. Split this way a maximal set, and consider each component as the union of an infinite recursive set and of a creative set, see III.6.7.e. The two creative sets thus obtained are as wanted.)

d) *There are two disjoint creative sets which are recursively separable.* (Hint: let $2x \in A \Leftrightarrow x \in \mathcal{K}$, and $2x + 1 \in B \Leftrightarrow x \in \mathcal{K}$. They are creative sets, separated by the even numbers.)

Exercises III.6.27 Conditions equivalent to effective inseparability. Let A and B be disjoint r.e. sets. *The following conditions are equivalent to effective inseparability of A and B.*

a) *For some recursive function f,*

$$A \subseteq \mathcal{W}_x \wedge B \subseteq \mathcal{W}_y \wedge \mathcal{W}_x \cup \mathcal{W}_{f(y)} = \omega \Rightarrow f(x,y) \in \mathcal{W}_x \cap \mathcal{W}_y.$$

This is the dual of the notion of recursive inseparability. (Muchnik [1958a])

b) *For some recursive function f,*

$$\mathcal{W}_x \subseteq \overline{A} \wedge \mathcal{W}_y \subseteq \overline{B} \wedge \mathcal{W}_x \cap \mathcal{W}_y = \emptyset \Rightarrow f(x,y) \in \overline{A \cup B} - (\mathcal{W}_x \cup \mathcal{W}_y).$$

This is the analogue of creativeness. (Muchnik [1958a], Smullyan [1961]) (Hint: the equivalence is trivial, and it does not use the Fixed-Point Theorem.)

c) *For some recursive function f,*

$$\overline{A} \subseteq \mathcal{W}_x \wedge \overline{B} \subseteq \mathcal{W}_y \Rightarrow f(x,y) \in (A \cap \mathcal{W}_x) \cup (B \cup \mathcal{W}_y).$$

This is the analogue of contraproductiveness.

d) *For every pair $(\mathcal{W}_x, \mathcal{W}_y)$ of disjoint r.e. sets, there is a recursive function f that simultaneously m-reduces them to (A, B), i.e.*

$$z \in \mathcal{W}_x \Leftrightarrow f(z) \in A \quad and \quad z \in \mathcal{W}_y \Leftrightarrow f(z) \in B.$$

This the analogue of m-completeness. (Muchnik [1958a], Smullyan [1961]) (Hint: see III.6.6, and use the Double Fixed-Point Theorem II.2.11.b.)

Further generalizations of creativeness (e.g. to infinite sequences of pairwise disjoint r.e. sets) are considered in Cleave [1961], Lachlan [1964], [1964a], [1965b], Malcev [1963], Vučkovich [1967], Carpentier [1968], [1969], [1970], Ershov [1977].

III.7   Recursive Isomorphism Types

In this section we introduce two natural generalizations of the notion of m-reducibility, obtained by considering special reducing functions. Interestingly, these two reducibilities are different, but the notions of degree induced by them coincide.

Mezoic sets and 1-degrees

A natural strengthening of m-reducibility is obtained by asking one-oneness of the reducing function.

Definition III.7.1 (Post [1944]) *A is **1-reducible** to B $(A \leq_1 B)$ if, for some one-one recursive function f, $x \in A \Leftrightarrow f(x) \in B$.*
   *A is **1-equivalent** to B $(A \equiv_1 B)$ if $A \leq_1 B$ and $B \leq_1 A$.*

Note that, since the proof of III.1.2 was obtained by the S_n^m-Theorem, which automatically provides one-one functions (II.1.7), *a set A is r.e. if and only if $A \leq_1 \mathcal{K}$.*

Exercises III.7.2 a) *If $A \leq_1 B$ and B is recursive, so is A.*
   b) *If $A \leq_1 B$ and B is r.e. then so is A.*
   c) *If A is r.e. then $A \leq_1 \overline{A}$ if and only if A is infinite, coinfinite and recursive.*
   d) *If $A \leq_1 B$ then $|A| \leq |B|$ and $|\overline{A}| \leq |\overline{B}|$. Thus, if $A \equiv_1 B$, A and B, as well as $\overline{A}$ and $\overline{B}$, must have the same cardinality.*
   e) *If A and B are recursive sets, $|A| \leq |B|$, and $|\overline{A}| \leq |\overline{B}|$, then $A \leq_1 B$.*
   f) *If A and B are infinite and coinfinite recursive sets, then $A \equiv_1 B$.*

Note that $\leq_1$ is a reflexive and transitive relation, and thus $\equiv_1$ is an equivalence relation.

Definition III.7.3 *The equivalence classes of sets w.r.t. 1-equivalence are called **1-degrees**, and $(\mathcal{D}_1, \leq)$ is the structure of 1-degrees, with the partial ordering $\leq$ induced on them by $\leq_1$.*
   *The 1-degrees containing r.e. sets are called **r.e. 1-degrees**, and two of them are:*

1. $\mathbf{0_1}$, *the 1-degree of the infinite and coinfinite recursive sets*

2. $\mathbf{0_1'}$, *the 1-degree of $\mathcal{K}$.*

*A set A is **1-complete** if it is r.e. and its 1-degree is $\mathbf{0_1'}$, i.e. $\mathcal{K} \leq_1 A$.*

Note that an r.e. 1-degree contains only r.e. sets. The 1-degrees containing recursive sets are infinitely many: one for each finite or cofinite cardinality, together with $\mathbf{0_1}$, ordered as follows:

$$0_1$$

$$a_2 \quad\quad b_2$$
$$a_1 \quad\quad b_1$$
$$a_0 \quad\quad b_0$$

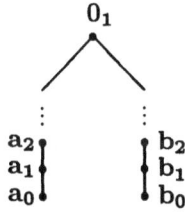

where a_n is the 1-degree of the finite sets with n elements, and b_n the 1-degree of the cofinite sets with n elements in the complement. Even if we consider only the 1-degrees of infinite and coinfinite sets, 0_1 is not the least 1-degree:

Proposition III.7.4 *If a set A has 1-degree above 0_1, then A is neither immune nor coimmune. In particular, if A is r.e. then it is not simple.*

Proof. If B is an infinite and coinfinite recursive set, and f is a one-one recursive function such that $x \in B \Leftrightarrow f(x) \in A$, then $f(B) \subseteq A$ and $f(\overline{B}) \subseteq \overline{A}$, and both $f(B)$ and $f(\overline{B})$ are infinite r.e. sets. $\square$

Simple sets are neither recursive nor 1-complete, and thus they solve a version of Post's problem for 1-degrees. But we could also formulate the problem as: are there (r.e.) 1-degrees strictly between 0_1 and $0_1'$? Then simple sets are of no help any more, since their 1-degrees are all incomparable with 0_1. To solve this version, we must first understand what the 1-complete sets are.

Theorem III.7.5 (Myhill [1955]) *A set is 1-complete if and only if it is creative.*

Proof. A 1-complete set is m-complete, and hence creative (by III.6.6). Let now A be creative, and f be a recursive function such that

$$W_x \subseteq \overline{A} \;\Rightarrow\; f(x) \in \overline{A} - W_x.$$

The same proof of III.6.6 would show that A is 1-complete, if we knew that f can be supposed to be one-one (recall that the reduction function of $\mathcal{K}$ to A is the composition of f and a function g obtained by the Fixed-Point Theorem, hence by the S_n^m-Theorem, and thus one-one by II.1.7). We then show how to build a one-one productive function h for $\overline{A}$, by induction. We can start by letting $h(0) = f(0)$. Suppose $h(0), \dots, h(n-1)$ have been defined: we want $h(n)$ such that

- $h(n) \notin \{h(0), \dots, h(n-1)\}$
- $W_n \subseteq \overline{A} \;\Rightarrow\; h(n) \in \overline{A} - W_n.$

Note that, by iteration, if $\mathcal{W}_{t(x)} = \mathcal{W}_x \cup \{f(x)\}$ then, when $\mathcal{W}_n \subseteq \overline{A}$, the elements $f(n), f(t(n)), f(t^2(n)), \ldots$ are all distinct and in $\overline{A} - \mathcal{W}_n$. Given n, generate this list, and see which of the following happens first.

- If we first find a repetition, we know that it cannot be $\mathcal{W}_n \subseteq \overline{A}$, and then $h(n)$ can be anything, as long as it does not interfere with the requirement that h be one-one. For example, let $h(n)$ be the least number different from $h(0), \ldots, h(n-1)$.

- If we first find an element not in $\{h(0), \ldots, h(n-1)\}$, then we can let $h(n)$ be the first such one. Then, as above, $h(n) \in \overline{A} - \mathcal{W}_n$ if $\mathcal{W}_n \subseteq \overline{A}$.
  $\square$

Exercises III.7.6 Special productive functions. a) *A set is productive if and only if it has a one-one, onto productive function.* (Rogers [1967]) (Hint: the proof above shows how to get a one-one productive function f. To get from it an onto one, let g be a recursive one-one enumeration of an infinite recursive set B of r.e. indices of ω, and let

$$h(x) = \begin{cases} g^{-1}(x) & \text{if } x \in B \\ f(x) & \text{otherwise.}) \end{cases}$$

b) *A set is contraproductive if and only if it has a one-one, onto contraproductive function.* (Horowitz [1978]) (Hint: symmetric to part a), using III.6.12.)

c) *A set is completely productive if and only if it has a one-one completely productive function.* (Horowitz [1978]) (Hint: if A is completely productive then, by part a), it has a one-one productive function f. By the Fixed-Point Theorem there is a one-one recursive function h such that $\mathcal{W}_{h(x)} = \mathcal{W}_x \cap \{f(h(x))\}$. Then fh is a one-one completely productive function for A.)

d) *If a set has an onto completely productive function, then its complement is r.e. (and hence creative).* (Horowitz [1978]) (Hint: if A is completely productive, there is a recursive function f such that $f(x) \in \overline{A} \Leftrightarrow f(x) \in \mathcal{W}_x$. If f is onto then, for any y, there is x such that $f(x) = y$, and y is in $\overline{A}$ if and only if it is in $\mathcal{W}_x$.)

e) *The complement of every creative set A has an onto completely productive function.* (Horowitz [1978]) (Hint: as in part a), starting with a completely productive function for $\overline{A}$, and using A in place of ω.)

Now that we know that the 1-complete sets are exactly the creative ones, we have a candidate for the solution to Post's problem for 1-degrees.

Definition III.7.7 (Dekker [1953]) *A is **mezoic** if it is an r.e. set which is neither recursive, nor creative, nor simple.*

We only have to show that such sets exist. Actually, they are quite abundant.

Proposition III.7.8 (Dekker [1953]) *Every nonrecursive r.e. T-degree contains a mezoic set.*

Proof. Let A be a simple set in the given r.e. nonrecursive T-degree (by III.2.14), and

$$\langle x, y \rangle \in B \iff x \in A.$$

Clearly B is an r.e. set in the same T-degree as A. Moreover B is mezoic, because:

- B is not recursive, because so is A.

- B is not simple, because the set $\{\langle a, y \rangle : y \in \omega\}$, for any $a \notin A$, is a recursive subset of $\overline{B}$.

- B is not creative, because if f were a productive function for $\overline{B}$, and $\mathcal{W}_{h(x)} = \{\langle z, y \rangle : z \in \mathcal{W}_x\}$, then

$$\mathcal{W}_x \subseteq \overline{A} \;\Rightarrow\; \mathcal{W}_{h(x)} \subseteq \overline{B} \;\Rightarrow\; f(h(x)) \in \overline{B} - \mathcal{W}_{h(x)}$$
$$\Rightarrow\; (f(h(x)))_1 \in \overline{A} - \mathcal{W}_x,$$

and A would be creative too. $\quad\square$

Exercises III.7.9 A classification of the r.e. sets (Uspenskii [1957]) An r.e. set A is **pseudocreative** if, for every r.e. subset B of $\overline{A}$, there is an infinite r.e. subset of $\overline{A}$ disjoint from B. The creative sets are exactly the effectively pseudocreative sets. An r.e. set A is **pseudosimple** if there is an infinite r.e. subset B of $\overline{A}$, such that $A \cup B$ is simple.

a) *The recursive, simple, pseudosimple, and pseudocreative sets are a partition of the class of the r.e. sets.*

a) *Every nonrecursive r.e. T-degree contains a pseudocreative set.* (Hint: see the proof of III.7.8.)

b) *Every nonrecursive r.e. T-degree contains a pseudosimple set.* (Hint: if A is simple, let $2x \in B \iff x \in A$.)

Exercises III.7.10 Splinters again. a) *Every splinter is recursive or pseudocreative, in particular is not simple.* (Ullian [1960]) (Hint: let $A = \{a, f(a), \ldots\}$ be nonrecursive and suppose that, for some r.e. set $B \subseteq \overline{A}$ disjoint from A, $A \cup B$ is simple. Given x, consider $\{x, f(x), \ldots\}$. If it is finite, then $x \notin A$, otherwise A is finite and hence recursive. If it is infinite, by simplicity of $A \cup B$ there must be n such that $f^{(n)}(x) \in A$ or $f^{(n)}(x) \in B$. In the second case, $x \notin A$. In the first, $f^{(n)}(x) = f^{(m)}(a)$ for some m, and x is in A if and only if $m \geq n$ and $x = f^{(m-n)}(a)$. Then A is recursive.)

b) *Every recursive set is a splinter.* (Ullian [1960])

c) *Every creative set is a splinter.* (Myhill [1959], Ullian [1960]) (Hint: let B be creative, and $\langle x, y \rangle \in A \iff x \in B$. A and B have the same m-degree and hence, by

III.7.5, the same 1-degree. By III.7.13, they are recursively isomorphic. Thus we can just prove that A is a splinter. Let $\{a_n\}_{n\in\omega}$ be a one-one enumeration of B. Picture the pairs of numbers as a double array: we have to generate the rows corresponding to pairs with first elements in B, i.e. the elements of the kind $\langle a_n, y\rangle$. Note that $A = A_0 \cup A_1$, where $A_0 = \{\langle a_n, y\rangle : n < y\}$, and $A_1 = \{\langle a_n, y\rangle : y \le n\}$. A_0 is recursive, and thus there is an enumeration $\{b_n\}_{n\in\omega}$ of it in increasing order. It is then enough to define f recursive that steps from one element of the following list to the following:

$$\langle a_0, 0\rangle, b_0$$
$$\langle a_1, 0\rangle, \langle a_1, 1\rangle, b_1$$
$$\langle a_2, 0\rangle, \langle a_2, 1\rangle, \langle a_2, 2\rangle, b_2$$
$$\cdots$$

Then let $f(\langle a_n, n\rangle) = b_n$, $f(b_n) = \langle a_{n+1}, 0\rangle$, and $f(\langle x, y\rangle) = \langle x, y+1\rangle)$ otherwise.)

Young [1967] has proved that *there are pseudocreative sets which are not splinters.*

Recursive isomorphism types

After having strengthened m-reducibility by requiring the reducing function to be one-one, we can make a further step and ask for ontoness as well.

Definition III.7.11 (Post [1944]) *A is **recursively isomorphic** to B $(A \equiv B)$ if, for some one-one onto recursive function f, $x \in A \Leftrightarrow f(x) \in B$.*

*The equivalence classes of sets w.r.t. recursive isomorphism are called **recursive isomorphism types**.*

Note that $\equiv$ is an equivalence relation, because the class of one-one and onto recursive functions (called **recursive permutations**) is closed under inverses.

Exercises III.7.12 (Rogers [1967]) a) *The recursive permutations form a group, which is not finitely generated.* (Hint: if the group were finitely generated, the recursive permutations could be recursively enumerated, and the diagonal method would produce a contradiction.)

b) *The group of the recursive permutations is not a normal subgroup of the group of all permutations of ω.* (Hint let h be a nonrecursive permutation, and

$$f(x) = \begin{cases} x & \text{if } x \text{ odd} \\ 2h(\frac{x}{2}) & \text{if } x \text{ even} \end{cases} \qquad g(x) = \begin{cases} x-1 & \text{if } x \text{ odd} \\ x+1 & \text{if } x \text{ even.} \end{cases}$$

Then $f^{-1}gf$ is not a recursive permutation.)

In Set Theory, the Cantor-Schröder-Bernstein Theorem shows that two sets which can be one-one mapped into one another must have the same cardinality. The next result is a constructive version of it.

Theorem III.7.13 (Isomorphism Theorem (Myhill [1955]) *1-degrees and isomorphism types coincide, i.e. for any pair of sets A and B,*

$$A \equiv_1 B \iff A \equiv B.$$

Proof. Let f and g be one-one recursive functions, such that

$$x \in A \iff f(x) \in B \text{ and } x \in B \iff g(x) \in A.$$

We want to define a recursive permutation h that interchanges A and B, and thus we will satisfy the condition $x \in A \iff h(x) \in B$. Since h has to be total, from time to time we ensure that the least element not yet in the domain gets into it (by defining h on it). Similarly, h has to be onto, and from time to time we ensure that the least element not yet in the range gets into it (by letting it be the value of h for some argument). We then have to show how to ensure that h is both one-one and a function. But h is a function when h^{-1} is one-one, and thus we will have a symmetric construction, that alternates steps to make h total and one-one, to steps to make it onto and a function.

We can easily start by letting $h(0) = f(0)$. Then

$$0 \in A \iff f(0) \in B \iff h(0) \in B.$$

Now take the first y not yet in the range (i.e. $y \neq f(0)$): we want an element x such that $h(x) = y$ (towards ontoness), and $x \neq 0$ (to ensure that h is a function). We obviously try $g(y)$: if this is not 0 then we have what we want, and can let $h(y) = g(y)$, since

$$y \in B \iff g(y) \in A \iff h(y) \in A.$$

But if $g(y) = 0$ then we have two different elements y and $f(0)$, and we know that $g(f(0))$ cannot be 0, since g is one-one. Then we can let $h(g(f(0))) = y$, since

$$y \in B \iff g(y) = 0 \in A \iff f(0) \in B \iff g(f(0)) = h(y) \in A.$$

The procedure is perfectly general. At any stage, having defined h on a set of elements $\{x_0, \ldots, x_n\}$ in such a way that

$$x_i \in A \iff h(x_i) \in B,$$

we know that, given $x \notin \{x_0, \ldots, x_n\}$ and $y \notin \{h(x_0), \ldots, h(x_n)\}$,

- one of $f(x), f(x_0), \ldots, f(x_n)$ is not in $\{h(x_0), \ldots, h(x_n)\}$

- one of $g(y), g(h(x_0)), \ldots, g(h(x_n))$ is not in $\{x_0, \ldots, x_n\}$.

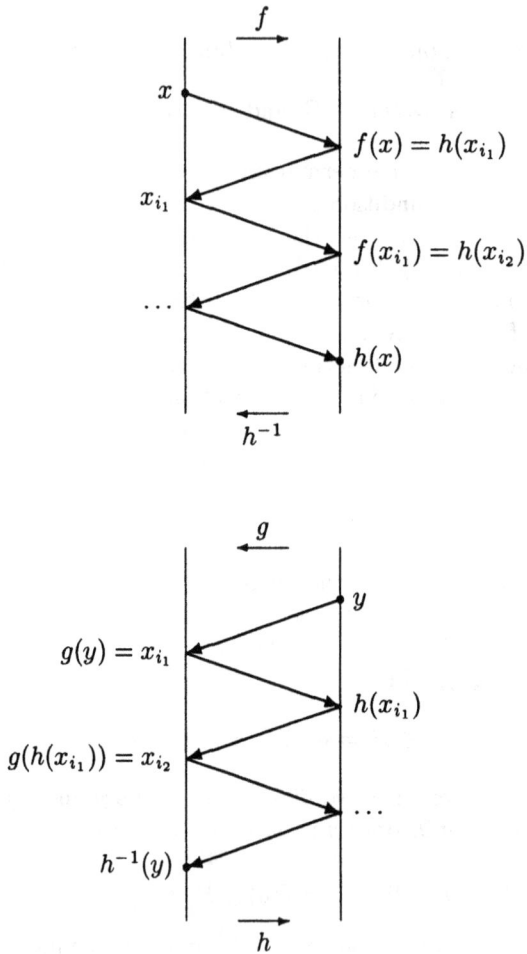

Figure III.1: Defining $h(x)$ and $h^{-1}(y)$

We can then proceed inductively, and extend h to a new element x_{n+1}, by letting x_{n+1} be the least element not yet in $\{x_0, \ldots, x_n\}$ if n is odd, and $h(x_{n+1})$ be the least element not yet in $\{h(x_0), \ldots, h(x_n)\}$ if n is even.

The procedure to find $h(x_{n+1})$ given x_{n+1} must succeed after at most $n+1$ trials. It is illustrated in Figure 1, and can be described recursively as follows. Given x not yet in the domain of h, let $z = x$, and consider $f(z)$: if it is not yet in the range of h, let $h(x) = f(z)$. Otherwise, change z into $h^{-1}(f(z))$, and try again. The procedure must stop after finitely many steps, as already noted, because the domain of h is finite, x is not in it, and f is one-one. And $x \in A \Leftrightarrow h(x) \in B$ because both f and h have this property, the latter by induction hypothesis.

The procedure to find x_{n+1} given $h(x_{n+1})$ is symmetric, by using g and h in place of f and h^{-1}. $\square$

Corollary III.7.14 *The creative sets are all recursively isomorphic.*

Proof. The creative sets are exactly the 1-complete ones, by III.7.5, and thus are all 1-equivalent. $\square$

Note that it also immediately follows that *the infinite, coinfinite recursive sets are all recursive isomorphic* (Post [1944], Dekker [1953]), although this can easily be proved directly.

Exercise III.7.15 *The effectively inseparable pairs of r.e. sets are all recursively isomorphic*, i.e. if (A, B) and (C, D) are effectively inseparable, there is a recursive permutation that simultaneously exchanges A and C, and B and D. (Muchnik [1958a], Smullyan [1961]) (Hint: use III.6.27.d.)

Rogers [1967] introduced Klein's approach in Recursion Theory, and stressed the importance of considering *properties invariant under recursive permutations*. We leave to the reader the verification that all concepts introduced so far (with only one exception, see III.7.16.c), as well as those that will be introduced in the next chapters, are indeed recursively invariant. Frequently, III.7.13 is a useful tool in such verifications.

Exercises III.7.16 a) *The property of being regressive is invariant under recursive permutations.* (Hint: if $A = \{a_0, a_1, \ldots\}$ is regressed by φ, then the partial recursive function ψ such that $\psi(h(x)) \simeq h(\varphi(x))$ regresses $h(A) = \{h(a_0), h(a_1), \ldots\}$. Note that the enumeration of the set has changed: thus, even if A is retraceable, in general $h(A)$ is only regressive.)

b) *The property of containing an infinite retraceable subset is invariant under recursive permutations.* (Dekker [1962]) (Hint: if A contains an infinite retraceable set B, and h is a recursive permutation, then $h(B)$ is an infinite regressive subset of $h(A)$, by part a), and it contains, by II.6.20, an infinite retraceable subset.)

c) *The property of being retraceable is not invariant under recursive permutations, even for co-r.e. sets.* (McLaughlin [1966]) (Hint: this requires the priority method. We fix the permutation in advance, and let h exchange $2x$ and $2x + 1$. We then build $A = \{a_0 < a_1 < \cdots\}$, and φ retracing it. The construction has to be effective to give φ partial recursive, and this produces A co-r.e. We ensure that $h(A)$ is not retraceable. Choose a_{2e} and a_{2e+1} to witness the fact that φ_e does not retrace $h(A)$. Also, let $\varphi(a_{n+1}) = a_n$ and $\varphi(a_0) = a_0$, so that φ retraces A. At stage $s + 1$, let φ_s be the approximation of φ obtained so far. Consider the smallest $e \leq s$ such that the condition 'φ_e does not retrace $h(A)$' has not yet been satisfied and not injured afterwards (in particular, $a_{2e+1}^s = a_{2e}^s + 1$), and $\varphi_{e,s}(a_{2e+1}^s) = a_{2e}^s$. Then $\varphi_{e,s}(h(a_{2e}^s)) = h(a_{2e+1}^s)$, and we want to destroy this. Let x be the smallest odd element which is greater than both a_{2e+1}^s and every element in the domain of φ_s, and let $x = a_{2e+1}^{s+1}$, and $\varphi_{s+1}(x) = a_{2e}^s$. The reason for doing this is that, if $a_{2e}^s = a_{2e}$, then $\varphi_e(h(a_{2e}))$ is not in $h(A)$, and so φ_e does not retrace $h(A)$.)

d) *There are r.e. sets coregressive, but not coretraceable.* (McLaughlin [1966]) (Hint: by the proof of part c), $h(\overline{A}) = \overline{h(A)}$ is such a set.)

Degtev [1971] and Soare [1972] have shown that every nonrecursive r.e. T-degree contains an r.e. set which is coregressive but not coretraceable.

Recursive equivalence types and isols ⋆

Dekker [1955] defines the relation of **recursive equivalence** as:

$$A \cong B \quad \Leftrightarrow \quad \text{for some one-one, partial recursive function } \varphi,$$
$$A \subseteq \text{dom } \varphi \text{ and } \varphi(A) = B.$$

$\cong$ is an equivalence relation, which may be seen as a constructivization of the property of having same cardinality, and its equivalence classes are called **recursive equivalence types** (r.e.t.), and are thus constructive analogues of cardinals.

In Set Theory there are two notions of finiteness: A is finite if it can be one-one mapped on a proper initial segment of ω, and is **Dedekind-finite** if it cannot be one-one mapped on a proper subset of itself. The two notions coincide if choice is assumed, but in choiceless Set Theory the latter is weaker, and some infinite sets may still be Dedekind-finite. A set A is recursively equivalent to a proper subset of itself if and only if it is infinite and not immune, and thus *the r.e.t.'s of finite or immune sets* (called **isols**, because such sets are isolated in the usual topology, see p. 186) *may be seen as a constructive version of Dedekind-finite cardinals* (even more appropriately they arise, via Kleene realizability, from standard cardinal arithmetic on subsets of ω, in Intuitionistic Set Theory, see McCarty [1986]).

The set Λ of isols can be given operations of sum and product, induced by

$A \oplus B$ and $A \cdot B$, and a partial order relation $\leq$ defined as

$$x \leq y \Leftrightarrow (\exists z)(x + z = y).$$

The structure of isols is quite rich: ω is embedded in Λ (by identifying n with the isol of finite sets of cardinality n), Λ has $2^{\aleph_0}$ elements, and $\langle \Lambda, \leq \rangle$ embeds every countable partial ordering (as well as many others, see Ellentuck [1973]).

Myhill [1958] has introduced a way to extend certain functions from ω to Λ. First note that each function $f : \omega \to \omega$ can be written as

$$f(n) = \sum_{i=0}^{\infty} c_i \left(\begin{array}{c} n \\ i \end{array} \right)$$

where the c_i's (Sterling coefficients) are positive or negative integers. If they are all positive or null, then f is called a **combinatorial function**. Thus every function on ω is the difference of two combinatorial ones. A combinatorial function is recursive if and only if the sequence $\{c_i\}_{i \in \omega}$ is. Combinatorial functions of many variables are defined similarly. The class of combinatorial functions is a rich one, closed under composition, and containing $\mathcal{I}_i^n$, the constant functions, sum, product, factorial and positive-base exponential. A **combinatorial set function** is a function on $\mathcal{P}(\omega)$ that maps each finite set to a finite set, and that respects cardinalities, intersections and countable unions (in particular, the function is generated by its behavior on the finite sets). It is called effective if the restriction to finite sets is recursive (on the characteristic indices). Combinatorial set functions induce functions on ω (since they preserve finite cardinalities), and effective combinatorial set functions induce functions on r.e.t.'s, in particular on Λ. The basic connection between the two notions is that the (recursive) combinatorial functions are exactly those induced by (effective) combinatorial set functions. Thus *for each recursive combinatorial function f on ω, there is a function f_Λ on Λ extending it.*

It is immediate to note that *for each recursive relation R on ω there is a relation R_Λ on Λ extending it*, since given R there are f, g recursive combinatorial functions such that

$$R(x) \Leftrightarrow f(x) = g(x),$$

and thus it is enough to define

$$R_\Lambda(x) \Leftrightarrow f_\Lambda(x) = g_\Lambda(x).$$

Note that if a recursive relation R admits an algebraic characterization on ω, R_Λ does not necessarily coincide with the interpretation of the characterization on Λ. E.g., if Pr is the set of prime numbers, Pr_Λ is different from the set of prime isols, defined as

$$x \text{ prime} \Leftrightarrow 2 \leq x \wedge \forall y \forall z (y \cdot z = x \Rightarrow y = 1 \vee z = 1).$$

Nerode [1961], [1962] has used these translations to show that *a substantial part of the theory of (finite cardinal) arithmetic can be extended to isols*. Consider a first-order language with equality, function constants for all recursive combinatorial functions, and relation constants for all recursive relations. Then *a universal Horn sentence* (see p. 39) *holds over ω if and only if its translation holds over Λ*. It immediately follows that, e.g., the following hold for isols:

$$x \neq x + 1, \quad nx = ny \Rightarrow x = y, \quad 2^x \cdot 2^y = 2^{x+y}.$$

The result holds in much greater generality than stated, but some restrictions are needed, because there are universal sentences true in ω that fail in Λ, e.g. the fact that every element is even or odd. Ellentuck [1967] has discovered a notion of **universal isol** (as an isol avoiding R_Λ, for each coinfinite recursive R, and thus having a sort of genericity): there are $2^{\aleph_0}$ such isols, and *each of them provides counterexamples to each universal sentence true in ω but false in Λ*. E.g. a universal isol is prime, but not a member of Pr_Λ.

The notion of isol can be regarded as an extension not only of the notion of integer, but also of nonstandard integer. Indeed, Nerode [1966] has proved that any countable nonstandard model of Arithmetic correct for diophantine equations (i.e. such that if an equation has a solution in the model, then it has a solution in the integers) can be embedded in the isols. Conversely, any subset of the isols whose differences generate an integral domain can be embedded in a nonstandard model of Arithmetic.

The theory of $\langle \Lambda, +, \cdot \rangle$ is at least as complicated as that of $\langle \omega, +, \cdot \rangle$ (in particular is not decidable), because the set of finite isols is definable over Λ, by the formula

$$\mathrm{Fin}(x) \Leftrightarrow (\forall y)(x \leq y \vee y \leq x)$$

(Dekker and Myhill [1960]). Actually, *the first-order theory of $\langle \Lambda, \leq \rangle$ is recursively isomorphic to the Second Order Arithmetic* (Nerode and Manaster [1971]), *and the same holds for the theory of $\langle \Lambda, +, \cdot \rangle$* (Ellentuck [1973a]).

For a treatment of the theory of isols (and r.e.t.'s in general) see Dekker and Myhill [1960], Dekker [1966], Crossley and Nerode [1974], McLaughlin [1982].

III.8   Variations of Truth-Table Reducibility $\star$

In this section we play on the theme of *tt*-reducibility, by first considering bounds on the number of elements used in the truth-tables, and then by relaxing the condition that we know in advance the effect of the answers to the queries made to the oracle. At the end we also introduce a number of other strong reducibilities.

Bounded truth-table degrees

The variation we consider first is, like c-reducibility or d-reducibility, a strengthening of tt-reducibility, but in a different direction. We impose restrictions not on the kind of truth tables we allow, but rather on their size.

Definition III.8.1 (Post [1944]) *A is **btt-reducible** to B ($A \leq_{btt} B$) if, for some recursive function f and some number m, called the **norm** of the reduction,*

> *1. $x \in A \Leftrightarrow B \models \sigma_{f(x)}$*
>
> *2. $\sigma_{f(x)}$ uses at most m elements.*

If m is the norm, we also write $A \leq_{btt(m)} B$.
  *A is **btt-equivalent** to B ($A \equiv_{btt} B$) if $A \leq_{btt} B$ and $B \leq_{btt} A$.*

Exercises III.8.2 a) *If A is recursive, then $A \leq_{btt} B$ for any set B.*
   b) *If $A \leq_{btt} B$ and B is recursive, so is A.*
   c) *$A \leq_{btt(1)} \overline{A}$.*

Note that $\leq_{btt}$ is a reflexive and transitive relation, and thus $\equiv_{btt}$ is an equivalence relation.

Definition III.8.3 *The equivalence classes of sets w.r.t. btt-equivalence are called **btt-degrees**, and $(\mathcal{D}_{btt}, \leq)$ is the structure of btt-degrees, with the partial ordering $\leq$ induced on them by $\leq_{btt}$.*
  *The btt-degrees containing r.e. sets are called **r.e. btt-degrees**, and two of them are:*

> *1. $\mathbf{0}_{btt}$, the btt-degree of recursive sets*
>
> *2. $\mathbf{0}'_{btt}$, the btt-degree of $\mathcal{K}$.*

*A set A is **btt-complete** if its r.e. and its btt-degree is $\mathbf{0}'_{btt}$, i.e. if $\mathcal{K} \leq_{btt} A$.*

In general $\leq_{btt(m)}$ is not transitive, but $\leq_{btt(1)}$ is. Since a set is always reducible to its complement by a bounded truth-table reduction of norm 1, but not always by m-reductions, $\leq_m$ and $\leq_{btt(1)}$ differ in general. But they agree on nontrivial r.e. sets.

Proposition III.8.4 *If A and B are r.e. sets, $B \neq \emptyset, \omega$ and $A \leq_{btt(1)} B$, then $A \leq_m B$.*

Proof. Since $A \leq_{btt(1)} B$, there is a recursive function f such that, depending on x,

$$x \in A \Leftrightarrow f(x) \in B \quad \text{or} \quad x \in A \Leftrightarrow f(x) \notin B.$$

We want a recursive function g such that $x \in A \Leftrightarrow g(x) \in B$. There are two cases:

- if $x \in A \Leftrightarrow f(x) \in B$, we just let $g(x) = f(x)$

- if $x \in A \Leftrightarrow f(x) \notin B$, then exactly one of $x \in A$ and $f(x) \in B$ happens. Generate A and B simultaneously, and find out which one. If $x \in A$ then we want $g(x) \in B$. If $f(x) \in B$ then $x \notin A$, and we want $g(x) \notin B$. Since $B \neq \emptyset, \omega$ we can pick up $a \in B$ and $b \notin B$, and let $g(x)$ be a in the first case, and b in the second. $\square$

Exercise III.8.5 *m-reducibility and btt-reducibility differ on the r.e. sets.* (Fischer [1963]) (Hint: Let $\langle x, y \rangle \in A \Leftrightarrow x \in B \wedge y \in B$, where B is the simple tt-complete set of III.3.5: then $A \leq_{btt(2)} B$. If $A \leq_m B$, there would be g recursive such that $x \in B \wedge y \in B \Leftrightarrow g(x, y) \in B$, and B would be creative because, if h is gotten by iteration of g, $x \in \mathcal{K} \Leftrightarrow F_x \subseteq B \Leftrightarrow h(x) \in B$.)

When $A \leq_{btt} B$, only a finite number m of elements are used in the reduction, and hence at most 2^{2^m} truth-tables may be used. Actually much more is true: a single truth-table is enough (although, in general, with a different norm).

Proposition III.8.6 (Fischer) *If $B \neq \emptyset, \omega$ and $A \leq_{btt} B$, then A is reducible to B by a fixed truth-table.*

Proof. Consider e.g. the case of norm 1, and let A be reducible to B via $\sigma_{f(x)}$. Fix $a \in B$ and $b \notin B$. Since $\sigma_{f(x)}$ is either $y \in X$ or $y \notin X$ for some y, let $\sigma_{g(x)}$ be the fixed formula

$$(y \in X \wedge z \in X) \vee (y \notin X \wedge z \notin X),$$

where y is the element appearing in $\sigma_{f(x)}$, and

$$z = a \quad \text{if} \quad \sigma_{f(x)} \text{ is } y \in X$$
$$z = b \quad \text{if} \quad \sigma_{f(x)} \text{ is } y \notin X.$$

Then $B \models \sigma_{f(x)} \Leftrightarrow B \models \sigma_{g(x)}$.

The general case is similar, although the addition of more than one variable is required, to be able to distinguish among many cases. $\square$

Thus a *btt*-reduction can actually be seen as the assignment, to every x, of a fixed number of elements $\{b_1^x, \ldots, b_n^x\}$, such that the answer to the question 'is x in A?' depends solely on the membership of the b_i^x's in B.

Proposition III.8.7 *A set is btt-reducible to $\mathcal{K}$ if and only if it is in the Boolean algebra generated by the r.e. sets.*

Proof. If a set A is in the Boolean algebra generated by the r.e. sets, it can be obtained from a finite number of r.e. sets $A_1, \ldots, A_n$ by the set-theoretical operations of union, intersection and complementation. Express $x \in A$ as a propositional formula involving the A_i's, and then substitute each atomic formula $x \in A_i$ with $f_i(x) \in \mathcal{K}$, where f_i is an m-reduction of A_i to $\mathcal{K}$, which exists because A_i is r.e. E.g. let $A = A_1 - (A_2 \cap A_3)$. Then

$$x \in A \;\Leftrightarrow\; x \in A_1 \wedge \neg(x \in A_2 \wedge x \in A_3)$$

and, if

$$\sigma_{g(x)} = f_1(x) \in X \wedge \neg(f_2(x) \in X \wedge f_3(x) \in X),$$

then $x \in A \Leftrightarrow \mathcal{K} \models \sigma_{g(x)}$, and $A \leq_{btt} \mathcal{K}$.

Conversely, if $A \leq_{btt} \mathcal{K}$ then, by the previous proposition, A can be reduced to $\mathcal{K}$ by a fixed truth-table, and the procedure just given can be inverted. $\square$.

The analogue of Post's Problem for *btt*-reducibility is: are there r.e. sets which are neither recursive, nor *btt*-complete? We already know that simple sets are not *m*-complete, and we now extend this result to show that they are also not *btt*-complete.

Theorem III.8.8 (Post [1944]) *A simple set is not btt-complete.*

Proof. Let $\mathcal{K} \leq_{btt} A$. We try to define a disjoint strong array $\{F_n\}_{n \in \omega}$ intersecting $\overline{A}$, with every element of fixed cardinality: then A is not simple, because an infinite r.e. subset of $\overline{A}$ can be obtained by the following procedure. Simultaneously generate A and the F_n's: as soon as all but one element of F_n have been generated in A, we know that the other is in $\overline{A}$. The only trouble is that there could be only finitely many n for which F_n has exactly one element in $\overline{A}$, and this procedure would not produce an infinite set. But if this happens then (by eliminating the finitely many exceptions) we may suppose that each F_n has at least two elements in $\overline{A}$. Again we can proceed as above, if there are infinitely many n for which F_n has exactly 2 elements in $\overline{A}$. In general, let z be the least number such that, for infinitely many n, $F_n \cap \overline{A}$ has exactly z elements. Then, for some n_0 and all $n \geq n_0$, $F_n \cap \overline{A}$ has at least z elements. Generate simultaneously A and the F_n's, for $n \geq n_0$: as soon as all but z elements of some F_n have been generated in A, the others must be in $\overline{A}$.

If $\mathcal{K} \leq_{btt} A$, then so too is $\overline{\mathcal{K}} \leq_{btt} A$. Let $x \in \overline{\mathcal{K}} \Leftrightarrow A \models \sigma_{f(x)}$, for some recursive function f, with a given reduction $\{b_1^x, \ldots, b_m^x\}$. Suppose we have already chosen

$$F_0 = \{b_1^{x_0}, \ldots, b_m^{x_0}\} \quad \cdots \quad F_{n-1} = \{b_1^{x_{n-1}}, \ldots, b_m^{x_{n-1}}\}$$

with the property that

$$F_i \cap \overline{A} \neq \emptyset \quad \text{and} \quad A \models \sigma_{f(x_i)}.$$

To define F_n, let C be the set consisting of all the x for which the formula $\sigma_{f(x)}$ is deducible, in the propositional calculus, from $\sigma_{f(x_0)}, \ldots, \sigma_{f(x_{n-1})}$ and the conditions '$z \in X$', for $z \in A$. By the inductive hypothesis (that $A \models \sigma_{f(x_i)}$) and logical properties,

$$x \in C \iff A \models \sigma_{f(x)} \iff x \in \overline{K},$$

and hence $C \subseteq \overline{K}$. If $C = W_a$, let $x_n = a$. From $a \in \overline{K} - C$ we have:

- $A \models \sigma_{f(a)}$, because $a \in \overline{K}$

- $\{b_1^a, \ldots, b_m^a\} \cap \overline{A} \neq \emptyset$. Otherwise, being $\sigma_{f(a)}$ true in A, and using only the elements $b_1^a, \ldots, b_m^a$ which are all in A, $\sigma_{f(a)}$ would be deducible from the conditions '$z \in X$' for $z \in A$. Then $a \in C$, contradiction.

However, we cannot prove that the F_n's are disjoint on $\overline{A}$. But this is unnecessary: it is enough to show that, given F_n, there are only finitely many sets of our sequence, with the same intersection on $\overline{A}$. Indeed, consider $\{b_1^x, \ldots, b_m^x\}$: since only membership in A or $\overline{A}$ matters, there are only 2^m possibilities, If two given conditions have same elements on $\overline{A}$, they are not only equivalent: they are also deducible one from the other from the conditions '$z \in X$' for $z \in A$. Thus in our sequence, by definition of C, at most 2^m conditions may have the same intersection on $\overline{A}$. Then the procedure described at the beginning still produces an infinite subset of $\overline{A}$. $\square$

Corollary III.8.9 *There are tt-complete, btt-incomplete sets.*

Proof. By III.3.5 there is a simple tt-complete set: by simplicity it cannot be btt-complete. $\square$

Exercises III.8.10 a) *A pseudosimple set is not btt-complete.* (Shoenfield [1957])

b) *If $B \leq_{btt} A$ and B is part of a recursively inseparable pair of r.e. sets, then A is not simple.* (Kobzev [1973]) (Hint: by induction on m, prove that if A is simple and $\{F_n\}_{n \in \omega}$ is a strong array of m-tuples intersecting $\overline{A}$, there is a finite set $D \subseteq \overline{A}$ such that, for all n, $F_n \cap D \neq \emptyset$. Then let $B \leq_{btt} A$ via $\{b_1^x, \ldots, b_m^x\}$, and consider the two cases $\{b_1^x, \ldots, b_m^x\} \subseteq A \Rightarrow x \in B$, and $\{b_1^x, \ldots, b_m^x\} \subseteq A \Rightarrow x \notin B$. If e.g. the first holds, and B and C are recursively inseparable, then $x \in C \Rightarrow \{b_1^x, \ldots, b_m^x\} \cap \overline{A} \neq \emptyset$. Let $D \subseteq \overline{A}$ be finite, and such that $x \in C \Rightarrow \{b_1^x, \ldots, b_m^x\} \cap D \neq \emptyset$. Moreover, let $x \in R \iff \{b_1^x, \ldots, b_m^x\} \cap D \neq \emptyset$. Then R is recursive, $C \subseteq R$, and $B \cap R$ and C are recursively inseparable. Split R into m recursive parts

$$x \in R_1 \iff x \in R \land b_1^x \in D \quad x \in R_2 \iff x \in R - R_1 \land b_2^x \in D \quad \cdots$$

For at least one i, $B \cap R_i$ and $C \cap R_i$ are recursively inseparable, and $B \cap R_i \leq_{btt} A$ with norm $m - 1$. So A is not simple by induction hypothesis. For $m = 1$ recall that $\leq_{btt(1)}$ coincides with $\leq_m$ on the r.e. sets, and see III.6.23.)

Theorem III.8.11 (Kobzev [1974], Lachlan [1975]) *A btt-complete set is d-complete.*

Proof. Let $K \leq_{btt} A$: for some recursive f

$$x \in K \Leftrightarrow A \models \sigma_{f(x)}$$

and, for some n, $\sigma_{f(x)}$ uses exactly n elements. We want to get

$$x \in K \Leftrightarrow D_{h(x)} \cap A \neq \emptyset,$$

for some recursive h. The obvious approach is to consider

$$G_x = \{\text{elements used in } \sigma_{f(x)}\}.$$

We can certainly modify the reduction in such a way to suppose that, whenever x is enumerated in K, then some element of G_x is enumerated in A at the same stage. This gives in particular

$$x \in K \Rightarrow G_x \cap A \neq \emptyset,$$

but we do not have the opposite: some element of G_x could go in A even if $x \notin K$.

We thus consider the set G_x dynamically: at every stage s we see the set $G_x \cap A_{s+1}$, and can consider the remaining elements of G_x. We thus have n r.e. sets F_i, and F_i consists of elements z_x^i which, at a certain stage $s + 1$, are put in F_i because $|G_{z_x^i} \cap A_{s+1}| = i$. We try to consider z_x^i in place of x and, as above,

$$\text{if } x \in K \Rightarrow z_x^i \in K \text{ then } x \in K \Rightarrow |G_{z_x^i} \cap A_{s+1}| > i.$$

If, for $j > i$ and at stage $s + 1$, we put in F_j only elements z not yet in K_{s+1}, and such that $|G_z \cap A_{s+1}| = j$, and we avoid putting in K elements z_x^i which are also in F_j (to avoid a situation in which z_y^i goes in K when $y \notin K$), then for the greatest i such that F_i is infinite, and for almost every x, we also have

$$|G_{z_x^i} \cap A| > i \Rightarrow x \in K.$$

Indeed, only finitely many z_x^i's can be in F_j for some $j > i$, and thus only for finitely many x's not in K it will be $|G_{z_x^i} \cap A| > i$.

There are however two problems.

1. We want $x \in \mathcal{K} \Rightarrow z_x^i \in \mathcal{K}$, but we do not have any control over $\mathcal{K}$, and thus we cannot directly force z_x^i into $\mathcal{K}$. What we *can* do, is to build an r.e. set D, and use a recursive function g such that $x \in D \Leftrightarrow g(x) \in \mathcal{K}$: and g can be used in the construction of D itself, by the Fixed-Point Theorem. Thus putting x in D will force $g(x)$ into $\mathcal{K}$. We will thus consider $G_{g(z_x^i)}$, in place of $G_{z_x^i}$.

   The *construction* is finally the following: at stage $s + 1$,

   - if $x \in \mathcal{K}_{s+1} - \mathcal{K}_s$ and $z_x^i \in F_{i,s}$ (i.e. the x-th element of F_i has already been generated), put z_x^i in D, unless $z_x^i \in F_j$, for some $j > i$.

   - if, for some i and some $x \leq s$,

   $$x \notin D_{s+1} \wedge g(x) \notin \mathcal{K}_{s+1} \wedge x \notin F_{i,s} \wedge |G_{g(x)} \cap A_{s+1}| = i,$$

   then choose i maximal, x minimal, and put x in F_i.

2. Let now i be the greatest such that F_i is infinite. Let

   $$D_{h(x)} = G_{g(z_x^i)} \cap \overline{A}_{s+1},$$

   where $s + 1$ is the stage in which z_x^i is generated in F_i. We would like to show

   $$x \in \mathcal{K} \Leftrightarrow D_{h(x)} \cap A \neq \emptyset,$$

   but here comes the second problem: by construction, $D_{h(x)} \cap A \neq \emptyset$ only for those x which are generated in $\mathcal{K}$ after z_x^i is generated in F_i. So consider these elements:

   $$x \in \mathcal{K}^* \Leftrightarrow \exists s(x \in \mathcal{K}_{s+1} - \mathcal{K}_s \wedge z_x^i \in F_{i,s}).$$

   Then $\mathcal{K} - \mathcal{K}^*$ is recursive, so $\mathcal{K}^*$ is creative, and to have A d-complete is enough to prove

   $$x \in \mathcal{K}^* \Leftrightarrow D_{h(x)} \cap A \neq \emptyset.$$

   If $x \in \mathcal{K}^*$ then $D_{h(x)} \cap A \neq \emptyset$ by construction, since an element of $G_{g(z_x^i)}$ is generated in A at the same stage in which $g(z_x^i)$ is generated in $\mathcal{K}$, hence after the stage z_x^i is generated in F_i.

   The opposite holds for almost every x, because i is the greatest such that F_i is infinite, and thus, except for finitely many x, we have $D_{h(x)} \cap A \neq \emptyset$ only if $x \in \mathcal{K}^*$, by the first part of the construction. $\square$

Since *btt*-complete sets are d-complete and hence quasicreative, and quasi-creative sets are not simple (III.6.18), we have a different proof of III.8.8.

Exercises III.8.12 a) *A semirecursive set is not btt-complete.* (Jockusch [1968a]) (Hint: a semirecursive set is not p-complete, by III.5.3.b.)

b) *A set is btt-complete if and only if, for some recursive function f and some n, $|D_{f(x)}| \leq n$, and $W_x \subseteq \overline{A} \Rightarrow D_{f(x)} \subseteq \overline{A} \wedge D_{f(x)} \not\subseteq W_x$.* (Kobzev [1974]) (Hint: a btt-complete set is bounded d-complete, and thus the bounded version of the quasi-creativeness criterion for d-completeness applies.)

Weak truth-table degrees

Recall that truth-table reducibility differs from Turing reducibility in that it is possible to foresee ahead of time, in a computation, both the elements on which the oracle is going to be queried, and the outcome of all possible answers. A natural intermediate reducibility is obtained by retaining the first characteristic, while relaxing the second.

Definition III.8.13 (Friedberg and Rogers [1959]) *A is **wtt-reducible** to B $(A \leq_{wtt} B)$ if $c_A \simeq \varphi_e^B$ for some number e and some recursive function f, and the calculation of $\varphi_e(x)$ requires only queries to the oracle B on elements less than $f(x)$.*

*A is **wtt-equivalent** to B $(A \equiv_{wtt} B)$ if $A \leq_{wtt} B$ and $B \leq_{wtt} A$.*

The main difference with truth-table reducibility is in the fact that *weak truth-table reductions may diverge*. If φ_e^X is a weak truth-table reduction of A to B, then φ_e^X needs to be total only for $X = B$, but not for oracles X different from B. Actually, by III.3.2, this must be the case if the reduction is not a truth-table one.

Exercise III.8.14 *A set A is wtt-reducible to $\mathcal{K}$ if and only if it is tt-reducible to it.* (Hint: if φ_e^X is a wtt-reduction to $\mathcal{K}$ with bound f, given x we may consider all the sets $X \subseteq \{0, \ldots, f(x)\}$. Each of them is recursive, and so we can ask, recursively in $\mathcal{K}$, if $\varphi_e^X(x)$ converges. This makes it possible to build the appropriate truth-table.)

Note that $\leq_{wtt}$ is a reflexive and transitive relation, and thus $\equiv_{wtt}$ is an equivalence relation.

Definition III.8.15 *The equivalence classes of sets w.r.t. wtt-equivalence are called **wtt-degrees**, and $(\mathcal{D}_{wtt}, \leq)$ is the structure of wtt-degrees, with the partial ordering $\leq$ induced on them by $\leq_{wtt}$.*

*The wtt-degrees containing r.e. sets are called **r.e. wtt-degrees**, and two of them are:*

1. *$\mathbf{0}_{wtt}$, the wtt-degree of recursive sets*

2. *$\mathbf{0}'_{wtt}$, the wtt-degree of $\mathcal{K}$.*

A set A is **wtt-complete** *if it is r.e. and its wtt-degree is* $0'_{wtt}$, *i.e. if* $\mathcal{K} \leq_{wtt} A$.

Proposition III.8.16 (Friedberg and Rogers [1959]) *A hypersimple set is not wtt-complete.*

Proof. We refer to the proof of III.3.10, which goes through practically unchanged, by letting
$$x \in C \Leftrightarrow \varphi_e^{A^*}(x) \simeq 0,$$
where $c_{\mathcal{K}} \simeq \varphi_e^A$, with bound f. If $C = W_a$ then $\varphi_e^{A^*}(a)$ and $\varphi_e^A(a)$ are both convergent and different, so there must be an element of $\overline{A}$ between the given n and $f(a)$. The rest proceeds as before.   □

A simple but useful observation is that *permitting preserves wtt-reducibility*. This allows us to extend many results from T-degrees to wtt-degrees. E.g. *every nonrecursive r.e. wtt-degree contains a simple set* (III.3.18), although not always a hypersimple one (since a hypersimple set is not wtt-complete). Jockusch [1981a] shows that *not every nonrecursive r.e. tt-degree contains a simple set*.

Another useful fact to notice is that the completeness criterion for T-reducibility (III.1.5) extends to wtt-reducibility as well, with the obvious definition of wtt-reducibility for functions, and a similar proof:

Proposition III.8.17 (Arslanov [1981]) *An r.e. set A is wtt-complete if and only if there is a function $f \leq_{wtt} A$ without fixed-points.*

Recall that an application of the original criterion showed that every effectively simple set is T-complete (III.2.18). Here the same results holds, for all the effectively simple sets not ruled out by the previous result.

Proposition III.8.18 (Kanovich [1970], [1970a], Arslanov [1981]) *Every effectively simple, not hypersimple set is wtt-complete.*

Proof. Let
$$W_e \subseteq \overline{A} \Rightarrow |W_e| \leq g(e).$$
Define $f \leq_T A$ such that
$$W_{f(e)} = \{\text{the first } g(e) + 1 \text{ elements of } \overline{A}\}.$$
Then f has no fixed-points, as in III.2.18. We show that $f \leq_{wtt} A$. Since A is not hypersimple, there is a strong array h intersecting $\overline{A}$. Thus the first $g(e)+1$ elements of $\overline{A}$ are below the maximum of $\bigcup_{z \leq g(e)} D_{h(z)}$, and this provides the needed recursive bound to the questions f has to answer.   □

In particular, *Post's simple set is wtt-complete* (Ladner). As we have already noted, the *tt*-completeness of Post's simple set depends on the acceptable system of indices for the r.e. sets (III.3.6, III.9.2).

Exercise III.8.19 *Not every strongly effectively simple set is wtt-complete.* (Hint: let A be any coinfinite r.e. set. If it is not hypersimple itself, it has a hypersimple superset, namely $A \cup \bigcup_{x \in B} D_{f(x)}$, where f is a disjoint strong array intersecting A, and B any hypersimple set. And if A is strongly effectively simple, so is any coinfinite r.e. superset of it.)

Exercises III.8.20 a) *An r.e. set A is wtt-complete if and only if*

$$\mathcal{W}_x \subseteq \overline{A} \Rightarrow D_{g(x)} \not\subseteq A \cup \mathcal{W}_x$$

for some recursive function g. (Kanovich [1969], [1970a]) (Hint: if such a g exists, let $\mathcal{W}_{f(x)} = D_{g(x)} \cap \overline{A}$. Then $f \leq_{wtt} A$, since it uses the oracle only for elements in $D_{g(x)}$. Suppose $\mathcal{W}_{f(x)} = \mathcal{W}_x$: then $\mathcal{W}_x \subseteq \overline{A}$, so $D_{g(x)} \not\subseteq A \cup \mathcal{W}_x$, contradicting $\mathcal{W}_x = D_{g(x)} \cap \overline{A}$. Then f has no fixed-points, and A is *wtt*-complete. Conversely, let $f \simeq \varphi_e^A$ with bound h be without fixed-points. Given x, let

$$\varphi_i(z) \simeq \begin{cases} 1 & \text{if } z \text{ shows up first in } A \\ 0 & \text{if } z \text{ shows up first in } \mathcal{W}_x. \end{cases}$$

Since φ_i is recursive, there is z such that $\mathcal{W}_{\varphi_e^{\varphi_i}(z)} = \mathcal{W}_z$ (by the Fixed-Point Theorem). Since $\mathcal{W}_{f(z)} \neq \mathcal{W}_z$, $\varphi_e^{\varphi_i}(z) \neq f(z)$. If $\mathcal{W}_x \subseteq \overline{A}$, then φ_i is correct on $A \cup \mathcal{W}_x$, hence there must be an element of $\overline{A \cup \mathcal{W}_x}$ below $h(z)$. Then let $D_{g(x)}$ be the set $\{0, \dots, h(z)\}$.)
    b) *For every nonrecursive r.e. set A there is f recursive such that*

$$\mathcal{W}_x \subseteq \overline{A} \Rightarrow \mathcal{W}_{f(x)} \text{ finite } \wedge \mathcal{W}_{f(x)} \not\subseteq A \cup \mathcal{W}_x.$$

(Soloviev [1976]) (Hint: put in $\mathcal{W}_{f(x)}$ the smallest element that does not appear to be in $A \cup \mathcal{W}_x$.)
    c) *The class of wtt-complete sets is properly included in the class of T-complete, not hypersimple sets.* (Soloviev [1976]) (Hint: let A be T-complete and hypersimple, and B be an infinite recursive subset of it. Consider $A - B$: it is T-complete because A is, and is not simple. It is also not *wtt*-complete, since $A - B \leq_{wtt} A$, and A is not *wtt*-complete.)

Other notions of reducibility $\star$

The reader certainly feels that we have introduced enough reducibilities, but many more have been considered. We just review some of them.
    A natural way to weaken reducibilities is by letting the finiteness condition in the sets defining them become an r.e. condition, like we did e.g. for *c*-reducibility, obtaining Q-reducibility. In a similar way, *d*-reducibility is weakened into

s-reducibility, defined as follows: $A \leq_s B$ if and only if, for some recursive function f,

$$x \in A \Leftrightarrow \mathcal{W}_{f(x)} \cap B \neq \emptyset.$$

Then the r.e. sets fall in just two s-degrees, one consisting of all the nonempty r.e. sets, and the other consisting of $\emptyset$ alone.

Similarly, positive reducibility is weakened in **enumeration reducibility**, introduced in Section II.3 (p. 197). Recall that $A \leq_e B$ if and only if, for some recursive function f,

$$x \in A \Leftrightarrow (\exists u)(D_u \subseteq B \wedge u \in \mathcal{W}_{f(x)}).$$

There is just one r.e. e-degree, which is also the least e-degree. As we have already seen, this reducibility is particularly suitable for the study of partial function (through their graphs), and we will deal with it in Chapter XIV.

A number of other reducibilities can be introduced by limiting the size of the sets used in defining known reducibilities, like we did for btt-reducibility. Thus we can define, in a natural way, notions of **bounded conjunctive, disjunctive, positive, weak truth-table** and **Q-reducibility**. Also T-reducibility can be restricted in a similar way, once we recall (III.1.4) that, on r.e. sets, $A \leq_T B$ if and only if there is a recursive function f such that

$$x \in \overline{A} \Leftrightarrow (\exists u)(D_u \subseteq \overline{B} \wedge u \in \mathcal{W}_{f(x)}).$$

Then A is **bounded Turing reducible** to B if, moreover, there is a fixed number n bounding the size of $\mathcal{W}_{f(x)}$.

There are other ways of imposing bounds on known reducibilities. E.g. wtt-reducibility is obtained by restricting the number of *elements* queried in a computation. A restriction on the number of *queries* can be imposed too. Jockusch [1972a] calls A **bounded search reducible** to B ($A \leq_{bs} B$) if A is Turing reducible to B with a recursive bound on the number of queries to the oracle. Since we may suppose that, once the oracle has been queried, the answer to the query is stored and remembered, a bound on the size of the queries implies a bound on their number. Thus

$$A \leq_{wtt} B \Rightarrow A \leq_{bs} B \Rightarrow A \leq_T B.$$

Jockusch [1972a] proves that $\leq_{bs}$ *is not transitive*, the intuitive reason being that if $A \leq_{bs} B \leq_{bs} C$ then, given x, we know that the oracle on B is queried a recursively bounded number of times, but we do not know for which elements, and thus we cannot use the fact that the queries on C are also recursively bounded. The special case of a constant recursive bound is obviously transitive. Jockusch has proved that *there are bs-complete sets which are not wtt-complete*, and Soloviev [1976] shows that *there are T-complete sets which are not bs-complete*.

A different way to weaken reducibilities is by considering partial function in place of total ones, thus obtaining **partial reducibilities**. E.g. Ershov [1977] calls a set A partially m-reducible to a set B ($A \leq_{pm} B$) if, for some partial recursive function φ,

$$x \in A \iff \varphi(x){\downarrow} \wedge \varphi(x) \in B.$$

This reducibility is particularly important for the study of the Δ_2^0 sets.

For results on some of the reducibilities quoted above, see Degtev [1979], [1981], [1982], [1983], Soloviev [1976a], Omanadze [1976a], [1980], Zakharov [1984], [1986].

III.9   The World of Complete Sets ⋆

We summarize here the work done in this chapter with respect to completeness properties. We have already proved most of the positive results, and only some counterexamples are missing. Some of them are constructed by using the priority method, and should therefore wait until Chapter X. Since however this is the right place for them, we sketch their proofs here anyway, for the reader already acquainted with the method.

Relationships among completeness notions

Our goal is to show that *the following implications hold, and no other one does*:

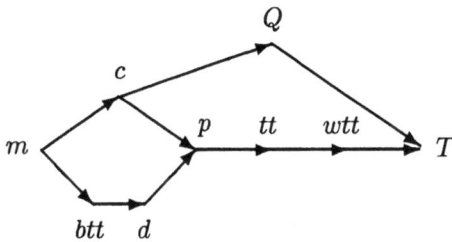

All the implications are trivially true, except for the fact, proved in III.8.11, that a btt-complete set is d-complete. We thus have only to provide counterexamples to the missing implications. They come from different sources, including Post [1944], Lachlan [1965], Young [1965], Jockusch [1968a], Gill and Morris [1974], Odifreddi [1981].

We first prove that no other arrow holds.

1. *There is a Q-complete, not wtt-complete set*
A hypersimple set is not *wtt*-complete, by III.8.16, and we proved on
p. 297 that there is a Q-complete, hypersimple set.

2. *There is a btt-complete, not Q-complete set*
We build two r.e. sets A and B such that $\mathcal{K} \leq_{btt} A$, and $B \nleq_Q A$. The
first condition ensures that A is *btt*-complete, the second that it is not
Q-complete.

To get $\mathcal{K} \leq_{btt} A$ we let

$$x \in \mathcal{K} \Leftrightarrow \{2x, 2x+1\} \cap A \neq \emptyset.$$

If at stage $s+1$ we see $x \in \mathcal{K}_s$ but $\{2x, 2x+1\} \cap A_s = \emptyset$, we put in A the
first element between $2x$ and $2x+1$ which is not restrained. If both are
restrained, we put in A the one restrained by the condition with lower
priority.

To get $B \nleq_Q A$, we want to spoil every reduction

$$x \in B \Leftrightarrow \mathcal{W}_{\varphi_e(x)} \subseteq A.$$

Pick up a witness a_e, and wait until $\varphi_e(a_e)$ converges. If it does not,
then φ_e was not a Q-reduction. Otherwise, wait for a later stage $s+1$
for which $\mathcal{W}_{\varphi_e(a_e)} \nsubseteq A_s$, i.e. $\mathcal{W}_{\varphi_e(a_e)} \cap \overline{A_s} \neq \emptyset$. If it never comes, then
$\mathcal{W}_{\varphi_e(a_e)} \subseteq A$, and a_e never gets into B, so φ_e does not Q-reduce B to
A. Otherwise, choose an element x_e of $\mathcal{W}_{\varphi_e(a_e)}$ which is not yet in A,
restrain it from entering A, and put a_e into B. If the condition is never
injured, i.e. x_e does not go into A for satisfaction of a requirement of the
first type (to make A *btt*-complete), then $\mathcal{W}_{\varphi_e(a_e)} \nsubseteq A$, while $a_e \in B$, and
again φ_e does not reduce B to A. Otherwise, a new attempt will have to
be made, to satisfy the requirement.

3. *There is a c-complete, not d-complete set*
Post's example III.3.5 of a simple set which is *tt*-complete, actually produces a *c*-complete set. But no simple set can be *d*-complete, by III.6.18.

4. *There is a btt-complete, not c-complete set*
The set built in part 2 above is *btt*-complete, and not Q-complete. It
cannot be *c*-complete, since *c*-reducibility implies Q-reducibility.

We now prove that no arrow can be reversed.

5. *There is a T-complete, not Q-complete set*
By III.4.23, there is a T-complete maximal set. Being hyperhypersimple,
it cannot be Q-complete (by III.4.10).

6. *There is a Q-complete, not c-complete set*
   We noted on p. 297 that a hypersimple set can be Q-complete, but by
   III.8.16 such a set cannot be *wtt*-complete, and in particular it cannot be
   c-complete.

7. *There is a c-complete, not m-complete set*
   There is a simple *tt*-complete set, by III.3.5, but a simple set is not
   m-complete, by III.2.7.

8. *There is a btt-complete, hence p-complete, set which is not c-complete,*
   hence not m-complete
   The set built in part 2 is p-complete but not Q-complete, and hence not
   c-complete.

9. *There is a d-complete, not btt-complete set*
   This is a modification of the proof of part 2. We build two r.e. sets A
   and B such that $\mathcal{K} \leq_d A$, and $B \not\leq_{btt} A$. Thus A is d-complete, but is
   not *btt*-complete.

   To get $\mathcal{K} \leq_d A$ we let

   $$x \in \mathcal{K} \Leftrightarrow I_x \cap A \neq \emptyset,$$

   where $I_0 = \{0\}$, $I_1 = \{1,2\}, \ldots$, and I_x has $x + 1$ elements. Of course
   we have to use tables with unbounded number of elements, otherwise
   we would have $\mathcal{K} \leq_{btt} A$, and there would be no hope of satisfying the
   remaining condition. To spoil *btt*-reductions with norm n we have to take
   care of n elements, and almost every I_x has more than n elements. This
   makes the argument work.

   To get $B \not\leq_{btt} A$, we want to spoil every reduction of bounded norm

   $$x \in B \Leftrightarrow A \models \sigma_{\varphi_e(x)}.$$

   Pick up a witness a_e, and wait until $\varphi_e(a_e)$ converges. If it does not,
   then φ_e was not a *btt*-reduction. Otherwise, wait for a later stage $s + 1$
   for which $A_s \models \sigma_{\varphi_e(x)}$. If it never comes, $A \models \sigma_{\varphi_e(a_e)}$ fails but a_e never
   gets into B, so φ_e does not *btt*-reduce B to A. Otherwise, restrain from
   entering A all the elements used in $\sigma_{\varphi_e(a_e)}$ and not yet in A, and put a_e
   into B. If the condition is never injured, i.e. no element used in $\sigma_{\varphi_e(a_e)}$
   and not in A_s goes into A for satisfaction of a requirement of the first
   type (to make A d-complete), then $A \models \sigma_{\varphi_e(a_e)}$ fails, while $a_e \in B$, and
   again φ_e does not reduce B to A. Otherwise, a new attempt will have to
   be made, to satisfy the requirement.

10. *There is a p-complete, not d-complete set*
    The simple *tt*-complete set of III.3.5 is actually *c*-complete, hence *p*-complete, but is not *d*-complete by III.6.18, being simple.

11. *There is a tt-complete, not p-complete set*
    A semirecursive set is not *p*-complete, by III.5.3.b, but may be *tt*-complete, by III.5.5.a.

12. *There is a T-complete, not wtt-complete set*
    A hypersimple set is not *wtt*-complete by III.8.16, but can be *T*-complete, by III.3.13.

As the reader will have noticed, only one result is missing. Its proof is more complicated than the ones given so far.

Theorem III.9.1 (Lachlan [1975]) *There is a set which is wtt-complete, but not tt-complete.*

Proof. We want to build A *wtt*-complete, but not *tt*-complete. We will reduce $\mathcal{K}$ to A, with a recursive bound. We will thus define a recursive function f (such that $f(0) = 0$), and a sequence of boxes

$$I_x = \{z : f(x) \leq z < f(x+1)\}.$$

- To get $\mathcal{K} \leq_{wtt} A$ is enough to ensure that, whenever $x \in \mathcal{K}_{s+1} - \mathcal{K}_s$, some element of $I_x \cap \overline{A}_s$ enters A_{s+1}. Then, to know if $x \in \mathcal{K}$, we look for s so big that all elements of $I_x \cap A$ have been generated in A_s, and then see if $x \in \mathcal{K}_s$.

- To get A not *tt*-complete, we define B r.e. such that $B \not\leq_{tt} A$. The strategy for this is the usual one: to spoil the *e*-th reduction, choose a witness a_e, wait for $\varphi_{e,s}(a_e)$ to converge, and then put a_e in B if and only if $\sigma_{\varphi_e(a_e)}$ fails on A.

Since requirements of the first kind have to be satisfied immediately (i.e. they have highest priority), they might interfere with the satisfaction of requirements of the second kind. To avoid this, we define a positive equivalence relation η which breaks down the boxes into pieces, i.e. such that, at any stage, its equivalence classes are subintervals of the boxes. Also, A_s is η-closed, and consists of initial segments of the boxes. In other words, at stage s a typical box I_x looks like this:

$$\boxed{\begin{array}{|c|c|c|c|c|c|} \hline A \cap I_x & & & \| & \\ \hline \end{array}}$$

$f(x)$ $f(x+1)$

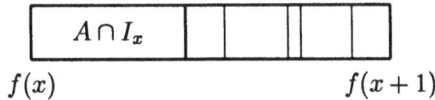

The *construction* is as follows. Suppose that $\mathcal{K}$ is enumerated in such a way that, for infinitely many stages, nothing is enumerated in it. At stage $s+1$ we do the following:

- if $x \in \mathcal{K}_{s+1} - \mathcal{K}_s$, take the smallest element of $I_x \cap \overline{A}_s$, and put its equivalence class into A. We will define f in such a way that $I_x \cap \overline{A}_s$ is always nonempty, so that this is always possible.

- if no element is generated in $\mathcal{K}$ at stage $s+1$, look for the smallest $e \leq s$ such that the requirement

$$R_e : \quad \neg(B \leq_{tt} A \text{ via } \varphi_e)$$

has not yet been satisfied and not injured afterwards, and $\varphi_{e,s}(a_e^s)$ converges (where a_e^s is the current witness for R_e). Consider the smallest $m \geq e$ such that all elements used in $\sigma_{\varphi_{e,s}}(a_e^s)$ are in $\bigcup_{i \leq m} I_i$. We act in such a way as to ensure that the truth value of $A \models \sigma_{\varphi_{e,s}}(a_e^s)$ depends, after stage $s+1$, only on $A \cap (\bigcup_{i < e} I_i)$, i.e. on the first e boxes. This requires action on $I_e, \ldots, I_m$. We show how to act on I_m: the same action will be taken on $I_{m-1}, \ldots, I_e$. Consider I_m at stage s and let $n_0, \ldots, n_q$ be the largest elements of the equivalence classes of I_m.

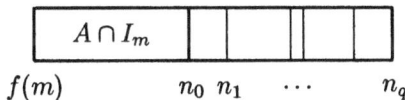

$$\boxed{\begin{array}{|c|c|c|c|c|c|} \hline A \cap I_m & & & \| & \\ \hline \end{array}}$$

$f(m)$ $n_0 \; n_1$ $\cdots$ n_q

Since, by construction, we put equivalence classes into A, the final value of $A \cap I_m$ will be one of $\{z : f(m) \leq z \leq n_i\}$, for some $i \leq q$. The final value of $A \cap (\bigcup_{i < m} I_i)$ is one of $2^{f(m)}$ many (since there are $f(m)$ elements in the first m boxes). For each n_i, there are thus $2^{2^{f(m)}}$ possibilities: it follows that, if $q \geq 2^{2^{f(m)}}$, then at least two n_i's determine the same truth-value of $\sigma_{\varphi_{e,s}}(a_e^s)$, for *any* possible choice of A below $f(m)$. In general, if $q \geq 2^{r \cdot 2^{f(m)}}$ then there are $r+1$ of the n_i's which determine the same truth-value of $\sigma_{\varphi_{e,s}}(a_e^s)$, for *any* possible choice of A below $f(m)$.

We will define f in such a way that r is big enough to ensure that, for any t, $I_m \cap \overline{A}_t \neq \emptyset$. Thus let $r \leq q$ be the maximum number such that there exists a subsequence $p_0, \ldots, p_r$ of $n_0, \ldots, n_q$ which determines the same truth-value of $\sigma_{\varphi_{e,s}(a_e^s)}$, for *any* possible choice of A below $f(m)$. Let

$$I_m \cap A_{s+1} = \{z : f(m) \leq z \leq p_0\},$$

and, for $i < r$, let $\{z : p_i < z \leq p_{i+1}\}$ be the new equivalence classes of η on I_m. Since in the future

$$I_m \cap A = \{z : f(m) \leq z \leq p_i\}$$

for some $i \leq r$ (by construction), the truth-value of $\sigma_{\varphi_{e,s}(a_e^s)}$ depends now only on the first m boxes.

We proceed similarly on $I_{m-1}, \ldots, I_e$ by descending induction, determining $I_i \cap A_{s+1}$ for $e \leq i \leq m$, and in the end we have

$$A_{s+1} = A_s \cup \bigcup_{e \leq i \leq m} (I_i \cap A_{s+1}).$$

Note that the truth-value of $A \models \sigma_{\varphi_{e,s}(a_e^s)}$ depends only on the first e boxes. Put a_e^s into B if and only if $A_{s+1} \models \sigma_{\varphi_{e,s}(a_e^s)}$ fails. The requirement R_e is now satisfied, and can be injured only if something changes on the first e boxes.

Of course we could have gone all the way down to I_0, to fix the truth-value of $\sigma_{\varphi_{e,s}(a_e^s)}$ once and for ever. The trouble with this is that the size of the boxes is finite, and we are not able to avoid a collapse, if every condition is free to interfere with every box. By letting R_e interfere with I_i only for $i \geq e$, we ensure that only finitely many conditions interfere with a given box I_x, and we are thus able to show that $I_x \cap \overline{A}_s$ is always nonempty (something which is needed to satisfy the first kind of requirements).

It only remains to determine f. Note that the internal situation of I_x can be changed only if either x gets into $\mathcal{K}$, or some R_e with $e \leq x$ is satisfied. Now R_e can be satisfied at two different stages only if it gets injured between them, and to injure R_e we must take action on $\bigcup_{i<e} I_i$. This is possible only if some $i < e$ gets into $\mathcal{K}$ (which can happen only e times), or some R_i with $i < e$ is satisfied. By induction, it follows that R_e can be satisfied less than 2^{e+1} times, hence I_x can change at most $\sum_{e \leq x} 2^{e+1} = 2^{x+2}$ times. It is thus enough to let:

$$f(0) = 0$$
$$f(x+1) = 2^{2^{x+2} \cdot 2^{f(x)}}. \quad \square$$

Exercise III.9.2 *There is an acceptable system of indices for which Post's simple set is tt-incomplete.* (Lachlan [1975]) (Hint: apply the method of III.3.6 to the above proof.)

We briefly discuss now bounded truth-table reducibilities (p. 340). The interesting fact is that *the notions of bd, bp, btt and bwtt-completeness coincide* (Lachlan [1975], Kobzev [1974], [1977]), the first three because a *btt*-complete set is *d*-complete and *p*-complete (III.8.11), and the last two with a proof similar to III.9.1. Moreover, from the next theorem it follows that *bc and m-completeness coincide*, and thus there are only two interesting notions of completeness for bounded truth-table reducibilities.

Theorem III.9.3 (Lachlan [1966]) *If $A \cdot B$ is m-complete, so is at least one of A and B.*

Proof. We want to build an r.e. set D such that one of the following holds:

1. $\mathcal{K} \leq_m D \leq_m A$

2. $\mathcal{K} \leq_m D \leq_m B$.

Note that, no matter how we define D, it will be an r.e. set. Since $A \cdot B$ is m-complete, there will be a recursive function h such that

$$x \in D \Leftrightarrow h(x) \in A \cdot B,$$

and thus there will be two recursive functions f and g such that

$$x \in D \Leftrightarrow f(x) \in A \wedge g(x) \in B.$$

By the Fixed-Point Theorem we can use D itself in its own definition, and hence we may suppose f and g given beforehand.
Consider the set

$$D^* = \{z : (\exists t)(z \notin D_t \wedge g(z) \in B_t\}.$$

For $z \in D^*$ we have $z \in D \Leftrightarrow f(z) \in A$, since $g(z) \in B$ already. There are then two cases:

- If D^* is infinite, let $\{z_0, z_1, \dots\}$ be a recursive enumeration of it. Then $z_x \in D \Leftrightarrow f(z_x) \in A$, and we can easily ensure condition 1 by letting $x \in \mathcal{K} \Leftrightarrow z_x \in D$.

- If D^* is finite we have $z \in D \Leftrightarrow g(z) \in B$, with at most finitely many exceptions. Indeed, if $z \in D$ then $g(z) \in B$; and there are only finitely many z such that $g(z) \in B \wedge z \notin D$, since they must be in D^*. Then we can easily ensure condition 2 by letting D and $\mathcal{K}$ differ only finitely.

The construction of D is the following. At stage $s+1$ we have $\mathcal{K}_s$, D_s, and an enumeration $\{z_0, \ldots, z_i\}$ of the elements of

$$D_s^* = \{z : (\exists t \leq s)(z \notin D_t \wedge g(z) \in B_t\}.$$

Then:

- if $x \in \mathcal{K}_{s+1}$, let $z_x \in D_{s+1}$ (if z_x exists, i.e. if enough elements have already been generated in D^*)

- if $x \in \mathcal{K}_{s+1}$ and $x \notin D_s^*$, let $x \in D_{s+1}$ (this kills x, in the sense that it ensures $x \notin D^*$, since x goes into D).

If D^* is infinite then $x \in \mathcal{K} \Leftrightarrow z_x \in D$: if $x \in \mathcal{K}$ then $z_x \in D$ by the first part of the construction; and if $z_x \in D$ then it must be $x \in \mathcal{K}$, since the second part of the construction puts in D only elements of $\overline{D^*}$, while $z_x \in D^*$.

If D^* is finite, then the second part of the construction applies, except for finitely many cases, and thus, for almost every z, $z \in \mathcal{K} \Leftrightarrow x \in D$. $\square$

Corollary III.9.4 *A bc-complete set is m-complete.*

Proof. A set A is bc-complete if there is a recursive function f such that

$$x \in \mathcal{K} \Leftrightarrow D_{f(x)} \subseteq A,$$

where the size of $D_{f(x)}$ is bounded by a fixed number n. By possibly adding elements of A to it, we can suppose that $D_{f(x)}$ always has exactly n elements. But then there are n recursive functions f_i such that

$$x \in \mathcal{K} \quad \Leftrightarrow \quad f_1(x) \in A \wedge \cdots \wedge f_n(x) \in A$$
$$\Leftrightarrow \quad \langle f_1(x), \ldots, f_n(x) \rangle \in A \cdots A.$$

But then $A \cdots A$ is m-complete and thus, by repeatedly applying the theorem, so is A. $\square$

Structural properties and completeness

We now summarize the connections between structural properties on one side, and completeness properties on the other.

We first look at properties implying completeness.

1. *A set is creative if and only if it is 1-complete (or m-complete)* (III.6.6, III.7.5).

2. *A set is quasicreative if and only if it is d-complete* (III.6.16).

3. *A set is subcreative if and only if it is Q-complete* (III.6.20).

4. *An effectively simple set is T-complete* (III.2.18), *but it may be Q-incomplete* (III.4.24.b, III.4.10) *or wtt-incomplete* (III.8.19).

5. *A strongly effectively simple set is Q-complete* (III.6.21.a), *but it may be wtt-incomplete* (III.8.19).

We then look at properties implying incompleteness.

6. *A simple set is not d-complete, and hence not btt-complete* (III.6.18.a), *but it may be c-complete* (III.3.5).

7. *A hypersimple set is not wtt-complete* (III.8.16), *but it may be Q-complete* (p. 297).

8. *A hyperhypersimple set is not Q-complete* (III.4.10), *but it may be T-complete* (III.4.23).

9. *A semirecursive set is not p-complete, and hence not btt-complete* (III.5.3.b), *but it may be tt-complete* (III.5.5.a) *or Q-complete* (p. 297).

We finally look at naturally defined sets with completeness properties.

10. $\mathcal{K}$ *is 1-complete and m-complete.*

11. *Post's simple set* (III.2.11) *is wtt-complete* (being effectively simple and not hypersimple, III.8.18) *and not m-complete* (being simple), *and can be tt-complete* (actually c-complete, III.3.6) *or not* (III.9.2), *depending on the acceptable system of indices.*

12. *The deficiency set of $\mathcal{K}$ is T-complete* (III.3.13) *but not wtt-complete* (being hypersimple).

III.10　Formal Systems and R.E. Sets ⋆

As we have seen (p. 253), the theory of r.e. sets arose from problems connected with the study of formal systems, and Post's problem was motivated by methodological considerations. But we have pushed the study of r.e. sets quite far, and it is natural to enquire whether we have lost touch with 'reality'. We will see here that, far from being so, many of the notions so far introduced have bearing on our original study of formal systems, and provide appropriate tools to describe natural phenomena.

Formal systems and r.e. sets $\star$

An abstract approach to formal systems, which isolates their basic properties, is the following. Fix a formal countable language, and identify (by arithmetization) the sets of well-formed formulas and of sentences (formulas with no free variable) with two recursive sets F and S, with $S \subseteq F$.

Definition III.10.1 *A* **formal system** $\mathcal{F}$ *in the given language is a pair* (T, R) *of r.e. sets contained in S and interpreted, respectively, as the sets of (codes of) theorems and of refutable formulas. $\mathcal{F}$ is said to be:*

1. **consistent** *if* $T \cap R = \emptyset$

2. **complete** *if* $T \cup R = S$

3. **decidable** *if T and R are recursive*

4. **undecidable** *if T is not recursive.*

By Post's Theorem, *a consistent and complete formal system is decidable* (Janiczak [1950]): given the code $n \in S$ of a sentence, enumerate T and R simultaneously, until n appears in one (and exactly one) of them.

Usually T is generated by isolating an r.e. subset Ax of S (set of **axioms**), and n-ary functions $r : S^n \to S$ (called **rules of deduction**). Then T is the closure of Ax under the rules. R can be generated in the same way, and independently of T, or there may be a function $n : S \to S$ (called **negation**) such that $R = n(T)$.

Exercise III.10.2 *Every formal system can be generated by means of finitely many recursive rules of deduction, from a finite set of axioms.* (Kleene [1952]) (Hint: since T is r.e., for some recursive R is $x \in T \Leftrightarrow \exists y R(x, y)$. As a system of axioms take the equations of a system which Herbrand-Gödel computes the characteristic function f of R. As rules, take R1 and R2, together with the rule that produces x from $f(\overline{x}, \overline{y}) = \overline{1}$.)

While the abstract notion of formal system is certainly broad enough to include all the common examples, the formal systems associated to r.e. sets (like in III.10.2) may look unnatural and ad hoc. In mathematics one usually uses **first-order formal systems**, whose axioms include all axioms of first-order logic with equality, the set of theorems is generated from the axioms by the first-order logical rules, and the refutable formulas are the negations of theorems. For first-order formal systems, *inconsistency implies that every sentence is a theorem*, because if φ and $\neg\varphi$ are both provable then so is $\neg\varphi \vee \psi$ (i.e. $\varphi \to \psi$) for any sentence ψ, and then ψ follows by modus ponens.

The next result shows that the r.e. sets do describe (at least up to T-equivalence) formal systems of common use.

Proposition III.10.3 (Feferman [1957]) *Every r.e. T-degree contains a first-order formal system.*

Proof. Let A r.e. be given. The idea is to consider a decidable theory, and code A into it. Consider the language of field theory (containing $=,0,1,+,\times$). First note that:

- *The theory of algebraically closed fields of any given characteristic is complete and decidable.*
  If neither φ nor $\neg\varphi$ are provable, both are consistent with the theory, and by the Completeness Theorem of first-order logic there are two algebraically closed fields of the given characteristic and same uncountable cardinality, satisfying one φ, and the other $\neg\varphi$. But this is against Steinitz Theorem, according to which any two such fields must be isomorphic, and in particular must satisfy the same formulas. Thus the theory is a complete and consistent formal system, and it is then decidable.

We use field characteristics to code the given r.e. set A: let T_A be the theory of algebraically closed fields whose characteristic is not p_n, for any $n \in A$ (i.e. add to the axioms of algebraically closed fields the statements $\underbrace{1 + \ldots + 1}_{p_n \text{ times}} \neq 0$,

for $n \in A$). Then:

- $A \leq_T T_A$
  Indeed $n \in A \Leftrightarrow \vdash_{T_A} \underbrace{1 + \ldots + 1}_{p_n \text{ times}} \neq 0$, for the following reasons.
  If $n \in A$ then $\underbrace{1 + \ldots + 1}_{p_n \text{ times}} \neq 0$ is obviously provable, being an axiom. And
  if $n \notin A$ then $\underbrace{1 + \ldots + 1}_{p_n \text{ times}} \neq 0$ cannot be provable, because in this case T_A
  is consistent with the theory of algebraically closed fields of characteristic p_n.

- $T_A \leq_T A$
  By definition T_A is r.e. in A. It is then enough to show also that its complement is r.e. in A. Since a formula φ is provable in T_A if and only if it is true in some algebraically closed field of characteristic 0 or p_n, for some $n \notin A$, by completeness of the theories of algebraically closed fields of given characteristic we have that φ is not provable in T_A if and only its negation is true in some field of characteristic either 0 or p_n, for some $n \notin A$. Then, by decidability of the same theories, $\overline{T_A}$ is r.e. in A. □

More results on the connections between degrees and formal systems are in Feferman [1957], Shoenfield [1958], and Hanf [1965]. In particular, *every r.e. tt-degree contains a finitely axiomatizable first-order formal system.* On the other hand, *not every r.e. m-degree contains a first-order formal system,* because closure under conjunction implies that if A is the set of theorems of a first-order formal system, then $A \cdot A \leq_m A$, and this fails in general for r.e. sets (see III.8.5).

Undecidability

To express in full generality the methods to prove undecidability, suppose that the language contains constants $\bar{n}$, for every n. Fix an effective enumeration $\{\psi_e\}_{e \in \omega}$ of the formulas of the language with one free variable, and suppose that there is a recursive **substitution** function s, such that $s(e, n)$ is the code of $\psi_e(\bar{n})$ as a sentence. The notion of representability of predicates (I.3.4) can be easily generalized to this abstract setting: a set A is **weakly representable** in $\mathcal{F}$ if, for some e,

$$x \in A \Leftrightarrow s(e, x) \in T.$$

Note that every set weakly representable in a formal system must be r.e.

The direct methods to prove the undecidability of a formal system are summarized in the following:

Theorem III.10.4 Direct undecidability proofs (Tarski, Mostowski and Robinson [1953], Bernays [1957], Putnam [1957], Smullyan [1958], Vaught) *A formal system $\mathcal{F}$ is undecidable if one of the following holds:*

1. *Every recursive set is weakly representable in $\mathcal{F}$.*

2. *Some nonrecursive r.e. set is weakly representable in $\mathcal{F}$.*

Proof. If $\mathcal{F}$ is decidable, the diagonal set

$$n \in T^* \Leftrightarrow s(n, n) \in T$$

is recursive, and then so is $\overline{T^*}$. If every recursive set is representable in $\mathcal{F}$, there is e such that

$$n \in \overline{T^*} \Leftrightarrow s(e, n) \in T.$$

For $n = e$ we get a contradiction, and this proves the first part.

The second method is trivial: if $\mathcal{F}$ is decidable, every set weakly representable in it must be recursive. $\square$

The two methods are substantially different in principle: the undecidability is forced in one case by the *quantity* of sets represented, despite the fact that all of them may be computable (the underlying reason being that there is no recursive enumeration of all the recursive sets), in the other by the (noncomputable) *quality* of some of them. It is not obvious however that they are distinct, since it could be that once all the recursive sets are weakly representable, also some nonrecursive r.e. set is. Actually, for the examples of II.2.17 much more is true: every r.e. set is weakly representable. In particular, all these examples weakly represent $\mathcal{K}$, and thus the sets of their theorems are all T-complete. Thus *the positive solution to Post's Problem implies that not every undecidable formal system can weakly represent every r.e. set*, but this does not solve our question yet. The answer has been obtained by Shoenfield [1961], with *a formal system in which the weakly representable sets are exactly the recursive ones*. The proof introduces a difficult extension of the priority method used for the solution of Post's Problem, and will be given in X.3.10.

There is also an indirect method to prove undecidability, and this was Post's direct concern. Formal systems can be interpreted one into another. The simplest way is having a recursive function f that translates formulas and preserves theorems: given $\mathcal{F}$ and $\mathcal{F}'$ with sets of theorems F and F',

$$x \in F \Leftrightarrow f(x) \in F'.$$

But this is only one possible way, and we can say in general that $\mathcal{F}$ is **interpretable** in $\mathcal{F}'$ if $F \leq_T F'$. Then the following is simply an observation:

Theorem III.10.5 Indirect undecidability proofs. *$\mathcal{F}$ is undecidable if an undecidable formal system is interpretable in it.*

The undecidability of the Predicate Calculus was obtained this way (II.2.18), as well as the undecidability of the formal systems of current use in mathematics (see e.g. Tarski, Mostowski and Robinson [1953], Ershov, Lavrov, Taimanov and Taitslin [1965]), using as starting point systems that (like $\mathcal{R}$) are all T-complete, and thus showing that they are all T-complete as well. Thus *the positive solution to Post's Problem implies that there are undecidable formal systems that cannot be proved undecidable by interpreting $\mathcal{R}$ in them.*

Essential undecidability

Since a consistent formal system $\mathcal{F}$ is described in terms of a pair of disjoint r.e. sets (T, R) contained in a recursive set S, and respectively coding the theorems and the refutable formulas, we may first of all expect that the theory of pairs of r.e. sets might be relevant.

Definition III.10.6 (Tarski [1949]) *A consistent formal system $\mathcal{F} = (T, R)$ is **essentially undecidable** if T and R are recursively inseparable.*

If we say that $\mathcal{F}' = (T', R')$ is an **extension** of $\mathcal{F} = (T, R)$ (in the same language) when $T', R' \subseteq S$, $T \subseteq T'$ and $R \subseteq R'$, then essential undecidability means that *not only the system itself, but also every consistent extension of it in the same language is undecidable.*

The notion of representability of sets can be generalized to pairs. Recall that s is the substitution function, and $s(e, n)$ is the code of the formula $\psi_e(\overline{n})$, for a given enumeration $\{\psi_e\}_{e \in \omega}$ of the formulas of the language with one free variable.

Definition III.10.7 (Kleene [1952], Putnam and Smullyan [1960])
*Given a formal system $\mathcal{F} = (T, R)$ and two disjoint sets A and B, we say that A and B are **separable** in $\mathcal{F}$ if there is e such that*

$$x \in A \Rightarrow s(e, x) \in T \quad and \quad x \in B \Rightarrow s(e, x) \in R.$$

As in the case of simple undecidability, we have two methods to prove essential undecidability.

Theorem III.10.8 Direct proofs of essential undecidability (Kleene [1952], Tarski, Mostowski and Robinson [1953], Smullyan [1958], Putnam and Smullyan [1960]) *A consistent formal system $\mathcal{F}$ is essentially undecidable if one of the following holds:*

1. *Every pair of disjoint recursive sets is separable in $\mathcal{F}$.*

2. *Some recursively inseparable pair is separable in $\mathcal{F}$.*

Proof. Note that everything separable in $\mathcal{F}$ remains separable in any consistent extension of it. And if every pair of disjoint recursive sets is separable in a consistent system (T', R'), every recursive set is weakly representable in it (given A recursive, let e be the number of the formula separating A and $\overline{A}$: then

$$\begin{aligned} x \in A &\Rightarrow s(e, x) \in T' \\ x \in \overline{A} &\Rightarrow s(e, x) \in R' \Rightarrow s(e, x) \notin T' \end{aligned}$$

by consistency, and hence $x \in A \Leftrightarrow s(e, x) \in T'$). Then every consistent extension of $\mathcal{F}$ is undecidable, by III.10.4 (because every recursive set is weakly representable in it), and $\mathcal{F}$ is essentially undecidable.

The second part holds because $\mathcal{F}$, if not essentially undecidable, has a decidable consistent extension (T', R'), in which the given recursively inseparable pair (A, B) is still separable:

$$x \in A \Rightarrow s(e, x) \in T' \quad and \quad x \in B \Rightarrow s(e, x) \in R'.$$

But then the recursive set $\{x : s(e, x) \in T'\}$ would separate A and B. $\square$

Exercises III.10.9 A set A is **representable** in $\mathcal{F} = (T, R)$ if, for some e,

$$x \in A \Rightarrow s(e, x) \in T \quad \text{and} \quad x \in \overline{A} \Rightarrow s(e, x) \in R,$$

i.e. if A and $\overline{A}$ are separable in $\mathcal{F}$.

a) *Every recursive set is representable in $\mathcal{F}$ if and only if every pair of disjoint recursive sets is separable in it.*

b) *If some nonrecursive set is representable in $\mathcal{F}$, then some recursively inseparable pair of sets is separable in it, but not conversely.* (Hint: in $\mathcal{R}$ only recursive sets are representable, by II.2.16, but every pair of disjoint r.e. sets is separable, see below.)

As already for the undecidability proofs, the two methods for essential undecidability are substantially different and distinct. We now see that they both apply, as usual, to consistent extensions of $\mathcal{R}$.

Theorem III.10.10 (Rosser [1936], Putnam and Smullyan [1960]) *In any consistent formal system $\mathcal{F}$ extending $\mathcal{R}$, every disjoint pair of r.e. sets is separable.*

Proof. If A and B are disjoint r.e. sets, there exist recursive relations R and Q such that

$$x \in A \Leftrightarrow \exists y R(x, y) \quad \text{and} \quad x \in B \Leftrightarrow \exists y Q(x, y).$$

We already know that all the recursive relations are representable in $\mathcal{F}$ (II.2.16), and then there are formulas ψ_1 and ψ_2 such that

$$R(x, y) \Rightarrow \vdash_{\mathcal{F}} \psi_1(\overline{x}, \overline{y}) \quad \text{and} \quad \neg R(x, y) \Rightarrow \vdash \neg \psi_1(\overline{x}, \overline{y})$$
$$Q(x, y) \Rightarrow \vdash_{\mathcal{F}} \psi_2(\overline{x}, \overline{y}) \quad \text{and} \quad \neg Q(x, y) \Rightarrow \vdash \neg \psi_2(\overline{x}, \overline{y}).$$

Then the formula

$$\varphi(x) \Leftrightarrow \exists y [\psi_1(x, y) \wedge (\forall z < y) \neg \psi_2(x, z)]$$

separates A and B:

- $x \in A \Rightarrow \vdash_{\mathcal{F}} \varphi(\overline{x})$

  If $x \in A$ then, for some y, $R(x, y)$ holds, and hence $\vdash_{\mathcal{F}} \psi_1(\overline{x}, \overline{y})$. Since A and B are disjoint, $x \notin B$: hence, for every z, $\neg Q(x, z)$ and so $\vdash_{\mathcal{F}} \neg \psi_2(\overline{x}, \overline{z})$. By the axioms of $\mathcal{R}$ we can treat bounded quantifiers, and hence we also have $\vdash_{\mathcal{F}} (\forall z < \overline{y}) \neg \psi_2(\overline{x}, z)$. Then $\vdash_{\mathcal{F}} \varphi(\overline{x})$.

- $x \in B \Rightarrow \vdash_{\mathcal{F}} \neg \varphi(\overline{x})$

  Note that $\neg \varphi(x) \Leftrightarrow \forall y [\neg \psi_1(x, y) \vee (\exists z < y) \psi_2(x, z)]$. If $x \in B$ then, for some z, $Q(\overline{x}, \overline{z})$ holds, and hence $\vdash_{\mathcal{F}} \psi_2(\overline{x}, \overline{z})$. Since A and B are disjoint, $x \notin A$ and so, for every y, $\neg R(x, z)$ holds, and $\vdash_{\mathcal{F}} \neg \psi_1(\overline{x}, \overline{y})$. By the axioms of $\mathcal{R}$ we know that, for every y and a fixed z, $y \leq \overline{z} \vee \overline{z} < y$. Thus we only have to consider finitely many cases because, again by the axioms of $\mathcal{R}$, $y \leq \overline{z}$ means that y is actually the numeral corresponding to some $y \leq z$. If $y \leq \overline{z}$ then $\vdash_{\mathcal{F}} \neg \psi_1(\overline{x}, \overline{y})$. If $\overline{z} < y$ then, from $\psi_2(\overline{x}, \overline{z})$, we have $(\exists z < y) \psi_2(\overline{x}, z))$. Thus $\vdash_{\mathcal{F}} \neg \varphi(\overline{x})$.   □

Corollary III.10.11 *If $\mathcal{F}$ is a consistent formal system extending $\mathcal{R}$ then $\mathcal{F}$ is essentially undecidable, and the sets of (codes of) theorems and of refutable formulas of $\mathcal{F}$ are effectively inseparable (and in particular creative).*

Proof. Both criteria for essential undecidability are applicable, since every pair of disjoint r.e. sets is separable in $\mathcal{F}$, in particular every pair of disjoint recursive sets, and any recursively inseparable pair.

Moreover, separability of A and B provides a simultaneous m-reduction of them to the sets of theorems and of refutable formulas. These are then effectively inseparable, because there is a pair A and B of effectively inseparable r.e. sets.   □

Smullyan [1961] calls $\mathcal{F} = (T, R)$ a **Gödel theory** if every r.e. set is weakly representable in it, and a **Rosser theory** if every pair of disjoint r.e. sets is separable in it. These two notions simply express m-completeness, respectively for r.e. sets and pairs of disjoint r.e. sets. Thus *if $\mathcal{F}$ is a Gödel theory then T is creative, and if $\mathcal{F}$ is a Rosser theory then T and R are effectively inseparable.* The fact that two disjoint creative sets may be recursively separable (III.6.26.d) can be seen as saying that *a Gödel theory is not necessarily essentially undecidable.* On the other hand, the existence of recursively inseparable pairs in any nonrecursive r.e. T-degree (III.6.22) shows that *an essentially undecidable formal system is not necessarily a Rosser theory.*

Pour El [1968] calls **effectively extensible** a theory $\mathcal{F} = (T, R)$ for which there is an effective procedure that produces, for any consistent extension $\mathcal{F}'$ of it, a formula which is neither provable nor refutable in $\mathcal{F}'$. This notion is clearly equivalent to the effective inseparability of T and R (see III.6.27.b), and stresses the effective content of the proof of Gödel's Theorem, which effectively exhibits undecidable sentences for any sufficiently strong consistent formal system.

The last results and comments show that the notions of creativeness and effective inseparability are useful tools for the description of common theories, for what concerns the description of undecidability and related phenomena. The limits of this usefulness are pointed out by III.7.13 and III.7.15 (and their extensions considered in Pour El and Kripke [1967], Pour El [1968]): *from a recursion-theoretic point of view, all effectively inseparable (and, in particular, a great number of common) formal systems are isomorphic, and thus mere*

variations of one another. To get a finer analysis the recursive, purely extensional approach (only considering the sets of theorems) has to give in to a
proof-theoretical, intensional one (accounting for the way the theorems are obtained). As Kreisel [1971] puts it, *Proof Theory begins where Recursion Theory
ends*.

Independent axiomatizability

Since consistent theories have disjoint sets of theorems and refutable formulas, if
we call a formal system nontrivial when it has at least infinitely many theorems
and infinitely many refutable formulas, then (the sets associated to) *consistent
nontrivial formal systems are not simple*. This does not mean that the notion
of simplicity, as well as its strengthenings, has no relevance for the description
of formal systems, as we now see.

Until now we have looked at decidability questions, and thus recursive enumerability was seen as opposed to recursiveness. This depended on the fact
that we considered formal systems only, that is r.e. sets of formulas. But in
mathematics theories need not be r.e. (and the results on the limitations of
formal systems show that there is no other way to override the incompleteness
phenomena, while preserving expressive power).

Definition III.10.12 *A set of formulas in a given language is called a* **first-
order theory** *if it is closed under logical consequence. A theory is:*

1. **axiomatizable** *if it is the closure under logical consequence of an r.e.
   set of formulas* $\{\alpha_e\}_{e \in \omega}$, *called the axioms*

2. **independently axiomatizable** *if moreover, for every e, the axiom* α_e
   is not a logical consequence of the remaining axioms

3. **finitely axiomatizable** *if it has a finite set of axioms.*

Note that *the axiomatizable theories are just the first-order formal systems*,
because the closure of an r.e. set of formulas under recursive rules (like the ones
of any standard complete formalization of the Predicate Calculus) is still an r.e.
set of formulas. Moreover, *axiomatizable theories always have a recursive set
of axioms* (Craig [1953]), because any formula is implied by the conjunction of
itself with other formulas, and thus we can substitute the n-th axiom with the
conjunction of the first n ones, obtaining a set enumerated in increasing order,
and hence recursive. Finally, *a finitely axiomatizable theory is independently
axiomatizable*.

Proposition III.10.13 (Kreisel [1957], Pour El [1968a]) *The following
are equivalent, for an axiomatizable first-order theory* $\mathcal{F}$:

1. $\mathcal{F}$ is independently axiomatizable

2. for any axiomatization $\{\alpha_e\}_{e \in \omega}$ of $\mathcal{F}$, the set

$$n \in A \Leftrightarrow \alpha_0 \wedge \cdots \wedge \alpha_n \models \alpha_{n+1}$$

   is not hypersimple

3. for some axiomatization $\{\alpha_e\}_{e \in \omega}$ of $\mathcal{F}$, the set

$$n \in A \Leftrightarrow \alpha_0 \wedge \cdots \wedge \alpha_n \models \alpha_{n+1}$$

   is not hypersimple.

Proof. Note that the conditions are trivially equivalent when $\mathcal{F}$ is finitely axiomatizable (because in this case the set A is finite, and certainly not hypersimple), and then we can suppose that $\mathcal{F}$ is not finitely axiomatizable. We first prove that if $\mathcal{F}$ is independently axiomatizable, and $\{\alpha_e\}_{e \in \omega}$ is any axiomatization of it, then the set

$$n \in A \Leftrightarrow \alpha_0 \wedge \cdots \wedge \alpha_n \models \alpha_{n+1}$$

is not hypersimple. Fix an independent axiomatization $\{\beta_n\}_{n \in \omega}$ of $\mathcal{F}$. It is enough to show that, given any n, we can effectively find an $m > n$ such that $\{n, \ldots, m - 1\} \cap \overline{A} \neq \emptyset$ (since this permits an inductive generation of a strong array intersecting $\overline{A}$). Given $\alpha_0, \cdots, \alpha_n$ we can find first of all a number p such that they are deducible from $\beta_0, \ldots, \beta_p$. Then we can find a number m such that β_{p+1} is deducible from $\alpha_0, \ldots, \alpha_m$. Now one of $n, \ldots, m - 1$ is not in A otherwise, inductively, all of $\alpha_{n+1}, \ldots, \alpha_m$ would be deducible from $\alpha_0, \ldots, \alpha_n$, and hence from $\beta_0, \ldots, \beta_p$. But then so would be β_{p+1}, contradicting the fact that the β's are an independent axiomatization.

We now show that if, for any axiomatization $\{\alpha_e\}_{e \in \omega}$ of $\mathcal{F}$, the set

$$n \in A \Leftrightarrow \alpha_0 \wedge \ldots \wedge \alpha_n \models \alpha_{n+1}$$

is not hypersimple, then $\mathcal{F}$ is independently axiomatizable. We do this in two steps:

- there is an effective procedure that produces, given any theorem γ of $\mathcal{F}$, another theorem γ' that is not deducible from it
  By hypothesis, we have a strong array $\{D_{f(x)}\}_{x \in \omega}$ that intersects $\overline{A}$. Given any theorem γ of $\mathcal{F}$, first find an m such that γ is deducible from $\alpha_0, \ldots, \alpha_m$, and then an x such that $D_{f(x)}$ contains only numbers greater than m. The conjunction γ' of the α_{n+1}'s such that n is in $D_{f(x)}$ is then a theorem of $\mathcal{F}$ (because a conjunction of axioms) that cannot be

deduced from γ. Indeed, one number n in $D_{f(x)}$ is not in A, i.e. α_{n+1} cannot be deduced from the α's with smaller indices, and in particular is not deducible from γ. But α_{n+1} is a conjunct of γ', and thus γ' is not deducible from γ either.

- there is an independent axiomatization of $\mathcal{F}$

Let $\{\alpha_n\}_{n\in\omega}$ be an enumeration of the theorems of $\mathcal{F}$. We want to generate this set with a set of independent axioms $\{\beta_n\}_{n\in\omega}$. The first part of the proof produces, given any formula γ, a formula γ' not deducible from it. The opposite is not necessarily true, since γ' could be too strong, and imply γ. To have independent formulas, we have to relax γ' a little, and this can be done by considering $\neg\gamma \vee \gamma'$, i.e. $\gamma \to \gamma'$. This is not deducible from γ as before, since otherwise (by modus ponens) so would be γ'. And if γ is not valid, it is not deducible from $\gamma \to \gamma'$, otherwise from $\neg\gamma$ (which is stronger than $\neg\gamma \vee \gamma'$) we would get γ, and then γ would be valid. More generally, the sequence

$$\gamma \quad \gamma \to \gamma' \quad \gamma' \to \gamma'' \quad \cdots$$

is independent. We still have to make sure that all the α_n's are going to be deducible from the axioms, and we simply add them one by one as conclusions, so that α_n can be obtained from the first $n+1$ axioms. Thus our set of independent axioms for $\mathcal{F}$ is a set of formulas inductively defined as follows. First we let

$$\beta_0 = \alpha_0'.$$

This makes sure that we start from a formula that is not valid. Then, if $\beta_n = \delta_n \to \gamma_n$,

$$\beta_{n+1} = \gamma_n \to \alpha_n \wedge (\alpha_n \wedge \gamma_n)'. \quad \square$$

Exercise III.10.14 *For any hypersimple set A there is a consistent first-order formal system $\mathcal{F}$ extending $\mathcal{R}$, and an axiomatization $\{\alpha_e\}_{e\in\omega}$ of it, such that*

$$n \in A \Leftrightarrow \alpha_0 \wedge \cdots \wedge \alpha_n \models \alpha_{n+1}$$

In particular, $\mathcal{F}$ is not independently axiomatizable. (Kreisel [1957]) (Hint: let α_0 be the conjunction of the finitely many axioms of $\mathcal{Q}$, see p. 23, and of the formula $\forall x(\varphi(x) \to P(x))$, where P is a new predicate, and φ weakly represents A in $\mathcal{R}$. Moreover, let α_{n+1} be $P(\overline{n})$. Then α_{n+1} can be deduced from α_0 only when $n \in A$.)

Pour El [1968a] shows that $\mathcal{Q}$, and more generally every theory with effectively inseparable sets of theorems and of refutable formulas, has a consistent extension which is not independently axiomatizable, and with the same language as the original theory.

For more recursion-theoretical results about formal systems see p. 510, as well as Smullyan [1961], Martin and Pour El [1970], and Downey [1987].

Chapter IV

Hierarchies and Weak Reducibilities

The theme of this chapter is **definability** in given languages, and a classification of sets and relations according to their best definition. Of course definability is not an absolute notion, and it depends on the given language. Here we will consider three natural ones, the first two for Arithmetic, the last one for Set Theory. The first two will differ in that we will allow only number quantifications in one case, but also function (or set) quantifications in the other, and will define, respectively, the **arithmetical** and **analytical sets**. The third approach will lead us to the **constructible sets**.

Definability is a linguistical, more than computational, notion. However, we already know that it is possible to characterize the recursive sets in a purely linguistical way, see I.3.6. This suggests the possibility of considering definability as an abstract version of computability, and we will see in later chapters that this program is indeed feasible, for some of the definability notions that will be introduced in this chapter.

We only scratch the surface of the subject in here, and refer to Volumes II and III for a detailed study of the arithmetical and the analytical sets, to which the two volumes are respectively dedicated. But we will prove a number of interesting results already in this chapter, providing some nontrivial characterizations of a number of classes. In particular, we deal with: the **limit sets**, that can be obtained as limits of recursive functions; the **hyperarithmetical sets**, that can be effectively computed modulo number-theoretical quantification; and the Σ_2^1 **sets**, that can be defined over the constructible universe in a particularly simple way.

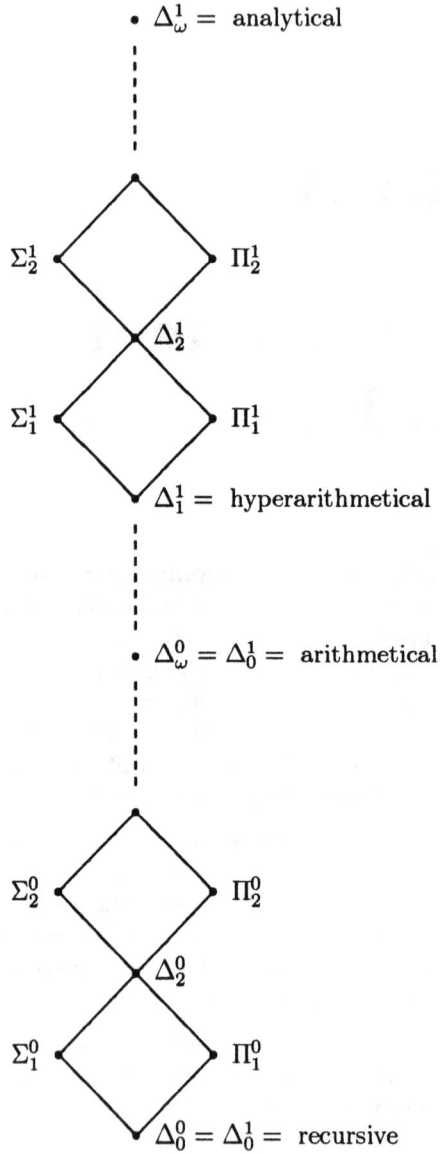

- $\Delta^1_\omega = $ analytical

Σ^1_2 Π^1_2

Δ^1_2

Σ^1_1 Π^1_1

$\Delta^1_1 = $ hyperarithmetical

- $\Delta^0_\omega = \Delta^1_0 = $ arithmetical

Σ^0_2 Π^0_2

Δ^0_2

Σ^0_1 Π^0_1

$\Delta^0_0 = \Delta^1_0 = $ recursive

Figure IV.1: The Arithmetical and Analytical Hierarchies

IV.1   The Arithmetical Hierarchy

We start by considering, as the simplest framework for the definition of sets of natural numbers, First-Order Arithmetic. We first introduce the notion of arithmetical definability, and then classify sets definable in Arithmetic by looking at their best possible definitions. We will thus obtain classes closely related, both in computational content and in structure theory, to the classes of recursive and recursively enumerable sets.

The definition of truth ⋆

> Pilate said unto him, What is truth?
> And when he had said this, he went out.
> (John, *Gospel*, XVIII, 38)

The idea for a definition of truth comes from Aristotle (*Metaphysica*, Γ 7, 1011b, 25–27):

> it is false to say that the being is not or that the non-being is; it is true to say that the being is and that the non-being is not.

Thus e.g.

> 'it is raining' is true if and only if it is raining,

where the quoted phrase is the one whose truth we are trying to establish, and it is considered as a purely syntactical object, while the unquoted phrase is taken semantically, for what it means. And the quoted phrase is true if and only if it reflects what happens in the world, as expressed by the unquoted phrase.

For natural languages this explains the meaning of truth, but it is not particularly manageable. The advantage of formal languages is that they are built by induction, and thus we can actually apply the ideas just introduced to produce an inductive definition of truth. The original definition will be applied directly only to atomic formulas, while for compound formulas we will rely on induction, and will force only the interpretation of the new symbols introduced. This was first done by Tarski [1936].

Truth in First-Order Arithmetic

There are two natural first-order languages that come to mind, for the purpose of classifying arithmetical definitions. They are in some sense extreme examples, allowing respectively for a minimal and a maximal set of nonlogical primitive functions and predicates. We consider both of them, and prove that their definitional power is the same.

Definition IV.1.1 Definition of truth in Arithmetic (Tarski [1936])
Let $\mathcal{L}$ be the first-order language with equality, augmented with constants $\bar{n}$ for each number n, and binary function symbols $+$ and $\times$, and let $\mathcal{A}$ be the intended structure for $\mathcal{L}$, i.e. the natural numbers, with the usual sum and product.

*Given a closed formula φ of $\mathcal{L}$, $\boldsymbol{\varphi}$ is **true** in $\mathcal{A}$ ($\mathcal{A} \models \varphi$) is inductively defined as follows:*

$$
\begin{aligned}
\mathcal{A} &\models \bar{m} + \bar{n} = \bar{p} &\Leftrightarrow&\quad m + n = p \\
\mathcal{A} &\models \bar{m} \times \bar{n} = \bar{p} &\Leftrightarrow&\quad m \cdot n = p \\
\mathcal{A} &\models \neg\psi &\Leftrightarrow&\quad not\ (\mathcal{A} \models \psi) \\
\mathcal{A} &\models \varphi_0 \wedge \varphi_1 &\Leftrightarrow&\quad \mathcal{A} \models \varphi_0\ and\ \mathcal{A} \models \varphi_1 \\
\mathcal{A} &\models \varphi_0 \vee \varphi_1 &\Leftrightarrow&\quad \mathcal{A} \models \varphi_0\ or\ \mathcal{A} \models \varphi_1 \\
\mathcal{A} &\models \exists x \psi(x) &\Leftrightarrow&\quad for\ some\ n,\ \mathcal{A} \models \psi(\bar{n}) \\
\mathcal{A} &\models \forall x \psi(x) &\Leftrightarrow&\quad for\ all\ n,\ \mathcal{A} \models \psi(\bar{n}).
\end{aligned}
$$

*An n-ary relation $\boldsymbol{P}$ is **definable in First-Order Arithmetic** (briefly, is **arithmetical**) if, for some formula φ with n free variables,*

$$
P(x_1, \ldots, x_n) \Leftrightarrow \mathcal{A} \models \varphi(\bar{x}_1, \ldots, \bar{x}_n).
$$

As we have already noted, this definition explicitly defines the meaning of any syntactical first-order formula over Arithmetic by first reducing the meaning of compound statements to the meaning of simpler ones, and then by forcing the meaning of the function constants $+$ and $\times$ to agree with the usual standard meaning of sum and product. After this, being arithmetical is then defined in the same way as being representable (I.3.4), but with the notion of provability in a formal system replaced by the notion of truth in First-Order Arithmetic.

This procedure of defining truth for formulas of a given language over some structure is quite general, and in the future we will simply indicate the changes needed to extend the above definition of truth to different languages and structures.

Definition IV.1.2 (Kleene [1943], Mostowski [1947]) *Let $\mathcal{L}^*$ be the first-order language with equality, augmented with constants $\bar{n}$ for each number n, and a relation symbol φ_R for each recursive relation R, and let $\mathcal{A}^*$ be the intended structure for $\mathcal{L}^*$, i.e. the natural numbers, with all the recursive relations.*

Given a closed formula φ of $\mathcal{L}^*$, $\mathcal{A}^* \models \varphi$ is defined inductively as above, starting from

$$
\mathcal{A}^* \models \varphi_R(\bar{x}_1, \ldots, \bar{x}_n) \Leftrightarrow R(x_1, \ldots, x_n).
$$

An n-ary relation P is in the **Arithmetical Hierarchy** *if, for some formula φ of $\mathcal{L}^*$ with n free variables,*

$$P(x_1, \ldots, x_n) \Leftrightarrow \mathcal{A}^* \models \varphi(\bar{x}_1, \ldots, \bar{x}_n).$$

Briefly stated, *the Arithmetical Hierarchy consists of the relations definable in First-Order Arithmetic, with the recursive relations as parameters.* This seems natural from our point of view, since we want to classify sets and relations according to their noneffectiveness, as expressed by the complexity of their definition, and thus the recursive relations may be given for free. The next result shows that we are classifying the same sets as before.

Theorem IV.1.3 (Gödel [1931]) *The Arithmetical Hierarchy contains exactly the arithmetical relations.*

Proof. One direction comes from the fact that the relations represented by the atomic formulas of the language $\mathcal{L}$ are recursive, being built up from plus and times only. The other direction follows from the fact, proved in I.3.6, that the recursive relations are representable in $\mathcal{R}$, and hence (since the axioms of $\mathcal{R}$ are true in $\mathcal{A}$) are arithmetical. $\square$

The Arithmetical Hierarchy

Before we can classify relations according to their definition in $\mathcal{L}^*$, we need to be able to put these definitions in some kind of normal form. This is easily accomplished in a standard way, by manipulation of quantifiers.

Proposition IV.1.4 *The following transformations of quantifiers are permissible (up to logical equivalence):*

1. *permutation of quantifiers of the same type*

2. *contraction of quantifiers of the same type*

3. *permutation of two quantifiers, one of which bounded*

4. *substitution of a bounded quantifier with an unbounded one of the same type.*

Proof. Part 1 is obvious. Part 2 can be accomplished by codifying the various quantified variables into a single one. E.g., given a formula

$$\forall x_1 \ldots \forall x_n R(x_1, \ldots, x_n),$$

this is equivalent to the formula

$$\forall x R((x)_1, \ldots, (x)_n),$$

where $x = \langle x_1, \ldots, x_n \rangle$.

Part 3 is obvious when the two quantifiers are of the same type. For the remaining cases, consider e.g.

$$(\forall x \leq a)(\exists y) R(x, y).$$

If we let y_x be a number such that $R(x, y_x)$, for $x \leq a$, then $y = \langle y_0, \ldots, y_a \rangle$ witnesses the truth of

$$(\exists y)(\forall x \leq a) R(x, (y)_{x+1}),$$

and thus the former formula implies the latter, while the converse implication holds trivially. The other case is treated similarly.

Part 4 is standard:

$$(\forall x \leq a) R(x) \quad \Leftrightarrow \quad \forall x (x \leq a \rightarrow R(x))$$
$$(\exists x \leq a) R(x) \quad \Leftrightarrow \quad \exists x (x \leq a \wedge R(x)). \quad \Box$$

Proposition IV.1.5 Prenex Normal Form (Kuratowski and Tarski [1931]). *Any relation in the Arithmetical Hierarchy is equivalent to one with a list of alternated quantifiers in the prefix, and a recursive matrix.*

Proof. The previous transformations allow to contract quantifiers of the same type, without changing the recursiveness of the matrix. Thus we only have to show how to push quantifiers in front. This is accomplished by the following well-known transformations, read from left to right, and which again work up to logical equivalence.

First of all note that bound variables can be renamed, according to the rules:

$$\exists x \alpha(x) \quad \Leftrightarrow \quad \exists y \alpha(y)$$
$$\forall x \alpha(x) \quad \Leftrightarrow \quad \forall y \alpha(y),$$

where y is any variable that does not occur free in α. Thus we can always suppose that, in the following rules, x does not occur free in β.

$$\neg(\exists x)\alpha \quad \Leftrightarrow \quad (\forall x)\neg\alpha$$
$$\neg(\forall x)\alpha \quad \Leftrightarrow \quad (\exists x)\neg\alpha$$
$$(\exists x \alpha) \wedge \beta \quad \Leftrightarrow \quad \exists x (\alpha \wedge \beta)$$
$$(\forall x \alpha) \wedge \beta \quad \Leftrightarrow \quad \forall x (\alpha \wedge \beta)$$
$$(\exists x \alpha) \vee \beta \quad \Leftrightarrow \quad \exists x (\alpha \vee \beta)$$
$$(\forall x \alpha) \vee \beta \quad \Leftrightarrow \quad \forall x (\alpha \vee \beta)$$

$$(\exists x \alpha) \to \beta \quad \Leftrightarrow \quad \forall x(\alpha \to \beta)$$
$$(\forall x \alpha) \to \beta \quad \Leftrightarrow \quad \exists x(\alpha \to \beta)$$
$$\beta \to (\exists x \alpha) \quad \Leftrightarrow \quad \exists x(\alpha \to \beta)$$
$$\beta \to (\forall x \alpha) \quad \Leftrightarrow \quad \forall x(\alpha \to \beta) \quad \square$$

It should be noted that the rules to bring a formula into prenex normal form can be applied in any order. In particular, *prenex normal forms are not unique, and may have different numbers of quantifiers*. However, since there are only finitely many possible manipulations, the prenex normal form with smallest number of quantifiers can always be found, starting from a given formula. Note that we are not claiming that there is an algorithm that gives the best prenex normal form to which the formula is equivalent, since better prenex normal forms might be produced by equivalent formulas.

The Prenex Normal Form suggests that we only have to count the number of alternations of quantifiers in the prefix, to measure the distance from recursiveness. And of course there are, for each such number, two possibilities, according to whether the first quantifier is existential or universal. It just remains to give everything a name.

Definition IV.1.6 The Arithmetical Hierarchy (Kleene [1943], Mostowski [1947])

1. Σ_n^0 *is the class of relations definable over* $\mathcal{A}^*$ *by a formula of* $\mathcal{L}^*$ *in prenex form with recursive matrix, and* n *quantifier alternations in the prefix, the outer quantifier being existential.*

2. Π_n^0 *is defined similarly, with the outer quantifier being universal.*

3. Δ_n^0 *is* $\Sigma_n^0 \cap \Pi_n^0$, *i.e. the class of relations definable in both the n-quantifier forms.*

4. Δ_ω^0 *is the class of the arithmetical relations.*

By extension, we will call a *formula* Σ_n^0 or Π_n^0, if it is in prenex normal form, with n quantifier alternations in the prefix, the outer one being, respectively, existential or universal.

Note also that, by contraction of quantifiers, n quantifier alternations are equivalent to n alternated quantifiers.

The levels of the Arithmetical Hierarchy

The first levels of the hierarchy are inhabited by old friends. First of all, the level 0 obviously consists of the recursive relations (because no quantifier is involved). More interestingly,

$$\Delta_1^0 = \text{recursive}$$
$$\Sigma_1^0 = \text{recursively enumerable.}$$

This obviously follows from II.1.10 and II.1.19.

If we wished to, we could define a similar hierarchy for formulas of $\mathcal{L}$. If we count quantifier alternations, and ask the matrix to be quantifier free (i.e. a Boolean combination of diophantine equations), we still get the r.e. sets at the first existential level, by Matiyasevitch result (see p. 135), and thus this hierarchy coincides, from the first level on, with the Arithmetical Hierarchy. By using $\mathcal{L}^*$ we simply avoid the proof of this representation theorem for the r.e. sets on one side, and have more freedom in the computations on the other.

If we are willing to compromise a little, we may allow the matrix to contain bounded quantifiers. Then the proof of IV.1.3 would suffice to show that the r.e. sets are at the first existential level of this hierarchy (because of the form the formulas that represent recursive relations in $\mathcal{R}$ have). The Δ_0^0 relations of this hierarchy (namely the sets definable with plus and times, by using connectives and bounded quantifiers) form the interesting class of **rudimentary predicates** (Smullyan [1961]), to which we will return in Chapter VIII.

Exercises IV.1.7 The Bounded Arithmetical Hierarchy (Davis [1958], Harrow [1978]) Let Δ_0^0 be the class of relations definable over $\mathcal{L}$ by using only connectives and bounded quantifiers. We can stratify it by counting the number of bounded quantifier alternations, getting classes $\Sigma_{0,n}^0$ and $\Pi_{0,n}^0$ in the natural way. Because of the collapse in d) below, $\Delta_{0,n}^0$ is defined as the class of sets A such that both A and $\overline{A}$ are in the n-th level of the hierarchy.

a) $\Pi_{0,n}^0$ and $\Sigma_{0,n}^0$ are closed under conjunction and disjunction.

b) For $n \geq 1$, $\Pi_{0,n}^0$ is closed under universal quantification bounded by a polynomial, and $\Sigma_{0,n}^0$ is closed under existential quantification bounded by a polynomial. (Hint: by induction, since e.g.

$$(\exists z \leq p(\vec{x}) + q(\vec{x}))P(z) \quad \Leftrightarrow \quad (\exists a \leq p(\vec{x}))(\exists b \leq q(\vec{x}))P(a+b)$$
$$(\exists z \leq p(\vec{x}) \cdot q(\vec{x}))P(z) \quad \Leftrightarrow \quad (\exists a \leq p(\vec{x}))(\exists b < q(\vec{x}))P(p(\vec{x}) \cdot b + a).)$$

c) The matrix can always be reduced to a diophantine equation. (Hint: diophantine equations $p(\vec{x}) = 0$, where p is a polynomial with integral coefficients, are closed under Boolean combinations. Precisely,

$$p(\vec{x}) = q(\vec{x}) \quad \Leftrightarrow \quad p(\vec{x}) - q(\vec{x}) = 0$$
$$p(\vec{x}) \neq 0 \quad \Leftrightarrow \quad p(\vec{x}) < 0 \vee 0 < p(\vec{x})$$
$$p(\vec{x}) = 0 \vee q(\vec{x}) = 0 \quad \Leftrightarrow \quad p(\vec{x}) \cdot q(\vec{x}) = 0$$
$$p(\vec{x}) = 0 \wedge q(\vec{x}) = 0 \quad \Leftrightarrow \quad p^2(\vec{x}) + q^2(\vec{x}) = 0$$
$$p(\vec{x}) < q(\vec{x}) \quad \Leftrightarrow \quad (\exists z \leq q(\vec{x}))(p(\vec{x}) + z + 1 = q(\vec{x})).$$

By part b), quantification bounded by a polynomial is allowed.)

d) $\Pi_{0,2n+1}^0 = \Pi_{0,2n}^0$ and $\Sigma_{0,2n+2}^0 = \Sigma_{0,2n+1}^0$, and thus the Bounded Arithmetical Hierarchy reduces to

$$\Pi_{0,0}^0 \subseteq \Sigma_{0,1}^0 \subseteq \Pi_{0,2}^0 \subseteq \Sigma_{0,3}^0 \ \ldots$$

(Hint: the bounded universal quantification of a diophantine equation is still diophantine. Indeed, consider

$$R(z, \vec{x}) \Leftrightarrow (\forall y \leq z)(p(y, \vec{x}) = 0) \Leftrightarrow (\sum_{y \leq z} p^2(y, \vec{x})) = 0.$$

This is a polynomial in y and $\vec{x}$, while we want it to be a polynomial in z and $\vec{x}$. Since p^2 is a polynomial, it can be written as $\sum_{n \leq k} q_n(\vec{x}) \cdot y^n$. Thus we have

$$R(z, \vec{x}) \Leftrightarrow \sum_{n \leq k} [q_n(\vec{x}) \cdot \sum_{y \leq z} y^n] = 0.$$

It only remains to note that, by induction, $\sum_{y \leq z} y^n$ is indeed a polynomial in z, of degree $n+1$, and with rational coefficients, that can be turned into integral coefficients by taking common denominators.)

e) The negation of a set in a given level belongs to the next level, and thus

$$\Pi_{0,2n}^0 \subseteq \Delta_{0,2n}^0 \subseteq \Sigma_{0,2n+1}^0 \subseteq \Delta_{0,2n+1}^0 \subseteq \Pi_{0,2n+2}^0.$$

f) If for any n two of the previous classes coincide, the hierarchy collapses. (Hint: if two classes coincide, then the common class is closed under both bounded quantifications, and thus is the whole hierarchy.)

It is not known whether the Bounded Arithmetical Hierarchy collapses or not.

The properties of the first level of the Arithmetical Hierarchy are inherited at higher ones, by very similar proofs, which we just sketch.

Proposition IV.1.8 Closure properties (Kleene [1943], Mostowski [1947])

1. R is Σ_n^0 if and only if $\neg R$ is Π_n^0
   R is Π_n^0 if and only if $\neg R$ is Σ_n^0

2. Δ_n^0 is closed under negations

3. Σ_n^0, Π_n^0 and Δ_n^0 are closed under conjunction, disjunction and bounded quantification

4. for $n \geq 1$, Σ_n^0 is closed under existential quantification, and Π_n^0 is closed under universal quantification

5. the universal quantification of a Σ_n^0 relation is Π_{n+1}^0, and the existential quantification of a Π_n^0 relation is Σ_{n+1}^0.

Proof. Everything easily follows from logical operations and quantifier manipulations. E.g., let $\exists x \forall y R$ and $\exists x \forall y Q$ be Σ_2^0 formulas. Then, if v, w, z, and w do not occur free in Q or R,

$$(\exists x \forall y R) \wedge (\exists x \forall y Q) \quad \Leftrightarrow \quad (\exists v \forall w R) \wedge (\exists z \forall t Q)$$
$$\Leftrightarrow \quad \exists v \exists z \forall w \forall t (R \wedge Q),$$

and this is a Σ_2^0 formula. $\square$

Theorem IV.1.9 Enumeration Theorem (Kleene [1943], Mostowski [1947]) *For each $n, m \geq 1$ there is an $m + 1$-ary Σ_n^0 relation that enumerates the m-ary Σ_n^0 relations. Similarly for Π_n^0.*

Proof. This is simply a consequence of the Normal Form Theorem for r.e. relations, II.1.10, according to which there is an $m + 1$-ary Σ_1^0 relation enumerating the m-ary Σ_1^0 relations. Then its negation enumerates the m-ary Π_1^0 relations.

Inductively, let $R(e, x, \vec{z})$ be an $m + 2$-ary Σ_n^0 relation enumerating the $m + 1$-ary Σ_n^0 relations. Then $\forall x R(e, x, \vec{z})$ is an $m + 1$-ary Π_{n+1}^0 relation enumerating the m-ary Π_{n+1}^0 relations, and its negation enumerates the m-ary Σ_{n+1}^0 relations. $\square$

Definition IV.1.10 *A set A is Σ_n^0-complete if it is Σ_n^0, and every Σ_n^0 set is m-reducible to it. Π_n^0-complete sets are defined similarly.*

The concept of Σ_n^0-completeness is the analogue, at level n, of the concept of m-completeness for r.e. sets. By induction we thus have:

Proposition IV.1.11 *For each $n \geq 1$, Σ_n^0-complete and Π_n^0-complete sets exist.*

In Chapter X we will see that many natural index sets (p. II.2.8) are complete at some level of the Arithmetical Hierarchy.

Exercises IV.1.12 Partial truth definitions. Let $\{\psi_e\}_{e \in \omega}$ be an effective enumeration of the closed formulas of $\mathcal{L}$.

a) *For each $n \geq 1$, the set*

$$e \in T_n \quad \Leftrightarrow \quad \psi_e \text{ is } \Sigma_n^0 \wedge \mathcal{A} \models \psi_e$$

is Σ_n^0-complete. Similarly for Π_n^0. (Hint: first note that to check, of a given formula, whether it is Σ_n^0 is a recursive procedure. To show that T_n is Σ_n^0 proceed inductively, using the definition of truth in $\mathcal{A}$: note that if there are no quantifiers then the formulas are recursive, and their truth can be effectively determined. To show completeness, use the existence of Σ_n^0-complete sets.)

b) *The set*
$$\langle e, n \rangle \in T \Leftrightarrow e \in T_n$$
is not arithmetical. (Tarski [1936]) (Hint: if it were, it would be Σ_n^0 for some n. This would contradict the Arithmetical Hierarchy Theorem given below.)

Theorem IV.1.13 Arithmetical Hierarchy Theorem (Kleene [1943], Mostowski [1947]) *The Arithmetical Hierarchy does not collapse. More precisely, for any $n \geq 1$ the following hold:*

1. $\Sigma_n^0 - \Pi_n^0 \neq \emptyset$, *and hence* $\Delta_n^0 \subset \Sigma_n^0$

2. $\Pi_n^0 - \Sigma_n^0 \neq \emptyset$, *and hence* $\Delta_n^0 \subset \Pi_n^0$

3. $\Sigma_n^0 \cup \Pi_n^0 \subset \Delta_{n+1}^0$.

Proof. The first two parts are equivalent, by taking negations. The proof of the first one is similar to the one of $II.2.3$: if $R(e, x)$ enumerates the unary Σ_n^0 relations, then
$$P(x) \Leftrightarrow R(x, x)$$
is Σ_n^0. But it cannot be Π_n^0, otherwise its negation would be Σ_n^0, and there would be e such that
$$R(e, x) \Leftrightarrow \neg P(x) \Leftrightarrow \neg R(x, x).$$

For $x = e$ a contradiction would follow.

For the last part, note that we can always add dummy quantifiers in front or at the end of the prefix, and thus $\Sigma_n^0 \cup \Pi_n^0 \subseteq \Delta_{n+1}^0$. To show that the inclusion is strict, let $P \in \Sigma_n^0 - \Pi_n^0$: then $\neg P \in \Pi_n^0 - \Sigma_n^0$. If
$$Q(x, z) \Leftrightarrow [P(x) \wedge z = 0] \vee [\neg P(x) \wedge z = 1]$$
then $Q \in \Delta_{n+1}^0$. But Q is not in Π_n^0, otherwise so would be
$$P(x) \Leftrightarrow Q(x, 0),$$
and similarly Q is not in Σ_n^0. $\square$

The results just proved justify the picture of the Arithmetical Hierarchy given in Figure 1 (p. 362), where upward connections mean strict inclusion, and no other inclusion holds.

We have defined the Arithmetical Hierarchy by iterating quantifiers and proved, by direct arguments, properties of the various levels similar to those of the first level. The next result gives a different, purely recursion-theoretical, definition of the Arithmetical Hierarchy, and explains the similarities of the various levels. It also suggests the possibility of different proofs for the results proved above, by relativization of the results of the first level.

Theorem IV.1.14 Post's Theorem (Post [1948], Kleene) *A relation is:*

1. Δ^0_{n+1} *if and only if it is recursive in a* Σ^0_n *or a* Π^0_n *relation*

2. Σ^0_{n+1} *if and only if it is recursively enumerable in a* Σ^0_n *or a* Π^0_n *relation.*

Proof. First note that, for what concerns relative recursive computations, Σ^0_n and Π^0_n are interchangeable, because an oracle and its complement are equivalent. We can thus use any of them.

We first prove part 2. Suppose P is Σ^0_{n+1}: then, for some R in Π^0_n,

$$P(\vec{x}) \Leftrightarrow \exists y R(\vec{x}, y).$$

By the relativized version of II.1.10, P is then r.e. in R, and hence r.e. in a Π^0_n relation.

Conversely, let P be r.e. in a Π^0_n relation Q. Then P is r.e. in a Π^0_n set A, obtained by coding Q:

$$\langle x_1, \ldots, x_m \rangle \in A \Leftrightarrow Q(x_1, \ldots, x_m).$$

By the relativized version of II.1.10, together with compactness and monotonicity (II.3.13), there is an r.e. relation R such that

$$P(\vec{x}) \Leftrightarrow \exists u \exists v (D_u \subseteq A \land D_v \subseteq \overline{A} \land R(\vec{x}, u, v)).$$

Note that

$$D_u \subseteq A \Leftrightarrow (\forall x \in D_u)(x \in A).$$

Since D_u is finite, the quantifier is bounded: but A is in Π^0_n, and then so is the whole expression $D_u \subseteq A$. Similarly,

$$D_v \subseteq \overline{A} \Leftrightarrow (\forall x \in D_v)(x \notin A)$$

is Σ^0_n, because the negation of A is used. Thus both expressions are Σ^0_{n+1}, and so is R, being r.e. (and hence Σ^0_1). But Σ^0_{n+1} is closed under existential quantification, and thus P is in it.

Part 1 follows from part 2 and the relativized version of II.1.19. If P is recursive in a Σ^0_n relation, then both P and $\neg P$ are r.e. in it, and hence Σ^0_{n+1}. But then P is both Σ^0_{n+1} and Π^0_{n+1}, and hence Δ^0_{n+1}. The converse is similar: suppose that P and $\neg P$ are each r.e. in some (not necessarily the same one) Σ^0_n relation. Then they are r.e. in the complete Σ^0_n set, and hence P is recursive in the complete Σ^0_n set. $\square$

Exercises IV.1.15 a) *For any $n \geq 1$, the union of two Σ^0_n sets can be reduced to the union of two disjoint Σ^0_n sets.* (Hint: see II.1.23.)

b) *For any $n \geq 1$, there are two disjoint Σ^0_n sets that cannot be separated by a Δ^0_n set.* (Hint: see II.2.5.)

c) *Any two Σ^0_n-complete sets are recursively isomorphic.* (Hint: use III.7.13.)

Δ_2^0 sets

As a special case of Post's Theorem, we get:

Proposition IV.1.16 *A set is Δ_2^0 if and only if it is recursive in $\mathcal{K}$.*

Proof. $\mathcal{K}$ is Σ_1^0-complete, and hence any Σ_1^0 or Π_1^0 relations is recursive in it. Thus, being recursive in some Σ_1^0 or Π_1^0 relation is equivalent to being recursive in $\mathcal{K}$. $\square$

This allows us to prove an interesting characterization, useful in constructions.

Proposition IV.1.17 The Limit Lemma (Shoenfield [1959]) *A is Δ_2^0 if and only if its characteristic function is the limit of a recursive function g, i.e.*

$$c_A(x) = \lim_{s \to \infty} g(x, s).$$

Proof. If g is given, then

$$
\begin{aligned}
x \in A \quad &\Leftrightarrow \quad \forall s \exists t (t \geq s \wedge g(x, t) = 1) \\
&\Leftrightarrow \quad \exists s \forall t (t \geq s \to g(x, t) = 1),
\end{aligned}
$$

and A is Δ_2^0.

Conversely, if A is Δ_2^0 then $A \leq_T \mathcal{K}$. Let e be such that $c_A \simeq \varphi_e^{\mathcal{K}}$, and

$$g(x, s) \simeq \varphi_e^{\mathcal{K}_s}(x).$$

Then c_A is the limit of g. $\square$

This characterization of Δ_2^0 suggests a possible hierarchy for it, obtained by bounding the number of times the approximating function g may change. The r.e. sets can be obtained by approximations that change at most once (let $g(x, s)$ be 0 until it is discovered that x is in the set, and 1 afterwards). If we consider the sets whose approximation may change only n times, we have the **Boolean Hierarchy**, which stratifies the set belonging to the smallest Boolean algebra generated by the r.e. sets (or, by III.8.7, the sets btt-reducible to $\mathcal{K}$) (Addison [1965], Gold [1965], Putnam [1965], Ershov [1968]). Obviously, Δ_2^0 is not exhausted by the Boolean Hierarchy (since there are sets recursive in $\mathcal{K}$ and not btt-reducible to it), but this can be achieved by an appropriate transfinite extension of the hierarchy (Ershov [1968a], [1970]).

Exercises IV.1.18 The Boolean Hierarchy (Ershov [1968]). For $n \geq 1$, let Σ_n^{-1} be the class of sets with a recursive approximation g which changes at most n times, and such that $g(x,0) = 0$. Π_n^{-1} is the class of complements of sets in Σ_n^{-1}, and $\Delta_n^{-1} = \Sigma_n^{-1} \cap \Pi_n^{-1}$. A set is called **$n$-r.e.** if it is Σ_n^{-1}, and **weakly n-r.e.** if it is Δ_{n+1}^{-1}.

a) Σ_1^{-1} *is the class of r.e. sets.*

b) *For $n \geq 1$, Σ_n^{-1} is the class of sets of the form $(A_1 - A_2) \cup (A_3 - A_4) \cdots$, with A_i r.e.*

c) *For each $n \geq 1$, there are Σ_n^{-1}-complete sets.* (Hint: let

$$\langle x,y \rangle \in A \cdot B \Leftrightarrow x \in A \wedge y \in B \quad \text{and} \quad \langle x,y \rangle \in A + B \Leftrightarrow x \in A \vee y \in B,$$

and consider the sequence of sets

$$\mathcal{K} \quad \mathcal{K} \cdot \overline{\mathcal{K}} \quad (\mathcal{K} \cdot \overline{\mathcal{K}}) + \mathcal{K} \quad \cdots)$$

d) *For each $n \geq 1$, $\Sigma_n^{-1} - \Pi_n^{-1} \neq \emptyset$, and $\Pi_n^{-1} - \Sigma_n^{-1} \neq \emptyset$.* (Hint: use the sets given in part c.)

e) *For each $n \geq 2$, there is a Δ_n^{-1}-complete set.* (Hint: if A and B are, respectively, Σ_n^{-1}-complete and Π_n^{-1}-complete, then $A \oplus B$ is Δ_{n+1}^{-1}-complete.)

f) *For any n, $\Sigma_n^{-1} \cup \Pi_n^{-1} \subset \Delta_{n+1}^{-1}$.* (Hint: use the set given in part e.)

g) *For $n \geq 1$, a set is Δ_{n+1}^{-1} if and only if it has a recursive approximation that changes at most n times.* (Hint: suppose A is both Σ_n^{-1} and Π_n^{-1}. Then both A and its complement can be approximated by recursive functions g and h that change at most $n + 1$ times, and with value 0 at stage 0. For any x, we then know that one of $g(x,0)$ and $h(x,0)$ is wrong. Look for the first s such that $g(x,s) \neq h(x,s)$, and let $g(x,s)$ be the first value of a new function f approximating A. Let f change only when g does, and g and h differ. Then f changes at most n times.)

Note that there is nothing special about the second level: the characterization of Δ_2^0 obviously extends, by relativization, to all levels.

Proposition IV.1.19 Shoenfield [1959]) *For $n \geq 1$, A is Δ_{n+1}^0 if and only if there is an $n + 1$-ary recursive function g such that*

$$c_A(x) = \lim_{s_1 \to \infty} \cdots \lim_{s_n \to \infty} g(x, s_1, \ldots, s_n).$$

Relativizations $\star$

We can relativize all the work done so far to a given oracle X, defining the classes $\Sigma_n^{0,X}$, $\Pi_n^{0,X}$ and $\Delta_n^{0,X}$, and obtaining similar results, simply by substituting 'recursive' with 'recursive in X'. In particular, we have

$$A \leq_T B \Leftrightarrow A \in \Delta_1^{0,B}.$$

We might thus think that a number of other reducibilities can be defined, by looking at higher levels. The next result shows that this is not the case.

Proposition IV.1.20 *The relation*

$$A \leq_{\Delta_n^0} B \Leftrightarrow A \in \Delta_n^{0,B}$$

is transitive only for $n = 1$.

Proof. The transitivity for $n = 1$ follows from the fact that in this case $\leq_{\Delta_1^0}$ is simply Turing reducibility. For any $n > 1$, we show a counterexample. Let $A \in \Sigma_n^0 - \Pi_n^0$, which exists by the Hierarchy Theorem. Then there is $B \in \Pi_{n-1}^0$ such that

$$x \in A \Leftrightarrow \exists y B(x, y).$$

By definition A is in $\Sigma_1^{0,B}$, and hence in $\Delta_n^{0,B}$ (since $n > 1$). Since B is in Δ_n^0 by its choice, if $\leq_{\Delta_n^0}$ were transitive we would have $A \in \Delta_n^0$, and hence $A \in \Pi_n^0$, contradiction. $\square$

The obvious reason why the transitivity of $\leq_{\Delta_n^0}$ fails for $n > 1$, is that quantifiers simply sum up: if B has n quantifiers, and we stick n more in front of it, one might collapse (if the leftmost quantifier of the prefix of B is of the same type of the rightmost added in front of it), but the others are going to remain. This ceases to be a problem, if we do not care anymore for a fixed number of quantifiers.

Definition IV.1.21 *A is **arithmetical** in B ($A \leq_a B$) if it is in the Arithmetical Hierarchy relativized to B.*
*A is **arithmetically equivalent** to B ($A \equiv_a B$) if $A \leq_a B$ and $B \leq_a A$.*

Exercises IV.1.22 a) *If A is arithmetical, then $A \leq_a B$ for any set B.*
b) *If $A \leq_a B$ and B is arithmetical, so is A.*

Note that $\leq_a$ is a reflexive and transitive relation, and thus $\equiv_a$ is an equivalence relation.

Definition IV.1.23 *The equivalence classes of sets w.r.t. arithmetical equivalence are called **a-degrees**, and $(\mathcal{D}_a, \leq)$ is the structure of a-degrees, with the partial ordering $\leq$ induced on them by $\leq_a$.*

The structure of the a-degrees will be studied in Chapter XIII.

IV.2   The Analytical Hierarchy

The Arithmetical Hierarchy certainly includes only a small portion of all possible sets, being countable (each set has a definition in $\mathcal{L}^*$, and there are only countably many definitions). Moreover, by Tarski's Theorem (see p. 166, or IV.1.12.b), there are natural examples of nonarithmetical sets (namely, the set of codes of sentences of $\mathcal{L}$ true in Arithmetic). We thus feel the need of enlarging the Arithmetical Hierarchy. One natural way to proceed is by extending the language and going to the second order, i.e. by considering quantifiers not only over numbers, but also over functions.

Truth in Second-Order Arithmetic

As already for First-Order Arithmetic, there are at least two possible languages of interest.

Definition IV.2.1 Definition of truth in Second-Order Arithmetic (Tarski [1936]). *Let $\mathcal{L}_2$ be the second-order language with first-order equality, augmented with number constants $\overline{n}$ for each number n, and function constants $\overline{f}$ for each total function f, among them $+$ and $\times$. Let also $\mathcal{A}_2$ be the intended structure for $\mathcal{L}_2$, i.e. the set of natural numbers and total functions over them.*
Given a closed formula φ of $\mathcal{L}_2$, $\mathcal{A}_2 \models \varphi$ is defined inductively as usual.

A relation P of n number variables and m function variables is **definable in Second-Order Arithmetic (from parameters)** *if, for some closed formula φ of $\mathcal{L}_2^*$ with n free number variables and m free function variables,*

$$P(x_1, \ldots, x_n, f_1, \ldots, f_m) \Leftrightarrow \mathcal{A}_2^* \models \varphi(\overline{x}_1, \ldots, \overline{x}_n, \overline{f}_1, \ldots, \overline{f}_m).$$

If φ contains no function constant except plus and times, then P is simply **analytical.**

Notice that 'analytical' is defined without the use of parameters (i.e. function constants other than $+$ and $\times$). Since the use of parameters is like the use of oracles (see Section II.3), the analytical relations are those that can be defined in Second-Order Arithmetic, without any help.

The reason for the name 'analytical' comes from the fact that *Second-Order Arithmetic allows for the formalization of elementary Analysis*, by coding real numbers as functions of integers.

Definition IV.2.2 (Lusin [1925], Sierpinski [1925], Kleene [1955]) *Let $\mathcal{L}_2^*$ be the second-order language with first-order equality, augmented with number constants $\overline{n}$ for each number n, function constants $\overline{f}$ for each total function f, and relation symbols φ_R for each restricted recursive relation of numbers and total functions. Let also $\mathcal{A}_2^*$ be the intended structure for $\mathcal{L}_2^*$, i.e. the set of natural numbers and total functions over them, with all the restricted recursive relations of numbers and total functions.*
Given a closed formula φ of $\mathcal{L}_2^*$, $\mathcal{A}_2^* \models \varphi$ is defined inductively as usual.

A *relation P of n number variables and m function variables is in the* **Projective Hierarchy** *if, for some closed formula φ of $\mathcal{L}_2^*$ with n free number variables and m free function variables,*

$$P(x_1, \ldots, x_n, f_1, \ldots, f_m) \Leftrightarrow \mathcal{A}_2^* \models \varphi(\bar{x}_1, \ldots, \bar{x}_n, \bar{f}_1, \ldots, \bar{f}_m).$$

If φ contains no function constant except plus and times, then P is in the **Analytical Hierarchy**.

It follows from IV.1.3 relativized that *the Analytical Hierarchy contains exactly the analytical relations.*

The Analytical Hierarchy

Function quantifiers can be manipulated in a way similar to number quantifiers.

Proposition IV.2.3 *The following transformations of function quantifiers are permissible (up to logical equivalence):*

1. *permutation of quantifiers of the same type*

2. *contraction of quantifiers of the same type*

3. *permutation of a number quantifier and a function quantifier*

4. *substitution of a number quantifier with a function quantifier of the same type.*

Proof. Part 1 is obvious. Parts 2 needs only a way to code and decode finitely many functions, e.g.

$$\langle f_1, \ldots, f_n \rangle(x) = \langle f_1(x), \ldots, f_n(x) \rangle$$
$$(f)_n(x) = (f(x))_n.$$

Then e.g. the formula

$$\forall f_1 \ldots \forall f_n R(f_1, \ldots, f_n),$$

is equivalent to

$$\forall f R((f)_1, \ldots, (f)_n).$$

Part 3 needs a way to code and decode infinitely many functions, e.g.

$$f(x) = \begin{cases} f_n(z) & \text{if } x = \langle n, z \rangle \\ 0 & \text{otherwise} \end{cases}$$
$$(f)_n(z) = f(\langle n, z \rangle).$$

Having this, and given e.g.

$$(\forall x)(\exists f)R(f,x),$$

let f_n be such that $R(f_n,n)$ holds. Then the given formula is equivalent to

$$(\exists f)(\forall x)R((f)_x,x).$$

Part 4 is easily obtained, noting that

$$\exists x R(x) \Leftrightarrow \exists f R(f(0)) \quad \text{and} \quad \forall x R(x) \Leftrightarrow \forall f R(f(0)). \quad \square$$

Proposition IV.2.4 Prenex Normal Form (Kuratowski and Tarski [1931]). *Any relation in the Analytical Hierarchy is equivalent to one with a list of alternated function quantifiers in the prefix, and an arithmetical matrix.*

Proof. Similar to IV.1.5. $\square$

Proposition IV.2.5 (Kleene [1955]) *The matrix of a relation in the Analytical Hierarchy can always be reduced to a single number quantifier, opposite in type to the rightmost function quantifier of the prefix, followed by a recursive predicate.*

Proof. Having a formula in prenex normal form, with a prefix consisting of alternated function quantifiers, and an arithmetical matrix, there are two cases. If there are at least two function quantifiers, then all the number quantifiers can be eliminated, by first moving near to the function quantifier of the same type, then rising to function quantifiers, and being contracted; afterwards, one dummy number quantifier can be reintroduced. If there is only one function quantifier, then this works only for the number quantifiers of its same type, while the other remain, and can be contracted to a single one.

More precisely, proceed as follows.

- Reduce the matrix in prenex normal form, with a prefix of alternated number quantifiers, and a recursive matrix.

- Take the leftmost number quantifier, and confront it with the rightmost function quantifier of the prefix. If they are of the same type, raise the number quantifier to a function quantifier, and then contract.

- If they are not of the same type, permute them. See if there is a function quantifier to the left. If so, it must be of the same type as the number quantifier, which can thus be eliminated as above. If not, proceed with the next number quantifier in the matrix, inductively.

- At the end, either all number quantifiers have been eliminated, or a sequence of them, of the same type, lies to the left of the (necessarily unique) function quantifier. Contract them, and permute the resulting number quantifier back to the matrix. □

Exercises IV.2.6 Set quantifiers. a) *The same analytical classes are obtained if we use set quantifiers in place of function quantifiers, and matrices with at most two number quantifiers.* (Kleene [1955]) (Hint: sets and their characteristic functions can be identified, and we can thus work with the latter. In one direction, to say that a function is a characteristic function only involves a number quantifier, e.g.

$$(\exists A)R(c_A) \Leftrightarrow \exists f[\forall x(f(x) \le 1) \wedge R(f)].$$

In the other direction, functions can be substituted by their graphs, and saying that f is the characteristic function of the graph of a total function is an arithmetical condition: f has values bounded by 1; whenever it is 1 for z, then $z = \langle x, y \rangle$, i.e. it codes a relation; whenever it is 1 for $\langle x, y \rangle$ and $\langle x, y' \rangle$ then $y = y'$, i.e. the relation is a function; and for every x there is y such that $f(\langle x, y \rangle) = 1$, i.e. the function is total. Note that this last condition introduces two number quantifiers.)

b) *Set quantifiers and matrices with only one number quantifier are not sufficient to generate the whole Analytical Hierarchy.* (Kreisel) (Hint: the predicate $\forall A \exists x R(\hat{c}_A(x))$, with R recursive, is r.e. Indeed, the set of sets is finitely branching. Thus, by König's Lemma, if for every A there is x such that $R(\hat{c}_A(x))$ holds, there must be an x that works for every A. This means that we only have to look for x such that $(\forall A)R(\hat{c}_A(x))$ holds. For a fixed x this is a recursive condition, because there are only finitely many possible sequence numbers of length x. Thus the whole condition is r.e.)

Exactly like we did for the Arithmetical Hierarchy, we can now count the number of alternations of function quantifiers, and stratify the Analytical Hierarchy.

Definition IV.2.7 The Analytical Hierarchy (Lusin [1925], Sierpinski [1925], Kleene [1955])

1. Σ_n^1 *is the class of relations definable over $\mathcal{A}_2^*$ by a formula of $\mathcal{L}_2^*$ (without parameters) in prenex form with arithmetical matrix, and n quantifier alternations in the prefix, the outer quantifier being existential.*

2. Π_n^1 *is defined similarly, with the outer quantifier being universal.*

3. Δ_n^1 *is $\Sigma_n^1 \cap \Pi_n^1$, i.e. the class of relations definable in both the n-quantifier forms.*

4. Δ_ω^1 *is the class of the analytical relations.*

By extension, we will call a *formula* without parameters Σ_n^1 or Π_n^1 if it is in prenex normal form with n function quantifier alternations in the prefix, the first of which, respectively, existential or universal.

Note also that, by contraction of quantifiers, n function quantifier alternations are equivalent to n alternated function quantifiers.

The levels of the Analytical Hierarchy

A good part of the theory of the Arithmetical Hierarchy goes through for the Analytical Hierarchy as well, with similar proofs, that we do not repeat.

Proposition IV.2.8 Closure properties (Kleene [1955])

1. *R is Σ_n^1 if and only if $\neg R$ is Π_n^1*
   R is Π_n^1 if and only if $\neg R$ is Σ_n^1

2. *Δ_n^1 is closed under negations*

3. *Σ_n^1, Π_n^1 and Δ_n^1 are closed under conjunction, disjunction and number quantification*

4. *for $n \geq 1$, Σ_n^1 is closed under existential function quantification, and Π_n^1 is closed under universal function quantification*

5. *the universal function quantification of a Σ_n^1 relation is Π_{n+1}^1, and the existential function quantification of a Π_n^1 relation is Σ_{n+1}^1.*

Proof. Everything easily follows from logical operations and quantifier manipulations. E.g., closure under number quantification follows by moving the number quantifier to the matrix, by successive permutations. $\square$

Theorem IV.2.9 Enumeration Theorem (Kleene [1955]) *For each n and $m \geq 1$, and any l, there is a Σ_n^1 relation with $m + 1$ number variables and l function variables that enumerates the Σ_n^1 relations with m number variables and l function variables. Similarly for Π_n^1.*

Proof. Similar to IV.1.9, using this time the Normal Form Theorem for r.e. relations of numbers and total functions, that follows from II.3.11. Note that, because of IV.2.5, the hierarchy is inductively built up by starting from the r.e. predicates, by negations and function quantifications. $\square$

Corollary IV.2.10 Normal Form Theorem for Π_1^1 sets (Lusin [1917], Suslin [1917], Kleene [1955]) *A is Π_1^1 if and only if, for some recursive predicate R,*

$$x \in A \Leftrightarrow \forall f \exists y R(x, \hat{f}(y)).$$

Definition IV.2.11 *A set A is* Σ_n^1*-complete if it is* Σ_n^1*, and every* Σ_n^1 *set is m-reducible to it.* Π_n^1*-complete sets are defined similarly.*

Proposition IV.2.12 *For each* $n \geq 1$*,* Σ_n^1*-complete and* Π_n^1*-complete sets exist.*

Theorem IV.2.13 Analytical Hierarchy Theorem (Lebesgue [1905], Lusin [1925], Sierpinski [1925], Kleene [1955]) *The Analytical Hierarchy does not collapse. More precisely, for any* $n \geq 1$ *the following hold:*

  1. $\Sigma_n^1 - \Pi_n^1 \neq \emptyset$, *and hence* $\Delta_n^1 \subset \Sigma_n^1$

  2. $\Pi_n^1 - \Sigma_n^1 \neq \emptyset$, *and hence* $\Delta_n^1 \subset \Pi_n^1$

  3. $\Sigma_n^1 \cup \Pi_n^1 \subset \Delta_{n+1}^1$.

We thus get a picture for the Analytical Hierarchy similar to the one obtained for the Arithmetical Hierarchy, see Figure 1 (p. 362).

Despite the similarities between the Arithmetical and Analytical Hierarchies, there are differences. The most striking one seems to be the fact that there is no recursion-theoretical generation of the Analytical Hierarchy from below. Thus *function quantifiers, unlike number quantifiers (that correspond to relative recursive enumerability) seem to have no computational content.* See p. 395 for a precise statement of this fact.

Π_1^1 sets

The Normal Form Theorem for Π_1^1 sets suggests some natural representations for them, in set-theoretical terms. To state them precisely, we first introduce some terminology.

Definition IV.2.14 *A set of sequence numbers T is a* **tree** *if it is closed under subsequences.* $\langle \emptyset \rangle$ *is the* **root**, *or* **vertex**, *of the tree. Elements of the tree are called* **nodes**. *A* **branch** *of T is a maximal linearly ordered subset of T. The terminal node of a branch is a* **leaf**. *A tree is* **well-founded** *if it has no infinite branch.*

Thus a tree is simply a subset of the tree of all sequence numbers ordered by extension, such that whenever a node is in T, so are all nodes between it and the vertex $\langle \emptyset \rangle$. Moreover, *an infinite branch of a tree can be thought of as a function f whose course-of-values determines the branch* (recall, see p. 89, that the course-of-value of f is the function $\hat{f}$, where $\hat{f}(y)$ is the sequence number coding the first y values of f).

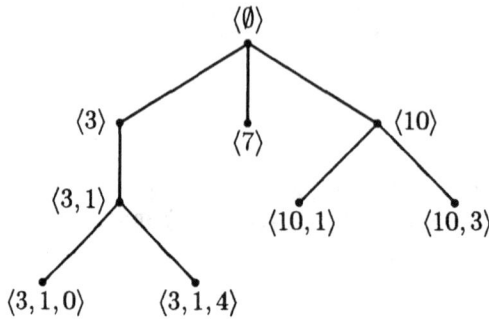

Figure IV.2: A tree

Well-foundedness of a partial ordering usually refers to lack of infinite descending sequences. Here a branch can be thought of as a descending sequence (if we picture the tree as growing downward, as in Figure 2): thus the terminology is consistent, and a tree is well-founded as a tree if and only if it is so as a partial ordering. What makes well-founded trees interesting is that it is possible to proceed by induction on them, from the leaves to the vertex. This procedure of backward induction is called **bar-recursion**. The next result is mostly a rephrasing of the Normal Form Theorem IV.2.10.

Theorem IV.2.15 First Representation Theorem for Π_1^1 sets (Lusin and Sierpinski [1923], Kleene [1955]) *A set A is Π_1^1 if and only if there is a recursive sequence $\{T_x\}_{x \in \omega}$ of recursive trees, such that*

$$x \in A \Leftrightarrow T_x \text{ is well-founded.}$$

Proof. If A is Π_1^1 then, for some recursive R,

$$x \in A \Leftrightarrow \forall f \exists y R(x, \hat{f}(y)).$$

It is now enough to define

$$u \in T_x \Leftrightarrow Seq(u) \wedge (\forall v \sqsubset u) \neg R(x, v),$$

where u and v range over sequence numbers and (see p. 89) $\sqsubset$ means proper initial subsequence. Clearly T_x is a recursive tree, uniformly in x. A leaf corresponds to a first place in which R holds. An infinite branch of T_x corresponds to a function f such that $\forall y \neg R(x, \hat{f}(y))$, and it exists if and only if $x \notin A$. Thus $x \in A$ if and only if T_x is well-founded.

Conversely, let $\{T_x\}_{x \in \omega}$ be a recursive sequence of recursive trees, and suppose

$$x \in A \Leftrightarrow T_x \text{ is well-founded.}$$

Then A is Π_1^1, since

$$x \in A \quad \Leftrightarrow \quad \neg \exists f \forall y (\hat{f}(y) \in T_x)$$
$$\Leftrightarrow \quad \forall f \exists y (\hat{f}(y) \notin T_x),$$

by the interpretation of infinite branches of trees as functions. Note that the relation $\hat{f}(y) \notin T_x$ is recursive, as a relation of f, x, and y, because T_x is uniformly recursive in x. □

Corollary IV.2.16 *The set T of characteristic indices of well-founded recursive trees is Π_1^1-complete.*

Proof. T is Π_1^1 because 'being the index of the characteristic function of a tree' is an arithmetical condition, and 'having no infinite branch' is Π_1^1:

$$e \in T \quad \Leftrightarrow \quad (\forall x)(\exists y)[y \leq 1 \wedge \varphi_e(x) \simeq y] \wedge$$
$$(\forall x)[\varphi_e(x) \simeq 1 \rightarrow Seq(x)] \wedge$$
$$(\forall x)(\forall y)[x \sqsubseteq y \wedge \varphi_e(y) \simeq 1 \rightarrow \varphi_e(x) \simeq 1] \wedge$$
$$(\forall f)(\exists y)[\varphi_e(\hat{f}(y)) \simeq 0].$$

T is Π_1^1-complete by the proof of the First Representation Theorem. More precisely, if

$$x \in A \Leftrightarrow \forall f \exists y R(x, \hat{f}(y))$$

for some recursive R then there is, by the S_n^m-Theorem, a recursive function g such that

$$\varphi_{g(x)}(u) \simeq \begin{cases} 1 & Seq(u) \wedge (\forall v \sqsubseteq u)\neg R(x, v) \\ 0 & \text{otherwise.} \end{cases}$$

Then $A \leq_m T$, because

$$x \in A \Leftrightarrow g(x) \in T. \quad \Box$$

Notice that it is possible to assign to each well-founded tree T an ordinal, as follows (by bar-recursion):

Definition IV.2.17 *If T is a well-founded tree, the **ordinal of a node** u on T is defined as:*

$$ord_T(u) = \begin{cases} sup\{1 + ord_T(u * \langle n \rangle)\} & \text{if } (\exists n)(u * \langle n \rangle \in T) \\ 1 & \text{otherwise.} \end{cases}$$

*The **ordinal of a well-founded tree** T is defined as:*

$$ord(T) = \begin{cases} ord_T(\langle \emptyset \rangle) & \text{if } T \neq \emptyset \\ 0 & \text{otherwise.} \end{cases}$$

Thus we assign 1 to the terminal nodes of each branch of a tree, and then we climb the tree by assigning to each nonterminal node the least ordinal not yet assigned to any of its successors. A nonempty tree has the ordinal of its vertex $\langle \emptyset \rangle$. Of course the whole procedure works because the tree is well-founded, and thus the inductive procedure of climbing the tree from the terminal nodes makes sense. The interesting fact is that this is really an equivalent way to look at countable ordinals.

Proposition IV.2.18 *An ordinal α is countable if and only if there is a well-founded tree T such that $\alpha = ord(T)$.*

Proof. Note that an ordinal is assigned to a node of T only if all the smaller ordinals have already been assigned to some of its successors. Since a tree has only countably many nodes, being made of sequence numbers, its ordinal can have only countably many predecessors, and it is thus countable.

Let now α be a countable ordinal: we define a well-founded tree T such that $\alpha = ord(T)$, by induction on the countable ordinals.

- $\alpha = 0$.
  Let $T = \emptyset$.

- $\alpha = \beta + 1$
  Let T_0 be a well-founded tree such that $\beta = ord(T_0)$. We only have to build a tree T that consists of a subtree isomorphic to T_0, and one vertex on top of it. Let
  $$T = \{\langle \emptyset \rangle\} \cup \{\langle 0 \rangle * u : u \in T_0\}.$$

- $\alpha = \sup_{n \in \omega} \alpha_n$
  Let T_n be well-founded trees such that $\alpha_n = ord(T_n)$. We only have to build a tree T that consists of subtrees isomorphic to the T_n's, and one vertex on top of them. Let
  $$T = \{\langle \emptyset \rangle\} \cup \bigcup_{n \in \omega} \{\langle n \rangle * u : u \in T_n\}. \quad \square$$

Since the notion of well-founded tree is easily constructivized, by considering recursive well-founded trees, we have a natural constructive analogue of the notion of ordinal: a **recursive ordinal** is an ordinal α such that $\alpha = ord(T)$, for some recursive well-founded tree T. Since the node of a tree is assigned an ordinal only if all smaller ordinals have already been assigned to some of its successors, and the subtree consisting of the extensions of a node of a recursive tree is isomorphic to a recursive tree, *the recursive ordinals form an initial segment of the ordinals*. The first nonrecursive ordinal is indicated by ω_1^{ck}, because it is the constructive analogue of the first uncountable ordinal ω_1, and it was introduced by Church and Kleene [1937]. Clearly ω_1^{ck} is countable, because there are only countably many recursive trees, and thus only countably many recursive ordinals. The theory of recursive ordinals will be developed in Volume III, and will be useful for a finer analysis of the hyperarithmetical (p. 391) and Π_1^1 sets.

Another way to introduce countable ordinals is by looking at countable well-orderings (equivalently, at well-orderings of ω), and thus we may guess that there is a result similar to IV.2.15, using well-orderings instead of well-founded trees.

Definition IV.2.19 *A* **(recursive) partial ordering** $\preceq$ *is a (recursive) binary relation on a (recursive) set A of numbers which is reflexive, antisymmetric and transitive, i.e. such that, for every $x, y, z \in A$,*

    1. $x \preceq x$

    2. $x \preceq y \wedge y \preceq x \Rightarrow x = y$

    3. $x \preceq y \wedge y \preceq z \Rightarrow x \preceq z$.

A **linear partial ordering** *is a partial ordering which is total on A, i.e. such that, for every $x, y \in A$,*

    4. $x \preceq y \vee y \preceq x$.

A **well-ordering** *is a linear partial ordering with no infinite descending sequence of elements.*

As usual, we will write $x \prec y$ for $x \preceq y \wedge x \neq y$.

Theorem IV.2.20 Second Representation Theorem for Π_1^1 sets (Lusin and Sierpinski [1923], Kleene [1955], Spector [1955]) *A set A is Π_1^1 if and only if there is a recursive sequence $\{\prec_x\}_{x \in \omega}$ of recursive linear orderings, such that*

$$x \in A \Leftrightarrow \prec_x \text{ is a well-ordering.}$$

Proof. By IV.2.15, it is enough to show that it is possible to go back and forth, in an effective way, from recursive trees to recursive linear orderings of ω, preserving well-foundedness and well-ordering.

Let $\prec$ be a recursive linear ordering of ω: we want a recursive tree $T_\prec$ such that

$$\prec \text{ is a well-ordering} \Leftrightarrow T_\prec \text{ is well-founded.}$$

It is enough to define

$$\langle x_1, \ldots, x_n \rangle \Leftrightarrow x_n \prec \cdots \prec x_1.$$

Thus branches of $T_\prec$ code maximal descending sequences of elements w.r.t. $\prec$, and there are infinite branches in $T_\prec$ if and only if there are infinite descending sequences in $\prec$.

Let T be a recursive tree: we want a recursive linear ordering $\prec_T$ such that

$$T \text{ is well-founded } \Leftrightarrow \prec_T \text{ is a well-ordering.}$$

A natural way to linearly order a tree is the lexicographical one: given two sequence numbers, we look at the branches on which they lie, and give the precedence to the one on the leftmost branch when the branches differ, and to the one on the higher level if they are on the same branch. But for our purposes this is overkilling: the resulting order is always a well-ordering (since each node has only finitely many predecessors at the same level, on different branches, and only finitely many predecessors at lower levels, on the same branch).

The reason why this ordering is not sensitive to infinite branches of the tree is because it follows each branch in increasing order: we then have only to modify it, and reverse the order on single branches. Thus we will still order lexicographically sequence numbers lying on different branches, but reverse the lexicographical order on single branches (and thus the only way to produce infinite descending sequences will be to have infinite branches). The wanted linear ordering is thus, for $x, y \in T$,

$$x \prec_T y \;\Leftrightarrow\; y \sqsubset x \vee (\exists n)[(x)_n < (y)_n \wedge (\forall i < n)((x)_i = (y)_i)].$$

This is sometimes called the **Kleene-Brouwer ordering** (Lusin and Sierpinski [1923], Brouwer [1924]) associated to T.   □

Exercises IV.2.21 a) *The set* W *of characteristic indices of recursive well-orderings of* ω *is* Π_1^1*-complete.* (Lusin and Sierpinski [1923], Spector [1955]) (Hint: the proof above shows that $T \leq_m W$, if $\prec_T$ is extended to a total ordering, e.g. by putting all numbers not in T before everything, with their usual ordering. Thus it is enough to show that W is Π_1^1.)

b) T *and* W *are recursively isomorphic.* (Hint: this holds in general, for Π_1^1-complete sets.)

Since well-orderings of ω are another way of looking at countable ordinals, we might think of defining the recursive ordinals by looking at recursive well-orderings, instead than recursive well-founded trees. Happily, by the proof just given, *the class of ordinals defined by recursive well-founded trees on one side and recursive well-orderings on the other coincide*, thus showing that the notion of recursive ordinal is natural and stable.

We have just scratched the surface of the theory of Π_1^1 sets, which will be developed in detail in Volume III. We just hint here at the fact that Π_1^1 *sets admit a natural interpretation in terms of infinitary computations.* Consider a recursive finitely branching tree: by König's Lemma, the process of determining whether it is well-founded or not is r.e. (see IV.2.6.b). Thus the process of determining whether a recursive tree is well-founded can be seen as an infinitary analogue of an r.e. question. This can be made precise, and it turns out that the Π_1^1 sets are exactly the r.e. sets relative to the ordinal ω_1^{ck} (see p. 385).

Δ_1^1 sets

The level 0 of the Analytical Hierarchy corresponds to the arithmetical sets, because we allow for arithmetical matrices. To describe Δ_1^1 we take a constructive stand, and will show how to obtain any member of it from below.

Recall that we constructed the smallest Boolean algebra generated by the r.e. sets by closing them under complements and finite unions, and by so doing we did not even exhaust Δ_2^0. The idea here is to consider not only finite unions, but infinite ones as well. Of course we have to be careful, since the singletons are r.e. sets, and if we allow arbitrary countable unions then we get any possible set: this shows that infinite unions are too powerful. We consider only constructive unions, i.e. unions of r.e. families of sets, and a great deal of the power is maintained, since we obtain this way all the Δ_1^1 sets.

Theorem IV.2.22 Suslin-Kleene Theorem (Suslin [1917], Kleene [1955a]) Δ_1^1 *is the smallest class of sets*

1. *containing every singleton set $\{n\}$*

2. *closed under complements*

3. *closed under effective unions.*

Proof. To be able to talk about effective unions, we are going to assign indices to sets in the class, and operate on them by using recursive functions. We thus call a class of sets $\mathcal{C} = \{C_x\}_{x \in C}$ an SK-class (for Suslin-Kleene) if there are recursive functions f_1, f_2 and φ such that

- $C_{f_1(n)} = \{n\}$

- $C_{f_2(n)} = \overline{C}_n$

- if φ_n is total then $\varphi(n)\downarrow$ and $C_{\varphi(n)} = \bigcup_{x \in \omega} C_{\varphi_n(x)}$.

The theorem thus amounts to proving that Δ_1^1 is the smallest SK-class. We do this in two steps, first explicitly defining the smallest SK-class, and then showing that it coincides with Δ_1^1.

1. *Definition of the smallest SK-class $\mathcal{G}$*

    We first define, by induction, the system of indices we are going to use, as the smallest set G such that

    * $\langle 0, n \rangle \in G$
    *   if $n \in G$ then $\langle 1, n \rangle \in G$
    *   if φ_n is total and $\forall x (\varphi_n(x) \in G)$ then $\langle 2, n \rangle \in G$.

    These indices obviously correspond to the three clauses of the definition of a SK-class. We can then define the class $\mathcal{G} = \{G_x\}_{x \in G}$ in the natural way:

    * $G_{\langle 0, n \rangle} = \{n\}$
    * $G_{\langle 1, n \rangle} = \overline{G}_n$
    * $G_{\langle 2, n \rangle} = \bigcup_{x \in \omega} G_{\varphi_n(x)}$.

    By definition, $\mathcal{G}$ is a SK-class. It is also the smallest such one, intuitively because we put in it only what is strictly required by the definition of SK-class, and formally because if $\mathcal{C} = \{C_x\}_{x \in \omega}$ is a SK-class, and f_1, f_2, φ are the recursive functions associated to it, then

    $$x \in G \Rightarrow G_x = C_{g(x)},$$

    i.e. each element of $\mathcal{G}$ is in $\mathcal{C}$, where g is any recursive function such that

    $$g(x) = \begin{cases} f_1(n) & \text{if } x = \langle 0, n \rangle \\ f_2(g(n)) & \text{if } x = \langle 1, n \rangle \\ \varphi(e) & \text{if } x = \langle 2, n \rangle \text{ and } \varphi_e(z) \simeq g(\varphi_n(z)) \\ 0 & \text{otherwise.} \end{cases}$$

    Note that such a function exists by the Fixed-Point Theorem.

2. *$\mathcal{G}$ is contained in Δ_1^1*

    We have to show that if $x \in G$ then $G_x \in \Delta_1^1$. It is enough to prove that $G_x \in \Pi_1^1$ for every $x \in G$: then, since $\overline{G}_x = G_{\langle 1, n \rangle}$, $\overline{G}_x \in \Pi_1^1$ and $G_x \in \Sigma_1^1$, hence $G_x \in \Delta_1^1$.

    The reason why $G_x \in \Pi_1^1$ is that $\mathcal{G}$ is the smallest class of sets satisfying an arithmetical condition, which means that every class satisfying the same condition must contain it: this can be expressed by a universal quantifier over classes of sets, and an arithmetical matrix, and it is 'almost' Π_1^1. We only have to talk about sets, rather than classes of sets, and this can easily be accomplished by considering an indexed class of sets as a binary relation, which can then be coded as a set. Precisely, we are going to show that $\mathcal{G}$ is uniformly Π_1^1, i.e. that there is a Π_1^1 set A such that

$$z \in G_x \Leftrightarrow \langle z, x \rangle \in A.$$

We will write an arithmetical predicate $P(z, x, X)$, meaning that X is a set coding a class satisfying the conditions of the definition of SK-class, and then we will say that A is the smallest such set, as follows:

$$\langle z, x \rangle \in A \Leftrightarrow (\forall X)(P(z, x, X) \rightarrow \langle z, x \rangle \in X).$$

Then A will be Π_1^1. It just remains to define P:

$$
\begin{aligned}
P(z, x, X) \quad \Leftrightarrow \quad & [x = \langle 0, n \rangle \wedge z = n] \vee \\
& [x = \langle 1, n \rangle \wedge \langle z, n \rangle \notin X] \vee \\
& [x = \langle 2, n \rangle \wedge \varphi_n \text{ total} \wedge (\exists t)(\langle z, \varphi_n(t) \rangle \in X)].
\end{aligned}
$$

3. Δ_1^1 *is contained in* $\mathcal{G}$

Suppose A is Δ_1^1: there are two recursive relations R and Q such that

$$
\begin{aligned}
x \in A \quad &\Leftrightarrow \quad \forall f \exists y R(x, \hat{f}(y)) \\
x \in \overline{A} \quad &\Leftrightarrow \quad \forall g \exists y Q(x, \hat{g}(y)).
\end{aligned}
$$

To show that $A \in \mathcal{G}$, it is enough to find a recursive function h such that

$$
G_{h(x)} = \begin{cases} \{x\} & \text{if } x \in A \\ \emptyset & \text{otherwise.} \end{cases}
$$

Then $A = \bigcup_{x \in \omega} G_{h(x)}$, i.e. A is an effective union of members of $\mathcal{G}$, and hence is in $\mathcal{G}$.

Since A is Π_1^1, consider the representation of A by uniformly recursive trees T_x given by the First Representation Theorem IV.2.15. If $x \in A$ then T_x is well-founded, and thus each branch is finite. We might label the leaves with the set $\{x\}$, and climb our way to the top of the tree, by giving the same label $\{x\}$ to any node whose successors all have that label. Then, if $x \in A$, the vertex (corresponding to the empty sequence number $\langle \emptyset \rangle$) also has that label. This is easily defined in recursive terms, by first giving the function

$$
\varphi_0(x, u) \simeq \begin{cases} \{x\} & \text{if } R(x, u) \wedge (\forall v \sqsubset u) \neg R(x, v) \\ \bigcup_{n \in \omega} \varphi_0(x, u * \langle n \rangle) & \text{otherwise,} \end{cases}
$$

and then considering $\varphi_0(x, \langle \emptyset \rangle)$. Note that there is a partial recursive function $\psi_0(x, u)$ such that $G_{\psi_0(x,u)} = \varphi_0(x, u)$, by the Fixed-Point Theorem:

$$\psi_0(x, u) \simeq \begin{cases} \langle 1, x \rangle & \text{if } R(x, u) \wedge (\forall v \sqsubset u) \neg R(x, v) \\ \langle 2, e \rangle & \text{if } \varphi_e(n) \simeq \psi_0(x, u * \langle n \rangle), \text{ otherwise.} \end{cases}$$

This almost produces what we want, and there is only one problem: $\varphi_0(x, \langle \emptyset \rangle)$ is undefined when $x \in \overline{A}$, because in this case some branch of the tree has no leaf, and thus we cannot climb our way to the top, since we never define φ_0 on that path. However, if $x \in \overline{A}$ then we can play a similar strategy on the other tree, corresponding to Q. We can thus get a function φ_1 such that $\varphi_1(x, \langle \emptyset \rangle)$ gives $\emptyset$ if $x \in \overline{A}$, and is undefined otherwise. Moreover, we can find a partial recursive function ψ_1 such that $G_{\psi_1(x,u)} = \varphi_1(x, u)$.

Since x is either in A or in $\overline{A}$, one and only one of $\varphi_0(x, \langle \emptyset \rangle)$ and $\varphi_1(x, \langle \emptyset \rangle)$ is defined, and we can thus let

$$h(x) = \begin{cases} \psi_0(x, \langle \emptyset \rangle) & \text{if } \psi_0(x, \langle \emptyset \rangle) \text{ converges} \\ \psi_1(x, \langle \emptyset \rangle) & \text{if } \psi_1(x, \langle \emptyset \rangle) \text{ converges.} \end{cases}$$

Then, as wanted,

$$G_{h(x)} = \begin{cases} \{x\} & \text{if } x \in A \\ \emptyset & \text{otherwise.} \quad \square \end{cases}$$

Corollary IV.2.23 *The arithmetical sets are properly included in Δ_1^1.*

Proof. We prove, by induction on n, that every Σ_n^0 set is Δ_1^1. Each r.e. set can be obtained as effective union of recursive sets. Precisely, if

$$x \in A \Leftrightarrow (\exists y) R(x, y)$$

then A is the union, over $y \in \omega$, of the sets

$$A_y = \{x : R(x, y)\}.$$

By relativization, each Σ_{n+1}^0 can be obtained as the effective union of Π_n^0 sets. By induction hypothesis, every Σ_n^0 set is Δ_1^1, and by closure under complements so is every Π_n^0 set. Thus every Σ_{n+1}^0 set is Δ_1^1.

Let now B_n be a Σ_n^0-complete set, for every $n \geq 1$. The set

$$C_n = \{\langle x, n \rangle : x \in B_n\}$$

is still Σ_n^0, and hence Δ_1^1. Then the set

$$C = \bigcup_{n \geq 1} C_n$$

is also Δ_1^1. If it were arithmetical, it would be Σ_n^0 for some n. But then e.g.

$$x \in B_{n+1} \Leftrightarrow \langle x, n+1 \rangle \in C,$$

and the Σ_{n+1}^0-complete set B_n would be Σ_n^0, contradiction. $\quad\square$

The characterization just given suggests the possibility of building a hierarchy for Δ_1^1, by looking at the stages in which a given relation is generated, in the inductive definition of the class. This will be done in Volume III, and will define the **Hyperarithmetical Hierarchy** (Davis [1950], Mostowski [1951], Kleene [1955a]), which can be seen as a transfinite extension (of length ω_1^{ck}, see p. 385) of the Arithmetical Hierarchy (by the proof of the corollary).

A suggestive reformulation of the Suslin-Kleene Theorem is that *the Δ_1^1 sets are exactly the sets computable with the help of an oracle that embodies number quantification* (disguised in the operation of union). Otherwise stated, the Δ_1^1 sets are the sets computable modulo a finite number of quantifications over ω. This can be made precise with the extension of the notion of relative recursiveness to higher types (see p. 199). Then the Δ_1^1 sets are the sets recursive in the type-2 object

$$E(f) \simeq \begin{cases} 0 & \text{if } \exists x(f(x) = 0) \\ 1 & \text{otherwise} \end{cases}$$

(Kleene [1959]).

Before too great expectations arise, we immediately show that *no result similar to the Suslin-Kleene Theorem may possibly hold for higher levels of the Analytical Hierarchy*, in a precise sense. Thus IV.2.22 is an exceptional characterization, working only because one function quantifier is needed to express the notion of 'smallest set satisfying a given condition': Δ_1^1 has no power to swallow it up, but all other higher Δ_n^1's do.

Proposition IV.2.24 (Addison and Kleene [1957]) *There is no $n > 0$ such that Δ_{n+1}^1 is the smallest indexed class of sets uniformly satisfying a Δ_n^1 condition.*

Proof. Let $P(z, x, X)$ be a Δ_n^1 formula meaning that X codes a class of sets indexed by numbers, as follows:

$$z \in G_x \Leftrightarrow \langle z, x \rangle \in X.$$

Then the smallest class of sets $\mathcal{G}$ defined by P is strictly contained in Δ_{n+1}^1. Indeed, let A be the set coding it:

$$\langle z, x \rangle \in A \Leftrightarrow (\forall X)(P(z, x, X) \to \langle z, x \rangle \in X).$$

Then A is Π_n^1 and hence, since $n > 0$, Δ_{n+1}^1. The set B defined as

$$x \in B \Leftrightarrow \langle x, x \rangle \notin A$$

is thus Δ_{n+1}^1 too, but it cannot be in the class coded by A, otherwise there would be an e such that

$$x \in B \Leftrightarrow \langle x, e \rangle \in A,$$

and for $x = e$ we would get a contradiction. $\square$

Descriptive Set Theory $\star$

The Arithmetical and Analytical Hierarchies resemble very much, both in notions and results, classical hierarchies that were studied at the beginning of the century in Descriptive Set Theory, which started as a way of generating from below interesting and graspable classes of functions and sets (on reals, not on natural numbers), as opposed to the definition of the continuum as a whole, by the power axiom. Part of the motivation was to prove the Continuum Hypothesis for larger and larger classes of sets, with the hope of finally getting to a complete solution of the problem.

The classical development has its first landmarks in Borel [1898], who introduces Borel sets (as the smallest class containing the open sets, and closed under complements and countable unions), Baire [1899], who introduces Baire sets (via Baire functions, defined as the smallest class containing the continuous functions, and closed under limits), and Lebesgue [1905], in which the classes of Borel and Baire sets are proved to coincide. Lebesgue also introduces the analytic sets (not to be confused with the analytical sets dealt with in this section) as projections of Borel sets, and falsely claims that an analytic set is Borel. This is corrected by Suslin [1917], where it is proved that a set is Borel if and only if it is both analytic and coanalytic. The Projective Hierarchy, obtained by iterating projections and complements, is defined and studied in Lusin [1925] and Sierpinski [1925]. The classical period ends in the late Thirties, when a stumbling block is reached, with the impossibility of extending the theory beyond the second level of the Projective Hierarchy. This will be explained later, after the introduction of the methods to prove the independence of the Continuum Hypothesis (Gödel [1940], Cohen [1963]): *the theory of higher projective classes is mostly independent of ZFC, and thus it cannot be pursued without additional set-theoretical hypothesis.*

A second period of development for the subject comes when the recursion theorists attempt to classify sets of natural numbers, from their own point of view. Kleene [1943] and Mostowski [1947] introduce the Arithmetical Hierarchy, and while the former works without awareness of the classical work, the latter explicitly develops the hierarchy as an analogue of the Projective Hierarchy. After the introduction by Kleene [1955] of the Analytical Hierarchy, the analogies begin to clarify. Addison [1954], [1959] not only makes them precise, but also sees that there is more: *classical Descriptive Set Theory can be obtained by relativization of the theory of recursion-theoretical hierarchies*, using the following translations (in which the left-hand side is the effective version of the right-hand one):

recursive function	continuous function
r.e. set	open set
hyperarithmetical set	Borel set
Σ_1^1 set	analytic set
analytical set	projective set.

In particular, the recursion-theoretical versions of classical results are stronger, and imply their classical counterparts. This explains the double assignment of credit to results, in this section.

After the classical and the effective periods, the subject has entered its modern era with the introduction of new set-theoretical axioms (which, as we have quoted, are necessary to go beyond the first two levels of the Analytical Hierarchy). The first axiom to imply results about analytical sets was the *Axiom of Constructibility* (IV.4.2): Gödel [1940] and Addison [1959a] showed that a coherent theory for all levels of the Analytical Hierarchy can be obtained from it. Although provably consistent with ZFC, the Axiom of Constructibility is however taken more as a useful technical tool than as a real additional axiom of Set Theory. The *existence of measurable cardinals* is taken more seriously, and it does provide for additional results about analytical sets (Solovay [1969], Martin and Solovay [1969]), but its influence does not seem to extend much beyond the fourth level of the hierarchy. The most successful axiom to date for the development of Descriptive Set Theory is the *Axiom of Projective Determinacy*, a restricted version of the full Axiom of Determinacy (V.1.15) saying that all projective games are determined. This can be seen as an axiom about the existence of very large cardinals, and it implies an extremely coherent picture of the Analytical Hierarchy (the first results in this direction have been obtained by Blackwell [1967], Addison and Moschovakis [1968] and Martin [1968]). We will come back to this subject in the last chapter of our book.

Classical references for Descriptive Set Theory are Hausdorff [1917], Lusin [1930a], Sierpinski [1950], Kuratowski and Mostowski [1968]. The definitive test

about the subject, treating classical, effective and modern theory, is Moschovakis [1980].

Relativizations ⋆

As for the Arithmetical Hierarchy, the work done for the Analytical Hierarchy
can also be relativized to a given oracle X, defining the classes $\Sigma_n^{1,X}$, $\Pi_n^{1,X}$ and
$\Delta_n^{1,X}$, and obtaining similar results. This time a number of new reducibilities
do arise.

Proposition IV.2.25 (Shoenfield [1962]) *The relation*

$$A \leq_{\Delta_n^1} B \Leftrightarrow A \in \Delta_n^{1,B}$$

is transitive, for any n.

Proof. For $n = 0$ this is obvious, since then $\leq_{\Delta_0^1}$ is just arithmetical reducibility. Let then $n > 0$, $A \in \Delta_n^{1,B}$, and $B \in \Delta_n^{1,C}$. There are $P \in \Pi_{n-1}^{1,B}$ and
$R \in \Sigma_{n-1}^{1,B}$ such that

$$x \in A \Leftrightarrow \exists f P(x, f, B) \Leftrightarrow \forall f R(x, f, B).$$

To show that $A \in \Sigma_n^{1,C}$, note that

$$x \in A \Leftrightarrow \exists D[D = B \land \exists f P(x, f, D)].$$

The expression $B = D$ can be rewritten as

$$\forall x[(x \in B \to x \in D) \land (x \in D \to x \in B)].$$

By using the $\Pi_n^{1,C}$ form for B in the first conjunct, and the $\Sigma_n^{1,C}$ form in the
second one we get, by manipulation of quantifiers, a $\Sigma_n^{1,C}$ form for $B = D$
which, substituted above, produces a $\Sigma_n^{1,C}$ form for A.

Similarly, noting that

$$x \in A \Leftrightarrow \forall D[D = B \to \forall f R(x, f, D)],$$

we get a $\Pi_n^{1,C}$ form for A, which is then $\Delta_n^{1,C}$. □

The relation $\leq_{\Delta_n^1}$ is obviously reflexive and thus it induces, as usual, an
equivalence relation.

Definition IV.2.26 *For $n \geq 1$, the equivalence classes of sets w.r.t. the equivalence relation*

$$A \equiv_{\Delta_n^1} B \Leftrightarrow A \leq_{\Delta_n^1} B \land B \leq_{\Delta_n^1} A$$

*are called $\mathbf{\Delta_n^1}$-degrees, and $(\mathcal{D}_n^1, \leq)$ is the structure of Δ_n^1-degrees, with the
partial ordering $\leq$ induced on them by $\leq_{\Delta_n^1}$.*

The Δ_1^1-degrees are also called **hyperdegrees**, and will be studied in Volume III, together with the other Δ_n^1-degrees. For an anticipation, see p. 441.

The nice closure properties that produce the transitivity of $\leq_{\Delta_n^1}$ also have some very negative consequences, to the effect that *no level of the Analytical Hierarchy can be thought of as built from below*. More precisely, no analogue of Post's Theorem IV.1.14 exists. Recall that, for the Arithmetical Hierarchy, Δ_{n+1}^0 was actually the class of sets recursive in some Σ_n^0 set. This result fails very badly to generalize here, even if recursive (i.e. Δ_1^0) is substituted by any Δ_n^1.

Proposition IV.2.27 (Addison and Kleene [1957], Shoenfield [1962])
There is no n such that Δ_{n+1}^1 is the class of sets Δ_n^1 in some Σ_n^1 set.

Proof. By the result proved above, the class of sets Δ_n^1 in some Σ_n^1 set is contained in Δ_{n+1}^1. We have to show that the inclusion is always proper. For $n = 0$ this is clear, since there are Δ_1^1 sets which are not arithmetic. Suppose then $n \geq 1$: we prove that there is $A \in \Delta_{n+1}^1$ such that, for any set $B \in \Sigma_n^1$, A is not Δ_n^1 in it. Let C be a Σ_n^1-complete set: since any other Σ_n^1 set is recursive in it, it is enough to get A not Δ_n^1 in C. Since we want to limit its complexity, the obvious first choice for A is a $\Sigma_n^{1,C}$-complete set, which is certainly not Δ_n^1 in C. Moreover, by the transitivity of $\leq_{\Delta_{n+1}^1}$, A is Δ_{n+1}^1, being Σ_n^1 in the Σ_n^1 set C. $\square$

Recall that we also proved another negative result (IV.2.24), to the effect that only Δ_1^1 can be constructed from below. Notice that even this case is not a real exception: the very notion of 'smallest set satisfying a given condition' is Π_1^1, and thus in some sense is more complicated than the class Δ_1^1 itself. And as soon as it becomes less complicated than a given class Δ_n^1 (i.e. for $n > 1$), no such characterization is possible anymore.

Post's Theorem in the Analytical Hierarchy ⋆

We have seen that, in a precise sense, Post's Theorem IV.1.14 has no analogue in the Analytical Hierarchy. But we should not forget that, in stepping up from the Arithmetical to the Analytical Hierarchy, the universe has changed quite radically: functions are now treated as objects, like we did with numbers all along, and we quantify over them. But there has been no corresponding increase in the power of computations: we have retained the notion of computability that we found suitable for numbers. In particular, the μ-operator is available as a search operator over numbers, but nothing of the kind exists over functions. The negative results obtained above could thus be interpreted as signs of the inadequacy of the notion of recursiveness as a notion of computability over the continuum.

Notions of computability over abstract domains have been proposed, see p. 202, and they all basically reduce to either prime or search computability (p. 203), which are abstract analogues of deterministic and nondeterministic computability. We have made a great use of searches over numbers in the proofs of the results of Section 1: computationally, the existential quantifiers are handled by searching for elements that satisfy the matrix. In particular, Post's Theorem is a relativization of II.1.19, whose proof amounted to a parallel search. This suggests that, in an abstract setting, it is search computability that we need. Moschovakis [1969] has indeed proved that, once the notion of recursiveness is replaced by that of search computability, and a Projective Hierarchy is defined in analogy to the Arithmetical Hierarchy (i.e. by allowing search computable matrices, and quantifications over the domain), then *the theory of the Arithmetical Hierarchy, including Post's Theorem, generalizes to abstract domains*.

It only remains to apply the general theory, and translate the notion of search computability over the continuum in the familiar terms of the Analytical Hierarchy. It turns out that *a relation is Δ^1_{n+2} if and only if it is search computable in Π^1_{n+1}*, and this provides for an analogue of Post's Theorem for the Analytical Hierarchy (Hinman and Moschovakis [1971]).

The idea of the proof for $n = 0$ (the other cases being similar by relativization) is the following. The main computation concerns the complexity of domains of functions search computable in Π^1_1, which are the analogues of (relativized) r.e. relations. One direction requires an arithmetization of the computation tree similar to that of I.7.3. Since the relevant objects can be coded only as functions (instead of as numbers), a search computable function converges if and only if there is a function coding a computation tree for it, and this introduces an outer existential function quantifier. Moreover, since computation trees are not finite objects anymore, they must well-founded, and to express this we need a universal function quantifier. Thus the whole relation is Σ^1_2. If the function is search computable in some oracle the complexity of the computation tree is accordingly increased, but Π^1_1 oracles only require a universal quantifier, which is present anyway because of well-foundedness, and in this case the total complexity is still Σ^1_2. Conversely, let A be Σ^1_2: then the function that searches for a witness of the Π^1_1 matrix is search computable in the characteristic function of the latter, and converges if and only if the witness exists. Thus A is the domain of a function search computable in Π^1_1.

By a refinement of the argument sketched above, Hinman and Moschovakis [1971] show that *a relation is Δ^1_2 if and only if it is search computable in the equality predicate for functions*.

These results point out that the proper analogue of recursiveness on the continuum seems to be Δ_2^1, rather than Δ_1^1. See also p. 442 on this.

IV.3   The Set-Theoretical Hierarchy

Arithmetic is a natural environment for the study of sets and functions of natural numbers, since these can arguably be seen as primitive objects. But modern mathematics is usually done (formally or informally) in the framework of Set Theory, in which the natural numbers can be defined. In this section we look at definability in the language of (formalized) Set Theory, with an eye to the applications to our subject.

Since this is not a book on Set Theory, we refer to Gödel [1944], [1964], and Kreisel and Krivine [1966] for general discussion, Fraenkel, Bar-Hillel, and Levy [1958] for foundations and history, Monk [1969] and Levy [1979] for the fundamentals of the subject, and Cohen [1966], Jech [1978], and Kunen [1980] for more advanced topics. However, this section and the next are mostly self-contained, and require only the working knowledge of elementary Set Theory used until now. We will also recall the needed concepts and results.

Truth in Set Theory

A primordial notion of set (Frege [1893]) corresponds to an *extensional perception of properties*. It can be formulated on two simple principles: **extensionality**, which ensures that a set is completely determined by its elements, and **full comprehension**, according to which one can consider the set of all the objects for which a given property holds, abstracting from the intensional way the given property is expressed. Unless one admits partial properties, that may possibly be undefined for a given argument, this naive point of view is shaken by Russell's paradox (p. 81), since the set corresponding to the property of not being a member of itself is contradictory.

One possible remedy to the situation is to switch to a different notion of set, corresponding to an *arbitrary collection of elements taken from a given collection*, like when we consider a set of natural numbers. This implies a cumulative conception of the universe, in which sets are obtained by stages, from simple to more complicated ones, and it suggests a number of properties of the membership relation $\in$. They fall into five different categories:

1. *extensionality*

   Membership being the only concept at hand, a set is completely determined by its elements, and thus two sets with the same elements are indistinguishable. As a particular instance of a more general principle going back to Leibniz (**identitas indiscernibilium**), according to which

two objects that cannot be distinguished are equal, we get the Axiom of
Extensionality:

$$(\forall z)(z \in x \leftrightarrow z \in y) \;\Rightarrow\; x = y.$$

2. *comprehension*

Full comprehension is simply false for the present notion of set (Gödel
[1958]), since now a set is a collection not of arbitrary elements, but only
of elements already belonging to a given collection. In this setting it
takes the form of the Axiom of **Separation**, according to which one can
consider the subset of a given set consisting of all the objects in it for
which a given property holds.

3. *set construction*

Since the Axiom of Separation only isolates subsets of given sets, without
additional axioms we could not produce sets bigger than the given ones.
On the other hand the paradoxes show that there are collections that are
simply too big to be sets, and thus some restrain is needed. We compromise by adopting the doctrine of **limitation of size** (Cantor [1899],
Russell [1906]), according to which a construction principle should produce only sets which are not too large, compared to sets on which they
build.

Typical examples of natural construction principles are the following:
Pairing, **Union**, **Power Set** (taking the set $\mathcal{P}(a)$ of all subsets of a set
a), and **Replacement** (taking the image of a set under a function). To
avoid the consideration of functions replacement can be substituted, in
presence of the other axioms, by **Collection** (taking a set b such that
$(\forall x \in a)(\exists y \in b)\varphi(x,y)$, provided $(\forall x \in a)(\exists y)\varphi(x,y)$ holds).

4. *set existence*

The axioms given until now do not ensure, by themselves, the existence
of any set: separation only produces subsets of given sets, while the
construction principles have to build on something. We thus need to
postulate the existence of at least a set. Even a single set, together with
the previous axioms, makes the subject nontrivial, as we will see in this
section. But to make possible the definition of the usual mathematical
objects, like the real numbers, we actually need an infinite set (say ω),
whose existence is ensured by the Axiom of **Infinity**.

The existence of a single infinite set, combined with the construction
principles, produces large sets. But the existence of *very* large sets (e.g.
of sets closed under replacement and power set) does not follow from the
axioms, and it must be postulated independently, if needed. See Drake

[1974] for a treatment of such axioms that, like the Axiom of Infinity, usually postulate the existence of large cardinals.

5. *foundation*

If the membership relation is well-founded, not only sets are obtained from simpler constituents: there are atomic constituents (**urelements**), which all sets are ultimately built of. Then the process of generation of sets can be seen as a well-ordered sequence of stages, through which the cumulative universe comes into being. This implies a form of the **vicious circle principle**, according to which a totality cannot contain members that presuppose it (Poincaré [1906], Russell [1908]). In particular, there cannot be sets x such that $x \in x$, otherwise they would give rise to an infinitely descending chain $\cdots \in x \in x$.

The well-foundedness of $\in$ is stated by the Axiom of **Foundation** (Mirimanoff [1917])

$$(\exists y)(y \in x) \to (\exists y)[y \in x \land (\forall z \in y)(z \notin x)],$$

which says that every nonempty set has a minimal element w.r.t. $\in$ (note the analogy with the Least Number Principle, p. 21).

The usefulness of foundation is that it allows for recursion on $\in$, also called $\in$-**induction**:

$$(\forall z)[(\forall y \in z)\varphi(y) \to \varphi(z)] \Rightarrow (\forall x)\varphi(x).$$

Suppose indeed that $(\exists x)\neg\varphi(x)$. We have to find a nonempty set to which apply the Axiom of Foundation. Fix x such that $\neg\varphi(x)$ holds, and consider

$$\hat{x} = \{x\} \cup \text{the downward closure of } x \text{ w.r.t. } \in,$$

which is a set (see IV.3.12). By separation, consider the set consisting of the elements of $\hat{x}$ that do not satisfy φ. By foundation, it has a minimal element z w.r.t. $\in$: by minimality, no element of $\hat{x}$ belonging to z satisfies φ. But every element of z is in $\hat{x}$, by its downward closure, and thus every element of z satisfies φ. Then, by hypothesis, also z satisfies it, contradiction.

The **Zermelo-Fraenkel system** ZF (Zermelo [1908], Fraenkel [1922]) is the first-order theory with equality, the relation symbol $\in$ as the only nonlogical symbol, and the axioms just stated. ZFC is obtained by adding to it the **Axiom of Choice** or, equivalently, the Well-Ordering Principle (every set can be well-ordered). We will mostly be concerned with the **Generalized Kripke-Platek system** GKP (Kripke [1964], Platek [1966]), which is obtained from

ZF by dropping Infinity and Power Set (see also p. 421), and with $\mathbf{ZF^-}$, which instead only drops the Power Set Axiom.

For convenience, the language of Set Theory is enriched of a term formation symbol $\{\,\}$, defined according to the following rule: whenever we can prove that

$$\exists y \forall x (x \in y \leftrightarrow \varphi(x)),$$

then we can denote such a y (which, by extensionality, is necessarily unique) by

$$\{x : \varphi(x)\}.$$

The existence of distinct urelements is not necessary for a development of Set Theory, since a single object is enough to start the whole process, and it is provided by the empty set (which has no elements). By extensionality, the empty set is the only urelement. The existence of other urelements is sometimes useful, and in these cases one can state the Axiom of Extensionality only for nonempty sets.

Definition IV.3.1 *The* **Cumulative Hierarchy** *is described, in terms of ordinals, as follows:*

$$
\begin{aligned}
V_0 &= \emptyset \\
V_{\alpha+1} &= \mathcal{P}(V_\alpha) \\
V_\beta &= \bigcup_{\alpha<\beta} V_\alpha \ (\beta \ limit) \\
V &= \bigcup_\alpha V_\alpha.
\end{aligned}
$$

In ZF we can prove, by $\in$–induction, that $\forall x \exists \alpha (x \in V_\alpha)$. Indeed, suppose $(\forall y \in x)(\exists \alpha)(x \in V_\alpha)$. By collection, $(\exists \alpha)(\forall y \in x)(x \in V_\alpha)$, and thus $x \subseteq V_\alpha$, i.e. $x \in V_{\alpha+1}$.

The notion of extension of a property, which axiomatic Set Theory meant to tame, can now be dealt with precisely. For any property φ, $\{x : \varphi(x)\}$ is a **class**. A class can be either a proper class, or a set. Sets are exactly those classes which are members of (or, equivalently, are contained in) some V_α, and are thus not too big. Proper classes are exactly those classes which are equinumerous to V.

V can be seen as the intuitive universe of Set Theory, and ZF as its axiomatization. To gain a better understanding of them we will consider different structures, that can be seen as 'models' of (parts of) ZF, in the sense that they satisfy some of the axioms of ZF. To make this more precise we need a notion of truth for formulas of the language of Set Theory in given structures, which we can define as usual (see IV.1.1).

Definition IV.3.2 Definition of truth in Set Theory (Tarski [1936])
Consider $\mathcal{A} = \langle A, \varepsilon \rangle$, where A is a class, and ε is a binary relation on it. Let

$\mathcal{L}_A$ be the first-order language with equality and $\in$, augmented with constants $\bar{a}$ for any a in A.

Given a closed formula φ of $\mathcal{L}_A$, φ **is true in** $\mathcal{A}$ ($\mathcal{A} \models \varphi$) is inductively defined as usual, starting with:

$$\mathcal{A} \models \bar{a} \in \bar{b} \quad \Leftrightarrow \quad a \,\varepsilon\, b$$
$$\mathcal{A} \models \bar{a} = \bar{b} \quad \Leftrightarrow \quad a = b.$$

An n-ary relation P **is definable over** $\mathcal{A}$ if, for some formula φ with n free variables and $x_1, \ldots, x_n \in A$,

$$P(x_1, \ldots, x_n) \quad \Leftrightarrow \quad \mathcal{A} \models \varphi(\bar{x}_1, \ldots, \bar{x}_n).$$

Intuitively, φ is true in $\langle A, \varepsilon \rangle$ if the formula obtained from φ by replacing $\in$ with ε, and restricting all quantifiers to A, is true (i.e. it holds in $\langle V, \in \rangle$).

Definition IV.3.3 *A structure $\mathcal{A}$ is a* **model** *of a theory T with axioms in the language of Set Theory, if all the axioms of T are true in $\mathcal{A}$.*

This notion is standard in logic, and so is the result that a first-order theory is consistent if and only if it has a model whose universe is a set (Gödel [1930]). By Gödel's Second Theorem (p. 169), we cannot then prove in ZF that there is a model of ZF with a set as universe. We will however find models whose universe is a class.

Standard structures

Among the various possible structures for Set Theory we isolate a particularly interesting class. The idea is that we would like our structures to satisfy some minimal conditions, namely extensionality and well-foundedness.

The first property is completely captured by the Axiom of Extensionality, which is satisfied by A if

$$\langle A, \in \rangle \models [(\forall z)(z \in x \leftrightarrow z \in y) \rightarrow x = y].$$

By the definition of truth this means that, whenever $x, y \in A$,

$$(\forall z \in A)(z \in x \leftrightarrow z \in y) \Rightarrow x = y,$$

that is

$$x \cap A = y \cap A \Rightarrow x = y.$$

A condition implying extensionality is thus **transitivity** of $\in$ (i.e. if $z \in x$ and $x \in A$ then $z \in A$), which amounts to saying that if $x \in A$ then $x \subseteq A$, and hence $x \cap A = x$.

For what concerns well-foundedness, there is some difficulty. A relation ε on A is well-founded (on A) if every nonempty subset of A has a minimal element with respect to ε. This can be written as:

$$(\exists y)(y \in x) \;\to\; (\exists y)[y \in x \wedge (\forall z \; \varepsilon \; y)(z \notin x)].$$

If we think of simply requesting $\langle A, \varepsilon \rangle$ to satisfy this, then we do not ensure well-foundedness of ε. The reason is that the intention is to consider every possible subset x of A, but either we leave the schema as such (by substituting x with any formula φ), and then we only talk about the countably many definable subsets of A, or we turn the schema into a second-order axiom, by universally quantifying x, and then we only talk about the members of A (because of the interpretation of quantifiers over a structure). In either case, a great deal of subsets of A are not considered, by Cantor's Theorem.

One reason to insist on well-foundedness of ε is that it almost justifies (in presence of separation) transfinite induction on ε, i.e.

$$(\forall y \in A)[(\forall z \; \varepsilon \; y)\varphi(z) \;\to\; \varphi(y)] \;\Rightarrow\; (\forall y \in A)\varphi(y).$$

The only thing that does not go through in the proof of recursion on $\in$ (p. 399) is that we cannot prove that $\hat{x}$ is a set. This is taken care by the additional assumption that the predecessors of any element of A w.r.t. ε form a set (in which case ε is called left-narrow).

One obvious way to have a well-founded membership relation is to let it be the usual $\in$. We are thus led to the following notion.

Definition IV.3.4 *A structure $\langle A, \varepsilon \rangle$ is* **standard** *if A is transitive, and ε is the membership relation restricted to A.*

Since ε is intended as $\in$, a standard structure is completely determined by its universe A, and we will usually refer only to it. Also, by definition of ordinal (set of smaller ordinals) and transitivity, the ordinals of a standard structure are closed downward. If A is a transitive set it thus makes sense to talk of the **ordinal of A**, meaning the smallest ordinal not in it or, equivalently, the set of ordinals in it.

The next result shows that standard structures really capture the essence of extensionality and well-foundedness, and it can be seen as a generalization of the fact that a well-ordered set is isomorphic, in a unique way, to an ordinal (which is obtained as a special case of it, when ε is a total ordering on the set A).

Theorem IV.3.5 The Collapsing Lemma (Gödel [1939], Mostowski [1949]) *If $\langle A, \varepsilon \rangle$ is an extensional, left-narrow, and well-founded structure, i.e.*

1. *for every* $x, y \in A$, *if* $(\forall z \in A)(z \varepsilon x \leftrightarrow z \varepsilon y)$ *then* $x = y$

2. *for every* $x \in A$, *the class* $\{y : y \in A \wedge y \varepsilon x\}$ *is a set*

3. *every nonempty subset of A has a minimal element w.r.t.* ε

then $\langle A, \varepsilon \rangle$ *is isomorphic, in a unique way, to a standard structure.*

Proof. Suppose $\langle A, \varepsilon \rangle$ is extensional and well-founded. The idea of the proof is quite simple. We picture the internal structure of A w.r.t. ε as a tree, which is well-founded because so is ε. We then relabel the tree, by assigning $\emptyset$ to the leaves, and the set of its predecessors (which, by left-narrowness, is really a set) to any nonterminal node. Thus we obtain a set which is transitive by construction (because we never skip a level, and proceed from the leaves in an orderly fashion), and is isomorphic to $\langle A, \varepsilon \rangle$ because their internal structures are represented by the same tree.

Formally, we define the collapsing function f as

$$f(x) = \{f(y) : y \in A \wedge y \varepsilon x\},$$

and show that f is the required isomorphism between $\langle A, \varepsilon \rangle$ and $\langle f(A), \in \rangle$.

1. *f is a function on A*
   By induction on ε it is easy to prove that, for each $x \in A$, $f(x)$ exist and is unique. Existence follows by collection and separation (or replacement), applied to the set of predecessors (w.r.t. ε) of x in A, which is a set by left-narrowness. Uniqueness follows by the Axiom of Extensionality, since a set is completely determined by its elements. For more details, see the proof of IV.3.10.

2. *$f(A)$ is transitive*
   Let $z \in u \in f(A)$: by definition of f, there exists $x \in A$ such that $u = f(x) = \{f(y) : y \in A \wedge y \varepsilon x\}$. Since $z \in u$, there is $y \in A$ such that $z = f(y)$, i.e. $z \in f(A)$.

3. *if $y \in A$ then $y \varepsilon x \Rightarrow f(y) \in f(x)$*
   Since $f(x) = \{f(y) : y \in A \wedge y \varepsilon x\}$, from $y \in A \wedge y \varepsilon x$ it follows that $f(y) \in f(x)$.

4. *f is one-one*
   We show, by induction on the well-founded relation ε, that

   $$\text{if } x, y \in A \text{ then } f(x) = f(y) \Rightarrow x = y.$$

   By induction hypothesis,

   $$(\forall s \varepsilon A)(\forall t \varepsilon A)(s \varepsilon x \wedge t \varepsilon y \wedge f(s) = f(t) \Rightarrow s = t).$$

By extensionality, we will have $x = y$ (since $x, y \in A$) if

$$(\forall s \in A)(s \; \varepsilon \; x \leftrightarrow s \; \varepsilon \; y).$$

Suppose $s \in A \wedge s \; \varepsilon \; x$. Then, by part 3, $f(s) \in f(x)$. If $f(x) = f(y)$ then $f(s) \in f(y) = \{f(t) : t \in A \wedge t \; \varepsilon \; y\}$. Thus $f(s) = f(t)$, for some $t \in A \wedge t \; \varepsilon \; y$. By the induction hypothesis $s = t$, and hence $s \; \varepsilon \; y$. The converse in similar.

5. if $y \in A$ then $f(y) \in f(x) \Rightarrow y \; \varepsilon \; x$
   Since $f(x) = \{f(t) : t \in A \wedge t \; \varepsilon \; x\}$, if $f(y) \in f(x)$ then $f(y) = f(t)$ for some $t \in A \wedge t \; \varepsilon \; x$. From $t \in A \wedge y \in A$ we have $y = t$ by part 4, and from $t \; \varepsilon \; x$ then $y \; \varepsilon \; x$.

6. f is unique
   Suppose there are two such isomorphisms: they induce an isomorphism of two transitive classes. We prove, by induction on $\in$, that if $g : B \to C$ is an isomorphism and B, C are transitive, then g is the identity. Suppose $x \in B$, and $g(y) = y$ for every $y \in x \cap B$: we show that $g(x) = x$.

   - $x \subseteq g(x)$
     If $y \in x$ then $y \in B$ by transitivity, and thus $y \in x \cap B$. By induction hypothesis, $g(y) = y$. Since g is an isomorphism, $g(y) \in g(x)$ because $y \in x$, and thus $y \in g(x)$.

   - $g(x) \subseteq x$
     If $z \in g(x)$ then $z \in C$ by transitivity and, since g is onto, there is $y \in B$ such that $g(y) = z$. Then $g(y) \in g(x)$ and, since g is an isomorphism, $y \in x$. From $y \in x \cap B$ and the inductive hypothesis, $y = g(y) = z$. Thus $z \in x$.

Since g is the identity, there can be only one isomorphism of $\langle A, \varepsilon \rangle$ with a transitive class. $\square$

Corollary IV.3.6 *An extensional class A can be collapsed to an isomorphic transitive class in a unique way, with the transitive subclasses of A left unchanged.*

Proof. The Collapsing Lemma can be applied to $\langle A, \in \rangle$, since extensionality holds by hypothesis, and $\in$ is automatically left-narrow and well-founded. Thus there is a unique isomorphism with a transitive class. Since now ε is $\in$, the definition of f is simply

$$f(x) = \{f(y) : y \in x \cap A\}.$$

Suppose $z \subseteq A$ is transitive. We want to show, by induction on $\in$, that $f(x) = x$ for $x \in z$, so that f does not change z. But if $x \in z$ then $x \subseteq z$ by transitivity, thus $x \subseteq A$ and $x \cap A = x$. Then

$$f(x) = \{f(y) : y \in x\}.$$

By induction hypothesis, $f(y) = y$ if $y \in x$ (since then $y \in z$, by transitivity), and thus

$$f(x) = \{y : y \in x\} = x. \quad \square$$

Note that, in particular, the transitive collapse of an extensional class does not change the ordinals of the class.

For any Set Theory with the Axiom of Extensionality, the Collapsing Lemma shows that there are only two kinds of models for it: the ones which are not well-founded and, up to isomorphism, the standard ones. We will thus identify the latter with the transitive models.

The Set-Theoretical Hierarchy

Manipulation of quantifiers similar to those already seen for number and function quantifiers in Arithmetic are possible also in our present context. Here we call **bounded quantifier** a quantifier of the type $(\forall x \in a)$ or $(\exists x \in a)$.

Proposition IV.3.7 *The following transformations of quantifiers are permissible (up to provable equivalence in GKP):*

1. *permutation of quantifiers of the same type*

2. *contraction of quantifiers of the same type*

3. *permutation of two quantifiers, one of which is bounded*

4. *substitution of a bounded quantifier with an unbounded one of the same type.*

Proof. Part 1 is obvious. Part 2 uses the Pairing Axiom:

$$\exists x \exists y \varphi(x,y) \iff \exists z (\exists x \in z)(\exists y \in z)[z = \{x,y\} \wedge \varphi(x,y)],$$

where z is a variable not occurring in φ.

Part 3 uses the Axiom of Collection:

$$(\forall x \in z)(\exists y)\varphi(x,y) \iff (\exists a)(\forall x \in z)(\exists y \in a)\varphi(x,y),$$

where a is a variable not occurring in φ.

Part 4 is just the definition of bounded quantifiers:

$$(\forall x \in a)\varphi(x) \iff \forall x(x \in a \to \varphi(x))$$
$$(\exists x \in a)\varphi(x) \iff \exists x(x \in a \wedge \varphi(x)). \quad \square$$

Proposition IV.3.8 Prenex Normal Form (Kuratowski and Tarski [1931]). *Any relation definable in the language of Set Theory is equivalent in GKP to one with a list of alternated quantifiers in the prefix, and a matrix without unbounded quantifiers.*

Proof. The previous transformations allow us to contract quantifiers of the same type, without introducing unbounded quantifiers in the matrix. And quantifiers can be pushed in front by the usual transformations (see the proof of IV.1.5). □

In the style of the Arithmetical and Analytical Hierarchies, we can now introduce a hierarchy for formulas of Set Theory.

Definition IV.3.9 The Set-Theoretical Hierarchy (Levy [1965])

1. Σ_n *is the class of relations definable over the language of Set Theory by a formula in prenex normal form with no unbounded quantifiers in the matrix and n quantifier alternations in the prefix, the outer quantifier being existential.*

2. Π_n *is defined similarly, with the outer quantifier being universal.*

3. Δ_n *is $\Sigma_n \cap \Pi_n$, i.e. the class of relations definable in both the n-quantifier forms.*

4. Σ_n^T, Π_n^T, Δ_n^T *are the classes of relations definable by formulas provably equivalent, in the theory T, to (respectively) $\Sigma_n, \Pi_n, \Delta_n$ formulas.*

5. $\Sigma_n^{\mathcal{A}}$, $\Pi_n^{\mathcal{A}}$, $\Delta_n^{\mathcal{A}}$ *are the classes of relations definable over the structure $\mathcal{A}$ by (respectively) $\Sigma_n, \Pi_n, \Delta_n$ formulas (also called $\Sigma_n, \Pi_n, \Delta_n$-definable over $\mathcal{A}$).*

By extension, we will call a *formula* Σ_n or Π_n if it is in prenex normal form with n quantifier alternations in the prefix, the outer one being, respectively, existential or universal.

We will also say that a *function* is in a given class if its graph is.

Δ_1^{GKP} functions

The level 0 of the Levy's Hierarchy, which consists of the relations definable by formulas without unbounded quantifiers, is quite interesting: *inclusion, set equality, ordered pair and relative projections, function, domain, range, graph, ordinal, ordinal successor, limit, n, natural number (i.e. $x \in \omega$), $x \subseteq \omega$, and $x = \omega$* are all Δ_0^{GKP} notions, as their natural definitions show. E.g., an ordinal is a transitive set which is totally ordered by the $\in$ relation (which induces the order relation on the ordinals); x is a natural number if and only if it is either

0 (i.e. $\emptyset$) or a successor ordinal (i.e. of the form $x \cup \{x\}$, with x ordinal), together with all its predecessors; x is ω if it is a limit ordinal (i.e. neither 0 nor a successor) such that all its elements are natural numbers. However, the existence of ω as a constant (is equivalent to, and hence) requires the Axiom of Infinity, and is thus not provable in GKP alone. Thus we can use individual natural numbers, but not their collection. Of course, ω is $\Delta_0^{ZF^-}$.

The level 1 contains various other interesting notions. For example, *well-foundedness is Δ_1^{GKP}*: a relation on a set x is well-founded if and only if every nonempty subset of x has a minimal element w.r.t. it (this provides the Π_1^{GKP} form), and if and only if there is a function with domain x and range contained in the ordinals which is order preserving (this provides the Σ_1^{GKP} form). Note that the last assertion exploits the possibility of defining a function by recursion on a well-founded relation (see IV.2.17 for details)

To prove a strong and useful closure property of Δ_1^{GKP}, we consider the analogue of primitive recursion for set functions. Recall that a nonzero natural number is inductively generated by its predecessor. A set is instead inductively generated by its elements. In both cases, primitive recursion gives the value of a function on a given element when its values are given for the elements that inductively generate it.

Theorem IV.3.10 Primitive recursion on $\in$ (Von Neumann [1923], [1928], Karp [1967]) *The class Δ_1^{GKP} is closed under primitive recursion on $\in$. Formally, let g be a total Δ_1^{GKP} function, and*

$$\hat{f}(\vec{x}, y) = \{f(\vec{x}, z) : z \in y\}.$$

Then the function

$$f(\vec{x}, y) = g(\vec{x}, y, \hat{f}(\vec{x}, y))$$

is total and Δ_1^{GKP}.

Proof. We omit the parameters $\vec{x}$, which are kept constant in the proof. We consider the functions that satisfy the recursion equation on their domain: if their domains are closed downward w.r.t. $\in$ (i.e. transitive), they can be seen as approximations to f. Formally, let

$$P(h) \quad \Leftrightarrow \quad h \text{ is a function } \wedge$$
$$Dom \, h \text{ is closed downward w.r.t. } \in \wedge$$
$$(\forall y \in Dom \, h)[h(y) = g(y, \hat{h}(y)],$$

where $Dom \, h$ is the domain of h. By the hypothesis on g, and the fact that the remaining notions used in the definition are Δ_0^{GKP}, P is Δ_1^{GKP}. If we let

$$f(y) = z \Leftrightarrow (\exists h)[P(h) \wedge h(y) = z],$$

then f is Σ_1^{GKP}. We now show that f is the required function.

1. f *is unique*
By induction on $\in$. Suppose $P(h_1) \wedge P(h_2)$, and let h_1 and h_2 be defined on y. Then

$$h_1(y) = g(y, \hat{h}_1(y)) \quad \text{and} \quad h_2(y) = g(y, \hat{h}_2(y)).$$

If $z \in y$ then, by closure downward of the domains of h_1 and h_2 w.r.t. $\in$, z is in them, and thus h_1 and h_2 are both defined on z. By induction hypothesis they agree on $z \in y$: thus $\hat{h}_1(y) = \hat{h}_2(y)$, and $h_1(y) = h_2(y)$.

2. f *is total*
By induction on $\in$ we show that f is defined on y. This means finding a function h defined on y and satisfying P. By induction hypothesis,

$$(\forall z \in y)(\exists h_z)[P(h_z) \wedge z \in Dom\, h_z].$$

By collection and separation, there is a set A containing the functions satisfying P, and with some $z \in y$ in their domains. The union of members of A is thus a function defined on all $z \in y$, and we can let h be this function extended to y by:

$$h(y) = g(y, \hat{h}(y)).$$

Note that $\hat{h}(y)$ is determined by the (union of) functions in A. It remains to be shown that h satisfies P, which can be easily verified:

- h is a function
  Because g and the members of A are.

- *Dom* h is transitive
  Because the domain of h is the union of $\{y\}$ and the domains of the functions in A, and the union of the latter is a transitive set containing all the predecessors of y w.r.t. $\in$.

- h satisfies the recursion equation on its domain
  On y this holds by the definition of h. On elements z of the domain different from y, it is enough to note that they are in the domain of some function h_1 in A, which satisfies the recursion equation on its domain by hypothesis (since it satisfies P). But the domain of h_1 is transitive, and thus h_1 is defined on all predecessors of z. Then $\hat{h}_1(z) = \hat{h}(z)$, and

$$h(z) = h_1(z) = g(z, \hat{h}_1(z)) = g(z, \hat{h}(z)).$$

3. f *is* Δ_1^{GKP}
We already know that f is Σ_1^{GKP}. Being total, it is also Π_1^{GKP}:

$$f(y) \neq z \Leftrightarrow (\exists u)[f(y) = u \wedge u \neq z].$$

4. *f satisfies the recursion equation*
   This is similar to what we have proved at the end of part 2. Given y, let h be a function satisfying P and defined on y. Its domain is transitive, and thus it contains all predecessors of y w.r.t. $\in$. By the uniqueness of f, $\hat{h}(y) = \hat{f}(y)$. But h satisfies the recursion equation on its domain, since it satisfies P, and thus

$$f(y) = h(y) = g(y, \hat{h}(y)) = g(y, \hat{f}(y)). \quad \Box$$

Course-of-values recursion on natural numbers gives the value of a function using any set of values for previous elements, and not only the value for the immediate predecessor. For sets, the set of previously generated elements can be seen as the transitive closure of the set, i.e. its downward closure under $\in$.

Definition IV.3.11 *The **transitive closure** $Tc(x)$ of a set x is the smallest transitive set containing all elements of x.*

Corollary IV.3.12 *The transitive closure is a Δ_1^{GKP} function.*

Proof. Let

$$Tc(x) = x \cup \left(\bigcup_{y \in x} Tc(y) \right).$$

By the theorem, $Tc(x)$ exists and it is Δ_1^{GKP}. Moreover, it is really the transitive closure:

1. $x \subseteq Tc(x)$
   By definition.

2. $Tc(x)$ *is transitive*
   By induction on $\in$. Suppose that $Tc(y)$ is transitive, for every $y \in x$. Let $z \in t \in Tc(x)$. By definition, there are two cases:

   - $t \in x$
     Then $Tc(t)$ is transitive, $z \in Tc(t) \subseteq Tc(x)$, and $z \in Tc(x)$.
   - $t \in \bigcup_{y \in x} Tc(y)$
     Then $t \in Tc(y)$ for some $y \in x$, and $Tc(y)$ is transitive. Thus $z \in Tc(y) \subseteq Tc(x)$, and $z \in Tc(x)$.

3. $x \subseteq z \wedge z$ *transitive* $\Rightarrow Tc(x) \subseteq z$
   By induction on $\in$. Suppose

$$y \subseteq z \wedge z \text{ transitive} \Rightarrow Tc(y) \subseteq z,$$

   for every $y \in x$. Given such a y, from $x \subseteq z$ we have $y \in z$, and $y \subseteq z$ by transitivity. Then the induction hypothesis applies, and $Tc(y) \subseteq z$. Thus $\left(\bigcup_{y \in x} Tc(y) \right) \subseteq z$. Since $x \subseteq z$ by hypothesis, $Tc(x) \subseteq z$. $\quad \Box$

We can think of primitive recursion on the transitive closure as being the analogue of course-of-values recursion. The next result is thus the analogue of I.7.1, and it shows that Δ_1^{GKP} has sufficiently strong closure properties to allow for the usual arithmetization results, in set-theoretical setting.

Corollary IV.3.13 Course-of-values recursion on $\in$ (Von Neumann [1923], [1928], Karp [1967]) *The class Δ_1^{GKP} is closed under course-of-values recursion over $\in$. Formally, let g be a total Δ_1^{GKP} function, and*

$$f(\vec{x}, y) = g(\vec{x}, y, \hat{f}(\vec{x}, Tc(y))).$$

Then f is total and Δ_1^{GKP}.

Proof. The proof of IV.3.10 is based on induction on $\in$, i.e.

$$(\forall z)[(\forall y \in z)P(y) \;\to\; P(z)] \Rightarrow (\forall x)P(x).$$

The same proof goes through, once we prove that a similar principle of induction on the transitive closure holds:

$$(\forall z)[(\forall y \in Tc(z))P(y) \;\to\; P(z)] \;\Rightarrow\; (\forall x)P(x).$$

Its hypothesis suggests to prove not $(\forall x)P(x)$ directly, but rather

$$(\forall z)(\forall x \in Tc(z))P(x)$$

(from which the former follows, since $x \in Tc(\{x\})$). We proceed by induction on $\in$. Given z we have, by induction hypothesis,

$$(\forall y \in z)(\forall x \in Tc(y))P(x),$$

and we want to prove $(\forall x \in Tc(z))P(x)$. If $x \in Tc(z)$, there are two cases:

- $x \in z$
  Then, by the induction hypothesis, $(\forall u \in Tc(x))P(u)$, and thus $P(x)$ holds by the hypothesis of the principle.

- $x \in \bigcup_{y \in z} Tc(y)$
  Then, by the induction hypothesis, $P(x)$ holds. $\square$

Exercises IV.3.14 a) *Sum and product on the ordinals are Δ_1^{GKP}.* (Hint: the order relation on the ordinals is induced by $\in$.)

b) *The Δ_1^{GKP} functions are closed under composition and case definition.* (Hint: by logical operations and quantifier manipulations.)

The levels of the Set-Theoretical Hierarchy

The classes of the Set-Theoretical Hierarchy share a number of properties with their analogues in the Arithmetical and Analytical Hierarchies.

Proposition IV.3.15 Closure properties (Levy [1965])

1. R is Σ_n^{GKP} if and only if $\neg R$ is Π_n^{GKP}
   R is Π_n^{GKP} if and only if $\neg R$ is Σ_n^{GKP}

2. Δ_n^{GKP} is closed under negations

3. Σ_n^{GKP}, Π_n^{GKP}, and Δ_n^{GKP} are closed under conjunction, disjunction, and bounded quantification

4. for $n \geq 1$, Σ_n^{GKP} is closed under existential quantification, and Π_n^{GKP} is closed under universal quantification

5. the universal quantification of a Σ_n^{GKP} relation is Π_{n+1}^{GKP}, and the existential quantification of a Π_n^{GKP} relation is Σ_{n+1}^{GKP}.

Proof. Everything easily follows from logical operations and quantifier manipulations, as in IV.1.8. □

The closure properties just proved can be dealt with in formal theories for Set Theory, like GKP, because they only require formulas manipulations. The remaining properties need instead the existence of elements, and hence they will be proved not for theories, but for models.

Theorem IV.3.16 Enumeration Theorem (Kripke [1964], Levy [1965], Platek [1966]) For any standard model A of GKP, and each $n, m \geq 1$, there is an $m + 1$-ary Σ_n^A relation that enumerates over A the m-ary Σ_n^A relations. Similarly for Π_n^A.

Proof. We prove the result for $n = 1$. The remaining cases follow from it by induction on n, as in IV.1.9. First we introduce some general tools:

1. a Δ_1^{GKP} coding and decoding mechanism
   We want to show how to code and decode finite sequences of sets. The idea is to use the Δ_0^{GKP} notions of ordered pair:

$$(x, y) = z \Leftrightarrow (\exists a \in z)(\exists b \in z)(a = \{x\} \wedge b = \{x, y\} \wedge z = \{a, b\}),$$

and relative projections:

$$(z)_1 = x \quad \Leftrightarrow \quad (\exists y \in z)(z = (x, y))$$
$$(z)_2 = y \quad \Leftrightarrow \quad (\exists x \in z)(z = (x, y)).$$

By course-of-values recursion (since the ordered pair of two sets is two levels higher than its components) we can define

$$\langle\rangle \;=\; \emptyset$$
$$\langle x_1,\ldots,x_{n+1}\rangle \;=\; (\langle x_1,\ldots,x_n\rangle, x_{n+1}).$$

It is then immediate to obtain, again by course-of-values recursion, a Δ_1^{GKP} predicate telling whether a set is a coding sequence, and Δ_1^{GKP} functions giving the length of a coding sequence, and the components of an n-tuple.

Note that the Axiom of Infinity is not needed in the coding procedure: we use the natural numbers individually, but we never need to consider their set. Recall that the expressions $x \in \omega$, $x \subseteq \omega$, $x = \omega$ are all Δ_0^{GKP}, even without the Axiom of Infinity.

2. a Δ_1^{GKP} satisfaction predicate for Δ_0^{GKP} formulas
   By using the coding and decoding mechanism, and the possibility of doing course-of-values recursions, we can now proceed as in usual arithmetizations. In particular, we can define a Δ_1^{GKP} satisfaction predicate $\mathcal{T}(e,x,y)$, meaning:

   > the formula coded by e, with inputs coded by x (i.e. when its free variables $v_1, v_2, \ldots$ are substituted, in an orderly fashion, by the components of x), is true in the transitive closure of y.

   $\mathcal{T}$ is Δ_1^{GKP} because the definition of truth in a standard structure is a course-of-values recursion over $\in$ (in general, it would only be a course-of-values recursion over the membership relation of the structure).

   Note that what we have here is a Δ_1^{GKP} satisfaction predicate for Δ_0^{GKP} formulas, since the effect of interpreting a formula over a (transitive) set is to bound all quantifiers over the set.

3. a Σ_1^{GKP} universal predicate
   The definition of a universal Σ_1^{GKP} predicate for Σ_1^{GKP} m-ary relations is now immediate:

   $$\mathcal{W}(e, x_1, \ldots, x_m) \Leftrightarrow$$
   $$(\exists a)[a \text{ is transitive } \wedge\, x_1, \ldots, x_m \in a \wedge \mathcal{T}(e, \langle x_1, \ldots, x_m\rangle, a)].$$

   Note that we only dealt with relations without parameters, but the case of relations with parameters can be treated similarly: all parameters can be coded into one, and thus m-ary relations with parameters are just $m+1$-ary relations without parameters, in which one variable has been substituted with a constant.

Let now A be a standard model of GKP, and consider

$$\mathcal{W}^A(e, x_1, \ldots, x_m) \Leftrightarrow \langle A, \in \rangle \models \mathcal{W}(\bar{e}, \bar{x}_1, \ldots, \bar{x}_m),$$

i.e. interpret $\mathcal{W}$ over A. This is obviously a Σ_1^A predicate, and we want to show that it enumerates the m-ary Σ_1^A predicates.

If $\varphi \in \Sigma_1^A$, let $v_1, \ldots, v_m$ be its free variables, and $x_1, \ldots, x_m$ be in A. It is enough to show that if $\varphi(x_1, \ldots, x_m)$ holds in A, it also holds in some transitive set $a \in A$ (this is called $\mathbf{\Sigma_1}$-**reflection**). Then, if e codes φ, A satisfies $\mathcal{W}(e, x_1, \ldots, x_m)$ by definition, as wanted.

Suppose $\varphi(x_1, \ldots, x_m)$ holds in A. Since φ is Σ_1^A, there is ψ without unbounded quantifiers, such that $\varphi(x_1, \ldots, x_m) \leftrightarrow (\exists y)\psi(x_1, \ldots, x_m, y)$. To say that, for $x_1, \ldots, x_n \in A$, $\varphi(x_1, \ldots, x_m)$ holds in A, means that there is $y \in A$ such that $\psi(x_1, \ldots, x_m, y)$ is true over A. But since ψ has no unbounded quantifiers, this is true in the transitive closure of $\{x_1, \ldots, x_m, y\}$ as well, which is a member of A (by the closure properties of A, which is a model of GKP). $\quad\square$

We can rephrase what we proved as follows. Since to interpret a formula of any complexity over a set turns it into Δ_0^{GKP} form, by bounding the quantifiers, the global satisfaction predicate over a *set* is Δ_1^{GKP}. The satisfaction predicate for Σ_{n+1}^{GKP} formulas over a *class* is Σ_{n+1}^{GKP}. Of course, there is no definable notion of global satisfiability over a class, by Tarski's Theorem (see p. 166).

Exercise IV.3.17 The Fixed-Point Theorem for Σ_1^A relations. *If A is a standard model of GKP, $\mathcal{W}^A$ is the enumeration predicate for Σ_1^A m-ary relations, and R is an $m+1$-ary relation, there is $e \in A$ such that*

$$\mathcal{W}^A(e, x_1, \ldots, x_m) \Leftrightarrow R(e, x_1, \ldots, x_m).$$

(Hint: there is a Δ_1^A analogue of the S_n^m-Theorem, by standard methods. Moreover Σ_1^A relations are closed under substitution of Δ_1^A functions, because

$$P(\vec{x}, f(\vec{x})) \Leftrightarrow (\exists y)[y = f(\vec{x}) \wedge P(\vec{x}, y)],$$

and Σ_1^A is closed under conjunctions and existential quantifications.)

Theorem IV.3.18 Set-Theoretical Hierarchy Theorem (Kripke [1964], Levy [1965], Platek [1966]) *For any standard model A of GKP, the Set-Theoretical Hierarchy over it does not collapse. More precisely, for any $n \geq 1$ the following hold:*

1. $\Sigma_n^A - \Pi_n^A \neq \emptyset$, and hence $\Delta_n^A \subset \Sigma_n^A$

2. $\Pi_n^A - \Sigma_n^A \neq \emptyset$, and hence $\Delta_n^A \subset \Pi_n^A$

3. $\Sigma_n^A \cup \Pi_n^A \subset \Delta_{n+1}^A$.

Proof. By diagonalization and IV.3.16, as in IV.1.13. $\quad\square$

$\mathcal{HF}$ and the Arithmetical Hierarchy

We have noted that the Axiom of Foundation allows a representation of the transitive closure of a set as a well-founded tree, describing the set-theoretical build-up of the set from the empty set. We now analyze the sets whose associated tree is finite.

Definition IV.3.19 $\mathcal{HF}$ *is the set of* **hereditarily finite sets,** *i.e. the smallest class* $\mathcal{A}$ *of sets such that:*

1. $\emptyset \in \mathcal{A}$

2. *if* $x_1, \ldots, x_n \in \mathcal{A}$ *then* $\{x_1, \ldots, x_n\} \in \mathcal{A}$.

Note the difference between being finite, i.e. having only finitely many elements (like $\{\omega\}$, that consists of only one element), and being hereditarily finite, i.e. having only finitely many elements, each of which is hereditarily finite (a definition by course-of-value recursion). In other words, a set is in $\mathcal{HF}$ if and only if its transitive closure is finite.

Proposition IV.3.20 $\mathcal{HF}$ *is the smallest transitive model of* GKP.

Proof. We first verify that $\mathcal{HF}$ is a transitive model of GKP:

1. *transitivity*
   The tree representation of an element of x is a subtree of the representation of x, and it is finite if this is.

2. *extensionality and foundation*
   Automatic from transitivity.

3. *pair*
   The tree representation of $\{x, y\}$ consists of a vertex on top of the tree representations of x and y, and thus it is finite if these are.

4. *union*
   The tree representation of $\bigcup x$ consists of a vertex on top of the tree representation of the elements of the elements of x, and thus it is finite if there are only finitely many of these trees, each of them finite.

5. *separation*
   The tree representation of a subset of x is a subtree of the representation of x, and it is finite if this is.

6. *collection*

Suppose $(\forall x \in a)(\exists y)\varphi(x,y)$ holds in $\mathcal{HF}$. Since a is finite, it has only finitely many elements $x_1, \ldots, x_n$. For each of them, there is a set y_i in $\mathcal{HF}$ such that $\varphi(x_i, y_i)$ holds. By pair and union, which we have already verified, $b = \{y_1, \ldots, y_n\}$ is in $\mathcal{HF}$, and thus $(\forall x \in a)(\exists y \in b)\varphi(x,y)$ holds in $\mathcal{HF}$.

We now verify that $\mathcal{HF}$ is the smallest transitive model of GKP. Suppose A is such a model: then it contains $\emptyset$, and it is closed under pair and union. We want to show that $\mathcal{HF} \subseteq A$. This is easily seen by induction, since each element $x \in \mathcal{HF}$ is obtained from the emptyset, by finitely many applications of pairing and union. $\square$

Note that the natural numbers, represented in set-theoretical terms (as finite ordinals), are all in $\mathcal{HF}$, by induction:

$$0 = \emptyset \quad \text{and} \quad n + 1 = n \cup \{n\}.$$

Exercises IV.3.21 a) $\mathcal{HF} = V_\omega$.

b) $\mathcal{HF}$ *is the smallest model of ZFC with the Axiom of Infinity replaced by its own negation.* (Hint: the power set of a finite set is finite. Choice is trivial, since the elements of $\mathcal{HF}$ are all finite. Since ω is not in $\mathcal{HF}$, the Axiom of Infinity does not hold. Its negation, that every set is finite, holds because if $x \in \mathcal{HF}$ then any function from x to some natural number is already in $\mathcal{HF}$.)

The reason why we are particularly interested in $\mathcal{HF}$ is that, for sets of natural numbers, set-theoretical definability over it coincides (and not only globally, but level by level) with arithmetical definability. This provides an alternative, set-theoretical way of seeing Recursion Theory, and it is the starting point for some interesting generalizations (see p. 421).

Theorem IV.3.22 Set-theoretical definability of the Arithmetical Hierarchy (Ackermann [1937]) *Let $A \subseteq \omega$. Then:*

1. *$A \in \mathcal{HF}$ if and only if A is finite*

2. *A is definable over $\mathcal{HF}$ if and only if A is arithmetical. More precisely, for $n \geq 1$:*

$$A \in \Delta_n^{\mathcal{HF}} \quad \Leftrightarrow \quad A \in \Delta_n^0$$
$$A \in \Sigma_n^{\mathcal{HF}} \quad \Leftrightarrow \quad A \in \Sigma_n^0.$$

Similarly for relations, of any number of variables.

Proof. The first assertion is easy to see: an hereditarily finite set is, in particular, finite; and a finite set of natural numbers is hereditarily finite, because so are the natural numbers (as set-theoretical objects).

The proof of the second assertion is more cumbersome, and it amounts to show that we can translate, by preserving the logical complexity, arithmetical assertions into set-theoretical ones, and conversely:

1. *translation from Arithmetic to Set Theory*

   We already know how to interpret natural numbers in set-theoretical terms. Then number quantifiers can be easily turned into set quantifiers:

   $$(\exists x)\varphi(x) \quad \Leftrightarrow \quad (\exists x)(x \in \omega \wedge \varphi(x))$$
   $$(\forall x)\varphi(x) \quad \Leftrightarrow \quad (\forall x)(x \in \omega \rightarrow \varphi(x)).$$

   The expression $x \in \omega$ is Δ_0^{GKP}, hence $\Delta_0^{\mathcal{HF}}$, and thus it does not increase the complexity of the matrix. It only remains to translate recursive matrices, i.e. graphs of recursive functions, into Δ_1^{GKP} predicates. We refer to the characterization of recursive functions given in I.1.8. Sum, product, and composition have already been dealt with in IV.3.14, while identities and equality are trivially Δ_1^{GKP}. For μ-recursion, let

   $$f(\vec{x}) = \mu y R(\vec{x}, y).$$

   Then

   $$f(\vec{x}) = y \quad \Leftrightarrow \quad R(\vec{x}, y) \wedge (\forall z < y)\neg R(\vec{x}, z).$$

   If R is Δ_1^{GKP} then so is the graph of f, by the closure properties of Δ_1^{GKP}, and the fact that bounded number quantifiers translate into bounded set quantifiers (because the order relation on ordinals is induced by the membership relation).

   Note that the whole argument does not require the existence of ω. If ω were present (i.e. if we worked with a model of GKP plus infinity), all the translations of arithmetical formulas would simply become Δ_0, because the number quantifiers would then be translated into bounded set quantifiers (see p. 419 for more on this point).

2. *translation from Set Theory to Arithmetic*

   First of all we have to interpret members of $\mathcal{HF}$ as natural numbers, and this can be done by induction on the construction of $\mathcal{HF}$:

   $$f(\emptyset) = 0$$
   $$f(\{x_1, \ldots, x_n\}) = 2^{f(x_1)} + \cdots + 2^{f(x_n)}$$

(we suppose all the x_i distinct, since a set is determined solely by its elements). This simply amounts to using canonical indices (II.5.13) hereditarily, by inductively decomposing the exponents in the binary decomposition of a number.

Now set-theoretical quantifiers can be turned into number quantifiers:

$$(\exists x)\varphi(x) \quad \Leftrightarrow \quad (\exists n)\varphi(f^{-1}(n))$$
$$(\forall x)\varphi(x) \quad \Leftrightarrow \quad (\forall n)\varphi(f^{-1}(n)).$$

It remains to be proved that the translations of $\Delta_0^{\mathcal{HF}}$ formulas are recursive. By the parallel closure properties of Δ_0^{GKP} formulas and recursive relations, this reduces to show how to deal with membership and constants. For the former, note that

$$x \in y \Leftrightarrow f(x) \in D_{f(y)},$$

which is a recursive relation. To deal with constants, note that among the values of f there are some that naturally correspond to the set-theoretical integers:

$$g(0) \;=\; 0$$
$$g(n+1) \;=\; g(n) + 2^{g(n)}$$

(because $0 = \emptyset$, and $n + 1 = n \cup \{n\}$). Now g can be thought of as a function both from $\mathcal{HF}$ to ω, and from ω to ω. In the latter case, it is a recursive function. And g can be used to substitute occurrences of set-theoretical natural numbers with occurrences of the corresponding natural numbers coding them as sets. $\square$

We have proved a correspondence between definability on Arithmetic and $\mathcal{HF}$, for relations on natural numbers. This suffices for our recursion-theoretical purposes, but there is more to it. It can actually be shown that, for statements about natural numbers, $\mathcal{PA}$ is equivalent to ZFC with the Axiom of Infinity replaced by its negation (which is equivalent to $V = \mathcal{HF}$) (Ackermann [1937]). In other words, the translations provided in the proof above are actually faithful interpretations of the stated theories into one another (where interpretation means that provable statements are translated into provable statements, and faithfulness that no translation is provable unless its original version was already provable). Note that a symmetric role is played by induction and foundation, which is the reason to consider Peano Arithmetic, and not weaker systems.

The absence of the Axiom of Infinity is enough for the faithfulness of the translation of $\mathcal{PA}$ into Set Theory. The substitution of the Axiom of Infinity with its negation is instead crucial to prove the faithfulness of the translation of Set Theory in $\mathcal{PA}$ (since otherwise $(\forall x)(x \in \mathcal{HF})$, which is equivalent to the negation of the Axiom of Infinity, is not provable in Set Theory, while its translation, which amounts to $(\forall x)(x \in \omega)$, is provable in $\mathcal{PA}$).

Absoluteness and the Analytical Hierarchy

Since there is no single privileged standard structure for Set Theory, we will have to interpret the formulas on the various structures. The problem is that the same formula could be true in some model of GKP and false in some other, thus not having an absolute meaning.

As an example, consider the set

$$x \in b \Leftrightarrow x \in a \wedge (\forall z)\varphi(x, z),$$

obtained from a by separation. Suppose φ has no quantifier. For any A such that $a \in A$, there is a set b_A obtained by interpreting the definition of b over A:

$$x \in b_A \Leftrightarrow x \in a \cap A \wedge (\forall z \in A)\varphi(x, z).$$

If A is transitive then $a \cap A = a$, so

$$x \in b_A \Leftrightarrow x \in a \wedge (\forall z \in A)\varphi(x, z),$$

But $(\forall z \in A)\varphi(x, z)$ could hold even if $(\forall z)\varphi(x, z)$ does not. Thus $b \subseteq b_A$, and 'A believes b_A is b', while this is not necessarily so.

Similarly, by changing the universal quantifier into an existential one, we could have $b_A \subseteq b$, while for more complicated formulas there is no simple relationship between the true b and the set b_A, that A believes to be b. This is of course an unpleasant situation, introducing an element of relativity in Set Theory: on one side we have the 'real' sets, on the other their interpretations over given models, with no apparent connection between them.

The situation is not as disruptive as it might seem at first sight, since a number of formulas turn out to have an absolute meaning, in the sense of defining the same set on every model.

Definition IV.3.23 (Gödel [1940]) *A formula is* **absolute for a class of structures** *if it has the same truth value in each structure of the given class.*

The next result isolates a class of formulas that are absolute, and it is quite useful in applications. The notion of standard model is used in a crucial way, thus providing another reason to restrict our attention to such models.

Proposition IV.3.24 Δ_1^T *formulas are absolute for the standard models of T.*

Proof. Fix a transitive model A of T: we show that the truth value of any Δ_1^T formula interpreted over A is independent of A, and it coincides with the truth value of the formula in the universe V of sets.

First of all note that, over elements of a transitive set A, membership is absolute: indeed, if $x \in A$ then $x \cap A = x$, and thus $z \in x$ has the same meaning over A and over V. This shows that, in particular, bounded quantifiers preserve absoluteness and thus, by induction on their complexity, Δ_0^T formulas are absolute for standard models of T.

Suppose now φ is Δ_1^T. We want to show that φ is true over A if and only if it is true (over V). Let ψ_1 and ψ_2 be Δ_0^T formulas such that, in T,

$$\varphi(\vec{x}) \leftrightarrow (\exists y)\psi_1(\vec{x}, y) \leftrightarrow (\forall y)\psi_2(\vec{x}, y).$$

Suppose $\varphi(\vec{x})$, i.e. $(\exists y)\psi_1(\vec{x}, y)$, is true over A. This means that, for some $y \in A$, $\psi_1(\vec{x}, y)$ is true over A. But this is a Δ_0^T formula, which is absolute. Then $\psi_1(\vec{x}, y)$, and hence $(\exists y)\psi_1(\vec{x}, y)$ and $\varphi(\vec{x})$, are true.

Suppose now $\varphi(\vec{x})$, i.e. $(\forall y)\psi_2(\vec{x}, y)$, is true. Then $\psi_2(\vec{x}, y)$ is true for all y, in particular for all $y \in A$. Thus $(\forall y)\psi_2(\vec{x}, y)$, and hence $\varphi(\vec{x})$, are true over A. $\square$

In the Arithmetical Hierarchy quantifiers range over ω. For any model A of GKP and the Axiom of Infinity, $\omega \in A$. Then number quantifiers can be interpreted as bounded set-theoretical quantifiers, and *arithmetical relations are absolute*, being translated into Δ_1^A formulas (when sum and product are replaced by their Δ_1^{GKP} definitions, see IV.3.14.a).

In the Analytical Hierarchy quantifiers range also over $\mathcal{P}(\omega)$, which is only a Π_1^{GKP} object:

$$x = \mathcal{P}(\omega) \Leftrightarrow \forall y(y \in x \leftrightarrow y \subseteq \omega).$$

$\mathcal{P}(\omega)$ *is not absolute*, and hence not Δ_1^{GKP}: by absoluteness of $y \subseteq \omega$, its interpretation over a transitive model A of GKP is $\mathcal{P}(\omega) \cap A$, and thus it varies with A (see also IV.4.27.c). This means that function quantifiers do not automatically translate into bounded quantifiers, and analytical relations are not automatically absolute. But relations in the first two levels are, as we now see.

Proposition IV.3.25 (Mostowski [1949]) Π_1^1 *relations are absolute for standard models of GKP containing ω (i.e. models of ZF^-).*

Proof. By the First Representation Theorem for Π_1^1 sets (IV.2.15), A is Π_1^1 if and only if there is a recursive sequence $\{T_x\}_{x \in \omega}$ of recursive trees, such that

$$x \in A \Leftrightarrow T_x \text{ is well-founded.}$$

But we have already noted that recursive relations (being arithmetical) and well-foundedness (being Δ_1^{GKP}) are absolute. Then so is A. □

Note that the proof shows that every Π_1^1 relation is actually Δ_1 over GKP plus infinity. Similarly for Σ_1^1 relations, by taking negations.

Exercise IV.3.26 Π_1^1 *formulas are not absolute for standard models of ZFC minus infinity.* (Hint: $\mathcal{HF}$ is a model for it, but the relations over ω definable over it are all arithmetical.)

Theorem IV.3.27 (Shoenfield [1961a]) Σ_2^1 *relations are absolute for standard models of GKP containing all countable ordinals.*

Proof. By the relativized version of the First Representation Theorem for Π_1^1 sets, A is Σ_2^1 if and only if there is a recursive sequence $\{T_{x,f}\}_{x \in \omega}$ of trees uniformly recursive in f, such that

$$x \in A \Leftrightarrow (\exists f)(T_{x,f} \text{ is well-founded}).$$

Since well-foundedness of $T_{x,f}$ is equivalent to the existence of an order-preserving map from $T_{x,f}$ to the countable ordinals (because the trees are countable),

$$x \in A \Leftrightarrow (\exists f)(\exists g)(g : T_{x,f} \to \omega_1 \text{ is order-preserving}).$$

Now we can reproduce the proof of the First Representation Theorem IV.2.15 (by looking at the first places in which the matrix fails), and get uniformly absolute trees $R_{x,f,g}$ on $\omega \times \omega_1$ such that

$$x \in A \Leftrightarrow R_{x,f,g} \text{ is not well-founded}$$

(since A is now defined by existential quantifiers, in place of universal ones). Thus, if M contains all countable ordinals, A is absolute for M. □

Note that the proof shows that every Σ_2^1 relation is actually of the form $(\exists \alpha \text{ countable})\varphi$, with α ranging over countable ordinals, and φ Δ_1 over GKP plus infinity.

Exercise IV.3.28 Σ_2^1 *formulas are not absolute for standard models of ZFC.* (Hint: the formula translating 'the set $X \subseteq \omega$ codes a countable transitive model of ZFC' is Δ_1^1, by arithmetization. 'There exists a set X coding a transitive model of ZFC' is thus a true Σ_2^1 formula, which is not true in the least countable transitive model of ZFC. Note that this reasoning requires the existence of a standard model of ZFC, and thus it is not formalizable in ZFC. For the weaker result relative to standard models of ZFC^- only, i.e. without the Power Set Axiom, such an assumption is not necessary.)

Admissible sets ⋆

We have worked with the theory GKP because we wanted to have structural results for all levels of the Set-Theoretical Hierarchy. But the full power of GKP is needed only for the full results, and it is possible to refine GKP, and isolate what is needed to get the structural results for the first (or, more generally, the n-th) level only.

The **Kripke-Platek system** KP (Kripke [1964], Platek [1966]) has the same axioms of GKP, with separation and collection limited to Δ_0 formulas, and it can be seen as a kind of constructive Set Theory. This theory is strong enough to prove (separation for Δ_1 formulas, collection for Σ_1 formulas, and) the closure properties of the first level of the Levy's Hierarchy.

The transitive sets which are models of KP (i.e. the standard models) are called **admissible sets**, and can be seen as domains suitable for a theory of Σ_1 relations and functions analogous to that of r.e. relations and partial recursive functions, and hence for a Generalized Recursion Theory on sets. Note that $\mathcal{HF}$ is the smallest admissible set, and thus the usual notion of recursiveness is a special case of recursion on an admissible set (by IV.3.22). Moreover, Gordon [1968] has proved that on an admissible set the notion of recursiveness coincides with that of search computability (see p. 204).

This should not be taken to mean that the notion of admissibility is either sufficient or necessary for an abstract analogue of all parts of Recursion Theory: Simpson [1974] and Harrington (see Chong [1984]) have shown that there are admissible sets which do not admit a positive solution to the analogue of Post's Problem, while Friedman and Sacks [1977] have extended a good deal of Recursion Theory, including a positive solution to the analogue of Post's Problem, to special nonadmissible sets.

The full power of GKP is not always avoidable, even in the study of the first level of the Set-Theoretical Hierarchy. A crucial example is the notion of well-foundedness, which is Δ_1^{GKP} but not Δ_1^{KP}, and thus is not absolute for admissible sets. What fails here, since well-foundedness is Π_1^{KP} by definition, is the possibility of carrying on recursion on well-founded relations (the so called β**-property**, Mostowski [1959]), and thus to provide the Σ_1^{KP} form: we have noted that the justification of recursion on well-founded relations requires some form of separation. Δ_1-separation, provided by admissibility, is not enough, although Σ_1-separation is. In particular, the Collapsing Lemma IV.3.5 is not provable in KP, although its corollary is (because its proof requires only a course-of-value recursion on $\in$).

For a development of the theory of admissibility see Barwise [1975] and Fenstad [1980]. The implications of the notion of admissibility for the study of $\mathcal{P}(\omega)$ will be dealt with in Volume III.

IV.4   The Constructible Hierarchy

The Analytical Hierarchy is immensely extended and it contains, already at low levels, all sets of natural numbers naturally occurring in practical considerations. Nevertheless it is still countable, and this has to be true of all the hierarchies that simply stratify the relations definable in some fixed countable language, including the universal language of Set Theory. If we want to overcome this defect, we have to allow for an extension of the notion of definability. One way to do this is by transfinitely iterating definitions, and we pursue this path here.

The Constructible Hierarchy

At the turn of the century various mathematicians began to feel uncomfortable with the development of Set Theory. The center of the dispute was the Power Set Axiom that, in one of its simplest applications, allowed consideration of the class of all sets of natural numbers as a completed totality. The discovery of paradoxes added ground to the objections, and one possible way out was seen in a strong form of the **vicious circle principle**, according to which a totality cannot contain members that are only definable in terms of it (Poincaré [1906], Russell [1908]). As a consequence, only **predicative definitions**, that define objects without referring to sets already containing them, would be accepted.

Contrary to the form of the vicious circle principle considered on p. 399, which is consistent with usual mathematical practice, and is actually a consequence of the Axiom of Foundation, this strong form would permit only a limited development of mathematics. See e.g. Weyl [1918] for what can be saved of analysis, and Gödel [1944], [1964] for discussion.

The idea of a predicative iterative approach, in which one would start with easily definable and graspable sets, and would add at each step only those sets that were definable by using the previously obtained ones, is however worth pursuing. The decision of when (i.e. at which ordinal) to stop the iteration process is crucial. If we really were interested only in the predicatively definable sets, then we should allow only for predicatively definable ordinals: this requires an analysis of the notion of predicativity, and will be considered in Volume III. Here we take a more generous stand, and allow for any number of iterations. By so doing we lose the property of predicativity, and *we can think of the constructible hierarchy as consisting of those sets which are predicatively definable modulo the ordinals*. This hierarchy should not be taken as exhausting the whole universe, but rather as a kind of minimal model (see IV.4.7).

The constructible hierarchy also extends ideas of Hilbert [1926], who tried to prove the Continuum Hypothesis by considering the generalized recursions (using higher-type objects) needed to generate all functions of natural numbers, and attempted a proof to show that they could be reduced to transfinite recursions on ordinals up to ω_1. Gödel's improvements on Hilbert's tentative are of two kinds: he uses all ordinals, instead of only countable ones, and first-order definitions, instead of recursions (which correspond only to one-quantifier definitions).

Definition IV.4.1 (Gödel [1939]) *The* **Constructible Hierarchy** *is described, in terms of ordinals, as follows:*

$$
\begin{aligned}
L_0 &= \emptyset \\
L_{\alpha+1} &= \mathrm{def}\,(L_\alpha) \\
L_\beta &= \bigcup_{\alpha<\beta} L_\alpha \ (\beta \ limit) \\
L &= \bigcup_\alpha L_\alpha,
\end{aligned}
$$

where $\mathrm{def}\,(x)$ *is the set of subsets* y *of* x *which are definable (with parameters) over* x. *If* $x \in L$ *then* x *is* **constructible**.

There are alternative ways of presenting the successor steps in the construction of L. First of all, we can allow parameters or not. The reason is that if a set is definable over L_α with parameters, it is also definable over it without parameters, by hereditarily substituting the parameters with their definitions (in the sense that the definition of a parameter may involve other parameters, which have to be discharged as well). Note that no circularity is involved, because the definition of an element of L_α can only use parameters from L_β with $\beta < \alpha$, and thus the ordinals are decreasing, although an ordinal greater than α may be needed to carry out the induction.

Another way to define L (Gödel [1940]) relies on an analysis of the operation $\mathrm{def}\,(x)$, and on the isolation of finitely many (eight) operations that explicitly produce, by composition over the transitive closure of x, any set definable over x. This approach effectively reduces the infinitely many formulas of the language of Set Theory to a finite set of operations, and it allows a finite axiomatization of Set Theory, based on the concept of class (the **Von Neumann-Gödel-Bernays system** NGB, see Fraenkel, Bar-Hillel, and Levy [1958]).

Finally, it has turned out that the levels L_α of the constructible hierarchy are well-behaved only for limit ordinals, while in general they are not closed under very natural functions (like ordered pairing) that, although obviously constructible, increase levels. For a finer analysis, the constructible hierarchy has been substituted by **Jensen Hierarchy** (Jensen [1972]), whose levels J_α are extensions of L_α, and possess the closure properties that only the limit levels of L have. The levels of the two hierarchies coincide exactly at those stages α such that $\omega \cdot \alpha = \alpha$. See Devlin [1984] for a treatment.

Axiom IV.4.2 *The* **Axiom of Constructibility** *is the assertion* $V = L$ *that every set is constructible, i.e.* $(\forall x)(x \in L)$.

Here we develop the study of L only for what concerns our immediate interest, i.e. the study of subsets of ω. General references for constructible sets are Gödel [1940], Mostowski [1969], and Devlin [1984].

The levels of the Constructible Hierarchy

The first few levels of L have already been considered: since $\mathrm{def}(x) = \mathcal{P}(x)$ for any finite set x, we have $L_\alpha = V_\alpha$ for $\alpha \leq \omega$. In particular, $L_\omega = \mathcal{HF}$.

Some of the properties of these first levels generalize to every level.

Proposition IV.4.3 (Gödel [1939]) L_α *is transitive, and it has finite cardinality if $\alpha < \omega$, and the same cardinality as α otherwise.*

Proof. L_α is transitive, by induction on α:

- $\alpha = 0$

  Obvious.

- $\alpha = \beta + 1$

  Let $x \in y \in L_\alpha$. By definition, $y \subseteq L_\beta$, and thus $x \in L_\beta$. By induction hypothesis L_β is transitive, and thus $x \subseteq L_\beta$. Then $x \in L_{\beta+1}$ because x is definable over L_β, by the formula $\varphi(z) \Leftrightarrow z \in x$ (in which x appears as a parameter).

- α *limit*

  The union of transitive sets is transitive.

L_n is finite, by induction on $n \in \omega$, because the power set of a finite set is finite. Suppose then $\alpha \geq \omega$: we show by induction on α that L_α has the same cardinality as α.

- $\alpha = \omega$

  L_ω is countable because all the L_n are finite.

- $\alpha = \beta + 1$

  The cardinality of $L_{\beta+1}$ is equal to the cardinality of the set of formulas with parameters in L_β, and hence to the cardinality of L_β (because the number of possible formulas is countable, and $\beta \geq \omega$). By induction hypothesis this is the cardinality of β, and hence of $\beta + 1 = \alpha$.

- α *limit*

  L_α is the union of α many sets, each of cardinality at most the cardinality of α. Thus it has cardinality at most equal to that of α (by well-known properties of cardinals, following from the Axiom of Choice). $\square$

Proposition IV.4.4 Hierarchy Theorem (Gödel [1939])

1. $\alpha \leq \beta \Rightarrow L_\alpha \subseteq L_\beta$

2. $\alpha < \beta \Rightarrow L_\alpha \subset L_\beta$.

Proof. Note that $L_\gamma \in L_{\gamma+1}$, being definable by $x = x$ over L_γ. To show that $L_\alpha \subseteq L_\beta$ if $\alpha \leq \beta$, we may suppose $\alpha < \beta$ (since the case $\alpha = \beta$ is trivial). And, if $\alpha < \beta$, it is enough to show that $L_\alpha \in L_\beta$, because L_β is transitive (and thus its elements are contained in it).

We thus prove, by induction on β, that $L_\alpha \in L_\beta$, for any $\alpha < \beta$:

- $\beta = 0$
  Obvious.

- $\beta = \gamma + 1$
  If $\alpha < \beta$ then either $\alpha = \gamma$, and thus $L_\alpha = L_\gamma \in L_{\gamma+1} = L_\beta$, or $\alpha < \gamma$, and thus, by induction hypothesis, $L_\alpha \in L_\gamma \in L_{\gamma+1}$: by transitivity, $L_\alpha \subseteq L_{\gamma+1}$.

- β limit
  If $\alpha < \beta$, there exists γ such that $\alpha < \gamma < \beta$. Then $L_\alpha \in L_\gamma$ by induction hypothesis, and $L_\gamma \subseteq L_\beta$ by definition of L_β for β limit. Thus $L_\alpha \in L_\beta$.

To show that the hierarchy is proper it is enough to show, by induction on α, that $\alpha \in L_{\alpha+1} - L_\alpha$. Suppose that, for every $\beta < \alpha$, $\beta \in L_{\beta+1} - L_\beta$: then $(\forall \beta < \alpha)(\beta \in L_\alpha)$ (because, by the first part of the proof, if $\beta < \alpha$ then $L_{\beta+1} \subseteq L_\alpha$), and $\alpha \subseteq L_\alpha$. Now:

- $\alpha \notin L_\alpha$
  Otherwise, by definition of L_α, $\alpha \subseteq L_\beta$ for some $\beta < \alpha$, and thus $\beta \in L_\beta$, contradicting the induction hypothesis.

- $\alpha \in L_{\alpha+1}$
  $\alpha \subseteq L_\alpha$, and $\alpha \notin L_\alpha$. Moreover, the ordinals in L_α are closed downward (because L_α is transitive). This means that α is the set of ordinals in L_α, and thus it is definable over it by the formula 'x is an ordinal'. Then $\alpha \in L_{\alpha+1}$. $\square$

The structure of L

The next result shows that the definition of L_α is absolute, and thus it has the same meaning in every model of ZF^-.

Proposition IV.4.5 (Takeuti [1960], Karp [1967])

1. L_α is a $\Delta_1^{ZF^-}$ function total on the ordinals

2. $x \in L_\alpha$ is $\Delta_1^{ZF^-}$, as a relation of x and α

3. $x \in L$ is $\Sigma_1^{ZF^-}$.

Proof. L_α is defined by recursion on the ordinals, and thus it is enough to verify that the recursion cases are $\Delta_1^{ZF^-}$. The only nontrivial case is the successor step. But to say that $x \subseteq L_\alpha$ is definable over L_α means that there is a formula (this introduces an existential quantifier, which is bounded over the $\Delta_1^{ZF^-}$ set of formulas, which is a set by the Axiom of Infinity[1]) such that x is the set defined by it over L_α (and the satisfaction predicate is Δ_1^{GKP}, as in IV.3.16).

Then $x \in L_\alpha$ is $\Delta_1^{ZF^-}$, and $x \in L$ is $\Sigma_1^{ZF^-}$ because

$$x \in L \Leftrightarrow (\exists \alpha)(\alpha \text{ ordinal} \wedge x \in L_\alpha). \quad \square$$

Note that L_α is absolute for standard models of ZF^-, being $\Delta_1^{ZF^-}$, but L is not, being only $\Sigma_1^{ZF^-}$. Given a standard model A of ZF^-, if $\alpha \in A$ then $L_\alpha \in A$, by replacement. Thus L in A becomes $\bigcup_{\alpha \in A} L_\alpha$, which is L itself if A contains all ordinals. In other words, *L is absolute for standard models of ZF^- containing all ordinals* (Shepherdson [1951]).

One important tool for later proofs is the following.

Theorem IV.4.6 Reflection Principle for L (Levy [1960], Montague [1961]) *Given a formula φ with n free variables, for every ordinal α there exists $\beta \geq \alpha$ such that, whenever $x_1, \ldots, x_n \in L_\beta$,*

$$L \models \varphi(x_1, \ldots, x_n) \Leftrightarrow L_\beta \models \varphi(x_1, \ldots, x_n).$$

Proof. Let φ have no occurrence of $\forall$ (by possibly replacing all occurrences of $\forall$ with $\neg \exists \neg$), and let $\psi_0, \ldots, \psi_m$ be the finitely many subformulas of φ. Define a sequence of ordinals by starting with $\beta_0 = \alpha$, and letting β_{n+1} be the least ordinal $\gamma > \beta_n$ such that, for every sentence $(\exists v_i)\psi_k(v_i)$ with constants in L_{β_n} and true in L, $\psi_k(a_i)$ is true in L for some $a_i \in L_\gamma$. Note that β_{n+1} exists, by the Axiom of Replacement. Let β be the l.u.b. of $\{\beta_n\}_{n \in \omega}$.

It is now immediate to check, by induction on the length, that if ψ is any instance of a subformula of φ (in particular φ itself) with constants from L_β, then ψ is true in L if and only if it is true in L_β. The atomic case holds by absoluteness, propositional connectives are trivially handled, and the case of

---

[1]The Axiom of Infinity could be avoided, thus replacing ZF^- by GKP throughout this section, if the alternative approach to constructibility by Gödel's functions, quoted after IV.4.1, were used.

(existential) quantification holds by construction. □

Note that the Reflection Principle has nothing much to do with L, and it works in general for any class which is the union of an increasing hierarchy of sets defined on all ordinals, such that the limit levels are defined as the union of the previous ones. In particular it works for V, and it shows that in ZF we can find a model (with a V_α, and hence a set, as universe) for any finite set of theorems of ZF, and thus prove the consistency of any finite part of ZF. By Gödel's Second Theorem (p. 169), ZF *is not finitely axiomatizable* (Mc-Naughton [1954], Montague [1961]). In other words, collection and separation are not reducible, like foundation was, to a finite set of statements. As we have quoted, there exists a finitely axiomatizable set of axioms NGB for Set Theory, based on the notion of class (see Fraenkel, Bar-Hillel, and Levy [1958]).

The next result characterizes L in set-theoretical terms. We consider the full theory ZF, instead of just ZF^-, not only because the stated result is stronger, but because we need to consider the Power Set Axiom in the next subsection.

Theorem IV.4.7 (Gödel [1939]) *L is the smallest standard model of ZF containing all the ordinals.*

Proof. We have already noted that every model of ZF^- containing all the ordinals must contain L, by absoluteness of the definition of L_α.

Since L contains all the ordinals, because $\alpha \in L_{\alpha+1}$ by the proof of IV.4.4, it remains to prove that L is a model of ZF. The argument is in four parts, exploiting different properties of L in the verifications of the axioms.

1. *extensionality and foundation*
   They are automatically satisfied, because L is transitive.

2. *infinity*
   Since ω is an absolute object, it is enough to show that it is in L. But $\omega \in L_{\omega+1}$.

3. *separation*
   Suppose $a \in L$, and φ is a formula with parameters in L. We want to show that the set x defined by

   $$z \in x \Leftrightarrow L \models z \in a \wedge \varphi(z)$$

   is in L. There exists α such that L_α contains both a and the parameters of φ, and hence the set x_α

   $$z \in x_\alpha \Leftrightarrow L_\alpha \models z \in a \wedge \varphi(z)$$

is in L, being in $L_{\alpha+1}$. But there is no reason to believe that $x = x_\alpha$, since φ may have quantifiers, that mean different things when interpreted over L and over L_α. To straighten this out, we apply the Reflection Principle IV.4.6: let $\beta \geq \alpha$ be such that, whenever $z \in L_\beta$,

$$L_\beta \models z \in a \wedge \varphi(z) \Leftrightarrow L \models z \in a \wedge \varphi(z).$$

Thus, if

$$z \in x_\beta \Leftrightarrow L_\beta \models z \in a \wedge \varphi(z),$$

we have $x_\beta \cap L_\beta = x \cap L_\beta$. But, since $a \in L_\alpha \subseteq L_\beta$, a is a subset of L_β by transitivity, and then so is x. Thus $x_\beta \cap L_\beta = x$, and $x \in L_{\beta+1}$.

4. *large sets existence*
The remaining axioms are all of the same type: given certain sets, they produce sets bigger than them (in contrast to the Separation Axiom, that isolates subsets of given sets). They are treated in the same way, and the proofs that they hold in L all follow from a single fact: that each subset of L is contained in some L_α. This is easily seen to be true: if $a \subseteq L$ then $(\forall x \in a)(\exists \alpha)(x \in L_\alpha)$, and by collection (or replacement) there is α such that $(\forall x \in a)(x \in L_\alpha)$, so that $a \subseteq L_\alpha$.

- *pair*
  The set b is the pair $\{x, y\}$ if

  $$(\forall z \in b)(z = x \vee z = y),$$

  which is an absolute definition. Given $x, y \in L$, we must then show that $\{x, y\} \in L$.
  $\{x, y\}$ exists by the Pairing Axiom in V, and it is contained in L. Then it is contained in L_α, for some α. Since it satisfies the definition above over L_α, it is in $L_{\alpha+1}$.

- *union*
  Similar.

- *power set*
  The set b is the power set $\mathcal{P}(a)$ if

  $$(\forall z)(z \in b \leftrightarrow z \subseteq a),$$

  which is not an absolute definition. Thus b is the power set of a in L if it satisfies this definition on L, i.e. (by absoluteness of inclusion)

  $$(\forall z \in L)(z \in b \leftrightarrow z \subseteq a),$$

which defines $\mathcal{P}(a) \cap L$ over V. Given $a \in L$, we must then show that $\mathcal{P}(a) \cap L$ is in L.

$\mathcal{P}(a) \cap L$ exists by the Power Set Axiom in V, and it is contained in L. Then it is contained in L_α, for some α. Since it satisfies the definition above over L_α, it is in $L_{\alpha+1}$.

- collection
  If $a \in L$, and
  $$L \models (\forall x \in a)(\exists y)\varphi(x,y),$$

we need to find $b \in L$ such that

$$L \models (\forall x \in a)(\exists y \in b)\varphi(x,y).$$

The hypothesis means that

$$(\forall x \in a \cap L)(\exists y \in L)(L \models \varphi(x,y)).$$

By collection in V, there is $c \subseteq L$ such that

$$(\forall x \in a \cap L)(\exists y \in c)(L \models \varphi(x,y)).$$

Then $c \subseteq L_\alpha$ for some α, and hence

$$(\forall x \in a \cap L)(\exists y \in L_\alpha)(L \models \varphi(x,y)).$$

Since $L_\alpha \in L$, it is enough to let $b = L_\alpha$ to have

$$(\forall x \in a \cap L)(\exists y \in b)(L \models \varphi(x,y)),$$

and hence

$$L \models (\forall x \in a)(\exists y \in b)\varphi(x,y). \quad \square$$

Corollary IV.4.8 L is a model of $V = L$.

Proof. We have to prove
$$L \models (\forall x)(x \in L).$$

Since L is absolute for models of ZF^- containing all ordinals, and L is such a model, this amounts to showing that

$$(\forall x \in L)(x \in L),$$

which is trivially true. $\square$.

The corollary is not only interesting for its own sake, but also in applications: to prove that some fact holds in L, it is enough to show that it is provable in ZF plus $V = L$.

Corollary IV.4.9 (Kreisel [1956], Shoenfield [1961a]) *If φ is a Σ_3^1 sentence of Second-Order Arithmetic provable in ZF plus $V = L$, then φ is already provable in ZF alone.*

Proof. Let $\varphi \leftrightarrow (\exists A)\psi(A)$, with $\psi \in \Pi_2^1$. If φ is provable in ZF plus $V = L$ then it is true in L, i.e. there is $A \in L$ such that $\psi(A)$ holds in L. By absoluteness of Π_2^1 formulas for L (IV.3.27) $\psi(A)$ must then be true, and thus φ holds. But the whole reasoning took place in ZF, since only ZF is needed to prove IV.4.12 and IV.3.27. Thus φ has been proved in ZF alone.   □

The result is best possible since, by IV.4.22, $(\forall A)(A \in L)$ is a Π_3^1 statement true in L (because L satisfies $V = L$), but not provable in ZFC (because its negation is consistent with ZFC).

The usefulness of the corollary is quite evident: $V = L$ does not prove any new Σ_3^1 sentence of Second-Order Arithmetic, and thus it can be used freely when trying to prove such a sentence in ZF alone. The same of course holds for any of the consequences of $V = L$, like the Axiom of Choice and the Continuum Hypothesis, proved below. Actually, in the latter cases the result is not the best possible, and it can be improved to hold for bigger classes of sentences (see Platek [1969]).

Theorem IV.4.10 (Gödel [1939], Karp [1967])

1. *There exists a $\Delta_1^{ZF^-}$ well-ordering $<_\alpha$ of L_α.*

2. *There exists a $\Sigma_1^{ZF^-}$ well-ordering $<_L$ of L.*

Proof. We define a $\Delta_1^{ZF^-}$ well-ordering $<_\alpha$ of L_α, by recursion on the ordinals.

1. $\alpha = 0$
   Since $L_0 = \emptyset$, $<_0 = \emptyset$.

2. $\alpha = \beta + 1$
   By induction hypothesis we have $<_\beta$ that well-orders L_β. Let

$$
\begin{aligned}
x <_\alpha y \quad \Leftrightarrow \quad &(x \in L_\beta \wedge y \in L_\beta \wedge x <_\beta y) \vee \\
&(x \in L_\beta \wedge y \in L_{\beta+1} - L_\beta) \vee \\
&(x \in L_{\beta+1} - L_\beta \wedge y \in L_{\beta+1} - L_\beta \wedge \\
&\quad \text{the first formula defining } x \text{ over } L_\beta \text{ precedes} \\
&\quad \text{the first formula defining } y \text{ over } L_\beta).
\end{aligned}
$$

3. α limit

$$
x <_\alpha y \quad \Leftrightarrow \quad (\exists \beta < \alpha)(x <_\beta y).
$$

By definition, if $\alpha < \beta$ then $<_\alpha$ is an initial segment of $<_\beta$. Since the definition is by recursion on the ordinals, it is enough to verify that the recursion is $\Delta_1^{ZF^-}$. Since L_α is $\Delta_1^{ZF^-}$, the only point to check is the case of $\alpha = \beta + 1$ and $x, y \in L_{\beta+1} - L_\beta$.

The formulas without parameters of the set-theoretical language are countably many, and there is a natural (lexicographical) $\Delta_1^{ZF^-}$ well-ordering of them, of length ω. By arithmetization everything can be coded in $\mathcal{HF} = L_\omega$, which exists by the Axiom of Infinity.

For what concerns parameters, they are in L_β and, by induction hypothesis, $<_\beta$ well-orders L_β. We can then say that φ with parameters $a_1, \ldots, a_n$ (ordered by $<_\beta$) precedes ψ with parameters $b_1, \ldots, b_m$ (ordered by $<_\beta$) if and only if:

- $n < m$

- $n = m$ and φ precedes ψ (as formulas)

- $n = m$, $\varphi = \psi$ and, for the first i such that $a_i \neq b_i$, $a_i <_\beta b_i$.

Again the conditions in this definition are $\Delta_1^{ZF^-}$, and thus the whole well-ordering is $\Delta_1^{ZF^-}$. Then

$$x <_L y \Leftrightarrow (\exists \alpha)(x \in L_\alpha \wedge y \in L_\alpha \wedge x <_\alpha y)$$

is $\Sigma_1^{ZF^-}$. Note that, for each α, $<_\alpha$ is an initial segment of $<_L$. $\quad\square$

Exercises IV.4.11 a) *If L_α is a model of ZF^-, the well-ordering $<_\alpha$ of L_α has length α.* (Hint: by replacement, L_α is closed under the rang function of $<_\alpha$

$$f(y) = \sup\{f(x) + 1 : x <_\alpha y\},$$

and thus $\sup\{f(x) : x \in L_\alpha\} \leq \alpha$. Conversely, since f is one-one, f^{-1} is a function with range L_α, and if the length of $<_\alpha$ were $\beta < \alpha$, then the range of f^{-1} would be in L_α, i.e. $L_\alpha \in L_\alpha$, contradiction.)

b) *If $V = L$ then $\leq_L$ is $\Delta_1^{ZF^-}$.* (Hint: in this case

$$\neg(x <_L y) \Leftrightarrow x = y \vee (\exists \alpha)(x \in L_\alpha \wedge y \in L_\alpha \wedge y <_\alpha x),$$

and thus $<_L$ is also $\Pi_1^{ZF^-}$.)

One of the original reasons to introduce L was to prove the following fact, whose consequence is that *the Axiom of Choice is consistent with ZF* (Gödel [1938]), since it holds in a model of ZF. Note that also V is a model of ZF, but there is no obvious way to show that it satisfies the Axiom of Choice, the difficulty being in well-ordering the power set of a given (well-ordered) set.

Corollary IV.4.12 *L is a model of ZFC plus $V = L$.*

Proof. By IV.4.7 and IV.4.8, it only remains to prove that the Axiom of Choice holds in L. The theorem just proved shows that L is well-orderable, and thus the Axiom of Choice holds if $V = L$. But L is a model of ZF plus $V = L$, and hence the Axiom of Choice holds in L. □

The fact that L is a model of ZFC has an important consequence: all theorems of ZFC are true in L. In particular, it is possible to carry on inside L the theory of cardinals, which is largely based on the Axiom of Choice.

Of particular interest for us is ω_1^L, the analogue of the first uncountable cardinal, which is the least ordinal α such that no one-one function $f : \alpha \to \omega$ exists in L. In other words, 'L believes ω_1^L is ω_1'. Clearly $\omega_1^L \le \omega_1$: if there is no function with certain properties in V, there is none in L either. *The assertion $\omega_1^L = \omega_1$ is consistent with ZFC*, since so is $V = L$ (by IV.4.8). But also *the assertion $\omega_1^L < \omega_1$ is consistent with ZFC* (Cohen [1966]), by cardinal collapsing (see Volume III). It also follows from large cardinals hypothesis (Rowbottom [1971]).

Constructible sets of natural numbers

We have introduced the Constructible Hierarchy not for its own sake, but to study subsets of ω. We thus turn now to our real interest, the set $\mathcal{P}(\omega) \cap L$ of constructible sets of natural numbers.

Since $\mathcal{P}(\omega) \cap L \subseteq L$, there is α such that $\mathcal{P}(\omega) \cap L \subseteq L_\alpha$, and thus there is a stage after which no subset of ω is ever generated. We now improve on this, by exhibiting such an α (which will turn out to be the best possible).

Theorem IV.4.13 (Gödel [1939]) $\mathcal{P}(\omega) \cap L \subseteq L_{\omega_1^L}$.

Proof. The theorem reduces to prove, in ZFC, that $\mathcal{P}(\omega) \cap L \subseteq L_{\omega_1}$. Since L is a model of ZFC, the result also holds in L. To see what this means, recall that $\mathcal{P}(\omega)$, the power set of ω, means $\mathcal{P}(\omega) \cap L$ in L (see the proof of IV.4.7). L is absolute for L, because L is a model of ZF^- containing all ordinals. And ω_1^L is, by definition, the ω_1 of L. Thus $\mathcal{P}(\omega) \cap L \subseteq L_{\omega_1}$ means $\mathcal{P}(\omega) \cap L \subseteq L_{\omega_1^L}$, when interpreted inside L, and this is the statement of the theorem.

We thus turn to the proof of $\mathcal{P}(\omega) \cap L \subseteq L_{\omega_1}$, which roughly consists of the following. Let $A \subseteq \omega$ be constructible, i.e. $A \in L$. Then $A \in L_\alpha$, for some α: we want to show that α may be supposed to be countable. To achieve this, we use the Löwenheim-Skolem Theorem of logic (for the reader not acquainted with it, we prove in 2 below the special case we need). Its essence is to cut out, from a model of a sentence, a *countable* model, leaving unchanged a given countable subset. Here we know that $A \in L_\alpha$, and thus we only have to find a sentence φ whose models are exactly the sets of the kind L_α (this is done in 1 below). Then L_α is a model of φ, and it contains $\{A\}$: the Löwenheim-Skolem Theorem produces a countable model of φ containing $\{A\}$, which must then be of the form L_α, with α countable (because, by IV.4.3, L_α is countable if and only if α is). It only remains to find φ, and prove the Löwenheim-Skolem Theorem.

1. *There is a sentence φ such that if a set M is a standard model for φ, then $M = L_\alpha$ for some (limit) α.*
   By absoluteness of L_α, if M is a standard model of ZF^- then $\bigcup_{\alpha \in M} L_\alpha$ is L in M, and it is contained in it. Let α_M be the smallest ordinal not in M: if α_M is limit then, by definition of L, $\bigcup_{\alpha \in M} L_\alpha = L_{\alpha_M}$, and thus $L_{\alpha_M} \subseteq M$. If $V = L$ also holds in M, i.e. M is equal to L in it, then $L_{\alpha_M} = M$.

   We can then let φ be the conjunction of the following:

   - the finitely many axioms needed to prove that the notions of ordinal and of L_α are absolute for standard models of them

   - the finitely many axioms needed to prove that any model of them is closed under L_α (as a function of α)

   - the finitely many axioms needed to prove that there is no greatest ordinal

   - $V = L$.

2. *Given a sentence φ, a set M which is a model for it, and a countable subset X of M, there is a countable model M' of φ such that $X \subseteq M' \subseteq M$* (Löwenheim [1915], Skolem [1920]).
   The proof is a variation of the proof of the Reflection Principle IV.4.6. As there, the only important thing is to provide the witnesses for the existential sentences true in M. Let φ have no occurrence of $\forall$ (by possibly replacing all occurrences of $\forall$ with $\neg\exists\neg$), and let $\psi_0, \ldots, \psi_m$ be the finitely many subformulas of φ. Define a sequence of subsets of M, starting with $M_0 = X$, and letting M_{n+1} be a countable subset of M containing M_n and such that, for every sentence $(\exists v_i)\psi_k(v_i)$ with constants in M_n and true in M, $\psi_k(a_i)$ is true in M for some $a_i \in M_{n+1}$. Let $M' = \bigcup_{n \in \omega} M_n$.

   First of all we have to prove, by induction on n, that M_{n+1} exists, i.e. that there is a countable structure as wanted. M_0 is countable by the hypothesis on X. Suppose M_n is countable: then there are only countably many possible sentences with constants in M_n to consider. Since we only need to add one witness for each of them, we can choose M_{n+1} as a countable subset of M.

It is now immediate to check, by induction on the length, that if ψ is any instance of a subformula of φ (in particular φ itself) with constants from M', then ψ is true in M if and only if it is true in M'. The atomic case holds by absoluteness, propositional connectives are trivially handled, and the case of (existential) quantification holds by construction.

Fact 1 refers to standard models, and this requires a small patch up of the sketch given at the beginning. Given $A \subseteq \omega$ such that $A \in L_\alpha$, we may suppose $\alpha \geq \omega$ limit: then we have a model L_α of φ, containing $\omega \cup \{A\}$. By the Löwenheim-Skolem Theorem (fact 2), there is a countable model with the same properties, but we cannot yet conclude that this is a countable L_α, because it is not necessarily a standard model (an assumption used in the proof of fact 1 above). But we can apply the (corollary of the) Collapsing Lemma IV.3.5, and find a standard structure isomorphic to it, and hence satisfying the same sentences, φ in particular. Now we do have a standard structure satisfying φ, still containing A (because $\omega \cup \{A\}$ is a transitive set, since $A \subseteq \omega$, and thus it is not changed by the collapsing function). Then this model must be a countable L_α, and α is countable itself by IV.4.3. $\square$

One of the original reasons to introduce L was to prove the following fact, whose consequence is that *the Continuum Hypothesis is consistent with ZFC* (Gödel [1938]), since it holds in a model of ZFC. The reason why the Continuum Hypothesis follows from $V = L$, while it is independent of ZFC, is that the former completely specifies the extension of the power set operation (left undetermined by the latter), by determining which subsets of a given set are available, namely only those that can be predicatively defined from the ordinals. By providing a sort of minimal interpretation for the power set operation, $V = L$ limits the number of subsets of a given set to the least possible value.

Corollary IV.4.14 *The Continuum Hypothesis holds in L.*

Proof. We have proved that $\mathcal{P}(\omega) \cap L \subseteq L_{\omega_1}$. If $V = L$ then this means that the power set of ω has at most the power of L_{ω_1}, which is ω_1 by IV.4.3. By Cantor's Theorem (II.2.1), the cardinality of the power set of ω must be uncountable, and hence at least ω_1. Thus it is exactly ω_1. This shows that the Continuum Hypothesis follows from ZFC plus $V = L$ (the Axiom of Choice is needed for cardinal arithmetic, in particular in the proof of IV.4.3), and by IV.4.12 it then holds in L. $\square$

We will see in Volume III that *all the following possibilities are consistent with ZFC*:

1. $\mathcal{P}(\omega) \subseteq L$, which follows from $V = L$.

2. $\mathcal{P}(\omega) \not\subseteq L$ and $\mathcal{P}(\omega) \cap L$ countable, by collapsing ω_1^L (this also follows from large cardinals hypothesis, by Rowbottom [1971]).

3. $\mathcal{P}(\omega) \not\subseteq L$ and $\mathcal{P}(\omega) \cap L$ uncountable, by starting with a model of the Continuum Hypothesis, and building a generic extension of it that preserves cardinals.

By the proof of the corollary, $\mathcal{P}(\omega) \cap L$ is not contained in L_α for $\alpha < \omega_1^L$, because otherwise the power set of ω would be countable in L, contradicting Cantor's Theorem in L. Thus $\{\mathcal{P}(\omega) \cap L_\alpha\}_{\alpha < \omega_1^L}$, as a hierarchy for $\mathcal{P}(\omega) \cap L$, has the smallest possible length. We now show that the hierarchy is not proper.

Proposition IV.4.15 (Putnam [1963]) *There exists $\alpha < \omega_1^L$ such that no new subset of ω is generated in L at stage $\alpha + 1$, i.e.*

$$\mathcal{P}(\omega) \cap (L_{\alpha+1} - L_\alpha) = \emptyset.$$

Proof. Let ψ be the formula

$$(\exists \alpha)[\mathcal{P}(\omega) \cap (L_{\alpha+1} - L_\alpha) = \emptyset].$$

By IV.4.13,

$$\mathcal{P}(\omega) \cap (L_{\omega_1^L+1} - L_{\omega_1^L}) = \emptyset$$

and thus, by absoluteness, ψ is true in some L_β with β limit (e.g. let β be $\omega_1^L + \omega$). By the same argument of the proof of IV.4.13 (using the formula φ whose models are all limit levels of the constructible hierarchy, the Löwenheim-Skolem Theorem, and the Collapsing Lemma), there is a countable β with the same properties. Since ψ holds in L_β, there is an ordinal $\alpha < \beta$ such that

$$\mathcal{P}(\omega) \cap (L_{\alpha+1} - L_\alpha) = \emptyset,$$

and α is countable, because so is β.

This is not enough yet, because we actually want $\alpha < \omega_1^L$, i.e. constructibly countable. To get this stronger version we just need to note that the Löwenheim-Skolem Theorem has been proved in ZFC, and hence it holds in L. Its version in L sounds as follows:

- Given a sentence φ, a set $M \in L$ which is a model for it, and a subset $X \in L$ of M which is constructibly countable, there is a constructibly countable model $M' \in L$ of φ such that $X \subseteq M' \subseteq M$.

By using the argument of IV.4.13 with this version, which is applicable because the starting model is L_β, which is constructible, we obtain a constructibly countable β as wanted. □

Since the hierarchy $\{L_\alpha\}_{\alpha<\omega_1^L}$ is obviously proper, by IV.4.4, the meaning of the last result is that, at the stages at which no new subset of ω is generated (called **gap ordinals**), new sets appear, that will later be used to define new subsets of ω. This shows that the definition of $\mathcal{P}(\omega) \cap L$ is intrinsically set-theoretical. Since $V = L$ is consistent with ZFC, and thus $\mathcal{P}(\omega) \cap L$ could be the true power set of ω, *set-theoretical methods are essential in the analysis of the continuum*, i.e. in Classical Recursion Theory as defined on p. 1. This adds to the remarks made by Sacks in the Foreword.

The study of gap ordinals has been pursued in various directions, some of which are treated in Volume III. On one side, the possible lengths of gaps have been investigated: it turns out that *there are gaps of any length less than ω_1^L, and the gaps are distributed in a very orderly fashion, according to their lengths* (Marek and Srebrny [1974]). On the other side, a characterization of gap ordinals has been obtained: α *is a gap ordinal if and only if*

$$L_{\alpha+1} \models \alpha \ \text{uncountable},$$

in the sense that there is no one-one function in $L_{\alpha+1}$ from α to ω (Boolos and Putnam [1968], Jensen [1972]). The condition is obviously sufficient, because such a function induces a well-ordering of ω of length α, that cannot be in L_α since α is not. The converse is a strengthening of IV.4.13, and it is obviously necessary, if $L_{\alpha+1}$ is a model of enough Set Theory: in this case IV.4.13 holds in it, and thus all constructible subsets of ω which are in $L_{\alpha+1}$ must be already in $L_{\omega_1^{L_{\alpha+1}}}$, i.e. must be constructed at stages countable in $L_{\alpha+1}$. Thus, by absoluteness,

$$\mathcal{P}(\omega) \cap (L_{\alpha+1} - L_{\omega_1^{L_{\alpha+1}}}) = \emptyset.$$

And if α is not countable in $L_{\alpha+1}$, $\omega_1^{L_{\alpha+1}} \leq \alpha$. Note that the gap could be quite large, depending on how much of Set Theory is modelled by $L_{\alpha+1}$, since $L_{\alpha+1}$ could prove the existence of various cardinals (inside it) greater than $\omega_1^{L_{\alpha+1}}$.

Exercises IV.4.16 a) *The gap ordinals are ω_1^L.* (Putnam [1963]) (Hint: there is a gap above any given constructibly countable ordinal.)

b) *For any ordinal γ admitting an absolute definition, there are gaps of length γ.* (Putnam [1963]) (Hint: consider $\alpha + \gamma$ instead of $\alpha + 1$ in the proof above.)

One might think of mimicking the definition of L in an autonomous way, by relying only on subsets of ω.

Definition IV.4.17 (Kleene [1959b]) *The* **Ramified Analytical Hierar-**

chy *is described, in terms of ordinals, as follows:*

$$
\begin{aligned}
RA_0 &= \text{ } \textit{the arithmetical sets} \\
RA_{\alpha+1} &= \text{ } \text{def}_2\,(RA_\alpha) \\
RA_\beta &= \text{ } \bigcup_{\alpha<\beta} RA_\alpha \text{ } (\beta \text{ } \textit{limit}) \\
RA &= \text{ } \bigcup_\alpha RA_\alpha,
\end{aligned}
$$

where $\text{def}_2(x)$ *is the set of subsets of* ω *which are definable in Second-Order Arithmetic with set quantifiers restricted to* x, *and parameters in* x.

RA will be studied in Volume III. Leeds and Putnam [1974] show that parameters can be avoided in its definition. By cardinality reasons, RA *breaks down at an ordinal* β_0 (since if each step adds a new subset, the subsets of ω run out in at most a continuum of steps). Boolos and Putnam [1968] show that, *when account is taken of the different starting points, the Ramified Analytical Hierarchy coincides, up to* β_0 *and level by level, with the Constructible Hierarchy restricted to* $\mathcal{P}(\omega)$. It is thus not surprising that β_0 *is the first gap ordinal*, i.e. the first point in which sets not in $\mathcal{P}(\omega)$ become essential for the definition of new subsets of ω.

Exercises IV.4.18 a) β_0 *is countable.* (Cohen [1963a]) (Hint: by cardinality considerations, there is an ordinal α such that $RA_{\alpha+1} = RA_\alpha$. As in IV.4.15, and by absoluteness of RA_α, there is a countable one.)

b) $RA \subset \Delta^1_2$. (Putnam [1964]) (Hint: there is a Δ^1_2 predicate enumerating the members of RA. Thus the members of RA are uniformly Δ^1_2. By diagonalizing the enumeration predicate, there is a Δ^1_2 set not in RA. To get the enumeration predicate, by part a) we may restrict attention to countable ordinals, and hence to well-orderings of ω. The methods of IV.2.22 and IV.4.5 show that $A \in RA_x$ is uniformly Δ^1_1, whenever x varies over the elements of a well-ordering of ω, used as parameter. Since $A \in RA$ if and only if there is a well-ordering of ω such that $A \in RA_x$ for some x in it, we have the Σ^1_2 form, because being a well-ordering is a Π^1_1 condition. Similarly, $A \in RA$ if and only if, for every well-ordering of ω which is sufficiently long, i.e. it has elements y and z such that z is the successor of y in it, and $RA_y = RA_z$, then $A \in RA_x$ for some x. This provides the Π^1_2 form, because being a well-ordering is now an hypothesis, and thus it becomes Σ^1_1.)

Σ^1_2 sets

We have studied the set $\mathcal{P}(\omega) \cap L$ as a totality, but we do not know yet which sets are in it. Of course, since $V = L$ is consistent with ZFC, it might be that all sets (of natural numbers) are constructible. But even if this were the case, this would not be provable in ZFC, because Cohen [1963] has shown that it is consistent with ZFC that $\mathcal{P}(\omega) \nsubseteq L$ (see Volume III).

It is thus a nontrivial problem to ask which subsets of ω are provably in L, in the sense of being provable in ZFC that they are constructible.

Theorem IV.4.19 (Mostowski [1949], Shoenfield [1961a]) $\Sigma_2^1 \cup \Pi_2^1 \subseteq L$.

Proof. Since $\omega \in L$, and L satisfies separation, for any formula φ there is a set $a \in L$ such that

$$x \in a \Leftrightarrow L \models x \in \omega \wedge \varphi(x).$$

This of course does not mean that every definable subset of ω (in the real world, i.e. in V) is in L, because φ does not need to be absolute, and thus the set it defines in V is not necessarily a. But this certainly holds for the formulas whose meanings are the same over L and in the real world. By IV.3.27, this is the case of Σ_2^1 formulas, because L is a standard model of GKP containing all the ordinals. Thus Σ_2^1 sets are all in L, and the same holds for their complements in ω, i.e. the Π_2^1 sets. $\square$.

This result is the best possible: Jensen and Solovay [1970] have shown that *it is consistent with ZFC that $\mathcal{P}(\omega) \cap L$ is properly contained in Δ_3^1*. This also follows from large cardinals assumptions (Solovay [1967]). On the other hand, Harrington [1974] has shown that *for any n such that $3 \leq n \leq \omega$, it is consistent with ZFC that $\mathcal{P}(\omega) \cap L = \Delta_n^1$*. See Volume III for all this.

Recall that the Arithmetical and Analytical Hierarchies classify sets of natural numbers according to the complexity of their definitions over the structure of Arithmetic. The difference between the two hierarchies is in the range of quantifiers, over ω in the former case, and over $\mathcal{P}(\omega)$ in the latter. Our original motivation for the introduction of the Levy's Hierarchy was to pursue this trend, and allow for more general quantifiers, over sets in a given model. Our first example was $\mathcal{HF}$, and definability over it coincided, for subsets of ω, with arithmetical definability (IV.3.22).

We turn now to definability over L. Of course, every set in L is definable over it, using itself as parameter, but this is not what we mean: in the Analytical Hierarchy we did quantify over sets, but did not allow them as parameters (no attention is needed over $\mathcal{HF}$, because the proof of IV.3.22 shows that elements of $\mathcal{HF}$ and natural numbers are, in a sense, the same thing). What we really look for is thus definability without parameters over L.

Theorem IV.4.20 Definability of Σ_2^1 sets (Takeuti and Kino [1962]) *Let $A \subseteq \omega$. Then*

$$A \in \Sigma_2^1 \Leftrightarrow A \in \Sigma_1^L \text{ without parameters.}$$

Similarly for relations, of any number of variables.

Proof. Let $A \in \Sigma_2^1$. By the proof of IV.3.27, A is of the form $(\exists \alpha \text{ countable})\varphi$, with α ranging over countable ordinals, and φ Δ_1 over GKP plus Infinity. Since L is a model of GKP, and it contains all the countable ordinals, A is Σ_1^L without parameters.

For the converse, suppose $A \subseteq \omega$ is Σ_1^L without parameters, i.e.

$$x \in A \iff L \models (\exists y)\psi(x, y),$$

with $\psi \in \Delta_1^L$. By definition of satisfaction,

$$x \in A \iff (\exists y \in L)(L \models \psi(x, y)).$$

By definition of L, and absoluteness of Δ_1^L formulas,

$$x \in A \iff (\exists \alpha)(\exists y \in L_\alpha)(L_\alpha \models \psi(x, y)),$$

and hence

$$x \in A \iff (\exists \alpha)(L_\alpha \models (\exists y)\psi(x, y)).$$

Using the Löwenheim-Skolem Theorem and the Collapsing Lemma as in the proof of IV.4.13, and with φ as there,

$$x \in A \iff (\exists a)[a \text{ countable transitive} \wedge a \models \varphi \wedge (\exists y)\psi(x, y)].$$

Suppose such an a exists: being countable, $\langle a, \in \rangle$ is isomorphic to a well-founded structure $\langle \omega, \varepsilon \rangle$, and the Collapsing Lemma applied to the latter reproduces the original a (because the transitive collapse is unique, and a is transitive). In particular, if f is the collapsing function, $x = f(f^{-1}(x))$, and $(\exists y)\psi(x, y)$ holds on a if and only if $(\exists y)\psi(z, y)$, with $f(z) = x$, holds on $\langle \omega, \varepsilon \rangle$. Thus

$$x \in A \iff (\exists \varepsilon)[\langle \omega, \varepsilon \rangle \text{ well-founded} \wedge$$
$$\langle \omega, \varepsilon \rangle \models \varphi \wedge (\exists y)(\exists z)(f(z) = x \wedge \psi(z, y))],$$

where f is the collapsing function of $\langle \omega, \varepsilon \rangle$.

It only remains to compute the complexity of the last expression for A:

- ε is a binary relation on ω, that can be coded by its characteristic function: the first existential quantifier is thus a function quantifier.

- Well-foundedness is Π_1^1, see e.g. IV.2.15.

- The graph of the collapsing function, when the values are restricted to ω, is Δ_1^1, because it can be defined by recursion with clauses arithmetical in ε, as follows. First of all, $f(z) = 0$ if and only if z is the (unique) element of ω which does not have predecessors w.r.t. ε. And $f(z) = n + 1$ if and only if, for some y, $f(y) = n$, and z is the successor of y w.r.t. ε (i.e. the set $y \cup \{y\}$, when membership in interpreted as ε).

- The satisfaction relation is Δ_1^1, by arithmetization.

The whole expression is thus Σ_2^1. $\square$

Corollary IV.4.21 *Let $A \subseteq \omega$. Then*

$$A \in \Sigma_2^1 \Leftrightarrow A \in \Sigma_1^{L_{\omega_1}} \text{ without parameters.}$$

Similarly for relations, of any number of variables.

Proof. The first part of the proof goes through because L_{ω_1} is a model of GKP, and it contains all the countable ordinals. The second part of the proof can be repeated as above, without appeal to the Löwenheim-Skolem Theorem, since every L_α with $\alpha < \omega_1$ is countable. $\square$

We have proved the result for relations on ω, but little work is needed to extend it to relations on ω and $\mathcal{P}(\omega)$: we just have to show how to take care of set variables, and for this it is enough to prove that the graph of the collapsing function f, when the values are restricted to $\mathcal{P}(\omega)$, is still Δ_1^1. Recall that, by definition of f (see IV.3.5),

$$f(z) = \{ f(y) : y \in \omega \wedge y \, \varepsilon \, z \}.$$

Thus

$$f(z) = A \Leftrightarrow (\forall x)[x \in A \leftrightarrow (\exists y)(x = f(y) \wedge y \, \varepsilon \, z)].$$

The use of f on the right-hand-side is restricted to number values, and thus it is Δ_1^1 by the proof of IV.4.20. Thus the whole expression is Δ_1^1, because only number quantifiers are used.

Corollary IV.4.22 (Gödel [1940], Mostowski, Addison [1959a]) *The relations $A \in L$ and $A \leq_L B$, for A and B subsets of ω, are Σ_2^1.*

Proof. We know from IV.4.5, IV.4.10, and IV.4.7 that $x \in L$ and $x \leq_L y$ are Σ_1^L. When restricted to $\mathcal{P}(\omega)$ they become relations of set variables, and thus they are Σ_2^1. $\square$

The complexity of $\mathcal{P}(\omega) \cap L$ computed above is best possible, since *if $A \in L$ is Π_2^1 then $\mathcal{P}(\omega) \subseteq L$*, i.e. every subset of ω is constructible: indeed, if $A \in L$ is Π_2^1 then $(\exists A)(A \notin L)$ is a Σ_2^1 formula false in L, and by absoluteness it is also false in V, which means that $\mathcal{P}(\omega) \subseteq L$.

By IV.4.11.b, if $V = L$ then $\leq_L$ is actually $\Delta_1^{ZF^-}$, and thus the well-ordering $\leq_L$ is actually Δ_2^1 on $\mathcal{P}(\omega)$. In particular, $A \leq_{\Delta_2^1} B$ if $A \leq_L B$. Thus *it is consistent with* ZFC *that the* Δ_2^1*-degrees are well-ordered*, and no result contradicting this can be proved in ZFC. The study of the structure of Δ_2^1-degrees can thus be pursued along two different paths: one is to prove consistency results, the other to introduce new axioms and study the structure under them. We will follow both paths, in Volume III. Of course similar considerations hold for the Δ_n^1-degrees, for any $n \geq 2$.

$\mathcal{HC}$ and the Analytical Hierarchy

We have seen in IV.3.22 that the Arithmetical Hierarchy can be viewed as a set-theoretical hierarchy over the hereditarily finite sets. We now provide a similar interpretation of the Analytical Hierarchy.

Definition IV.4.23 $\mathcal{HC}$ *is the set of* **hereditarily countable sets,** *i.e. the smallest class* $\mathcal{A}$ *of sets such that:*

1. $\emptyset \in \mathcal{A}$

2. *if* $(\forall n \in \omega)(x_n \in \mathcal{A})$ *then* $\{x_n : n \in \omega\} \in \mathcal{A}$.

A number of properties of $\mathcal{HF}$ can be extended to $\mathcal{HC}$, by similar proofs. E.g., a set is hereditarily countable if and only if its transitive closure is countable.

Proposition IV.4.24 $\mathcal{HC}$ *is a transitive model of* GKP *containing all countable ordinals.*

Proof. The axioms of GKP can be checked in a way similar to IV.3.22. Moreover, note that an ordinal is transitively closed, and thus it is in $\mathcal{HC}$ if and only if it is countable. □

Since a countable subset of $\mathcal{HC}$ is a countable set with hereditarily countable members, it belongs to $\mathcal{HC}$. In particular, since $\omega \in \mathcal{HC}$, $\mathcal{P}(\omega) \subseteq \mathcal{HC}$.

Note that the smallest transitive model of GKP containing all countable ordinals is L_{ω_1}, and thus $L_{\omega_1} \subseteq \mathcal{HC}$ (directly, if $x \in L_{\omega_1}$ then, for some countable α, x is in L_α, which is transitive: then the transitive closure of x is contained in L_α, which is countable). However, since it is consistent with ZFC that some subset of ω is not constructible, we cannot prove that $L_{\omega_1} = \mathcal{HC}$.

Exercises IV.4.25 a) *If* $V = L$ *then* $\mathcal{HC} = L_{\omega_1}$. (Hint: by $\in$-induction. Suppose that $(\forall y \in x)(y \in L_{\omega_1})$, and x is countable. Then $x \subseteq L_{\omega_1}$, and $x \subseteq L_\alpha$ for some countable α. As in IV.4.13, by starting from $L_\alpha \cup \{x\}$, if $V = L$ then $x \in L_{\omega_1}$.)

b) $\mathcal{HC}$ *is a model of* ZFC^- *plus* $V = \mathcal{HC}$. (Hint: choice follows from the following facts: a subset of $\mathcal{HC}$ is well-orderable in V by the Axiom of Choice; a wellordering of an hereditarily countable set is still such; well- foundedness is absolute for standard models of GKP. $V = \mathcal{HC}$ holds because if $x \in \mathcal{HC}$ then any function from x into ω is already in $\mathcal{HC}$.)

The main reason for us to consider $\mathcal{HC}$ is the following analogue of IV.3.22, which provides an alternative set-theoretical way of seeing the Analytical Hierarchy. One should note that each level of the hierarchy on $\mathcal{HC}$ corresponds to the next level of the Analytical Hierarchy. This is not accidental, and it has already been observed in the discussion on p. 397.

Theorem IV.4.26 Set-theoretical definability of the Analytical Hierarchy. *Let $A \subseteq \omega$. Then A is definable over $\mathcal{HC}$ without parameters if and only if A is analytical. More precisely, for any n:*

$$A \in \Delta^1_{n+2} \quad \Leftrightarrow \quad A \in \Delta^{\mathcal{HC}}_{n+1} \ \text{without parameters}$$
$$A \in \Sigma^1_{n+2} \quad \Leftrightarrow \quad A \in \Sigma^{\mathcal{HC}}_{n+1} \ \text{without parameters}$$

Similarly for relations, of any number of variables.

Proof. For $n = 0$ the proof of IV.4.21 goes through without changes, using absoluteness of $\Delta^{\mathcal{HC}}_1$ formulas. Moreover, as noted after the proof of IV.4.21, the result holds for relations not only over ω, but also over $\mathcal{P}(\omega)$. The cases for $n > 0$ then follow by adding quantifiers, because $\mathcal{P}(\omega) \subseteq \mathcal{HC}$. □

To generalize the result for Σ^1_2 to any level we have used in a crucial way the fact that $\mathcal{P}(\omega) \subseteq \mathcal{HC}$. The analogue could not be proved for L, and this was the reason why we had to content ourselves with the absolute result for Σ^1_2, in IV.4.20. Of course, the full result can be obtained also for L if we assume $V = L$ (Takeuti and Kino [1962]), by the same proof (or by IV.4.25.a).

As it was already the case for IV.3.22, embedded in the proof of IV.4.26 are translations of Second-Order Arithmetic and Set Theory on $\mathcal{HC}$ into one another. The translation of Second-Order Arithmetic to Set Theory is standard, while the other consists of seeing hereditarily countable sets as countable trees (and then coding them as subsets of ω), and by interpreting equality as tree isomorphism. Peano Arithmetic was equivalent to ZFC with the Axiom of Infinity replaced by $V = \mathcal{HF}$: we now have that *Second-Order Arithmetic is equivalent to ZFC with the Power Set Axiom replaced by $V = \mathcal{HC}$* (Kreisel [1968], Zbierski [1971]).

It should be noted that by Second-Order Arithmetic we mean here the second-order version of $\mathcal{PA}$ plus the Axiom of Comprehension for analytical formulas (asserting the existence of the analytical sets), and an Axiom of Choice which asserts that if $(\forall x)(\exists A)\varphi(x, A)$ then there is a subset A of $\omega \times \omega$ such that, if A_x is the section of A w.r.t. x, then $(\forall x)\varphi(x, A_x)$. This obviously corresponds to collection for sets, and it is needed (Gandy [1967a]) to model the Axiom of Collection.

Second-Order Arithmetic, as well as subsystems of it obtained by variously restricting the Axioms of Comprehension and Choice, will be studied in Volume III.

Exercises IV.4.27 a) **Levy Absoluteness Lemma.** *A Σ_1^{ZF} formula with parameters in $\mathcal{HC}$ true in V is already true in $\mathcal{HC}$.* (Levy [1965]) (Hint: by Löwenheim-Skolem, Collapsing Lemma, and absoluteness, as in IV.4.13.)

b) *$\mathcal{P}(\omega) \cap L \subseteq L_{\omega_1}$ follows from Levy's Absoluteness Lemma and Δ_1^{ZF}-definability of L_α* (Karp [1967]). *This provides a slightly different and easier proof of IV.4.13.* (Hint: if $A \subseteq \omega$ and $A \in L$ the Σ_1^{ZF} formula $(\exists \alpha)(A \in L_\alpha)$ with parameter $A \in \mathcal{HC}$ is true in V. By Levy Absoluteness it is true for some countable ordinal.)

c) *$\mathcal{P}(\omega)$ is not Δ_1^{ZF}.* (Hint: otherwise the formula asserting its existence would be true in $\mathcal{HC}$.)

Recursion Theory on the ordinals ⋆

The whole idea of constructibility rests on the fact that L is definable by recursion on the ordinals. Since usually the ordinals are defined within Set Theory, so is L. While investigating the theory of ordinals, Takeuti [1957] discovered that it could be developed independently of Set Theory. It thus became clear that if one could also develop independently a theory of recursion on the ordinals, this would allow a different approach to L. Takeuti [1960] carried out the task, by defining the notion of recursive function on the ordinals by schemata, in a way similar to recursiveness on the integers. He then discovered that Gödel's result IV.4.13 could be recast in recursion-theoretical terms by saying that ω_1 (and, more generally, any uncountable ordinal) was **stable**, in the sense of being closed under the recursive functions on all the ordinals. With this, Recursion Theory was generalized both to the class of all ordinals, and to cardinals.

Independently, and motivated by needs related to the theory of infinitary languages, Machover [1961] developed an equivalent approach to recursion on cardinals, using systems of equations. The circle was closed by Takeuti and Kino [1962], when it was realized that recursion on the ordinals was actually equivalent to Σ_1^L definability.

After discovering that cardinals were appropriate domains for Recursion Theory, it was natural to wonder whether the strong closure properties of cardinals were somehow needed. Kripke [1964] and Platek [1966] answered the question by reversing the attack. They relativized the previous approaches to any ordinal α, by defining the α-recursive functions (e.g. by schemata, using a search operator on ordinals less than α). Then they defined **admissible**

ordinals as the α's closed under the α-recursive functions, and showed that for them all approaches are equivalent. In particular, for an admissible ordinal α, α-recursiveness means $\Sigma_1^{L_\alpha}$ definability, and L_α is the smallest standard model of KP of ordinal α. This relates effective Set Theory to Recursion Theory on the ordinals, and provides finer versions of various results of this section. The first two admissible ordinals are ω and ω_1^{ck} (see p. 385).

While admissible sets turn out to be nice domains only for elementary Recursion Theory (see p. 421), many deeper parts of Recursion Theory carry over to any admissible ordinal (see Chong [1984] for a detailed treatment). In particular, Post's Problem always admits a positive solution (Sacks and Simpson [1972]).

Admissible ordinals will provide, in Volume III, a uniform way of describing a number of classes of subsets of ω, and will also be useful from a methodological point of view, for a better understanding of which properties of ω are used in proofs of single results.

Relativizations $\star$

As for the Arithmetical and Analytical Hierarchies, the work done for the Constructible Hierarchy can also be relativized to a given set A. There are two natural ways of doing this, corresponding to adding A as a constant or as a predicate.

Definition IV.4.28 (Hajnal [1956], Levy [1960a])

1. *The class $L[A]$ is defined like L, by starting with $A \cup Tc(A)$ in place of $\emptyset$.*

2. *The class $L(A)$ is defined like L by allowing, at successor stages, also parameters over A.*

From a set-theoretical point of view, the two ways are not equivalent. The first produces the smallest standard model of ZF containing A and all the ordinals, and it does not necessarily satisfy the Axiom of Choice, without further assumptions on A. The second produces the smallest standard model M of ZFC containing $M \cap A$ and all the ordinals. From our point of view, however, they are equivalent, since we only consider relativizations to sets $A \subseteq \omega$.

Definition IV.4.29 *A is **constructible from** B $(A \leq_L B)$ if it is in $L[B]$. A is **constructibly equivalent** to B $(A \equiv_L B)$ if $A \leq_L B$ and $B \leq_L A$.*

Exercises IV.4.30 a) *If A is constructible, then $A \leq_L B$ for any B.*
  b) *If $A \leq_L B$ and B is constructible, so is A.*

Note that $\leq_L$ is reflexive and transitive, and thus $\equiv_L$ is an equivalence relation.

Definition IV.4.31 *The equivalence classes of sets w.r.t. constructibility equivalence are called* **L-degrees**, *and* $(\mathcal{D}_L, \leq)$ *is the structure of L-degrees, with the partial ordering $\leq$ induced on them by $\leq_L$.*

Of course, not much can be said about the structure of L-degrees in ZFC alone: *the assertion that there is exactly one L-degree (containing all subsets of ω) is consistent with ZFC*, since so is $V = L$, and thus no result contradicting it can be proved in ZFC. The study of the structure of L-degrees can thus be pursued along two different paths: one is to prove consistency results, the other to introduce new axioms, and study the structure under them. We will follow both paths, in Volume III.

Chapter V

Turing Degrees

The first four chapters of this book introduced the basic notions and methods of Recursion Theory. It is now time to put all this machinery to good use, and begin a systematic study of the continuum from a recursion-theoretical point of view. While a great deal of Recursion Theory, as seen in previous and following chapters, has a more limited scope and analyzes increasingly bigger, but always countable, subsets of $\mathcal{P}(\omega)$, this and the next chapters attempt a global attack, by trying to characterize the structure of $\mathcal{P}(\omega)$ in terms of degrees. Here we study Turing degrees, by developing a paradigm that will later be followed for the study of many other notions of degrees.

The main results we will obtain are of two kinds:

Algebraic. We will look for results that describe the algebraic structure of degrees, as a partially ordered set. We will ask natural questions about this structure, concerning: linearity, density, embeddability of partial orderings, ideals, nontrivial automorphisms, and so on. Despite the fact that (contrary to the case of m-degrees) there is no complete characterization yet, in Section 7 we will be able to derive a number of interesting global results.

Computational. The first natural question here is the decidability of theory of degrees, i.e. the existence of an effective method that would tell, of any given sentence in the language of degrees, whether it is true or not. We will be able to show in Section 7 that such a method does not exist, and that actually it is as difficult to decide first-order sentences about the order of degrees as it is to decide analytical sentences of arithmetic. In precise words, *the theory of degrees is recursively isomorphic to Second-Order Arithmetic.* Knowing that the theory is undecidable, we may look for partial decidability results: we will discuss the fact that it is possible to decide the theory of two-quantifier sentences, but not that of three quantifiers.

The division in sections is methodological and it reflects the different, increasingly more powerful tools used in the proofs. We start in Section 2 with the **finite extension method**, which is just a version of Baire Category, the relationship being analyzed in Section 3. We discuss the fact that some results, like the existence of minimal degrees, cannot be proved by finite extensions. We then introduce more powerful tools, namely the **coinfinite extension method** in Section 4, and the **tree method** in Section 5. The work culminates in Section 7, where the global structure of T-degrees is investigated, and the quoted undecidability results are proved.

V.1   The Language of Degree Theory

We have defined the structure of Turing degrees in II.3.3, and here we begin a close look at it. To improve readability, as well as to follow common use notations, we adopt (in this chapter) a number of conventions:

1. *degree will always mean T-degree*

2. *sets will be used as representatives for the degrees*, and this is possible because a function and its graph are T-equivalent, and thus each degree contains a set

3. *a set will be identified with its characteristic function*

4. *partial recursive functions with oracle A will sometimes be denoted by $\{e\}^A$, in place of φ_e^A*

5. *we will use lowercase Greek letters for strings*, i.e. for partial functions with finite domains (see V.2.1)

6. *degrees will be denoted by lowercase boldface letters*

7. $a < b$ *will mean* $a \leq b \wedge a \neq b$

8. $a|b$ *will mean* $a \nleq b \wedge b \nleq a$, i.e. that a and b are incomparable.

The join operator

Since $(\mathcal{D}, \leq)$ is a partially ordered set, it makes sense to talk about the l.u.b. and the g.l.b. of a pair of degrees.

Definition V.1.1 *Given a and b, then:*

1. $a \cup b$ (also called the **join** of a and b) is their least upper bound

2. $a \cap b$ is their greatest lower bound.

We now show that the l.u.b. always exists, and it is induced by the disjoint sum of two sets. The g.l.b. of two degrees may instead exist or not (V.2.16 and V.4.7).

Recall that the disjoint sum of two sets is defined as

$$A \oplus B \Leftrightarrow \{2x : x \in A\} \cup \{2x + 1 : x \in B\}.$$

Notice that $A \oplus B$ is obviously invariant under T-equivalence, and thus it makes sense to consider it as an operator on T-degrees.

Proposition V.1.2 *Given a and b, $a \cup b$ is the degree of $A \oplus B$, for any $A \in a$ and $B \in b$.*

Proof. Clearly $A, B \leq_T A \oplus B$, since

$$x \in A \Leftrightarrow 2x \in A \oplus B \quad \text{and} \quad x \in B \Leftrightarrow 2x + 1 \in A \oplus B.$$

And if $A, B \leq_T C$, let $A \simeq \varphi_a^C$ and $B \simeq \varphi_b^C$. Then $A \oplus B \simeq \varphi(x)$, where φ is the function partial recursive in C such that

$$\varphi(z) \simeq \begin{cases} \varphi_a^C(x) & \text{if } z = 2x \\ \varphi_b^C(x) & \text{if } z = 2x + 1. \end{cases}$$

Thus $A \oplus B$ is the l.u.b. of A and B w.r.t. $\leq_T$. $\quad\square$

Notice that *the join operator is definable in $(\mathcal{D}, \leq)$*, as

$$a \cup b = c \Leftrightarrow a \leq c \wedge b \leq c \wedge (\forall d)(a \leq d \wedge b \leq d \rightarrow c \leq d).$$

We can thus freely add $\cup$ to the structure $(\mathcal{D}, \leq)$ simply recalling, when needed, that one universal quantifier is needed to express it.

We can also introduce a more general notion of join operation, as follows:

Definition V.1.3 *Given a countable family $\{A_n\}_{n \in I}$, with $I \subseteq \omega$, then*

$$\oplus_{n \in I} A_n = \{\langle n, x \rangle : x \in A_n \wedge n \in I\}.$$

This notion is going to be useful, but we should be aware of the fact that it is not invariant under T-equivalence.

Exercises V.1.4 (Kleene and Post [1954]) a) *If $I = \{a, b\}$ then $(\oplus_{n \in I} A_n)$ is recursively equivalent to $A_a \oplus A_b$.*

b) *If I is finite then the degree of $(\oplus_{n \in I} A_n)$ is uniquely determined by the degrees of the A_n's, but this fails if I is infinite.* (Hint: let A be any set, and $x \in A_n \Leftrightarrow n \in A$. Then the A_n's are all recursive, but $(\oplus_{n \in \omega} A_n)$ has the degree of A.)

c) *If $A_n \leq_T B_n$ uniformly in $n \in I$, then $(\oplus_{n \in I} A_n) \leq_T (\oplus_{n \in I} B_n)$.* (Hint: the hypothesis means that there is a recursive function f such that $A \simeq \varphi_{f(n)}^{B_n}$ if $n \in I$.)

The jump operator

The fact that $\mathcal{K}$ is a complete r.e. set, and thus it embodies one existential number quantifier, is now generalized with the introduction of a degree operation that corresponds to number quantification.

Definition V.1.5 *The* **jump of a set** A *is the relativization of* $\mathcal{K}$ *to* A, *defined as*

$$x \in A' \Leftrightarrow x \in \mathcal{W}_x^A \Leftrightarrow \{x\}^A(x)\downarrow \ .$$

By relativization of the properties of $\mathcal{K}$ (recall the general comments on p. 177), we get the following properties of the jump operator:

1. *A is r.e. in B if and only if $A \leq_1 B'$*
   From the fact that A is r.e. if and only if $A \leq_1 \mathcal{K}$ (p. 320).

2. *$A \leq_1 A'$*
   This is just a particular case of 1.

3. *A' is not recursive in A*
   From the fact that $\mathcal{K}$ is not recursive (II.2.3).

The next result exhibits a connection between Turing and one-one reducibilities, through the jump operator.

Proposition V.1.6 *$A \leq_T B$ if and only if $A' \leq_1 B'$.*

Proof. If $A \leq_T B$, from A' r.e. in A we have A' r.e. in B: by fact 1 above, then $A' \leq_1 B'$.
   To show $A \leq_T B$ we just have to prove that both A and $\overline{A}$ are r.e. in B or, by fact 1 above, that $A \leq_1 B'$ and $\overline{A} \leq_1 B'$. Both follow from $A' \leq_1 B'$ by transitivity, the former because $A \leq_1 A'$, the latter because $\overline{A} \leq_1 A'$ (since $\overline{A}$ is r.e. in A). $\square$

In particular this shows that the jump operator is invariant under T-equivalence, and thus it induces an operator on degrees.

Definition V.1.7 (Kleene and Post [1954]) *The* **jump** $\mathbf{a}'$ *of* $\mathbf{a}$ *is the degree of A', for any $A \in \mathbf{a}$, and $(\mathcal{D}, \leq,')$ is the structure of Turing degrees, with the partial ordering $\leq$ and the jump operator* $'$.

Usually we simply write $\mathcal{D}$ when we refer to the structure of degrees without jump operator, and $\mathcal{D}'$ when we include the jump operator. We will see in XI.5.17 that *the jump operator is definable in* $\mathcal{D}$.

Recall that we defined $\mathbf{0}$ as the degree of the recursive sets, and $\mathbf{0}'$ as the degree of $\mathcal{K}$. This is consistent with the present notation, since $\mathcal{K}$ is by definition the jump of the emptyset, and $\emptyset \in \mathbf{0}$.

The jump operator can be iterated, both on sets and on degrees.

Definition V.1.8 *For any set A, the n-th jump $A^{(n)}$ of A is defined inductively as follows:*

$$A^{(0)} = A \qquad A^{(n+1)} = (A^{(n)})'.$$

The n-th jump $a^{(n)}$ of a degree a is the degree of $A^{(n)}$, for any set $A \in a$.

Of course $a^{(n)}$ can also be directly defined by induction, as follows:

$$a^{(0)} = a \qquad a^{(n+1)} = (a^{(n)})'.$$

We will use $\emptyset^{(n)}$ as our usual representative for $\mathbf{0}^{(n)}$.

Exercises V.1.9 a) $\emptyset^{(n+1)}$ *is Σ_{n+1}^0-complete.* (Hint: by iteration of the fact that $\emptyset' = \mathcal{K}$ is Σ_1^0-complete.)

b) *A set A is arithmetical if and only if it is T-reducible to $\emptyset^{(n)}$ for some n, i.e. if its degree is bounded by some $\mathbf{0}^{(n)}$.*

There is also a transfinite operation of jump, that can be defined using the infinite join of the finite iterations.

Definition V.1.10 (Kleene and Post [1954]) *For any set A, the ω-jump $A^{(\omega)}$ of A is defined as*

$$A^{(\omega)} = \oplus_{n \in \omega} A^{(n)}.$$

For any degree a, the ω-jump $a^{(\omega)}$ of a is the degree of $A^{(\omega)}$, for any set $A \in a$.

Exercises V.1.11 a) *If $A \leq_T B$ then $A^{(\omega)} \leq_1 B^{(\omega)}$. In particular, the ω-jump is well-defined on degrees.* (Hint: if $A \leq_T B$ then $A \leq_1 B'$.)

b) *The converse fails.* (Hint: A and A' have the same ω-jump.)

First properties of degrees

We now state some simple but basic facts about $\mathcal{D}$ and $\mathcal{D}'$.

Proposition V.1.12 (Kleene and Post [1954]) *As a partially ordered structure $\mathcal{D}$ is an uppersemilattice of cardinality $2^{\aleph_0}$ with a least but no maximal element. Moreover, each element has $2^{\aleph_0}$ successors and at most countably many predecessors.*

Proof. We have already noted that the join operator provides with a least upper bound operator, and thus $\mathcal{D}$ is an uppersemilattice. The least element is the degree 0 of the recursive sets. There can be no maximal element, since the jump operator is a strictly increasing operator.

Given a set A, for any set B such that $B \leq_T A$ there is a number $e \in \omega$ such that $B \simeq \varphi_e^A$, and thus there can be at most countably many sets T-reducible to A. This means that a degree has at most countably many predecessors, and it contains at most countably many sets. In particular, there are at least $2^{\aleph_0}$ degrees.

There cannot be more than $2^{\aleph_0}$ degrees, because we can define a one-one map from the degrees into $\mathcal{P}(\omega)$, by choosing a set from each degree.

On the other hand, the map $B \mapsto A \oplus B$ is a one-one map from $\mathcal{P}(\omega)$ into $\{X : A \leq_T X\}$: this means that there are at least $2^{\aleph_0}$ sets to which A is T-reducible, and hence there are at least $2^{\aleph_0}$ successors of any degree. Then there are exactly $2^{\aleph_0}$ ones. $\quad\square$

Note that 0, being the least degree, is definable in D as

$$c = 0 \Leftrightarrow (\forall x)(c \leq x).$$

Thus we can freely add 0 to the structure $\mathcal{D}$ simply recalling, when needed, that one universal quantifier is needed to express it.

Definition V.1.13 *The* **cone above** a *is the set*

$$\mathcal{D}(\geq a) = \{b : b \geq a\}.$$

The properties just proved for $\mathcal{D}$ are obviously true for $\mathcal{D}(\geq a)$ as well, and we may wonder whether the structure $\mathcal{D}$ is **homogeneous**, in the sense of being isomorphic (or at least elementarily equivalent, i.e. satisfying the same first-order properties) to all of its cones. If it were so, this would justify the principle of relativization (p. II.3). As it happens, this actually fails (V.7.13), and thus relativization of a result has to be verified in each case.

A connection between the join and jump operators is given by the following result.

Proposition V.1.14 (Kleene and Post [1954]) *For any pair of degrees* a *and* b,
$$a \leq b \Rightarrow a' \leq b' \quad and \quad a' \cup b' \leq (a \cup b)'.$$

Proof. We know that if $A \leq_T B$ then $A' \leq_1 B'$, and hence $A' \leq_T B'$: this proves the first part.

Since $a \cup b$ is the l.u.b. of a and b, we have $a \leq a \cup b$ and hence $a' \leq (a \cup b)'$. Similarly $b' \leq (a \cup b)'$. Since $a' \cup b'$ is the l.u.b. of a' and b', it follows that $a' \cup b' \leq (a \cup b)'$.   □

We will see (V.2.27) that the result is best possible, since every possibility compatible with the properties above can be realized.

The Axiom of Determinacy ⋆

We have argued above that the cardinality of $\mathcal{D}$ is $2^{\aleph_0}$, by providing a one-one map from $\mathcal{D}$ to $\mathcal{P}(\omega)$, but this required the Axiom of Choice (we chose a set from each degree). Yates [1970] has proved that the use of this axiom is not avoidable, since it is consistent with ZF (plus the Axiom of Dependent Choices) that there is no one-one map from $\mathcal{D}$ into $\mathcal{P}(\omega)$. We will see the proof in Volume III, but now we introduce a new axiom that implies the same result.

Suppose I and II are two players, that alternately play integers:

$$\begin{array}{lcccc} \text{I} & a_0 & & a_2 & & \cdots \\ \text{II} & & a_1 & & a_3 & & \cdots \end{array}$$

Putting together their moves, we get a function $f(n) = a_n$. To decide who is going to win the play, we choose ahead of time a set $\mathcal{A}$ of total functions, and we say that I wins if, at the end of the game, the function f is in $\mathcal{A}$, and II wins otherwise. Thus any set $\mathcal{A}$ defines a game $G(\mathcal{A})$.

One of the two players might not only win, but even have a **winning strategy**, i.e. a way to decide his or her moves so that, independently of how the other player moves, he or she will win. E.g., if $\mathcal{A}$ is the set of functions with value 0 for odd arguments a winning strategy for player II simply consists of playing 0 when it is his or her turn. Of course only one of the two players can have a winning strategy, but it is conceivable that there are games for which none has one.

Definition V.1.15 (Gale and Stewart [1953], Mycielski and Steinhaus [1962]) *A* **winning strategy** *for player I in the game $G(\mathcal{A})$ is a function $w_I : Seq \to \omega$ such that I wins the game if he consistently plays following the strategy. In other words, any function f such that $f(2n) = w_I(\hat{f}(2n-1))$ is in $\mathcal{A}$, independently of the values of f for odd arguments. A winning strategy for player II is defined similarly.*

The game $G(\mathcal{A})$ is **determined** *if one of the two players has a winning strategy in it, and the* **Axiom of Determinacy** *is the assertion AD that every game $G(\mathcal{A})$ is determined.*

The interest of the Axiom of Determinacy for Degree Theory lies in the next result.

Theorem V.1.16 (Martin [1968]) *If AD holds, then every set of degrees either contains a cone or is disjoint from a cone.*

Proof. Given a set $\mathcal{A}$ of degrees, let $\mathcal{A}^*$ consist of the sets whose degrees are in $\mathcal{A}$. Suppose there is a winning strategy for I in $G(\mathcal{A}^*)$. The strategy, being a function from (sequence) numbers to numbers, has a certain degree a: we show that every degree $b \geq a$ must be in $\mathcal{A}$. Choose any function g in b, and let player II play according to g (i.e. the n-th move of II is $g(n)$). Let also player I play according to the winning strategy. Since the moves of I are determined by a function of degree a, and those of II by a function of degree b, and $a \leq b$, the final outcome will be a function of degree b. But since I was following a winning strategy he wins the game, and thus the outcome must be in $\mathcal{A}^*$: this means that its degree, i.e. b, must be in $\mathcal{A}$.

Similarly, if player II has a winning strategy then all degrees above the one in which the strategy lies must be in $\overline{\mathcal{A}}$, since II wins the game.   □

This results has a number of interesting consequences for degrees. An astonishing one is the following.

Theorem V.1.17 (Martin) *If AD holds, every map from degrees to sets is constant on a cone.*

Proof. Let $F : \mathcal{D} \to \mathcal{P}(\omega)$ be given. $F(x)$ is a set of numbers, for each degree x. Let $\mathcal{A}_n = \{x : n \in F(x)\}$: this is a set of degrees, and thus there is a degree a_n which is the base of a cone contained either in $\mathcal{A}_n$ or in $\overline{\mathcal{A}}_n$. This means that n is in $F(x)$ either for all x's above a_n or for none, and thus the behavior of $F(x)$ on n is fixed on the cone above a_n. Consider a degree a above all the a_n's: it exists, because there are only countably many of them (e.g. take the infinite join of sets $A_n \in a_n$). Then F must be constant on the cone with a as a base, since this time its behavior is fixed for any n.   □

Corollary V.1.18 *If AD holds, there is no one-one map from $\mathcal{D}$ into $\mathcal{P}(\omega)$.*

This contradicts the Axiom of Choice, which allows to choose a set from each degree, thus producing a one-one map from $\mathcal{D}$ into $\mathcal{P}(\omega)$. In particular, *the Axiom of Determinacy is inconsistent with the Axiom of Choice* (Gale and Stewart [1953]).

It should be noted that the proof of the theorem above used a weak form of the Axiom of Choice, by actually picking up representatives in countably many degrees. This weak form is actually a consequence of AD, and can thus be freely used. To see why, note that we know that $(\forall n)(\exists A)(A \in a_n)$. Consider the game in which I constantly plays n while II plays the characteristic function of a set A, and such that II wins if $A \in a_n$. Player I does not have a winning strategy, since there is an $A \in a_n$ that can be played by II. Then, by AD, II has a winning strategy, which produces a function choosing, for every initial move n of I, a set in a_n.

We might read the corollary as saying that *if AD holds there are more degrees than sets*, but we probably should not, when Choice is not present, compare cardinalities by using one-one maps. If we use instead onto maps, these peculiarities disappear.

Exercise V.1.19 *Without using the Axiom of Choice, there are onto maps between $\mathcal{D}$ and $\mathcal{P}(\omega)$. (Hint: in one direction, send a set to its degree. In the other direction, build a tree of sets with pairwise incomparable degrees, see V.2.11, then send the degree of each branch to the branch itself, and all other degrees to a fixed set. This suffices, because $\mathcal{P}(\omega)$ is isomorphic to a tree.)*

Another interesting consequence of V.1.16 is the following.

Proposition V.1.20 (Martin) *If AD holds, there is a countably additive measure on $\mathcal{D}$.*

Proof. Recall that a countable additive measure is any function μ defined on subsets of $\mathcal{D}$ such that

1. $\mu(\emptyset) = 0$ and $\mu(\mathcal{D}) = 1$

2. if $X \subseteq Y$ then $\mu(X) \le \mu(Y)$

3. $\mu(\{a\}) = 0$

4. if $\{X_n\}_{n \in \omega}$ is a collection of pairwise disjoint sets of degrees, then

$$\mu\left(\bigcup_{n \in \omega} X_n\right) = \sum_{n \in \omega} \mu(X_n).$$

It is then enough to define

$$\mu(\mathcal{A}) = \begin{cases} 1 & \text{if } \mathcal{A} \text{ contains a cone} \\ 0 & \text{otherwise.} \end{cases}$$

The only condition that requires some arguing is the last one. There are two cases:

- *some X_{n_0} contains a cone*
  Then so does $\bigcup_{n \in \omega} X_n$, and $\mu(\bigcup_{n \in \omega} X_n) = 1$. Moreover no X_n with $n \ne n_0$ contains a cone, because the intersection of two cones is not empty, while the X_n are disjoint. Then $\sum_{n \in \omega} \mu(X_n) = \mu(X_{n_0}) = 1$.

- *no X_n contains a cone*

Then $\sum_{n \in \omega} \mu(X_n) = 0$. Moreover $\overline{X}_n$ contains a cone for each n, by V.1.16, and then so does their intersection, as in V.1.17. But

$$\bigcap_{n \in \omega} \overline{X}_n = \overline{\bigcup_{n \in \omega} X_n},$$

and thus $\bigcup_{n \in \omega} X_n$ does not contain a cone: hence $\mu(\bigcup_{n \in \omega} X_n) = 0$. □

Since the countable ordinals are represented by well-ordered sets of natural numbers, it immediately follows that *if AD holds then $\aleph_1$ is a measurable cardinal* (Solovay), by giving a set $\mathcal{A}$ of countable ordinals the same measure as the set of degrees containing well-orderings with ordinal in $\mathcal{A}$.

One might wonder why we should accept an axiom like AD. The fact is that, as we will prove in following chapters, various special cases of it can be proved in ZFC, and thus some versions of the results proved above simply hold. More precisely, Martin [1975] has showed in ZFC that *every Borel game is determined*: thus every Borel set of degrees either contains a cone or it is disjoint from a cone, and every Borel map from $\mathcal{D}$ to sets is constant on a cone. We will investigate the effect of restricted versions of AD on Recursion Theory in Volume III.

V.2   The Finite Extension Method

Degrees are represented by sets, and thus existential properties of degrees can be proved by building appropriate sets. Since building sets by successive approximations will be our main concern in this chapter, we set up an efficient notation to deal with the problem. Recall that a set is identified with its characteristic function, which is just a sequence of 0's and 1's.

Definition V.2.1 *A* **string** *is a partial function $\sigma : \omega \to \{0,1\}$ with finite domain. If the domain of σ is an initial segment of ω, we call σ an* **initial segment**, *and let $|\sigma|$ be its length. Given two strings σ and σ', then:*

1. σ' is an **extension** *of σ if it extends it as a partial function, i.e.*

$$\sigma(x) \downarrow \; \Rightarrow \; \sigma'(x) \downarrow \wedge \sigma(x) \simeq \sigma'(x).$$

2. σ and σ' are **incompatible** *if they differ on some argument on which they are both defined.*

Initial segments will be identified, when needed, with sequences of 0's and 1's, or with their sequence numbers. Then $\sigma * \tau$ will be the juxtaposition of the two sequences σ and τ, and $ln(\sigma)$ the length of σ.

Strings are finite objects, and thus they can be coded by natural numbers. In particular, *there is a canonical ordering of strings*, to which we will implicitly refer when talking of 'smallest string satisfying a given property'.

Recall that, by compactness and monotonicity (II.3.13), convergent computations with oracle A can be approximated by convergent computations having as oracle some finite approximation of A, which can then be coded by a string. In other words, using the notations for this chapter,

$$\{e\}^A(x) \simeq y \Leftrightarrow (\exists \sigma \subseteq A)(\{e\}^\sigma(x) \simeq y).$$

It is exactly because of these continuity properties that we will be able to prove results about degrees.

Incomparable degrees

Looking at the partially ordered structure $\mathcal{D}$, the first problem that comes to mind is whether the order is really partial. The next result shows that this is the case, and it introduces our first method of proof.

Theorem V.2.2 (Kleene and Post [1954]) *There are two incomparable degrees.*

Proof. We want to build sets A and B such that $A \not\leq_T B$ and $B \not\leq_T A$. The proof consists of two steps: first we break down these two global conditions into infinitely many local conditions, and then satisfy each of them by a finite action.

Note that $A \leq_T B$ means that, for some e, $A \simeq \{e\}^B$. Similarly for $B \leq_T A$. We can thus rewrite the global conditions as the sequence:

$$R_{2e} \quad : \quad A \not\simeq \{e\}^B$$
$$R_{2e+1} \quad : \quad B \not\simeq \{e\}^A.$$

The construction of A and B is by finite initial segments. We will let $A = \bigcup_{s \in \omega} \sigma_s$ and $B = \bigcup_{s \in \omega} \tau_s$. At each stage of the construction we will take care of one requirement, once and for all. We begin by setting $\sigma_0 = \tau_0 = \emptyset$. At stage $s + 1$, suppose σ_s and τ_s are given.

- If $s = 2e$ then we satisfy R_{2e}
  Let x be the first element such that $\sigma_s(x)$ does not converge: this means that we have not yet decided whether x has to be in A or not. We will decide this now, and we will use x to witness that $A \not\simeq \{e\}^B$. In other

words, we will arrange that $A(x) \not\simeq \{e\}^B(x)$: the idea is obviously to diagonalize, i.e. to make A on x different from $\{e\}^B$ on x. But since we have not yet defined B, we do not even know whether $\{e\}^B(x)$ converges. What we do know, by compactness, is that if it will converge in the end so will $\{e\}^\tau(x)$, for some $\tau \subseteq B$. Since $\tau_s \subseteq B$ by construction, if such a τ will exist then it will be compatible with τ_s (since B will extend both). We may also suppose, by monotonicity, that τ will actually extend τ_s. At this point we may turn things around, and see whether there is any string $\tau \supseteq \tau_s$ such that $\{e\}^\tau(x)$ converges.

If such a string does not exist, we know that $\{e\}^B(x)$ will be undefined, and since $A(x)$ will certainly be defined (being a total function), it does not matter what we do. E.g., let σ_{s+1} be the smallest initial segment extending σ_s and defined on x. Since nothing has to be done on B, let $\tau_{s+1} = \tau_s$.

If however such a τ exists, then it might be the case that $\{e\}^B(x)$ will converge, and we want to dispose of this case too. What we can do, since we are also building B, is to insure that B will extend τ: then $\{e\}^B(x) \simeq \{e\}^\tau(x)$ by monotonicity. To insure this, we only have to let $\tau_{s+1} = \tau$. But now we have to be careful with A: we know that $\{e\}^B(x)$ converges, and thus we want $A(x)$ different from it. Then we let σ_{s+1} be the smallest initial segment extending σ_s, and such that

$$\sigma_{s+1}(x) = 1 - \{e\}^B(x).$$

- If $s = 2e + 1$ then we satisfy R_{2e+1}
  The construction is the same in this case, simply with the roles of A and B interchanged.

Notice that there is no reason to define the strings as initial segments, except for being sure that a set is obtained as their union. If we did not do this, then we could just say that A is any set extending $\bigcup_{s \in \omega} \sigma_s$, and similarly for B. Also, the only reason to specify that we choose particular x's and particular σ's, when many possibilities are open, is simply to have the construction as effective as possible: this is going to be useful in the corollary below, where an evaluation on the complexity of the construction is needed. □

We can distinguish between **global results**, that hold for the whole structure $\mathcal{D}$, and **local results**, that hold instead in the degrees $\mathcal{D}(\leq a)$ below a given degree a. Not all results can be localized below a given nonrecursive degree, and those that do may require more sophisticated proofs. But the simple analysis of a proof will usually provide an upper bound to the noneffective parts in it, and hence a local version of the global result (although not necessarily the sharpest one).

Corollary V.2.3 *There are incomparable degrees below $0'$.*

Proof. We only have to show that the sets A and B built above are recursive in $\mathcal{K}$. This follows from the fact that the only nonrecursive step in the construction is to decide questions of the form: 'is there a string σ' extending a given string σ and such that $\{e\}^{\sigma'}(x)$ converges, for a given x?' This is easily seen to be recursively enumerable: it is enough to dovetail the computations of $\{e\}^{\sigma'}(x)$ for all strings σ' extending σ (note that this last condition can be effectively checked, because strings are finite functions). Since $\mathcal{K}$ is m-complete, each of these questions can thus be answered recursively in $\mathcal{K}$. $\square$

Exercises V.2.4 a) *The existence of incomparable degrees follows from $2^{\aleph_0} \neq \aleph_1$, by cardinality considerations.* (Myhill [1961]) (Hint: each degree has only countably many predecessors, and thus a linear ordering of degrees must have cardinality at most $\aleph_1$. But there are $2^{\aleph_0}$ degrees.)

b) *The existence of incomparable degrees can be proved by set-theoretical considerations.* (Kreisel) (Hint: the existence of incomparable degrees is absolute for standard models of ZF^- by IV.3.25, being a Σ_1^1 statement. Thus it is enough to show that it holds in a model of ZF^-. By forcing, there is such a model in which the Continuum Hypothesis fails and thus, by part a), in which there are incomparable degrees.)

c) *The Continuum Hypothesis is equivalent to the assertion that there is a cofinal chain of degrees of order type $\aleph_1$, where cofinal means that each degree is bounded by some element of the chain.* (Hint: the downward closure of such a chain still has cardinality $\aleph_1$, because each degree has at most countably many predecessors.)

Embeddability results

Having shown that $\mathcal{D}$ is not linear, one immediately wonders about its possible complexity as a partial order. This can be measured by the quantity of partial orderings that can be embedded in the structure. We show in this subsection that, in this sense, $\mathcal{D}$ is quite complicated.

We could proceed by brute force and build, given any countable partial ordering, a set of degrees isomorphic to it. The constructions of these sets would not differ very much, and most of the work would be to reproduce the given partial orderings. We will instead isolate the recursion theoretical ideas in just one result, which will produce a kind of universal countable partial ordering, in which all the others are embedded. The idea consists in building degrees which are very independent, one from the others, in the sense that none is recoverable from the remaining ones: the universal partial ordering will then be given by all their possible combinations. The notion of independence is captured by the following definition.

Definition V.2.5 (Kleene and Post [1954]) *A set* $\{A_n\}_{n \in I}$ *is* **recursively independent** *if, for each* $n \in I$,

$$A_n \not\leq_T \oplus\{A_m : m \in I \wedge m \neq n\}.$$

Exercises V.2.6 a) *The set* $\{A\}$ *is recursively independent if and only if* A *is not recursive.*

b) $\{A, B\}$ *is recursively independent if and only if* A *and* B *are incomparable.*

c) *If* $\{A_n\}_{n \in I}$ *is recursively independent then the* A_n *are mutually incomparable, but not conversely.* (Hint: if $\{A, B, C\}$ is recursively independent, then $A \oplus B$, $B \oplus C$, and $A \oplus C$ are mutually incomparable but not independent, since e.g. $A \oplus B$ is reducible to $(A \oplus C) \oplus (B \oplus C)$.)

Proposition V.2.7 (Kleene and Post [1954]) *There exists a countable, recursively independent set.*

Proof. The proof differs only in the details from that of V.2.2. We have to build a countable sequence $\{A_n\}_{n \in \omega}$ of sets, and we just build a giant set A that puts them all together. We will then let

$$x \in A_n \Leftrightarrow \langle n, x \rangle \in A,$$

so that A can be thought of as the infinite join of the A_n's, and each A_n can be thought of as the n-th column of A. The requirements are:

$$R_{\langle e,n \rangle} \; : \; A_n \not\cong \{e\}^{\oplus_{m \neq n} A_m}.$$

We will build A by finite initial segments σ_s. We start with $\sigma_0 = \emptyset$. At stage $s + 1$, let σ_s be given. If $s = \langle e, n \rangle + 1$, then we attack $R_{\langle e,n \rangle}$. We choose a number $\langle n, x \rangle$ such that σ_s is not yet defined on it, so that in particular membership of x in A_n has not yet been decided. Note that in the end $\oplus_{m \neq n} A_m$ will be equal to A except for the n-th column, which will not contain any elements. Then we look for a string σ such that:

- σ is 0 on the elements of the n-th column, i.e. those of the form $\langle n, z \rangle$

- σ extends σ_s when this is defined, on elements which are not on the n-th column

- $\{e\}^{\sigma}(\langle n, x \rangle)\downarrow$.

If such a string does not exist then $\{e\}^{\oplus_{m \neq n} A_m}$ cannot converge on $\langle n, x \rangle$, and we let σ_{s+1} be any initial segment extending σ_s and defined on $\langle n, x \rangle$.

Otherwise, we take one such string σ, and define σ_{s+1} as any initial segment which extends σ_s, extends σ on elements not on the n-th column, and

$$\sigma_{s+1}(\langle n, x \rangle) = 1 - \{e\}^{\sigma}(\langle n, x \rangle). \quad \square$$

As we announced, the result is enough to prove all the embeddability results we need.

Exercise V.2.8 *There is a recursive partial ordering in which all the countable partial ordering are embeddable.* (Mostowski [1938]) (Hint: take the pairs of rationals, ordered as

$$(x,y) < (x',y') \Leftrightarrow x < x' \wedge y < y'.$$

This a planar version of the fact that every countable linear ordering is embeddable in the rationals. A direct construction is also possible, by having at any finite stage a finite approximation, and extending it by adding a new element in each possible way consistent with the partial ordering requirement.)

Theorem V.2.9 (Sacks [1963]) *Any countable partial ordering is embeddable in the degrees (below $0'$).*

Proof. By the exercise, it is enough to show that any recursive countable partial ordering $\preceq$ is embeddable in the degrees. We use the recursively independent set $\{A_n\}_{n \in \omega}$, and associate to each element a in the domain of $\preceq$ the set $\oplus_{n \preceq a} A_n$:

$$\langle n, x \rangle \in B_a \Leftrightarrow n \preceq a \wedge x \in A_n.$$

It remains to show that

$$a \preceq b \Leftrightarrow B_a \leq_T B_b,$$

so that the structure of the degrees of the B_a's is isomorphic to $\preceq$:

- if $a \preceq b$ then $B_a \leq_T B_b$
  Note that, since $\preceq$ is transitive, $B_a \subseteq B_b$ in this case, and thus

$$\langle n, x \rangle \in B_a \Leftrightarrow n \preceq a \wedge \langle n, x \rangle \in B_b.$$

  Thus $B_a \leq_T B_b$, because $\preceq$ is recursive.

- if $B_a \leq_T B_b$ then $a \preceq b$
  Suppose $a \not\preceq b$: then each element $n \preceq b$ is different from a. It follows that $B_b \leq_T (\oplus_{n \neq a} A_n)$:

$$\langle n, x \rangle \in B_b \Leftrightarrow n \preceq b \wedge \langle n, x \rangle \in (\oplus_{n \neq a} A_n).$$

  But $A_a \leq_T B_a$, since $a \preceq a$, and thus it cannot be $B_a \leq_T B_b$ otherwise, by transitivity, $A_a \leq_T (\oplus_{n \neq a} A_n)$, contradicting the recursive independence of the A_n's.

If we let A_n be the set constructed in V.2.9, then $\oplus_{n \in \omega} A_n$ is recursive in $\mathcal{K}$, and hence so are all the B_a's. Thus $\preceq$ is actually embeddable in the degrees below $\mathbf{0}'$.   □

A sentence in the language of $\mathcal{D}$ is a **one-quantifier sentence** if it is logically equivalent to a sentence in prenex normal form with a prefix consisting of quantifiers of the same type, and a matrix consisting of a Boolean combination of atomic formulas of the form $x \leq y$.

Corollary V.2.10 (Lerman [1972]) *The one-quantifier sentences of $\mathcal{D}$ and $\mathcal{D}(\leq 0')$ admit a decision procedure.*

Proof. It is enough to decide the existential sentences. But such a sentence simply asserts the existence of finitely many elements $x_1, \ldots, x_n$ in a certain order relationship. Since any countable (and hence any finite) partial ordering is embeddable in $\mathcal{D}$, the sentence is true in $\mathcal{D}$ if and only if it is consistent with the fact that $\leq$ is a partial ordering.   □

Exercises V.2.11 Independent sets of degrees. Since the infinite join is not a degree-theoretical operation, a set of degrees $\mathcal{A}$ is called independent if no element in it is bounded by the l.u.b. of some finite subset of $\mathcal{A}$.

a) *There is a countable independent set of degrees.* (Kleene and Post [1954]) (Hint: see V.2.7.)

b) *There is an independent set of degrees of cardinality $2^{\aleph_0}$.* (Sacks [1961]) (Hint: build a tree of sets, each branch of which is not recursive in any finite join of different branches.)

Sacks [1961] has proved that *every partial ordering with one of the following properties is embeddable in $\mathcal{D}$*:

1. *cardinality $2^{\aleph_0}$ and finite predecessor property*

2. *cardinality $\aleph_1$ and countable predecessor property*

3. *cardinality $2^{\aleph_0}$, countable predecessor property and $\aleph_1$ successor property.*

In particular, the last two conditions are equivalent and best possible, if the Continuum Hypothesis is assumed. The best absolute result would clearly be to prove the embeddability of any partial ordering with

4. cardinality $2^{\aleph_0}$ and countable predecessor property

but it is unknown whether this holds.

The splitting method

We have completely decided the one-quantifier theory of $\mathcal{D}$. We thus start to deal with two-quantifier sentences, that will be discussed in general on p. 490. A typical problem is to get a degree incomparable with a given one. Here the game is more subtle than the one in V.2.2: there we could build both sets, while here one of them is given to us and we have no control over it. The technique used in the solution to the problem is going to be extremely useful, and will be exploited in many different situations. The idea is natural, and it is a refinement of the one already used: since we control only one of the two sides, but we still need two possibilities to be able to diagonalize, we look for both of them on the side we control.

Definition V.2.12 *Two strings σ_1 and σ_2 are an **e-splitting** if, for some x, $\{e\}^{\sigma_1}(x)$ and $\{e\}^{\sigma_2}(x)$ both converge, and are different. In this case we say that σ_1 and σ_2 **e-split** on x.*

Theorem V.2.13 (Kleene and Post [1954]) *Given any nonrecursive degree $\mathbf{b}$, there is a degree $\mathbf{a}$ incomparable with it.*

Proof. Let B be nonrecursive: we want A such that

$$
\begin{aligned}
R_{2e} &\quad : \quad A \not\equiv \{e\}^B \\
R_{2e+1} &\quad : \quad B \not\equiv \{e\}^A.
\end{aligned}
$$

The two types of requirements look the same, but are very different: B is given ahead of time, while A is constructed. Thus they require different actions.

As usual, we will let $A = \bigcup_{s \in \omega} \sigma_s$. We start with $\sigma_0 = \emptyset$. At stage $s + 1$, let σ_s be given.

- If $s = 2e$ then we satisfy R_{2e}
  This is done by the method of the last subsection: we choose x on which σ_s is not yet defined, and we extend σ_s to a σ_{s+1} such that

$$
\sigma_{s+1}(x) = \begin{cases} 1 - \{e\}^B(x) & \text{if } \{e\}^B(x)\downarrow \\ 0 & \text{otherwise.} \end{cases}
$$

  Note that we simply ask whether $\{e\}^B(x)$ converges, instead of asking, as before, whether we may make it converge (because B is given).

- If $s = 2e + 1$ then we satisfy R_{2e+1}
  We now see if there are e-splitting extensions of σ_s, i.e. if we have two possible choices for $\{e\}^A$, on some element x. If this the case, choose τ_1 and τ_2 extending σ_s and e-splitting on some x. Then $\{e\}^{\tau_1}(x)$ and

$\{e\}^{\tau_2}(x)$ are both convergent, and one of them must be different from $B(x)$. Let σ_{s+1} be τ_i, where

$$B(x) \not\simeq \{e\}^{\tau_i}(x).$$

If such strings do not exist, let $\sigma_{s+1} = \sigma_s$.

We still have to argue that R_{2e+1} is really satisfied, also in the case that e-splitting extensions of σ_s do not exist. Here is where the nonrecursiveness of B comes into the game: we claim that, in this case, $\{e\}^A$ is either not total or recursive, and thus it must be different from B, which is total and nonrecursive.

Suppose $\{e\}^A$ is total: by compactness, given any x there must be a string $\tau \subseteq A$ such that $\{e\}^\tau(x)$ gives the right value. We can obviously suppose, by monotonicity, that τ extends σ_s. But since there are no e-splitting extending σ_s it must be the case that, as long as $\{e\}^\tau(x)$ converges, the value is unique. This then suggests a recursive method to compute $\{e\}^A$: given x, dovetail the possible computations $\{e\}^\tau(x)$, for all strings τ extending σ_s. The first converging one gives the right value. $\square$

From a topological point of view, the difference between the previous proof and the pure extension method of the last subsection is that *the splitting method requires something more than simple continuity*, since we use the fact that when a functional is constant on an open set (i.e. there is no splitting above a given string), then its value is recursive.

Notice also that a simple analysis of the proof does *not* give the result for degrees below $\mathbf{0}'$, since if $\mathbf{0} < \mathbf{b} < \mathbf{0}'$ we only get $\mathbf{a} \leq \mathbf{b}' \leq \mathbf{0}''$. The result still holds below $\mathbf{0}'$, but a different proof is needed. The theory of degrees below $\mathbf{0}'$ thus requires a separate study, which we will undertake in Chapter XI.

Exercises V.2.14 a) *Given a countable set of nonrecursive degrees, there is a degree incomparable with every element of the set.* (Shoenfield [1960]) (Hint: the requirements to satisfy are still countably many.)

b) *Every maximal antichain of degrees is uncountable.* (Hint: use part a) and Zorn's Lemma.)

c) *Every maximal independent set of degrees is uncountable.* (Hint: every countable independent set of degrees can be extended.)

The next definition introduces two notions of minimality, one for pairs and one for single degrees, which are going to be useful in the future.

Definition V.2.15

1. *Two degrees $\mathbf{a}$ and $\mathbf{b}$ form a* **minimal pair** *if they are nonrecursive and their g.l.b. is $\mathbf{0}$, i.e.*

$$(\forall \mathbf{c})(\mathbf{c} \leq \mathbf{a} \wedge \mathbf{c} \leq \mathbf{b} \Rightarrow \mathbf{c} = \mathbf{0}).$$

2. *A degree a is* **minimal** *if it is nonrecursive and there is no degree between 0 and a, i.e.*

$$(\forall c)(c \leq a \Rightarrow c = 0 \vee c = a).$$

Clearly two distinct minimal degrees form a minimal pair, but a minimal pair might consist of nonminimal degrees. The existence of minimal degrees is proved in Section 5, and it requires a different method of proof. With the present tools we can however prove the existence of minimal pairs.

Proposition V.2.16 *There exists a minimal pair of degrees. Actually, each nonrecursive degree is part of a minimal pair.*

Proof. Let B be nonrecursive: we want A such that

$$
\begin{array}{lll}
R_{2e} & : & A \not\simeq \{e\} \\
R_{2\langle e,i\rangle+1} & : & \{e\}^A \simeq \{i\}^B \simeq C \Rightarrow C \text{ recursive.}
\end{array}
$$

The first kind of requirement ensures that A is not recursive, while the second ensures that any set recursive in both A and B is recursive. They are satisfied as usual, the first by diagonalization, the second by the splitting method.

The idea to satisfy $R_{2\langle e,i\rangle+1}$ is the following. First one tries to make the requirement vacuously true, by having convergent computations such that

$$\{e\}^A(x) \not\simeq \{i\}^B(x).$$

This can be done, at a given stage n, by looking for e-splitting extensions of σ_s. If they do exist, we choose the one that produces a disagreement with $\{i\}^B$ on the x for which the two strings e-split. If they do not exist, we can argue that then $\{e\}^A$ is recursive if total, as in V.2.13. Note that in this case the hypothesis that B is nonrecursive plays no role in the proof, and it is needed only to satisfy the definition of minimal pair. $\square$

Corollary V.2.17 *There exists a pair of degrees with greatest lower bound.*

Exercises V.2.18 a) *Two nonrecursive sets A and B form a minimal pair if, for each a,*

$$\{a\}^A \simeq \{a\}^B \simeq C \Rightarrow C \text{ recursive.}$$

(Posner) This slightly simplifies the presentation of requirements. (Hint: choose an element z in one of the two sets but not in the other, say $z \in A - B$. Consider

$$\{a\}^X \simeq \begin{cases} \{e\}^X & \text{if } z \in X \\ \{i\}^X & \text{otherwise.} \end{cases}$$

Then $\{a\}^A \simeq \{e\}^A$ and $\{a\}^B \simeq \{i\}^B$.)

b) *There is a set of $2^{\aleph_0}$ degrees, such that each pair from it is a minimal pair.*
This will be greatly improved in V.5.12, which however requires a more difficult proof.
(Hint: build a tree of sets such that any two branches are a minimal pair.)

The proof of the next result is quite clever, and it manages to diagonalize
against uncountable sets of degrees by a counting trick. It is an example of the
method used by Sacks [1961] to obtain the results on uncountable embeddings
quoted on p. 462.

Definition V.2.19 *A set of degrees is:*

1. *a* **chain** *if it is linearly ordered*

2. *an* **antichain** *if its members are mutually incomparable nonzero degrees.*

The condition of nonrecursiveness for members of an antichain is imposed
only to avoid the trivial case $\{0\}$.

Proposition V.2.20 (Sacks [1961])

1. *Every countable chain of degrees is extendable, and thus every maximal
   chain has cardinality $\aleph_1$.*

2. *Every antichain of cardinality less than $2^{\aleph_0}$ is extendable, and thus every
   maximal antichain has cardinality $2^{\aleph_0}$.*

Proof. The first part is easy: given any countable set $\{A_n\}_{n\in\omega}$, the set
$(\oplus_{n\in\omega} A_n)'$ has degree strictly above all the degrees of the A_n's. Since every degree has at most countably many predecessors, the maximal length of a
chain is $\aleph_1$.

For the second part, let $\mathcal{A}$ be a set of degrees of cardinality $2^{\aleph_0}$, and such
that any pair in it is a minimal pair (it exists, by the last exercise above). Let
$\mathcal{B}$ be any set of nonrecursive degrees of cardinality less than $2^{\aleph_0}$. Note that

- the downward closure of $\mathcal{B}$ has cardinality less than $2^{\aleph_0}$, since each member of $\mathcal{B}$ has at most countably many predecessors

- for each member of $\mathcal{B}$ there is at most one member of $\mathcal{A}$ above it, since
  any degree below two distinct members of $\mathcal{A}$ must be recursive, and hence
  not in $\mathcal{B}$.

Then less than $2^{\aleph_0}$ members of $\mathcal{A}$, and hence not all of them, are comparable
with some member of $\mathcal{B}$. $\square$

We have noticed in V.2.14 that maximal antichains and maximal independent sets of degrees are uncountable. We have just strengthened the first result,

by showing that maximal antichains actually have cardinality $2^{\aleph_0}$. A similar strengthening of the second result is impossible, since *the assertion that maximal independent sets of degrees have cardinality $2^{\aleph_0}$ is independent of ZFC* (Sacks [1961], Groszek and Slaman [1983]).

Forcing the jump

After a first study of some elementary properties of $\mathcal{D}$, we now we turn our attention to $\mathcal{D}'$. To control the behavior of the jump operator a new idea is needed, which will be variously mixed with different methods. We first present it by itself, in the proof of the next result.

Proposition V.2.21 (Spector [1956]) *The jump operator is not one-one.*

Proof. It is enough to get $a > 0$ such that $a' = 0'$. To get A nonrecursive we diagonalize, while to have $A' \leq_T \mathcal{K}$ we want to decide, for each e, whether $\{e\}^A(e)\downarrow$. The requirements are then:

$$R_{2e} \quad : \quad A \not\cong \{e\}$$
$$R_{2e+1} \quad : \quad \text{decide whether } \{e\}^A(e)\downarrow .$$

Note that we do not consider the condition $\mathcal{K} \leq_T A'$, which is automatically satisfied because $\emptyset \leq_T A$, and hence $\mathcal{K} = \emptyset' \leq_T A'$.

As usual we build A by finite initial segments, starting with $\sigma_0 = \emptyset$. At stage $s + 1$, let σ_s be given.

- If $s = 2e$ then we satisfy R_{2e}
  Let x be the first element on which σ_s is undefined, and let σ_{s+1} be the smallest initial segment extending σ_s and such that

$$\sigma_{s+1}(x) = \begin{cases} 1 - \{e\}(x) & \text{if } \{e\}(x)\downarrow \\ 0 & \text{otherwise.} \end{cases}$$

- If $s = 2e + 1$ then we satisfy R_{2e+1}
  See if there is an initial segment σ extending σ_s such that $\{e\}^\sigma(e)\downarrow$. If so, let σ be the smallest one, and $\sigma_{s+1} = \sigma$. Otherwise, let $\sigma_{s+1} = \sigma_s$.

By construction A is not recursive. Moreover,

$$e \in A' \Leftrightarrow \{e\}^A(e)\downarrow \Leftrightarrow \{e\}^{\sigma_{2e+2}}(e)\downarrow$$

and, since the construction is recursive in $\mathcal{K}$, $A' \leq_T \mathcal{K}$. $\square$

Corollary V.2.22 *The jump operator is never one-one on its range.*

Proof. By relativization, given any C we get A such that $A \not\leq_T C$ and $(A \oplus C)' \leq_T C'$. But then $A \oplus C$ is a set with degree different from C, and jump C'. $\square$

Despite the absolute simplicity of the proof just given, the ideas involved in it are quite deep. The argument used is a forerunner and miniaturized version of the **forcing method**, introduced by Cohen [1963] to prove the independence of the Continuum Hypothesis, and which will play a major role in the next volumes. Its main idea is to *approximate truth by finite information*, although here we only consider one-quantifier sentences (through the jump operator which, as we know, corresponds to one quantifier). Note that it is always the case that, whenever a computation with oracle A converges, then it does because of a finite amount of information about A. But there is no reason to believe that the same holds for divergent computations as well, as the following example shows:

$$\{e\}^A(x) \simeq \begin{cases} 0 & \text{if } (\exists x)(x \in A) \\ \text{undefined} & \text{otherwise.} \end{cases}$$

What the proof given above accomplishes is to build a set A such that *not only the convergence, but also the divergence of any computations using A as an oracle can be determined by a finite amount of information on A*. Such sets are called **1-generic**, and will be studied in Section XI.2. They are particularly useful in the examination of $\mathcal{D}(\leq 0')$.

Exercises V.2.23 a) *Even without diagonalization, A is automatically nonrecursive.* (Posner and Epstein [1978]) (Hint: let σ_e determine whether $\{e\}^A(e)$ converges for all $A \supseteq \sigma_e$, or diverges for all $A \supseteq \sigma_e$. Let $A = \bigcup_{e \in \omega} \sigma_e$: if A is recursive, for some e,

$$\{e\}^\sigma(x) \simeq \sigma(x) \Leftrightarrow \sigma \not\subseteq A.$$

Now some extension of σ_e defined on e is contained in A, and hence makes $\{e\}$ converge on e, and some is not, and makes $\{e\}$ diverge on e, contradicting the choice of σ_e.)

b) *There is a recursively independent set of degrees with jump $\mathbf{0}'$.* (Hint: combine V.2.7 with the proof above.)

A natural question about the jump operator concerns its range, and it is answered by the next result.

Theorem V.2.24 Jump Inversion Theorem (Friedberg [1957b]) *The range of the jump operator is the cone $\mathcal{D}(\geq \mathbf{0}')$.*

Proof. We have already noted that $a' \geq 0'$, for any a. To get the converse, let C be a set such that $K \leq_T C$: we want to get A such that $A' \equiv_T C$. This splits into two separate conditions: $A' \leq_T C$, which will be satisfied as above, and $C \leq_T A'$, which will be achieved by coding C into A. The two strategies will be pursued alternately.

As usual we build A by finite initial segments, starting with $\sigma_0 = \emptyset$. At stage $s + 1$, let σ_s be given.

- If $s = 2e$ then we see if there is an initial segment $\sigma \supseteq \sigma_s$ such that $\{e\}^\sigma(e) \downarrow$. If so, we let $\sigma_{s+1} = \sigma$ for the smallest such, and otherwise we let $\sigma_{s+1} = \sigma_s$.

- If $s = 2e + 1$, we code the e-th element of C into A:

$$\sigma_{s+1} = \sigma_s * \langle C(e) \rangle.$$

The construction is recursive in C: the first step is recursive in $\mathcal{K}$, which is recursive in C by hypothesis, and the second uses C directly. Thus, since

$$e \in A' \Leftrightarrow \{e\}^{\sigma_{2e+1}}(e) \downarrow,$$

we have $A' \leq_T C$.

The construction is also recursive in $\mathcal{K}$ and A: the second step simply determines the value of A for the next undefined element, which is $|\sigma_{2e+1}|$. Then

$$e \in C \Leftrightarrow \sigma_{2e+2}(|\sigma_{2e+1}|) = 1$$

and $C \leq_T A \oplus \mathcal{K}$. But $A \oplus \mathcal{K} \leq_T A'$ (because $A \leq_T A'$ and $\mathcal{K} \leq_T A'$), and hence $C \leq_T A'$. $\square$

Notice that the **least possible jump** of a degree a is $a' = a \cup 0'$, since $a \cup 0' \leq a'$ always holds. Not every degree realizes the least possible jump, e.g. any degree $c \geq 0'$ does not (since then $c \cup 0' = c$), but the proof above actually shows that *every degree $c \geq 0'$ is the jump of a degree realizing the least possible jump*.

Exercise V.2.25 *For any $c \geq 0'$ there are infinitely many degrees with jump c.* (Hint: relativize V.2.23.b, and apply the Jump Inversion Theorem.)

The jump operator by itself is not very problematic: using the previous exercise, it can be shown that *the first-order theory with equality of $(\mathcal{D}, ')$ is decidable* (Jockusch and Soare [1970]). But adding the jump to $\mathcal{D}$ is a different story, and even the one-quantifier sentences of $\mathcal{D}'$ are not known to be decidable.

Our last immediate goal is to determine the possible behavior of the jump operator. To provide some necessary counterexamples, we first prove a result.

Proposition V.2.26 (Spector [1956]) *For any $c \geq 0'$ there exist two degrees* a *and* b *such that* $a \cup b = a' = b' = c$.

Proof. We modify the proof of the Jump Inversion Theorem. Let C such that $\mathcal{K} \leq_T C$ be given: we want two sets A and B such that $A' \equiv_T B' \equiv_T C$, and $A \oplus B \equiv_T C$. The last condition involves the only new idea in the proof, and consists of building A and B by initial segments of the same length, and to let the two sides have the same value only when coding elements of C, so that C will be recoverable from the two sets together.

We let $A = \bigcup_{s \in \omega} \sigma_s$ and $B = \bigcup_{s \in \omega} \tau_s$, and we start with $\sigma_0 = \tau_0 = \emptyset$. At stage $s + 1$, let σ_s and τ_s be given and of the same length.

- If $s = 3e$ then we see if there is an initial segment $\sigma \supseteq \sigma_s$ such that $\{e\}^\sigma(e) \downarrow$, and let σ_{s+1} be the smallest such string if one exists, and σ_s otherwise. Moreover, we also extend τ_s on the new elements x on which σ_{s+1} has been defined, if any, by letting

$$\tau_{s+1}(x) = 1 - \sigma_{s+1}(x),$$

  so that σ_{s+1} and τ_{s+1} differ on these elements.

- If $s = 3e + 1$ we do the same, interchanging the roles of A and B (and hence of the σ's and τ's).

- If $s = 3e + 2$ then we code $C(e)$, by letting

$$\sigma_{s+1} = \sigma_s * \langle C(e) \rangle \qquad \tau_{s+1} = \tau_s * \langle C(e) \rangle.$$

  Note that this is the only step where the two sides receive the same value.

As before, we have

$$A \oplus \mathcal{K} \equiv_T B \oplus \mathcal{K} \equiv_T B' \equiv_T A' \equiv_T C.$$

To recover $C(e)$, we only have to look at the e-th place in which A and B agree, and this is recursive in $A \oplus B$. So

$$C \leq_T A \oplus B \leq_T A' \oplus B' \leq_T C,$$

and hence the sets $A \oplus B$ and C are T-equivalent. $\quad\Box$

Recall that V.1.14 provided some necessary conditions for the behavior of the jump operator. We can show now that they are best possible.

Proposition V.2.27 (Spector [1956], Shoenfield [1959]) *Every possibility compatible with any of the properties*

$$a \le b \Rightarrow a' \le b' \quad and \quad a' \cup b' \le (a \cup b)'$$

is realized.

Proof. The following are the possible cases:

1. *jumps of comparable degrees*
   Let $a \le b$. Then we can have:

   - $a' = b'$ by the proof of V.2.21.
   - $a' < b'$, if $a = 0$ and $b = 0'$.

2. *jumps of incomparable degrees*
   Let $a | b$. Then we can have:

   - $a' = b'$ by V.2.26, since two degrees with jump c are strictly below it, and if they join to it then they must be incomparable (otherwise their join would be one of them).
   - $a' < b'$, by having a degree b incomparable with $a = 0'$ and with jump greater than $0''$. Note that a degree realizing the least possible jump cannot be above $0'$, and thus we only need to avoid $b < 0'$. This is automatic if $b' > 0''$, and thus it is enough to apply the Jump Inversion Theorem, and get b such that $b' = 0'''$.
   - $a' | b'$, by taking two degrees whose jumps are incomparable, which exist by the Jump Inversion Theorem (since incomparable degrees above $0'$ exist, e.g. by relativization of V.2.2).

3. *distributivity of jump over join*

   - $a' \cup b' = (a \cup b)' = 0''$ if $a = 0$ and $b = 0'$.
   - $a' \cup b' < (a \cup b)'$ for the degrees in V.2.26. $\square$

V.3   Baire Category $\star$

In Section II.1 we looked at the class $\mathcal{P}$ of partial functions, from a topological point of view. Here we do the same for the class of total functions, and we will distinguish the class of functions (ω^ω, the **Baire space**) from the class of characteristic functions or sets ($\mathcal{P}(\omega)$, the **Cantor space**). There are various ways of introducing a topology on these sets, but the induced topology will be unique, and thus particularly stable. As a reference on topology the reader can consult Kelley [1955].

Topologies on total functions

Natural topologies on ω^ω and $\mathcal{P}(\omega)$ are the following:

1. *the topology induced by the topology of $\mathcal{P}$*
   The open sets are the intersection of open sets of $\mathcal{P}$ with ω^ω or $\mathcal{P}(\omega)$. Thus a basic open set is determined by a finite initial segment, see p. 186.

2. *the product topology of the discrete topology on ω or $\{0,1\}$*
   Here every subset of ω is open, and a basic open set on ω^ω is a product

$$A_0 \times A_1 \times \cdots \times A_n \times \omega \times \omega \times \cdots$$

   where the A_i's are nonempty open sets. Thus the basic open sets are unions of sets determined by finite initial segments, namely the finite initial segments σ of length $n+1$ such that $\sigma(i) \in A_i$.

3. *the topology induced by the following metric:*

$$d(f,g) = \begin{cases} 0 & \text{if } f = g \\ \frac{1}{\mu x\,[(f(x) \neq g(x)] + 1} & \text{otherwise.} \end{cases}$$

   Thus the distance between two functions is determined by the smallest argument on which they differ, and the factor '+1' takes care of the possibility that they differ on 0. The basic open sets are the balls relative to the metric, and again they are determined by finite initial segments.

4. *the topology induced by the order topology on the reals*
   Here we use the canonical homeomorphism between ω^ω and the irrationals between 0 and 1, which can be produced in two different ways. One is to give the direct map

$$f \longmapsto \cfrac{1}{f(0) + \cfrac{1}{f(1) + \cdots}}.$$

   For the other, first note that there is a one-one map from ω^ω to $\mathcal{P}(\omega)$, given by

$$f \longmapsto \underbrace{1 \cdots 1}_{f(0)+1} \underbrace{0 0 \cdots 0}_{f(1)+1} \underbrace{1 1 \cdots 1}_{f(2)+1} 0 0 \cdots$$

   Then consider the reals between 0 and 1, in binary representation. Each such real may be written as $\sum_{i \in I} 2^{-(i+1)}$, for some $I \subseteq \omega$. The binary irrationals are those reals between 0 and 1 that cannot be so represented with a finite I. The composition of these two maps gives a homeomorphism between ω^ω and the binary irrationals. These and the irrationals are homeomorphic, because binary rationals and rationals are both countable and dense in $[0, 1]$, so there is a one-one order preserving correspondence between them, which may be extended in a unique way to a homeomorphism of $[0, 1]$ with itself. The restriction of it to the binary irrationals is a homeomorphism with the irrationals.

Since all these topologies are equivalent, we can use them interchangeably, but we will mostly continue to use the topology induced by the topology of $\mathcal{P}$, whose basic open sets are determined by finite strings.

Proposition V.3.1 ω^ω *is separated, but neither compact nor connected.* $\mathcal{P}(\omega)$ *is separated and compact, but not connected.*

Proof. For separation, given two different functions f and g, let x be the smallest point on which they differ. Then the two open sets determined by $\hat{f}(x + 1)$ and $\hat{g}(x + 1)$ are disjoint, and separate f and g. This also proves that the two spaces are not connected.

$\mathcal{P}(\omega)$ is compact by König's Lemma (see V.5.23), but ω^ω obviously is not since, e.g., the open sets determined by the finite initial segments $\langle n \rangle$ form a disjoint covering, from which no finite one can be extracted. □

Corollary V.3.2 ω^ω, $\mathcal{P}(\omega)$, *and* $[0, 1]$ *are pairwise not homeomorphic.*

Proof. The first two are not homeomorphic because one is compact but the other is not. And they are not homeomorphic to $[0, 1]$, because they are not connected, but the reals are. □

Exercises V.3.3 a) *A set* $\mathcal{A} \subseteq \omega^\omega$ *is compact if and only it is closed and bounded,* i.e. *there is a function* f *such that, for any* $g \in \mathcal{A}$, $(\forall n)(g(n) \leq f(n))$. (Hint: use König's Lemma for finitely generated trees.)

b) *A set* $\mathcal{A} \subseteq \omega^\omega$ *is clopen if and only if, for some well-founded tree,* $\mathcal{A}$ *is the union of the open sets determined by the branches of the tree.*

c) *A set* $\mathcal{A} \subseteq \mathcal{P}(\omega)$ *is clopen if and only if, for some tt-condition* σ, $\mathcal{A}$ *is the class of sets satisfying* σ.

d) *Both* ω^ω *and* $\mathcal{P}(\omega)$ *are dimensionless, in the sense that any finite power of them is homeomorphic to the original space, but* $[0, 1]$ *is not.* (Hint: by taking one internal point away we disconnect $[0, 1]$ but not $[0, 1] \times [0, 1]$.)

Comeager sets

The reason we introduce the topological approach to total functions is given by the next notions, which will have an immediate bearing on the methodology of Degree Theory.

Definition V.3.4 (Baire [1899]) *A set* $\mathcal{A}$ *in the Baire or Cantor spaces is:*

1. **dense** *if its closure, i.e. the smallest closed set containing it, is the whole space (alternately, if its complement does not contain any open sets)*

2. **comeager** *if it contains the intersection of a countable family of open dense sets*

3. **meager** *if its complement is comeager.*

Exercises V.3.5 a) *The intersection of a countable family of comeager sets is still comeager.*
   b) *The superset of a comeager set is comeager.*
   c) *The comeager sets form a filter in the lattice of subsets of the Baire or Cantor spaces under inclusion, and the meager sets form an ideal.*

Proposition V.3.6 *Given a set A in the Baire or Cantor spaces, then*

1. *A is dense if and only if*

$$(\forall \sigma)(\exists f \supseteq \sigma)(f \in A)$$

2. *A contains an open dense set if and only if*

$$(\forall \sigma)(\exists \tau \supseteq \sigma)(\forall f \supseteq \tau)(f \in A).$$

Proof. In the first part, the right-hand side says that no basic open set is contained in $\overline{A}$. This is clearly equivalent to density, which means that no open set is contained in $\overline{A}$.

For the second part, let $A \supseteq B$, and B be open dense. We just proved that, given σ, there is $g \supseteq \sigma$ in B. Since B is open, there is a basic open set containing g, and contained in B. Let σ_1 be the string determining it: then $g \supseteq \sigma_1$, and any function extending σ_1 is in B. If τ is the smallest string containing both σ and σ_1, which exists because they are comparable (being both contained in g), then any function $g \supseteq \tau$ must be in B, because $g \supseteq \sigma_1$. The condition is thus satisfied for B, and hence for A.

Conversely, suppose

$$(\forall \sigma)(\exists \tau_\sigma \supseteq \sigma)(\forall f \supseteq \tau_\sigma)(f \in A).$$

Then the open dense set

$$\bigcup_\sigma \{f : f \supseteq \tau_\sigma\}$$

is contained in A.   □

We now prove the basic result of this subsection, by the same finite extension method used to prove the results on degrees in the last section.

Theorem V.3.7 Baire Category Theorem (Baire [1899]) *A comeager set is not empty.*

Proof. If $\mathcal{A}$ is comeager there is a sequence $\{\mathcal{A}_n\}_{n\in\omega}$ of open dense sets such that $\mathcal{A} \supseteq \bigcap_{n\in\omega} \mathcal{A}_n$. It is enough to prove that $(\bigcap_{n\in\omega} \mathcal{A}_n) \neq \emptyset$.

We build a function $f \in \bigcap_{n\in\omega} \mathcal{A}_n$, by initial segments. Let $\sigma_0 = \emptyset$. At stage $n + 1$, let σ_n be given: we ensure that $f \in \mathcal{A}_n$. Since $\mathcal{A}_n$ is open dense, there is a string $\sigma \supseteq \sigma_n$ such that all functions extending it are in $\mathcal{A}_n$: it is then enough to let σ_{n+1} be any initial segment extending σ. Since in the end $f = \bigcup_{n\in\omega} \sigma_n$, we will have $f \supseteq \sigma_{n+1}$ and hence $f \in \mathcal{A}_n$. $\quad\square$

Corollary V.3.8 *A comeager set is not meager.*

Proof. Note that the intersection of two comeager sets is comeager. If a comeager set $\mathcal{A}$ were meager, $\mathcal{A} \cap \overline{\mathcal{A}}$ would be comeager and hence not empty, contradiction. $\quad\square$

Note that it is essential that we are considering only countably many dense open sets: *the intersection of $2^{\aleph_0}$ open dense sets is not necessarily nonempty* ($\overline{\{f\}}$ is open dense, being the union of the open sets determined by the strings not contained in f, but $\bigcap_{f\in\omega^\omega} \overline{\{f\}}$ is obviously empty). If the Continuum Hypothesis is assumed, this settles the question. But if the Continuum Hypothesis fails then there are uncountable cardinals below $2^{\aleph_0}$, and we may ask whether *the intersection of less than $2^{\aleph_0}$ open dense sets is nonempty*. The positive answer is known as **Martin's Axiom**, and it is independent of ZFC (Martin and Solovay [1970]). For some of its interesting set-theoretical consequences see, e.g., Jech [1978] and Levy [1979].

Exercises V.3.9 a) *A comeager set is dense.* (Hint: it is enough to show that it intersects any nonempty basic open set. The proof is as above, only starting with σ_0 determining the given basic open set.)

b) *A comeager set has cardinality $2^{\aleph_0}$.* (Hint: build a binary tree, each time extending a given string in two incomparable ways, and then continuing separately on each of them.)

c) *A countable set is meager.* (Hint: it is enough to show that any singleton $\{f\}$ is meager. Since comeager sets are closed under supersets, it is enough to find a comeager set not containing f, so that $\overline{\{f\}}$ is comeager. For every σ there is $\tau_\sigma \supseteq \sigma$ such that every $g \supseteq \tau_\sigma$ is different from f. Let $\mathcal{A}$ be the union of the basic open sets determined by τ_σ. By definition $\mathcal{A}$ is open dense, hence comeager, and does not contain f.)

Exercise V.3.10 Banach–Mazur games. Given $\mathcal{A}$, consider the following game. The two players play strings, and each must play a proper extension of the last move.

Player I wins if and only if the union of the moves is in $\mathcal{A}$. Then $\mathcal{A}$ *is comeager if and only if player I has a winning strategy, and $\mathcal{A}$ is meager if and only if player II has a winning strategy.* (Oxtoby [1957]) (Hint: suppose I has a winning strategy, and let $\mathcal{A}_0$ be the basic open set determined by the first string played by I, according to the strategy. For any possible response τ of II consider the response σ_τ of I, according to the strategy, and let $\mathcal{A}_1$ be the open set obtained as the union of the basic open sets determined by the σ_τ, and so on. The $\mathcal{A}_n$'s are open dense and $\mathcal{A} \supseteq \bigcap_{n \in \omega} \mathcal{A}_n$, because I uses a winning strategy.)

The intuition about comeager and meager sets is that they are, respectively, very large and very small. The next result accords with this intuition, and tells that almost all sections of a large (small) set in a product space must be large (small).

Theorem V.3.11 Kuratowski-Ulam Theorem. *Given* $\mathcal{A} \subseteq \omega^\omega \times \omega^\omega$ *or* $\mathcal{A} \subseteq \mathcal{P}(\omega) \times \mathcal{P}(\omega)$, *let*

$$f \in \mathcal{A}_g \Leftrightarrow (f, g) \in \mathcal{A}$$

be the section of $\mathcal{A}$ at g. Then:

1. *if $\mathcal{A}$ is comeager, the set $\{g : \mathcal{A}_g$ comeager$\}$ of its comeager sections is comeager*

2. *if $\mathcal{A}$ is meager, the set $\{g : \mathcal{A}_g$ meager$\}$ of its meager sections is comeager.*

Proof. The second part follows from the first, by taking complements. Let then $\mathcal{A}$ be comeager: $\mathcal{A}$ contains $\bigcap_{n \in \omega} \mathcal{A}_n$, where the $\mathcal{A}_n$'s are open dense in the product space. The comeager sets are closed upward under inclusion, and thus it is enough to find a comeager set of functions g for which $\mathcal{A}_g$ is comeager. Since $\mathcal{A}_g \supseteq \bigcap_{n \in \omega}(\mathcal{A}_n)_g$, we look for g's such that $(\mathcal{A}_n)_g$ is open dense. By definition of product topology $(\mathcal{A}_n)_g$ is open for any g, and we would want

$$(\forall \sigma)(\exists \tau \supseteq \sigma)(\forall f \supseteq \tau)[f \in (\mathcal{A}_n)_g],$$

and hence

$$(\forall \sigma)(\exists \tau \supseteq \sigma)(\forall f \supseteq \tau)[(f, g) \in \mathcal{A}_n].$$

If this holds for every n, then $\mathcal{A}_g$ is comeager. Let us consider the set of g's for which this holds:

$$g \in \mathcal{B} \Leftrightarrow (\forall n)(\forall \sigma)(\exists \tau \supseteq \sigma)(\forall f \supseteq \tau)[(f, g) \in \mathcal{A}_n].$$

We see whether $\mathcal{B}$ is comeager. If

$$g \in \mathcal{B}_{(n, \sigma)} \Leftrightarrow (\exists \tau \supseteq \sigma)(\forall f \supseteq \tau)[(f, g) \in \mathcal{A}_n],$$

then $\mathcal{B} = \bigcap_{n,\sigma} \mathcal{B}_{(n,\sigma)}$, but the $\mathcal{B}_{(n,\sigma)}$ are not necessarily open, because we use all of g in their definition. Thus $\mathcal{B}$ is not necessarily comeager, but a small variation of it will do. Let

$$g \in \mathcal{C}_{(n,\sigma)} \Leftrightarrow (\exists y)(\exists \tau \supseteq \sigma)(\forall f \supseteq \tau)(\forall h \supseteq \hat{g}(y))[(f,h) \in \mathcal{A}_n].$$

Now $\mathcal{C}_{(n,\sigma)}$ is open (because union of basic open sets) and dense (because so is $\mathcal{A}_n$). If

$$C = \bigcap_{n,\sigma} \mathcal{C}_{(n,\sigma)}$$

then $\mathcal{C}$ is comeager and, as above, $\mathcal{A}_g$ is comeager whenever $g \in \mathcal{C}$. ☐

Exercises V.3.12 $\mathcal{A}$ has the **Baire property** if there is an open set $\mathcal{U}$ such that $(\mathcal{A} - \mathcal{U}) \cup (\mathcal{U} - \mathcal{A})$ is meager, i.e. $\mathcal{A}$ differs from an open set by a meager set. *If $\mathcal{A}$ has the Baire property, then the converse implications in the Kuratowski-Ulam Theorem hold.* (Hint: suppose $\mathcal{A}$ is not meager, but $\{g : \mathcal{A}_g \text{ meager}\}$ is comeager. Since $\mathcal{A}_g$ differs from the open set $\mathcal{U}_g$ by a meager set, $\mathcal{U}_g$ must be meager for a comeager set of g's. But $\mathcal{U}$ is not empty, and it contains a basic open set: hence there is also a comeager set of g's such that $\mathcal{U}_g$ contains an open set, and thus is not meager. The intersection of these two comeager sets is not empty by the Baire Category Theorem, contradiction.)

Baire Category and Degree Theory

Going back to the proofs of the results in Section 2, we notice that they all shared the following characteristic features. We had a set of requirements R_n to satisfy, which may be identified with the class of sets satisfying them. The general pattern of the proofs was to show that, for each n,

$$(\forall \sigma)(\exists \tau)(\forall A \supseteq \tau)(A \in R_n).$$

Thus we were actually showing that, for each n, there is an open dense set $\mathcal{A}_n$ contained in R_n (by V.3.6). The Baire Category Theorem could then be applied to claim that the intersection of requirements is not empty, and hence that there is a set satisfying all the requirements, without further constructions. More precisely:

Proposition V.3.13 The Finite Extension Method (Myhill [1961], Sacks [1963]) *Given a countable collection of requirements R_n such that*

$$(\forall \sigma)(\exists \tau \supseteq \sigma)(\forall A \supseteq \tau)(A \in R_n),$$

the set $\bigcap_{n \in \omega} R_n$ is comeager (and hence nonempty).

The categorical approach is useful for a methodological analysis of the finite extension method. Precisely, from it we obtain that:

1. *requirements can be taken care of separately, by showing that each of them is dense*

2. *we can freely combine constructions known to be performable separately, as long as the global list of requirements remains countable*
   For example, suppose we want to prove that, given a countable sequence $\{A_0, A_1, \ldots\}$ of nonrecursive sets, there is a set B incomparable with all of them. Then we only have to prove that, for a given e and a given nonrecursive set C,

   - $(\forall \sigma)(\exists \tau)(\forall A \supseteq \tau)(A \not\simeq \{e\}^C)$
   - $(\forall \sigma)(\exists \tau)(\forall A \supseteq \tau)(C \not\simeq \{e\}^A)$.

   We know that we can do this, by the work done in Section 2. It thus follows that, for a fixed nonrecursive set C, the set

   $$\{A : A \not\leq_T C \wedge C \not\leq_T A\}$$

   is comeager, and hence so is

   $$\{A : (\forall n)(A \not\leq_T A_n \wedge A_n \not\leq_T A)\},$$

   being a countable intersection of comeager sets. In particular, the set is not empty by Baire Category Theorem. Note that we could even avoid the consideration of the condition $A \not\leq_T C$, because there are only countably many sets recursive in C: the set $\{A : A \not\leq_T C\}$ is thus automatically comeager.

3. *we cannot use the finite extension method to produce sets belonging to a meager class*
   This follows from the fact that the finite extension method produces sets in a comeager class, and the intersection of two comeager classes is nonempty, by the Baire Category Theorem. In particular, if we can use the finite extension method to build a set satisfying certain requirements, then we cannot use the same method to build a set not satisfying the same requirements.

4. *a game-theoretical approach to Degree Theory is possible, via Banach-Mazur games*
   This follows from V.3.10, and has been developed by Yates [1976]. See also p. 495.

We can reformulate a number of results proved before in terms of category notions, with the convention that a set of degrees $\mathcal{A}$ is comeager or meager if such is the class of sets whose degree is in $\mathcal{A}$. This makes sense because a degree contains only countably many sets, and thus it is a meager class.

Proposition V.3.14 *The following sets of degrees are comeager:*

1. $\{a : a$ *is incomparable with a fixed nonrecursive degree*$\}$

2. $\{a : a$ *is the l.u.b. of two incomparable degrees*$\}$

3. $\{a : a$ *is the l.u.b. of a minimal pair*$\}$

4. $\{a : a$ *realizes the least possible jump*$\}$.

The following sets of degrees are meager:

5. $\{a : a$ *is comparable with a fixed nonrecursive degree*$\}$

6. $\{a : a$ *is a minimal degree*$\}$.

Proof. The proof of V.2.2 builds two sets A and B with incomparable degree: this can be seen as the construction of a single set $A \oplus B$ which is the least upper bound of two incomparable sets, and proves 2. Part 6 follows from this, since a minimal degree cannot be the l.u.b. of two incomparable degrees.

Similarly, the other parts follow from V.2.13, V.2.16 and V.2.24. □

Note that this shows, in particular, that *degrees comparable with a given nonrecursive degree, as well as minimal degrees, cannot be built by the finite extension method.*

The work done so far also allows to decide whether any quantifier-free question about jumps, l.u.b.'s and g.l.b.'s of degrees holds for a comeager set of degrees or not.

Proposition V.3.15 (Stillwell [1972]) *The theory of degrees*

$$(\mathcal{D}, \leq, \cup, \cap, ', 0),$$

with subformulas containing a term $t_1 \cap t_2$ thought of as prefixed by '$t_1 \cap t_2$ exists', and with the quantifiers $\forall$ and $\exists$ interpreted, respectively, as meaning 'for a comeager set of degrees' and '$\neg\forall\neg$', is decidable.

Proof. For simplicity, we will say 'almost always' to mean 'for a comeager set of n-tuples of degrees'.

Note that terms are obtained from variables and 0 by using $\cap$, $\cup$, and $'$. By induction, we show that for every term t there are degrees a_i occurring in t such that, for some m,

$$t = a_1 \cup \cdots \cup a_n \cup 0^{(m)}.$$

is almost always true. This reduces every term to a normal form.

1. $(a_1 \cup \cdots \cup a_m \cup 0^{(p)}) \cup (b_1 \cup \cdots \cup b_m \cup 0^{(q)}) =$
$$a_1 \cup \cdots \cup a_m \cup b_1 \cup \cdots \cup b_m \cup 0^{(\max(p,q))}.$$
   This always holds.

2. $(a_1 \cup \cdots \cup a_m \cup 0^{(p)}) \cap (b_1 \cup \cdots \cup b_m \cup 0^{(q)}) = c_1 \cup \cdots \cup c_s \cup 0^{(\min(p,q))}$
   almost always, where

$$\{c_1, \ldots, c_s\} = \{a_1, \ldots, a_m\} \cap \{b_1, \ldots, b_n\}.$$

   First note that it is always possible to rearrange terms, and possibly introduce a new $0^{(\min(p,q))}$ (which has no influence on $\cup$) in such a way to have, for some d and e,

$$a_1 \cup \cdots \cup a_m \cup 0^{(p)} = (c_1 \cup \cdots \cup c_s \cup 0^{(\min(p,q))}) \cup d$$
$$b_1 \cup \cdots \cup b_n \cup 0^{(q)} = (c_1 \cup \cdots \cup c_s \cup 0^{(\min(p,q))}) \cup e.$$

   It is then enough to show that almost always, given a and d, is

$$(a \cup d) \cap (a \cup e) = a.$$

   And this is just the relativization to a of the fact that almost every degree d is part of a minimal pair (V.2.16).

3. $(a_1 \cup \cdots \cup a_m \cup 0^{(p)})' = a_1 \cup \cdots \cup a_m \cup 0^{(p+1)}$ almost always.
   First, almost always $a' = a \cup 0'$ and, by relativization to b, almost always $(a \cup b)' = a \cup b'$. Then, by induction, almost always $a \cup 0^{(p)} = a^{(p)}$. Finally, almost always

$$a_1 \cup \cdots \cup a_m \cup 0^{(p)} = (a_1 \cup \cdots \cup a_m)^{(p)},$$

   and hence

$$(a_1 \cup \cdots \cup a_m \cup 0^{(p)})' = (a_1 \cup \cdots \cup a_m)^{(p+1)}$$
$$= a_1 \cup \cdots \cup a_m \cup 0^{(p+1)}.$$

The decision procedure now follows easily, since every formula with free variables is 0,1-valued, being satisfied either by a comeager set of degrees or by a meager one. Precisely:

1. $t_1 \leq t_2$, with t_1 and t_2 terms
   After putting the two terms in normal form, $t_1 \leq t_2$ holds if and only if all the variables of t_1 appear in t_2, and the exponent of 0 in t_1 is not bigger than the exponent of 0 in t_2.

2. ¬ψ

   The complement of a meager set is comeager, and the complement of a comeager set is meager.

3. ψ ∧ φ

   The intersection of two comeager sets is comeager, and the intersection of any set with a meager set is meager.

4. ∀xψ(x)

   This follows from the Kuratowski-Ulam Theorem, by the interpretation of the universal quantifier.   □

We should not expect too much from this decision procedure, since practically every interesting question we may ask will involve real quantifiers as well. E.g., it is true that almost every degree has no minimal predecessor (see V.3.17), but we cannot express this sentence in the above language, since the notion of minimal degree requires a true universal quantifier.

Meager sets of degrees

The results quoted above were simply old facts rephrased in categorical terms. We now prove a theorem whose very statement genuinely requires the notions introduced in this section. The plan of the proof can be read independently, but its implementation relies on methods and notations that will be introduced in Section 5.

Theorem V.3.16 (Martin [1967]) *If $\mathcal{A}$ is a downward closed, meager class of degrees then the upward closure of $\mathcal{A} - \{0\}$ is still meager.*

Proof. We want to build A by finite extensions, in such a way to have

$$(\forall C \leq_T A)(C \notin \mathcal{A} - \{0\}).$$

This condition can be broken down in the requirements

$$R_e \ : \ \{e\}^A \text{ total } \Rightarrow \{e\}^A \notin \mathcal{A} - \{0\}.$$

Let $\sigma_0 = \emptyset$. At stage $e + 1$, let σ_e be given: we attack R_e.

1. If there is $\sigma \supseteq \sigma_e$ such that

$$(\exists x)(\forall \tau \supseteq \sigma)(\{e\}^\tau(x)\uparrow)$$

then we let $\sigma_{e+1} = \sigma$. This ensures that if $A \supseteq \sigma_{e+1}$ then $\{e\}^A$ is not total.

2. If there is a $\sigma \supseteq \sigma_e$ with no e-splitting extensions, again let $\sigma_{e+1} = \sigma$. This ensures that if $A \supseteq \sigma_{e+1}$ then $\{e\}^A$ is recursive.

3. Suppose now that:

   - $(\forall \sigma \supseteq \sigma_e)(\forall x)(\exists \tau \supseteq \sigma)(\{e\}^\tau(x) \downarrow)$
   - for all $\sigma \supseteq \sigma_e$ there are e-splitting extensions of σ.

It will be useful in the following to picture $\{e\}^\sigma$ as a finite string, with the property that

$$\sigma \subseteq \sigma' \Rightarrow \{e\}^\sigma \subseteq \{e\}^{\sigma'}.$$

The proof goes roughly as follows. We build an order-preserving recursive map Φ from strings to strings such that

$$(\forall \tau \supseteq \Phi(\{e\}^{\sigma_e}))(\exists \sigma \supseteq \sigma_e)[\tau = \Phi(\{e\}^\sigma)].$$

This gives a homeomorphism between the set $\{\Phi(\{e\}^\sigma) : \sigma \supseteq \sigma_e\}$ and the whole space of strings. Since $\mathcal{A}$ is meager, there is an open set contained in $\overline{\mathcal{A}}$, hence there is $\sigma \supseteq \sigma_e$ such that

$$(\forall A \supseteq \sigma)(\Phi(\{e\}^A) \notin \mathcal{A}).$$

But Φ is recursive, and so $\Phi(\{e\}^A) \leq_T \{e\}^A$. Moreover, $\mathcal{A}$ is closed downward, hence

$$(\forall A \supseteq \sigma)(\{e\}^A \notin \mathcal{A}),$$

and R_e is satisfied.

To finish the proof, it remains to build Φ. We first build an admissible triple (i.e. a uniform tree, see V.6.2) with the following properties: if $T_0(n)$ is the set of strings extending σ_e, of length $g(n+1)$, and agreeing with f_L on the interval $[g(n), g(n+1))$, and $T_1(n)$ is defined similarly using f_R, then whenever $\sigma_0 \in T_0(n)$ and $\sigma_1 \in T_1(n)$, σ_0 and σ_1 e-split. The construction is the same as in V.6.5, only we have to consider every string extending σ_e and of length $g(n)$, when we build the $n+1$ level (instead of only the strings of length $g(n)$ which are on the tree, as we did in V.6.5).

We now define Φ as follows. If $\Phi(\mu)(m)$ converges for all $m < n$ and $n < |\mu|$, then:

$$\Phi(\mu)(n) \simeq \begin{cases} 0 & \text{if } \mu \supseteq \{e\}^\sigma \text{ for some } \sigma \in T_0(n) \\ 1 & \text{if } \mu \supseteq \{e\}^\sigma \text{ for some } \sigma \in T_1(n) \\ 1 & \text{if } \mu | \{e\}^\sigma \text{ for every } \sigma \in T_0(n) \cup T_1(n) \\ \text{undefined} & \text{otherwise,} \end{cases}$$

where $\mu|\{e\}^\sigma$ means that the two strings are incompatible. By definition $\Phi(\mu)$ is an initial segment, and Φ is order-preserving and single-valued (if $\sigma \in T_0(n)$ and $\sigma' \in T_1(n)$ then $\{e\}^\sigma$ and $\{e\}^{\sigma'}$ are incompatible). The third condition in the definition ensures that $\Phi(\mu)(n)$ is defined whenever μ is long enough. More precisely, if $\Phi(\mu)(n)$ converges for all $m < n$, then so does $\Phi(\mu)(n)$, whenever

$$|\mu| \geq \sup\{|\{e\}^\sigma| : \sigma \in T_0(n) \cup T_1(n)\}.$$

Hence, in general, $\Phi(\mu)(n)$ converges if

$$|\mu| \geq \sup\{|\{e\}^\sigma| : \sigma \in \bigcup_{m \leq n}(T_0(m) \cup T_1(m))\}.$$

This also shows that, if we choose our admissible triple in such a way to have $|\{e\}^\sigma| \geq g(n)$ when $\sigma \in T_0(n) \cup T_1(n)$ (which is possible because, for each $A \supseteq \sigma_e$, $\{e\}^A$ is total and so $\sup_\sigma |\{e\}^\sigma| = \infty$). Then $\Phi(\{e\}^\sigma)(n) \downarrow$ for all $\sigma \in T_0(n) \cup T_1(n)$, and

$$\Phi(\{e\}^\sigma)(n) = \begin{cases} 0 & \text{if } \sigma \in T_0(n) \\ 1 & \text{if } \sigma \in T_1(n). \end{cases}$$

Fix now $\sigma \supseteq \sigma_e$ in some $T_0(n) \cup T_1(n)$. We show that Φ gives the homeomorphism we wanted between $\{\Phi(\{e\}^\mu) : \mu \supseteq \sigma\}$ and the whole space. Given $\tau \supseteq \Phi(\{e\}^\sigma)$, let μ be the string of length $g(|\tau|)$ such that

$$\mu \in T_{\tau(m)}(m) \text{ for } |\sigma| \leq m < |\tau|.$$

Then $\Phi(\{e\}^\mu)(m) = \tau(m)$ by the considerations above, and $|\Phi(\{e\}^\mu)| = |\tau|$. This proves the claim. $\square$

The theorem may be seen as a further step in the determination of the classes $\mathcal{A}$ of degrees for which we can always find a degree incomparable with every element of $\{\mathcal{A}\} - 0$. We showed in V.2.20 that this is possible if $\mathcal{A}$ has cardinality less than $2^{\aleph_0}$, and the theorem just proved gives a solution also for some classes of cardinality $2^{\aleph_0}$.

Corollary V.3.17 *The class of degrees without minimal predecessors is comeager, and hence nonempty. In particular, the structure $\mathcal{D}$ is not atomic.*

Proof. The class of minimal degrees plus 0 is meager, and obviously closed downward. Thus the upward closure of the class of minimal degrees, i.e. the class of degrees with minimal predecessors, is still meager. $\square$

The existence of degrees without minimal predecessors can also be proved by initial segments results, see V.6.16.c.

Measure Theory and Degree Theory $\star$

In place of using Baire Category we might have used **measure theory**, by considering Lebesgue measure on $\mathcal{P}(\omega)$ or, equivalently, the product measure of the measure on $\{0, 1\}$ given by $\mu(\{0\}) = \mu(\{1\}) = \frac{1}{2}$. There is a well-known analogy, according to which meager and comeager sets correspond, respectively, to sets with measure 0 and 1 (although a comeager set can have measure 0). The measure-theoretical approach does not seem to have the same natural bearing on Degree Theory that Baire Category does, and this is reflected in the fact that proofs of results tend to be more complicated. Usually, however, the picture obtained by the two points of view is consistent, i.e. a set of degrees looks large in one case if it does in the other (Sacks [1963], Martin [1967], Stillwell [1972]). A notable exception is provided by the analogue of the theorem just proved: Paris [1977] has shown that *the upward closure of a measure 0 downward closed set of nonzero degrees is not necessarily of measure 0* (see also Kurtz [1983] for a natural example, namely the set of 1-generic degrees), although *the upward closure of the minimal degrees still has measure 0*.

V.4   The Coinfinite Extension Method

We have proved in the last section that construction principles more powerful than the finite extension method V.3.13 are needed, if we wish to prove results like the existence of minimal degrees. We introduce one of them here, by approximating a set not by finite, but by coinfinite extensions.

Definition V.4.1 *A* **coinfinite condition** *is a partial function* $\theta : \omega \to \omega$ *with coinfinite recursive domain. A coinfinite condition is* **recursive** *if it is recursive as a partial function.*

The notions of extension and compatibility for coinfinite conditions are the usual ones for partial functions.

Recursive coinfinite conditions are the next logical step after finite strings (which are just particular recursive coinfinite conditions, with finite domain). It thus makes sense to see whether we can build, by using them, sets which we cannot build by the finite extension method V.3.13. For example, we will see in Section 6 that we can build a minimal degree by recursive coinfinite extensions.

Nonrecursive coinfinite conditions are also useful because they allow coding a given set in a set we are building, in just one step. In this section we will deal with applications of this kind. This is also one direct way to prove relativized results above given nonrecursive degrees, which cannot be proved by the finite extension method V.3.13 (by V.3.14).

Exact pairs and ideals

By building a minimal pair we have proved that some pairs of degrees have g.l.b. We will now prove that there are pairs without g.l.b., by generalizing the notion of minimal pair.

Definition V.4.2 (Kleene and Post [1954]) *Two degrees **a** and **b** form an* **exact pair** *for a set of degrees C if*

1. *both **a** and **b** are above all degrees in C, i.e.*

$$(\forall c \in C)(c \leq a \wedge c \leq b)$$

2. *any degree below both **a** and **b** is also below some degree in C, i.e.*

$$x \leq a \wedge x \leq b \rightarrow (\exists c \in C)(x \leq c).$$

Theorem V.4.3 Spector Theorem (Kleene and Post [1954], Lacombe [1954], Spector [1956]) *Every countable set of degrees in which every pair of elements is bounded has an exact pair.*

Proof. Given $\{B_n\}_{n \in \omega}$ we want to build two sets A and B such that

1. $(\forall n)(B_n \leq_T A, B)$

2. $C \leq_T A, B \Rightarrow C \leq_T B_n$, for some n.

Actually, since any pair of B_m's, and hence any finite set of them, is bounded by some B_n, we can replace the second condition with the weaker one:

3. $C \leq_T A, B \Rightarrow C \leq_T (\oplus_{m \leq n} B_m)$, for some n.

The second condition looks like the requirement for minimal pairs, except that here we require a set recursive in both A and B to be not outright recursive, but only recursive in some element of the given set. We thus extend the proof of V.2.16 in the natural way. First we get one upper bound for free, simply by letting $B = \oplus_{n \in \omega} B_n$. We then build A by coinfinite approximations θ_s.

To ensure that the first condition is satisfied, we require

$$C_e : B_e \leq_T A.$$

This is easily obtained, by coding B_e into the e-th column A_e of A, with at most finitely many exceptions. Using coinfinite conditions will make this step trivial, and it will be possible to satisfy C_e in just one step. Since A_e and B_e will differ only finitely they will have the same degree, and then $A = \oplus_{n \in \omega} A_n$ will also be an upper bound for $\{B_n\}_{n \in \omega}$.

The finite modifications are needed to ensure the third condition, with the relative requirements

$$R_e \; : \; \{e\}^A \simeq \{e\}^B \simeq C \; \Rightarrow \; C \leq_T (\oplus_{m \leq n} B_m), \; \text{for some } n$$

(note that we use V.2.18.a to simplify the presentation of the requirements). They will be satisfied as in the minimal pair construction.

We start with $\theta_0 = \emptyset$. At each step $e + 1$, given θ_e (see Figure 1), we simultaneously attack R_e and C_e. As in V.2.16, we see if there are e-splitting strings σ_0 and σ_1 which are compatible with θ_e. Note that we still use finite strings, because they are always enough to determine convergent computations, although now we cannot look for extensions of θ_e, which is infinite, and thus we only look for strings compatible with it. If they exist, choose the one σ_i that produces a disagreement with $\{e\}^B$ on the element for which σ_0 and σ_1 split. We then define

$$\theta_{e+1}(x) \simeq \begin{cases} \theta_e(x) & \text{if } \theta_e(x)\downarrow \\ \sigma_i(x) & \text{if } \sigma_i(x)\downarrow \\ B_e(z) & \text{if } x = \langle e, z \rangle, \text{otherwise.} \end{cases}$$

If no e-splitting compatible with θ_e exists, we let

$$\theta_{e+1}(x) \simeq \begin{cases} \theta_e(x) & \text{if } \theta_e(x)\downarrow \\ B_e(z) & \text{if } x = \langle e, z \rangle, \text{otherwise.} \end{cases}$$

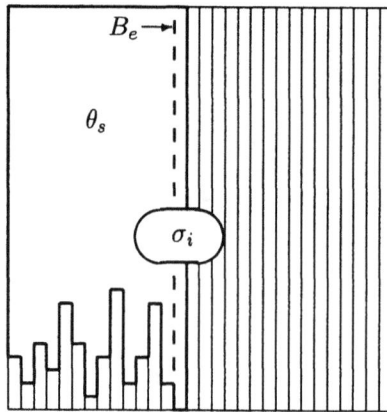

Figure V.1: Step $e + 1$ for Spector's Theorem

Let $A = \bigcup_{e \in \omega} \theta_e$. By construction, θ_{e+1} extends θ_e infinitely on the e-th column, and finitely elsewhere. Thus, by induction, we always code B_e in the e-th column of A, except for finitely many points. Hence, for almost every z,

$$z \in B_e \Leftrightarrow \langle e, z \rangle \in A.$$

Thus $B_e \leq_T A$ holds, and the coding requirement C_e is satisfied.

It remains to see that R_e has been satisfied by the action taken at step $e+1$. This is vacuously true if there were e-splitting extensions as required, because we then defined θ_{e+1}, and hence A, in such a way to have $\{e\}^A \not\cong \{e\}^B$. And if they did not exist, then any string σ compatible with θ_e and making $\{e\}^\sigma(x)$ converge must produce the right value. This gives a procedure recursive in θ_e to compute the function $\{e\}^A$, if total. But θ_e is defined only finitely outside the first e-columns, and the m-th column, for $m < e$, codes B_m except on finitely many points. Thus θ_e is recursive in $\bigoplus_{m \leq e} B_m$, and hence R_e is satisfied. □

Note that *the exact pair for* $\{B_n\}_{n \in \omega}$ *given by the proof is actually recursive in* $(\bigoplus_{n \in \omega} B_n)'$. More precisely, step $e+1$ is r.e. in the join of the columns coded in θ_e, and hence the exact pair is recursive in $\bigoplus_{e \in \omega} (\bigoplus_{m \leq e} B_m)'$.

In particular, even if $(\bigoplus_{n \in \omega} B_n) \leq_T \mathcal{K}$, i.e. if the set is uniformly recursive in $\mathbf{0}'$, then the exact pair just obtained is only below $\mathbf{0}''$. We will prove a partial version of the theorem for degrees below $\mathbf{0}'$ in XI.3.18. However, *the full analogue of Spector's Theorem fails below* $\mathbf{0}'$ for cardinality reasons (basically, there are many ideals but few possible exact pairs, see V.4.6).

Note also that the only instance of the theorem that is provable by the finite extension method V.3.13 is the existence of exact pairs for $\{\mathbf{0}\}$, when the theorem reduces to the existence of minimal pairs, because if the set contains a nonzero degree then the theorem produces degrees comparable with it, and hence a member of a meager class.

If $\mathcal{C}$ is downward closed then an exact pair $\mathbf{a}$ and $\mathbf{b}$ for $\mathcal{C}$ obviously determines it, since then

$$c \in \mathcal{C} \Leftrightarrow c \leq a \wedge c \leq b.$$

There are thus two conditions that have some bearing for exact pairs, and we collect them in the next definition.

Definition V.4.4 *A set of degrees is:*

1. *an **initial segment** if is closed downward*

2. *an **ideal** if it is closed downward and under joins*

3. *a **principal ideal** if it is an ideal determined by a degree $\mathbf{a}$, i.e. it is of the form $\mathcal{D}(\leq a) = \{c : c \leq a\}$.*

Corollary V.4.5 *Every countable ideal is the intersection of two principal ideals.*

Proof. An ideal is closed under joins and thus, if countable, it has an exact pair. Being closed downward, it is determined by it.   □

Exercise V.4.6 *There are ideals of degrees below $0'$, without exact pairs below $0'$.*
(Hint: let $\{A_n\}_{n\in\omega}$ be a recursively independent set recursive in $\mathcal{K}$, see V.2.9. Then the ideals generated by $\{A_n\}_{n\in X}$, for any X, are distinct for distinct X's, by recursive independence. Thus there are $2^{\aleph_0}$ ideals, but only countably many degrees recursive in $\mathcal{K}$, and thus only countably many possible exact pairs.)

Greatest lower bounds and least upper bounds

The notion of exact pair is useful for studying l.u.b.'s of sets of degrees, as well as g.l.b.'s of pairs.

Theorem V.4.7 (Kleene and Post [1954], Spector [1956]) *There are two degrees below $0'$ without g.l.b.*

Proof. Consider an infinite ascending sequence of degrees: by Spector's Theorem it has an exact pair. The degrees forming such a pair cannot have g.l.b., since any lower bound admits an element of the chain above it (by definition of exact pair), and thus is not the greatest lower bound (because the sequence is increasing).

An infinite ascending sequence certainly exists (e.g. iterate the jump operator, starting from any degree). But if we wish to get the exact pair below $0'$, we have to choose a chain $\{B_n\}_{n\in\omega}$ such that $(\oplus_{n\in\omega}B_n)' \leq_T \mathcal{K}$, because this is the bound we obtained from the proof of Spector's Theorem. For this it is enough to build (see V.2.7 and V.2.21) a recursively independent set $\{A_n\}_{n\in\omega}$ such that $(\oplus_{n\in\omega}A_n)' \leq_T \mathcal{K}$: then the sets $B_n = \oplus_{m\leq n}A_m$ form a strictly ascending sequence, by recursive independence of the A_n's, and

$$(\oplus_{n\in\omega}B_n)' \leq_T (\oplus_{n\in\omega}A_n)' \leq_T \mathcal{K}. \quad \square$$

Corollary V.4.8 $\mathcal{D}$ *and* $\mathcal{D}(\leq 0')$ *are not lattices.*

Note that the finite extension method V.3.13 produced a minimal pair, hence a pair with g.l.b., and thus it cannot produce a pair without g.l.b. However, *it is possible to prove by the finite extension method that $\mathcal{D}$ is not a lattice*, as shown in the exercises.

Exercises V.4.9 a) *Any pair without g.l.b. is the exact pair of an infinite ascending chain.* Thus the proof given above is in a sense the only possible one. (Hint: given a pair of degrees, consider an enumeration of the countable set of degrees below both of them and, for each n, consider the join of the first n degrees in the list.)

b) *A set bounding a pair without g.l.b. can be built by the finite extension method.* (Jockusch [1981]) (Hint: build simultaneously an infinite ascending chain and an exact pair for it. More precisely, build a recursively independent set $\{B_n\}_{n \in \omega}$, so that the sets $B_0 \oplus \cdots \oplus B_n$ form an ascending sequence. Build also another set C, which will provide finite modifications of the B_n's, as follows:

$$A_n(x) = \begin{cases} C(x) & \text{if } x \leq n \\ B_n(x) & \text{otherwise.} \end{cases}$$

Then the degree of A_n and B_n is the same. Let

$$A = \oplus_{n \in \omega} A_n \quad \text{and} \quad B = \oplus_{n \in \omega} B_n,$$

and make sure that A and B are an exact pair for the chain $\{B_n\}_{n \in \omega}$.)

c) *The set of degrees a such that $\mathcal{D}(\leq a)$ is not a lattice is comeager.* (Jockusch [1981])

We now turn to l.u.b.'s of sets of degrees.

Proposition V.4.10 Compactness for l.u.b.'s (Spector [1956])

1. *A chain of degrees has l.u.b. if and only if it is eventually constant.*

2. *A set of degrees has l.u.b. if and only if it there is a finite subset of it whose join provides an upper bound for the whole set.*

Proof. Given a chain, consider an exact pair for the ideal generated by it. The l.u.b. of the chain is the g.l.b. of the exact pair, and if the chain is not eventually constant it cannot exist, as in V.4.7. Given a set $\mathcal{A}$, it is enough to reduce it to a chain that has l.u.b. if and only if $\mathcal{A}$ has it. First note that we may suppose $\mathcal{A}$ countable, otherwise it cannot have upper bounds. Let then $\{a_0, a_1, \ldots\}$ be an enumeration of $\mathcal{A}$, and consider the chain

$$\begin{aligned} b_0 &= a_0 \\ b_{n+1} &= b_n \cup a_{n+1}. \end{aligned}$$

Now the chain $\{b_0 \leq b_1 \leq \cdots\}$ has l.u.b. if and only if it is eventually constant, and thus $\mathcal{A}$ has l.u.b. if and only if it has an upper bound which is the l.u.b. of a finite subset of it. $\square$

In particular, *no infinite ascending sequence of degrees has l.u.b.*

Exercise V.4.11 *The existence of a set of degrees without l.u.b. can be proved by the finite extension method V.3.13.* (Kleene and Post [1954]) (Hint: since any countable partial ordering can be embedded in the degrees below $\mathbf{0}'$, by V.2.9, there is a set of degrees below $\mathbf{0}'$ isomorphic to the rationals. If a subset of it has l.u.b. then this must be below $\mathbf{0}'$, and hence only countably many subsets can have l.u.b. But there are $2^{\aleph_0}$ such subsets, since they correspond to the reals, as Dedekind sections.)

Extensions of embeddings

We have shown in V.2.10 how the truth of one-quantifier questions reduces to embedding problems. Similarly, Shore [1978] and Lerman [1983] have noticed that *the truth of two-quantifier questions reduces to problems of extension of embeddings*. The idea is that the truth of $(\forall \vec{x})(\exists \vec{y})\varphi(\vec{x}, \vec{y})$ can be decided if one knows whether

$$(\forall \vec{x})[D(\vec{x}) \;\rightarrow\; (\exists \vec{y})\varphi(\vec{x}, \vec{y})],$$

for each of the finitely many D's describing a possible order relationship among the x's (i.e. a conjunction of atomic statements $x_i \leq x_j$ or negations of them). Moreover, φ can be reduced (by writing it into disjunctive normal form) to a finite disjunction of descriptions of possible order relationships among the x's and y's.

The general form of the problem we want to study is thus the following: given an embedding of a partial ordering P in $\mathcal{D}$, and an extension R of P, can we extend the embedding of P to an embedding of R? We cannot expect to be able to do this in general, since there are three restrictions we have already encountered:

1. $\mathcal{D}$ *has cardinality* $2^{\aleph_0}$ *and countable predecessor property.*
   Thus appropriate bounds on the cardinality of R are needed.

2. $\mathcal{D}$ *is an uppersemilattice.*
   Thus there is no hope to embed in $\mathcal{D}$ any partial ordering which does not respect the uppersemilattice structure of $\mathcal{D}$. If P is already embedded then we do not have to worry about it, but we certainly have to ask this of R. For example, given a_1, a_2 and a_3 such that $a_3 < a_1 \cup a_2$, we cannot introduce a new degree b which is above a_1 and a_2 but not above a_3.

3. *There are minimal pairs.*
   Thus we cannot expect to extend any given partial ordering by inserting new elements below given ones. For example, if a_1 and a_2 are a minimal pair, we cannot introduce a new degree b strictly between them and $\mathbf{0}$.

We are thus led to the following definition.

Definition V.4.12 *The partial ordering* $(R, \sqsubseteq_R)$ *is a* **consistent extension** *of the uppersemilattice* $(P, \sqsubseteq_P, \sqcup_P)$ *if:*

1. *R respects the uppersemilattice structure of P, i.e.*

$$a_1, a_2 \in P \wedge b \in R \wedge a_1, a_2 \sqsubseteq_R b \Rightarrow a_1 \sqcup_P a_2 \sqsubseteq_R b$$

2. *R is an end-extension of P, i.e.*

$$b \in R - P \wedge a \in P \Rightarrow b \not\sqsubseteq_R a.$$

The possible extensions of embeddings will always refer to consistent extensions.

Theorem V.4.13 (Kleene and Post [1954]) *Any embedding of a finite partial ordering P in $\mathcal{D}$ can be extended to an embedding of any finite consistent extension R of P.*

Proof. Since R is finite, it is possible to find a sequence of finitely many successive consistent extensions that add one element at a time, starting from P and ending with R. Thus we only need to know how to treat one-element extensions.

This amounts to proving that, given two finite sets $\{B_n\}_{n \in F}$ and $\{C_m\}_{m \in G}$ such that no C_m is recursive in $\oplus_{n \in F} B_n$, we can build A such that

1. for each $n \in F$, $B_n \leq_T A$

2. for each $m \in G$, $C_m \not\leq_T A$.

This can easily be done, by the same methods of the proof of Spector's Theorem. Since there are only finitely many sets to code, the first conditions can be satisfied ahead of time, simply by setting

$$\theta_0(x) \simeq B_e(z) \quad \text{if} \quad e \in F \wedge x = \langle e, z \rangle.$$

To satisfy the second condition let θ_s be given, and suppose we want to satisfy

$$C_m \not\simeq \{e\}^A$$

for some $m \in G$ and e. We look for two strings which are compatible with θ_s and e-split. If they exist, then we choose the one σ that gives a value different from C_m on the element on which they split, and by letting $\theta_{s+1} = \theta_s \cup \sigma$ we satisfy the requirement. If they do not exist then $\{e\}^A$, if total, will be recursive in θ_s as usual, and thus recursive in $\oplus_{n \in F} B_n$ because, by induction, θ_s is a finite modification of it. But since C_m is not recursive in $\oplus_{n \in F} B_n$, it cannot be equal to $\{e\}^A$. $\square$.

In the opposite direction, it is possible to show that the condition on R of being a consistent extension of P is necessary, not only for some P (as we have already argued), but also for *any* P. One part is trivial (since $\mathcal{D}$ an uppersemilattice), but it holds also for the other one, since any finite uppersemilattice with least element P is isomorphic to an initial segment of $\mathcal{D}$ (see p. 528), and thus some embedding of P cannot be extended by inserting new elements below given ones. This completely characterizes the possible finite extensions of embeddings of finite uppersemilattices into $\mathcal{D}$, and it allows to give *a decision procedure for the two-quantifier sentences of $\mathcal{D}$* (Shore [1978], Lerman [1983]). Schmerl has instead shown that *the three-quantifier theory of $\mathcal{D}$ is undecidable* (see Lerman [1983]).

Note that only notational changes are needed in the proof given above (coding one set at a time), to obtain one-element consistent extensions of countable embeddings (the requirements on the B_n's and C_m's being now that no C_m is recursive in any finite join of the B_n's). But not every consistent countable extension can be obtained by a sequence of one-element consistent extensions (e.g. when an infinite descending chain is added above a given element), and some more work (sketched in the exercise) is needed for the general case.

Exercise V.4.14 *Any embedding of a countable partial ordering P in $\mathcal{D}$ can be extended to an embedding of any countable consistent extension R of P.* (Kleene and Post [1954], Sacks [1961]) (Hint: this extends V.2.9. Given $\{C_n\}_{n \in \omega}$, build sets $\{A_n\}_{n \in \omega}$ recursively independent over the C_m's, and not introducing any new relationships among them, i.e. such that

$$A_n \not\leq_T (\oplus_{m \in \omega} C_m) \oplus (\oplus_{m \neq n} A_m)$$
$$C_n \leq_T (\oplus_{m \in F} C_m) \oplus (\oplus_{m \in \omega} A_m) \Rightarrow C_n \leq_T (\oplus_{m \in F} C_m),$$

for any finite set F. The second condition is ensured by the e-splitting method. If now $P = \{c_n\}_{n \in \omega}$ and $R - P = \{b_n\}_{n \in \omega}$, and the degrees of the C_n's are isomorphic to the ordering of the c_n's, we let

$$B_n = (\oplus_{c_m \sqsubseteq b_n} C_m) \oplus (\oplus_{b_m \sqsubseteq b_n} A_m).$$

Note that the second condition above, together with the fact that R respects the uppersemilattice structure of P, proves the crucial part, namely that

$$c_k \not\sqsubseteq b_n \Rightarrow C_k \not\leq_T B_n.$$

Indeed, suppose $C_k \leq_T B_n$. Since no new relationship among the C's is introduced, $C_k \leq_T (\oplus_{c_m \sqsubseteq b_n} C_m)$. Since R extends P, $c_k \sqsubseteq (\bigsqcup\{c_m : c_m \sqsubseteq b_n\})$. Since R respects joins in P, $(\bigsqcup\{c_m : c_m \sqsubseteq b_n\}) \sqsubseteq b_n$. So $c_k \sqsubseteq b_n$.)

Sacks [1961] shows that any embedding of a partial ordering P with cardinality less than $2^{\aleph_0}$ in $\mathcal{D}$ can be extended to an embedding of any consistent extension R of P such that $R - P$ is countable. He then uses this result to get the uncountable embeddings quoted on p. 462.

V.5   The Tree Method

Our knowledge of the structure $\mathcal{D}$ is beginning to shape up, but we have not answered yet an important question, regarding the density of the structure. We will now prove that not only is $\mathcal{D}$ not dense, it even has minimal elements. To get this result we cannot use the finite extension method V.3.13, as we know, because the minimal degrees are a meager class. It would be possible to use recursive coinfinite extensions (see V.6.9), but first we prove the existence of minimal degrees in a simpler way, using a more powerful method.

The idea of the finite extension method was to build an increasing sequence of strings σ_n, and then take their union $\bigcup_{n \in \omega} \sigma_n$. We may think of this as building a decreasing sequence of open sets $T_n = \{X : X \supseteq \sigma_n\}$, and then taking their intersection $\bigcap_{n \in \omega} T_n$. This naturally leads us to consider more general sets T_n.

Definition V.5.1 (Shoenfield [1966]) *A* **tree** *is a function T from initial segments to initial segments (which we will identify with sequences of 0's and 1's), with the following properties:*

1. $T(\sigma) \!\downarrow \wedge \tau \subseteq \sigma \Rightarrow T(\tau) \!\downarrow \wedge T(\tau) \subseteq T(\sigma)$

2. if one of $T(\sigma * 0), T(\sigma * 1)$ is defined, both are defined and incompatible.

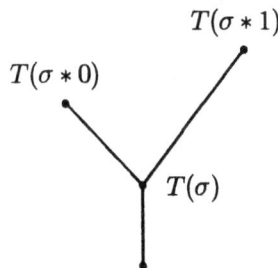

This way of looking at trees in not different from the one of IV.2.14: it only has a different emphasis. What really matters in a tree T is always its range (which is still a tree in the sense of IV.2.14), but to think of it in terms of a function is a useful tool: we can simply talk of $T(\sigma * 0)$ or $T(\sigma * 1)$, instead of 'the two smallest incompatible extensions of $T(\sigma)$ on the tree'. Similarly, to state that both or neither of $T(\sigma * 0), T(\sigma * 1)$ are defined is no loss of generality, since if e.g. we are only interested in the branch extending $T(\sigma * 0)$, we may define $T(\sigma * 1)$ as well, and then let the tree grow afterwards only above $T(\sigma * 0)$. A **total tree**, i.e. a tree which is total as a function from strings to strings, is nothing more than a closed perfect subset of $\{0,1\}^\omega$, in the terminology of Cantor [1883] (according to which a set $\mathcal{A} \subseteq \{0,1\}^\omega$ is perfect if there is no $f \in \mathcal{A}$ and an open set $\mathcal{O}$ such that $\mathcal{O} \cap \mathcal{A} = \{f\}$, i.e. it has no isolated points). Intuitively, *a total tree is just a set of branches or functions, none of which is isolated*.

We say that

1. A is on T, or A is a **branch** of T, if $T(\sigma) \subseteq A$ for infinitely many σ's

2. σ is on T if it is in the range of T

3. T^* is a **subtree** of T $(T^* \subseteq T)$ if every σ on T^* is also in T

4. T^* is the **full subtree** of T above σ if it consists of every string on T extending σ.

The concept of subtree is for trees what the concept of extension is for strings. The finite extension method consists of building a decreasing sequence $\{T_n\}_{n \in \omega}$ of trees, where T_0 is the **identity tree** (consisting of all strings, also called the full binary tree), and T_{n+1} is a full subtree of T_n. The method of trees is more general because it allows T_{n+1} to be any subtree of T_n.

The simplest application of the tree method will be with **recursive trees**, i.e. trees that are recursive as (total) functions from strings to strings (appropriately coded as numbers). We are really interested in the range of a tree, but the terminology is consistent: if a tree is recursive as a function, then so is its range as a set (to know if $\sigma \in T$ we generate the tree up to a level in which all branches have length at least equal to the length of σ, and see if σ is on T at that level). In later applications we will also deal with **partial recursive trees**, which need not be total: in this case the range will be r.e.

The method of trees used above can be recast in the same categorical framework of Section 3, by using a different topology than the one generated by basic open sets determined by finite strings. More precisely, consider any set $\mathcal{C}$ of recursive trees, with the following properties:

- the identity tree is in $\mathcal{C}$

- if $T \in \mathcal{C}$ and $\sigma \in T$, then the full subtree of T above σ is in $\mathcal{C}$.

Then $\mathcal{C}$ is the base of a topology on $\mathcal{P}(\omega)$ finer than the original one dealt with in Section 3. We can then say that:

1. $\mathcal{A} \subseteq \mathcal{C}$ is $\mathcal{C}$-**dense** if for any $T \in \mathcal{C}$ there is $S \subseteq T$ such that $S \in \mathcal{A}$

2. $\mathcal{A} \subseteq \mathcal{P}(\omega)$ is $\mathcal{C}$-**comeager** if it contains the intersection of countably many $\mathcal{C}$-dense sets.

A good deal of the theory of Baire Category can be developed in this generalized framework (Yates [1976]), in particular the Baire Category Theorem holds. Basic examples of $\mathcal{C}$ are given by the set of full trees (in which case we obtain the same notions as in Section 3), the set of recursive coinfinite conditions, and the set of recursive trees.

Hyperimmune-free degrees

Dealing with the next notion will provide a useful warm up with the tree method, before the real applications come. Moreover, the notion will be useful in extending results from T-degrees to other types of degrees (see VI.6.18).

Definition V.5.2 (Martin and Miller [1968]) *A degree* $a > 0$ *is* **hyper-immune-free** *if for every* $A \in a$ *and* $f \leq_T A$, f *is majorized by a recursive function, i.e. there is a recursive function* g *such that* $(\forall x)(f(x) \leq g(x))$.

The name 'hyperimmune-free' is justified below. The definition has been given in a form which makes it easier to deal with it in applications.

Exercises V.5.3 Hyperimmune degrees. (Martin and Miller [1968]) A degree is hyperimmune if it contains a hyperimmune set, see III.3.7.

a) *A degree is hyperimmune-free if and only if it is not hyperimmune.* (Hint: one direction follows from III.3.8. The other from the fact that if $\{a_n\}_{n\in\omega}$ enumerates A in increasing order, and $f \leq_T A$ is not majorized by any recursive function, then the set B defined as $A \oplus \{a_{f(n)} : n \in \omega\}$ is in the same degree as A, because $f \leq_T A$, and is hyperimmune. Otherwise, there would be a recursive function majorizing the elements of B in increasing order, and thus a recursive function majorizing f too, because $f(n) \leq a_{f(n)}$.)

b) *The hyperimmune degrees are closed upwards, and the hyperimmune-free degrees are closed downward.* (Hint: the second assertion holds by definition, the first follows from it by part a).)

c) *The set of hyperimmune-free degrees is meager.* (Hint: we have to build a hyperimmune set by the finite extension method V.3.13, which is immediate by definition: given any string σ and an infinite disjoint strong array, there is a string extending σ such that all sets extending it intersect one element of the strong array.)

d) *Every nonzero degree comparable with* $\mathbf{0}'$ *is hyperimmune.* (Hint: since $\mathbf{0}'$ contains a hyperimmune degree, by part b) so do all degrees above it. To show that every nonzero degree below it does too, modify III.3.13 by using the limit lemma IV.1.17. Precisely, given $A \leq_T \mathcal{K}$ nonrecursive, let A be the limit of g recursive. If $f(x)$ is the smallest stage where g gives the right value of $A(y)$, for all $y \leq x$, then f is increasing, and hence its range B is recursive in A. By part b), it is enough to show that B is hyperimmune. If it were not then A would be recursive, because to know the real value of $A(x)$ we just have to compute g at x, at the stage given by the x-th element of B.)

Since the set of hyperimmune-free degrees is meager, we cannot expect to build one member of it by the finite extension method V.3.13. By using trees we can instead easily obtain the result. For later use we isolate the steps needed in the proof.

Proposition V.5.4 Diagonalization Lemma. *Given e and a recursive tree T, there is a recursive tree $Q \subseteq T$ such that, for every A on Q, $A \not\leq \{e\}$.*

Proof. Since $T(0)$ and $T(1)$ are incompatible, at least one of them must disagree with $\{e\}$ on some x. If $T(i)$ is such, let Q be the full subtree of T above it. $\square$

Proposition V.5.5 Totality Lemma (Martin and Miller [1968]) *Given e and a recursive tree T, there is a recursive tree $Q \subseteq T$ such that one of the following holds:*

1. *for every A on Q, $\{e\}^A$ is not total*

2. *for every A on Q, $\{e\}^A$ is total and*

$$(\forall n)(\forall \sigma)(|\sigma| = n \Rightarrow \{e\}^{Q(\sigma)}(n)\downarrow),$$

   where $|\sigma|$ is the length of σ.

Proof. See if

$$(\exists \sigma \in T)(\exists x)(\forall \tau \supseteq \sigma)(\tau \in T \Rightarrow \{e\}^\tau(x)\uparrow).$$

If so, choose such a σ: for any string τ extending it, $\{e\}^\tau$ is undefined on a fixed element. It is then enough to let Q be the full subtree of T above σ. Then case 1 holds.

Otherwise, for any string σ and any x there is an extension τ of σ that makes $\{e\}^\tau$ defined on x. We can then build a tree by successive levels, and make $\{e\}^A$ converge on more and more elements. Precisely, let Q be defined as follows. First we start with

$$Q(\emptyset) = \text{least } \tau \in T \text{ such that } \{e\}^\tau(0)\downarrow.$$

Inductively, given $Q(\sigma)$ on T we know that there is an extension of it on T, say $T(\tau)$ for some τ, such that $\{e\}^{T(\tau)}(|\sigma|)\downarrow$. Then let

$$Q(\sigma * i) = T(\tau * i),$$

for $i = 0, 1$, so that they are two incomparable extensions of $Q(\sigma)$. $\square$

Proposition V.5.6 (Martin and Miller [1968]) *There are hyperimmune-free degrees.*

Proof. The requirements on A are:

$$R_{2e} \quad : \quad A \not\equiv \{e\}$$
$$R_{2e+1} \quad : \quad \{e\}^A \text{ total} \; \Rightarrow \; \text{for some recursive } g, (\forall n)(\{e\}^A(n) \leq g(n)).$$

Define a sequence of recursive trees T_n such that $T_{n+1} \subseteq T_n$, and all branches of T_{n+1} satisfy R_n. Precisely:

$$
\begin{aligned}
T_0 \quad &= \quad \text{identity tree} \\
T_{2e+1} \quad &= \quad \text{the } Q \text{ of the Diagonalization Lemma, for } T = T_{2e} \\
T_{2e+2} \quad &= \quad \text{the } Q \text{ of the Totality Lemma, for } T = T_{2e+1}.
\end{aligned}
$$

Let now $A \in \bigcap_{n \in \omega} T_n$: it exists because $T_n(\emptyset) \subseteq T_{n+1}(\emptyset)$, since $T_n \subseteq T_{n+1}$, and thus we just have to consider $A = \bigcup_{n \in \omega} T_n(\emptyset)$.

A is not recursive because it is on T_{2e+1}, and hence it is different from $\{e\}$. Suppose now $\{e\}^A$ is total. Since A is on T_{2e+2}, it must be that the second case holds in the Totality Lemma, hence

$$\{e\}^A(n) \leq \max_{|\sigma|=n} \{e\}^{T_{2e+2}(\sigma)}(n),$$

since $\{e\}^A(n)$ is already defined at the n-th level of the tree, for each possible branch A. Thus $\{e\}^A$ is majorized by a recursive function. $\quad\Box$

Exercises V.5.7 (Martin and Miller [1968]) a) *There is a hyperimmune-free degree below $\mathbf{0}''$.* (Hint: in the construction above, only two-quantifier questions were asked, and thus the degree is less than or equal to $\mathbf{0}''$. To show it is strictly below it, note that $\mathbf{0}''$ is hyperimmune, because comparable with $\mathbf{0}'$.)

b) *There is an infinite ascending sequence of hyperimmune-free degrees.* (Hint: by relativization, there is a hyperimmune-free degree above any hyperimmune-free degree.)

c) *There are $2^{\aleph_0}$ hyperimmune-free degrees.* (Hint: build a tree of sets, all of whose branches are of hyperimmune-free degree. This can be achieved by building a **tree of trees**, i.e. a function T from pairs of strings to strings, such that for each σ the function $T_\sigma(\tau) = T(\sigma, \tau)$ is a tree. The desired tree will be given by the function $T_\sigma(\emptyset)$. Start by $T_\emptyset$ being the identity tree. At any stage, instead of just building one subtree of the given tree, build two, above two incomparable strings. Precisely, given T_σ, apply the Diagonalization or the Totality Lemma, depending on whether $|\sigma|$ is $2e$ or $2e + 1$, to the full subtree of T_σ above $T_\sigma(0)$ to get $T_{\sigma*0}$, and above $T_\sigma(1)$ to get $T_{\sigma*1}$.)

Jockusch [1969b] shows that a hyperimmune degree contains a bi-immune set (while the converse does not necessarily hold), and strengthens V.5.6 by producing a degree without bi-immune sets. See Jockusch [1972] and Simpson [1977] for more on this topic.

The results of this subsection show that a characterization of the degrees containing hyperimmune sets cannot be simple. With regard to other immunity properties introduced in Chapter II, we quote the following results:

1. *a immune* $\Leftrightarrow$ *$a > 0$*
   (Dekker and Myhill [1958], see II.6.13)

2. *$a' \geq 0''$* $\Rightarrow$ *a cohesive* $\Rightarrow$ *a hyperhyperimmune* $\Rightarrow$ *$a' > 0'$*
   (Jockusch [1969a], [1973a])

3. *If $a \leq 0'$, then a hyperhyperimmune* $\Leftrightarrow$ *a cohesive* $\Leftrightarrow$ *$a' = 0''$*
   (Cooper [1972]).

Minimal degrees

We turn now to a most important application of the tree method. We want to build a minimal degree, i.e. a nonrecursive set A such that

$$C \leq_T A \Rightarrow C \text{ recursive or } C \equiv_T A.$$

The proof we give is a simplification, due to Shoenfield [1966], of the original construction of Spector, which is given in Section 6.

When we constructed a minimal pair, we used the e-splitting method: if there was no e-splitting then we got a recursive set, while in the opposite case we fixed one side, and diagonalized against it on the other, thus taking full advantage of the fact of being able to work with two sets. Here we will have to get by with just one set, and this will complicate our life a bit, but the ideas will always be the same. The needed change is to use e-splittings in a global way.

Definition V.5.8 *T is an **e-splitting tree** if, for every σ, $T(\sigma*0)$ and $T(\sigma*1)$ e-split.*

The interest of the notion comes from the following lemma.

Proposition V.5.9 (Spector [1956]) *Given e, a recursive tree T, and A on T, if $\{e\}^A$ is total then:*

1. *if there is no e-splitting on T, $\{e\}^A$ is recursive*

2. *if T is e-splitting, $A \leq_T \{e\}^A$.*

Proof. Since $\{e\}^A$ is total, given x we know that $\{e\}^A(x)$ converges, and thus there must be $\sigma \subseteq A$ such that $\{e\}^\sigma(x)$ converges, and gives the right value. Since A is on T, by monotonicity we may suppose $\sigma \in T$. If there is no e-splitting on T, to compute $\{e\}^A(x)$ is then enough to look for any string τ on T such that $\{e\}^\tau(x)$ converges, since its value must be equal to $\{e\}^\sigma(x)$ (otherwise σ and τ would e-split on x), and hence to $\{e\}^A(x)$.

Suppose now that T is instead e-splitting. We show how to generate increasingly long segments of A recursively in $\{e\}^A$. Given $T(\sigma) \subseteq A$, since A is on T either $T(\sigma * 0)$ or $T(\sigma * 1)$ will be included in A, and we have to decide which one. Being T e-splitting, there is an x such that

$$\{e\}^{T(\sigma * 0)}(x) \not\simeq \{e\}^{T(\sigma * 1)}(x),$$

with both sides converging. Only one of them can agree with $\{e\}^A(x)$, and this determines which of the two strings is contained in A. Precisely, $T(\sigma * i) \subseteq A$ if $\{e\}^{T(\sigma * i)}(x) \simeq \{e\}^A(x)$. $\square$

Proposition V.5.10 Minimality Lemma (Spector [1956]) *Given e and a recursive tree T, there is a recursive tree $Q \subseteq T$ such that one of the following holds:*

1. for every A on Q, $\{e\}^A$ total $\Rightarrow \{e\}^A$ recursive

2. for every A on Q, $\{e\}^A$ total $\Rightarrow A \leq_T \{e\}^A$.

Proof. The result follows from the previous lemma, if we just build Q with either no e-splitting on it, or as an e-splitting tree.

If there is a string on T with no e-splitting above it, take Q as the full subtree above it: then Q has no e-splitting.

If any string on T has two e-splitting extensions (i.e. two strings on T which extend it and e-split), then we can build an e-splitting subtree Q of T, by induction: given $Q(\sigma)$, let $Q(\sigma * 0)$ and $Q(\sigma * 1)$ be two e-splitting extensions of it. $\square$

Note that the last lemma is a modification of the minimal pair construction. If there is a string on T with no e-splitting extensions, we just take the full subtree above it, and have $\{e\}^A$ recursive for any A on it, as in the minimal pair construction. If any string on T has e-splitting extensions, take two of them, say σ_0 and σ_1. Then, for some x, $\{e\}^{\sigma_0}(x) \not\simeq \{e\}^{\sigma_1}(x)$, with both sides converging. Certainly $\{e\}^A(x)$, whatever A may be, disagrees with one of them, say $\{e\}^{\sigma_i}$. Then A does not agree with σ_i. If σ_0 and σ_1 were the only possible beginnings of A, then A would have to agree with the other string. And we can force this to happen, by just discharging every other possibility, i.e. by making σ_0 and σ_1 the first level of our subtree. And we can continue this, thereby building an e-splitting tree such that any set A on it is then recursive in $\{e\}^A$.

Theorem V.5.11 (Spector [1956]) *There exists a minimal degree.*

Proof. We only have to satisfy the requirements

$$R_{2e} \quad : \quad A \not\simeq \{e\}$$
$$R_{2e+1} \quad : \quad C \simeq \{e\}^A \Rightarrow C \text{ recursive or } A \leq_T C.$$

We build a sequence of recursive trees, as follows:

$$
\begin{aligned}
T_0 \quad &= \quad \text{identity tree} \\
T_{2e+1} \quad &= \quad \text{the } Q \text{ of the Diagonalization Lemma, for } T = T_{2e} \\
T_{2e+2} \quad &= \quad \text{the } Q \text{ of the Minimality Lemma, for } T = T_{2e+1}.
\end{aligned}
$$

Then $A = \bigcup_{n \in \omega} T_n(\emptyset)$ satisfies the requirements. $\square$

Having obtained minimal degrees by the tree method, and knowing that we cannot obtain them by the finite extension method V.3.13, it is natural to investigate the situation more thoroughly. More precisely, we would like to know how much of the machinery just introduced is really needed, in particular which methods are sufficient, and which are necessary. To the first question we will answer in V.6.9, where it will be shown that *recursive coinfinite extensions are sufficient*, giving a sort of best possible answer. The second question is less precise, but we can argue that *total e-splitting trees are not necessary*, because they can be combined with totality requirements and automatically produce hyperimmune-free minimal degrees (see the exercises below). Now, hyperimmune-free degrees cannot be comparable with $\mathbf{0'}$, but in XI.4.5 we will show that there are minimal degrees below $\mathbf{0'}$, which must then be built by partial trees. In XI.4.3 we will also show that, for the construction of minimal degrees below $\mathbf{0'}$, partial e-splitting trees are not only sufficient, but also necessary (Chong [1979]).

Exercises V.5.12 a) *There exists a minimal degree below* $\mathbf{0''}$. (Spector [1956]) (Hint: in the proof above, only two-quantifier questions were asked, and thus the degree is less than or equal to $\mathbf{0''}$. To show it is strictly below it, note that $\mathbf{0''}$ is obviously not minimal.)

b) *There are $2^{\aleph_0}$ minimal degrees.* (Lacombe) (Hint: build a tree of minimal degrees, as in V.5.7.c.)

c) *The diagonalization steps in the construction of minimal degrees are not needed.* (Posner and Epstein [1978]) (Hint: similar to V.2.23, this time using trees T_e which are either e-splitting or without e-splittings, and A on all of them. Another way is given by part b), because we can build a tree of sets satisfying all the minimality conditions, and only countably many of these sets can be recursive.)

Exercises V.5.13 Hyperimmune-free degrees. a) *There is a minimal, hyperimmune-free degree.* (Martin and Miller [1968]) (Hint: combine the proofs of V.5.11 and V.5.6.)

b) *There is a hyperimmune-free degree that is not minimal.* (Martin and Miller [1968]) (Hint: by V.5.7.b.)

That *there is a minimal degree which is not hyperimmune-free* follows from the existence of minimal degrees below $\mathbf{0}'$, see XI.4.5.

Exercises V.5.14 Jumps of minimal degrees. a) *Not every minimal degree realizes the least possible jump.* (Sasso [1974]) (Hint: build a set A of minimal degree such that $A' \not\leq_T A \oplus \mathcal{K}$. The requirements are treated by the following lemma: given e and a recursive tree T, there is a recursive tree $Q \subseteq T$ such that, for any A on Q, $A' \not\simeq \{e\}^{A \oplus \mathcal{K}}$. The idea is that the question: 'does A always go left at even levels of T?' can be phrased as a question about A', since it involves only one quantifier over A. Let a be such that

$$\{a\}^\sigma(x) \simeq 0 \Leftrightarrow \sigma \text{ branches right at some even level of } T.$$

By definition of jump, $a \in A' \Leftrightarrow \{a\}^A(a)\downarrow$. See if, for some $\tau \in T$, $\{e\}^{\tau \oplus \mathcal{K}}(a) \simeq 0$. If so, to diagonalize we want $a \in A'$, i.e. for every A we require $\{a\}^A(a)\downarrow$, and hence A must branch right at some even level of T. Take Q as the full subtree of T above $\tau * 11$, since $\tau * 11$ branches right twice, and one of $\tau * 1$ and $\tau * 11$ is at an even level. Otherwise, let Q be the subtree of T that branches left at every even level. Note that we use only even levels, because we still need to get a tree, and this is taken care of at odd levels.)

b) *Every degree above $\mathbf{0}''$ is the double jump of a minimal degree.* (Cooper [1973]) (Hint: note that if we also use the Totality Lemma in the construction of a set of minimal degree we can decide, recursively in the construction and hence in $\emptyset'''$, whether $\{e\}^A$ is total or not, depending on which case applies in the lemma. Any question about the second jump of A or, equivalently, with two quantifiers over A, can be rephrased as a question about the totality of a given function recursive in A, and thus $A'' \leq_T A \oplus \emptyset''$. We can now build a tree of $T \leq_T \emptyset''$, all of whose branches A have this property. If $\emptyset'' \leq_T C$ let A be the branch of T determined by C, i.e. $A = \bigcup_{\sigma \subseteq C} T(\sigma)$. Then

$$A \leq_T C \oplus T \leq_T C \oplus \emptyset'' \leq_T C,$$

and hence $A \oplus \emptyset'' \leq_T C$. Conversely,

$$C \leq_T A \oplus T \leq_T A \oplus \emptyset'',$$

and thus $C \equiv_T A \oplus \emptyset'' \equiv_T A''$.)

Jockusch and Posner [1978] show that, *for any minimal degree a, $a' = (a \cup \mathbf{0}')'$*, see XI.2.15. Cooper [1973] proves that actually *every degree above $\mathbf{0}'$ is the jump of a minimal degree*, but this is much more difficult to prove, see Epstein [1975] or Posner [1981] for a proof by recursive approximations, and Lerman [1983] for one by approximations recursive in $\emptyset'$.

Exercises V.5.15 Autoreducible sets. A set A is **autoreducible** if, for each x, the question 'is x in A?' can be answered recursively in A, without ever asking the oracle about x. Thus every single element encodes redundant information, retrievable from the rest of the set. A nonrecursive degree containing only autoreducible sets is called **completely autoreducible**.

a) *Every m-degree contains autoreducible sets.* (Trakhtenbrot [1970]) (Hint: consider $A \oplus A$. Then $2x$ and $2x + 1$ give the same information.)

b) *An introreducible set (II.6.7) is autoreducible.* (Jockusch and Paterson [1976]) (Hint: let B be nonautoreducible. We build, by the finite extension method V.3.13, an infinite subset A of B such that $B \not\leq_T A$. To get $B \neq \{e\}^A$ at stage n, given σ_n look for an extension σ of it such that $\sigma^{-1}(1) \subseteq B$, and there is x such that every $\tau \supseteq \sigma$ such that $\tau^{-1}(1) \subseteq B$ satisfies $\{e\}^\tau(x) \not\simeq B(x)$. Such a string must exist, since B is not autoreducible. Thus let $\sigma_{n+1} = \sigma_n$.)

c) *The set of completely autoreducible degrees is meager.* (Trakhtenbrot [1970]) (Hint: build, by the finite extension method, a set A such that, for every e, there is x such that $A(x) \not\simeq \{e\}^{A-\{x\}}(x)$.)

d) *There exists a completely autoreducible degree.* (Jockusch and Paterson [1976]) (Hint: build a set as in the minimal degree construction, with double e-splittings taking the place of e-splittings, i.e. using trees T such that, for any σ, there are two distinct elements on which $T(\sigma * 0)$ and $T(\sigma * 1)$ e-split. If T is doubly e-splitting, A is on T, and $\{e\}^A$ is a characteristic function, then $\{e\}^A$ is autoreducible: to compute $\{e\}^A(x)$ with oracle $\{e\}^A$ without using this value, first build large enough segments of A, with oracle $\{e\}^A$ but avoiding the particular value $\{e\}^A(x)$, which is possible thanks to the double e-splittings. Then compute $\{e\}^A(x)$ with oracle A.)

Obviously, the completely autoreducible degree just built is minimal. The existence of minimal, not completely autoreducible degrees and of completely autoreducible, not minimal degrees will be proved in V.6.10.d and V.6.16.e. Jockusch and Paterson [1976] show that nonzero r.e. degrees and degrees above $\mathbf{0}'$ are not completely autoreducible.

Minimal upper bounds $\star$

Exactly as we generalized the notion of minimal pair to that of exact pair, we can extend the notion of minimal degree to that of minimal upper bound.

Definition V.5.16 *A degree a is a* **minimal upper bound** *for a set of degrees C if*

1. *it is a strict upper bound to C, i.e. $(\forall c \in C)(c < a)$*

2. *there is no strict upper bound to C below it, i.e.*

$$(\forall b)[b \leq a \wedge (\forall c \in C)(c < b) \Rightarrow b = a].$$

A **minimal cover** *for a degree b is a minimal upper bound for $\{b\}$, i.e. a degree $a > b$ such that there is no degree strictly between a and b.*

A **strong minimal cover** *for* b *is a degree* $a > b$ *such that anything strictly below it is bounded by* b, *i.e.*

$$(\forall c)(c \le a \Rightarrow c \le b \vee c = a).$$

Both notions of minimal cover are plausible generalizations of the notion of minimal degree above a given set. Relativizations of the results about minimal degrees produce results about minimal covers, but not about strong minimal covers. For example, V.5.11 relativized shows that *every degree has a minimal cover*. On the other hand, *not every degree has a strong minimal cover* (Shoenfield [1959]), e.g. $0'$ does not (by V.2.26 any degree above $0'$ is the join of two strictly smaller degrees, and thus it cannot be a strong minimal cover of it, otherwise the two degrees would be below $0'$, and so would be their join). Jockusch [1981] shows that *the set of degrees without strong minimal covers is comeager*, and thus the class of degrees with a strong minimal cover is meager. It is not known whether the minimal degrees are included in it, i.e. if every minimal degree has a strong minimal cover.

Exercises V.5.17 *There is a cone of minimal covers.* (Jockusch [1973]) (Hint: every minimal degree has a minimal cover, and thus the set of degrees which are not minimal covers cannot contain a cone. By V.1.16, if AD holds then the set of degrees which are minimal covers must contain a cone. But this set is arithmetical, and thus only the Axiom of Determinacy for arithmetical sets is needed, which is provable in ZFC by Martin [1975].) A direct proof of the result, not using Determinacy, will be given in XII.3.13.

Spector's Theorem showed that any countable ideal has an exact pair. The proof was an extension of the construction of minimal pairs, plus a coding method that allowed us to push the set constructed above given degrees. We now prove a similar result for minimal upper bounds, by extending the construction of minimal degrees. We also need a coding method, and this is provided by the next notion, implicit in the proof of Spector's Theorem.

Definition V.5.18 (Sacks [1971]) *A* **recursively pointed tree** T *is a tree which is recursive in all of its branches, i.e.* $T \le_T A$ *whenever A is on T.*

Exercises V.5.19 a) *If T is recursively pointed, then T has branches of every degree above the degree of T.* (Sacks [1971]) (Hint: given $T \le_T A$, consider $B = \bigcup_{\sigma \subseteq A} T(\sigma)$.)

b) *If T is recursively pointed, $Q \subseteq T$, and $Q \le_T T$, then Q is also recursively pointed, and $Q \equiv_T T$.* (Sacks [1971]) (Hint: Q is pointed because if $A \in Q \subseteq T$ then $Q \le_T T \le_T A$. If A is the leftmost branch of Q then $T \le_T A$ by pointedness, and $A \le_T Q$ by definition, so $T \le_T Q$.)

Proposition V.5.20 (Sacks [1971]) *If T is recursively pointed and $T \leq_T A$, then there is some recursively pointed tree $Q \subseteq T$ such that $Q \equiv_T A$.*

Proof. We have to code A in Q, and at every even level $2n$ we thin the tree down by just taking the right or left branch, according to whether n is in A or not. Precisely, we define Q by induction. Given $Q(\sigma)$ on T, let τ be a string such that $Q(\sigma) = T(\tau)$, which exists because $Q(\sigma)$ is on T, by induction. Let

$$Q(\sigma * i) = T(\tau * A(|\sigma|) * i).$$

Then $Q \leq_T T \oplus A$, so $Q \leq_T A$ (because $T \leq_T A$). From any path B of Q and T itself we can recover A, and so $A \leq_T B$ (since $T \leq_T B$ by pointedness). To have $A \leq_T Q$ it is enough to choose any $B \leq_T Q$, e.g. the leftmost branch.

Finally, Q is pointed: given $B \in Q$ we can recover Q itself from B and T, by the uniformity of the construction, and again $T \leq_T B$, so $Q \leq_T B$.   □

Proposition V.5.21 (Sacks [1963]) *Every countable set of degrees has a minimal upper bound.*

Proof. Let $\mathcal{A} = \{a_0, a_1, \dots\}$ be a countable set of degrees. We may suppose that $\mathcal{A}$ is a chain, otherwise we can just consider the chain defined as

$$b_0 = a_0 \qquad b_{n+1} = b_n \cup a_{n+1},$$

whose minimal upper bounds are exactly those of $\mathcal{A}$. We can also assume that the chain has no greatest element, otherwise the result is obtained by taking any minimal cover of it. This assumption is made only to avoid direct diagonalization against all the degrees in $\mathcal{A}$.

Let then $A_n \in a_n$. We will build a set $A \in \bigcap_{n \in \omega} T_n$, where T_n is a recursively pointed tree of degree a_n. This automatically implies that A is an upper bound for $\mathcal{A}$.

To get minimality, we use the following extension of the Minimality Lemma. Given e and a tree T, there is a tree $Q \subseteq T$ such that one of the following holds:

1. *for every A on Q, $\{e\}^A$ total $\Rightarrow \{e\}^A \leq_T Q$*

2. *for every A on Q, $\{e\}^A$ total $\Rightarrow A \leq_T \{e\}^A \oplus Q$.*

This is proved by the proof of V.5.10, simply because when the relevant trees are not recursive then we have to take them into account in our computations.

The construction is as follows. Let T_0 be a recursively pointed tree of the same degree as A_0 (which can be obtained by starting from the identity tree, and applying V.5.20). Given T_e recursively pointed of the same degree as A_e, first get $T \subseteq T_e$ recursively pointed of the same degree as A_{e+1}, by V.5.20, since $A_e \leq_T A_{e+1}$. Then let T_{e+1} be the Q of the Minimality Lemma stated above, which is still recursively pointed, and of the same degree as A_{e+1}.

If $A \in \bigcap_{n \in \omega} T_n$, suppose $\{e\}^A$ is total. By construction, there are two cases:

- $\{e\}^A \leq_T T_{e+1}$
  Then $\{e\}^A \leq_T A_{e+1}$ (since T_{e+1} has the same degree as A_{e+1}), A is below some element of the chain, and it is not an upper bound to it.

- $A \leq_T \{e\}^A \oplus T_{e+1}$
  If $\{e\}^A$ is itself an upper bound to the chain then $T_{e+1} \leq_T \{e\}^A$, and hence $A \leq_T \{e\}^A$.

Thus no upper bound to the chain is strictly below A, and A is a minimal upper bound. $\square$

Exercises V.5.22 a) *Every countable set of degrees has $2^{\aleph_0}$ minimal upper bounds.* (Sacks [1963]) (Hint: build a tree of minimal upper bounds.)

b) *Every countable ascending chain of hyperimmune-free degrees has a hyperimmune-free minimal upper bound* (Martin and Miller [1968]). Note that the result fails for countable sets in place of chains, because the join of hyperimmune-free degrees in not necessarily hyperimmune-free, see V.6.10.c.

König's Lemma and Π_1^0 classes $\star$

In this subsection we consider again general trees as in IV.2.14 (i.e. as sets of sequence numbers closed under subsequences), and study their infinite branches. The classical result in this field is the following.

Theorem V.5.23 König's Lemma (König [1926]) *An infinite tree in which every node has only finitely many immediate successors has an infinite branch.*

Proof. Let T be such a tree. We define an infinite branch by induction. We start with σ_0. Given σ_n with infinitely many extensions on T, let σ_{n+1} be an immediate successor of σ_n with infinitely many extensions on T. It exists because σ_n has infinitely many extension on T, but only finitely many immediate successors. Thus at least one of them must have infinitely many extensions on T. $\square$

The rest of this subsection is devoted to an analysis of constructive versions of this result. We first investigate **binary trees** which, by definiteness, we define as sets of sequences of 0's and 1's.

Definition V.5.24 *A class of sets is a Π_1^0 class if it is the set of infinite branches of some infinite recursive binary tree.*

The reason for the name is obvious: if T is a recursive tree, then A is an infinite branch of it if and only if $(\forall x)(\hat{c}_A(x) \in T)$. More generally, if P is a recursive predicate then the class of sets A such that $(\forall x)P(\hat{c}_A(x))$ is a Π_1^0 class, since it is the set of infinite branches of the tree obtained by closing P under subsequences.

Note that, by König's Lemma, a Π_1^0 class is never empty. As a fundamental example, *the sets separating two disjoint r.e. sets A and B form a Π_1^0 class*:

$$C \in \mathcal{S}_{A,B} \Leftrightarrow (\forall x)[(x \in A \to x \in C) \wedge (x \in B \to x \notin C)].$$

Explicitly, a recursive tree whose branches are exactly the members of $\mathcal{S}_{A,B}$ is the following:

$$x \in T_{A,B} \Leftrightarrow Seq(x) \wedge x \text{ is correct up to stage } ln(x),$$

where 'x is correct up to stage s' means that for every $i \le ln(x)$, if $i \in A_s$ then $(x)_i = 1$, and if $i \in B_s$ then $(x)_i = 0$. In other words, we seal off a branch of $T_{A,B}$ as soon as we discover that it is incorrect.

Note that $T_{A,B}$ has an infinite branch if and only if A and B are disjoint. Moreover, an infinite branch of $T_{A,B}$ is the characteristic function of a set separating A and B. We thus immediately have:

Proposition V.5.25 Failure of the recursive version of König's Lemma (Kleene [1952]) *There is an infinite recursive binary tree without infinite recursive branches.*

Proof. If A and B are recursively inseparable (II.2.5), then $T_{A,B}$ has no infinite recursive branches. $\square$

It is the existence of finite branches, i.e. the consideration of partial recursive trees in the sense of this section, that produces the possibility of recursive trees without infinite recursive branches. If we restricted our attention to total recursive trees, then the leftmost branch would be recursive. Instead, the leftmost infinite branch of a partial recursive tree need not even be r.e., although it always has r.e. degree (see V.5.26 and V.5.33.a).

Exercise V.5.26 *There is an infinite recursive binary tree which has no infinite r.e. branch.* (Jockusch and Soare [1972a]) (Hint: let S be Post's simple set (III.2.11), and consider a disjoint strong array $\{D_{f(x)}\}_{x \in \omega}$ intersecting $\overline{S}$. Let

$$A \in \mathcal{C} \Leftrightarrow A \subseteq \overline{S} \wedge (\forall x)(D_{f(x)} \cap A \ne \emptyset).$$

$\mathcal{C}$ is a nonempty Π_1^0 class because it contains $\overline{S}$, and it has no r.e. member because its members are immune.)

Infinite binary trees without infinite recursive branches must be quite fat:

Proposition V.5.27 (Jockusch and Soare [1972]) *A Π_1^0 class without recursive members has cardinality $2^{\aleph_0}$.*

Proof. It is enough to show that every isolated infinite branch of a recursive tree is recursive: then if there are no infinite recursive branches every branch splits, and the number of infinite branches is $2^{\aleph_0}$.

Suppose A is the unique branch of T above $\sigma \in T$. To decide whether $x \in A$, consider all strings of T above σ and of length greater than x. Generate T until all of them except one die out. The only surviving one is contained in A, and is defined on x. Thus $\sigma(x)$ tells whether x is in A or not. □

Exercise V.5.28 *A Π_1^0 class without recursive members is meager.* (Jockusch and Soare [1972]) (Hint: a Π_1^0 class is closed, since membership in $\overline{A}$ is determined by a finite initial segment, and thus by an open set contained in $\overline{A}$. If $(\forall \sigma)(\exists A \supseteq \sigma)(A \notin \mathcal{A})$ then $(\forall \sigma)(\exists \tau \supseteq \sigma)(\forall A \supseteq \tau)(A \notin \mathcal{A})$, and by V.3.6 $\overline{A}$ is comeager. Otherwise, $(\exists \sigma)(\forall A \supseteq \sigma)(A \in \mathcal{A})$, and one such A is recursive.)

Since the recursive sets do not provide, in general, witnesses for branches of any infinite binary tree, we introduce the following notion:

Definition V.5.29 *A class of sets $\mathcal{A}$ is a **basis** for Π_1^0 classes if every Π_1^0 class has an element in $\mathcal{A}$. A class of degrees is a basis if so is the class of sets with degrees in it.*

We now start a search of bases for Π_1^0 classes. First we show that it will be impossible to find a single best answer.

Proposition V.5.30 *The intersection of all bases for Π_1^0 classes is the class of recursive sets, which is not a basis. In particular, there is no least basis.*

Proof. Every recursive set must be in every basis, because given B recursive the condition $A = B$ defines a Π_1^0 class with B as the only member.

To show that the intersection of all bases is the class of recursive sets, it is enough to show that given A nonrecursive there is a basis $\mathcal{B}$ not containing A. We build $\mathcal{B}$ by putting an element of each Π_1^0 class in it. Given a Π_1^0 class, there are two cases: either it has recursive elements, in which case we can put one in $\mathcal{B}$, or (by the proposition above) it has $2^{\aleph_0}$ elements, and thus we can choose one different from A. □.

A simple analysis of the proof of König's Lemma provides the first result.

Proposition V.5.31 Kreisel's Basis Lemma (Kreisel [1950]) *An infinite recursive binary tree has a Δ_2^0 infinite branch.*

Proof. Let T be a given infinite recursive binary tree. Given $\sigma_n \in T$, we have to decide whether to choose σ_{n+1} as $\sigma_n * \langle 0 \rangle$ or $\sigma_n * \langle 1 \rangle$. We see if

$$(\forall m > n)(\exists \tau \in T)(|\tau| = m \wedge \tau \supseteq \sigma_n * \langle 0 \rangle).$$

If yes, we let $\sigma_{n+1} = \sigma_n * \langle 0 \rangle$. Otherwise, $\sigma_{n+1} = \sigma_n * \langle 1 \rangle$. Let $A = \bigcup_{n \in \omega} \sigma_n$. Then $A \in T$.

Since the quantifier on τ is bounded, because there are only finitely many strings of length m, and T is recursive, the question is Π_1^0. It can thus be answered recursively in $\mathcal{K}$. Then $A \leq_T \mathcal{K}$, and $A \in \Delta_2^0$. $\quad\square$

Kreisel's Basis Lemma has been improved by Shoenfield [1960a], who proved that there is always a branch of degree less than $\mathbf{0}'$. The next result is much stronger.

Theorem V.5.32 Low Basis Theorem (Jockusch and Soare [1972])
The degrees $\mathbf{a}$ such that $\mathbf{a}' = \mathbf{0}'$, called low degrees, form a basis for Π_1^0 classes.

Proof. We proceed as in V.2.21, on trees. Let T be a given infinite recursive binary tree. We want to build A on T such that $e \in A'$ can be decided recursively in $\mathcal{K}$.

Let $T_0 = T$. Given T_e, to decide whether $e \in A'$ we consider the set

$$U_e = \{\sigma \in T_e : \{e\}^\sigma(e)\uparrow\}.$$

There are two cases:

1. *U_e is infinite*
   Then we can let $T_{e+1} = U_e$, and for every A on T_{e+1} we will have $e \notin A'$. Note that U_e is indeed a tree, being closed under subsequences (since if a string does not decide a computation, neither does any substring of it).

2. *U_e is finite*
   Then we can let $T_{e+1} = T_e$. For every A on T_{e+1} we will have $e \in A'$, since $\{e\}^\sigma(e)$ is undefined for at most finitely many strings on T_{e+1}, and thus it must converge for any string which is big enough.

Since the case distinction is recursive in $\mathcal{K}$, if $A \in \bigcap_{e \in \omega} T_e$ then $A' \leq_T \mathcal{K}$. And clearly $A \in T$, because $T_0 = T$. $\quad\square$

Exercises V.5.33 (Jockusch and Soare [1972], [1972a]) a) *The r.e. degrees form a basis for Π_1^0 classes.* (Hint: given an infinite recursive binary tree T, consider its leftmost infinite branch A. If

$$\sigma \in B \leftrightarrow \sigma \in T \wedge \sigma \text{ is on the left of } A$$

then $A \equiv_T B$. Moreover, B is r.e. because if $\sigma \in T$ is in B then we discover it by generating all strings of T of length up to n, for n big enough.)

b) *The r.e. degrees strictly below $\mathbf{0}'$ do not form a basis for Π_1^0 classes.* (Hint: the example given in V.5.26 consists only of effectively immune sets. They, by the proof of III.2.18, cannot have r.e. degree strictly below $\mathbf{0}'$.)

c) *If $b \geq \mathbf{0}'$ then $\mathbf{0}$ and the degrees with jump b form a basis for Π_1^0 classes.* (Hint: let T be an infinite recursive binary tree without infinite recursive branches. Since it has $2^{\aleph_0}$ branches, it is possible to build a total subtree $Q \leq_T \mathcal{K}$ all of whose branches force the jump, as in V.5.32. If $B \leq_T \mathcal{K}$ then the branch determined by B, i.e. turning right at level n if $n \in B$ and left otherwise, has jump B.)

Proposition V.5.34 (Jockusch and Soare [1972]) *The hyperimmune-free degrees form a basis for Π_1^0 classes.*

Proof. We proceed as in V.5.6. Let T be a given infinite recursive binary tree. We want to build A on T such that $\{e\}^A$, whenever is total, is majorized by a recursive function.

Let $T_0 = T$. Given T_e, consider

$$U_e^x = \{\sigma \in T_e : \{e\}^\sigma(x)\uparrow\}.$$

There are two cases:

1. U_e^x *is infinite for some x*
   Then we can let $T_{e+1} = U_e^x$ for one such x, and for every A on T_{e+1} we will have $\{e\}^A$ not total. Note that U_e^x is indeed a tree, being closed under subsequences (since if a string does not decide a computation, neither does any substring of it).

2. U_e^x *is finite for every x*
   Then we can let $T_{e+1} = T_e$. For every A on T_{e+1} it is enough to find recursively a level n such that, for all strings σ of length n, $\{e\}^\sigma(x)\downarrow$. Then

$$\{e\}^A(x) \leq \max_{\sigma \in T \wedge |\sigma|=n} \{e\}^\sigma(x). \quad \square$$

The results just proved for binary recursive trees can easily be seen to hold also for **recursively bounded recursive trees**, i.e. recursive trees such that the size (i.e., the greatest component) of nodes on T of length n is bounded by $f(n)$, for some recursive f. E.g., a binary tree is recursively bounded by 2.

The results for recursively bounded recursive trees can be extended to **finitely branching recursive trees**, i.e. recursive trees such that the number of nodes on T of length n is finite. The reason is that such trees are bounded by a function recursive in $\mathbf{0}'$: to know a bound on the size of strings of level

n on T we can ask, for any m, whether the size of every string of level n on T is bounded by m. This can be answered recursively in $\mathcal{K}$, and we can inductively determine the smallest m for which such a sentence holds (which exists, because T is finitely branching). Thus *the results of this subsection hold for finitely branching recursive trees, when relativized to* $\mathbf{0}'$. E.g., V.5.32 shows that a finitely branching infinite recursive tree has an infinite branch with jump recursive in $\mathbf{0}''$.

   If the condition of being finitely branching is dropped, then the situation changes radically (e.g., V.5.23 fails). The theory of infinite **recursive trees** with infinite branches can be developed in a way largely parallel to the one for binary recursive trees, when the notions of recursive set and Turing degree are replaced by those of hyperarithmetical set and hyperdegree. See Volume III for a treatment.

Complete extensions of Peano Arithmetic $\star$

Since the class of sets separating two disjoint r.e. sets is a nonempty Π_1^0 class, so are:

1. *the set of consistent extensions of a given consistent theory*

2. *the set of complete extensions of a given consistent theory.*

Jockusch and Soare [1972] and Hanf [1975] provide converses to these examples in the spirit of III.10.3, showing that *the class of degrees of members of a given* Π_1^0 *class coincides with the class of degrees of complete extensions of some (finitely axiomatizable) first-order theory.*

   In the following we will restrict our attention to complete extensions of Peano Arithmetic, because they provide a possible description of the arithmetical world in accord with the partial but fundamental picture given by $\mathcal{PA}$. In particular, we will thus be able to take advantage of the power of $\mathcal{PA}$ in proofs, including the Induction Principle.

   The basic link with the subject of the last subsection is another basis result.

Theorem V.5.35 Scott Basis Theorem (Scott [1962]) *If $\mathcal{F}$ is a consistent extension of $\mathcal{PA}$, the sets recursive in $\mathcal{F}$ form a basis for Π_1^0 classes.*

Proof. Let T be an infinite recursive tree. To be able to choose an infinite branch recursively in $\mathcal{F}$, we proceed inductively. Let $\sigma_0 = \emptyset$. Given σ_n of length n, consider all its extensions of length $n + 1$ on T. Given any two of them, say τ_0 and τ_1, we have to decide which one looks better as an initial segment of an infinite branch of T. The statement

$$\psi_0 \Leftrightarrow (\exists m)(\tau_0 \text{ has an extension of length } m \text{ on } T, \text{ but } \tau_1 \text{ does not})$$

is Σ_1^0 (note that the quantifiers on strings are restricted to strings of a given length). If it is true then, for some m, so is

$$\tau_0 \text{ has an extension of length } m \text{ on } T, \text{ but } \tau_1 \text{ does not.}$$

But this is a true recursive sentence, which is then provable in $\mathcal{PA}$, and hence in $\mathcal{F}$. Then so is ψ_0. Similarly for

$$\psi_1 \Leftrightarrow (\exists m)(\tau_1 \text{ has an extension of length } m \text{ on } T, \text{ but } \tau_0 \text{ does not}).$$

Now ψ_0 and ψ_1 cannot both be provable, because $\mathcal{F}$ is consistent (otherwise there would be a number m such that one of τ_0 and τ_1 has at the same time an extension of length m, and no extension of length m). Thus τ_i looks better if ψ_i is provable in $\mathcal{F}$, and otherwise τ_0 and τ_1 look the same.

We can now compare all pairs of strings on T of length $n+1$ extending σ_n. Since there is an infinite branch on T, all strings which do not extend to an infinite branch are eliminated (when compared to a string that does). All the remaining ones do extend to an infinite branch, and we can choose any of them as σ_{n+1}. $\square$

The next result provides a converse to Scott Basis Theorem.

Theorem V.5.36 Characterization of the degrees of complete extensions of $\mathcal{PA}$ (Solovay) *The following conditions are equivalent:*

1. *$\boldsymbol{a}$ is the degree of a consistent extension of $\mathcal{PA}$*

2. *$\boldsymbol{a}$ is the degree of a complete extension of $\mathcal{PA}$*

3. *$\mathcal{D}(\leq \boldsymbol{a})$ is a basis for Π_1^0 classes.*

Proof. Clearly, 2 implies 1. Scott Basis Theorem shows that 1 implies 3. If 3 holds then there is a complete extension of $\mathcal{PA}$ recursive in $\boldsymbol{a}$ (because the set of complete extensions of $\mathcal{PA}$ is a Π_1^0 class): that 3 implies 2 thus follows from the upward closure of the degrees of complete extensions of $\mathcal{PA}$, which we now prove.

Let $\mathcal{F}$ be a complete extension of $\mathcal{PA}$ recursive in a set C. It is enough to build a tree of complete extensions of $\mathcal{PA}$, recursively in $\mathcal{F}$. Then the branch determined by C has the same degree of C. Let $\{\varphi_n\}_{n \in \omega}$ be an enumeration of the sentences in the language of arithmetic. We start with $\mathcal{F}_\emptyset = \mathcal{PA}$. Given $\mathcal{F}_\sigma$, we proceed in two steps:

1. *completeness*
   Given φ_n, with $n = |\sigma|$, we decide how to consistently add to $\mathcal{F}_\sigma$ one of

φ_n and $\neg\varphi_n$. As in V.5.35, we let

$$\psi_0 \Leftrightarrow (\exists m)(m \text{ codes a proof of } \varphi_n \text{ in } \mathcal{F}_\sigma,$$
$$\text{but no smaller } m' \text{ codes a proof of } \neg\varphi_n)$$
$$\psi_1 \Leftrightarrow (\exists m)(m \text{ codes a proof of } \neg\varphi_n \text{ in } \mathcal{F}_\sigma,$$
$$\text{but no smaller } m' \text{ codes a proof of } \varphi_n).$$

Since $\mathcal{F}_\sigma$ is a finite extension of $\mathcal{PA}$, it is a formal system, and thus ψ_0 and ψ_1 are Σ_1^0, and hence provable in $\mathcal{F}$ if true. Moreover, by consistency of $\mathcal{F}$, they are not both provable. Recursively in $\mathcal{F}$ we can see which of them, if any, is provable.

If ψ_1 is provable in $\mathcal{F}$ then we know that ψ_0 is not provable in $\mathcal{F}_\sigma$, and hence that $\neg\varphi_n$ is consistent with it. We can thus let

$$\mathcal{F}'_\sigma = \mathcal{F}_\sigma \cup \{\neg\varphi_n\}.$$

Otherwise, we can let

$$\mathcal{F}'_\sigma = \mathcal{F}_\sigma \cup \{\varphi_n\}.$$

2. branching

Since the sets of provable and refutable formulas of $\mathcal{PA}$ are an effectively inseparable pair of r.e. sets (see III.10.11), there is a recursive function that produces, given a disjoint pair (A, B) of r.e. sets extending them, an element not in $A \cup B$. This applies in particular to the sets of provable and refutable formulas of $\mathcal{F}'_\sigma$ which is, by construction, a finite extension of $\mathcal{PA}$. In other words, there is an effective way to find a sentence ψ which is neither provable nor refutable from $\mathcal{F}'_\sigma$. Then we can let

$$\mathcal{F}_{\sigma*(0)} = \mathcal{F}'_\sigma \cup \{\psi\}$$
$$\mathcal{F}_{\sigma*(1)} = \mathcal{F}'_\sigma \cup \{\neg\psi\}.$$

If $\mathcal{F} \leq_T C$ then $\bigcup_{\sigma \subseteq C} \mathcal{F}_\sigma$ is a complete extension of $\mathcal{PA}$ of the same degree as C. $\square$

The theorem shows that the degrees of consistent or complete extensions of $\mathcal{PA}$ describe particularly simple bases for Π_1^0 classes. A complete degree-theoretical characterization of them is not known, but from the basis results we already have we can easily derive a number of consequences, both positive and negative. The latter show that a complete extension of $\mathcal{PA}$ cannot be too simple, in various ways. They generalize II.2.17, which stated only that a consistent extension of $\mathcal{PA}$ cannot be recursive.

Proposition V.5.37 (Scott and Tennenbaum [1960], Jockusch and Soare [1972], [1972a]) *A consistent extension of $\mathcal{P}A$ can have neither incomplete r.e. degree, nor minimal degree. But there are complete extensions of $\mathcal{P}A$ of low degree, as well as of degree $0'$.*

Proof. If there were a consistent extension of $\mathcal{P}A$ of r.e. incomplete degree, by Scott Basis Theorem the r.e. incomplete degrees would be a basis, contradicting V.5.33.b.

Similarly, if there were a consistent extension of $\mathcal{P}A$ of minimal degree then the minimal degrees together with the recursive sets would be a basis, contradicting the fact that the members of the following nonempty Π_1^0 class are neither recursive nor of minimal degree:

$$A \oplus B \in \mathcal{C} \quad \Leftrightarrow \quad (\forall e)(B(e) \not\simeq \{e\}(e)) \wedge$$
$$(\forall e)(A(2e) \not\simeq \{e\}(2e)) \wedge$$
$$(\forall e)(A(2e+1) \not\simeq \{e\}^B(2e+1)).$$

The first two conditions imply that both A and B are not recursive, since they diagonalize against every recursive function (on fixed arguments). The last condition implies that $A \not\leq_T B$. Thus $A \oplus B$ cannot be recursive or of minimal degree. And $\mathcal{C}$ is Π_1^0 because, e.g.,

$$B(e) \not\simeq \{e\}(e) \quad \Leftrightarrow \quad \{e\}(e){\uparrow} \vee B(e) \neq \{e\}(e),$$

and to diverge on a given argument is a Π_1^0 condition.

The existence results follow from V.5.32 and V.5.33.a: the latter implies the existence of a complete extension of $\mathcal{P}A$ of r.e. degree, which must be $0'$ because all the r.e. incomplete degrees are ruled out by the first part of the proof. $\square$

Jockusch and Soare [1972a] show that *the complete extensions of an essentially undecidable formal system can have both incomplete r.e. degree and minimal degree.* Thus the role of $\mathcal{P}A$ (through the fact, used in the proof of V.5.36, that the sets of theorems and of refutable formulas are not only recursively, but also effectively inseparable) is crucial.

An interesting way to analyze the behavior of consistent extensions of $\mathcal{P}A$ is to see which sets are weakly representable in them. Clearly, in the standard model exactly the arithmetical sets are. By II.2.16, in every consistent formal system extending $\mathcal{R}$ exactly the r.e. sets are weakly representable. But for consistent extensions $\mathcal{F}$ of $\mathcal{R}$ which are not formal systems the following might happen. If φ weakly represents a set A in $\mathcal{R}$, $\varphi(\overline{x})$ might become provable in $\mathcal{F}$ for some $x \in \overline{A}$, and thus φ represents only a superset of A in $\mathcal{F}$, possibly a cofinite (and hence recursive) one. Moreover, since $\mathcal{F}$ is not a formal system, the set represented by φ is not necessarily r.e.

Proposition V.5.38 *For every consistent extension $\mathcal{F}$ of $\mathcal{PA}$, the class of sets weakly representable in $\mathcal{F}$ properly includes the recursive sets.*

Proof. Since, by II.2.16, the recursive sets are actually strongly representable in $\mathcal{R}$, they remain strongly representable in every extension of it, and hence weakly representable in every consistent extension.

Given a recursive enumeration $\{\varphi_n\}_{n\in\omega}$ of the sentences in the language of arithmetic let, as in V.5.35 and V.5.36,

$$\psi_0(n) \iff (\exists m)(m \text{ codes a proof of } \varphi_n \text{ in } \mathcal{PA},$$
$$\text{but no smaller } m' \text{ codes a proof of } \neg\varphi_n)$$
$$\psi_1(n) \iff (\exists m)(m \text{ codes a proof of } \neg\varphi_n \text{ in } \mathcal{PA},$$
$$\text{but no smaller } m' \text{ codes a proof of } \varphi_n)$$

By consistency of $\mathcal{F}$, ψ_0 weakly represents a set A. If φ_n is a theorem of $\mathcal{PA}$ then $\psi_0(n)$ is a true Σ_1^0 formula, which is then provable in $\mathcal{F}$. If $\neg\varphi_n$ is a theorem of $\mathcal{PA}$ then $\psi_1(n)$ is provable in $\mathcal{F}$ and hence, by consistency of $\mathcal{F}$, $\psi_0(n)$ is not provable. Thus A separates theorems and refutable formulas of $\mathcal{PA}$, and thus it cannot be a recursive set (by III.10.11). $\square$

We now provide some examples of complete extensions of $\mathcal{PA}$, pathological from the point of view of weakly representable sets.

Proposition V.5.39 (Jockusch and Soare [1972])

1. *There is a complete extension of $\mathcal{PA}$ in which no hypersimple set is weakly representable.*

2. *There is a complete extension of $\mathcal{PA}$ in which only Δ_2^0 sets are weakly representable.*

3. *For any $n \geq 2$ there is a complete extension of $\mathcal{PA}$ in which only Δ_{n+1}^0 sets are weakly representable, but no nonrecursive Σ_n^0 or Π_n^0 set is.*

4. *There is a complete extension of $\mathcal{PA}$ in which no arithmetical nonrecursive set is weakly representable.*

Proof. By V.5.34 there is a complete extension $\mathcal{F}$ of $\mathcal{PA}$ of hyperimmune-free degree. If A is weakly representable in $\mathcal{F}$, it is recursive in it. Thus it has hyperimmune-free degree, because the hyperimmune-free degrees are downward closed (V.5.3). Thus $\overline{A}$ cannot be hyperimmune, and A is not hypersimple.

Similarly, 2 follows from the Low Basis Theorem, which provides a complete extension of $\mathcal{PA}$ recursive in $\mathcal{K}$, in which all weakly representable sets must then be Δ_2^0.

For the remaining results, we first show that *given countably many nonre-cursive sets A_n and an infinite binary recursive tree, there is A on it in which no A_n is recursive.* This can be obtained inductively as usual, once we know how to get, given e, n, and an infinite recursive tree T, an infinite recursive subtree T' of T such that $A_n \not\simeq \{e\}^A$, for every A on T. There are three cases:

1. *for some x, there are infinitely many $\sigma \in T$ such that $\{e\}^\sigma(x)\uparrow$*
   Then the set
   $$\{\sigma \in T : \{e\}^\sigma(x)\uparrow\}$$
   is an infinite subtree of T (being closed under subsequences), and we can let it be T'. If $A \in T'$ then $\{e\}^A$ is not total, and hence is different from A_n.

2. *for some x, there is $\sigma \in T$ such that $\{e\}^\sigma(x)$ is defined and different from $A_n(x)$, and σ has infinitely many extensions on T*
   Then we can let T' be the full subtree of T above σ. If $A \in T'$ then $\{e\}^A$ is different from A_n.

3. *otherwise*
   Then we let $T' = T$, and show that $\{e\}^A$ is recursive for any A on T, and hence different from A_n. Given x, go to a level n of T such that $\{e\}^\sigma(x)\downarrow$ for every $\sigma \in T$ of length n: this is possible because we are not in the first case. If two strings give different values on x, then they cannot both have infinitely many extensions on T, otherwise one of them would give a value different from A_n, and we would be in the second case. We can thus generate enough of the tree to discover which of the strings belong to finite branches, until only strings with the same value remain. This must be the value of $\{e\}^A(x)$, for any $A \in T$.

By letting $\{A_n\}_{n\in\omega}$ be a list of the arithmetical nonrecursive sets, we get part 3. For part 2 we only have to let $\{A_n\}_{n\in\omega}$ be a Δ_{n+1}^0 list of $\Sigma_n^0 \cup \Pi_n^0$, and to compute the complexity of the construction. The division in cases is based on two-quantifier questions, and thus it is recursive in $\emptyset''$, which accounts for the bound $n \geq 2$. $\square$

Kučera [1986] (see X.4.8) has shown that part 2 cannot be improved as in part 3, because *there is no complete extension of $\mathcal{PA}$ in which only Δ_2^0 sets are weakly representable, but no nonrecursive r.e. set is.*

For more on Π_1^0 classes and their applications to complete extensions of $\mathcal{PA}$, see Jockusch and Soare [1971], [1972], [1972a], Jockusch [1974], [1989], Kučera [1985], [1986], [1988], [1989].

V.6   Initial Segments ⋆

Initial segments more complicated than minimal degrees have been used in the original proofs of many of the global results of the next section. They are now obsolete in this respect, since much simpler proofs have been obtained. Initial segments are still necessary for a complete algebraic characterization of the algebraic structure of $\mathcal{D}$, as well as for some advanced parts of Degree Theory (like the results quoted on p. 492). The techniques involved in the proofs are however mostly not recursion-theoretical, and outside the scope of our book. The reader is referred to Epstein [1979] and Lerman [1983] for detailed treatments. We will introduce only techniques which have other uses as well.

Uniform trees

To obtain initial segments we need trees that have more flexibility than the simple ones used in the previous section. The following notion introduces some uniformities in our trees.

Definition V.6.1 T *is a* **uniform tree** *if, for every* $i = 0, 1$ *and* σ,

1. $|T(\sigma)|$ *depends only on* $|\sigma|$

2. *there is a unique* τ_i, *depending only on* $|\sigma|$, *such that* $T(\sigma * i) = T(\sigma) * \tau_i$.

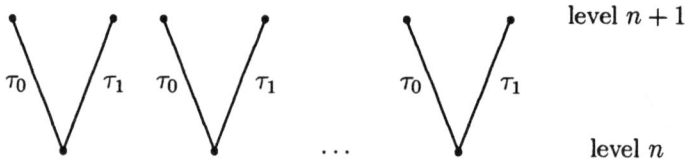

A uniform tree is nothing more than a tree in which, at each level, the strings immediately following each node are independent of the node itself, and have the same length.

A useful way of representing the situation is the following: there are three functions g (strictly increasing), and f_L, f_R (left and right functions) such that f_L and f_R take values in $\{0, 1\}$ and are incompatible (i.e. they differ on at least one argument) in every interval $[g(n), g(n+1))$. Thus the levels of the tree are determined by g, and a branch of the tree is simply a path which at every node follows one of f_L and f_R, up to the next node.

Definition V.6.2 (Spector [1956]) *An* **admissible triple** *is a triple* g, f_L, f_R *of functions from* ω *into* $\{0, 1\}$, *such that:*

1. $(\forall n)[g(n) < g(n+1)]$

2. $(\forall n)(\exists x)[g(n) \le x < g(n+1) \ \wedge \ f_L(x) \ne f_R(x)]$.

The triple is **recursive** *if* g, f_L, f_R *are.*

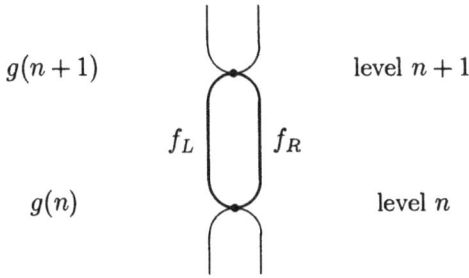

Our present task is to build minimal degrees by uniform trees. By doing so we will actually reproduce Spector's original proof. What we do is to reprove the lemmas of the last section, this time building uniform trees instead of simple trees.

Proposition V.6.3 Diagonalization Lemma for uniform trees. *Given e and a recursive uniform tree T, there is a recursive uniform tree $Q \subseteq T$ such that, for every A on Q, $A \not\ge \{e\}$.*

Proof. The same proof of the Diagonalization Lemma works, because any full subtree of a uniform tree is still uniform. ☐

Proposition V.6.4 Totality Lemma for uniform trees (Martin and Miller [1968]) *Given e and a recursive uniform tree T, there is a recursive uniform tree $Q \subseteq T$ such that one of the following holds:*

1. *for every A on Q, $\{e\}^A$ is not total*

2. *for every A on Q, $\{e\}^A$ is total and*

$$(\forall n)(\forall \sigma)(|\sigma| = n \ \Rightarrow \ \{e\}^{Q(\sigma)}(n)\downarrow),$$

where $|\sigma|$ is the length of σ.

Proof. See if

$$(\exists \sigma \in T)(\exists x)(\forall \tau \supseteq \sigma)(\tau \in T \Rightarrow \{e\}^\tau(x)\uparrow).$$

If so, choose such a σ: as in the Totality Lemma, we let Q be the full subtree of T above σ, which is uniform since T was. Then case 1 holds.

Otherwise, we define Q by induction as follows.

$$Q(\emptyset) = \text{ least } \tau \in T \text{ such that } \{e\}^{\tau}(0)\downarrow .$$

Given $Q(\sigma_i)$, for $1 \le i \le 2^n$ and σ_i string of length n, take:

$$\begin{array}{ll}
Q(\sigma_1) * \tau_1 \in T & \text{such that } \{e\}^{Q(\sigma_1)*\tau_1}(n)\downarrow \\
Q(\sigma_2) * \tau_1 * \tau_2 \in T & \text{such that } \{e\}^{Q(\sigma_2)*\tau_1*\tau_2}(n)\downarrow \\
\text{etc.}
\end{array}$$

Let $\tau = \tau_1 * \cdots * \tau_{2^n}$: for each i we then have $\{e\}^{Q(\sigma_i)*\tau}(n)\downarrow$. It is now enough to take two incomparable extensions μ_0 and μ_1 of τ with the same length and such that $Q(\sigma_i) * \mu_0$ and $Q(\sigma_i) * \mu_1$ are on T for every i, which is possible because T is uniform, and let

$$Q(\sigma_i * 0) = Q(\sigma_i) * \mu_0 \quad \text{and} \quad Q(\sigma_i * 1) = Q(\sigma_i) * \mu_1.$$

By definition Q is then uniform, since we extend all strings of a given level in the same way. $\square$

At this point we know how to build hyperimmune-free degrees by uniform trees. To obtain the same result for minimal degrees it is enough to show that we can handle, by uniform trees, the only case of the Minimality Lemma in which we are not taking full subtrees.

Proposition V.6.5 (Spector [1956]) *Given e and a recursive uniform tree T, if*

1. *every σ on T has e-splitting extensions on T*

2. $(\forall \sigma \in T)(\forall x)(\exists \tau \supseteq \sigma)(\tau \in T \wedge \{e\}^{\tau}(x)\downarrow)$

then T has a recursive e-splitting uniform subtree Q.

Proof. Define Q inductively by the following procedure. Suppose $Q(\sigma_i)$ is given, for any $1 \le i \le 2^n$ and σ_i string of length n.

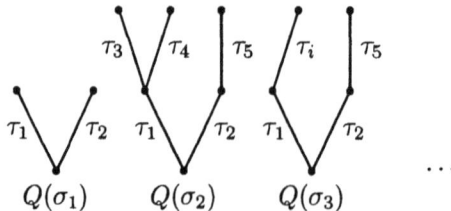

By 1 there are τ_1 and τ_2 such that $Q(\sigma_1) * \tau_1$ and $Q(\sigma_1) * \tau_2$ are on T and e-split. Reproduce them above $Q(\sigma_2)$: since T is uniform, the new strings are still on T. By 1 again, there are τ_3 and τ_4 such that $Q(\sigma_2) * \tau_1 * \tau_3$ and $Q(\sigma_2) * \tau_1 * \tau_4$ are on T and e-split, say on x. By 2 there is some τ_5 such that $\{e\}^{Q(\sigma_2)*\tau_2*\tau_5}(x)\downarrow$. Thus, for $i = 3$ or $i = 4$, $Q(\sigma_2) * \tau_1 * \tau_i$ and $Q(\sigma_2) * \tau_2 * \tau_5$ e-split on x. Reproduce $\tau_1 * \tau_i$ and $\tau_2 * \tau_5$ above $Q(\sigma_3)$, and so on. We thus get, at the end, two big strings τ and τ' such that, for each i, $Q(\sigma_i) * \tau$ and $Q(\sigma_i)*\tau'$ are on T and e-split. By possibly extending one of them to a string on T, we can actually find two such strings of the same length, and they provide the new level in Q. □

Theorem V.6.6 (Spector [1956]) *It is possible to build minimal degrees by recursive uniform trees.*

Proof. We start with
$$T_0 = \text{ identity tree,}$$
which is clearly a recursive uniform tree. Given T_{2e}, we let

$$T_{2e+1} = \text{ the } Q \text{ of the Diagonalization Lemma, for } T = T_{2e}.$$

To define T_{2e+2}, let $T = T_{2e+1}$ and see if the condition 2 of the above proposition holds. If not, choose a string σ on T for which it fails, and let

$$T_{2e+2} = \text{ full subtree of } T \text{ above } \sigma.$$

Then $\{e\}^A$ is not total, for any A on it. If condition 2 holds, see if 1 does. If not, there is a string σ with no e-splitting extensions, and we let

$$T_{2e+2} = \text{ full subtree of } T \text{ above } \sigma.$$

Then $\{e\}^A$ is recursive if total, for any A on it. Otherwise, let

$$T_{2e+2} = \text{ the } Q \text{ of the proposition above.}$$

Then $A \leq_T \{e\}^A$ if $\{e\}^A$ is total, for any A on it. If A is on all the T_n's, then it has minimal degree. □

One might wonder why we should want to build minimal degrees by uniform trees, since the proof is more complicated than the one given in V.5.11. There are two independent answers to this. The first is that this is a first step toward the construction of minimal degrees by recursive coinfinite conditions. The second is that the proof just given can be modified and, taking advantage of the uniformities, turned into a proof of the existence of more complicated initial segments. These two applications are treated in the next subsections.

Minimal degrees by recursive coinfinite extensions

We close now the circle started in Section 4, by showing how coinfinite conditions are nothing else than particular uniform trees.

Definition V.6.7 (Lachlan [1971]) T *is a* **strongly uniform tree** *(or a* **1-tree***) if it is uniform and, for all* σ, $T(\sigma * 0)$ *and* $T(\sigma * 1)$ *differ only on one argument (we say they are* **adjacent***).*

Equivalently, we could define strongly uniform trees as admissible triples (V.6.2) satisfying the stronger condition that the functions f_L and f_R differ, at each level, on exactly one argument. We can see the arguments on which the two sides differ as the uncommitted ones, and thus a strongly uniform tree defines a coinfinite condition

$$\theta(x) \simeq \begin{cases} 0 & \text{if } f_L(x) = f_R(x) = 0 \\ 1 & \text{if } f_L(x) = f_R(x) = 1 \\ \text{undefined} & \text{if } f_L(x) \neq f_R(x) \end{cases}$$

Of course the same translation would work for uniform trees as well. What makes strongly uniform trees special is that any set A extending θ is on the tree (since in each interval there is only one uncommitted point, the two branches f_L and f_R take care of all the possibilities). This is not true if the tree is only uniform (if in an interval there are n uncommitted points then there are 2^n possible extensions, but only two of them are on the tree).

Conversely, a coinfinite condition θ defines a strongly uniform tree as follows. Let g enumerate, in increasing order, the elements on which θ is undefined, and

$$f_L(x) = \begin{cases} \theta(x) & \text{if } \theta(x)\downarrow \\ 0 & \text{otherwise} \end{cases} \qquad f_R(x) = \begin{cases} \theta(x) & \text{if } \theta(x)\downarrow \\ 1 & \text{otherwise.} \end{cases}$$

Moreover, the translations just given preserve recursiveness (recall that a coinfinite condition always has recursive domain, and so g is always recursive), and thus *recursive strongly uniform trees and recursive coinfinite conditions are interchangeable*.

It is immediate to note that the Totality Lemma proved for uniform trees also works for strongly uniform ones (since we just split a string at the very end), and thus *it is possible to build hyperimmune-free degrees by recursive coinfinite extensions*. To prove the analogue of the Minimality Lemma requires instead much more work.

Proposition V.6.8 (Lachlan [1971]) *Given* e *and a strongly uniform tree* T, *if*

    1. every $\sigma \in T$ *has e-splitting extensions on* T

2. $(\forall \sigma \in T)(\forall x)(\exists \tau \supseteq \sigma)(\tau \in T \wedge \{e\}^\tau(x)\downarrow)$

3. T *does not have strongly uniform subtrees without e-splittings*

then T *has a strongly uniform e-splitting subtree* Q.

Proof. Clearly condition 1 is redundant, and it follows from 3. We make it explicit only to show where the new hypothesis is used. Since 2 holds, we may suppose that

$$(\forall n)(\forall \sigma)(|\sigma| = n \Rightarrow \{e\}^{T(\sigma)}(n)\downarrow),$$

otherwise we apply V.6.4 first. Again we proceed by induction, showing the first two steps.

Given σ, by 1 there are τ and τ' extending it and e-splitting, say on x. We may suppose they are of the same length (otherwise extend the shortest), and that for all strings μ of that length, $\{e\}^\mu(x)\downarrow$ (otherwise go to a level high enough, by the initial observation). Since T is strongly uniform, there is a sequence $\tau_0, \ldots, \tau_i$ of strings on it of the same length, each adjacent to the following in the list, and such that $\tau = \tau_0$ and $\tau' = \tau_i$. Since $\{e\}^{\tau_j}(x)\downarrow$ for all $j \le i$, and $\{e\}^{\tau_0}(x) \not\simeq \{e\}^{\tau_i}(x)$, two of these adjacent strings e-split on x. So $\sigma \in T$ has e-splitting adjacent extensions. This shows how to build the first level of the e-splitting uniform subtree.

Now let σ_1 and σ_2 be given. We set up to build the second level.

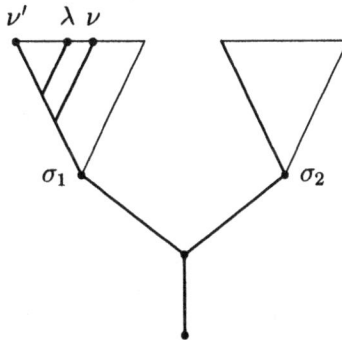

1. We first build a strongly uniform subtree of T above σ_2, as follows. Take a pair of e-splitting adjacent extensions of σ_2, and let it be the first level. Then consider the leftmost branch, take a pair of e-splitting adjacent extensions of it, and reproduce them on the rightmost one, and this is the second level. We go on by considering the leftmost branch at each level, finding an e-splitting adjacent pair above it, and reproducing it on every node of the same level.

2. Then we take this subtree and reproduce it, as it is, above σ_1. There must be an e-splitting on this strongly uniform subtree of T, by condition 3. We want to find one such that the two branches only differ on one element (since we are building a strongly uniform tree), and they also e-split above σ_2. Note that we know only that the leftmost branches e-split above σ_2, and we thus make our search by staying on the leftmost branch of our subtree above σ_1.

By methods we know, we may choose an e-splitting ν and ν' such that $|\nu| = |\nu'|$ and:

- ν' is on the leftmost branch of the strong uniform subtree (given any e-splitting, say on x, it is enough to wait until, for a long enough segment ν' on the leftmost branch, $\{e\}^{\nu'}(x) \downarrow$: one of the original branches and ν' must e-split on x.)

- ν goes right on the tree as late as possible (i.e. the common part of ν and ν' is maximal)

- if x is the element on which ν and ν' e-split, and $\mu \in T$, then $|\mu| = |\nu| = |\nu'| \Rightarrow \{e\}^{\mu}(x) \downarrow$.

Now take λ which is like ν, only it goes right one level after ν does: λ and ν are obviously adjacent, and $\{e\}^{\lambda}(x) \simeq \{e\}^{\nu'}(x)$ by the choice of ν (since the common part of ν and ν' is maximal). Then $\{e\}^{\lambda}(x) \not\simeq \{e\}^{\nu}(x)$, and λ and ν are an adjacent e-splitting above σ_1. By definition they are also an (adjacent) e-splitting above σ_2 (since ν' lies on the leftmost branch).

This shows how to build the second level of the e-splitting uniform subtree. The remaining levels can be built inductively, in the same way. $\square$

Theorem V.6.9 (Lachlan [1971]) *It is possible to build minimal degrees by recursive coinfinite extensions.*

Proof. We use strongly uniform recursive trees, which we know to be interchangeable with recursive coinfinite conditions. Let

$$T_0 = \text{identity tree},$$

which is clearly a strongly uniform recursive tree. Given T_{2e}, let

$$T_{2e+1} = \text{the } Q \text{ of the Diagonalization Lemma, for } T = T_{2e}.$$

To define T_{2e+2}, let $T = T_{2e+1}$, and see if condition 2 of the above proposition holds. If not, choose a string σ on T for which it fails, and let

$$T_{2e+2} = \text{full subtree of } T \text{ above } \sigma.$$

Then $\{e\}^A$ is not total, for any A on it. If condition 2 holds, see if 3 holds. If not, let

$$T_{2e+2} = \text{a strongly uniform subtree of } T \text{ without } e\text{-splittings.}$$

Then $\{e\}^A$ is recursive if total, for any A on it. Otherwise, let

$$T_{2e+2} = \text{the } Q \text{ of the proposition above.}$$

Then $A \leq_T \{e\}^A$ if $\{e\}^A$ is total, for any A on it. If A is on all the T_n's, then it has minimal degree. □

Exercises V.6.10 a) *The minimal degree built above is below* **0″**. (Lachlan [1971]) (Hint: to ask whether there is a strongly uniform subtree without e-splitting is too complicated. But we can ask if the inductive process of building the e-splitting subtree in the proposition above does terminate or not. If yes, take the e-splitting subtree. If not, we know that there is a strongly uniform subtree without e-splitting, and thus we can search for an index of it.)

b) *There is a cone of degrees such that every element in it is the join of two minimal degrees.* (Cooper [1972a]) (Hint: build a strongly uniform tree of minimal degrees recursive in $\emptyset''$. Given C such that $\emptyset'' \leq_T C$, choose A and B on the tree as follows. At level n, let x be the unique element on which the branches differ. Let A follow the branch that on x agrees with C, and B follow the other branch. Then $A, B \leq_T C \oplus \emptyset'' \leq_T C$, so $A \oplus B \leq_T C$. And $C \leq_T A \oplus B$, since $C(n)$ can be recovered as the value of A on the n-th element on which A and B differ.)

c) *The hyperimmune-free degrees are not closed under join.* (Martin and Miller [1968]) (Hint: build a strongly uniform tree of hyperimmune-free degrees recursive in $\emptyset''$. It follows as in part b) that the degrees above **0″**, which are hyperimmune because comparable with **0′**, are joins of two hyperimmune-free degrees.)

d) *There is a minimal, not completely autoreducible degree.* (Jockusch and Paterson [1976]) (Hint: build a set A of minimal degree, by recursive coinfinite extensions. Insert steps to insure that A is not autoreducible. The two constructions are compatible.)

The three-element chain

Recall that an initial segment is simply a set of degrees closed downward. The set $\{\mathbf{0}, \mathbf{a}\}$, with $\mathbf{a}$ a minimal degree, is thus the simplest nontrivial initial segment, the two-element chain. The set of minimal degrees and $\mathbf{0}$ is a nontrivial initial segment of power $2^{\aleph_0}$.

Since every degree has a minimal cover, there are degrees $\mathbf{a}$ and $\mathbf{b}$ such that $\mathbf{0} < \mathbf{b} < \mathbf{a}$, and there is no degree strictly between either $\mathbf{0}$ and $\mathbf{b}$, or $\mathbf{b}$ and $\mathbf{a}$. But there could be some degree $\mathbf{c}$ incomparable with $\mathbf{b}$, and still below $\mathbf{a}$. We now want to build $\mathbf{a}$ and $\mathbf{b}$ as above, but with the additional property that

$\{0, b, a\}$ is an initial segment, i.e. the only degrees below a are 0 and b. This is called a three-element chain.

The idea is to build A such that the odd part of A, defined as

$$Od(A) = \{x : 2x + 1 \in A\},$$

plays the role of the intermediate degree. Since $Od(A) \leq_T A$ automatically, the requirements on A are thus:

1. $Od(A)$ nonrecursive

2. $A \not\leq_T Od(A)$

3. $\{e\}^A$ total $\Rightarrow$ $\{e\}^A$ recursive or $\{e\}^A \equiv_T Od(A)$ or $\{e\}^A \equiv_T A$.

These are just generalizations of the conditions for A being minimal, and in many respects so is the construction, which employs uniform trees. The most crucial parts are the idea of forcing $Od(A)$ to be minimal (which allows us to build just one set A, instead of the separate sets A and B), and V.6.12 (which allows us to make A not recursive in its odd part, and uses uniformity in a crucial way, thus forcing us to use uniform trees in the construction of initial segments).

Proposition V.6.11 *Given e and a recursive uniform tree T, if for some σ*

$$T(\sigma * 0), T(\sigma * 1) \text{ disagree on their odd parts}$$

then there is a recursive uniform tree $Q \subseteq T$ such that, for every A on Q, $Od(A) \neq \{e\}$.

Proof. Since $T(\sigma * 0), T(\sigma * 1)$ disagree on their odd parts, the odd part of one of them, say $T(\sigma * i)$, disagrees with $\{e\}$. Let Q be the full subtree of T above $T(\sigma * i)$. □

Proposition V.6.12 (Titgemeyer [1962]) *Given e and a recursive uniform tree T, if for some σ*

$$T(\sigma * 0), T(\sigma * 1) \text{ agree on their odd parts}$$

then there is a recursive uniform tree $Q \subseteq T$ such that, for every A on Q, $A \neq \{e\}^{Od(A)}$.

Proof. Let x be such that $T(\sigma * 0)(x) \neq T(\sigma * 1)(x)$. Such an x must exist, because $T(\sigma * 0)$ and $T(\sigma * 1)$ are incompatible. By the hypothesis on σ, it is not on their odd parts. See if

$$(\exists \tau \supseteq T(\sigma * 0))(\tau \in T \wedge \{e\}^{Od(\tau)}(x) \downarrow).$$

If not, let Q be the full subtree above $T(\sigma * 0)$: if A is on Q, $\{e\}^{Od(A)}$ is not total (hence it differs from A).

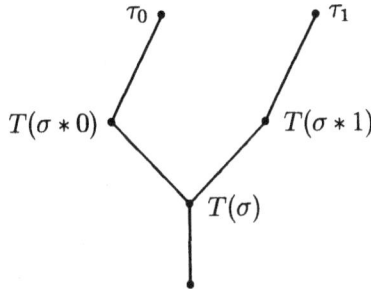

Otherwise, let τ_0 be such a string, and let τ_1 extend $T(\sigma * 1)$ in the same way as τ_0 extends $T(\sigma * 0)$. Since T is uniform, $\tau_1 \in T$. Now τ_0 and τ_1 have the same odd parts, since they extend in the same way $T(\sigma * 0)$ and $T(\sigma * 1)$, which have the same odd parts. Then $\{e\}^{Od(\tau_0)}(x) \simeq \{e\}^{Od(\tau_1)}(x)$. But $\tau_0(x) \neq \tau_1(x)$, and for $i = 0$ or $i = 1$ we have

$$\tau_i(x) \neq \{e\}^{Od(\tau_i)}(x).$$

Then let Q be the full subtree of T above τ_i. □

From the two lemmas we know that we can diagonalize if we use uniform trees such that $T(\sigma * 0)$ and $T(\sigma * 1)$ agree on the odds (i.e. on their odd parts) for infinitely many σ's, and disagree on the odds for infinitely many σ's. We will work, from now on, with trees which alternate agreements and disagreements on the odds. The lemmas just proved work for them too, since we are just taking full subtrees in their proofs.

Of course lemma V.5.9 still holds, and we thus have conditions ensuring that $\{e\}^A$ is recursive (A on a tree T with no e-splitting), or $A \leq_T \{e\}^A$ (A on a e-splitting tree). With a similar proof we get a similar lemma, that takes care of the remaining case:

Proposition V.6.13 (Titgemeyer [1962], Hugill [1969]) *Given e, a recursive uniform tree T, and A on T, if $\{e\}^A$ is total then:*

1. *if on T there is no e-splitting which agrees on the odd, then*

$$\{e\}^A \leq_T Od(A)$$

2. *if whenever $T(\sigma * 0)$ and $T(\sigma * 1)$ disagree on the odds they e-split,*

$$Od(A) \leq_T \{e\}^A.$$

Proof. In the first case we compute $\{e\}^A$. Given x, we know that $\{e\}^A(x)\downarrow$, and to get the right value it is enough to search for σ on T such that $\{e\}^\sigma(x)\downarrow$ and $Od(\sigma) \subseteq Od(A)$, and this is recursive in $Od(A)$.

In the second case we compute initial segments of $Od(A)$, by induction. Suppose $T(\sigma)$ such that $Od(T(\sigma)) \subseteq Od(A)$ is given. If $T(\sigma * 0)$ and $T(\sigma * 1)$ agree on the odds, then both work (i.e. their odd parts are contained in the odd part of A). Otherwise, they disagree on the odds, and then they e-split, say on x. If $\{e\}^{T(\sigma * i)}(x) \simeq \{e\}^A(x)$, then $T(\sigma * i) \subseteq Od(A)$.   □

Now we can prove the analogue of the Minimality Lemma.

Proposition V.6.14 (Hugill [1969]) *Given e and a recursive uniform tree T alternating agreements and disagreements on the odds, there is a recursive uniform tree $Q \subseteq T$ alternating agreements and disagreements on the odds, such that one of the following holds:*

1. *for every A on Q, $\{e\}^A$ is not total*

2. *for every A on Q, $\{e\}^A$ is recursive*

3. *for every A on Q, $\{e\}^A \equiv_T Od(A)$*

4. *for every A on Q, $A \leq_T \{e\}^A$.*

Proof. If there are σ on T and x such that $(\forall \tau \supseteq \sigma)(\tau \in T \Rightarrow \{e\}^\tau(x)\uparrow)$, then let Q be the full subtree above σ. If A is on Q, then $\{e\}^A$ is not total.

Otherwise, see if there is σ with no e-splitting above it. If so, let Q be the full tree above it. If A is on Q, then $\{e\}^A$ is recursive (by V.5.9).

Otherwise, there are e-splittings above any strings. See however if there is σ with no e-splitting extensions agreeing on the odds. If so, let Q be a uniform subtree of T above σ, alternating branches agreeing on the odds and e-splitting branches. By hypothesis the e-splittings do not agree on the odds, and on Q there are no e-splittings agreeing on the odds. The conditions of V.6.13 are satisfied, and if A is on Q then $A \equiv_T \{e\}^A$.

Otherwise, we can always find e-splitting agreeing on the odds. We want Q e-splitting uniform tree, alternating agreements and disagreements on the odds. The level agreeing on the odds can be made e-splitting by hypothesis. The level disagreeing on the odds can also be done, because we can take e-splittings, and then extend them to disagree on the odds. Let Q be such an e-splitting subtree of T, which can be made uniform by the techniques of V.6.5. If A is on Q then $A \leq_T \{e\}^A$, because Q is e-splitting (V.5.9).   □

Theorem V.6.15 (Titgemeyer [1962]) *The three-element chain is embeddable as initial segment of $\mathcal{D}$.*

Proof. We already have all the ingredients, and the result follows by starting with

$$T_0 = \text{identity tree,}$$

which is a recursive uniform tree alternating branches agreeing on the odds and disagreeing on the odds. Then we let:

$$
\begin{array}{rcl}
T_{3e+1} & = & \text{the } Q \text{ of V.6.11, applied to } T = T_{3e} \\
T_{3e+2} & = & \text{the } Q \text{ of V.6.12, applied to } T = T_{3e+1} \\
T_{3e+3} & = & \text{the } Q \text{ of V.6.14, applied to } T = T_{3e+2}.
\end{array}
$$

The first two steps make $Od(A)$ not recursive and A not recursive in $Od(A)$, so that we do have one nontrivial degree below A. The last step makes sure that this is the only one. □

Note that in particular the proof produces a minimal degree with a strong minimal cover (V.5.16). Actually, with terminology as on p. 495, *the set of minimal degrees with a strong minimal cover is comeager, for the topology induced by recursive uniform trees* (Simpson [1977]): it is enough to consider, given such a tree T, a recursive uniform tree T^* that alternates branches agreeing on the odds with branches disagreeing on the odds, and such that the odd parts of the branches are exactly the branches of T. The proof above shows that the set of degrees which are top elements of three-element chains is comeager, for the topology induced by the trees T^*. The intermediate degrees, which are minimal degrees with a strong minimal cover, are then comeager for the topology induced by the trees T. It can then be argued, as in Section 3, that *minimal degrees without a strong minimal cover cannot be built by recursive uniform trees*. The existence of such degrees is an open problem, and the construction of one of them would then require new methods of proof.

Exercises V.6.16 More initial segments. a) *The diamond is embeddable as initial segment of* $\mathcal{D}$, i.e. there is a degree with exactly two nontrivial degrees below it, and they are incomparable. (Sacks [1963]) (Hint: build A such that $Ev(A) = \{x : 2x \in A\}$ and $Od(A)$ do the job. The only modification in the proof of the three-element chain is that $Od(A)$ and $Ev(A)$ have to be treated symmetrically and, whenever we had in V.6.15 branches that simply disagreed on the odds, now we actually want them to agree on the evens. Note that on trees on which all branches either agree on the odds or agree on the evens, whenever there are e-splittings then we may find them agreeing on the odds or on the evens, by going by adjacent paths as in the first part of V.6.8, where now adjacent means agreeing on the odds or on the evens.)

  b) *Every recursive linear ordering is embeddable as initial segment of* $\mathcal{D}$. (Hugill [1969]) (Hint: we may restrict ourselves to infinite orderings $\preceq$ of ω in which 0 is the least element, and 1 the greatest. Choose a set $\{X_n\}_{n\in\omega}$ of disjoint, uniformly recursive sets such that

$$m \preceq n \quad \Rightarrow \quad X_m \subseteq X_n$$
$$m \prec n \quad \Rightarrow \quad X_n - X_m \text{ infinite,}$$

and $X_0 = \emptyset$, $X_1 = \omega$. We build A such that, if we define

$$x \in X_n(A) \Leftrightarrow \text{the } x\text{-th element (in order of magnitude) of } X_n \text{ is in } A,$$

then the degrees of $\{X_n(A)\}_{n \in \omega}$ form an initial segment isomorphic to the partial ordering. Note that $X_0(A) = \emptyset$, $X_1(A) = A$, and $X_n(A) \leq_T A$. E.g., in V.6.15 we took $X_2 =$ odd numbers. The idea is simply to approximate the ordering, by considering at stage s the subordering induced by it on $\{0, \ldots, s\}$, and ensure conditions on A that make the degrees of $\{X_n(A)\}_{n \leq s}$ isomorphic to it. The lemmas above can be rewritten with reference to any X_n, instead of the odds. To be able to diagonalize we need to have, whenever $m \prec n$, infinitely many branches agreeing on X_m and disagreeing on X_n. For this we enumerate these pairs $\langle m, n \rangle$ with infinitely many repetitions, and ask that the i-th level of the tree agrees on X_m and disagrees on X_n, if $\langle m, n \rangle$ is the i-th pair in the enumeration. The reason why dealing with only finitely many conditions at each stage is sufficient, is that every function recursive in A has infinitely many indices and, for all but finitely many such, our trees will have the desired properties.)

c) *There is a degree with no minimal predecessor.* (Martin [1967], Hugill [1969]) (Hint: embed an ordering with order type $1 + \omega^*$, i.e. the reverse ordering of the natural numbers with a least element added to it. For a different proof see V.3.17.)

d) *There is an ascending sequence of degrees with an upper bound below which there is no minimal upper bound.* (Yates [1970]) (Hint: embed an ordering with order type $\omega + \omega^*$.)

e) *There is a completely autoreducible, not minimal degree.* (Jockusch and Paterson [1976]) (Hint: use double e-splittings in the proof of V.6.15, as in V.5.15.d. This makes the top degree completely autoreducible.)

The initial segments of the degrees ⋆

The ideas involved in the proofs of the embeddings of the three-element chain and the diamond can be pushed further, to prove that every finite distributive lattice is embeddable as initial segment of $\mathcal{D}$. This relies on the fact that if a finite lattice is distributive, then it is isomorphic to a sublattice of the power set of some finite set. It is then enough, given a finite distributive lattice, to find a partition of ω into infinite, coinfinite recursive sets whose lattice structure under inclusion mirrors the given lattice (like $\emptyset, Ev, Od$ and ω did for the diamond), and build the top degree of the initial segment. What is needed is some kind of representation theorem for the lattices, but no new recursion-theoretical ideas are involved. This theorem is due to Lachlan [1968a], but the approach just sketched comes from Yates [1972], who did it for Boolean algebras (which require only the full power set of some set), and Epstein [1979], who got the representation.

The same approach can be adopted for countable bottomed distributive lattices, with a new ingredient: if the lattice is not recursive, its representation via recursive subsets cannot be chosen ahead of time, and has to be built along the way. This gives the result of Lachlan [1968a], and it is as far as the method can go: only distributive lattices can be treated this way.

Nondistributive lattices are much more difficult to deal with, and have been treated by Shoenfield, Lerman [1969], [1971], and Lachlan and Lebeuf [1976]. The final result for countable initial segments is the following: *a countable partial ordering is isomorphic to an initial segment of $\mathcal{D}$ if and only if it is an uppersemilattice with a least element.* This completely characterizes the possible isomorphism types of $\mathcal{D}(\leq a)$, since any degree has at most countably many predecessors. Namely, the principal ideals of $\mathcal{D}$ realize every possible countable uppersemilattice with least and greatest element. We will prove that $\mathcal{D}$ is not distributive without using initial segments, as a corollary of an embedding result in the r.e. degrees, see X.6.27.b.

Treatments of the embeddings quoted so far can be found in Epstein [1979] and Lerman [1983].

Uncountable initial segments have been considered by Thomason [1970] and Rubin. Here it is not possible to assume the existence of a greatest element, and hence to construct only the top degree. Thus the given uppersemilattice has to be approximated, and a further obstacle is introduced by the fact that not every initial segment is extendible (e.g. not every degree has a strong minimal cover), and thus the approximations cannot simply be end-extensions. The method of forcing is used to control the amalgamation of the various approximations, and Abraham and Shore [1986] have proved that *a partial ordering with power at most $\aleph_1$ is isomorphic to an initial segment of $\mathcal{D}$ if and only if it is an uppersemilattice with a least element, and countable predecessor property.* This completely characterizes the possible isomorphism types of linearly ordered initial segments, because a linear ordering with countable predecessor property must have power at most $\aleph_1$.

This result settles the initial segments problem if the Continuum Hypothesis holds, and it is the best possible result in ZFC, since Groszek and Slaman [1983] have shown that *it is consistent with ZFC to have an uncountable uppersemilattice (even of power $\aleph_2$ less than the continuum) with countable (even finite) predecessor property, which is not embeddable in $\mathcal{D}$ as an uppersemilattice (and hence as an initial segment).*

The characterization of initial segments is just one step toward a complete characterization of the algebraic structure of $\mathcal{D}$. The next step would be to know which initial segments are extendable, but not even the very first step has been answered: it is still not known if every minimal degree has a strong minimal cover.

V.7   Global Properties

Until now we have studied first-order or local properties of degrees, mainly telling that given configurations exist in the degrees, possibly extending given ones. We now take a different stand, and look at second-order or global properties of degrees: on the one hand we analyze the difficulty of the first-order theory of $\mathcal{D}$ (proving its undecidability and much more), and on the other hand we study its global algebraic properties. We will capitalize here on the work done so far, and a cascade of interesting results will be obtained.

Definability from parameters

What Spector's Theorem accomplishes is to define any countable ideal, by using two parameters. Indeed, if a and b are an exact pair for a countable ideal $\mathcal{I}$, then

$$c \in \mathcal{I} \Leftrightarrow c \leq a \wedge c \leq b.$$

As we can see, the formula that defines the ideal is always the same, and only the parameters change, for different ideals. This uniformity is quite important, because then we can avoid talking about ideals, replacing them by their exact pairs (since the dependence of the ideal on the exact pair is fixed). This shows that *the first-order theory of $\mathcal{D}$ includes an interpretation of quantification over countable ideals*.

We are now going to extend this result by showing that, in a little more complicated way, any countable relation is uniformly definable from a fixed number of parameters, which depends only on the number of arguments of the relation. We first prove a weaker result, which provides all the technical work.

Proposition V.7.1 (Slaman and Woodin [1986]) *Every countable antichain is definable from finitely many parameters in $\mathcal{D}$, in a uniform way.*

Proof. Since an antichain $\mathcal{C} = \{c_n\}_{n \in \omega}$ is made of pairwise incomparable degrees, the idea to define it is to find a property P for which the given c_n's are the minimal solutions. We will use an approach symmetric to the one adopted for Spector's Theorem, namely we define two degrees a and b and we look at the degrees x satisfying the property P so defined:

$$P(x) \Leftrightarrow x \neq (x \cup a) \cap (x \cup b).$$

If we can make sure that the elements of $\mathcal{C}$ are the minimal solutions of P, then

$$x \in \mathcal{C} \Leftrightarrow P(x) \wedge \neg(\exists z)(z < x \wedge P(z)).$$

The trouble with this approach is that we are asking too much, because being a minimal solution of P is a condition involving all degrees, and only countable information can be coded into a pair of sets. We thus relax the requirement a bit, and consider not all degrees, but only the countably many ones below any given degree c bounding C. Then we only need to have:

$$x \in C \iff x \leq c \wedge P(x) \wedge \neg(\exists z \leq c)(z < x \wedge P(z)).$$

We will thus define C using three parameters a, b, and c, of which c is simply any degree bounding every c_n.

Given $\{C_n\}_{n \in \omega}$ with $C_n \in c_n$, we can simply let $C = \bigoplus_{n \in \omega} C_n$. The conditions on A and B are the following:

1. c_n *is a solution of* P

   Since $x \leq (x \cup a) \cap (x \cup b)$ always holds, we ensure $(x \cup a) \cap (x \cup b) \not\leq x$, for x in C. This will be achieved by having, for every n, a set D_n such that

   $$D_n \leq_T C_n \oplus A, C_n \oplus B \quad \text{but} \quad D_n \not\leq_T C_n.$$

2. c_n *is a minimal solution of* P

   For this we need, for any $x \leq c$,

   $$x \neq (x \cup a) \cap (x \cup b) \implies c_n \leq x, \text{ for some } n.$$

   Hence, for any $X \leq_T C$,

   $$(\exists D)(D \leq_T X \oplus A, X \oplus B \wedge D \not\leq_T X) \implies C_n \leq_T X, \text{ for some } n.$$

As in Spector's Theorem, A and B will be built by coinfinite extensions θ_s and ϑ_s. We will have $A = \bigcup_{s \in \omega} \theta_s$ and $B = \bigcup_{s \in \omega} \vartheta_s$. As usual, A and B will be thought of as made of columns A_n and B_n, and corresponding columns in them will differ only finitely (so that their degrees will be the same). This time however it is D_n, rather than C_n, that gets coded in the n-th columns A_n and B_n.

Since D_n has to be recoverable from each of $C_n \oplus A$ and $C_n \oplus B$, we will use only the elements of C_n to code information in A and B. Precisely,

$$D_n(x) = \begin{cases} 0 & \text{if } x \notin C_n \\ A_n(x) & \text{otherwise.} \end{cases}$$

Then $D_n \leq_T C_n \oplus A$ by definition, and $D_n \leq_T C_n \oplus B$ because A_n and B_n will differ only finitely, by construction. It remains to ensure, by diagonalization, that $D_n \not\leq_T C_n$, and hence $D_n \not\leq_T \{e\}^{C_n}$, for every e.

Since we use coinfinite conditions, we will be able to take care of all these requirements, for a fixed n, in just one step. Indeed, we can define

$$D_n^*(x) = \begin{cases} 1 - \{e\}^{C_n}(x) & \text{if } x \text{ is the } e\text{-th element of } C_n, \text{ and } \{e\}^{C_n}(x){\downarrow} \\ 0 & \text{otherwise.} \end{cases}$$

D_n^* obviously is not recursive in C_n, and neither is any set differing only finitely from it (because each recursive function has infinitely many indices, and thus D_n^* differs infinitely often from every set recursive in C_n). It will then be enough to have A_n, and hence D_n, differ only finitely from D_n^*.

We now consider the minimality requirements. We will work with a fixed list $\{X_m\}_{m \in \omega}$ of the sets recursive in C. Recall that we want, for each m,

$$D \leq_T X_m \oplus A, X_m \oplus B \wedge D \not\leq_T X_m \Rightarrow C_n \leq_T X_m, \text{ for some } n.$$

The requirements are:

$$R_{e,m} : \{e\}^{X_m \oplus A} \simeq \{e\}^{X_m \oplus B} \simeq D \wedge D \not\leq_T X_m \Rightarrow C_n \leq_T X_m, \text{ for some } n$$

(note that we use V.2.18.a to simplify the presentation of the requirements). They will be satisfied as in the proof of Spector's Theorem.

We start with $\theta_0 = \vartheta_0 = \emptyset$. At each step $s + 1$ we are given θ_s and ϑ_s, defined on the first s columns and on finitely many other points, and differing on each of the first s columns only finitely. We take care of the requirement $R_{e,m}$ if $s = \langle e, m \rangle$, and in any case we will ensure the needed diagonalizations on D_s. There are three possible cases:

- we can make $\{e\}^{X_m \oplus A}$ not total

  In other words, there is a string σ compatible with θ_s, such that

  $$(\exists y)(\forall \tau \supseteq \sigma)(\tau \text{ compatible with } \theta_s \Rightarrow \{e\}^{X_m \oplus \tau}(y){\uparrow}).$$

  Then take one such string, and let:

  $$\theta_{s+1}(x) = \begin{cases} \theta_s(x) & \text{if } \theta_s(x){\downarrow} \\ \sigma(x) & \text{if } \sigma(x){\downarrow} \\ D_s^*(z) & \text{if } x = \langle s, z \rangle, \text{ otherwise.} \end{cases}$$

  ϑ_{s+1} is obtained similarly, by using ϑ_s in place of θ_s.

- otherwise, but we can force a special kind of disagreement

  In general we would only look for two strings σ and τ, respectively compatible with θ_s and ϑ_s, such that $\{e\}^{X_m \oplus \sigma}(x) \not\simeq \{e\}^{X_m \oplus \tau}(x)$ on some x. For reasons that will appear clear later, here we also request that they agree on elements of C_n, on the n-th column, for any $n < s$. If two such strings exist, take any of them and define θ_{s+1} and ϑ_{s+1} as above, using σ and θ_s for the former, and τ and ϑ_s for the latter.

- *otherwise*

Then simply let

$$\theta_{s+1}(x) = \begin{cases} \theta_s(x) & \text{if } \theta_s(x)\downarrow \\ D_s^*(z) & \text{if } x = \langle s, z \rangle, \text{ otherwise.} \end{cases}$$

ϑ_{s+1} is obtained by using ϑ_s in place of θ_s.

It is clear that in the first two cases the requirement $R_{e,m}$ is vacuously satisfied. We now have to argue that it is so also in the last case. Suppose that in the end we have

$$\{e\}^{X_m \oplus A} \simeq \{e\}^{X_m \oplus B} \simeq D \text{ and } D \not\leq_T X_m.$$

We have to prove that

$$C_n \leq_T X_m, \text{ for some } n.$$

Suppose there is no e-splitting compatible with θ_s: then, as usual, we would have $\{e\}^{X_m \oplus A} \simeq D \leq_T X_m$, but this is impossible by hypothesis. Thus there are e-splittings compatible with θ_s, and actually infinitely many such (because the same reasoning works above any given string). But we also supposed

$$\{e\}^{X_m \oplus A} \simeq \{e\}^{X_m \oplus B},$$

which means that the construction did not succeed in forcing a disagreement. Certainly the first case of the construction cannot have happened, because $\{e\}^{X_m \oplus A} \simeq D$ is a total function. Then it must have been the second step that failed. Since θ_s and ϑ_s disagree only finitely, there are infinitely many e-splitting pairs, with one side compatible with θ_s, and the other compatible with ϑ_s. Since we did not pick up any of them, it must be because none satisfies the additional condition of case two, and thus the two sides do not always agree on the elements of C_n, on the n-th columns, for some $n < s$.

We now show that $C_n \leq_T X_m$, for some $n < s$, in three steps:

1. *for a fixed $n < s$, there must be infinitely many pairs of e-splitting strings, compatible with θ_s and differing on some element of C_n*
   This is simply because the same holds for some $n < s$, and there are only finitely many such n's. Thus infinitely many disagreements must hold for the same n. We now fix this n.

2. *there is an infinite subset of C_n which is recursive in X_m*
   We show how to generate an infinite ascending sequence of elements of C_n. Suppose we have an initial segment μ of the characteristic function of such a subset. Simply look for an e-splitting σ, σ' extending μ, and compatible with θ_s. As argued above, σ and σ' must disagree on some element of

C_n (unless their pair is one of finitely many exceptions, a case that is effectively testable because the exceptions can be given in advance). But they can also disagree on some other element, not necessarily in C_n. To be able to sort out the right element, we only have to interpolate the two strings, by a sequence of strings $\sigma_0, \ldots, \sigma_i$ all extending μ, making $\{e\}^{\sigma_j}$ converge on the element on which σ and σ' e-split (which is possible, because we are not in the first case of the construction) and differing, each from the following one, on just one element. Moreover, $\sigma_0 = \sigma$ and $\sigma_i = \sigma'$. Since σ_0 and σ_i e-split, there must be $j < i$ such that σ_j and σ_{j+1} e-split. These strings can effectively be found, and now they differ only on one element, which must then be in C_n.

3. $C_n \leq_T X_m$

   We now have an infinite subset of C_n recursive in X_m. To be able to have all of C_n recursive in X_m, we play a trick: we choose, from the very beginning, sets C_n's which are introreducible (see II.6.7), i.e. recursive in each of their infinite subsets. This is possible, because every degree contains such a set (II.6.13). If C_n is introreducible, then the previous part already shows that $C_n \leq_T X_m$. $\square$

Note that *the parameters needed to define $\{C_n\}_{n \in \omega}$ can be obtained arithmetically in $\oplus_{n \in \omega} C_n$.*

Theorem V.7.2 Definability from parameters (Slaman and Woodin [1986]) *Every countable relation is definable from finitely many parameters in $\mathcal{D}$, in a uniform way.*

Proof. We start by dealing with sets. Given $\mathcal{C} = \{c_n\}_{n \in \omega}$, we would like to spread it out to get a set of incomparable degrees, since we know how to define the latter. Choose c above every element of $\mathcal{C}$ (which is possible because $\mathcal{C}$ is countable). Define a set $\mathcal{A} = \{a_n\}_{n \in \omega}$ of pairwise incomparable degrees not introducing any new relation on degrees below c, i.e. such that, for $x \leq c$,

$$x \cup a_m \leq y \cup a_n \Leftrightarrow x \leq y \wedge m = n.$$

$\mathcal{A}$ can be constructed by the methods of Section 4 (see V.2.9). We now spread out $\mathcal{C}$ by using any infinite subset $\mathcal{A}^* = \{a_{f(n)}\}_{n \in \omega}$ of $\mathcal{A}$, for any one-one total function f (for this part of the proof the set $\mathcal{A}$ itself would be perfectly sufficient, but the added generality will be useful later on). Clearly $\mathcal{A}^*$ is an antichain (since so is $\mathcal{A}$), and so is the set

$$\mathcal{C}^* = \{c_n \cup a_{f(n)}\}_{n \in \omega},$$

by the choice of $\mathcal{A}$ (because $c_n \cup a_{f(n)} \leq c_m \cup a_{f(m)}$ only if $f(m) = f(n)$, and hence if $m = n$, f being one-one). Then both $\mathcal{A}^*$ and $\mathcal{C}^*$ are definable with parameters, by the previous proposition. It follows that:

- C is definable
  Indeed
  $$x \in C \Leftrightarrow x \le c \wedge (\exists a \in A^*)(x \cup a \in C^*).$$

  Only the right-to-left implication has to be checked. Suppose that $x \le c$, and $x \cup a_{f(n)}$ is in C^*: then

  $$x \cup a_{f(n)} = c_m \cup a_{f(m)},$$

  for some m. By the properties of A it must be

  $$x \cup a_{f(n)} = c_n \cup a_{f(n)}.$$

  Then from $x \cup a_{f(n)} \le c_n \cup a_{f(n)}$ we have $x \le c_n$, and similarly $c_n \le x$, from which $x = c_n$. Thus $c \in C$.

- the map f^* from C to C^* induced by f is definable
  As for C,
  $$f^*(x) = y \Leftrightarrow x \in C \wedge x \cup y \in C^*.$$

We now turn to the definability of countable relations. Given an n-ary relation R on degrees, we can first of all define the n projections of R, since the degrees in them are the relevant ones:

$$x \in R_i \Leftrightarrow (\exists y_1) \cdots (\exists y_{n-1}) R(y_1, \ldots, y_{i-1}, x, y_i, \ldots, y_{n-1}).$$

We can now take as c a degree bounding all the elements of $\bigcup_{1 \le i \le n} R_i$, as above. The new problem that we have to face here is a coding problem: we would like to reduce R to a set of (incomparable) degrees, but we actually have n-tuples of degrees, one in each of the projections. What we do is to define A as above, with the additional requirement that the finite joins of its elements are uniquely determined by the elements themselves, so that the join of n elements uniquely codes them. Since we are going to use finite joins of elements of A, we want all of them to be incomparable. And since we are going to code the individual projections of R as well, the finite joins will have to not introduce new relationships among the degrees below c. All these conditions follow automatically, when A is a set of degrees recursively independent over the degrees below c: such a set can be constructed by the methods of Section 4 (see V.2.9). Having A, we can first of all pick up n disjoint subsets of it (by using one-one functions f_i) to code the projections R_i. As noted above, not only the sets R_i, but also the functions f_i^*, choosing the elements needed to spread out R_i, are definable. Now we can simply let

$$B = \{f_1^*(x_1) \cup \cdots \cup f_n^*(x_n) : R(x_1, \ldots, x_n)\}.$$

$\mathcal{B}$ is definable, because the degrees $f_1^*(x_1) \cup \cdots \cup f_n^*(x_n)$ are finite joins of members of $\mathcal{A}$, and hence pairwise incomparable. And R is recoverable from $\mathcal{B}$, because the joins of elements of $\mathcal{A}$ are uniquely determined by their elements, and hence

$$R(x_1, \ldots, x_n) \iff f_1^*(x_1) \cup \cdots \cup f_n^*(x_n) \in \mathcal{B}.$$

Moreover, R is definable because the maps f_i^* are definable, and so is $\mathcal{B}$. □

Note that in a sense the theorem is the best possible, since *the parameters cannot be eliminated in general*: the relations definable in $\mathcal{D}$ are only countably many, but there are uncountably many countable relations. There are however many relations that can be definable without parameters, as we will see.

The complexity of the theory of degrees

The definability result of the last subsection allows for a quick proof of the following theorem, which completely characterizes the complexity of the theory of degrees.

Theorem V.7.3 Simpson's Theorem (Simpson [1977]) *The first-order theory of $\mathcal{D}$ has the same degree (and actually the same isomorphism type) as the theory of Second-Order Arithmetic.*

Proof. We prove that the two theories have the same m-degree by interpreting each in the other, thus providing faithful translations that will preserve theorems. Since the translations will actually be one-one, the theories will have the same 1-degree, and hence will be recursively isomorphic by III.7.13.

One direction is clear, since every formula about the ordering of degrees can be interpreted, in the natural way, as a formula about sets of integers. Thus the theory of degrees is interpretable in Second-Order Arithmetic.

For the converse, we want to show that Second-Order Arithmetic is interpretable in $\mathcal{D}$. A *model of arithmetic* is a structure $\langle A, R, f_1, f_2, f_3, a \rangle$ such that:

1. A is a countable set

2. R is a total ordering on A with first element a, successor given by f_1, and such that every element different from a has an immediate predecessor

3. f_2 and f_3 satisfy the axioms of Robinson Arithmetic Q (p. 23), when interpreted as sum and product, with f_1 interpreted as successor.

A *standard model of arithmetic* is a model in which R is a well-ordering.

A standard model exists in the degrees, since any countable partial ordering is embeddable in boldmath $\mathcal{D}$ (by V.2.9). Moreover, given degrees $\vec{a}$ coding a countable set of degrees (intended to be the universe A of a standard model of Arithmetic in $\mathcal{D}$), we can say in a first-order statement of $\mathcal{D}$ that given degrees $\vec{c}$ code a relation and (the graphs of) three functions on the set coded by $\vec{a}$, satisfying the requirements for being a standard model of Arithmetic. This is because Q has only finitely many axioms, and well-ordering can be expressed by replacing quantification over subsets of A by quantification over parameters that define them (since a subset is defined by a fixed number of parameters, in a uniform way).

Then a sentence φ of Second-Order Arithmetic is true if and only if the sentence φ^* of $\mathcal{D}$ is, where φ^* says that there are degrees $\vec{a}$ and $\vec{c}$ coding a standard model of arithmetic in which the translation of φ (obtained, as above, by replacing quantification over subsets of A by quantification over parameters that define them) holds. Since φ^* can be effectively obtained from φ, we thus have an m-reduction of Second-Order Arithmetic to the theory of $\mathcal{D}$. $\square$

Corollary V.7.4 (Lachlan [1968]) *The first-order theory of $\mathcal{D}$ is undecidable and not axiomatizable.*

Proof. So is the theory of (first-order) arithmetic. $\square$

Corollary V.7.5 (Jockusch and Simpson [1976], Simpson [1977]) *The first-order theory of $\mathcal{D}$ is not absolute with respect to models of ZFC containing all the ordinals.*

Proof. There is a sentence about the ordering of degrees which is true in $\mathcal{D}$ if and only if $\mathcal{P}(\omega) \subseteq L$, since the latter can be translated into a sentence of Second-Order Arithmetic by IV.4.22. $\square$

Lachlan obtained the undecidability of the theory of degrees as a corollary of the embeddability of every countable distributive lattice as an initial segment of $\mathcal{D}$ (see p. 528), by using the undecidability of the theory of distributive lattices (and the Löwenheim-Skolem Theorem, to be able to restrict attention to countable distributive lattices). Subsequently, Thomason [1970] observed that the result follows from the embeddability of the finite distributive lattices only, this time using the fact that the set of sentences true in all distributive lattices and the set of sentences false in some finite distributive lattice are recursively inseparable.

In the first proofs of Simpson's Theorem, when only Spector's Theorem was available, only ideals (and hence initial segments) could be used to code sets and relations in $\mathcal{D}$ in a definable way, and more ingenuity was required. The original coding of Arithmetic by Simpson [1977] was very direct, and interpreted the natural number n as the degree $\mathbf{0}^{(n)}$, $m+n$ as $\mathbf{0}^{(m+n)}$, and $m \cdot n$ as $\mathbf{0}^{(m \cdot n)}$. To define the set $\{\mathbf{0}^{(n)}\}_{n \in \omega}$ Simpson hooked it up to an initial segment with controlled (double) jump, and the same method was used for sum and product. This required only simple initial segments (namely chains), but the proof that the coding worked was not straightforward (in particular, the jump operator had to be eliminated). An exposition of this method is in Epstein [1979].

A much simpler coding, requiring more initial segments (the countable distributive lattices) but less direct work, was devised by Nerode and Shore [1980]. They noticed that second-order logic on countable sets could be translated into the theory of countable distributive lattices with quantification over ideals, by first coding relations by graphs, and then graphs by ideals of distributive lattices. The advantages of this method were simplicity, which allowed for improved calculations and sharper consequences, and generality, which permitted the extension of Simpson's Theorem to a variety of different degree structures. An exposition of this method for the study of Turing degrees is in Lerman [1983], while the general result is proved in VI.4.7.

The coding method we used is due to Slaman and Woodin [1986]. It is easier than all of the above, since it codes sets and relations directly, and thus it makes the translation of Arithmetic straightforward. It also has the advantage that the only needed result (V.7.2) can be proved directly, and with very little machinery (only the coinfinite extension method), in particular avoiding initial segment constructions.

Some additional work provides a substantial generalization of Simpson's Theorem.

Theorem V.7.6 (Nerode and Shore [1980], [1980a]) *If C is an ideal of $\mathcal{D}$ closed under jump, the first-order theory of C has the same degree (and actually the same isomorphism type) as the theory of Second-Order Arithmetic with set quantifiers restricted to sets with degree in C.*

Proof. We refine the proof of V.7.3. As there, the translation of the theory of C into Second-Order Arithmetic with set quantifiers restricted to sets with degrees in C is immediate.

For the converse, we first need to show that we can pick up a standard model of arithmetic in C. Certainly the needed configurations exist in the degrees below $0'$ (by V.2.9), and hence in C. Moreover, also parameters $\vec{a}$ and $\vec{c}$ defining a standard model of arithmetic exist in C, because the proof of V.7.2 provides parameters coding a given set arithmetically in its infinite join, and the methods of V.4.7 allow to choose the needed configurations not only below $0'$, but actually with infinite join below it.

A simple translation of the relevant notions shows that there is a first-order sentence of the theory of C that says that degrees $\vec{a}$ and $\vec{c}$ code a model of arithmetic in C. But the method used in V.7.3 to show that the same holds for standard models as well does not work here, since by quantifying over parameters in $\mathcal{C}$ we do not take care of all possible subsets, and hence we do not define a real well-ordering.

To be able to handle standard models we proceed as follows. Consider the standard part (corresponding to the set of integers) of a model. If the model is standard then this part exhausts its universe, and thus every proper initial segment of the universe is finite and has a least upper bound. Conversely, if the model is not standard then the standard part is a proper initial segment with no least upper bound. Thus the model is standard if and only if every proper initial segment of the universe has least upper bound. This is still a second-order sentence, but it really needs only the standard part of the model and the finite subsets of the universe. If we show that these subsets are all coded by parameters in $\mathcal{C}$ when the universe is, then we can quantify only over parameters in $\mathcal{C}$ and still take care of all the needed subsets.

Finite subsets are easily handled, since $\mathcal{C}$ is an ideal closed under jump, and the parameters coding a finite subset are arithmetical in its join. The standard part can be enumerated arithmetically in $\vec{a}$ and $\vec{c}$, because an index relative to them of the first element (corresponding to 0) can be given for free, while the procedure to pick up the index of the successor of a given element only involves a few quantifiers (needed to express the order relation $\leq_T$, and the successor relation). The parameters coding the standard part are arithmetical in the enumerating function (which gives the infinite join) and, by transitivity, arithmetical in $\vec{a}$ and $\vec{c}$.

This already provides a translation of first-order arithmetic into the theory of $\mathcal{C}$. To be able to handle set quantifiers as well, we need to know which sets can be coded by parameters in $\mathcal{C}$, with respect to standard models of arithmetic in $\mathcal{C}$. We claim they are exactly the sets whose degrees are in $\mathcal{C}$, and this will finish the proof.

First of all we can identify sets of natural numbers and sets of degrees contained in the standard part of the model, by identifying the number n with the n-th element of the standard part. This preserves membership of the degree of the set in $\mathcal{C}$, since the standard part of the model can be recovered arithmetically from the parameters coding it. Given a set B, parameters $\vec{b}$ coding it can be obtained arithmetically in it, by the proof of V.7.2. And given parameters $\vec{b}$, the set B coded by them can be recovered arithmetically from them, from the definition provided by the proof of V.7.2. Since $\mathcal{C}$ is closed under jump, B has degree in $\mathcal{C}$ if and only if there are parameters $\vec{b}$ coding it in $\mathcal{C}$, which is what we wanted to prove. $\square$

Note that two ideals can be different but isomorphic, as the case of two different minimal degrees illustrates. This cannot happen for ideals closed under jump.

Corollary V.7.7 *Two isomorphic ideals closed under jump are identical.*

Proof. The proof above shows that an ideal closed under jump consists exactly of the degrees of sets which are coded by parameters in the ideal. Then two isomorphic ideals closed under jump code exactly the same sets, and must then be identical.   □

The corollary shows that two different ideals closed under jump are not isomorphic. The theorem implies that whenever the quantifications over sets with degrees in the two ideals have different power, then the two ideals are also not elementarily equivalent. We provide immediately a first example, and many others will be given in following chapters.

Corollary V.7.8 (Jockusch [1973], Nerode and Shore [1980a]) $\mathcal{D}$ *is not elementarily equivalent to the degrees of Δ_n^1 sets, for any $n \geq 0$.*

Proof. Clearly the degrees of Δ_n^1 sets are an ideal closed under jump, since the Δ_n^1 sets are closed under number quantification. We can thus apply the theorem. The sets definable in Second-Order Arithmetic with set quantification restricted to Δ_n^1 sets are all arithmetical in any enumeration of the Δ_n^1 sets, because set quantifiers can be replaced by number quantifiers over the indices of the enumeration. Since there is such an enumeration which is Δ_{n+1}^1 such sets are all Δ_{n+1}^1, by IV.2.25. The complete Σ_{n+1}^1 is thus an example of a set that can be defined in Second-Order Arithmetic, but not by set quantifiers restricted to Δ_n^1 sets.   □

This is as far as we can go, since it will be proved in Volume III that *the assertion that $\mathcal{D}$ is elementarily equivalent to the degrees of analytical sets is independent of ZFC.*

Absolute definability

We know that every countable set or relation is definable in $\mathcal{D}$ from parameters. We are not completely satisfied with this, because each set or relation needs different parameters. We may think of uniformly defining sets and relations which are definable in Arithmetic simply by using their own definitions in Arithmetic, and interpreting them over a fixed standard model of arithmetic in $\mathcal{D}$ (as in the previous subsection). This uses only the fixed parameters needed to define the model and it is more satisfactory, in particular the parameters can be fixed in advance. In this section we show how to get rid of these parameters too, for a wide class of sets and relations.

The main results of this subsection, as well as many others proved below and depending on it, will be stated in a strong form, without any assumption. We will freely use the following result, whose proof will be given only in XII.3.14, because it relies on methods and ideas that do not belong to this chapter (the construction of a minimal degree below $0'$ among them).

Theorem V.7.9 Definability of the arithmetical degrees (Jockusch and Shore [1984]) *The set of arithmetical degrees is definable in $\mathcal{D}$.*

However, the result just quoted is needed only to get the sharpest formulations of the main theorems of this section. The work done here is fruitful even without ever getting to the proof of V.7.9, because what we prove can be reinterpreted in the following ways:

1. The set of the arithmetical degrees can be used as a parameter. The results so obtained are then not absolute, but the parameter used is fixed and natural.

2. The set of the arithmetical degrees can be defined in $\mathcal{D}'$ in a trivial way (see V.8.1). The results obtained can then be interpreted as results about $\mathcal{D}'$, and as such they are absolute, although weaker than when stated for $\mathcal{D}$ alone.

The next result is the key to obtain definability and other results.

Proposition V.7.10 (Shore [1982]) *For every standard model of arithmetic in the arithmetical degrees, the map taking a degree x above all the arithmetical degrees into a set (of natural numbers in the standard model) of degree x is definable in $\mathcal{D}$, with parameters the degrees coding the standard model.*

Proof. Fix arithmetical degrees coding a standard model of arithmetic. We want to define in $\mathcal{D}$ the relation

X is a set (of natural numbers in the standard model) of degree x.

Consider the set of degrees

$\mathcal{X} = \{z : \text{the sets coded by parameters below } z \text{ are all recursive in } X\}$.

Being definable in Second-Order Arithmetic with the degrees coding the standard model as parameters, $\mathcal{X}$ is definable in $\mathcal{D}$ with the same parameters. If we showed that $deg(X)$ is its least upper bound, then

$$x = deg(X) \Leftrightarrow x \text{ is the l.u.b. of } \mathcal{X}$$

would be the definition we are looking for.

To be the least upper bound of $\mathcal{X}$, $deg(X)$ must be an upper bound. But if we take a degree z and parameters below it coding a set, then we can recover the set recursively not in z, but only in $z \cup a$ for some arithmetical degree a, the reason being that we also need to recover the standard part of the model, and this involves a few quantifiers over the (arithmetical) parameters coding it, and hence an arithmetical degree. To be sure that z always stays below $deg(X)$, we then modify the definition of $\mathcal{X}$ into:

$$\mathcal{X} = \{z : (\forall a \text{ arithmetical})(\text{the sets coded by}$$
$$\text{parameters below } z \cup a \text{ are recursive in } X)\}.$$

Given z, there is a set of degree z coded by parameters below $z \cup a$, for some arithmetical degree a. If $z \in \mathcal{X}$ then z is recursive in $deg(X)$, and thus $deg(X)$ is now an upper bound of $\mathcal{X}$.

To show that it is the least upper bound, it is enough to show that there are two degrees in $\mathcal{X}$ which join to $deg(X)$. Note that there is a fixed n such that if

$$(\forall a \text{ arithmetical})[(z \cup a)^{(n)} \leq deg(X)]$$

then z is in $\mathcal{X}$. Indeed, below $z \cup a$ one can code only sets of degree at most $(z \cup a)^{(n)}$, for a fixed n depending only on the coding procedure. Then it is enough to find two degrees z_1 and z_2 such that

$$z_1^{(n)} \cup z_2^{(n)} = z_1 \cup z_2 = deg(X).$$

That such degrees exist follows, by induction on n, from a relativization of V.2.26, together with the fact that $deg(X)$ is above all the arithmetical degrees, and in particular above $0^{(n)}$. $\square$

Theorem V.7.11 Absolute definability (Harrington and Shore [1981], Shore [1981], Jockusch and Shore [1984]) *A relation on degrees above all the arithmetical ones is definable in $\mathcal{D}$ if and only if it is definable in Second-Order Arithmetic.*

Proof. As usual, a relation can be definable in $\mathcal{D}$ only if it is definable in Second-Order Arithmetic. Let then R be such a relation on degrees x above all the arithmetical ones. First consider the relation R^* on sets that says that R holds for the degrees of its arguments:

$$R^*(X_1, \ldots, X_n) \Leftrightarrow R(deg(X_1), \ldots, deg(X_n)).$$

R^* is still analytical and hence, given a standard model of arithmetic in $\mathcal{D}$, it is faithfully translatable into a first-order formula φ^* of $\mathcal{D}$ with parameters the degrees coding the standard model.

Then $R(x_1, \ldots, x_n)$ holds if and only if for some standard model of arithmetic coded by arithmetical degrees, φ^* holds for some sets $X_1, \ldots, X_n$ such that $deg(X_i) = x_i$. Then R is definable in $\mathcal{D}$ because so are the set of the arithmetical degrees, the property of coding a standard model of arithmetic, and the map taking a degree x above all the arithmetical ones into a set X (of natural numbers in the standard model) of degree x. The quantification over the degrees coding the standard model eliminates the explicit reference to them. □

We can actually avoid the restriction on degrees above all the arithmetical ones, with no additional work.

Corollary V.7.12 *Let R be a relation on degrees which is invariant under join with arithmetical degrees, i.e. such that for every degree x_i and every arithmetical degree a_i*

$$R(x_1, \ldots, x_n) \Leftrightarrow R(x_1 \cup a_1, \ldots, x_n \cup a_n).$$

Then R is definable in $\mathcal{D}$ if and only if it is definable in Second-Order Arithmetic.

Proof. It is enough to modify the definition of R given above by stating that X_i is a set of degree $x_i \cup a_i$, for some arithmetical degree a_i. □

We now have a number of interesting examples of definable sets and relations. E.g., *the sets of degrees of Δ_n^1, analytical, and constructible sets are all definable in $\mathcal{D}$*. By relativization, also *the notions of Δ_n^1-degrees and constructibility degrees are definable in $\mathcal{D}$*.

Similarly, individual degrees above all the arithmetical ones and definable in Second-Order Arithmetic are definable in $\mathcal{D}$. An example is $0^{(\omega)}$. By relativization, *the ω-jump operation is definable in $\mathcal{D}$*. Lerman and Shore [1988] have proved that no degree $a > 0$ is definable in $\mathcal{D}$ by a $\exists\forall$–formula.

Of course the definitions provided by the proof above are not very natural from a recursion-theoretical point of view, because they simply translate definitions from Second-Order Arithmetic. In the next section we will start a path that will be pursued all along the rest of the book, by finding natural recursion theoretical definitions of particular classes of (and relations on) degrees.

Homogeneity

The fact that every particular result about $\mathcal{D}$ seems to relativize above any given degree, led to the following conjectures:

1. **strong homogeneity** (Rogers [1967])
   For every degree a, the structures $\mathcal{D}$ and $\mathcal{D}(\geq a)$ are isomorphic.

2. **homogeneity** (Yates [1970])
   For every degree a, the structures $\mathcal{D}$ and $\mathcal{D}(\geq a)$ are elementarily equivalent, i.e. they satisfy the same first-order formulas.

The same relativization phenomenon which led to the homogeneity conjectures is the key to their disproval.

Theorem V.7.13 Failure of homogeneity (Shore [1982], Harrington and Shore [1981], Jockusch and Shore [1984]) *If $\mathcal{D}(\geq a)$ is elementarily equivalent to $\mathcal{D}$, then a is arithmetical.*

Proof. Consider the formula $\varphi(x)$ defining the arithmetical degrees: when interpreted in $\mathcal{D}$ it defines the set $\mathcal{A}$ of the arithmetical degrees, while when interpreted in $\mathcal{D}(\geq a)$ it defines (by relativization of the proof of V.7.9) the set $\mathcal{A}_a$ of the degrees arithmetical in a and above it.

If $\mathcal{D}$ and $\mathcal{D}(\geq a)$ are elementarily equivalent, then so are $\mathcal{A}$ and $\mathcal{A}_a$ and hence, by V.7.6, the theories of Second-Order Arithmetic with set quantifiers restricted, respectively, to sets arithmetical and arithmetical in a. Then a is arithmetical, otherwise the sentence saying that there is a nonarithmetical set would distinguish them (being false in the former and true in the latter).   □

Corollary V.7.14 $\mathcal{D}$ *and* $\mathcal{D}(\geq 0^{(\omega)})$ *are not elementarily equivalent.*

Since homogeneity fails, relativized versions of results that hold in $\mathcal{D}$ are not automatically true, and require proofs. Also, there are results that simply fail to relativize. However this very result (saying that not everything relativizes) does relativize to degrees b which are definable in Second-Order Arithmetic (because then we can define in Second-Order Arithmetic the formula saying that all degrees satisfying φ are arithmetical in b). The exercises show that it is consistent, but unlikely, that the relativization holds in general.

Exercises V.7.15 A **cone of elementarily equivalent cones** is a cone such that, for any a and b in it, the cones $\mathcal{D}(\geq a)$ and $\mathcal{D}(\geq b)$ are elementarily equivalent. The existence of such a cone would provide a homogeneous substructure of the degrees.

a) *If Projective Determinacy holds then there is a cone of elementarily equivalent cones.* (Martin [1968]) (Hint: as in V.7.18, using V.1.16 and cones instead of comeager sets. Note that the set of degrees a such that $\mathcal{D}(\geq a)$ satisfies φ is an analytical set, and thus only Projective Determinacy is needed, in place of full Determinacy.)

b) *If $V = L$ then there is no cone of elementarily equivalent cones.* (Shore [1982]) (Hint: if $V = L$ then the degrees above a and b are elementarily equivalent only if a and b are arithmetically equivalent. Indeed, with notations as in V.7.13, an exact pair for the set of degrees satisfying φ defines $\mathcal{A}_C$ in $\mathcal{D}(\geq c)$. The least such exact pair w.r.t. $\leq_L$, which is definable in Second-Order Arithmetic, defines uniformly the same set in elementarily equivalent structures.)

c) *In a cone of elementarily equivalent cones, no degree can be definable in Second-Order Arithmetic.* (Hint: since V.7.13 relativizes to degrees definable in Second-Order Arithmetic, in any cone with such a base there is a cone not elementarily equivalent to it.)

To get a result that fully relativizes we need to look at isomorphism, rather than elementary equivalence.

Theorem V.7.16 Failure of strong homogeneity (Shore [1979], [1981], Harrington and Shore [1981], Jockusch and Shore [1984]) *If $\mathcal{D}(\geq a)$ is isomorphic to $\mathcal{D}(\geq b)$, then a and b are arithmetically equivalent.*

Proof. Consider an isomorphism carrying $\mathcal{D}(\geq a)$ into $\mathcal{D}(\geq b)$. The image of a copy of the standard model of Arithmetic defined in the degrees arithmetical in and above a is carried into a structure isomorphic to it, in the degree arithmetic in and above b. Indeed, an isomorphism preserves definable properties, and thus degrees satisfying the formula defining the arithmetical degrees in $\mathcal{D}(\geq a)$ are sent into degrees satisfying the same formula in $\mathcal{D}(\geq b)$. Now the sets arithmetical in a are coded by degrees arithmetical in a, and their images in $\mathcal{D}(\geq b)$ must code the same set (by the isomorphism), which is now arithmetical in b. Thus a must be arithmetical in b. The converse holds similarly, and thus a and b have the same arithmetical degree. $\square$

A **cone of isomorphic cones** is a cone such that, for any a and b in it, the cones $\mathcal{D}(\geq a)$ and $\mathcal{D}(\geq b)$ are isomorphic. The existence of such a cone would provide a strongly homogeneous substructure of the degrees.

Corollary V.7.17 *There is no cone of isomorphic cones.*

Proof. By the previous result, two cones can be isomorphic only if their bases are arithmetical one in the other. Then there are at most countably many cones isomorphic to a given one, and there cannot be any cone of isomorphic cones (because a cone has uncountably many elements). $\square$

We turn now to positive cases of homogeneity. We do not know whether there is a cone with nontrivial base which is elementarily equivalent to $\mathcal{D}$, or whether there are two isomorphic cones, but certainly there are lots of elementarily equivalent cones.

Proposition V.7.18 (Jockusch [1981]) *There is a comeager set of degrees which are bases of elementarily equivalent cones, i.e. a comeager set such that if a and b are in it, the cones $\mathcal{D}(\geq a)$ and $\mathcal{D}(\geq b)$ are elementarily equivalent.*

Proof. Consider the first-order sentences of the language of partial orderings. For each such sentence φ, consider the set of degrees a such that $\mathcal{D}(\geq a)$ satisfies φ. It is easy to prove (see V.3.15) that this set is either meager or comeager. Let $\mathcal{A}_\varphi$ be this set if it is comeager, and its complement otherwise. Then $\mathcal{A}_\varphi$ is always comeager, and φ holds either in every $\mathcal{D}(\geq a)$ or in none of them, for a in $\mathcal{A}_\varphi$. Since there are only countably many sentences φ, the intersection $\mathcal{A}$ of all the $\mathcal{A}_\varphi$ is still comeager. Moreover, the truth-value of φ in $\mathcal{D}(\geq a)$ is independent of a in $\mathcal{A}$, for every φ. This means that the first-order theory of $\mathcal{D}(\geq a)$ is independent of a in $\mathcal{A}$. □

Exercises V.7.19 a) *There is a comeager set of degrees of elementarily equivalent principal ideals $\mathcal{D}(\leq a)$.* (Jockusch [1981]) (Hint: as for cones.)

b) *There is a cone of elementarily equivalent principal ideals.* (Martin [1968]) (Hint: as for cones, with the added fact that the set $\mathcal{A}$ of degrees a such that $\mathcal{D}(\leq a)$ satisfies a given formula φ is an arithmetical set, because set quantifications over sets recursive in a fixed set A can be replaced by number quantifications over the indices relative to A. Then only Arithmetical Determinacy, which is provable in *ZFC* by Martin [1975], is needed to show that either $\mathcal{A}$ or its complement contains a cone.)

c) *The same results hold for jump intervals $\mathcal{D}([a, a'])$.*

Note that Simpson's Theorem has not been fully relativized to any degree a: it is obvious that *the first-order theory of $\mathcal{D}(\geq a)$ is 1-reducible to the theory of Second-Order Arithmetic with an added predicate for a, uniformly in a, but it is not known whether the converse holds* (the proof of V.7.3 shows only that the theory of Second-Order Arithmetic is 1-reducible to the theory of $\mathcal{D}(\geq a)$, uniformly in a). If this were provable, and in a uniform way, then it would follow that $\mathcal{D}(\geq a)$ and $\mathcal{D}(\geq b)$ are elementarily equivalent if and only if $a = b$, thus giving a final answer to the homogeneity problem.

Automorphisms

The questions of the last subsection, about homogeneity and strong homogeneity, can be asked about the relationships of $\mathcal{D}$ not only with some cone, but also with itself. The relevant notions are the following.

Definition V.7.20 *A map $f : \mathcal{D} \to \mathcal{D}$ is called:*

1. *an* **automorphism** *if it is an isomorphism that preserves the order, i.e.*

$$x \leq y \ \Leftrightarrow \ f(x) \leq f(y)$$

2. an **elementary map** *if it is preserves the first-order formulas, i.e. for any first-order formula φ,*

$$\mathcal{D} \models \varphi(x_1, \ldots, x_n) \Leftrightarrow \mathcal{D} \models \varphi(f(x_1), \ldots, f(x_n)).$$

The analogues of homogeneity and strong homogeneity then ask about the existence of nontrivial elementary maps and automorphisms. First of all, the two questions are equivalent.

Proposition V.7.21 (Slaman and Woodin [1986]) *A map from $\mathcal{D}$ to $\mathcal{D}$ is elementary if and only if it is an automorphism.*

Proof. An automorphism obviously is an elementary map. For the converse, an elementary map automatically preserves the order, and it is one-one because it preserves the equality relation. It remains to prove that any elementary map f is onto, i.e. that for any y there is x such that $y = f(x)$.

The proof of part 2 of the next theorem will show that f is the identity on the cone above a degree a (e.g. 0^ω). By V.7.2 there are degrees $\vec{c}$ coding a standard model of arithmetic, and $\vec{d}$ coding the graph of a function g enumerating (on the natural numbers of the model) the degrees below $y \cup a$. Moreover, there is a first-order sentence with parameters y and a stating that $\vec{c}$ and $\vec{d}$ have the desired properties. Since f is elementary, this statement is true of $f(y)$, $f(a)$, $f(\vec{c})$, and $f(\vec{d})$. In particular, $f(\vec{c})$ code a standard model of arithmetic, and $f(\vec{d})$ code a function that enumerates (on the natural numbers of this model) the degrees below $f(y \cup a)$. But $f(y \cup a) = y \cup a$ because f is the identity above a, and hence y is one of the degrees enumerated by the function coded by $f(\vec{d})$, say the n-th in the enumeration. Since f is elementary, it preserves all the relevant properties. In particular, it must be that y is the image via f of the n-th degree x enumerated by the function coded by $\vec{d}$. This shows that y is in the range of f, as wanted. $\square$

We can formulate the analogue of (strong) homogeneity as follows: the only automorphism of $\mathcal{D}$ is the identity. Algebraic structures without nontrivial automorphisms are called **rigid**. Although the rigidity of $\mathcal{D}$ has not been proved, all known results point in that direction, and at least show that *the automorphisms of $\mathcal{D}$ are severely restricted.*

Theorem V.7.22 Restrictions on automorphisms (Nerode and Shore [1980a], Harrington and Shore [1981], Shore [1981], Jockusch and Shore [1984]) *Every automorphism of $\mathcal{D}$:*

1. *sends any degree into a degree which is arithmetically equivalent to it*

2. *is the identity on every degree above all the arithmetical ones, in particular on the cone above* 0^ω.

Proof. Consider an automorphism $f : \mathcal{D} \to \mathcal{D}$. The first part follows from V.7.16 and the fact that f induces an isomorphism between the cones above x and $f(x)$, for any x.

For the second part, note that there is a copy of the standard model of arithmetic such that the relation 'the degrees $\vec{b}$ code a set of degree x' is analytical, and hence first-order definable for degrees x above all the arithmetical ones, by V.7.10. Then this relation is preserved by f and hence, for any $\vec{b}$ coding a set of degree x above all the arithmetical ones, $f(\vec{b})$ code a set of degree $f(x)$. But f is an automorphism, and hence the degrees $\vec{b}$ and $f(\vec{b})$ must actually code the same set. Then $x = f(x)$. $\quad\square$

Actually something better can be achieved, by more direct calculations: *any automorphism of $\mathcal{D}$ is the identity on a cone having an arithmetical degree as a base* (Nerode and Shore [1980a], Harrington and Shore [1981], Shore [1981], Jockusch and Shore [1984]).

We now introduce a useful tool for the study of automorphisms.

Definition V.7.23 *An* **automorphism basis** *for $\mathcal{D}$ is any set of degrees $\mathcal{A}$ such that the behavior of any automorphism is completely determined by its behavior on elements of $\mathcal{A}$.*

Producing many automorphism bases is one way to show that there are few automorphisms. Another one is to show that there are small bases. We will exhibit results in both directions.

The first way to obtain automorphism bases is to consider sets of degrees that generate $\mathcal{D}$ under $\cup$ and $\cap$.

Definition V.7.24 *Given a set of degrees $\mathcal{A}$, the* **set generated by $\mathcal{A}$ in $\mathcal{D}$** *is the smallest set:*

1. *containing $\mathcal{A}$*

2. *closed under joins*

3. *closed under g.l.b.'s, whenever they exist.*

Any automorphism of a partially ordered structure must preserve l.u.b.'s and g.l.b.'s, whenever they exist, and thus the behavior of an automorphism on a set $\mathcal{A}$ completely determines its behavior on the set generated by it. In particular, *if $\mathcal{A}$ generates $\mathcal{D}$ then $\mathcal{A}$ is an automorphism basis.*

Proposition V.7.25 (Jockusch and Posner [1981]) *If $\mathcal{A}$ is a comeager set of degrees, then $\mathcal{A}$ generates $\mathcal{D}$ under $\cup$ and $\cap$. More precisely, any degree can be represented in the form*

$$(a_1 \cup a_2) \cap (a_3 \cup a_4),$$

with $a_i \in \mathcal{A}$.

Proof. Let a be a given degree: by relativization of the minimal pair construction (V.2.16), given b we can get c such that

$$(a \cup b) \cap (a \cup c) = a.$$

Fix $b \in \mathcal{A}$: then the set of such degrees c is comeager, and hence so is the intersection of this set with $\mathcal{A}$. Thus we have b and c in $\mathcal{A}$ such that the above equation holds. We only have to represent any degree of the form $a \cup d$, with d in $\mathcal{A}$, as the join of two degrees in $\mathcal{A}$. Note that, in general,

$$A \oplus D \equiv_T D \oplus (A \triangle D)$$

for any A and D, where $A \triangle D$ is the symmetric difference $(A - D) \cup (D - A)$. Moreover, if $\mathcal{A}$ is comeager then so is the set

$$\mathcal{A}^{\triangle} = \{A \triangle D : D \text{ has degree in } \mathcal{A}\}.$$

We may then suppose that the degrees b and c above are not only in $\mathcal{A}$, but also in $\mathcal{A}^{\triangle}$, and then the result follows. $\square$

Clearly a countable set cannot generate $\mathcal{D}$, which is uncountable. But *there are uncountable meager sets of degrees that do generate $\mathcal{D}$*, as the next exercise shows.

Exercise V.7.26 *The minimal degrees generate $\mathcal{D}$.* (Jockusch and Posner [1981]) (Hint: given two sets A and B such that $A \leq_T B$, we show that there are sets of minimal degree M_1 and M_2 such that B and $M_1 \oplus M_2$ have A as g.l.b. We then apply this to any B above A and $\emptyset''$, which by V.6.10.b is the join of two minimal degrees, and have that any degree is generated by four minimal degrees. We extend the proof of V.6.10.b and build M_1 and M_2 by recursive coinfinite conditions. There are two additional requirements:

$$A \leq_T M_1 \oplus M_2$$
$$C \leq_T B, M_1 \oplus M_2 \ \Rightarrow \ C \leq_T A.$$

The first one is satisfied by building a strongly uniform tree of minimal degrees, and letting M_1 be the branch that, at level n, agrees with $A(n)$ on the unique element on which the two branches disagree, while M_2 follows the other branch. The tree is built by stages, and infinitely many times a new level will be added. The second condition is satisfied by the usual minimal pair construction, with coinfinite conditions that satisfy the coding requirement: this accounts for the fact that, in absence of e-splittings, not outright recursiveness, but only recursiveness in A is obtained.)

A proper cone cannot generate the degrees, being closed under l.u.b.'s and g.l.b's, but can nevertheless be an automorphism basis.

Proposition V.7.27 (Jockusch and Posner [1981]) *There is a comeager set of degrees which are bases of cones that are automorphism bases.*

Proof. It is enough to find a comeager set $\mathcal{A}$ such that, for any $\boldsymbol{a}$ in it, $\mathcal{D}(\geq \boldsymbol{a}) \cup \mathcal{D}(\geq 0^{(\omega)})$ generates $\mathcal{D}$. Since every automorphism is the identity above $0^{(\omega)}$, then $\mathcal{D}(\geq \boldsymbol{a})$ must be an automorphism basis.

First note that, for any $\boldsymbol{b}$, the set

$$\{\boldsymbol{a} : (\boldsymbol{b} \cup \boldsymbol{a}) \cap (\boldsymbol{b} \cup 0^{(\omega)}) = \boldsymbol{b}\}$$

is comeager (by relativization of the fact that the set of degrees which form a minimal pair together with $0^{(\omega)}$ is comeager, see V.2.16). Then the set of pairs $\boldsymbol{a}$ and $\boldsymbol{b}$ such that $(\boldsymbol{b} \cup \boldsymbol{a}) \cap (\boldsymbol{b} \cup 0^{(\omega)}) = \boldsymbol{b}$ is comeager, and so is the set of degrees $\boldsymbol{a}$ such that $\mathcal{D}(\geq \boldsymbol{a}) \cup \mathcal{D}(\geq 0^{(\omega)})$ generates a comeager set (and hence $\mathcal{D}$ itself, by V.7.25). $\square$

V.8 Degree Theory with Jump $\star$

Although the jump operator is known to be definable in $\mathcal{D}$ (see XI.5.17), here we briefly discuss what happens when we add it to $\mathcal{D}$, thus obtaining the structure $\mathcal{D}'$. Some of the results are simply implied by results on the structure without jump, e.g. $\mathcal{D}'$ is obviously still undecidable, and recursively equivalent to Second-Order Arithmetic. Other results exploit the extra power provided by the jump operator, and produce both easier proofs and sharper definability results. For example, the definability of the arithmetical degrees, which we had to postpone to Volume II if working in $\mathcal{D}$ alone, is easily obtained in $\mathcal{D}'$.

Proposition V.8.1 (Jockusch and Soare [1970]) *The set $\mathcal{A}$ of the arithmetical degrees is definable in $\mathcal{D}'$.*

Proof. The natural definition of $\mathcal{A}$ is:

$$\mathcal{A} = \text{ the smallest jump-ideal.}$$

This is easily expressed in $\mathcal{D}'$, as:

$$x \in \mathcal{A} \Leftrightarrow (\forall a)(\forall b)[(\forall z)(z \leq a, b \Rightarrow z' \leq a, b) \Rightarrow x \leq a, b]. \quad \Box$$

As we have already noted on p. 541, this result can be used to turn the proofs of the results (about definability, homogeneity and automorphisms) given for $\mathcal{D}$ under the assumption of the definability of the arithmetical degrees in $\mathcal{D}$, to proofs of the same results for $\mathcal{D}'$, with the advantage of avoiding the proof of V.7.9. This use of the jump operator is conservative: the advantage of having easier proofs is paid by obtaining weaker results (since the same results are proved in a stronger structure).

A more genuine use of the jump operator would be to take full advantage of the added strength, and prove stronger results for the stronger structure. These require new methods and ideas, some of which we will introduce later on. Here we just quote the statements of the improved results:

1. **definability**
   Every relation on degrees above $\mathbf{0}^{(3)}$ *which is definable in Second-Order Arithmetic is definable in* $\mathcal{D}'$ (Simpson [1977], Nerode and Shore [1980a], Shore [1982]).

2. **homogeneity**
   If $\mathcal{D}'(\geq a)$ *is elementarily equivalent to* $\mathcal{D}'$, *then* a *has triple jump* $\mathbf{0}^{(3)}$ (Simpson [1977], Nerode and Shore [1980a], Shore [1981]).

   If $\mathcal{D}'(\geq a)$ *is isomorphic to* $\mathcal{D}'(\geq b)$, *then* a *and* b *have the same triple jump* (Feiner [1970], Yates [1972], Nerode and Shore [1980a], Shore [1981]).

3. **automorphisms**
   Every automorphism preserving the jump is the identity on the cone above $\mathbf{0}^{(3)}$ (Jockusch and Solovay [1977], Richter [1979], Epstein [1979]).

Some of the results are actually stronger than we stated, and do not require the full power of the jump operator, but only a small fraction of it. For example, *every automorphism fixing* $\mathbf{0}'$ *is the identity on the cone above* $\mathbf{0}^{(3)}$ (Nerode and Shore [1980a]).

A word is perhaps in order, to explain why there is a factor of three jumps in the results quoted above. The reason is that the methods of proofs use embeddings of partial orderings into the degrees, and three jumps are already needed to express the relation $\leq_T$, and hence the order relation. To obtain better results, different methods are needed (see Nies, Shore and Slaman [1998]).

Improving bounds is only one way to improve results. Another way is to improve the explicit definitions of degrees or relations, making them more intelligible or more natural. As an example, we provide a simple and natural definition in $\mathcal{D}'$ of the ω-jump operator, which we already know to be definable in $\mathcal{D}$ by V.7.11. We first extend the notion of least upper bound.

Definition V.8.2 (Sacks [1971]) *A degree a is an n-least* **upper bound** *for a set of degrees C if it is the least element of*

$$\{x^{(n)} : (\forall c \in C)(c \leq x)\}.$$

Note that 0-l.u.b.'s are the usual l.u.b.'s. We now consider the chain $\{0^{(n)}\}_{n \in \omega}$ which, being an increasing chain, has no l.u.b. (by V.4.10).

Exercises V.8.3 a) $0^{(\omega)}$ *is not a minimal upper bound for* $\{0^{(n)}\}_{n \in \omega}$. (Kleene and Post [1954]) (Hint: the proof of Spector's Theorem produces an exact pair for the chain, below $0^{(\omega)}$.)

b) *There is a minimal upper bound for* $\{0^{(n)}\}_{n \in \omega}$ *below* $0^{(\omega)}$. (Sacks [1963]) (Hint: analyze the proof of V.5.21.)

Theorem V.8.4 (Enderton and Putnam [1970], Sacks [1971]) $0^{(\omega)}$ *is the 2-least upper bound of* $\{0^{(n)}\}_{n \in \omega}$.

Proof. There are two things to prove:

1. *if a is an upper bound of* $\{0^{(n)}\}_{n \in \omega}$, *then* $0^{(\omega)} \leq a^{(2)}$
   The hypothesis is that $0^{(n)} \leq a$, for every n. The problem is that we do not have a uniform reduction in general, and thus we cannot conclude anything about the relationship between $0^{(\omega)}$ and a. We however show that $0^{(n)} \leq a^{(2)}$ does hold uniformly in n, so that $0^{(\omega)} \leq a^{(2)}$. To do this, we have to show how to effectively get an index of $\emptyset^{(n)}$, relative to $A^{(2)}$, for a fixed $A \in a$. This can be done inductively, because $\emptyset^{(n+1)} = (\emptyset^{(n)})'$, and thus there is an e such that

$$(\forall x)(x \in \emptyset^{(n+1)} \Leftrightarrow \{e\}^{\emptyset^{(n)}}(x)\downarrow)$$

   Since, by induction hypothesis, we can express $\emptyset^{(n)}$ recursively in A, both sides are recursive in A', and the quantifier adds one more jump. Thus we can obtain such an e recursively in $A^{(2)}$.

2. *there is an upper bound a of* $\{0^{(n)}\}_{n \in \omega}$ *such that* $0^{(\omega)} = a^{(2)}$
   To get an upper bound A of $\{\emptyset^{(n)}\}_{n \in \omega}$, we let $A \in \bigcap_{n \in \omega} T_n$, where T_n is a recursively pointed tree of degree $0^{(n)}$. We use the following extension of the Totality Lemma V.5.5. Given e and a tree T, there is a tree $Q \subseteq T$ such that one of the following holds:

   • for every A on Q, $\{e\}^A$ is not total

- for every A on Q, $\{e\}^A$ is total.

This is proved by the proof of V.5.5, simply because when the relevant trees are not recursive, then we have to take them into account in our computations.

The construction is as follows. Let T_0 be the identity tree (which is recursively pointed). Given T_e recursively pointed of the same degree as $\emptyset^{(e)}$, first get $T \subseteq T_e$ recursively pointed of the same degree as $\emptyset^{(e+1)}$, by V.5.20, which is possible because $\emptyset^{(e)} \leq_T \emptyset^{(e+1)}$. Then let T_{e+1} be the Q of the Totality Lemma stated above, which is still recursively pointed, and of the same degree as $\emptyset^{(e+1)}$. The construction is recursive in $\emptyset^{(\omega)}$ and we can, recursively in it, decide whether $\{e\}^A$ is total or not. Then $A^{(2)} \leq_T \emptyset^{(\omega)}$. □

Corollary V.8.5 *The ω-jump is definable in $\mathcal{D}'$.*

Proof. By relativization, $a^{(\omega)}$ is the 2-least upper bound of $\{a^{(n)}\}_{n \in \omega}$. Clearly, since $\{a^{(n)}\}_{n \in \omega}$ is an increasing chain, $a^{(\omega)}$ is the 2-l.u.b. of the ideal generated by it, which is the smallest ideal containing a and closed under jump. Since the ideals are first-order definable in $\mathcal{D}'$ (by Spector's Theorem), so is then the ω-jump. □

Exercises V.8.6 a) $0^{(\omega)}$ *is the least element of the set of double jumps of degrees which are minimal upper bounds of $\{0^{(n)}\}_{n \in \omega}$.* (Sacks [1971]) (Hint: in the second part of the previous proof, make A also a minimal upper bound of $\{0^{(n)}\}_{n \in \omega}$, using the methods of V.5.21.)

b) $\{0^{(n)}\}_{n \in \omega}$ *has no 1-l.u.b.* (Sacks [1971]) (Hint: given any upper bound B of it, build another one A such that $B' \not\leq_T A'$, so that B is not the 1-l.u.b. To make A upper bound, use recursively pointed trees. To satisfy the requirements $B' \neq \{e\}^{A'}$, we have to control two-quantifier sentences over A. Start from a tree of sets such that the two-quantifier sentences over them are all decided by finite strings: such sets can be obtained by extending the proof of V.2.21, and the tree can be made recursive in $\emptyset''$. Thus all the trees of the construction will be arithmetical. Given one such tree T, to satisfy $B' \neq \{e\}^{A'}$ see if there are two strings on T that would decide, on the same element, $\{e\}^{A'}$ in two different way. If so, choose the full tree above the one that makes $\{e\}^{A'}$ different from B'. If not, $\{e\}^{A'}$ will be recursive in the tree, hence arithmetical and different from B', since B is not arithmetical.)

One trend of next chapters will be to prove a number of results with a similar flavor, producing nice definitions for interesting sets of degrees and operations on them.

Chapter VI

Many-One and Other Degrees

In this chapter we study the structure of **m-degrees**, with an approach similar to that used for T-degrees in Chapter V. The main difference between the two cases is the fact, proved in Section 1, that the structure of m-degrees is distributive. This is a major regularity, and the main reason allowing for a nice structure theory. This time the material is organized structurally, toward a characterization theorem that will be given in Section 4. We begin in Section 2 by an observation of Lachlan, a simple grain of sand that inserted in the oyster of distributivity will produce the final pearl. Having these two ingredients everything really becomes natural: layer after layer we build all the countable initial segments in Section 2, and the uncountable ones in Section 3. We disclose the oyster in Section 4, by giving Ershov's characterization of the structure of m-degrees up to isomorphism, as a strongly universal uppersemilattice. From it a number of other global results will follow, after which we will have a fairly complete understanding of the algebraic structure of m-degrees, a unique case in Recursion Theory.

We conclude with some additional topics, among them a comparison of the structures of **1-degrees**, **tt-degrees**, and **wtt-degrees** in Section 5, and the **structure inside degrees** of a given kind with respect to stronger reducibilities in Section 6.

VI.1 Distributivity

Our subject is the structure $\mathcal{D}_m$ introduced in Section III.2. Recall that there are three m-degrees containing recursive sets, namely $\mathbf{0}_m$, $\{\emptyset\}$, and $\{\omega\}$. Since the last two are incomparable and smaller than 0_m, but all other m-degrees are greater than or equal to 0_m, by convention we will always consider nontrivial sets, and thus 0_m may be considered as the least m-degree. We also borrow conventions and notations from Chapter V, e.g. the notations for degrees (as boldface letters) and cones.

Proposition VI.1.1 *As a partially ordered structure, $\mathcal{D}_m$ is an uppersemilattice of cardinality $2^{\aleph_0}$ with a least but no maximal element. Moreover, each element has $2^{\aleph_0}$ successors and at most countably many predecessors.*

Proof. See V.1.12, and note that 0_m is the least m-degree, $A \leq_m A'$, and $A \oplus B$ is actually the least upper bound of A and B w.r.t. m-reducibility. □

Exercises VI.1.2 The jump operator. a) *The jump operator is well-defined on m-degrees, and preserves the ordering. In particular, $0_m \leq a'$ for any m-degree a.* (Hint: if $A \leq_T B$ then $A' \leq_m B'$.)

b) *The only m-degree with jump $0'_m$ is 0_m.* (Hint: if $A' \leq_m \mathcal{K}$, from $A \leq_m A'$ we have that A is r.e. Since $A \equiv_T \overline{A}$ we have $A' \equiv_m \overline{A}'$, and thus also $\overline{A}$ is also r.e.)

c) *The jump operator is not one-one on m-degrees.* (Hint: $\mathcal{K}$ and $\overline{\mathcal{K}}$ are m-incomparable but, being T-equivalent, their jumps have the same m-degree.)

d) *There are m-degrees above $0'_m$ which are not jumps of any m-degree. Thus the analogue of the Jump Inversion Theorem fails.* (Hint: the m-degree of $\mathcal{K} \oplus \overline{\mathcal{K}}$ is above the m-degree of $\mathcal{K}$, but it cannot be a jump because it is in the same m-degree of its complement while A', being a complete Σ_1^0 set relative to A, cannot have this property.)

Distributive uppersemilattices

The next concept is going to be crucial for the study of m-degrees.

Definition VI.1.3 $(P, \sqsubseteq, \sqcup)$ *is a* **distributive uppersemilattice** *if it is an uppersemilattice such that if $a \sqsubseteq b \sqcup c$ there are $b_0 \sqsubseteq b$ and $c_0 \sqsubseteq c$ such that $a = b_0 \sqcup c_0$.*

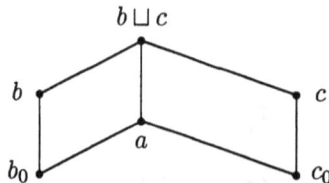

The notion of distributive uppersemilattice is consistent with the usual notion of distributivity. Recall that a lattice $(P, \sqsubseteq, \sqcup, \sqcap)$ is distributive if the distributive laws hold in it, i.e.

$$a \sqcap (b \sqcup c) = (a \sqcap b) \sqcup (a \sqcap c)$$
$$a \sqcup (b \sqcap c) = (a \sqcup b) \sqcap (a \sqcup c).$$

By duality, each of the two laws implies the other.

Proposition VI.1.4 *A lattice is distributive as a lattice if and only if it is distributive as an uppersemilattice.*

Proof. If the lattice is distributive and $a \sqsubseteq b \sqcup c$ then

$$a = a \sqcap (b \sqcup c) = (a \sqcap b) \sqcup (a \sqcap c).$$

Then it is enough to let $b_0 = a \sqcap b$ and $c_0 = a \sqcap c$ to have distributivity as an uppersemilattice.

If the lattice is distributive as an uppersemilattice, since

$$(a \sqcap b) \sqcup (a \sqcap c) \sqsubseteq a \sqcap (b \sqcup c)$$

always holds, it is enough to show that the converse holds too. Since

$$a \sqcap (b \sqcup c) \sqsubseteq b \sqcup c$$

holds, for some $b_0 \sqsubseteq b$ and $c_0 \sqsubseteq c$ is $a \sqcap (b \sqcup c) = b_0 \sqcup c_0$. But then $b_0, c_0 \sqsubseteq a$, and hence

$$a \sqcap (b \sqcup c) = b_0 \sqcup c_0 \sqsubseteq (a \sqcap b) \sqcup (a \sqcap c). \quad \square$$

Definition VI.1.5 *An element of an uppersemilattice is* **indecomposable,** *or an* **atom,** *if whenever $a = b \sqcup c$ then $a = b$ or $a = c$.*

Proposition VI.1.6 *In a distributive uppersemilattice the indecomposable elements below $b \sqcup c$ are exactly the indecomposable elements below b or c.*

Proof. One direction is trivial. Conversely, if $a \sqsubseteq b \sqcup c$ then, by distributivity, $a = b_0 \sqcup c_0$ for some $b_0 \sqsubseteq b$ and $c_0 \sqsubseteq c$. If a is indecomposable then $a = b_0$ or $a = c_0$, and hence $a \sqsubseteq b$ or $a \sqsubseteq c$. $\quad \square$

Exercises VI.1.7 a) *In a Boolean algebra the indecomposable elements are exactly the atoms.*

b) *In a linear ordering every element is indecomposable.*

c) *In a finite uppersemilattice every element is the l.u.b. of the indecomposable elements below it.*

Our reason to introduce distributive uppersemilattices is the following.

Proposition VI.1.8 (Lachlan [1970]) $\mathcal{D}_m$ *is a distributive uppersemilattice.*

Proof. Given $A \leq_m B \oplus C$ we want $B_0 \leq_m B$ and $C_0 \leq_m C$ such that $A \equiv_m B_0 \oplus C_0$. Given f recursive such that $x \in A \Leftrightarrow f(x) \in B \oplus C$, we let

$$x \in B_0 \quad \Leftrightarrow \quad f(x) \text{ even and } \frac{f(x)}{2} \in B$$

$$x \in C_0 \quad \Leftrightarrow \quad f(x) \text{ odd and } \frac{f(x)-1}{2} \in C.$$

Then, e.g., $B_0 \leq_m B$ via

$$g(x) = \begin{cases} b & \text{if } f(x) \text{ not even} \\ \frac{f(x)}{2} & \text{otherwise,} \end{cases}$$

where b is a fixed element of B. Similarly for $C_0 \leq_m C$. Moreover, it is easily checked that $A \equiv_m B_0 \oplus C_0$. $\quad\square$

Ideals of distributive uppersemilattices

The special properties of distributive uppersemilattices severely limit the possible ideals, and thus produce necessary conditions for the ideals of $\mathcal{D}_m$.

Proposition VI.1.9 *A finite distributive uppersemilattice with least element is a distributive lattice.*

Proof. By VI.1.4 it is enough to show that a finite uppersemilattice with least element is a lattice. But the greatest lower bound of two elements a and b is the least upper bound of the set of elements below them, which exists because the latter is a nonempty finite set. $\quad\square$

Corollary VI.1.10 (Lachlan [1970]) *Every finite ideal of $\mathcal{D}_m$ is a distributive lattice.*

To get conditions for any topped ideal we must develop some of the theory of distributive uppersemilattices.

Proposition VI.1.11 (Lachlan [1972a]) *Given a distributive uppersemilattice with least element, its ideals ordered by inclusion form a distributive lattice.*

Proof. We denote by 0 the least element of $(P, \sqsubseteq, \sqcup)$, and by $\mathcal{I}$ the set of its ideals. The g.l.b. of two ideals is their set-theoretical intersection. The l.u.b. is the least ideal containing their union. Since P is distributive, given $I_1, I_2 \in \mathcal{I}$ their l.u.b. is actually the set

$$I_1 \oplus I_2 = \{b \sqcup c : b \in I_1 \wedge c \in I_2\}.$$

Indeed, $I_1, I_2 \subseteq I_1 \oplus I_2$, because 0 is in every ideal. Every ideal containing I_1 and I_2 must clearly contain $I_1 \oplus I_2$. $I_1 \oplus I_2$ is closed under l.u.b. because I_1 and I_2 are. It is closed downward because if $a \sqsubseteq b \sqcup c$ then $a = b_0 \sqcup c_0$, with $b_0 \sqsubseteq b$ and $c_0 \sqsubseteq c$: if $b \sqcup c \in I_1 \oplus I_2$ then we may suppose $b \in I_1$ and $c \in I_2$, and thus $b_0 \in I_1$ and $c_0 \in I_2$, because I_1 and I_2 are closed downward. Then $a \in I_1 \oplus I_2$. Thus $I_1 \oplus I_2$ is an ideal.

Finally, $\mathcal{I}$ is distributive as an uppersemilattice: if $I \subseteq I_1 \oplus I_2$ then $I = I_1^* \oplus I_2^*$, where $I_i^* = I_i \cap I$. Indeed, $I_1^*, I_2^* \subseteq I$ and thus $I_1^* \oplus I_2^* \subseteq I$ because $\oplus$ is the l.u.b. operation. Conversely, if $a \in I$ then $a \in I_1 \oplus I_2$, so $a = b \sqcup c$ with $b \in I_1$ and $c \in I_2$: but $b, c \sqsubseteq a$ and $a \in I$, hence $b \in I_1^*$ and $c \in I_2^*$, from which $a \in I_1^* \oplus I_2^*$. $\square$

Proposition VI.1.12 (Ershov [1975]) *In a distributive uppersemilattice P every finite subset closed under l.u.b. is embedded as an uppersemilattice (i.e. with l.u.b. preserved) in a finite distributive sublattice of P.*

Proof. There is a natural embedding from P to $\mathcal{I}$, given by

$$a \longmapsto \hat{a} = \text{ the principal ideal generated by } a = \{b : b \sqsubseteq a\}.$$

Given a finite uppersemilattice $F \subseteq P$, consider $\mathcal{F} \subseteq \mathcal{I}$ so defined:

$$\mathcal{F} = \{\hat{a} : a \in F\}.$$

Let $\mathcal{L}$ be the finite distributive lattice generated by $\mathcal{F}$ in $\mathcal{I}$ (see VI.1.11). We want to pull back $\mathcal{L}$ in P, but the obvious trouble is that not every element of $\mathcal{L}$ is a principal ideal. Thus we need a one-one homomorphism of uppersemilattices $\varphi : \mathcal{L} \to P$ such that $F \subseteq \varphi(\mathcal{L})$. The obvious approach would be to take, for $I \in \mathcal{L}$,

$$\varphi(I) = \text{ l.u.b. of } I,$$

but not every ideal needs to have a l.u.b. (unless it is finite). So we define

$$G_I = \text{ a finite subset of } I$$
$$\varphi(I) = \text{ l.u.b. of } G_I.$$

In particular this gives $\varphi(I) \in I$, since an ideal is closed under l.u.b.

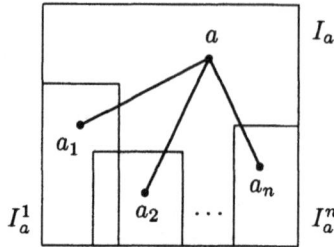

To have φ one-one, we pick up one element in each nonempty $I_1 - I_2$, for every $I_1, I_2 \in \mathcal{L}$. Let H be the set containing these elements, as well as the elements of F. For any $a \in H$, let I_a be the smallest element of $\mathcal{L}$ containing a. Then I_a is the l.u.b. of indecomposable elements $I_a^i \in \mathcal{L}$, and by VI.1.11 it must be $I_a = I_a^1 \oplus \cdots \oplus I_a^n$. Then there are elements $a_i \in I_a^i$ such that $a = a_1 \sqcup \cdots \sqcup a_n$. Let

$$
\begin{aligned}
G_a &= \{a_1, \ldots, a_n\} \\
G &= \bigcup_{a \in H} G_a \\
G_I &= \bigcup \{G \cap J : J \subseteq I \text{ indecomposable element of } \mathcal{L}\}.
\end{aligned}
$$

We now show that the function defined as

$$\varphi(I) = \text{ the l.u.b. of } G_I$$

has the needed properties:

1. φ is a homomorphism

   This follows from the fact that the indecomposable elements contained in $I_1 \oplus I_2$ are exactly those contained in I_1 or I_2. So $G_{I_1 \oplus I_2} = G_{I_1} \cup G_{I_2}$, and

   $$\varphi(I_1 \oplus I_2) = \varphi(I_1) \sqcup \varphi(I_2).$$

2. φ is one-one

   If I_1 and I_2 are different suppose, e.g., $I_1 - I_2 \neq \emptyset$, and let $a \in H \cap (I_1 - I_2)$, which exists by definition. Then $\varphi(I_1) \neq \varphi(I_2)$, because $\varphi(I_2) \in I_2$ (always), but $\varphi(I_1) \notin I_2$. Indeed, $I_a \subseteq I_1$ because I_a is the smallest element of $\mathcal{L}$ containing a, and thus $\varphi(I_a) \sqsubseteq \varphi(I_1)$, because φ is a homomorphism. $G_a \subseteq G_{I_a}$ by definition, and hence

   $$a = (\text{ l.u.b. of } G_a) \sqsubseteq (\text{ l.u.b. of } G_{I_a}) = \varphi(I_a).$$

   Thus $a \sqsubseteq \varphi(I_1)$, and if $\varphi(I_1) \in I_2$ then $a \in I_2$ by downward closure.

3. $F \subseteq \varphi(\mathcal{L})$
   It is enough to show that $\varphi(\hat{a}) = a$, for $a \in F$. Indeed, $I_a = \hat{a}$ if $a \in F$: so $a \sqsubseteq \varphi(I_a) = \varphi(\hat{a})$, by the previous part of the proof. Conversely, $\varphi(\hat{a}) \sqsubseteq a$ because $\varphi(\hat{a}) \in \hat{a}$ always holds. □

Corollary VI.1.13 (Lachlan [1970]) *Every topped ideal of $\mathcal{D}_m$ is the direct limit (in the sense of uppersemilattices) of an ascending sequence of finite distributive lattices.*

Proof. Consider an enumeration $\{a_0, a_1, \ldots\}$ of a topped ideal $I = \mathcal{D}_m\,(\leq a)$. Start with $D_0 = \{0_m, a\}$. Given $D_n \subseteq I$ consider the finite uppersemilattice generated by $D_n \cup \{a_n\}$, which has a as greatest element, and let D_{n+1} be the finite distributive lattice obtained as in the proof above. Since the former is embedded in the latter as an uppersemilattice, a is also the greatest element of D_{n+1}, and thus $D_{n+1} \subseteq I$. □

We now have a necessary condition for topped ideals of $\mathcal{D}_m$. In the next section we will show that this condition is also sufficient, thus completely characterizing the countable initial segments of $\mathcal{D}_m$.

VI.2   Countable Initial Segments

The following observation is going to be crucial for our later development.

Proposition VI.2.1 (Lachlan [1970]) *Let A be a set, and $\mathbf{a}$ its m-degree. For any r.e. set $\mathcal{W}_e$ and any recursive function f with range $\mathcal{W}_e$, let:*

$$x \in \Phi(\mathcal{W}_e) \iff f(x) \in A.$$

Then the map Φ induces an onto homomorphism of uppersemilattices between the r.e. sets modulo the finite sets (ordered by inclusion) and $\mathcal{D}_m\,(\leq a)$.

Proof. We have to check the following:

1. the m-degree of $\Phi(\mathcal{W}_e)$ does not depend on f
   Let g be another recursive function with range $\mathcal{W}_e$. If

   $$h(x) = \mu z[g(z) = f(x)] \quad \text{and} \quad t(x) = \mu z[g(x) = f(z)]$$

   then

   $$f(x) \in A \iff gh(x) \in A \quad \text{and} \quad g(x) \in A \iff ft(x).$$

2. $\Phi(\mathcal{W}_e) \leq_m A$
By definition.

3. if B and C are r.e. sets and $B \subseteq C$ then $\Phi(B) \leq_m \Phi(C)$
Let B and C be the ranges of f and g, and $h(x) = \mu z[g(z) = f(x)]$. Then h is total because $B \subseteq C$, and

$$x \in \Phi(B) \Leftrightarrow f(x) \in A \Leftrightarrow gh(x) \in A \Leftrightarrow h(x) \in \Phi(C).$$

4. if B and C are r.e. sets which differ finitely then $\Phi(B) \equiv_m \Phi(C)$
Let B and C be the ranges of f and g. Fix $a \in \Phi(C)$ and $b \notin \Phi(C)$, and let

$$h(x) = \begin{cases} a & \text{if } f(x) \in A \cap \overline{C} \\ b & \text{if } f(x) \in \overline{A} \cap \overline{C} \\ \mu z[g(z) = f(x)] & \text{if } f(x) \in C. \end{cases}$$

h is recursive because $f(x) \in B$ by definition, and $B \cap \overline{C}$ is finite (and hence so are $B \cap A \cap \overline{C}$ and $B \cap \overline{A} \cap \overline{C}$). Then $\Phi(B) \leq_m \Phi(C)$ via h. Symmetrically, $\Phi(C) \leq_m \Phi(B)$.

5. if $C \leq_m A$ then there is an r.e. set B such that $\Phi(B) \equiv_m C$
Let $x \in C \Leftrightarrow f(x) \in A$: if B is the range of f, $\Phi(B) = C$. □

The result shows that instead of controlling directly the m-degrees below A we can control the m-degrees of $\Phi(\mathcal{W}_e)$, for every e. This is the technique that we are going to use to build initial segments.

Finite initial segments

Our goal is VI.3.6, which completely characterizes the initial segments of $\mathcal{D}_m$. But its proof is quite complicated and it involves a number of separate ideas, taking care of different problems. We thus actually prove a sequence of results, solving one problem at the time. We keep the same notations throughout, and indicate how to adapt previous proofs to the new needs.

To illustrate the technique of construction of initial segments of m-degrees we start with the simplest case.

Proposition VI.2.2 (Lachlan [1970]) *Every finite ordinal is isomorphic to an initial segment of $\mathcal{D}_m$.*

Proof. We want to build a set A such that the m-degrees below its m-degree are isomorphic to $\{0, 1, \ldots, n\}$. We use the homomorphism of VI.2.1 to control the m-degrees below the m-degree of A, and thus we only have to control each $\Phi(\mathcal{W}_e)$.

We build $n+1$ infinite recursive sets P_i, for $i \leq n$, as follows:

$$P_i = \{\{(n+1)x + i\} : x \in \omega\}.$$

Note that P_i is a set of singletons. In the construction we will build equivalence classes as boxes, and these singletons are our initial equivalence classes.

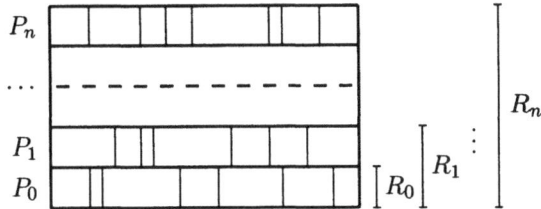

Consider the r.e. sets $R_x = \bigcup_{i \leq x} P_i$, for $x \leq n$. Since $R_0 \subseteq R_1 \subseteq \cdots \subseteq R_n$, we automatically have

$$\Phi(R_0) \leq_m \Phi(R_1) \leq_m \cdots \leq_m \Phi(R_n).$$

Thus Φ already gives a homomorphism between $\{R_0, \ldots, R_n\}$ and $\{0, \ldots, n\}$, for any set A. We set up to build A in such a way that the homomorphism is onto and one-one. We simultaneously build, by stages, two disjoint sets A and B, with the intention that B is contained in $\overline{A}$ (i.e. it contains elements we want to leave out of A). At stage $s+1$ we have finite approximations A_s and B_s of A and B, and sets $R_{x,s}$ such that $\Phi(R_{x,s}) \equiv_m \Phi(R_x)$. As above, $R_{x,s} = \bigcup_{i \leq x} P_{i,s}$, and $P_{i,s}$ is going to be a strong array: we look at the elements of the strong array as equivalence classes (boxes), with the intention that at following stages we put into A or B not single elements, but equivalence classes.

At stage $s+1$ we will do (according to the order decided by an exhaustive list of requirements) one of the following two types of action, respectively intended to make Φ onto and one-one:

1. to ensure that for each e there is x such that $\Phi(W_e) \equiv_m \Phi(R_x)$
   Since $\Phi(R_{x,s}) \equiv_m \Phi(R_x)$, it is enough to ensure that $\Phi(W_e) \equiv_m \Phi(R_{x,s})$, for some x. We may suppose W_e infinite, otherwise $\Phi(W_e)$ is automatically recursive. We look at the greatest $x \leq n$ such that $W_e \cap P_{x,s}$ is infinite and define, for each $i \leq x$, a new $P_{i,s+1}$ with the property that each equivalence class of it intersects W_e. We do this by moving boxes intersecting W_e from $P_{x,s}$ to $P_{i,s+1}$, when needed. For $i > x$ we let $P_{i,s+1} = P_{i,s}$.

- $\Phi(\mathcal{W}_e) \equiv_m \Phi(R_{x,s+1})$

  Note that $R_{n,s} \cup A_s \cup B_s = \omega$, and $A_s \cup B_s$ is finite. Then, by construction and the choice of x, $\mathcal{W}_e$ is contained in $R_{x,s}$, except for at most finitely many elements. Since Φ is invariant under finite differences, we may actually suppose that $\mathcal{W}_e \subseteq R_{x,s} = R_{x,s+1}$. Thus $\Phi(\mathcal{W}_e) \leq_m \Phi(R_{x,s+1})$ automatically, by the properties of Φ.

  Conversely, every box of $R_{x,s+1}$ contains an element of $\mathcal{W}_e$ by construction. If $\mathcal{W}_e$ and $R_{x,s+1}$ are respectively the ranges of f and g, let

$$h(z) = \mu y[f(y) \text{ and } g(z) \text{ are in the same box of } R_{x,s+1}].$$

  Then

$$z \in \Phi(R_{x,s+1}) \quad \Leftrightarrow \quad g(z) \in A$$
$$\Leftrightarrow \quad fh(z) \in A \qquad \text{(being in the same box)}$$
$$\Leftrightarrow \quad h(z) \in \Phi(\mathcal{W}_e),$$

  and $\Phi(R_{x,s+1}) \leq_m \Phi(\mathcal{W}_e)$. Note that it is crucial here that in the following we put boxes into A, and not only single elements.

- $\Phi(R_{i,s+1}) \equiv_m \Phi(R_{i,s})$

  We only have to prove this for $i \leq x$, since for $i > x$ we have $R_{i,s} = R_{i,s+1}$. If $i \leq x$ then $R_{i,s} \subseteq R_{i,s+1}$, since we only enlarge boxes, so $\Phi(R_{i,s}) \leq_m \Phi(R_{i,s+1})$.

  Since every new box of $R_{i,s+1}$ contains an old box of $R_{i,s}$, by associating to every element of $R_{i,s+1}$ an element of $R_{i,s}$ in the same box (as above) we get $\Phi(R_{i,s+1}) \leq_m \Phi(R_{i,s})$.

2. *to ensure that* $\Phi(R_x) \not\equiv_m \Phi(R_y)$ *if* $x \neq y$

   Since $x, y \leq n$ we may suppose, e.g., $x < y$. Then $R_x \subseteq R_y$, and automatically $\Phi(R_x) \leq_m \Phi(R_y)$. To ensure $\Phi(R_y) \not\leq_m \Phi(R_x)$, suppose that for some recursive function h

$$z \in \Phi(R_y) \quad \Leftrightarrow \quad h(z) \in \Phi(R_x).$$

Let R_y and R_x be the ranges of f and g: then

$$f(z) \in A \quad \Leftrightarrow \quad gh(z) \in A.$$

Note that there is a partial recursive function φ with domain R_y and range contained in R_x, and such that $w \in A \Leftrightarrow \varphi(w) \in A$ for each w on which φ is defined. For example,

$$\varphi(w) \simeq gh(z) \text{ for the smallest } z \text{ such that } f(z) = w.$$

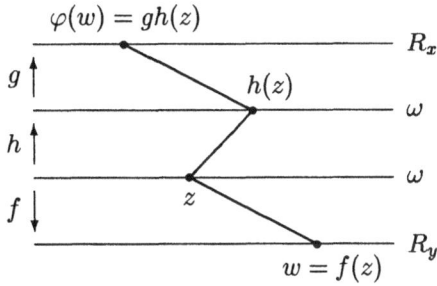

$$\varphi(w) = gh(z)$$

To ensure $\Phi(R_y) \not\leq_m \Phi(R_x)$ is thus enough to look at the partial recursive functions φ with domain containing R_y and range contained in R_x. Since $R_x \subseteq R_y$, there is a such that P_a is not contained in R_x, but it is contained in R_y. Consider $P_{a,s}$: the construction is such that every box of $P_{a,s}$ contains a box of from P_a. Choose $w \in P_a \cap P_{a,s}$: then $\varphi(w) \in R_x$. We ensure that $w \in A$ if and only if $\varphi(w) \notin A$, thus diagonalizing against φ:

- if $\varphi(w) \in A_s$ then put the box of w into B_{s+1}
- if $\varphi(w) \in B_s$ then put the box of w into A_{s+1}
- if $\varphi(w) \notin A_s \cup B_s$ then, e.g., put the box of w into A_{s+1}, and the box of $\varphi(w)$ into B_{s+1}.

In this case $\Phi(R_{i,s}) \equiv_m \Phi(R_{i,s+1})$ is automatic, since we just produce a finite modification, moving the box of w and, possibly, the box of $\varphi(w)$ too. $\quad \Box$

The construction just given is the basis for all the embeddings to be obtained. In the following we will limit ourselves to indicate the modifications needed to take care of the increasingly more complicated situations.

Proposition VI.2.3 (Lachlan [1970]) *The topped finite initial segments of* $\mathcal{D}_m$ *are exactly the finite distributive lattices.*

Proof. The condition is necessary by VI.1.10. For the converse, let D be a finite distributive lattice. We have to define R_x for $x \in D$ in such a way that $\{\Phi(R_x) : x \in D\}$ is isomorphic to D. This could appear to be complicated, since we have to preserve the relationships among the members of D, but a small trick makes everything trivial. Let H be the set $\{a_0, \ldots, a_n\}$ of the indecomposable elements of D (in the case of linear orderings we had $D = H$). Since H generates D (VI.1.7.a), we only have to define:

$$\begin{aligned} P_{a_i} &= \{\{(n+1)x + i\} : x \in \omega\} \\ R_x &= \bigcup \{P_a : a \leq x \wedge a \in H\}. \end{aligned}$$

The map Φ is a homomorphism from D to $\mathcal{D}_m(\le a)$, where a is the m-degree of the set A we build, because D is distributive: if $x = y \sqcup z$ then the indecomposable elements below x are exactly the indecomposable elements below y or z (VI.1.6), hence $R_x = R_y \cup R_z$ and $\Phi(R_x) \equiv_m \Phi(R_y) \oplus \Phi(R_z)$, by the properties of Φ.

As before we have to ensure that Φ is onto and one-one, and this is obtained in the same way as in the previous result, with the following modifications.

1. *ontoness*

   Given $\mathcal{W}_e$ infinite, consider all the i's such that $\mathcal{W}_e \cap P_{a_i,s}$ is infinite, and make all the classes of $P_{a_j,s+1}$ intersect $\mathcal{W}_e$, for every j such that $a_j \sqsubseteq a_i$ for some such i. If $i_0, \ldots, i_m$ is a list of all such i's, we have as before $\Phi(\mathcal{W}_e) \equiv_m \Phi(R_{x,s+1})$, where $x = a_{i_0} \sqcup \cdots \sqcup a_{i_m}$.

   And $\Phi(R_x) \equiv_m \Phi(R_{x,s+1})$ for all x, since this holds for indecomposable elements as before, and for the other elements because Φ is a homomorphism.

2. *one-oneness*

   If $x \not\sqsubseteq y$ we want $\Phi(R_x) \not\le_m \Phi(R_y)$. This is obtained as before, because if $x \not\sqsubseteq y$ then there is an indecomposable element a such that $a \sqsubseteq x$ and $a \not\sqsubseteq y$ (otherwise all the indecomposable elements below x are also below y, and hence so is their l.u.b. x). But then we can choose an element of $P_{a,s} \cap P_a$ to diagonalize. $\square$

Countable initial segments

The proofs given above may be easily extended to handle infinite initial segments whose representations can be chosen ahead of time (like recursive distributive lattices), but some new ideas are needed to treat the general case. Again we start with the linear orderings.

Proposition VI.2.4 (Lachlan [1970]) *Every countable linear ordering with least element is isomorphic to an initial segment of $\mathcal{D}_m$.*

Proof. When we consider a countable linear ordering L we cannot define R_x ahead of time for all $x \in L$, unless L is recursive. Then we approximate L by finite pieces. We may suppose that L has a greatest element, otherwise we can add one to it, and we call 0 and 1 the least and greatest elements. Let $L = \bigcup_{n \in \omega} L_n$, where $L_0 = \{0, 1\}$, and each L_n is finite and contained in L_{n+1}. Let $L_n = \{a_0 \sqsubset \cdots \sqsubset a_m\}$, with $a_0 = 0$ and $a_m = 1$. At stage $s + 1$ we have $P_{a_i,s}$. If b is a new element of L_{n+1}, i.e. b is not one of the a_i's, we have to create a new row P_b for it. Since 0 and 1 are already in L_0, there must be i such that $a_i \sqsubset b \sqsubset a_{i+1}$. Then we can simply let $P_{a_j,s+1} = P_{a_j,s}$ if $j \ne i+1$, and we create P_b by taking its boxes from $P_{a_{i+1},s}$. It is clear that for $x \in L_n$ it is still $R_{x,s+1} = R_{x,s}$, and thus $\Phi(R_{x,s+1}) \equiv_m \Phi(R_{x,s})$.   □

Proposition VI.2.5 (Lachlan [1970]) *The topped initial segments of $\mathcal{D}_m$ are exactly the direct limits of ascending sequences of finite distributive lattices.*

Proof. The condition is necessary by VI.1.10. For the converse, let D be the limit of $\{D_n\}_{n \in \omega}$, with D_n a finite distributive lattice, and $D_n \subseteq D_{n+1}$. Suppose also that D_0 already contains the least and greatest elements 0 and 1 of D, otherwise start from the first D_n that does. As in VI.2.3 we just have to consider the indecomposable elements, but now D is only approximated by D_n, and hence at a given stage we only have

$$H_n = \{a_0, \ldots, a_m\} = \text{set of the indecomposable elements of } D_n.$$

When we step from D_n to D_{n+1} two new things might happen:

1. *a new atom b appears*
   As in VI.2.4 we have to create a new P_b for it, and this time the atoms are not linearly ordered. We may simply build P_b as a sequence of boxes, each one being the union of one box from $P_{a_i,s}$ for each i such that $b \sqsubset a_i$ (some such i must exist, because the greatest element of D is already in D_0). This allows us to get $\Phi(R_{x,s+1}) \equiv_m \Phi(R_{x,s})$ as follows.

   If $b \not\sqsubset x$ then there is nothing to prove, since $R_{x,s+1} = R_{x,s}$. If $b \sqsubset x$ we may suppose that x is an atom of D_n, since we only have to prove the property for atoms (the rest follows automatically from the fact that Φ is a homomorphism). But then:

   - $\Phi(R_{x,s}) \leq_m \Phi(R_{x,s+1})$
     This follows from $R_{x,s} \subseteq R_{x,s+1}$ (note that P_a may be in $R_{x,s+1}$ now).

   - $\Phi(R_{x,s+1}) \leq_m \Phi(R_{x,s})$
     As usual it is enough to get effectively from every element of $R_{x,s+1}$ an element of $R_{x,s}$ in the same box. The only nontrivial case is when we consider elements in boxes of P_b, but each such box contains a box of $P_{x,s}$ by construction.

2. *an atom $a \in H_n$ becomes decomposable in D_{n+1}*
   In this case we do not want to consider $P_{a,s}$ anymore, and we simply throw it into B_{s+1}, which then becomes an infinite recursive set (as opposed to the finite set of the previous propositions). To check

   $$\Phi(R_{x,s+1}) \equiv_m \Phi(R_{x,s})$$

we simply note that for $a \sqsubset x$ is $R_{x,s+1} = R_{x,s} - P_a$, which is still r.e. since P_a is recursive. But then $R_{x,s} = R_{x,s+1} \cup P_{a,s}$ and, by the properties of Φ, $\Phi(R_{x,s}) \equiv_m \Phi(R_{x,s+1}) \oplus \Phi(P_{a,s})$. Since $P_{a,s} \subseteq \overline{A}$ by construction, because we threw it into B_{s+1}, $\Phi(P_a)$ has m-degree $\mathbf{0}_m$ and thus $\Phi(R_{x,s}) \equiv_m \Phi(R_{x,s+1})$.

Case 2 introduces the need of small adjustments in the proof.

1. We have to suppose that the top element of D is indecomposable in every D_n, so that we always have a top line to use when a new atom appears, and a new line has to be created. For this it is enough to add, if necessary, an element on top of the greatest one, and suppose it is already in D_0.

2. To show that Φ is onto we prove as before that $\Phi(\mathcal{W}_e) \equiv_m \Phi(R_x)$ for some x, noting that for an appropriate x we have $\mathcal{W}_e \subseteq A_s \cup B_s \cup R_{x,s}$, so that

$$\Phi(\mathcal{W}_e) \equiv_m \Phi(\mathcal{W}_e \cap A_s) \oplus \Phi(\mathcal{W}_e \cap B_s) \oplus \Phi(\mathcal{W}_e \cap R_{x,s}).$$

But

- $\Phi(\mathcal{W}_e \cap A_s)$ has m-degree $\mathbf{0}_m$ because $\mathcal{W}_e \cap A_s$ is finite
- $\Phi(\mathcal{W}_e \cap B_s)$ has m-degree $\mathbf{0}_m$ because $B_s \subseteq \overline{A}$
- $\Phi(\mathcal{W}_e \cap R_{x,s}) \equiv_m \Phi(R_{x,s})$ as usual. $\square$

Corollary VI.2.6 *Every countable distributive lattice is isomorphic to an initial segment of $\mathcal{D}_m$.*

Proof. If D is a countable distributive lattice, let $D_0 = \{0,1\}$. Given D_n take $a \in D - D_n$, if it exists, and consider the finite distributive sublattice D_{n+1} of D generated by $D_n \cup \{a\}$. Then D is the direct limit of $\{D_n\}_{n \in \omega}$. $\square$

Exercise VI.2.7 $\mathcal{D}_m$ *is not a lattice.* (Hint: let D_n be the finite distributive lattice consisting of an ascending sequence of n elements, with a diamond on top: the direct limit of the D_n's is an infinite chain with two incomparable elements and their join above it. These two elements have no l.u.b.)

A different method of proving VI.2.5 has been given by Lachlan [1971], and consists of building the same initial segments for T-degrees using strongly uniform trees (V.6.7). This produces common initial segments of Turing and m-degrees.

Exercise VI.2.8 *If A is a set of minimal Turing degree constructed by using strongly uniform trees, then A has minimal m-degree.* (Lachlan [1971]) (Hint: it is enough to prove that if f_L, f_R, g are an e-splitting strongly uniform tree, A is on it, and $\{e\}^A$ is an m-reduction, then $A \leq_m \{e\}^A$. But if $B = \{e\}^A$ is an m-reduction, there is h recursive such that $x \in B \Leftrightarrow h(x) \in A$. To define f recursive such that $x \in A \Leftrightarrow f(x) \in B$, see if f_L and f_R agree on x. If so, then let $f(x)$ be a fixed element of B if the common value is 1, and a fixed element of $\overline{B}$ otherwise. If not, x is in the range of h, because the tree is e-splitting. Then let $f(x) = h^{-1}(x)$.)

VI.3   Uncountable Initial Segments

The obvious trouble with uncountable initial segments is that we cannot approximate them by finite pieces in countably many steps. We are able to overcome this difficulty by extending given countable pieces.

Strong minimal covers

Recall (V.5.16) that a is a strong minimal cover of b when, for every c, if $c < a$ then $c \leq b$.

Proposition VI.3.1 (Lachlan [1972]) *Every m-degree has a strong minimal cover.*

Proof. The construction is an extension of the one for minimal m-degrees, which is the special case of VI.2.2 for $n = 1$. We build two sets

$$P_a = \{\{2x\} : x \in \omega\}$$
$$P_b = \{\{2x+1\} : x \in \omega\},$$

and let $R_a = P_a \cup P_b$ and $R_b = P_b$. Since $R_b \subseteq R_a$, we automatically have $\Phi(R_b) \leq_m \Phi(R_a)$. We want to ensure that:

1. $\Phi(R_b) \equiv_m B$, for a given set $B \in b$.

2. $\Phi(W_e) \equiv_m \Phi(R_a)$ or $\Phi(W_e) \leq_m \Phi(R_b)$, for every e

3. $\Phi(R_a) \not\leq_m \Phi(R_b)$.

We indicate the modifications to make in the construction of a minimal degree to ensure these conditions.

1. Let $B \in b$: we ensure $\Phi(R_b) \equiv_m B$ by putting the x-th box of P_b into A if and only if $x \in B$. This is the main new idea, to be fully exploited in this subsection. Let $R_b = P_b$ be the range of f. Then $\Phi(R_b) \leq_m B$ via $g(z) = \mu x[f(z) = 2x + 1]$, since

$$z \in \Phi(R_b) \Leftrightarrow f(z) \in A \Leftrightarrow g(z) \in B,$$

and $B \leq_m \Phi(R_b)$ via $h(x) = \mu z[f(z) = 2x + 1]$, since

$$x \in B \Leftrightarrow 2x + 1 \in A \Leftrightarrow fh(x) \in A \Leftrightarrow h(x) \in \Phi(R_b).$$

This part of the construction makes the approximation of A at a given stage infinite. But we can still think of A at stage s as the union of a finite part A_s on a side (obtained by the other two parts of the construction, as in VI.2.2) and of the infinite part of A contained in $P_{b,s}$. It is understood that anything falling into the x-th box of P_b during the construction goes into A, if $x \in B$.

2. This is ensured as before. If $W_e \cap P_{a,s}$ is infinite then we use appropriate boxes from $P_{a,s}$ to force all boxes of $R_{a,s+1}$ to intersect W_e, thus obtaining $\Phi(W_e) \equiv_m \Phi(R_{a,s+1})$. If $W_e \cap P_{a,s}$ is finite then, except for finitely many elements, $W_e \subseteq P_{b,s}$: then, automatically, $\Phi(W_e) \leq_m \Phi(R_{b,s})$.

3. This is also ensured as before, since we can use elements of $P_a \cap P_{a,s}$ to diagonalize against $\varphi : R_a \to R_b$.   □

Recall (V.5.17) that there is a cone of minimal covers in the Turing degrees. A similar proof shows here that *there is a cone of strong minimal covers in* $\mathcal{D}_m$. This stronger result fails for Turing degrees, and actually (in accord with V.1.16) *there is a cone without strong minimal covers in* $\mathcal{D}$, since by V.2.26 no degree above $\mathbf{0}'$ can be a strong minimal cover.

Uncountable linear orderings

Linearly ordered initial segments are especially simple because of the countable predecessor property, which implies that any such linear ordering can be obtained by a (possibly uncountable) sequence of countable extensions. We thus have only to learn how to do countable extensions.

Proposition VI.3.2 (Ershov [1975]) *Every countable linearly ordered initial segment* $\mathcal{I}$ *of* $\mathcal{D}_m$ *can be extended to an initial segment* $\mathcal{L}$ *isomorphic to* L, *for any countable linear ordering* L *having an initial segment* I *isomorphic to* $\mathcal{I}$.

Proof. The proof combines the ideas of VI.2.4 and VI.3.1. The only step to be added to VI.2.4 is the case when a new atom b appearing in L_{n+1} is in I. Then we build the corresponding P_b by using the row $P_{a,s}$, for the smallest $a \in L_n - I$ (which is automatically above b): we may always suppose it exists, by possibly topping L and putting the greatest element in L_0. We then order the classes of P_b, e.g. by ordering in the natural way their least elements, and put the x-th class into A if and only if $x \in B$, for some fixed B in the m-degree b of $\mathcal{I}$ corresponding to $b \in I$.

- $\Phi(R_b) \equiv_m B$
  $\Phi(P_b) \equiv_m B$ as in VI.3.1, and $R_b = \bigcup_{c \sqsubseteq b} P_c$, so $\Phi(R_b) \equiv_m \oplus_{c \sqsubseteq b} \Phi(P_c)$.
  But $b \in I$ and I is downward closed, so if $c \sqsubseteq b$ is $c \in I$ and P_c was introduced in the same way as P_b, in particular $\Phi(P_c)$ has m-degree $c \le b$. Thus $\Phi(R_b)$ has m-degree b.

- $\Phi(R_{x,s+1}) \equiv_m \Phi(R_{x,s})$
  For $x \in L_n - I$, which is the only interesting case, $R_{x,s+1} = R_{x,s}$.

- *ontoness*
  This is obtained as in VI.3.1, by considering the greatest $a \in L_n - I$, if it exists, such that $W_e \cap P_{a,s}$ is infinite.

- *one-oneness*
  This is obtained as in VI.3.1 and VI.2.4, by ensuring $\Phi(R_y) \not\le_m \Phi(R_x)$ whenever $x \sqsubset y$ and either $(x \in I \wedge y \notin I)$ or $(x \notin I \wedge y \notin I)$. Indeed, for $x, y \in I$ this is automatically achieved by the previous strategy. □

At this point we can already give some final results.

Corollary VI.3.3 *The order-types of the linearly ordered initial segments of $\mathcal{D}_m$ are exactly the linear orderings with least element and countable predecessor property.*

Proof. The conditions are obviously necessary. Given a linear ordering L with least element 0 and countable predecessor property, let $I_0 = \{0\}$ and $\mathcal{I}_0 = \{0_m\}$. For any ordinal $\alpha < \omega_1$ let I_α and $\mathcal{I}_\alpha$ be isomorphic countable initial segments of L and $\mathcal{D}_m$. Take $a \in L - I_\alpha$, if it exists, and let $I_{\alpha+1}$ be the downward closure of if $I_\alpha \cup \{a\}$ in L: $I_{\alpha+1}$ is countable because L has countable predecessor property, and $I_\alpha \subseteq I_{\alpha+1}$ since I_α is closed downward. By VI.3.2 there is an initial segment $\mathcal{I}_{\alpha+1}$ of $\mathcal{D}_m$ isomorphic to $I_{\alpha+1}$ and extending $\mathcal{I}_\alpha$. At limit stages take unions. This procedure gives an initial segment isomorphic to L in at most ω_1 steps, since L has countable predecessor property. □

In particular, *the ordinals of the well-ordered initial segments of $\mathcal{D}_m$ are exactly the ordinals $\le \omega_1$.*

Abraham and Shore [1986] have shown that the linearly ordered initial segments of $\mathcal{D}$ are the same as those of $\mathcal{D}_m$.

Uncountable initial segments

We are finally at the last step of our long journey. In the next result all the ingredients so far introduced, and some new ones, will be employed to produce all possible countable extensions of given initial segments. We will then show

how the result can be modified and used to characterize the initial segments of
$\mathcal{D}_m$.

Proposition VI.3.4 (Ershov [1975]) *Every countable ideal $\mathcal{I}$ of $\mathcal{D}_m$ can be
extended to an initial segment $\mathcal{L}$ isomorphic to L, for any countable distributive
uppersemilattice L having an ideal I isomorphic to $\mathcal{I}$.*

Proof. First note that, by possibly doing countably many successive extensions, we may simply restrict our attention to topped countable distributive
uppersemilattices L. As in VI.1.13 we can see that L is a direct limit of an
ascending sequence of finite distributive lattices. The proof then extends the
one of VI.2.5.

When new atoms appear, we treat them as in VI.2.5 and VI.3.2. Namely,
we create P_b as a sequence of boxes, each one being the union of one box from
$P_{a,s}$ for each atom a not in I and above b. Two new things might happen here:

1. *an atom $a \in I$ becomes decomposable*
   As in VI.2.5 we throw $P_{a,s}$ away, i.e. we put it into B_{s+1}. This now
   causes a little trouble, since $P_{a,s}$ contains boxes which were in A and
   that go now into $\overline{A}$, and A is not built monotonically. We still want to
   have $\Phi(R_{x,s+1}) \equiv_m \Phi(R_{x,s})$. Suppose $a \sqsubseteq x$: since $R_{x,s+1} = R_{x,s} - P_{a,s}$
   we have, as in VI.2.5, $\Phi(R_{x,s}) \equiv_m \Phi(R_{x,s+1}) \oplus \Phi(P_{a,s})$. It is thus enough
   to show that $\Phi(P_{a,s}) \leq_m \Phi(R_{x,s+1})$. We may suppose we already have
   in $R_{x,s+1}$ the P_b's for the atoms that make a decomposable. Now $b \sqsubseteq x$
   for all these b's, so $P_{b,s} \subseteq R_{x,s+1}$ and $\Phi(P_{b,s}) \leq_m \Phi(R_{x,s+1})$. But $b \in I$,
   since $a \in I$ and I is closed downward: by construction then $\Phi(P_{b,s})$
   has the m-degree $\boldsymbol{b}$ in $\mathcal{I}$ corresponding to b in I, and $\Phi(P_{a,s})$ has the
   m-degree $\boldsymbol{a}$ corresponding to a. But $\boldsymbol{a}$ is the l.u.b. of these $\boldsymbol{b}$'s, and so
   $\Phi(P_{a,s}) \leq_m \Phi(R_{x,s+1})$.

2. *one-oneness*
   We want $\Phi(R_y) \not\leq_m \Phi(R_x)$ whenever $y \not\sqsubseteq x$. This is automatic if x and
   y are in I.

   If $x \in I$ but $y \notin I$ then we have no trouble: there is an atom $a \sqsubseteq y$ such
   that $a \notin I$, otherwise $y \in I$ because I is an ideal, hence closed under
   l.u.b., and y is the l.u.b. of the atoms below it. But $a \not\sqsubseteq x$ since $x \in I$
   and I is closed downward, so we can use P_a to diagonalize.

   Let us thus suppose that $x \notin I$, $y \in I$, and $y \not\sqsubseteq x$. If we simply consider
   $\varphi : R_y \to R_x$ then we might not be able to diagonalize, since φ might
   have range contained in R_z, where $z \sqsubseteq x$ is the l.u.b. of the atoms of I
   below x ($z \in I$ because I is an ideal): thus φ sends rows corresponding
   to atoms of I to similar rows, and we do not want to touch them because

they code information. But we now show that if $\Phi(R_y) \leq_m \Phi(R_x)$ then there is a partial recursive function φ such that:

- φ has domain R_y and range contained in R_x
- $w \in A$ if and only $\varphi(w) \in A$, whenever $\varphi(w)$ is defined
- the range of φ intersects P_a, for some $P_a \subseteq R_x$ and $a \notin I$.

Consider indeed the usual φ (VI.2.2), and let $\mathcal{W}$ be its range. If φ does not have the required properties then $\mathcal{W} \subseteq R_z$, for the $z \in I$ considered above. So $\Phi(\mathcal{W}) \leq_m \Phi(R_z)$. Since $w \in A \Leftrightarrow \varphi(w) \in A$ and R_y is the domain of φ, $\Phi(R_y) \equiv_m \Phi(\mathcal{W})$ and hence $\Phi(R_y) \leq_m \Phi(R_z)$. But y and z are in I, and Φ is by definition an isomorphism on I, hence $y \sqsubseteq z \sqsubseteq x$, contradiction. We can then always consider φ with the stated properties, and use a row P_a for some $a \notin I$ to diagonalize.   □

The result already has a number of important consequences, e.g. *every countable ideal of $\mathcal{D}_m$ has a strong minimal cover, and it is the intersection of two principal ideals*. To derive the most important consequence of all we however need a strengthening of it, stated in the exercise.

Exercise VI.3.5 *For every countable ideal $\mathcal{I}$ of $\mathcal{D}_m$, and any countable distributive uppersemilattice L having an ideal I isomorphic to $\mathcal{I}$, there is a continuum of initial segments of $\mathcal{D}_m$ isomorphic to L, and such that their parts isomorphic to $L - I$ are pairwise disjoint.* (Paliutin [1975]) (Hint: the properties of the construction above are not affected if countably many times we take a box from the line relative to the top element, and put it into A_s or B_s. We can thus build a tree of sets whose degrees are top degrees of extension of $\mathcal{I}$ as wanted.)

We can now give the promised characterization of the initial segments of $\mathcal{D}_m$. Note that the exercise allows us to avoid any use of the Continuum Hypothesis, and thus the result is final.

Theorem VI.3.6 Characterization of the ideals of $\mathcal{D}_m$ (Ershov [1975], Paliutin [1975]) *The ideals of $\mathcal{D}_m$ are exactly, up to isomorphism, the distributive uppersemilattices with least element, countable predecessor property, and power at most that of the continuum.*

Proof. The conditions are clearly necessary. Conversely, given an uppersemilattice L with the stated conditions, we want to define an ideal $\mathcal{L}$ of $\mathcal{D}_m$ isomorphic to L. Let 0 be the smallest element of L, and define $I_0 = \{0\}$, $\mathcal{I}_0 = \{0_m\}$, and $\varphi_0(0) = 0_m$.

For $\alpha < 2^{\aleph_0}$ let I_α and $\mathcal{I}_\alpha$ be ideals of L and $\mathcal{D}_m$ isomorphic via φ_α, and of power less than the continuum. Take $a \in L - I_\alpha$, if it exists: $\hat{a}$ (the set of the predecessors of a in L) is a countable distributive uppersemilattice (because L has countable predecessor property) with $\hat{a} \cap I_\alpha$ as a countable ideal (since both $\hat{a}$ and I_α are ideals). By VI.3.4 we can extend the isomorphism φ_α of domain I_α to an isomorphism φ of domain $\hat{a} \cup I_\alpha$. Let $I_{\alpha+1}$ be the ideal generated by $\hat{a} \cup I_\alpha$: we want to show that φ is extendable to an isomorphism $\varphi_{\alpha+1}$ of domain $I_{\alpha+1}$. By VI.1.11 the elements of $I_{\alpha+1}$ are of the form $x \sqcup y$, for $x \in \hat{a}$ and $y \in I_\alpha$. We then let

$$\varphi_{\alpha+1}(x \sqcup y) = \varphi(x) \cup \varphi(y).$$

To check that $\varphi_{\alpha+1}$ is well-defined, consider $x' \sqcup y' = x \sqcup y$. Since $x' \sqsubseteq x \sqcup y$, there are $x_0' \sqsubseteq x$ and $y_0' \sqsubseteq y$ such that $x' = x_0' \sqcup y_0'$. But φ is an isomorphism on $\hat{a} \cup I_\alpha$, and thus $\varphi(x_0') \leq \varphi(x)$ and $\varphi(y_0') \leq \varphi(y)$. Then

$$\varphi_{\alpha+1}(x') = \varphi_{\alpha+1}(x_0' \sqcup y_0') = \varphi(x_0') \cup \varphi(y_0') \leq \varphi(x) \cup \varphi(y) = \varphi_{\alpha+1}(x \sqcup y).$$

Similarly, $\varphi_{\alpha+1}(y') \leq \varphi_{\alpha+1}(x \sqcup y)$, and hence

$$\varphi_{\alpha+1}(x' \sqcup y') = \varphi_{\alpha+1}(x') \cup \varphi_{\alpha+1}(y') \leq \varphi_{\alpha+1}(x \sqcup y).$$

The converse holds similarly, and then $\varphi_{\alpha+1}$ is well-defined.

But if we only use VI.3.4 as stated then there is no reason to believe that $\varphi_{\alpha+1}$ is also one-one. This is where the strengthening VI.3.5 comes into play: it provides enough choices to make the degrees below $\varphi(a)$ but not in $\varphi_\alpha(\hat{a} \cap I_\alpha)$ disjoint from $\varphi_\alpha(I_\alpha)$, and thus $\varphi_{\alpha+1}$ one-one.

At limit stages we take unions. The procedure gives an initial segment isomorphic to L in at most $2^{\aleph_0}$ steps, since L has at most the power of the continuum. $\square$

Refinements of the results of this section have been found by Malc'ev [1981], [1984], respectively on localization of initial segments and relativization to the uppersemilattice of immune sets.

VI.4   Global Properties

We now turn to the study of global properties of $\mathcal{D}_m$, following the path set up in Section V.7 for Turing degrees, with two major differences. First of all, we will be completely successful in characterizing the algebraic structure of $\mathcal{D}_m$, while an analogous characterization is not known for Turing degrees, and might even be independent of ZFC (see the results of Grozsek and Slaman quoted on pp. 467 and 529). Secondly, the properties of definability, homogeneity, and automorphisms for m-degrees are exactly the opposite of those for Turing degrees: no nontrivial countable set of m-degrees is definable, strong homogeneity holds, and there are lots of automorphisms. However, the complexity of the theories of Turing and m-degrees is the same.

Characterization of the structure of many-one degrees

We now characterize $\mathcal{D}_m$ as an algebraic structure. This result is the only known example of an absolute characterization of a degree theory, and it provides an alternative, recursion-theoretical description of the continuum.

The results proved in this chapter, which culminated in VI.3.6, already provide the tools needed to characterize $\mathcal{D}_m$ with a two-line proof.

> Aguzza qui, lettor, ben li occhi al vero,
> ché 'l velo è ora ben tanto sottile,
> certo che 'l trapassar dentro è leggero.[1]
> (Dante, *Purgatorio*, VIII)

Theorem VI.4.1 Characterization of $\mathcal{D}_m$ (Ershov [1975], Paliutin [1975]) *Up to isomorphism, $\mathcal{D}_m$ is the only structure with the following properties:*

1. *it is a distributive uppersemilattice with least element*

2. *every element has at most countably many predecessors*

3. *it has the power of the continuum*

4. *every ideal $\mathcal{I}$ with power less than the continuum can be extended to an ideal $\mathcal{L}$ isomorphic to L, for any distributive uppersemilattice L with power less than the continuum having an ideal I isomorphic to $\mathcal{I}$.*

Proof. A back-and-forth argument in the style of VI.3.6 easily gives an isomorphism between any two structures with the given properties. □

Note that no use is made of extra set-theoretical assumptions, like the Continuum Hypothesis, and thus the result is absolute.

Definability, homogeneity, and automorphisms

The algebraic characterization of $\mathcal{D}_m$ contains all the information about the structure, and it is thus not surprising that we can easily derive from it solutions to a number of algebraic problems. They should be contrasted with the opposite ones obtained for Turing degrees in Section V.7.

---

[1] Sharpen thy sight now, reader, to regard
the truth, for so transparent grows the veil,
to pass within will surely not be hard.

Theorem VI.4.2 Strong homogeneity. *Any two cones of m-degrees are isomorphic.*

Proof. Given a, it is enough to show that $\mathcal{D}_m (\geq a)$ satisfies the properties of VI.4.1. It then follows that any cone is isomorphic to $\mathcal{D}_m$, and hence any two cones are isomorphic.

The only nontrivial property is the fourth one. Let $\mathcal{I}$ be an ideal of $\mathcal{D}_m (\geq a)$ of power less than the continuum, and L be a distributive uppersemilattice of power less than the continuum and with an ideal I isomorphic to $\mathcal{I}$. Let $\hat{\mathcal{I}}$ be the downward closure of $\mathcal{I}$ in $\mathcal{D}_m$: we can extend I to a distributive uppersemilattice $\hat{I}$ isomorphic to $\hat{\mathcal{I}}$. If we show that $\hat{I} \cup L$ is an uppersemilattice with $\hat{I}$ as an ideal, then we can apply the methods of VI.3.6 to extend $\hat{\mathcal{I}}$ to an ideal isomorphic to $\hat{I} \cup L$ in $\mathcal{D}_m$, thus extending $\mathcal{I}$ to an ideal isomorphic to L in $\mathcal{D}_m (\geq a)$.

Since I is an ideal, so is $\hat{I}$ by its definition. To show that $\hat{I} \cup L$ is an uppersemilattice it is enough, by VI.1.11, to define $x \sqcup y$ when $x \in \hat{I}$ and $y \in L$. There are two cases:

- if $x \in L$ then define $x \sqcup y$ as the l.u.b. of x and y in L

- if $x \notin L$ then first consider the l.u.b. of x and a (the element of $\hat{I}$ corresponding to a) in $\hat{I}$: this is now an element of L, being above a. Then take the l.u.b. of it and y in L.   □

The next proposition allow us to derive information about definability and automorphisms.

Proposition VI.4.3 $\mathcal{D}_m (\leq a)$ *and* $\mathcal{D}_m (\leq b)$ *are isomorphic if and only if there is an automorphism of* $\mathcal{D}_m$ *carrying* a *into* b.

Proof. One direction is trivial, since any automorphism carrying a into b induces an isomorphism of $\mathcal{D}_m (\leq a)$ and $\mathcal{D}_m (\leq b)$. Conversely, given an isomorphism of $\mathcal{D}_m (\leq a)$ and $\mathcal{D}_m (\leq b)$ we can extend it to an automorphism of $\mathcal{D}_m$ by a back-and-forth argument as in VI.4.1.   □

Corollary VI.4.4 0_m *is the only m-degree fixed under every automorphism, as well as the only definable m-degree.*

Proof. Given $a > 0_m$, $\mathcal{D}_m (\leq a)$ is isomorphic to a countable distributive uppersemilattice I. $L = I \times I$ is still a countable distributive uppersemilattice, containing two distinct copies of I as ideals. Extend $\mathcal{D}_m (\leq a)$ to an ideal isomorphic to L, by VI.3.4, and let $b \neq a$ be the top m-degree corresponding to the second copy of I in L. Then $\mathcal{D}_m (\leq a)$ and $\mathcal{D}_m (\leq b)$ are isomorphic, and there is an automorphism of $\mathcal{D}_m$ carrying a into b. Then a is not fixed under every automorphism of $\mathcal{D}_m$, and in particular it cannot be definable. Thus 0_m is the only m-degree fixed under every automorphism, and the only definable m-degree.   □

Corollary VI.4.5 *Every definable set of m-degrees different from $\{0_m\}$ has power of the continuum.*

Proof. Given a set $S \neq \{0_m\}$ of m-degrees of power less than the continuum, choose $a \in S - \{0_m\}$. The m-degree b obtained as in the previous corollary can be taken to be not in S, because S has power less than the continuum, while there are (by VI.3.5) $2^{\aleph_0}$ possible choices for b. Since there is an automorphism of $\mathcal{D}_m$ carrying a into b, S is not closed under automorphisms and hence it cannot be definable.   □

It follows that many natural classes of m-degrees are not definable, e.g. the r.e. and the arithmetical m-degrees. On the other hand, there are nontrivial definable sets of m-degrees, e.g. the minimal m-degrees, and thus the result is the best possible.

From VI.4.3 it easily follows that there are $2^{\aleph_0}$ automorphisms of $\mathcal{D}_m$: given a minimal m-degree a, for any other minimal m-degree b there is an automorphism carrying a into b (because $\mathcal{D}_m(\leq a)$ and $\mathcal{D}_m(\leq b)$ are isomorphic), and there are $2^{\aleph_0}$ minimal m-degrees. This is not the best possible result, since there are $2^{2^{\aleph_0}}$ possible maps from $\mathcal{D}_m$ to $\mathcal{D}_m$. We now show that this bound is attained.

Proposition VI.4.6 (Shore) *There are $2^{2^{\aleph_0}}$ automorphisms of $\mathcal{D}_m$.*

Proof. Note that VI.4.2 produces an automorphism of $\mathcal{D}_m$, if both cones are $\mathcal{D}_m$ itself. Moreover, the back-and-forth argument of VI.4.1, on which the proof of VI.4.2 relies, takes $2^{\aleph_0}$ steps (we have to ensure that each m-degree is in both the domain and the range). The only new step here is to actually build a tree of height $2^{\aleph_0}$ of automorphisms of $\mathcal{D}_m$, by extending every partial automorphism in two different ways (by using VI.3.5) at successor stages, and taking unions at limit stages. Each branch of the tree is now an automorphism of $\mathcal{D}_m$, and different branches produce different automorphisms by construction. Thus there are $2^{2^{\aleph_0}}$ automorphisms.   □

The complexity of the theory of many-one degrees

We have characterized the complexity of the first-order theory of $\mathcal{D}$ in V.7.3. If we try to adapt the proof used there to $\mathcal{D}_m$ we run into trouble. The main point is that we are unable to prove the analogue of V.7.1, because its proof uses in an essential way the fact that every Turing degree contains an introreducible

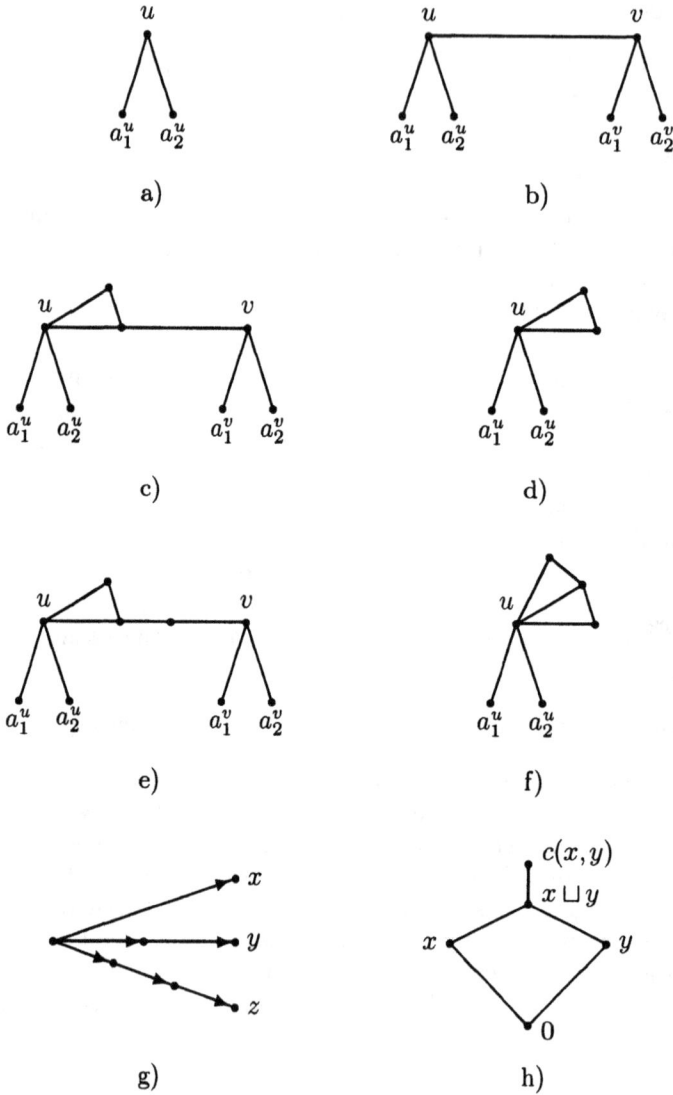

Figure VI.1: Coding by graphs

set (II.6.7), and this is false for m-degrees: a nonrecursive, not immune set is not recursive in each of its infinite subsets (since some of them are recursive), and there are m-degrees containing only nonrecursive, not immune sets (e.g. any m-degree above the m-degree of $\overline{\mathcal{K}}$, see III.6.10.b).

The next result comes to the rescue and provides a different, slightly less direct way of coding arithmetic. We state it in more generality than needed because of its interest.

Theorem VI.4.7 (Nerode and Shore [1980]) *Let* $(P, \sqsubseteq, \sqcup)$ *be an upper-semilattice with least element such that:*

1. *every countable ideal is the intersection of two principal ideals*

2. *every countable distributive lattice is isomorphic to an initial segment of* P.

Then the theory of Second-Order Arithmetic is 1-reducible to the first-order theory of P.

Proof. There are many steps toward the result:

1. *translate Second-Order Arithmetic into second-order logic on countable sets*
   This is a standard and well-known procedure, based on the fact that Peano Axioms actually define ω up to isomorphism, in second-order logic (see p. 22).

2. *translate second-order logic on countable sets into the theory of countable distributive lattices with quantification over ideals*
   This is the crucial step, which we split into two parts. We refer to the various parts of Figure 1.

   - *code relations by graphs (Lavrov [1963], Rabin and Scott)*
     We start with a binary relation R. Recall that a graph is a symmetric, irreflexive, binary relation, which we may picture as a set of points related by lines. First of all we have to put down the elements of the domain. A simple way is the following: for each element u add two points a_1^u and a_2^u, and relate them as in part a) of the picture. Thus in the graph the elements of the domain of the given relation are the points in which two lines coming from end points arrive.

     We now have to relate points u and v when $R(u, v)$ holds. A simple-minded solution as in part b) of the picture is not enough, since R might be in general not symmetric, while the proposal is. So we add two elements, and relate them as in part c) to show that $R(u, v)$, as opposed to $R(v,u)$, holds. This is still not enough since if, e.g., $R(u,u)$ holds then we would have the situation of part d). But this is ambiguous, because we might then think that $R(u,u)$ holds whenever we see a triangle with a vertex in u, while the triangle might come from the coding of $R(u,v)$ for some $v \neq u$. Our final choice is then the following: given u and v we add three new elements, and relate them as in part e), to show that $R(u,v)$ holds. If $R(u,u)$ holds then we get the unambiguous situation of part f).

This technique codes binary relations. But n-ary relations are easily reduced to binary ones, and thus to graphs as above. For example, if R is ternary we can introduce a nonsymmetric binary relation as follows. When $R(x,y,z)$ holds introduce four new elements, and relate them as in part g). An arrow $u \to v$ shows that the new binary relation holds for (u,v). The old elements are those from which no arrow comes out.

If there is more than one relation, we simply have to arrange for their domains to coincide.

- *code graphs by ideals of a distributive lattice*
  Given graphs on a countable domain we build a countable distributive lattice as follows. The atoms of the lattice correspond to the elements of the domain. Add l.u.b.'s $x \sqcup y$ for every pair of atoms x and y, and on top of $x \sqcup y$ add an element $c(x,y)$ as a code for $\{x,y\}$, see part h) of the picture (the reason why we do not simply take $c(x,y) = x \sqcup y$ is that we want the codes to be indecomposable elements, for reason to be explained shortly). Then add the necessary elements to get a distributive lattice.

A graph on the elements is simply a set of unordered pairs, and can be translated as a set of codes. There is thus a natural correspondence between ideals of the lattice and graphs on the atoms, as follows: an ideal I defines a graph R as

$$R(x,y) \iff c(x,y) \in I,$$

and a graph R defines the ideal generated by the codes $c(x,y)$, for every x and y such that $R(x,y)$ holds.

The crucial fact is that the correspondence is one-one: if R_1 and R_2 are different graphs, they generate different ideals. Indeed, the only obstacle to this could be that, given a graph R, the ideal generated by R as above also contains codes $c(x,y)$ for x and y such that $R(x,y)$ does not hold, so that decoding the ideal would not produce the original graph. That this is impossible follows from VI.1.6 and the fact that the codes are indecomposable elements.

3. *translate the theory of countable distributive lattices with quantification over ideals into the first-order theory of P*

   A formula φ with quantification over elements and ideals can be translated into a formula φ^* by replacing the ideals by exact pairs coding them, and quantification over ideals by quantification over exact pairs.

   By the initial segment assumption on P, φ is satisfiable in the theory of countable distributive lattices if and only if there is an element $a \in P$ such that the initial segment determined by a in P is a distributive lattice, and φ^* holds in it. □

Corollary VI.4.8 (Nerode and Shore [1980]) *The first-order theory of $\mathcal{D}_m$ is recursively isomorphic to the theory of Second-Order Arithmetic.*

Proof. We prove that the two theories have the same m-degree by interpreting each in the other, thus providing faithful translations that will preserve theorems. Since the translations will actually be one-one, the theories will have the same 1-degree, and hence will be recursively isomorphic by III.7.13.

One direction is clear, since every formula about the ordering of m-degrees can be interpreted, in the natural way, as a formula about sets of integers. Thus the theory of m-degrees is interpretable in Second-Order Arithmetic.

For the converse, we want to show that Second-Order Arithmetic is interpretable in $\mathcal{D}_m$. It is enough to show that $\mathcal{D}_m$ satisfies the conditions of the theorem, which it does: the required initial segments exist by VI.2.6, while the existence of exact pairs follows from VI.3.4, although it is also essentially implied by the proof of Spector's Theorem for Turing degrees (since m-degrees are closed under finite differences). □

Corollary VI.4.9 (Lachlan [1970) *The first-order theory of $\mathcal{D}_m$ is undecidable and not axiomatizable.*

For what concerns the extent of decidability, as for Turing degrees (see p. 490) we have that *the two-quantifier theory of $\mathcal{D}_m$ is decidable* (Degtev [1979]), and *the three-quantifier theory of $\mathcal{D}_m$ is undecidable* (Nies [1996]). Both proofs exploit the distributivity of the many-one degrees, and are thus quite different from the proofs of the same results for the Turing degrees given in Lerman [1983].

The next result can now be proved as in Section V.7 (using the present coding for arithmetic), and has the same consequences as there.

Theorem VI.4.10 (Nerode and Shore [1980a]) *If $\mathcal{C}$ is an ideal of $\mathcal{D}_m$ closed under jump, the first-order theory of $\mathcal{C}$ has the same degree (and actually the same isomorphism type) as the theory of Second-Order Arithmetic with set quantifiers restricted to sets with degree in $\mathcal{C}$.*

VI.5   Comparison of Degree Theories ⋆

In this section we consider other notions of degree introduced in Chapter III, namely **1-degrees**, **tt-degrees**, and **wtt-degrees**. We will mostly quote results about them, and will content ourselves to develop their theories only to the point needed to show that they are not elementarily equivalent among themselves and with Turing and m-degrees (with the only exception of tt-degrees and wtt-degrees, for which it is not known whether this holds).

1-degrees

We have already seen in Section III.7 that $\mathcal{D}_1$ is a special case among all the degree structures we have introduced, because we cannot talk of a least 1-degree in any natural way (III.7.4). The next result shows that the differences are even deeper.

Proposition VI.5.1 (Young [1964]) *$\mathcal{D}_1$ is neither an upper nor a lower semilattice, i.e. l.u.b. and g.l.b. do not always exist.*

Proof. By III.2.14 every nonrecursive r.e. T-degree contains a simple set. We will prove in X.1.7 that there are incomparable r.e. T-degrees, and thus there are two incomparable simple sets A and B. Suppose they have a l.u.b. D w.r.t. 1-reducibility:

1. *for some $z \in \overline{D}$, $A \leq_1 D \cup \{z\}$ and $B \leq_1 D \cup \{z\}$*
   Since D is an upper bound for A and B, there are recursive one-one functions f and g such that

   $$x \in A \Leftrightarrow f(x) \in D \quad \text{and} \quad x \in B \Leftrightarrow g(x) \in D.$$

We show that $\overline{D} \cap \overline{\text{range of } f} \neq \emptyset$ and $\overline{D} \cap \overline{\text{range of } g} \neq \emptyset$. Suppose, e.g., that $\overline{D} \cap \overline{\text{range of } f} = \emptyset$: we show $D \leq_1 A$ and thus $B \leq_1 D \leq_1 A$, contradicting the fact that A and B are incomparable. Simultaneously enumerate D and the range of f, and define h by induction as follows:

- if x shows up first in D, let $h(x)$ be the smallest element of A not yet in $\{h(0), \dots, h(x-1)\}$, so that $h(x) \in A$.

- if x shows up first in the range of f, let y be the unique element such that $f(y) = x$. If $y \notin \{h(0), \ldots, h(x-1)\}$ let $h(x) = y$, so that

$$x \in D \Leftrightarrow f(y) \in D \Leftrightarrow y \in A \Leftrightarrow h(x) \in A.$$

Otherwise, y must have been defined by the first clause, so let $h(x)$ be the smallest element of A not yet in $\{h(0), \ldots, h(x-1)\}$.

Thus $\overline{D} \cap \overline{\text{range of } f} \neq \emptyset$, and $\overline{D} \cap \overline{\text{range of } g} \neq \emptyset$ can be shown similarly. Let then $z \in \overline{D} \cap \text{range of } f$ and $z^* \in \overline{D} \cap \text{range of } g$. $A \leq_1 D \cup \{z\}$ via f itself (since z is not in the range of f), and $B \leq_1 D \cup \{z\}$ via g^* so defined:

$$g^*(x) = \begin{cases} g(x) & \text{if } g(x) \neq z \\ z^* & \text{otherwise.} \end{cases}$$

Note that g^* is still one-one, because z^* is not in the range of g.

2. $D \cup \{z\} <_1 D$, and hence D is not the l.u.b. of A and B
First note that $A, B \leq_1 A \oplus B$, so if D is the l.u.b. of A and B then $D \leq_1 A \oplus B$. But $A \oplus B$ is simple because so are A and B, and then so is D. Suppose $D \leq_1 D \cup \{z\}$. There is f recursive such that

$$x \in D \Leftrightarrow f(x) \in D \cup \{z\}.$$

Then (since $z \in \overline{D}$) $\overline{D}$ has an infinite r.e. subset $\{z, f(z), f^{(2)}(z), \ldots\}$, contradiction. Thus $D \cup \{z\} <_1 D$, since $D \cup \{z\} \leq_1 D$ clearly holds.

By a symmetrical argument (using $z \in D$ and $D - \{z\}$) one can show that A and B have no g.l.b.   □

Even if there is no least 1-degree, one can consider segments above $\mathbf{0}_1$. Lachlan [1969] proves that *every distributive uppersemilattice which is the direct limit of an ascending sequence of finite distributive lattices is isomorphic to a segment of* $\mathcal{D}_1$ *above* $\mathbf{0}_1$. The proof consists in forcing all the m-degrees of the initial segment of $\mathcal{D}_m$ built in VI.2.5 to contain only cylinders (see VI.6.1), so that they are actually 1-degrees.

Exercise VI.5.2 *If A is a set of minimal Turing degree constructed by using strongly uniform trees which is neither immune nor coimmune, then A has minimal 1-degree.* (Hint: as in VI.2.8, using recursive subsets of A and $\overline{A}$ to make the m-reductions one-one.)

A complete characterization of the segments of 1-degrees above $\mathbf{0}_1$ is not known, even for the finite ones, and the following results of Lachlan [1969] show that it might be complicated:

1. *every finite segment of $\mathcal{D}_1$ is a lattice* (this is not as trivial as VI.1.10, since $\mathcal{D}_1$ is not an uppersemilattice)

2. *some finite segment of $\mathcal{D}_1$ is nondistributive*

3. *not all finite lattices are isomorphic to a finite segment of $\mathcal{D}_1$.*

What is known is however enough for the analogue of Simpson's Theorem, proved by Nerode and Shore [1980]: *the first-order theory of $\mathcal{D}_1$ is recursively isomorphic to the theory of Second-Order Arithmetic.* The proof uses VI.4.7, once some problems are solved.

The first problem is that $\mathcal{D}_1$ is not an uppersemilattice, and thus VI.4.7 has to be rephrased for directed sets, in which every pair of elements has an upper bound.

The second problem is that the segments we have for $\mathcal{D}_1$ are only above 0_1. This does not introduce complications, because 0_1 *is definable in $\mathcal{D}_1$*: $\{\emptyset\}$ and $\{\omega\}$ are the only minimal 1-degrees (in the sense of being degrees with no smaller degree), and 0_1 is the smallest degree above both of them. Thus we can work only above 0_1.

The final problem is the existence of exact pairs. The same proof of Spector's Theorem V.4.3 shows that for any countable set of 1-degrees in which every pair of elements is bounded there is a pair such that every set 1-reducible to it is also 1-reducible to the disjoint union of finitely many finite modifications of representatives of the given 1-degrees. By the restriction above we only work with 1-degrees above 0_1, which are closed under finite modifications because a set A whose 1-degree is above 0_1 is neither immune nor coimmune, and thus infinite recursive subsets of A and $\overline{A}$ can be used to patch up finite modifications. Closure under disjoint union is needed only for the ideals generated by subsets of the codes of the distributive lattices used for VI.4.7, and it is proved in Nerode and Shore [1980].

Truth-table degrees and weak truth-table degrees

We will treat the two structures $\mathcal{D}_{tt}$ and $\mathcal{D}_{wtt}$ simultaneously, in the sense that *we will state our results for tt-degrees only, but note here that they all hold for wtt-degrees as well*, either by the same proofs or by minor changes that we will indicate when needed.

To get an elementary difference between $\mathcal{D}$ and $\mathcal{D}_{tt}$ we must develop some theory of the latter. We use for it the usual notation for the jump operator, which is well-defined on *tt*-degrees by V.1.6. The next result provides the analogue of the Jump Inversion Theorem V.2.24.

Theorem VI.5.3 Jump Inversion Theorem for $\mathcal{D}_{tt}$ (Mohrherr [1984])
The range of the jump operator on $\mathcal{D}_{tt}$ is the cone $\mathcal{D}_{tt}(\geq 0'_{tt})$.

Proof. By V.1.6, for any tt-degree a we have $a' \geq 0'_{tt}$. To get the converse, let C be a set such that $\mathcal{K} \leq_{tt} C$: we want to get A such that $A' \equiv_{tt} C$. Consider the construction of V.2.24, with the understanding that 'the least string $\sigma \supseteq \sigma_s$ such that $\{e\}^\sigma(e) \downarrow$' means 'the least string $\sigma \supseteq \sigma_s$ such that the search of a pair (σ, t) (in an exhaustive recursive list of them) for which $\{e\}^\sigma_t(e) \downarrow$ succeeds'. Then to be such a σ is an r.e. predicate.

1. $A' \leq_{tt} C$

   By induction on e we want to determine a truth-table which is satisfied by C if and only if $e \in A'$. Recall that $e \in A'$ is decided at stage $2e + 1$ of the construction, since

   $$ e \in A' \Leftrightarrow \{e\}^{\sigma_{2e+1}}(e) \downarrow \Leftrightarrow (\exists \sigma \supseteq \sigma_{2e})(\{e\}^\sigma(e) \downarrow), $$

   where σ_{s+1} is inductively defined as follows:

   $$ \sigma_{s+1} = \begin{cases} \sigma_s & \text{if } s = 2i \wedge i \notin A' \\ \mu\sigma(\sigma \supseteq \sigma_s \wedge \{i\}^\sigma(i) \downarrow) & \text{if } s = 2i \wedge i \in A' \\ \sigma_s * \langle C(i) \rangle & \text{if } s = 2i + 1. \end{cases} $$

   Since we only need to determine σ_{2e}, we only have to use $i < e$. We can thus fix two initial segments τ_0 and τ_1 of A' and C of length e.

   Recalling the initial observation, the following is an r.e. predicate:

   > there is a string σ_{2e} that satisfies the above inductive definition with τ_0 and τ_1 in place of A' and C, and a string $\sigma \supseteq \sigma_{2e}$ such that $\{e\}^\sigma(e) \downarrow$.

   It can thus be reduced to a question on $\mathcal{K}$ and hence, using the fact that $\mathcal{K} \leq_{tt} C$, to a tt-condition which is satisfied by C if and only if the predicate is true.

   We still have to express the fact that τ_0 and τ_1 really are, respectively, initial segments of A' and C. By induction hypothesis, for each $i < e$ we already have a tt-condition which is satisfied by C if and only if $i \in A'$. Thus there is a tt-condition that is satisfied by C if and only if τ_0 of length e is an initial segment of A'. And it is trivial to find a tt-condition satisfied by C if and only if τ_1 of length e is an initial segment of C itself.

   Thus we find a tt-condition, depending on τ_0 and τ_1 of length e, which is satisfied by C if and only if $e \in A'$. We still have to eliminate the reference to τ_0 and τ_1, which is easily done by considering all possible pairs of strings τ_0 and τ_1 of length e, and the disjunction of the tt-conditions relative to them.

2. $C \leq_{tt} A'$

By induction on e we want to determine a truth-table which is satisfied by A' if and only if $e \in C$. Recall that $e \in C$ is decided at stage $2e + 2$ of the construction, since

$$e \in C \Leftrightarrow \sigma_{2e+2}(|\sigma_{2e+1}|) = 1 \Leftrightarrow |\sigma_{2e+1}| \in A.$$

We thus have to find $|\sigma_{2e+1}|$, which can be explicitly defined as

$$|\sigma_{2e+1}| = e + \sum_{i \leq e} use\,(i, A),$$

where

$$use\,(i, A) = \left\{ \begin{array}{ll} |\mu\sigma(\sigma \supseteq \sigma_{2i} \wedge \{i\}^\sigma(i)\downarrow)| & \text{if } i \in A' \\ 0 & \text{otherwise.} \end{array} \right.$$

Indeed, at stage $2i + 1$ we see if we can make $\{i\}^A(i)$ converge, and if so we take a string that does it, otherwise we leave $\sigma_{2i+1} = \sigma_{2i}$. Thus $use\,(i, A)$ determines the length increase due to forcing the jump, while coding C always produces a one-point extension, and this accounts for the factor e in the expression for $|\sigma_{2e+1}|$.

Unraveling the definition of $|\sigma_{2e+1}|$ (and using the first e values of C, which we suppose to know by induction hypothesis) we can write down $|\sigma_{2e+1}|$, and thus a tt-condition on A' whose truth-value is equivalent to $e \in C$. The trouble is that we have used A' explicitly, while tt-reducibility allows only recursive procedures.

The first step is to consider, as above, approximations τ_0 and τ_1 of A' and C, respectively of length $e + 1$ and e. This however might cause a problem when computing $use\,(i, A)$, since we might look for a string σ such that $\{i\}^\sigma(i) \downarrow$ because the approximation to A' tells us that $i \in A'$, and we thus believe that such a string exists, while this might not be the case. But we only need to look for a string σ of length bounded by the true $use\,(i, A)$, since we may ask whether $(\exists \sigma \subseteq A)(\{i\}^\sigma(i) \downarrow)$: this is a question r.e. in A, which can be translated into a tt-condition on A'.

The fact that τ_0 and τ_1 are, respectively, initial segments of A' and C can be dealt with as above, this time using the induction hypothesis on C. And reference to τ_0 and τ_1 can also be eliminated as above.   □

The Jump Inversion Theorem actually holds for any reducibility $\leq_r$ between $\leq_{tt}$ and $\leq_T$, because if $\mathcal{K} \leq_r C$ then $\mathcal{K} \leq_{tt} \mathcal{K} \oplus C \equiv_r C$. By the theorem there is A such that $A' \equiv_{tt} \mathcal{K} \oplus C$, and hence $A' \equiv_r C$.

Kallibekov [1973] showed that $0'_{tt}$ is not a minimal cover in the r.e. tt-degrees, and a simple modification of his proof actually shows that $0'_{tt}$ is not a minimal cover in the tt-degrees. This result relativizes, and shows that no tt-degree which contains a jump is a minimal cover. By the Jump Inversion Theorem we then get the next result, which provides an elementary difference between $\mathcal{D}$ and $\mathcal{D}_{tt}$. The proof uses methods and notations typical of the study of r.e. degrees (priority, coding, and Sacks agreement strategy), and will be best understood with some knowledge of Chapter X.

Theorem VI.5.4 (Mohrherr [1984]) $\mathcal{D}_{tt}(\geq 0'_{tt})$ *is dense.*

Proof. By the Jump Inversion Theorem for $\mathcal{D}_{tt}$, it is enough to show that given sets A and C such that $A <_{tt} C'$ there is a set B such that $A <_{tt} B <_{tt} C'$. We build B by columns, as in Section V.4. First of all, we let

$$\langle 0, x \rangle \in B \Leftrightarrow x \in A.$$

This codes A into B, and produces $A \leq_{tt} B$. The other columns of B will be globally built as a set r.e. in C, hence tt-reducible to C'. Being B the join of its columns, $B \leq_{tt} C'$ follows from $A \leq_{tt} C'$.

The requirements for the construction of B are:

$$P_e \quad : \quad B \not\leq_{tt} A \text{ via } \varphi_e$$
$$N_e \quad : \quad C' \not\leq_{tt} B \text{ via } \varphi_e.$$

Fix an enumeration $\{C'_s\}_{s \in \omega}$ of C' recursive in C (since C' is r.e. in C), and an approximation $\{A_s\}_{s \in \omega}$ of A recursive in C and correct in the limit (by the Limit Lemma IV.1.17). Let

$$l_p(e, s) \quad = \quad \max\{z : (\forall y < z)(\varphi_{e,s}(y)\downarrow \wedge y \in B_s \Leftrightarrow A_s \models \sigma_{\varphi_e(y)})\}$$
$$l_n(e, s) \quad = \quad \max\{z : (\forall y < z)(\varphi_{e,s}(y)\downarrow \wedge y \in C'_s \Leftrightarrow B_s \models \sigma_{\varphi_e(y)})\}.$$

The construction is the following. At stage $s + 1$, for each $e \leq s$,

$$\langle e + 1, x \rangle \in B_{s+1} \quad \text{if} \quad x \in C'_s \wedge x < l_p(e, s)$$
$$\langle i + 1, x \rangle \in B_{s+1} \quad \text{if} \quad e < i \leq s \wedge x \leq s \wedge x \leq l_n(e, s).$$

Thus P_e takes action only in the e-th column, while N_e takes action on the $i + 1$ column, for every $i > e$.

We show that P_e and N_e are satisfied, by induction on e. Suppose P_i and N_i are satisfied for every $i < e$. Then, for $i < e$,

$$\lim_{s \to \infty} l_p(i, s) < \infty \quad \text{and} \quad \lim_{s \to \infty} l_n(i, s) < \infty.$$

Thus P_i puts only finitely many elements on the $(i + 1)$-th column for $i < e$, and so does N_i for $i \leq e$. Then:

- P_e is satisfied
  Suppose $B \leq_{tt} A$ via φ_e. Then $\lim_{s \to \infty} l_p(e, s) = \infty$ and, by construction,

$$\langle e + 1, x \rangle \in B \Leftrightarrow x \in C'$$

  except for at most finitely many elements, and $C' \leq_{tt} B \leq_{tt} A$, contradicting the hypothesis $A <_{tt} C'$. It is important to note that the strategy succeeds in this case because no negative condition, except N_i for $i < e$, can interfere with the $e + 1$ column.

- N_e is satisfied
  Suppose $C' \leq_{tt} B$ via φ_e. Then $\lim_{s \to \infty} l_n(e, s) = \infty$, and each $(i+1)$-th column with $i > e$ is contained in B. But then $B \equiv_{tt} A$, since the 0-th column codes A, the $(i + 1)$-th column for $i \leq e$ is finite, and the other columns are contained in B. Then $C' \leq_{tt} B \leq_{tt} A$, again contradicting the hypothesis $A <_{tt} C'$. □

The proof actually shows that *for any reducibility caught in between $\leq_1$ and $\leq_{wtt}$ no degree containing a jump is a minimal cover*. The crucial property is that if a computation converges at infinitely many stages then it converges. The property and the result both fail for Turing reducibility (every Turing degree has a minimal cover, and every degree above $0'$ is a jump).

For reducibilities (like $\leq_{tt}$ and $\leq_{wtt}$) for which the Jump Inversion Theorem holds it follows that no degree above $0'$ is a minimal cover. This fails for m-degrees, and we thus have a different proof of the fact that the Jump Inversion Theorem fails for m-degrees (VI.1.2.d).

The next result is a typical example of a transfer method that allows us to carry embedding results from Turing degrees to tt-degrees.

Proposition VI.5.5 (Martin) *Any hyperimmune-free minimal Turing degree is also a minimal tt-degree. In particular, there is a minimal tt-degree.*

Proof. Let $\boldsymbol{a}$ be a hyperimmune-free Turing degree, and $A \in \boldsymbol{a}$. We show that if $B \leq_T A$ then $B \leq_{tt} A$. If $B \simeq \{e\}^A$ then the function

$$f(x) = \mu s(\{e\}_s^A(x) \downarrow)$$

is recursive in A and, being $\boldsymbol{a}$ hyperimmune-free, it is majorized by a recursive function g. Then $B \leq_{tt} A$ by III.3.2, since if

$$\{e\}^C \simeq \begin{cases} \{e\}^C(x) & \text{if it converges in less than } g(x) \text{ steps} \\ 0 & \text{otherwise} \end{cases}$$

then $\{e\}^C$ is total for every C, and $\{e\}^A \simeq B$.

If also $B \in a$, by symmetry $B \equiv_{tt} A$ if $B \equiv_T A$. It follows that, if A has minimal Turing degree,

$$B \leq_T A \;\Rightarrow\; B \text{ recursive or } B \equiv_T A \;\Rightarrow\; B \text{ recursive or } B \equiv_{tt} A.$$

Thus A has minimal tt-degree as well. □

For future reference, note that the method just used is not useful below $\mathbf{0}'$, because no nonzero Turing degree comparable with $\mathbf{0}'$ is hyperimmune-free (V.5.3.d). The method however gives the stronger result that *the Turing degrees below a hyperimmune-free Turing degree are all tt-degrees* (recall that the hyperimmune-free degrees are downward closed).

The proof of VI.5.5 can be relativized, and shows that for any set A there is a set B which is a minimal cover of it with respect to tt-reductions via functions recursive in A. This falls short of being a minimal cover with respect to tt-reductions via recursive functions. In particular, it is not even true in general that $A \leq_{tt} B$, although $x \in A \Leftrightarrow B \models \sigma_{f(x)}$ for some function $f \leq_T A$.

Since every hyperimmune-free Turing degree has a hyperimmune-free minimal cover, *many tt-degrees have minimal covers*. But we cannot improve much on this, by VI.5.4.

Corollary VI.5.6 Failure of homogeneity. $\mathcal{D}_{tt}$ *and* $\mathcal{D}_{tt}(\geq \mathbf{0}'_{tt})$ *are not elementarily equivalent.*

Proof. The sentence asserting the existence of a minimal degree is true in $\mathcal{D}_{tt}$, but false in $\mathcal{D}_{tt}(\geq \mathbf{0}'_{tt})$. □

$\mathcal{D}_{tt}$ is obviously an uppersemilattice with least element $\mathbf{0}_{tt}$, and the proof of Spector's theorem V.4.3 shows that any countable ideal of $\mathcal{D}_{tt}$ is the intersection of two principal ideals (because tt-degrees are closed under finite modifications). Thus VI.4.7 allows one to prove that *the first-order theories of* $\mathcal{D}_{tt}$ *and* $\mathcal{D}_{wtt}$ *are recursively isomorphic to the theory of Second-Order Arithmetic* (Nerode and Shore [1980]), provided one has enough initial segments. The proofs of the initial segment results for Turing degrees can easily be modified to get the same results for hyperimmune-free degrees (for topped initial segments it is enough to ensure this for the top degree, since the hyperimmune-free Turing degrees are downward closed), and hence for tt-degrees, as in VI.5.5. Then *any uppersemilattice with a least element, countable predecessor property, and power at most $\aleph_1$ is isomorphic to an initial segment of* $\mathcal{D}_{tt}$ (Abraham and Shore [1986]). In particular, this holds for countable distributive lattices, and thus VI.4.7 applies.

Other global results about $\mathcal{D}_{tt}$ and $\mathcal{D}_{wtt}$ have been obtained by Nerode and Shore [1980a] and Mohrherr [1984], although the full analogues of the results on absolute definability and automorphisms proved for Turing degrees in Section V.7 are not known to hold.

Elementary inequivalences

What we have proved until now allows us to state the main result of this section, on the comparison of degree theories.

Theorem VI.5.7 (Young [1964], Lachlan [1970], Shore [1982a]) *The theories of $\mathcal{D}_1$, $\mathcal{D}_m$, $\mathcal{D}_{tt}$, $\mathcal{D}_{wtt}$, and $\mathcal{D}$ are pairwise not elementarily equivalent, with the only possible exception of $\mathcal{D}_{tt}$ and $\mathcal{D}_{wtt}$.*

Proof. $\mathcal{D}_1$ differs from all the remaining structures because it is not an uppersemilattice, by VI.5.1, while all the other ones are.

$\mathcal{D}_m$ differs from all the remaining structures because every m-degree has a strong minimal cover, by V.5.16, while this fails in the other cases, by VI.5.4 and V.2.26 (the latter implies that no Turing degree above $0'$ has a strong minimal cover).

$\mathcal{D}_{tt}$ and $\mathcal{D}_{wtt}$ differ from $\mathcal{D}$ because not every tt-degree or wtt-degree has a minimal cover, by VI.5.4, while every Turing degree does, by relativization of V.5.11. $\square$

We do not know whether $\mathcal{D}_{tt}$ and $\mathcal{D}_{wtt}$ are elementarily equivalent. In any case, the following result shows that their relationship is a special one.

Proposition VI.5.8 (Shore [1982a]) $\mathcal{D}_{tt}$ *and* $\mathcal{D}_{wtt}$ *have isomorphic cones. Precisely,* $\mathcal{D}_{tt}(\geq 0'_{tt})$ *and* $\mathcal{D}_{wtt}(\geq 0'_{wtt})$ *are isomorphic.*

Proof. Consider the natural map $\Phi : \mathcal{D}_{tt} \to \mathcal{D}_{wtt}$ defined as follows, for any tt-degree a:

$$\Phi(a) = \text{the } wtt\text{-degree of any } A \in a.$$

In particular, $\Phi(0'_{tt}) = 0'_{wtt}$.

Since tt-reducibility is stronger than wtt-reducibility, we have

$$a \leq b \text{ in } \mathcal{D}_{tt} \ \Rightarrow \ \Phi(a) \leq \Phi(a) \text{ in } \mathcal{D}_{wtt}.$$

In particular, $0'_{tt} \leq b \Rightarrow 0'_{wtt} \leq \Phi(b)$, and hence Φ is a homomorphism from $\mathcal{D}_{tt}(\geq 0'_{tt})$ to $\mathcal{D}_{wtt}(\geq 0'_{wtt})$.

To show that Φ is onto let $\Phi(a) \leq b$, and choose A in the tt-degree a, and B in the wtt-degree b. Then $A \leq_{wtt} B$ and $A \leq_{tt} A \oplus B \equiv_{wtt} B$, and thus b is the image of the tt-degree of $A \oplus B$ under Φ.

To show that Φ is one-one we prove that if b is a tt-degree above $0'_{tt}$, then

$$\Phi(a) \leq \Phi(a) \text{ in } \mathcal{D}_{wtt} \ \Rightarrow \ a \leq b \text{ in } \mathcal{D}_{tt}.$$

It is enough to show that if $\mathcal{K} \leq_{tt} B$ and $A \leq_{wtt} B$ then $A \leq_{tt} B$. Let $A \simeq \varphi_e^B$, with recursive bound f. Given x there are $2^{f(x)+1}$ sets $X \subseteq \{0, \ldots, f(x)\}$, each of them recursive. Then recursively in $\mathcal{K}$ we may know if $\varphi_e^X(x) \downarrow$. Since $\mathcal{K} \leq_{tt} B$, we can then build a truth-table reduction of A to B.   □

Note that *no other pair of structures among* $\mathcal{D}_m$, $\mathcal{D}_{tt}$, $\mathcal{D}_{wtt}$, *and* $\mathcal{D}$ *admits isomorphic cones*, because the properties used in VI.5.7 to show elementary inequivalence actually hold on a cone. Thus, even if $\mathcal{D}_{tt}$ and $\mathcal{D}_{wtt}$ were elementarily inequivalent, they would resemble each other more than any other of these pairs.

Bulitko [1980] and Selivanov [1982] have shown that the only possible truth-table-like reducibilities are: $\leq_m$, $\leq_{btt(1)}$, $\leq_l$, $\leq_d$, $\leq_c$, $\leq_p$, and $\leq_{tt}$ (see pp. 268 and 331 for definitions). Among them, the structures induced by $\leq_c$ and $\leq_d$ are isomorphic, since

$$A \leq_c B \iff \overline{A} \leq_d \overline{B},$$

and so are the structures induced $\leq_m$ and $\leq_{btt(1)}$ (Malc'ev [1985]). The remaining ones are pairwise elementarily inequivalent (Degtev [1979], [1985]), with the only possible exception of the structures induced by $\leq_p$ and $\leq_{tt}$.

VI.6   Structure Inside Degrees ⋆

In the last section we have compared various notions of degree from the point of view of the structures they induce. Despite the elementary differences proved in VI.5.7, we have noted a resemblance of methods of proofs in the study of 1-degrees and m-degrees on one side, and tt-degrees, wtt-degrees, and Turing degrees on the other.

In the present section we take a different perspective, and analyze the possible structure of degrees of one type inside degrees of another. We will again discover that 1-degrees and m-degrees are close, in the sense that they coincide in a large number of cases, and the same will hold for tt-degrees and Turing degrees (and hence for wtt-degrees). On the other hand, m-degrees and tt-degrees never coincide, and this shows where the real demarcation among degree notions lies.

Cylinders

Recall that the jump operator provides a homomorphism from $\mathcal{D}$ to $\mathcal{D}_1$, since

$$A \leq_T B \iff A' \leq_1 B'.$$

We now define canonical homomorphisms from $\mathcal{D}_m$ and $\mathcal{D}_{tt}$ to $\mathcal{D}_1$.

Definition VI.6.1 (Myhill [1959], Rogers [1967]) *The cylindrification of a set A is the set*

$$A \cdot N = \{\langle x, n \rangle : x \in A\}.$$

A is a **cylinder** *if $A \equiv A \cdot N$, i.e. A and $A \cdot N$ are recursively isomorphic.*

Proposition VI.6.2 (Rogers [1967]) $A \equiv_m A \cdot N$. *Moreover,*

$$A \leq_m B \quad \Leftrightarrow \quad A \cdot N \leq_1 B \cdot N.$$

Proof. $A \leq_1 A \cdot N$ via $f(x) = \langle x, x \rangle$, and $A \cdot N \leq_m A$ via $g(x) = (x)_1$. Thus $A \equiv_m A \cdot N$.

Let $A \leq_m B$: since $A \cdot N \leq_m A$, then $A \cdot N \leq_m B$ via some recursive function f, and $A \cdot N \leq_1 B \cdot N$ via $g(x) = \langle f(x), x \rangle$, which is one-one. Conversely, let $A \cdot N \leq_1 B \cdot N$: since $A \leq_m A \cdot N$ and $B \cdot N \leq_m B$, then $A \leq_m B$. □

We now give some conditions for a set to be a cylinder.

Proposition VI.6.3 (Young [1966a], Rogers [1967]) *The following conditions are equivalent:*

1. *A is a cylinder*

2. *for every set B, $B \leq_m A \Rightarrow B \leq_1 A$*

3. *there is a recursive function f such that, for every x,*

   - $\mathcal{W}_{f(x)}$ *is infinite*
   - $x \in A \Rightarrow \mathcal{W}_{f(x)} \subseteq A$
   - $x \in \overline{A} \Rightarrow \mathcal{W}_{f(x)} \subseteq \overline{A}$.

Proof. 1 implies 3 because if $A \equiv A \cdot N$ then there is a one-one recursive function g reducing $A \cdot N$ to A, and thus it is enough to let

$$\mathcal{W}_{f(x)} = \{g(\langle x, n \rangle) : n \in \omega\}.$$

3 implies 2 because we can use $\mathcal{W}_{f(x)}$ to turn a many-one reduction g of B to A into a one-one reduction, as follows. Let $h(0) = g(0)$. Given $h(x)$, let $h(x+1)$ be the first element generated in $\mathcal{W}_{f(g(x+1))}$ and not in $\{h(0), \ldots, h(x)\}$.

2 implies 1 because $A \leq_1 A \cdot N$ and $A \cdot N \leq_m A$ always hold, and by 2 the latter implies $A \cdot N \leq_1 A$. Thus $A \equiv_1 A \cdot N$, and $A \equiv A \cdot N$ by III.7.13. □

Exercises VI.6.4 a) *A set A is a cylinder if and only if, for some recursive function g and every $D_x \neq \emptyset$,*

$$D_x \subseteq A \Rightarrow g(x) \in A - D_x \quad \text{and} \quad D_x \subseteq \overline{A} \Rightarrow g(x) \in \overline{A} - D_x.$$

(Rogers [1967]) (Hint: the condition is equivalent to VI.6.3.)

b) *Logical theories are cylinders.* (Hint: show that condition VI.6.3.3 is satisfied, using the fact that φ and $\neg\neg\varphi$ are equivalent.)

c) *Every cylinder is a splinter.* (Myhill [1959]) (Hint: see the proof of III.7.10.c.)

d) *A cylinder is either recursive or pseudocreative.* (Hint: by c) and III.7.10.a.)

e) *The only recursive sets which are cylinders are* $\emptyset$, ω, *and the infinite coinfinite sets.* (Hint: by VI.6.3.3.) It follows that not every splinter is a cylinder. Young [1966] has shown that there are infinite coinfinite splinters which are not cylinders.

f) *Every creative set is a cylinder.* (Myhill [1959])

g) *Not every pseudocreative set is a cylinder.* (Young [1964a]) (Hint: let B be a simple set. Then $B \cdot B$ is pseudocreative, since if $x \in \overline{B}$ then $\{x\} \cdot B$ is an infinite r.e. subset of $\overline{B \cdot B}$. Choose B such that $B \cdot B \not\leq_m B$, see III.8.5. Then $B \cdot B$ is not a cylinder, otherwise there is f like in VI.6.3.3, and since $\mathcal{W}_{f(x)}$ is infinite and B is simple there is $\langle a, b \rangle \in \mathcal{W}_{f(x)}$ such that one of a and b is in B. Let $g(x)$ be the other. Then $x \in B \cdot B \Leftrightarrow g(x) \in B$, contradiction.)

We now turn to the analogue of the notion of cylinder for tt-reducibility.

Definition VI.6.5 (Rogers [1967]) *The tt-cylindrification of a set A is the set*

$$A^{tt} = \{x : A \models \sigma_x\}.$$

A is a tt-cylinder if $A \equiv A^{tt}$.

Proposition VI.6.6 (Rogers [1967]) $A \equiv_{tt} A^{tt}$. *Moreover,*

$$A \leq_{tt} B \quad \Leftrightarrow \quad A^{tt} \leq_1 B^{tt}.$$

Proof. $A \leq_1 A^{tt}$ via f such that $\sigma_{f(x)}$ is the tt-condition '$x \in X$', and $A^{tt} \leq_{tt} A$ via the identity function, since $x \in A^{tt} \Leftrightarrow A \models \sigma_x$ by definition.

Let $A \leq_{tt} B$: since $A^{tt} \leq_{tt} A$ then $A^{tt} \leq_{tt} B$, and for some recursive f

$$x \in A^{tt} \Leftrightarrow B \models \sigma_{f(x)} \Leftrightarrow f(x) \in B^{tt}.$$

Thus $A^{tt} \leq_m B^{tt}$. But f can be made one-one by induction, by substituting tt-conditions with equivalent ones if necessary (obtained by adding redundant clauses, like $n \in X \vee n \notin X$). Thus $A^{tt} \leq_1 B^{tt}$. Conversely, if $A^{tt} \leq_1 B^{tt}$ then $A \leq_{tt} B$, since $A \leq_{tt} A^{tt}$ and $B^{tt} \leq_{tt} B$. $\square$

Proposition VI.6.7 (Rogers [1967]) *The following conditions are equivalent:*

1. *A is a tt-cylinder*

2. *for every set B, $B \leq_{tt} A \Rightarrow B \leq_1 A$*

Proof. 1 implies 2 because if $B \leq_{tt} A$ then $B \leq_m A^{tt}$ by definition, and hence $B \leq_1 A^{tt}$ as in the proof of VI.6.6. If A is a tt-cylinder then $A^{tt} \leq_1 A$, and hence $B \leq_1 A$.

2 implies 1 because $A \leq_1 A^{tt}$ and $A^{tt} \leq_{tt} A$ always hold, and by 2 the latter implies $A^{tt} \leq_1 A$. Thus $A \equiv_1 A^{tt}$, and $A \equiv A^{tt}$ by III.7.13.   □

In particular, *a tt-cylinder is a cylinder*.

Inside many-one degrees

Since $\leq_1$ is stronger than $\leq_m$, an m-degree can be thought of as consisting of 1-degrees.

Definition VI.6.8 *An m-degree is* **irreducible** *if it consists of only one 1-degree.*

Obviously, *an m-degree is irreducible if and only if it contains only cylinders*, since $A \equiv_m A \cdot N$ always holds.

Trivial examples of irreducible m-degrees are $\{\emptyset\}$ and $\{\omega\}$. The first example of a nontrivial irreducible m-degree was given by Myhill [1955], who showed that the m-degree of $\mathcal{K}$ is such a degree (see III.6.6 and III.7.5). By relativization, *the m-degree of a jump set A' is irreducible*, and thus every Turing degree above $0'$ contains an irreducible m-degree.

Proposition VI.6.9 (Kobzev [1975]) *If A is r.e. and nonrecursive, the m-degree of A^{tt} is irreducible.*

Proof. Consider $B \equiv_m A^{tt}$. Since A^{tt} is a cylinder, from $B \leq_m A^{tt}$ we have $B \leq_1 A^{tt}$. For the converse, let $A^{tt} \leq_m B$ via f: we want to show that we have infinitely many equivalent choices for each value of f, so that f can be turned into a one-one reduction, and $A^{tt} \leq_1 B$. Consider

$$\sigma_z = (\sigma_x \wedge \neg\sigma_y) \vee (\neg\sigma_x \wedge \sigma_y),$$

for a given y:

- if $x \in A^{tt}$ then $A \models \sigma_x$, and thus

$$z \in A^{tt} \Leftrightarrow A \models \sigma_z \Leftrightarrow A \models \neg\sigma_y \Leftrightarrow y \notin A^{tt}$$

- if $x \notin A^{tt}$ then $A \models \neg\sigma_x$, and thus

$$z \in A^{tt} \Leftrightarrow A \models \sigma_z \Leftrightarrow A \models \sigma_y \Leftrightarrow y \in A^{tt}.$$

These are still only equivalences, but if we choose $y \notin A^{tt}$ we explicitly have:

- if $x \in A^{tt}$ then $z \in A^{tt}$ and $f(z) \in B$
- if $x \notin A^{tt}$ then $z \notin A^{tt}$ and $f(z) \notin B$.

Fix then an r.e. subset C of $\overline{A^{tt}}$. There is a recursive function g such that

$$f(z) \in W_{g(x)} \Leftrightarrow (\exists y \in C)[\sigma_z = (\sigma_x \wedge \neg \sigma_y) \vee (\neg \sigma_x \wedge \sigma_y)].$$

Moreover, $W_{g(x)}$ is an r.e. subset of B if $x \in A^{tt}$, and of $\overline{B}$ otherwise.

It remains to choose C in such a way that $W_{g(x)}$ is always infinite. Note that it is enough to have C and A^{tt} recursively inseparable, because then if $W_{g(x)}$ were finite the set

$$y \in R \Leftrightarrow f(z) \in W_{g(x)}$$

(with σ_z as above) would be a recursive set separating C and A^{tt}. Indeed, $C \subseteq R$ by definition. Moreover, once x is fixed the only thing that matters for z, and hence for $f(z)$, is whether y is in A^{tt} or not. Since for $y \in C$ we have $y \in R$, and $C \subseteq \overline{A^{tt}}$, it cannot be that $y \in A^{tt} \cap R$, and hence $R \subseteq \overline{A^{tt}}$.

To find C as wanted, note that $A \leq_1 A^{tt}$ and $A^{tt} \leq_1 \overline{A^{tt}}$ (the latter because $\overline{A^{tt}} \leq_{tt} A^{tt}$, and A^{tt} is a tt-cylinder). Thus $A \leq_1 \overline{A^{tt}}$. Let C be the image of A under this 1-reduction: it is an r.e. set because A is, and it cannot be separable from $\overline{A^{tt}}$ by a recursive set, otherwise A would be the inverse image of this set, and it would then be recursive. □

Corollary VI.6.10 *Every nonrecursive r.e. tt-degree contains an irreducible m-degree.*

Proof. $A \equiv_{tt} A^{tt}$. □

While not every nonrecursive tt-degree contains an irreducible m-degree (Degtev [1979]), every nonrecursive Turing degree contains such an m-degree (Downey [1993]). We only prove a weaker result, namely that *almost* every nonrecursive Turing degree contains an irreducible m-degree.

Proposition VI.6.11 (Degtev [1979]) *Every Turing degree not below $\mathbf{0}'$ contains an irreducible m-degree.*

Proof. We modify the proof of the previous result to show that if A is non-recursive and the m-degree of A^{tt} is not irreducible, then $A \in \Delta_2^0$. The result follows from the Limit Lemma IV.1.17, and the fact that $A \equiv_T A^{tt}$.

Since $A^{tt} \leq_1 \overline{A^{tt}}$ (because $\overline{A^{tt}} \leq_{tt} A^{tt}$, and A^{tt} is a tt-cylinder), it is enough to show that $A^{tt} \in \Sigma_2^0$. Since A^{tt} is a cylinder, if its m-degree is not irreducible then it must contain a set B which is not a cylinder, and thus such that $A^{tt} \leq_m B$ but $A^{tt} \not\leq_1 B$ (by VI.6.3.2). Let f be a recursive function such that $x \in A^{tt} \Leftrightarrow f(x) \in B$. Given n, let

$$f(z) \in W_{g(n,x)} \quad \Leftrightarrow \quad (\exists y)[\sigma_z = \sigma_x \wedge (\ \sigma_n \vee \sigma_y)] \vee$$
$$(\exists y)[\sigma_z = \sigma_x \vee (\neg \sigma_n \wedge \sigma_y)].$$

We show that

$$n \in A^{tt} \quad \Leftrightarrow \quad (\exists x)(W_{g(n,x)} \text{ finite}).$$

Then $A^{tt} \in \Sigma_2^0$, because finiteness of an r.e. set is a Σ_2^0 condition, since it can be expressed as saying that there is a number such that all greater ones are not in the set.

1. If $n \in A^{tt}$ then $A \models \sigma_n$, and hence

$$A \models \sigma_x \wedge (\sigma_n \vee \sigma_y) \quad \Leftrightarrow \quad A \models \sigma_x \vee (\neg \sigma_n \wedge \sigma_y) \quad \Leftrightarrow \quad A \models \sigma_x.$$

If $W_{g(n,x)}$ is infinite for every x then we can easily get $A^{tt} \leq_1 B$, contradicting the hypothesis.

2. If $n \notin A^{tt}$ then $A \models \neg \sigma_n$, and hence

$$x \in A^{tt} \quad \Rightarrow \quad A \models \sigma_x \wedge (\ \sigma_n \vee \sigma_y) \text{ iff } A \models \sigma_y$$
$$\Rightarrow \quad \{f(z) : \sigma_z = \sigma_x \wedge (\ \sigma_n \vee \sigma_y)\} \text{ is infinite}$$
$$x \notin A^{tt} \quad \Rightarrow \quad A \models \sigma_x \vee (\neg \sigma_n \wedge \sigma_y) \text{ iff } A \models \sigma_y$$
$$\Rightarrow \quad \{f(z) : \sigma_z = \sigma_x \vee (\neg \sigma_n \wedge \sigma_y)\} \text{ is infinite},$$

otherwise A is recursive (because $A \leq_m A^{tt} \leq_m B$). $\square$

Exercises VI.6.12 A set A is **perfect** if it is η-closed, for some nontrivial r.e. equivalence relation η whose only recursive η-closed sets are $\emptyset$ and ω. Note that every equivalence class of η is infinite and r.e.

a) *If A is perfect then its m-degree is irreducible.* (Ershov [1971]) (Hint: if $B \leq_m A$ via f, define g one-one reducing B to A by letting $f(x)$ and $g(x)$ be in the same equivalence class: this works because A is η-closed. If $A \leq_m B$ via f, consider the new equivalence $x\eta^*y \Leftrightarrow x\eta y \vee f(x) = f(y)$. A is still closed and perfect w.r.t. η^*. f m-reduces $[x]_{\eta^*}$ to $f([x]_{\eta^*})$, and thus the latter is still infinite. Define g one-one reducing B to A by letting $g(x) \in f([x]_{\eta^*})$.)

b) *Not every irreducible m-degree contains a perfect set.* (Denisov [1974]) (Hint: modify III.6.23.c to prove that if $B \leq_{tt} A$ and B is perfect then A is not hypersimple. Then use VI.6.10.)

We have now examples of irreducible m-degrees, as well as of m-degrees containing infinitely many 1-degrees (like $\mathbf{0}_m$). These are the only possible cases.

Proposition VI.6.13 (Young [1966a]) *An m-degree contains either only one or infinitely many 1-degrees.*

Proof. If there is more than one 1-degree in a given m-degree, there is a set A which is not a cylinder. Note that

$$A \oplus A \quad (A \oplus A) \oplus (A \oplus A) \quad \cdots$$

are all in the same m-degree, since $\oplus$ induces the l.u.b. for m-degrees. Since $B \leq_1 B \oplus B$ always holds, the two following facts produce, by induction, an infinite ascending chain of 1-degrees in the given m-degree:

1. $B \oplus B \leq_1 B \Rightarrow B \oplus B$ cylinder
   Given f one-one such that

   $$z \in B \oplus B \Leftrightarrow f(z) \in B \Leftrightarrow 2f(z) \in B \oplus B,$$

   the set

   $$\mathcal{W}_{g(z)} = \{z, 2f(z), 2f(2f(z)), \ldots\}$$

   satisfies condition VI.6.3.3.

2. $B \oplus B$ cylinder $\Rightarrow B$ cylinder
   If g satisfies VI.6.3.3 then

   $$x \in B \Rightarrow 2x \in B \oplus B \Rightarrow \mathcal{W}_{g(2x)} \subseteq B \oplus B$$
   $$x \notin B \Rightarrow 2x \notin B \oplus B \Rightarrow \mathcal{W}_{g(2x)} \subseteq \overline{B \oplus B}.$$

   It is thus enough to let

   $$\mathcal{W}_{h(x)} = \{z : 2z \in \mathcal{W}_{g(2x)} \vee 2z + 1 \in \mathcal{W}_{g(2x)}\}. \quad \Box$$

The proof shows that if an m-degree contains infinitely many 1-degrees then it contains an infinite chain. Young [1966a] shows that actually every countable linear ordering is embeddable in the 1-degrees of such an m-degree. It is not known whether there must always be an infinite antichain as well.

Exercises VI.6.14 a) *Every m-degree contains a greatest 1-degree (consisting exactly of the cylinders in the given m- degree).* (Rogers [1967]) (Hint: given A consider $A \cdot N$.)

b) *There are m-degrees without least 1-degree.* (Dekker and Myhill [1960]) (Hint: if A is simple and $z \in \overline{A}$ then $A - \{z\} \equiv_m A$ but $A - \{z\} <_1 A$, by the proof of VI.5.1.)

Inside truth-table degrees

We now look at m-degrees inside tt-degrees.

Proposition VI.6.15 (Jockusch [1969]) *Every nonrecursive tt-degree contains infinitely many m-degrees.*

Proof. Let C be a nonrecursive set. Consider the tree of binary sequence numbers, and the branch A defined by C:

$$x \in A \iff Seq(x) \wedge (\forall n)_{1 \leq n \leq ln(x)}[(x)_n = C(n-1)].$$

By definition $A \equiv_{tt} C$ (since there are only finitely many binary sequence numbers of a given length). Moreover, A is retraceable via

$$f((\langle x_1, \ldots, x_n, x_{n+1}\rangle)) = \langle x_1, \ldots, x_n \rangle.$$

Being nonrecursive and retraceable, A is immune (by II.6.5). It is also not hyperimmune, because the strong array whose elements are the sets of binary sequence numbers of a given length intersects A.

For $n \geq 1$, let $(\overline{A})^n$ be the recursive product of $\overline{A}$ with itself n times:

$$\langle x_1, \ldots, x_n \rangle \in (\overline{A})^n \iff (\forall i)_{1 \leq i \leq n}(x_i \in \overline{A}).$$

Clearly, $(\overline{A})^n \equiv_{tt} A$, and $(\overline{A})^n \leq_m (\overline{A})^{n+1}$. We now prove $(\overline{A})^{n+1} \not\leq_m (\overline{A})^n$, so that the tt-degree of A (and hence that of B) contains infinitely many m-degrees.

The plan of the proof is the following. Suppose $(\overline{A})^{n+1} \leq_m (\overline{A})^n$. We want to find an r.e. subset B of $\overline{A}$ so big that at most $2n$ elements at any level of the tree are in $\overline{B}$. We then get a contradiction as follows. From the fact that A is not hyperimmune we have a recursive function g such that $g(x)$ majorizes the x-th element (i.e. the one of level x) a_x of A. Then we can define S_x such that $|S_x| \leq 2n$ as follows: take all elements less than $g(x)$ which are on level x of the tree, and eliminate those that are generated in B, until at most $2n$ remain. Clearly, a_x is not eliminated, because $B \subseteq \overline{A}$. We thus get a strong array of bounded cardinality intersecting A, contradicting the fact that A is immune (II.6.10.b).

It remains to find B. Note that

$$(\overline{A})^n \equiv_m \overline{\{y : |D_y| \leq n \wedge D_y \cap A \neq \emptyset\}},$$

so that the hypothesis $(\overline{A})^{n+1} \leq_m (\overline{A})^n$ can be reformulated as:

$$\{y : |D_y| \leq n+1 \wedge D_y \cap A \neq \emptyset\} \leq_m \{y : |D_y| \leq n \wedge D_y \cap A \neq \emptyset\}.$$

Thus there is a recursive function h such that

$$|D_y| \le n+1 \;\Rightarrow\; |D_{h(y)}| \le n \wedge (D_y \cap A = \emptyset \leftrightarrow D_{h(y)} \cap A = \emptyset).$$

Let

$$x \in B_1 \;\Leftrightarrow\; (\exists y)(|D_y| \le n+1 \wedge x \in D_y \wedge x|D_{h(y)})$$
$$x \in B_2 \;\Leftrightarrow\; (\exists y)(|D_y| \le n+1 \wedge x|D_y \wedge x \mapsto D_{h(y)}),$$

where $x|D_z$ means that x and any element of D_z are not sent by iterations of f on a same element, and $x \mapsto D_z$ that x is sent by iterations of f on an element of D_z.

Clearly, $B_1 \cup B_2$ is r.e. Moreover:

1. $B \subseteq \overline{A}$

   Suppose $x \in B_1 \cap A$. For some y, $x \in D_y \cap A$ and $x|D_{h(y)}$, so that $D_y \cap A \ne \emptyset$ and hence $D_{h(y)} \cap A \ne \emptyset$. But the latter contradicts $x|D_{h(y)}$, because the elements of A are all sent by iteration of f on a same element (e.g., a_0).

   Suppose $x \in B_2 \cap A$. For some y, $x|D_y$ and $x \mapsto D_{h(y)}$, so that $D_y \cap A = \emptyset$ (because $x \in A$ and $x|D_y$) and hence $D_{h(y)} \cap A = \emptyset$. But the latter contradicts $x \mapsto D_{h(y)}$, because $x \in A$ and it is thus sent only on elements of A by iterations of f.

2. $\overline{B}$ has at most $2n$ elements at any level of the tree

   For the sake of contradiction, suppose there is $\{x_1, \dots, x_{2n+1}\} \subseteq \overline{B}$, whose elements we may suppose to be ordered from the left on the tree. Consider

   $$D_y = \{x_1, x_3, x_5, \dots, x_{2n+1}\},$$

   which has $n + 1$ elements. By hypothesis $D_y \subseteq \overline{B}$, and in particular $D_y \subseteq \overline{B_1}$. Since $|D_y| \le n + 1$ and $x_{2i+1} \in D_y$, then it cannot be $x_{2i+1}|D_{h(y)}$. Since $|D_{h(y)}| \le n$ and $|D_y| = n + 1$, there must be $i < j$ such that x_{2i+1} and x_{2j+1} are sent on a same element of $D_{h(y)}$. Since x_{2i+1} and x_{2j+1} are on the same level of the tree, also x_{2i+2} (which is between them and on the same level) must be sent on the same element, by the definition of f. But this is impossible, since then $x_{2i+2}|D_y$ but $x_{2i+2} \mapsto D_{h(y)}$, and hence $x_{2i+2} \in B_2$, contradicting the hypothesis that $x_{2i+2} \in \overline{B}$. □

The proof shows that every nonrecursive tt-degree contains an infinite chain of m-degrees, and Stephan [199?] has proved the same for infinite antichains.

Exercises VI.6.16 a) *Every tt-degree contains a greatest m-degree and a greatest 1-degree.* (Rogers [1967]) (Hint: consider A^{tt}.)

b) *No nonrecursive tt-degree contains a least m-degree.* (Jockusch) (Hint: let A be an immune retraceable set in the given tt-degree, by II.6.13. Then A and $\overline{A}$ form a minimal pair of m-degrees. Indeed, let $x \in C \Leftrightarrow f(x) \in A \Leftrightarrow g(x) \in \overline{A}$. If $f(x)$ and $g(x)$ are sent by the retracing function into the same element, this is in A. The set of such elements is finite, being r.e. Let $a \in A$ be greater than its maximum. If C is not recursive the ranges of f and g are unbounded, and $A - \{0, \ldots, a-1\}$ is r.e., contradicting immunity: if $z \geq a$ choose x such that $f(x), g(x) > z$. See if one is sent into z by the retracing function. At most one of them can be, by the choice of a, and if $z \in A$ one does.)

c) *Every nonrecursive tt-degree contains incomparable m-degrees.* (Jockusch [1968a]) (Hint: let A be a semirecursive set in the given tt-degree, by III.5.5. Then A and $\overline{A}$ are m-incomparable.)

Inside Turing degrees

We now look at tt-degrees inside Turing degrees.

Definition VI.6.17 *A Turing degree is **irreducible** if it consists of only one tt-degree.*

The irreducible Turing degrees are old friends.

Proposition VI.6.18 (Jockusch [1969], Martin) *A Turing degree is irreducible if and only if it is hyperimmune-free.*

Proof. We have already proved in VI.5.5 that if A has hyperimmune-free degree and $B \equiv_T A$ then $B \equiv_{tt} A$. Thus a hyperimmune-free degree contains only one tt-degree.

Suppose now that the Turing degree of A contains only one tt-degree. We may suppose that A has the greatest m-degree in it (otherwise we can consider A^{tt}). Suppose $f \leq_T A$ is not recursively majorized. With no loss of generality we may suppose f increasing. Let

$$e \in B \Leftrightarrow \{e\}(e) \text{ converges in less than } f(e) \text{ steps, and } \{e\}(e) \notin A.$$

Then $B \leq_T A$, and $A \equiv_T A \oplus B$. We now show that $B \not\leq_m A$: thus $A <_m A \oplus B$, contradicting the fact that A has the greatest m-degree.

Suppose $B \leq_m A$ via g. Let $e_0, e_1, \ldots$ be indices of g, with $e_x > x$. Then

$$e_x \in B \Leftrightarrow g(e_x) \in A \Leftrightarrow \{e_x\}(e_x) \in A.$$

Thus $\{e_x\}(e_x)$ must converge in more than $f(e_x) > f(x)$ steps (recall that f is increasing), and the number of steps needed to compute $\{e_x\}(e_x)$ is thus a recursive function majorizing f, contradiction. □

As in the case of m-degrees, there are only two possibilities.

Corollary VI.6.19 *A Turing degree contains either only one or infinitely many tt-degrees.*

Proof. The proof given above works also in the case that a Turing degree has a greatest tt-degree, by choosing A in the greatest m-degree of the greatest tt-degree. Since $A{\oplus}B \equiv_T A$, $A{\oplus}B \leq_{tt} A$ because A is in the greatest tt-degree, and hence actually $A \equiv_{tt} A \oplus B$. Thus $A <_m A \oplus B$ produces a contradiction, because A has greatest m-degree in its tt-degree.

But if there are finitely many tt-degrees, there is a greatest one (their l.u.b.). Thus the Turing degrees consisting of finitely many tt-degrees are still the hyperimmune-free ones, i.e. the irreducible ones. □

Exercises VI.6.20 a) *A Turing degree has a greatest tt-degree if and only if it is irreducible.* (Hint: by the proof of the corollary.)

b) *Non-irreducible Turing degrees contain an infinite chain of tt-degrees.* (Hint: by part a.)

c) *Non-irreducible Turing degrees contain an infinite antichain of tt-degrees.* (Degtev [1972]) (Hint: this uses the priority method. Let $\overline{A} = \{a_0 < a_1 < \cdots\}$, A hyperimmune. We build A_z in the Turing degree of A so that, for $m \neq n$, $A_m \not\leq_{tt} A_n$. Code A into A_z by letting $x \in A \Leftrightarrow 2x+1 \in A_z$: thus $A \leq_T A_z$. To ensure $A_m \not\leq_{tt} A_n$ via φ_e, we want some x such that $x \in A_m \Leftrightarrow A_n \not\models \sigma_{\varphi_e(x)}$. Wait until a fresh witness x appears, such that $\varphi_e(x)$ converges in less than a_x steps. Then see if $A_{n,s} \models \sigma_{\varphi_e(x)}$. If so, restrain x out of A_m. Otherwise, put x into A_m. In both cases, restrain out of A_n the elements used in the computation and not yet in it. Note that if φ_e is total there are infinitely many x such that $\varphi_e(x)$ converges in less than a_x steps, otherwise $\overline{A}$ would be majorized by the least number of steps needed to compute $\varphi_e(x)$. The construction is recursive in A, and thus $A_z \equiv_T A$.)

Bibliography

We include only references quoted in the book. A complete bibliography of Recursion Theory has been edited by Hinman as Volume IV of the Ω-*Bibliography of Mathematical Logic*, Springer Verlag, 1987. On-line bibliographies can be found at the following Web addresses:

- http://www-logic.uni-kl.de/bibl/index.html
- http://www.nd.edu:80/~cholak/computability/bib/bib.html

We indicate publications by abbreviating their original names. The following is a list of full names of Russian publications and of their official translations.

1. *Algebra i Logika* (Algebra and Logic)
2. *Doklady Akademii Nauk S.S.S.R.* (Soviet Mathematics, Dokladi)
3. *Izvestiya Vysshikh Uchebnykh Zavedenij Matematika* (Soviet Mathematics, Izvestiya)
4. *Kibernetika* (Cybernetics)
5. *Matematicheskii Sbornik* (Mathematics of the U.S.S.R.)
6. *Matematicheskie Zametki* (Mathematical Notes of the Academy of Science U.S.S.R.)
7. *Problemy Kibernetiki* (Problems of Cybernetics)
8. *Problemi Peredachi Informatsii* (Problems of Information Transmission)
9. *Sibirskij Matematicheskii Zhurnal* (Siberian Mathematical Journal)
10. *Uspekhi Matematicheskikh Nauk* (Russian Mathematical Surveys).

Aberth, O.
[1971]   The failure in computable analysis of a classical existence theorem for differential equations, *Proc. Am. Math. Soc.* 30 (1971) 151–156.
[1980]   *Computable analysis*, McGraw Hill, 1980.

Abian, S., and Brown, A.B.
[1961]   A theorem on partially ordered sets with applications to fixed-point theorems, *Can. J. Math.* 13 (1961) 78–83.

Abraham, U., and Shore, R.A.
[1986]   Initial segments of the degrees of size $\aleph_1$, *Isr. J. Math.* 53 (1986) 1–51.

Ackermann, W.
[1928]   Zum Hilbertschen Aufbau der reellen Zahlen, *Math. Ann.* 9 (1928) 118–133, transl. in Van Heijenoort [1967], pp. 493–507.
[1937]   Die Widerspruchsfreiheit der allegemeinen Mengenlehere, *Math. Ann.* 114 (1937) 305–315.
[1954]   *Solvable cases of the decision problem*, North Holland, 1954.

Addison, J.W.
[1954] *On some points of the Theory of Recursive Functions*, Ph.D. Thesis, University of Wisconsin, 1954.
[1959] Separation principles in the hierarchies of classical and effective descriptive set theory, *Fund. Math.* 46 (1959) 123–135.
[1959a] Some consequences of the axiom of constructibility, *Fund. Math.* 46 (1959) 337–357.
[1965] The undefinability of the definable, *Notices Am. Math. Soc.* 12 (1965) 347.

Addison, J.W., and Kleene, S.C.
[1957] A note on function quantification, *Proc. Am. Math. Soc.* 8 (1957) 1002–1006.

Addison, J.W., and Moschovakis, Y.N.
[1968] Some consequences of the axiom of definable determinateness, *Proc. Nat. Acad. Sci.* 59 (1968) 708–712.

Adler, A.
[1969] Some recursively unsolvable problems in analysis, *Proc. Am. Math. Soc.* 22 (1969) 523–526.

Agerwala, T.
[1974] A complete model for representing the coordination of asynchronous processes, *Report*, John Hopkins University, 1974.

Anderson, A.R.
[1964] *Minds and machines*, Prentice Hall, 1964.

Appel, K.I.
[1967] There exist two regressive sets whose intersection is not regressive, *J. Symb. Log.* 32 (1967) 322–324.

Appel, K.I., and McLaughlin, T.G.
[1965] On properties of regressive sets, *Trans. Am. Math. Soc.* 115 (1965) 83–93.

Arbib, M.A.
[1964] *Brains, machines and mathematics*, Wiley, 1964.
[1969] *Theories of abstract automata*, Prentice Hall, 1969.
[1969a] Self-reproducing automata - some implications for theoretical biology, in *Towards a Theoretical Biology*, Waddington ed., Edinburgh University Press, 1969, pp. 204–226.
[1972] *The metaphorical brain*, Wiley, 1972.
[1973] Automata theory in the context of theoretical neurophysiology, in *Foundations of mathematical biology*, Rosen ed., Academic Press, 1973, vol. 3, pp. 191–282.

Arslanov, M.M.
[1969] On effectively hypersimple sets, *Alg. Log.* 8 (1969) 143–153, transl. 8 (1969) 79–85.
[1970] On complete hypersimple sets, *Sov. Math.* 95 (1970) 30–35.
[1981] On some generalizations of the Fixed-Point Theorem, *Sov. Math.* 228 (1981) 9–16, transl. 228 (1981) 1–10.
[1985] Effectively hyperimmune sets and majorization, *Mat. Zam.* 38 (1985) 302–309, transl. 38 (1985) 677–680.

Arslanov, M.M., and Soloviev, V.D.
[1978] Effectivizations of definitions of classes of simple sets, *Alg. Avtom.* 10 (1978) 95–99.

Ashby, W.R.
[1952] *Design for a brain*, Wiley, 1952.

Babbage, C.
[1837]   On the mathematical powers of the calculating engine, in *The origins of digital computers*, Randell ed., Springer, 1973, pp. 19–54.

Backus, J.W.
[1981]   Is computer science based on the wrong fundamental concept of 'program'? An extended concept, in *Algorithmic languages*, de Bakker et al. eds., North Holland, 1981, pp. 133–165.

Backus, J.W., et al.
[1957]   The FORTRAN automatic coding system, *Proc. West. Joint Comp. Conf.* 11 (1957) 188–198.

Baire, R.
[1899]   Sur les fonctions de variables reelles, *Ann. Mat. Pura Appl.* 3 (1899) 1–122.

Banach, S.
[1922]   Sur les operations dans les ensembles abstraits et leurs applications aux equations integrales, *Fund. Math.* 3 (1922) 7–33.

Barendregt, H.
[1975]   Normed uniformly reflexive structures, *Springer Lect. Not. Comp. Sci.* 37 (1975) 272–286.
[1981]   *The lambda calculus, its syntax and semantics*, North Holland, 1981.

Barna, B.
[1956]   Über die Divergenzpunkte des Newtonschen Verfahrens zur Bestimmung von Wurzeln algebraischer Gleichungen, II, *Publ. Math. Debr.* 4 (1956) 384–397.

Barwise, J.
[1975]   *Admissible sets and structures*, Springer, 1975.

Beeson, M.J.
[1975]   The nonderivability in intuitionistic formal systems of theorems on the continuity of effective operations, *J. Symb. Log.* 40 (1975) 321–346.
[1985]   *Foundations of constructive mathematics*, Springer, 1985.

Beeson, M.J., and Ščedrov, A.
[1984]   Church's thesis, continuity, and set theory, *J. Symb. Log.* 49 (1984) 630–643.

Beigel, R.J., Gasarch, W.I., Gill, J., and Owings, J.C.
[1993]   Terse, superterse, and verbose sets, *Inf. Comp.* 103 (1993) 68–85.

Bell, J.S.
[1964]   On the Einstein-Podolski-Rosen paradox, *Phys.* 1 (1964) 195–200.
[1966]   On the problem of hidden variables in quantum mechanics, *Rev. Mod. Phys.* 38 (1966) 447–452.

Benioff, P.A.
[1980]   The computer as a physical system: a microscopical quantum mechanical Hamiltonian model of computers as represented by Turing machines, *J. Stat. Phys.* 22 (1980) 563–591.
[1981]   Quantum mechanical Hamiltonian models of discrete processes, *J. Math. Phys.* 22 (1981) 494–507.
[1982]   Quantum mechanical Hamiltonian models of discrete processes that erase their own history: application to Turing machines, *Int. J. Theor. Phys.* 21 (1982) 177–201.

Bennett, C.H.
[1973]   Logical reversibility of computations, *I.B.M. J. Res. Dev.* 6 (1973) 525–532.
[1982]   The thermodynamics of computation - a review, *Int. J. Theor. Phys.* 21 (1982) 905–940.

Berlekamp, E., Conway, J., and Guy, R.
[1982]   *Winning ways*, Academic Press, 1982.

Bernays, P.
[1957]   Review of Myhill [1955], *J. Symb. Log.* 22 (1957) 73–76.

Beth, E.W.
[1947]   Semantical considerations on intuitionistic mathematics, *Indag. Math.* 9 (1947) 572–577.

Bezboruah, A., and Shepherdson, J.C.
[1976]   Gödel's Incompleteness Theorem for Q, *J. Symb. Log.* 41 (1976) 503–512.

Bienestock, E., Fogelman Soulié, F., and Weisbuch, G., eds.
[1985]   *Disordered systems and biological organization*, Springer, 1985.

Birkoff, G., and Von Neumann, J.
[1936]   The logic of quantum mechanics, *Ann. Math.* 37 (1936) 823–843.

Bishop, E.
[1967]   *Foundations of constructive analysis*, McGraw Hill, 1967.

Blackwell, D.
[1967]   Infinite games and analytic sets, *Proc. Nat. Acad. Sci.* 58 (1967) 1836–1837.

Blum, M., and Marques, I.
[1973]   On complexity properties of recursively enumerable sets, *J. Symb. Log.* 38 (1973) 579–593.

Böhm, C., and Jacopini, G.
[1966]   Flow diagrams, Turing machines and languages with only two formation rules, *Comm. Ass. Comp. Mach.* 9 (1966) 366–371.

Boolos, G., and Putnam, H.
[1968]   Degrees of unsolvability of constructible sets of integers, *J. Symb. Log.* 33 (1968) 497–513.

Boone, W.W.
[1959]   The word problem, *Ann. Math.* 70 (1959) 207–265.

Borel, E.
[1898]   *Leçons sur la théorie des fonctions*, Gauthier-Villars, 1898.
[1912]   Le calcul des intégrales definies, *J. Math. Pur. Appl.* 8 (1912) 159–210.

Bremermann, H.J.
[1962]   Optimization through evolution and recombination, in *Self organizing systems*, Yovits et al. eds., Spartan Books, 1962, pp. 93–106.
[1982]   Minimum energy requirements of information transfer and computing, *Int. J. Theor. Phys.* 21 (1982) 203–217.

Brouwer, L.E.
[1919]   Intuitionistische Mengenlehre, *Jahr. Deutsch. Math. Verein.* 28 (1919) 203–208.
[1924]   Beweiss, dass jede volle Funktion gleichmässig stetig ist, *Konin. Neder. Akad. Weten. Proc.* 27 (1924) 189–193.
[1927]   Über Definitionsbereiche von Funktionen, *Math. Ann.* 97 (1927) 60–75, transl. in Van Heijenoort [1967], pp. 446–463.

Bulitko, V.K.
[1980]   Reducibility by Zhegalkin linear tables, *Sib. Math. J.* 21 (1980) 23–31, transl. 21 (1980) 332–339.

Burks, A.W., ed.
[1970]   *Essays on cellular automata*, University of Illinois Press, 1970.

Burks, C., and Farmer, D.
[1984] Toward modelling DNA sequences as automata, in Farmer, Toffoli and Wolfram [1984], pp. 157–167.

Byerly, R.B.
[1982] Recursion theory and the lambda calculus, *J. Symb. Log.* 47 (1982) 67–83.
[1985] Mathematical aspects of recursive function theory, in Harrington, Morley, Sčedrov and Simpson [1985], pp. 339–352.

Cantor, G.
[1874] Über eine Eigenschaft des Inbegriffes aller reellen algebraischen Zahlen, *J. Math.* 77 (1874) 258–262.
[1883] Über unendiche lineare Punktmannigfaltigkeiten, *Math. Ann.* 21 (1883) 545–586.
[1899] Letter to Dedekind, in Van Heijenoort [1967], pp. 113–117.

Carnap, R.
[1934] *Logische syntax der sprache*, Wien, 1934.

Carpentier, A.
[1968] Creative sequences and double sequences, *Notre Dame J. Form. Log.* 9 (1968) 35–61.
[1969] Complete enumerations and double sequences, *Zeit. Math. Log. Grund. Math.* 15 (1969) 1–6.
[1970] Uniformly precomplete enumerations, *Zeit. Math. Log. Grund. Math.* 16 (1970) 463–468.

Casalegno, P.
[1985] On the *T*-degrees of partial functions, *J. Symb. Log.* 50 (1985) 580–588.

Cayley, A.
[1879] The Newton-Fourier imaginary problem, *Am. J. Math.* 2 (1879) 97.

Čeitin, G.S.
[1959] Algorithmic operators in constructive complete separable metric spaces, *Dokl. Acad. Nauk* 128 (1959) 49–52.
[1962] Algorithmic operators in constructive metric spaces, *Trud. Mat. Inst. Steklov* 67 (1962) 295–361, A.M.S. transl. 64 (1967) 1–80.

Chaitin, G.J.
[1966] On the length of programs for computing finite binary sequences, *J. Ass. Comp. Mach.* 13 (1966) 547–569.
[1974] Information-theoretic limitations of formal systems, *J. Ass. Comp. Mach.* 21 (1974) 403–424.
[1979] Toward a mathematical definition of 'life', in *The maximum entropy formalism*, Levine at al. eds., M.I.T. Press, 1979, pp. 477–498.

Chomsky, N.
[1956] Three models for the description of languages, *I.R.E. Trans. Inf. Th.* 2 (1956) 113–124.
[1959] On certain formal properties of grammars, *Inf. Contr.* 2 (1959) 137–167.

Chong, C.T.
[1979] Generic sets and minimal α-degrees, *Trans. Am. Math. Soc.* 254 (1979) 157–169.
[1984] *Techniques of admissible Recursion Theory*, Springer Lect. Not. Math. 1106, 1984.

Church, A.
[1933] A set of postulates for the foundation of logic (second paper), *Ann. Math.* 34 (1933) 839–864.

[1936]   An unsolvable problem of elementary number theory, *Am. J. Math.* 58 (1936) 345–363, also in Davis [1965], pp. 89–107.

[1936a] A note on the Entscheidungsproblem, *J. Symb. Log.* 1 (1936) 40–41, also in Davis [1965], pp. 110–115.

[1940]   On the concept of random sequence, *Bull. Am. Math. Soc.* 46 (1940) 130–135.

[1941]   *The calculi of lambda conversion*, Princeton University Press, 1941.

[1957]   Application of recursive arithmetic to the problem of circuit synthesis, *Talks Cornell Summ. Inst. Symb. Log.*, Cornell, 1957, pp. 3–50.

Church, A., and Kleene, S.C.

[1937]   Formal definitions in the theory of ordinal numbers, *Fund. Math.* 28 (1937) 11–21.

Church, A., and Rosser, B.J.

[1936]   Some properties of conversion, *Trans. Am. Math. Soc.* 39 (1936) 472–482.

Cleave, J.P.

[1961]   Creative functions, *Zeit. Math. Log. Grund. Math.* 7 (1961) 205–212.

Codd, E.F.

[1968]   *Cellular automata*, Academic Press, 1968.

Cohen, P.J.

[1963]   The independence of the continuum hypothesis, *Proc. Nat. Acad. Sci.* 50 (1963) 1143–1148.

[1963a] A minimal model for Set Theory, *Bull. Am. Math. Soc.* 69 (1963) 537–540.

[1966]   *Set Theory and the Continumm Hypothesis*, Benjamin, 1966.

Cohen, P.F., and Jockusch, C.

[1975]   A lattice property of Post's simple sets, *Ill. J. Math.* 19 (1975) 450–453.

Colmerauer, A., Kanoui, H., Pasero, R., and Roussel, P.

[1972]   Un système de communication homme-machine en français, *Rapport preliminaire*, Groupe d'Intelligence Artificielle, Université d'Aix-Marseille, Luminy, 1972.

Cook, S.A.

[1982]   Toward a complexity theory of synchronous parallel computations, *Eins. Math.* 27 (1982) 99–124.

Cooper, S.B.

[1972]   Jump equivalence of the Δ_2^0 hyperhyperimmune sets, *J. Symb. Log.* 37 (1972) 598–600.

[1972a] Degrees of unsolvability complementary between r.e. degrees, *Ann. Math. Log.* 4 (1972) 31–73.

[1973]   Minimal degrees and the jump operator, *J. Symb. Log.* 38 (1973) 249–271.

Craig, W.

[1953]   On axiomatizability within a system, *J. Symb. Log.* 18 (1953) 30–32.

Crick, F., and Asanuma, C.

[1986]   Certain aspects of the anatomy and physiology of the cerebral cortex, in McClelland and Rumelhart [1986], vol. 2, pp. 333–371.

Crossley, J.N., ed.

[1981]   *Aspects of effective algebra*, Upside Down, 1981.

Crossley, J.N., and Nerode, A.

[1974]   *Combinatorial functors*, Springer, 1974.

Curry, H.B.

[1930]   Grundlagen der kombinatorischen Logik, *Am. J. Math.* 52 (1930) 509–536, 789–

834.

[1942] The inconsistency of certain formal logics, *J. Symb. Log.* 7 (1942) 115–117.

Curry, H.B., and Feys, R.

[1958] *Combinatory logic*, North Holland, 1958.

Curry, H.B., Hindley, J.R., and Seldin, J.P.

[1972] *Combinatory logic*, volume II, North Holland, 1972.

Curry, J., Garnett, L., and Sullivan, D.

[1983] On the iteration of rational functions: computer experiments with Newton's method, *Comm. Math. Phys.* 91 (1983) 267–277.

Dahl, O.J., Dijkstra, E.W., and Hoare, C.A.R.

[1972] *Structured programming*, Academic Press, 1972.

Darwin, C.

[1859] *On the origin of species by means of natural selection, or the preservation of favored races in the struggle for life*, 1859.

Davis, M.

[1950] *On the theory of recursive unsolvability*, Ph.D. Thesis, Princeton University, 1950.

[1956] A note on universal Turing machines, in Shannon and McCarthy [1956], pp. 167–175.

[1957] The definition of universal Turing machine, *Proc. Am. Math. Soc.* 8 (1957) 1125–1126.

[1958] *Computability and unsolvability*, McGraw Hill, 1958.

[1959] Computable functionals of arbitrary finite types, in Heyting [1959], pp. 281–284.

[1965] *The undecidable*, Raven Press, 1965.

[1973] Hilbert's tenth problem is unsolvable, *Am. Math. Month.* 80 (1973) 233–269.

[1974] *Computability*, Courant Institute Lecture Notes, New York University, 1974.

[1982] Why Gödel didn't have Church's thesis, *Inf. Contr.* 54 (1982) 3–24.

Davis, M., Matijasevich, Y., and Robinson, J.

[1976] Hilbert's tenth problem. Diophantine equations: positive aspects of a negative solution, *Proc. Symp. Pure Math.* 28 (1976) 323–378.

Dedekind, R.

[1888] *Was sind und was sollen die Zahlen?*, Braunschweig, 1888.

Degtev, A.N.

[1970] Remarks on retraceable, regressive and pointwise-decomposable sets, *Alg. Log.* 9 (1970) 651–660, transl. 9 (1970) 390–395.

[1971] Hypersimple sets with retraceable complements, *Alg. Log.* 10 (1971) 235–246, transl. 10 (1971) 147–154.

[1972] Hereditary sets and tabular reducibility, *Alg. Log.* 11 (1972) 257–269, transl. 11 (1972) 145–152.

[1973] tt and m-degrees, *Alg. Log.* 12 (1973) 143–161, transl. 12 (1973) 78–89.

[1979] Some results on uppersemilattices and m-degrees, *Alg. Log.* 18 (1979) 664–679, transl. 18 (1979) 420–430.

[1981] Relationships among complete sets, *Sov. Math.* 228 (1981) 50–55, transl. 228 (1981) 53–61.

[1982] Comparison of linear reducibility with other reducibilities truth-table like, *Alg. Log.* 21 (1982) 511–529, transl. 21 (1982) 339– 353.

[1983] Relationships between truth-table like reducibilities, *Alg. Log.* 22 (1983) 243–259, transl. 22 (1983) 173–185.

[1985] On the uppersemilattice of disjunctive and linear degrees, *Mat. Zam.* 38 (1985) 310–316, transl. 38 (1985) 681–684.

Dekker, J.C.E.

[1953] Two notes on r.e. sets, *Proc. Am. Math. Soc.* 4 (1953) 495–501.

[1953a] The constructivity of maximal dual ideals in certain Boolean algebras, *Pac. J. Math.* 3 (1953) 73–101.

[1954] A theorem on hypersimple sets, *Proc. Am. Math. Soc.* 5 (1954) 791–796.

[1955] A non-constructive extension of the number system, *J. Symb. Log.* 20 (1955) 204–205.

[1955a] Productive sets, *Trans. Am. Math. Soc.* 78 (1955) 129–149.

[1962] Infinite series of isols, *Proc. Symp. Pure Appl. Math.* 5 (1962) 77–96.

[1966] *Les fonctions combinatoires et les isols*, Gauthier-Villars, 1966.

Dekker, J.C.E., and Myhill, J.

[1958] Retraceable sets, *Can. J. Math.* 10 (1958) 357–373.

[1958a] Some theorems on classes of recursively enumerable sets, *Trans. Am. Math. Soc.* 89 (1958) 25–69.

[1960] Recursive equivalence types, *Univ. Cal. Publ. Math.* 3 (1960) 67–214.

De Leeuw, K., Moore, E.F., Shannon, C.E., and Shapiro, N.

[1956] Computability and probabilistic machines, in Shannon and McCarthy [1956], pp. 183–212.

Demongeot, J., Golès, E., and Tchuente, M., eds.

[1985] *Dynamical systems and cellular automata*, Academic Press, 1985.

Denisov, S.D.

[1974] Three theorems on elementary theories and *tt*-reducibility, *Alg. Log.* 13 (1974) 5–8, transl. 13 (1974) 1–2.

Descartes, R.

[1637] *Discours de la méthode*, 1637.

[1664] *L'homme et un traité de le formation du phoetus*, Paris, 1664.

Detlovs, V.K.

[1953] Normal algorithms and recursive functions, *Dokl. Akad. Nauk* 90 (1953) 723–725.

[1958] The equivalence of normal algorithms and recursive functions, *Trud. Math. Inst. Stekl.* 52 (1958) 75–139, A.M.S. transl. 23 (1963) 15–81.

Devlin, K.J.

[1984] *Constructibility*, Springer, 1984.

De Witt, B., and Graham, N., eds.

[1973] *The many-worlds interpretation of quantum mechanics*, Princeton University Press, 1973.

Dijksterhuis, E.J.

[1961] *The mechanization of the world picture*, Oxford University Press, 1961.

Dijkstra, E.W.

[1968] 'Go to' statements considered harmful, *Comm. Ass. Comp. Mach.* 11 (1968) 417–418.

Di Paola, R.A., and Heller, A.

[1987] Dominical categories: Recursion Theory without elements, *J. Symb. Log.* 52 (1987) 594–635.

Downey, R.G.

[1987] Maximal theories, *Ann. Pure Appl. Log.* 33 (1987) 245–282.

[1993] On irreducible *m*-degrees, *Rend. Sem. Mat. Univ. Pol. Torino* 51 (1993) 109–112.

Drake, F.R.

[1974] *Set Theory*, North Holland, 1974.

Dreben, B., and Goldfarb, W.D.

[1979]   *The decision problem. Solvable cases of quantificational formulas*, Addison Wesley, 1979.

Eccles, J.C.

[1953]   *The neurophysiological basis of mind*, Clarendon Press, 1953.

[1973]   *The understanding of the brain*, McGraw Hill, 1973.

Ehrenfeucht, A., and Feferman, S.

[1960]   Representability of r.e. sets in formal theories, *Arch. Math. Log. Grund.* 5 (1960) 37–41.

Eilenberg, S.

[1974]   *Automata, languages and machines*, Academic Press, 1974.

Eilenberg, S., and Elgot, C.C.

[1970]   *Recursiveness*, Academic Press, 1970.

Einstein, A., Podolski, B., and Rosen, N.

[1935]   Can quantum mechanical description of physical reality be considered complete?, *Phys. Rev.* 47 (1935) 777–780.

Elgot, C.C., and Robinson, A.

[1964]   Random-access stored program machines, an approach to programming languages, *J. Ass. Comp. Mach.* 11 (1964) 365–399.

Elgot, C.C., Robinson A., and Rutledge, J.D.

[1967]   Multiple control computer models, in *Systems and computers science*, Hart et al. eds., Toronto University Press, 1967, pp. 60–76.

Ellentuck, E.

[1967]   Universal isols, *Math. Zeitschr.* 98 (1967) 1–8.

[1973]   Uncountable suborderings of the isols, *Compos. Math.* 26 (1973) 277–282.

[1973a] Degrees of isolic integers, *Notre Dame J. Form. Log.* 14 (1973) 331–340.

Enderton, H.B., and Putnam, H.

[1970]   A note on the hyperarithmetical hierarchy, *J. Symb. Log.* 35 (1970) 429–430.

Epstein, R.L.

[1975]   *Minimal degrees of unsolvability and the full approximation construction*, Memoirs A.M.S. 163, 1975.

[1979]   *Degrees of unsolvability: structure and theory*, Springer Lect. Not. Math. 759, 1979.

Ershov, A.P.

[1960]   Algorithmic operators, *Probl. Kib.* 3 (1960) 5–48, transl. 3 (1960) 696–763.

[1981]   Abstract computability on algebraic structures, *Springer Lect. Not. Comp. Sci.* 122 (1981) 397–420.

Ershov, Y.L.

[1968]   A hierarchy of sets, *Alg. Log.* 7 (1968) 47–74, transl. 7 (1968) 25–43.

[1968a] A hierarchy of sets II, *Alg. Log.* 7 (1968) 15–47, transl. 7 (1968) 212–232.

[1968b] On recursive enumerations, *Alg. Log.* 7 (1968) 71–99, transl. 7 (1968) 330–346.

[1970]   A hierarchy of sets III, *Alg. Log.* 9 (1970) 34–51, transl. 9 (1970) 20–31.

[1971]   Positive equivalences, *Alg. Log.* 10 (1971) 620–650, transl. 10 (1971) 378–394.

[1972]   Computable functionals of finite type, *Alg. Log.* 11 (1972) 367–437, transl. 11 (1972) 203–242.

[1973]   The theory of A-spaces, *Alg. Log.* 12 (1973) 369–416, transl. 12 (1973) 209–232.

[1974]   Maximal and everywhere defined functionals, *Alg. Log.* 13 (1974) 374–397, transl. 13 (1974) 210–225.

[1975]   The uppersemilattice of enumerations of a finite set, *Alg. Log.* 14 (1975) 258–284, transl. 14 (1975) 159–175.

[1977]   *The theory of enumerations*, Nauka, 1977.

[1980]   *Decidability problems and constructive models*, Nauka, 1980.

Ershov, Y.L., Lavrov, I.A., Taimanov A.D., and Taislin, M.A.

[1965]   Elementary theories, *Usp. Mat. Nauk* 124 (1965) 37–108, transl. 20 (1965) 35–105.

Farber, D.J., Griswold, R.E., and Polonsky. I.P.

[1964]   SNOBOL, a string manipulation language, *J. Ass. Comp. Mach.* 11 (1964) 21–30.

Farmer, D., Toffoli, T., and Wolfram, S., eds.

[1984]   *Cellular automata*, North Holland, 1984.

Feferman, S.

[1957]   Degrees of unsolvability associated with classes of formalized theories, *J. Symb. Log.* 22 (1957) 161–175.

[1964]   Systems of predicative analysis, *J. Symb. Log.* 29 (1964) 1–30.

[1977]   Inductive schemata and recursively continuous functionals, in *Logic Colloquium '76*, Gandy et al. eds., North Holland, 1977, pp. 373–392.

Feferman, S., and Spector, C.

[1962]   Incompleteness along paths in progressions of theories, *J. Symb. Log.* 27 (1962) 383–390.

Feiner, L.

[1970]   The strong homogeneity conjecture, *J. Symb. Log.* 35 (1970) 375–377.

Fenstad, J.E.

[1974]   On axiomatizing recursion theory, in *Generalized Recursion Theory*, Fenstad et al. eds., North Holland, 1974, pp. 385–404.

[1980]   *General recursion theory*, Springer, 1980.

Feynman, R.P.

[1982]   Simulating physics with computers, *Int. J. Theor. Phys.* 21 (1982) 467–488.

[1996]   *Feynman's Lectures on computation*, Addison Wesley, 1996.

Feynman, R.P., Leighton, R.B., and Sands, M.

[1963]   *The Feynman lectures on Physics*, Addison Wesley, 1963.

Fischer, P.C.

[1963]   A note on bounded truth-table reducibility, *Proc. Am. Math. Soc.* 14 (1963) 875–877.

Florence, J.B.

[1967]   Infinite subclasses of recursively enumerable classes, *Proc. Am. Math. Soc.* 18 (1967) 633–639.

[1969]   Strong enumeration properties of recursively enumerable sets, *Zeit. Math. Log. Grund. Math.* 15 (1969) 181–192.

[1975]   On splitting an infinite recursively enumerable class, *Can. J. Math.* 27 (1975) 1127–1140.

Fraenkel, A.A.

[1922]   Zu den Grundlagen der Cantor-Zermeloschen Mengenlehre, *Math. Ann.* 86 (1922) 230–137.

Fraenkel, A.A., Bar-Hillel, Y., and Levy, A.

[1958]   *Foundations of Set Theory*, North Holland, 1958.

Fredkin, E., and Toffoli, T.

[1982]   Conservative logic, *Int. J. Theor. Phys.* 21 (1982) 219–253.

Frege, G.
[1893]   *Grundgesetze der Arithmetik, begriffsschriftlich abgeleitet*, Pohle, 1893.

Freivalds, R.V.
[1978]   Effective operations and functionals computable in the limit, *Zeit. Math. Log. Grund. Math.* 24 (1978) 193–206.

Freud, S.
[1917]   *Vorlesungen zur Einführung in die Psychoanalyse*, 1917.

Freyd, P., and Sčedrov, A.
[1987]   Some semantic aspects of polymorphic lambda calculus, *I.E.E.E. Log. Comp. Sci.* 2 (1987) 315–319.

Friedberg, R.M.
[1957]   The fine structure of degrees of unsolvability of recursively enumerable sets, *Talks Cornell Summ. Inst. Symb. Log.*, Cornell, 1957, pp. 404–406.

[1957a] Two recursively enumerable sets of incomparable degrees of unsolvability, *Proc. Nat. Acad. Sci.* 43 (1957) 236–238.

[1957b] A criterion for completeness of degrees of unsolvability, *J. Symb. Log.* 22 (1957) 159–160.

[1958]   Three theorems on recursive enumeration, *J. Symb. Log.* 23 (1958) 309–316.

[1958a] Four-quantifier completeness: a Banach-Mazur functional not uniformly partial recursive, *Bull. Acad. Pol. Sci.* 6 (1958) 1–5.

[1958b] Un contre-exemple relatif aux fonctionelles récursives, *Compt. Rend. Acad. Sci.* 247 (1958) 852–854.

Friedberg, R.M., and Rogers, H.
[1959]   Reducibilities and completeness for sets of integers, *Zeit. Math. Log. Grund. Math.* 5 (1959) 117–125.

Friedman, H.
[1971]   Axiomatic recursive function theory, in *Logic Colloquium '69*, Gandy et al. eds., North Holland, 1971, pp. 113–137.

[1971a] Algorithmic procedures, generalized Turing algorithms, and elementary recursion theory, in *Logic Colloquium '69*, Gandy et al. eds., North Holland, 1971, pp. 361–389.

Friedman, S., and Sacks, G.E.
[1977]   Inadmissible Recursion Theory, *Bull. Am. Math. Soc.* 83 (1977) 255–256.

Gale, D., and Stewart, F.M.
[1953]   Infinite games with perfect information, *Ann. Math. Stud.* 28 (1953) 245–266.

Galilei, G.
[1923]   *Il saggiatore*, 1623.
[1638]   *Discorsi intorno a due nuove scienze*, 1638.

Gandy, R.O.
[1967]   General recursive functionals of finite type and hierarchies of functions, *Ann. Fac. Sci. Univ. Clermont-Ferrand* 35 (1967) 5–24.

[1967a] Relations between analysis and set theory, *J. Symb. Log.* 32 (1967) 434.

[1980]   Church's thesis and principles for mechanisms, in *Kleene Symposium*, Barwise et al. eds., North Holland, 1980, pp. 123–148.

Gandy, R.O., and Hyland, J.M.E.
[1977]   Computable and recursively countable functions of higher type, in *Logic Colloquium '76*, Gandy et al. eds., North Holland, 1977, pp. 407–438.

Georgieva, N.
[1976]   Another simplification of the recursion schema, *Arch. Math. Log. Grund.* 18 (1976) 1–3.

Gill, J., and Morris, P.
[1974]   On subcreative sets and *s*-reducibility, *J. Symb. Log.* 39 (1974) 669–677.

Gladstone, M.D.
[1967]   A reduction of the recursion scheme, *J. Symb. Log.* 32 (1967) 505–508.
[1971]   Simplification of the recursion scheme, *J. Symb. Log.* 36 (1971) 653–665.

Gödel, K.
[1930]   Die Vollständigkeit der Axiome des logischen Funktionenkalküls, *Monash. Math. Phys.* 37 (1930) 349–360, transl. in Van Heijenoort [1967], pp. 583–591.
[1931]   Über formal unentscheidbare Sätze der Principia Mathematica und verwandter Systeme I, *Monash. Math. Phys.* 38 (1931) 173–198, transl. in Davis [1965], pp. 5–38, also in Van Heijenoort [1967], pp. 595–616.
[1933]   Eine Interpretation des intuitionistischen Aussagenkalküls, *Ergeb. Math. Koll.* 4 (1933) 39–40.
[1934]   On undecidable propositions of formal mathematical systems, *mimeographed notes*, 1934, in Davis [1965], pp. 41–81.
[1936]   Über die Lange der Beweise, *Ergeb. Math. Koll.* 7 (1936) 23–24, transl. in Davis [1965], pp. 82–83.
[1938]   The consistency of the axiom of choice and of the generalized continuum hypothesis, *Proc. Nat. Acad. Sci.* 24 (1938) 556–557.
[1939]   Consistency proof for the generalized continuum hypothesis, *Proc. Nat. Acad. Sci.* 25 (1939) 220–224.
[1940]   *The consistency of the axiom of choice and of the generalized continuum hypothesis*, Princeton University Press, 1940, 2nd ed. 1951.
[1944]   Russell's mathematical logic, in *The philosophy of Bertrand Russell*, Schlipp ed., Northwestern University Press, 1944, pp. 123–153.
[1946]   Remarks before the Princeton bicentennial conference on problems in mathematics, in Davis [1965], pp. 84–88.
[1958]   Über eine bisher noch nicht benütze Erweiterung des finiten Standpunktes, *Dial.* 12 (1958) 280–287.
[1964]   What is Cantor's continuum problem, in *Philosophy of mathematics*, Benacerraf et al. eds., Prentice-Hall, 1964, pp. 258–273.
[1986]   *Collected works*, Volume I, Oxford University Press, 1986.

Gold, E.M.
[1965]   Limiting recursion, *J. Symb. Log.* 30 (1965) 28–48.
[1967]   Language identification in the limit, *Inf. Contr.* 10 (1969) 447–474.

Goldstine, H.H., and Von Neumann, J.
[1947]   Planning and coding of problems for an electronic computing instrument, part II, *Report for U.S. Army Ord. Dept.*, 1947, also in Von Neumann's *Collected Works*, 5 (1963) 80–151.

Goodstein, R.L.
[1961]   *Recursive analysis*, North Holland, 1961.

Gordon, C.E.
[1968]   *A comparison of abstract computability theories*, Ph.D. Thesis, U.C.L.A., 1968.
[1970]   Comparison between some generalizations of recursion theory, *Comp. Math.* 22 (1970) 333–346.
[1974]   Prime and search computability, characterized as definability in certain sublanguages of constructible $\mathcal{L}_{\omega_1,\omega}$, *Trans. Am. Math. Soc.* 197 (1974) 391–407.

Gordon, G.
[1961]   A general purpose system simulation program, *Proc. E.J.C.C.* (1961) 87–104.

Grassmann, H.
[1861]   *Lehrbuch der Arithmetik für höhere Lehranstalten*, Berlin, 1861.

Greenspan, D.
[1973]   *Discrete models*, Addison Wesley, 1973.
[1980]   *Arithmetic applied mathematics*, Pergamon Press, 1980.
[1982]   Deterministic computer physics, *Int. J. Theor. Phys.* 21 (1982) 505–523.

Greibach, S.A.
[1981]   Formal languages: origins and directions, *Ann. Hist. Comp.* 3 (1981) 14–41.

Grilliott, T.J.
[1971]   On effectively discontinuous type-2 objects, *J. Symb. Log.* 36 (1971) 245–248.
[1971a] Inductive definitions and computability, *Trans. Am. Math. Soc.* 158 (1971) 309–317.

Groszek, M.S., and Slaman, T.A.
[1983]   Independence results on the global structure of the Turing degrees, *Trans. Am. Math. Soc.* 277 (1983) 579–588.

Grzegorczyk, A.
[1953]   Some classes of recursive functions, *Rozpr. Mat.* 4 (1953) 1–45.
[1955]   Computable functionals, *Fund. Math.* 42 (1955) 168–202.
[1957]   On the definitions of computable real continuous functions, *Fund. Math.* 44 (1957) 61–71.
[1959]   Some approaches to constructive analysis, in Heyting [1959], pp. 43–61.

Grzegorczyk, A., Mostowski, A., and Ryll-Nardzewski, C.
[1958]   The classical and ω-complete arithmetic, *J. Symb. Log.* 23 (1958) 188–206.

Hack, M.
[1975]   *Decidability question for Petri nets*, Ph.D. Thesis, M.I.T., 1975.

Haddon, R.C., and Lamola, A.A.
[1985]   The molecular electronic device and the biochip computer: present status, *Proc. Nat. Acad. Sci.* 82 (1985) 1874–1878.

Hajnal, A.
[1956]   On a consistency theorem connected with the generalized continuum problem, *Zeit. Math. Log. Grund. Math.* 2 (1956) 131–136.

Hanf, W.
[1965]   Model-theoretic methods in the study of elementary logic, in *The theory of models*, Addison et al. eds., North Holland, 1965, pp. 132–145.
[1975]   The Boolean algebra of logic, *Bull. Am. Math. Soc.* 81 (1975) 587–589.

Hardy, G.H.
[1910]   *Orders of infinity*, Cambridge, 1910.

Harrington, L.A.
[1974]   The constructible reals can be (almost) anything, *Unpublished manuscript*, 1974.

Harrington, L.A., Morley, M.D., Sčedrov, A., and Simpson, S.G., eds.
[1985]   *Harvey Friedman's research on the foundations of mathematics*, North Holland, 1985.

Harrington, L.A., and Shore, R.A.
[1981]   Definable degrees and automorphisms of $\mathcal{D}$, *Bull. Am. Math. Soc.* 4 (1981) 97–100.

Harrow, K.
[1978]   The bounded arithmetical hierarchy, *Inf. Contr.* 36 (1978) 102–117.

Hartmanis, J.
[1982]   A note on natural complete sets and Gödel numberings, *Theor. Comp. Sci.* 17 (1982) 75–89.

Hartmanis, J., and Baker, T.P.
[1975]   On simple Gödel numberings and translations, *S.I.A.M. J. Comp.* 4 (1975) 1–11.

Hartmanis, J., and Stearns, R.E.
[1965]   On the computational complexity of algorithms, *Trans. Am. Math. Soc.* 117 (1965) 285–306.
[1966]   *Algebraic structure theory for sequential machines*, Prentice Hall, 1966.

Hausdorff, F.
[1917]   *Grundzüge der Mengenlehre*, Leipzig, 1917, reprinted Chelsea, 1949.

Hebb, D.O
[1949]   *The organization of behavior*, Wiley, 1949.

Helm, J.P.
[1971]   On effectively computable operators, *Zeit. Math. Log. Grund. Math.* 17 (1971) 231–244.

Herbrand, J.
[1931]   Sur la non-contradiction de l'Arithmétique, *J. Reine Angew. Math.* 166 (1931) 1–8, transl. in Van Heijenoort [1967], pp. 620–628.

Hermes, H.
[1965]   *Enumerability, decidability, computability*, Springer, 1965.

Heyting, A., ed.
[1959]   *Constructivity in mathematics*, North Holland, 1959.

Higman, G.
[1961]   Subgroups of finitely presented groups, *Proc. Roy. Soc.* 262 (1961) 455–474.

Hilbert, D.
[1900]   Mathematische Probleme, *Proc. Int. Congr. Math.* (1900) 58–114.
[1904]   Über die Grundlagen der Logik und der Arithmetik, *Proc. Int. Congr. Math.* (1904) 174–185, transl. in Van Heijenoort [1967], pp. 129–138.
[1926]   Über das Unendliche, *Math. Ann.* 95 (1926) 161–190, transl. in Van Heijenoort [1967], pp. 367–392.

Hilbert, D., and Bernays, P.
[1934]   *Grundlagen der Mathematik*, Berlin, 1934.
[1939]   *Grundlagen der Mathematik*, vol. II, Berlin, 1939.

Hindley, J.R., and Seldin, J.P.
[1986]   *Introduction to combinators and λ-calculus*, London Mathematical Society, 1986.

Hinman, P.G.
[1973]   Degrees of continuous functionals, *J. Symb. Log.* 38 (1973) 393–395.
[1978]   *Recursion-theoretical hierarchies*, Springer, 1978.

Hinman, P.G., and Moschovakis, Y.N.
[1971]   Computability over the continuum, in *Logic Colloquium '69*, Gandy et al. eds., North Holland, 1971, pp. 77–105.

Hobbes, T.
[1655]   *De corpore*, 1655.

Hodges, A.
[1983]   *Alan Turing: the enigma*, Simon and Schuster, 1983.

Hofstadter, D.R.
[1979]   *Gödel, Escher, Bach: an Eternal Golden Braid*, Basic Books, 1979.
[1985]   *Metamagical themas: questing for the essence of mind and pattern*, Basic Books, 1985.

Hofstadter, D.R., and Dennett, D.C., eds.
[1981]   *The mind's I*, Basic Books, 1981.

Hopcroft, J.E., and Ullman, J.D.
[1979]   *Introduction to automata theory, languages and computations*, Addison Wesley, 1979.

Hopfield, J.J
[1982]   Neural networks and physical systems with emergent collective computational abilities, *Proc. Nat. Acad. Sci.* 79 (1982) 2554–2558.
[1984]   Neurons with graded response have collective computational properties like those of two-state neurons, *Proc. Nat. Acad. Sci.* 81 (1984) 3088–3092.

Horowitz, B.M.
[1978]   Sets completely creative via recursive permutations, *Zeit. Math. Log. Grund. Math.* 24 (1978) 445–452.

Hugill, D.
[1969]   Initial segments of Turing degrees, *Proc. Lond. Math. Soc.* 19 (1969) 1–15.

Hyland, J.M.E.
[1979]   Filter spaces and continuous functionals, *Ann. Math. Log.* 36 (1979) 101–143.
[1982]   The effective topos, in *The L.E.J. Brouwer centenary symposium*, Troelstra et al. eds., North Holland, 1982, pp. 165–216.

Ianov, Y.I.
[1958]   The logical schemes of algorithms, *Probl. Kib.* 1 (1958) 75–127, transl. 1 (1960) 82–140.

Janiczak, A.
[1950]   A remark concerning decidability of complete theories, *J. Symb. Log.* 15 (1950) 277–279.

Jech, T.
[1978]   *Set theory*, Academic Press, 1978.

Jensen, R.B.
[1972]   The fine structure of the constructible hierarchy, *Ann. Math. Log.* 4 (1972) 229–308.

Jensen, R.B., and Solovay, R.M.
[1970]   Some applications of almost disjoint sets, in *Mathematical logic and foundations of set theory*, Bar-Hillel ed., North Holland, 1970, pp. 84–104.

Jeroslow, R.G.
[1973]   Redundancies in the Hilbert-Bernays derivability conditions for Gödel's second incompleteness theorem, *J. Symb. Log.* 38 (1973) 359–367.

Jockusch, C.G.
[1966]   *Reducibilities in Recursive Function Theories*, Ph.D. Thesis, M.I.T., 1966.
[1968]   Uniformly introreducible sets, *J. Symb. Log.* 33 (1968) 521–536.
[1968a] Semirecursive sets and positive reducibility, *Trans. Am. Math. Soc.* 131 (1968) 420–436.
[1969]   Relationships between reducibilities, *Trans. Am. Math. Soc.* 142 (1969) 229–237.
[1969a] The degrees of hyperhyperimmune sets, *J. Symb. Log.* 34 (1969) 489–493.
[1969b] The degrees of bi-immune sets, *Zeit. Math. Log. Grund. Math.* 15 (1969) 135–140.

[1972] Upward closure of bi-immune degrees, *Zeit. Math. Log. Grund. Math.* 18 (1972) 285–287.

[1972a] A reducibility arising from Boone groups, *Math. Scand.* 31 (1972) 262–266.

[1973] An application of Σ_4^0-determinacy to the degrees of unsolvability, *J. Symb. Log.* 38 (1973) 293–294.

[1973a] Upward closure and cohesive degrees, *Isr. J. Math.* 15 (1973) 332–335.

[1974] Π_1^0 classes and Boolean combinations of recursively enumerable sets, *J. Symb. Log.* 39 (1974) 95–96.

[1981] Degrees of generic sets, *Lond. Math. Soc. Lect. Not.* 45 (1981) 110–139.

[1981a] Three easy constructions of recursively enumerable degrees, *Springer Lect. Not. Math.* 859 (1981) 83–91.

[1989] Degrees of functions with no fixed-points, *Log. Phil. Meth. Sci.* 8 (1989) 191–201.

Jockusch, C., Lerman, M., Soare, R.I., and Solovay, R.M.

[1989] Recursively enumerable sets modulo iterated jumps and extensions of Arslanov's completeness criterion, *J. Symb. Log.* 54 (1989) 1288–1323.

Jockusch, C., and Paterson, M.

[1976] Completely autoreducible degrees, *Zeit. Math. Log. Grund. Math.* 22 (1976) 571–575.

Jockusch, C., and Posner, D.

[1978] Double jumps of minimal degrees, *J. Symb. Log.* 43 (1978) 715–724.

[1981] Automorphism bases for degrees of unsolvability, *Isr. J. Math.* 40 (1981) 150–164.

Jockusch, C., and Shore, R.A.

[1984] Pseudo-jump operators II: transfinite iterations, hierarchies and minimal covers, *J. Symb. Log.* 49 (1984) 1205–1236.

Jockusch, C., and Simpson, S.G.

[1976] A degree-theoretic definition of the ramified analytical hierarchy, *Ann. Math. Log.* 10 (1976) 1–32.

Jockusch, C., and Soare, R.I.

[1970] Minimal covers and arithmetical sets, *Proc. Am. Math. Soc.* 25 (1970) 856–859.

[1971] A minimal pair of Π_1^0 classes, *J. Symb. Log.* 36 (1971) 66–78.

[1972] Π_1^0 classes and degrees of theories, *Trans. Am. Math. Soc.* 173 (1972) 33–56.

[1972a] Degrees of members of Π_1^0 classes, *Pac. J. Math.* 40 (1972) 605–616.

[1973] Post's problem and his hypersimple set, *J. Symb. Log.* 38 (1973) 446–452.

Jockusch, C., and Solovay, R.M.

[1977] Fixed-points of jump-preserving automorphisms of degrees, *Isr. J. Math.* 26 (1977) 91–94.

Jones, N.D., and Matijasevich, Y.V.

[1984] Register machine proof of the theorem on exponential diophantine representation of enumerable sets, *J. Symb. Log.* 49 (1984) 818–829.

Jones, N.D., and Shepherdson, J.C.

[1983] Variants of Robinson's essentially undecidable theory $\mathcal{R}$, *Arch. Math. Log. Grund.* 23 (1983) 61–64.

Julia, G.

[1918] Sur l'iteration des fonctions rationelles, *J. Math. Pur. Appl.* 8 (1918) 47–245.

Kallibekov, S.

[1973] On degrees of recursively enumerable sets, *Sib. Math. J.* 14 (1973) 421–426, transl. 14 (1973) 290–293.

Kalmar, L.
[1955]   Über ein problem, betreffend die definition des Begriffes der allegemein-rekursiven Funktion, *Zeit. Math. Log. Grund. Math.* 1 (1955) 93–96.
[1959]   An argument against the plausibility of Church's thesis, in Heyting [1959], pp. 72–80.

Kandel, E.R., and Schwartz, J.H.
[1981]   *Principles of neural science*, North Holland, 1981.

Kanovich, M.I.
[1969]   On the decision complexity of algorithms, *Dokl. Acad. Nauk* 186 (1969) 1008–1009, transl. 10 (1969) 700–701.
[1970]   On the decision complexity of r.e. sets, *Dokl. Acad. Nauk* 192 (1970) 721–723, transl. 11 (1970) 704–706.
[1970a] On the decision complexity of a recursively enumerable set as a criterion for its completeness, *Dokl. Acad. Nauk* 194 (1970) 500–503, transl. 11 (1970) 1224–1228.
[1975]   Dekker's construction and effective non-recursiveness, *Dokl. Acad. Nauk* 222 (1975) 1028–1030, transl. 16 (1975) 719–721.

Karp, C.R.
[1967]   A proof of the relative consistency of the continuum hypothesis, in *Sets, models, and Recursion Theory*, Crossley ed., North Holland, 1967, pp. 1–32.

Kechris, A.S., and Moschovakis, Y.N.
[1977]   Recursion in higher types, in *Handbook of Mathematical Logic*, Barwise ed., North Holland, 1977, pp. 681–737.

Kelley, J.
[1955]   *General topology*, Springer, 1955.

Kfoury, D.
[1974]   Translatability of schemas over restricted interpretations, *J. Comp. Syst. Sci.* (1974) 387–408.

Khutorezkii, A.B.
[1969]   On the reducibility of computable enumerations, *Alg. Log.* 8 (1969) 251–262, transl. 8 (1969) 145–151.
[1969a] Two existence theorems for recursive enumerations, *Alg. Log.* 8 (1969) 483–492, transl. 8 (1969) 277–282.

Kleene, S.K.
[1935]   A theory of positive integers in formal logic, *Am. J. Math.* 57 (1935) 153–173, 219–244.
[1936]   General recursive functions of natural numbers, *Math. Ann.* 112 (1936) 727–742, also in Davis [1965], pp. 237–252.
[1936a] A note on recursive functions, *Bull. Am. Math. Soc.* 42 (1936) 544–546.
[1936b] λ-definability and recursiveness, *Duke Math. J.* 2 (1936) 340–353.
[1938]   On notations for ordinal numbers, *J. Symb. Log.* 3 (1938) 150–155.
[1943]   Recursive predicates and quantifiers, *Trans. Am. Math. Soc.* 53 (1943) 41–73, also in Davis [1965], pp. 255–287.
[1945]   On the interpretation of intuitionistic number theory, *J. Symb. Log.* 10 (1945) 109–124.
[1950]   A symmetric form of Gödel's theorem, *Ind. Math.* 12 (1950) 244–246.
[1952]   *Introduction to metamathematics*, North Holland, 1952.
[1955]   Arithmetical predicates and function quantifiers, *Trans. Am. Math. Soc.* 79 (1955) 312–340.
[1955a] Hierarchies of number theoretic predicates, *Bull. Am. Math. Soc.* 61 (1955) 193–

213.

[1959] Recursive functionals and quantifiers of finite types I, *Trans. Am. Math. Soc.* 91 (1959) 1–52.

[1959a] Countable functionals, in Heyting [1959], pp. 81–100.

[1959b] Quantification of number-theoretic functions, *Comp. Math.* 14 (1959) 23–40.

[1963] Recursive functionals and quantifiers of finite type II, *Trans. Am. Math. Soc.* 108 (1963) 106–142.

[1969] *Formalized recursive functionals and formalized realizability*, A.M.S. Memoirs 89, 1969.

[1978] Recursive functionals and quantifiers of finite types revisited I, in *Generalized Recursion Theory II*, Fenstad et al. eds., North Holland, 1977, pp. 185–222.

[1981] Origins of recursive function theory, *Ann. Hist. Comp.* 3 (1981) 52–67.

[1981a] The theory of recursive functions approaching its centennial, *Bull. Am. Math. Soc.* 5 (1981) 43–61.

[1985] Unimonotone functions of finite types, *Proc. Symp. Pure Math.* 42 (1985) 119–138.

[1987] Reflections on Church's Thesis, *Notre Dame J. Form. Log.* 28 (1987) 490–498.

Kleene, S.C., and Post, E.L.

[1954] The uppersemilattice of degrees of recursive unsolvability, *Ann. Math.* 59 (1954) 379–407.

Klop, J.W.

[1982] Extending partial combinatory algebras, *Bull. Europ. Ass. Theor. Comp. Sci.* 16 (1982) 30–34.

Knaster, B.

[1928] Un théorème sur les fonctions d'ensembles, *Ann. Soc. Polon. Math.* 6 (1928) 133–134.

Knorr, W.

[1983] 'La croix des mathématiciens': the Euclidean theory of irrational lines, *Bull. Am. Math. Soc.* 9 (1983) 41–69.

Kobzev, G.N.

[1973] *btt*-reducibility, *Alg. Log.* 12 (1973) 190–204, transl. 12 (1973) 107–115.

[1974] On the complete *btt*-degree, *Alg. Log.* 13 (1974) 22–25, transl. 13 (1974) 10–12.

[1975] On r-separable sets, in *Studies of Mathematical Logic and the Theory of Algorithms*, Tbilisi, 1975, pp. 19–30.

[1977] On recursively enumerable *bw*-degrees, *Mat. Zam.* 21 (1977) 839–846, transl. 21 (1977) 473–477.

Kolmogorov, A.

[1932] Zur Deutung der Intuitionistischen Logik, *Math. Zeit.* 35 (1932) 58–65.

[1963] On tables of random numbers, *Ind. J. Stat.* 25 (1963) 369–376.

[1965] Three approaches to the definition of the concept 'amount of information', *Probl. Pered. Inf.* 1 (1965) 3–11, transl. 1 (1965) 1–7.

Kolmogorov, A., and Uspenskii, V.A.

[1958] On the definition of an algorithm, *Usp. Mat. Nauk* 13 (1958) 3–28, A.M.S. transl. 29 (1963) 217–245.

König, D.

[1926] Sur les correspondances multivoques des ensembles, *Fund. Math.* 8 (1926) 114–134.

Kowalski, R., and Van Emden, M.

[1976] The semantics of predicate logic as programming language, *J. Ass. Comp. Mach.* 23 (1976) 733–743.

Koymans, K.
[1982]   Models of λ-calculus, *Inf. Contr.* 52 (1982) 306–332.

Kozmnikh, V.V.
[1968]   On primitive recursive functions of one argument, *Alg. Log.* 7 (1968) 75–90, transl. 7 (1968) 44–53.

Kreisel, G.
[1950]   Note on arithmetical models for consistent formulae of the predicate calculus, *Fund. Math.* 37 (1950) 265–285.
[1956]   Some uses of metamathematics, *Brit. J. Phil. Sci.* 7 (1956) 161–173.
[1957]   Independent recursive axiomatization, *J. Symb. Log.* 22 (1957) 109.
[1959]   Interpretation of analysis by means of constructive functionals of finite types, in Heyting [1959], pp. 101–128.
[1960]   La prédicativité, *Bull. Soc. Math. France* 88 (1960) 371–391.
[1962]   On weak completeness of intuitionistic predicate logic, *J. Symb. Log.* 27 (1962) 139–158.
[1965]   Mathematical logic, in *Lectures on modern mathematics*, Saaty ed., Wiley, 1965, vol. 3, pp. 95–195.
[1965a] Model-theoretic invariants: applications to recursive and hyperarithmetic operations, in *The theory of models*, Addison et al. eds., North Holland, 1965, pp. 190–205.
[1966]   Mathematical logic: what has it done for the philosophy of mathematics?, in *Bertrand Russell. Philosopher of the century*, Schoemann ed., Allen and Unwin, 1966, pp. 201–272.
[1968]   A survey of proof theory, *J. Symb. Log.* 33 (1968) 321–388.
[1970]   Church's thesis: a kind of reducibility axiom for constructive mathematics, in *Intuitionism and proof theory*, Kino et al. eds., North Holland, 1970, pp. 121–150.
[1970a] Hilbert's programme and the search for automatic proof procedures, *Springer Lect. Not. Math.* 125 (1970) 128–146.
[1971]   Some reasons for generalizing recursion theory, in *Logic Colloquium '69*, Gandy et al. eds., North Holland, 1971, pp. 139–198.
[1972]   Which number-theoretic problems can be solved in recursive progressions on Π_1^1 paths through $\mathcal{O}$?, *J. Symb. Log.* 37 (1972) 311–334.
[1974]   A notion of mechanistic theory, *Synth.* 29 (1974) 11–26.
[1982]   Review of Pour El and Richards [1979] and [1981], *J. Symb. Log.* 47 (1982) 900-902.
[1987]   Church's Thesis and the ideal of informal rigour, *Notre Dame J. Form. Log.* 28 (1987) 499–519.

Kreisel, G., and Krivine, J.K.
[1966]   *Elements de logique mathématique*, Durad, 1966, transl. North Holland, 1967.

Kreisel, G., Lacombe, D., and Shoenfield, J.R.
[1957]   Fonctionnelles récursivement définissable et fonctionnelles récursives, *Compt. Rend. Acad. Sci.* 245 (1957) 399–402.
[1959]   Partial recursive functionals and effective operations, in Heyting [1959], pp. 195–207.

Kreisel, G., and Tait, W.W.
[1961]   Finite definability of number-theoretical functions and parametric completeness of equation calculi, *Zeit. Math. Log. Grund. Math.* 7 (1961) 28–38.

Kreisel, G., and Troelstra, A.S.
[1970]   Formal systems for some branches of intuitionistic analysis, *Ann. Math. Log.* 1 (1970) 229–387.

Kreitz, C., and Weihrauch, K.
[1984] A unified approach to constructive and recursive analysis, *Springer Lect. Not. Math.* 1104 (1984) 259–278.

Kripke, S.
[1964] Transfinite recursion on admissible ordinals, *J. Symb. Log.* 29 (1964) 161–162.

Kronecker, L.
[1887] Über den Zahlenbegriff, *J. für Math.* 101 (1887) 337–355.

Kučera, A.
[1985] Measure, Π_1^0 classes and complete extensions of $\mathcal{P}\mathcal{A}$, *Springer Lect. Not. Math.* 1141 (1985) 245–259.
[1986] An alternative, priority-free solution to Post's problem, *Springer Lect. Not. Comp. Sci.* 233 (1986) 493–500.
[1988] On the role of $0'$ in recursion theory, in *Logic Colloquium '86*, Drake et al. eds., North Holland, 1988, pp. 133–141.
[1989] On the use of diagonally nonrecursive functions, in *Logic Colloquium '87*, Ebbinghaus et al. eds., North Holland, 1989, pp. 219–239.

Kunen, K.
[1980] *Set Theory*, North Holland, 1980.

Kuratowski, C.
[1936] Sur les théorèmes de separation dans la théorie des ensembles, *Fund. Math.* 26 (1936) 183–191.
[1958] *Topologie*, Warsaw, 1958, transl. Academic Press, 1966.

Kuratowski, C., and Mostowski, A.
[1968] *Set Theory*, North Holland, 1968.

Kuratowski, C., and Tarski, A.
[1931] Les opérations logiques et les ensembles projectifs, *Fund. Math.* 17 (1931) 240–248.

Kurtz, S.A.
[1983] Notions of weak genericity, *J. Symb. Log.* 48 (1983) 764–770.

Kušner, B.A.
[1973] *Lectures on constructive mathematical analysis*, Nauka, 1973.

La Budde, R.A.
[1980] Discrete Hamiltonian mechanics, *Int. J. Gen. Syst.* 6 (1980) 3–12.

Lachlan, A.H.
[1964] Effective operations in a general setting, *J. Symb. Log.* 29 (1964) 163–178.
[1964a] Standard classes of recursively enumerable sets, *Zeit. Math. Log. Grund. Math.* 10 (1964) 23–42.
[1965] Some notions of reducibility and productiveness, *Zeit. Math. Log. Grund. Math.* 11 (1965) 17–44.
[1965a] On recursive enumerations without repetitions, *Zeit. Math. Log. Grund. Math.* 11 (1965) 209–220.
[1965b] Effective inseparability for sequences of sets, *Proc. Am. Math. Soc.* 16 (1965) 647–653.
[1966] A note on universal sets, *J. Symb. Log.* 31 (1966) 573–574.
[1967] On recursive enumerations without repetitions: a correction, *Zeit. Math. Log. Grund. Math.* 13 (1967) 99–100.
[1968] Complete recursively enumerable sets, *Proc. Am. Math. Soc.* 19 (1968) 99–102.
[1968a] Distributive initial segments of the degrees of unsolvability, *Zeit. Math. Log. Grund. Math.* 14 (1968) 457–472.

[1969]   Initial segments of one-one degrees, *Pac. J. Math.* 29 (1969) 351–366.
[1970]   Initial segments of many-one degrees, *Can. J. Math.* 22 (1970) 75–85.
[1971]   Solution to a problem of Spector, *Can. J. Math.* 23 (1971) 247–256.
[1972]   Two theorems on many-one degrees of recursively enumerable sets, *Alg. Log.* 11 (1972) 216–229, transl. 11 (1972) 127–132.
[1972a] Recursively enumerable many-one degrees, *Alg. Log.* 11 (1972) 326–358, transl. 11 (1972) 186–202.
[1975]   *wtt*-complete sets are not necessarily *tt*-complete, *Proc. Am. Math. Soc.* 48 (1975) 429–434.

Lachlan, A.H., and Lebeuf, R.
[1976]   Countable initial segments of the degrees of unsolvability, *J. Symb. Log.* 41 (1976) 289–300.

Lacombe, D.
[1954]   Sur le semi-réseau constitué par les degrès d'indecidabilité récursive, *Compt. Rend. Ac. Sci.* 239 (1954) 1108–1109.
[1955]   Extension de la notion de fonction récursive aux fonctions d'une ou plusieurs variables réelles I, II, III, *Compt. Rend. Acad. Sci.* 240 (1955) 2478–2480, and 241 (1955) 13–14, 151–153.
[1957]   Les ensembles récursivement ouverts ou fermès et leurs applications a l'analyse récursive, *Compt. Rend. Acad. Sci.* 244 (1957) 838–840, 996–997, and 245 (1957) 1040–1043.
[1960]   La théorie des fonctions récursives et ses applications, *Bull. Soc. Math. Fran.* 88 (1960) 393–468.
[1964]   Deux généralisations de la notion de récursivité relative, *Compt. Rend. Acad. Sci.* 258 (1964) 3410–3413.
[1965]   *Propriétés récursives des structures énumérées*, Dissertation, Université de Paris, 1965.

Lambek, J.
[1961]   How to program an infinite abacus, *Can. Math. Bull.* 4 (1961) 295–302.

La Mettrie, J.O. de
[1748]   *L'homme machine*, Leida, 1748.

Landauer, R.
[1961]   Irreversibility and heat generation in the computing process, *I.B.M. J. Res. Dev.* 5 (1961) 183–191.
[1976]   Fundamental limitations in the computational process, *Ber. Bunseng. Phys. Chem.* 80 (1976) 1041–1056.
[1982]   Uncertainty principle and minimal energy dissipation in the computer, *Int. J. Theor. Phys.* 21 (1982) 183–297.
[1985]   Fundamental physical limitations of the computational process, *Ann. New York Acad. Sci.* 426 (1985) 161–170.

Landin, P.J.
[1963]   The mechanical evaluation of expressions, *Comp. J.* 6 (1963) 308–320.

Langton, C. G.
[1984]   Self-reproduction in cellular automata, in Farmer, Toffoli and Wolfram [1984], pp. 135–144.

Laventrieff, M.
[1925]   Sur les sous-classes de la classification de M. Baire, *Compt. Rend. Acad. Sci.* 180 (1925) 111-114.

Lavrov, I.A.
[1963]   Effective inseparability of the sets of identically true formulae and finitely refuta-

ble formulae for certain theories, *Alg. Log.* 2 (1963) 5–18.

[1967] Use of arithmetical progressions of order k for the construction of basis of the algebra of primitive recursive functions, *Dokl. Acad. Nauk* 172 (1967) 279–282, transl. 8 (1967) 83–86.

[1968] Answer to a question of Young, *Alg. Log.* 7 (1968) 48–54, transl. 7 (1968) 98–101.

Lebesgue, H.

[1905] Sur les fonctions représentables analytiquement, *J. de Math.* 1 (1905) 139–216.

Lee, C.H.

[1963] A Turing machine which prints its own code script, *Proc. Symp. Math. Th. Aut.* (1963) 155–164.

Leeds, S., and Putnam, H.

[1974] Solution to a problem of Gandy, *Fund. Math.* 81 (1974) 99–106.

Leibniz, G.W.

[1666] *Dissertatio de arte combinatoria*, 1666.

[1903] *Opuscules et fragments inédits*, Couturat ed., Paris, 1903.

Lerman, M.

[1969] Some nondistributive lattices as initial segments of the degrees of unsolvability, *J. Symb. Log.* 34 (1969) 85–98.

[1971] Initial segments of the degrees of unsolvability, *Ann. Math.* 93 (1971) 365–389.

[1972] On suborderings of the α-recursively enumerable α-degrees, *Ann. Math. Log.* 4 (1972) 369–392.

[1983] *The degrees of unsolvability*, Springer, 1983.

Lerman, M., and Shore, R.A.

[1988] Decidability and invariant classes for degree structures, *Trans. Am. Math. Soc.* 310 (1988) 669–692.

Levy, A.

[1960] Axiom schemata of strong infinity in axiomatic set theory, *Pac. J. Math.* 10 (1960) 223–238.

[1960a] A generalization of Gödel's notion of constructibility, *J. Symb. Log.* 25 (1960) 147–155.

[1965] *A hierarchy of formulas in set theory*, Memoirs A.M.S. 57, 1965.

[1979] *Basic set theory*, Springer, 1979.

Lewis, H.R.

[1979] *Unsolvable cases of quantificational formulas*, Addison Wesley, 1979.

Li, Xiang

[1983] Effectively immune sets, program index sets and effectively simple sets, in *Southeastern Asian conference on logic*, Chong et al. eds., North Holland, 1983, pp. 97–105.

Lipschitz, L., and Rubel, L.A.

[1987] A differential algebraic replacement theorem, and analog computability, *Proc. Am. Math. Soc.* 99 (1987) 367–372.

Löb, M.H.

[1955] Solution of a problem of Leon Henkin, *J. Symb. Log.* 20 (1955) 115–118.

[1970] A model-theoretic characterization of effective operations, *J. Symb. Log.* 35 (1970) 217–222, correction ibid. 39 (1974) 225.

Longo, G., and Moggi, E.

[1984] The hereditary partial effective functionals and recursion theory in higher types, *J. Symb. Log.* 49 (1984) 1319–1332.

Löwenheim, L.
[1915]   Uber Möglichkeiten im Relativkalkül, *Math. Ann.* 76 (1915) 447–470, transl. in Van Heijenoort [1967], pp. 232–251.

Luckham, D., Park, D.
[1964]   The undecidability of the equivalence problem for program schemata, *Mimeogr. notes*, 1964.

Lusin, N.
[1917]   Sur la classification de M. Baire, *Compt. Rend. Acad. Sci.* 164 (1917) 91–94.
[1925]   Sur les ensembles projectifs de M. Henri Lebesgue, *Compt. Rend. Acad. Sci.* 180 (1925) 1318–1320, 1572–1574, 1817–1819.
[1930]   Sur le problème de M.J. Hadamard d'uniformization des ensembles, *Compt. Rend. Acad. Sci.* 190 (1930) 349–351.
[1930a] *Leçons sur les ensembles analytiques et leurs applications*, Gauthier-Villar, 1930.

Lusin, N., and Sierpinski, W.
[1923]   Sur un ensemble non mesurable, *J. Math.* 2 (1923) 53–72.

Lynch, N.A.
[1974]   Approximations to the halting problem, *J. Comp. Syst. Sci.* 9 (1974) 143–150.

Machover, M.
[1961]   The theory of transfinite recursion, *Bull. Am. Math. Soc.* 67 (1961) 575–578.

Mac Lane, S.
[1971]   *Categories for the working mathematician*, Springer, 1971.

Machtey, M., Winklmann, K., and Young, P.R.
[1978]   Simple Gödel numberings, isomorphisms, and programming properties, *S.I.A.M. J. Comp.* 7 (1978) 39–60.

Malc'ev, A.A.
[1981]   Construction of the m-jump, *Sib. Math. J.* 22 (1981) 129–135, transl. 22 (1981) 583–589.
[1984]   On the structure of the families of immune, hyperimmune, and hyperhyperimmune sets, *Mat. Sborn.* 124 (1984) 307–319, transl. 52 (1985) 301–313.
[1985]   The structure of the uppersemilattice of btt_1-degrees, *Sib. Math. J.* 26 (1985) 132–139, transl. 26 (1985) 264–270.

Malc'ev, A.I.
[1961]   Constructive algebras, *Usp. Math. Nauk.* 16 (1961) 3–60, transl. in [1971], pp. 148–214.
[1963]   Sets with complete numberings, *Alg. Log.* 2 (1963) 4–29, transl. in [1971], pp. 287–312.
[1964]   Toward a theory of computable families of objects, *Alg. Log.* 3 (1964) 5–31, transl. in [1971], pp. 353–378.
[1965]   *Algorithms and recursive functions*, Nauka, 1965.
[1971]   *The metamathematics of algebraic systems*, North Holland, 1971.

Mandelbrot, B.
[1980]   Fractal aspects of the iteration of $z \mapsto \lambda(1-z)$ for complex λ, *Ann. N.Y. Acad. Sci.* 357 (1980) 249–259.
[1982]   *The fractal geometry of nature*, Freeman, 1982.

Manna, Z.
[1974]   *Mathematical theory of computation*, McGraw Hill, 1974.

Marchenkov, S.S.
[1971]   On minimal enumerations of the class of recursively enumerable sets, *Dokl. Acad.*

Nauk 198 (1971) 530–532, transl. 12 (1971) 843–846.

[1976]   One class of partial sets, *Mat. Zam.* 20 (1976) 473–478, transl. 20 (1976) 823–825.

[1976a] On the comparison of the uppersemilattice of r.e. *m*-degrees and *tt*-degrees, *Mat. Zam.* 20 (1976) 19–26, transl. 20 (1976) 567–570.

Marek, W., and Srebrny, M.

[1974]   Gaps in the constructible universe, *Ann. Math. Log.* 6 (1974) 359–394.

Markov, A.A.

[1947]   Impossibility of certain algorithms in the theory of associative systems, *Dokl. Acad. Nauk* 55 (1947) 587–590, transl. 55 (1947) 583–586.

[1951]   The theory of algorithms, *Trud. Math. Ist. Stekl.* 38 (1951) 176–189, A.M.S. transl. 15 (1960) 1–14.

[1954]   *The theory of algorithms*, Trud. Math. Inst. Stekl. 42, 1954.

Martin, D.A.

[1963]   A theorem on hyperhypersimple sets, *J. Symb. Log.* 28 (1963) 273–278.

[1966]   Completeness, the recursion theorem and effectively simple sets, *Proc. Am. Math. Soc.* 17 (1966) 838–842.

[1966a] Classes of recursively enumerable sets and degrees of unsolvability, *Zeit. Math. Log. Grund. Math.* 12 (1966) 295–310.

[1967]   Measure, category and degrees of unsolvability, *Mimeographed Notes*, 1967.

[1968]   The axiom of determinateness and reduction principles in the analytical hierarchy, *Bull. Am. Math. Soc.* 74 (1968) 687–689.

[1975]   Borel determinacy, *Ann. Math.* 102 (1975) 363–371.

Martin, D.A., and Pour El, M.B.

[1970]   Axiomatizable theories with few axiomatic extensions, *J. Symb. Log.* 35 (1970) 205–209.

Martin, D.A., and Solovay, R.M.

[1969]   A basis theorem for Σ_3^1 sets of reals, *Ann. Math.* 89 (1969) 138–160.

[1970]   Internal Cohen extensions, *Ann. Math. Log.* 2 (1970) 143–178.

Martin, R.L., ed.

[1978]   *The paradox of the liar*, Ridgeview, 1978.

[1984]   *Recent essays on truth and the liar paradox*, Oxford University Press, 1984.

Marx, K., and Engels, F.

[1848]   *Manifest der Kommunistichen Partei*, London, 1848.

Matijasevich, Y.

[1970]   Enumerable sets are diophantine, *Dokl. Acad. Nauk* 191 (1970) 279–282, transl. 11 (1970) 354–357.

[1972]   Diophantine sets, *Usp. Mat. Nauk* 27 (1972) 185–222, transl. 27 (1972) 124–164.

McCarthy, J.

[1960]   Recursive functions of symbolic expressions and their computation by machine, *Comm. Ass. Comp. Mach.* 3 (1960) 184–195.

[1963]   A basis for a mathematical theory of computation, in *Computer programming and formal systems*, Braffort et al. eds., North Holland, 1963, pp. 33–70.

McCarty, C.

[1986]   Realizability and recursive set theory, *Ann. Pure Appl. Log.* 32 (1986) 153–183.

[1987]   Variations on a thesis: intuitionism and computability, *Notre Dame J. Form. Log.* 28 (1987) 536–580.

McClelland, J.L., and Rumelhart, D.E., eds.

[1986]   *Parallel distributed processing*, M.I.T. Press, 1986.

McCulloch, W., and Pitts, W.

[1943] A logical calculus of the ideas immanent in nervous activity, *Bull. Math. Biophys.* 5 (1943) 115–133.

McLaughlin, T.G.

[1962] A note on contraproductive domains, *Math. Scand.* 11 (1962) 175–178.

[1962a] On an extension of a theorem of Friedberg, *Notre Dame J. Form. Log.* 3 (1962) 270–273.

[1964] On contraproductive sets which are not productive, *Zeit. Math. Log. Grund. Math.* 10 (1964) 49–52.

[1965] On a class of complete sets, *Can. Math. Bull.* 8 (1965) 33–37.

[1966] Retraceable sets and recursive permutations, *Proc. Am. Math. Soc.* 17 (1966) 427–429.

[1973] On retraceable sets with rapid growth, *Proc. Am. Math. Soc.* 40 (1973) 573–576.

[1982] *Regressive sets and the theory of isols*, Dekker, 1982.

McNaughton, R.

[1954] A non-standard truth definition, *Proc. Am. Math. Soc.* 5 (1954) 505–509.

Medvedev, Y.T.

[1955] On non-isomorphic recursively enumerable sets, *Dokl. Acad. Nauk* 102 (1955) 211–214.

Melzak, Z.A.

[1961] An informal arithmetical approach to computability and computation, *Can. Math. Bull.* 4 (1961) 279–293.

Mermin, D.N.

[1985] Is the moon there when nobody looks? Reality and quantum theory, *Phys. Today* 38 (1985) 38–47.

Metakides, G., and Nerode, A.

[1979] Effective content of field theory, *Ann. Math. Log.* 17 (1979) 289–320.

[1982] The introduction of non-recursive methods into mathematics, in *The L.E.J. Brouwer centenary symposium*, Troelstra et al. eds., North Holland, 1982, pp. 319–334.

Meyer, A.R.

[1982] What is a model of the lambda calculus?, *Inf. Contr.* 52 (1982) 87–122.

Meyer, A.R., and Ritchie, D.M.

[1967] The complexity of loop programs, *Proc. Ass. Comp. Mach. Conf.* 22 (1967) 465–469.

Miller, W., and Martin, D.A.

[1968] The degrees of hyperimmune sets, *Zeit. Math. Log. Grund. Math.* 14 (1968) 159–166.

Minsky, M.L.

[1961] Recursive unsolvability of Post's problem of tag and other topics in the theory of Turing machines, *Ann. Math.* 74 (1961) 437–454.

[1967] *Computation: finite and infinite machines*, Prentice Hall, 1967.

Mirimanoff, D.

[1917] Les antinomies de Russell et de Burali-Forti et le problème fondamental de la théorie des ensembles, *Eins. Math.* 19 (1917) 37–52.

Mitchell, R.

[1966] A generalization of productive sets, *J. Symb. Log.* 31 (1966) 455–459.

Mohrherr, J.
[1984]   Density of a final segment of the truth-table degrees, *Pac. J. Math.* 115 (1984)
         409–419.

Moldestad, J., Stoltenberg-Hansen, V., Tucker, J.V.
[1980]   Finite algorithmic procedures and inductive definability, *Math. Scand.* 46 (1980)
         62–76.
[1980a] Finite algorithmic procedures and computation theories, *Math. Scand.* 46 (1980)
         77–94.

Monk, D.J.
[1969]   *Introduction to set theory*, McGraw Hill, 1969.

Montague, R.M.
[1961]   Semantical closure and non-finite axiomatizability I, in *Infinitistic methods*,
         Pergamon Press, 1961, pp. 45–69.

Moore, E.F.
[1962]   Machine models of self-reproduction, *Proc. Symp. Appl. Math.* 14 (1962) 17–
         23.

Moschovakis, Y.N.
[1964]   Recursive metric spaces, *Fund. Math.* 55 (1964) 215–238.
[1965]   Notations systems and recursive ordered fields, *Comp. Math.* 17 (1965) 40-71.
[1969]   Abstract first-order computability I, *Trans. Am. Math. Soc.* 138 (1969) 427–464.
[1969a] Abstract computability and invariant definability, *J. Symb. Log.* (1969) 605–633.
[1971]   Axioms for computation theories - First draft, in *Logic Colloquium '69*, Gandy et
         al. eds., North Holland, 1971, pp. 199–255.
[1980]   *Descriptive Set Theory*, North Holland, 1980.

Mostowski, A.
[1938]   Über gewisse universelle Relationen, *Ann. Soc. Pol. Math.* 17 (1938) 117–118.
[1947]   On definable sets of positive integers, *Fund. Math.* 34 (1947) 81–112.
[1949]   An undecidable arithmetical statement, *Fund. Math.* 36 (1949) 143–164.
[1951]   A classification of logical systems, *Studia Phil.* 4 (1951) 237–274.
[1959]   A class of models for second-order arithmetic, *Bull. Acad. Pol. Sci.* 7 (1959) 401–
         404.
[1962]   Representability of sets in formal systems, *Proc. Symp. Pure Appl. Math.* 5
         (1962) 29–48.
[1969]   *Constructible sets with applications*, North Holland, 1969.

Muchnik, A.A.
[1956]   Negative answer to the problem of reducibility in the theory of algorithms, *Dokl.
         Acad. Nauk* 108 (1956) 194–197.
[1956a] On the separability of recursively enumerable sets, *Dokl. Acad. Nauk* 109 (1956)
         29–32.
[1958a] Solution of Post's reduction problem and of certain other problems in the theory
         of algorithms, *Trud. Mosk. Math. Obsc.* 7 (1958) 391–401, A.M.S. transl. 29
         (1963) 197–215.

Mulry, P.S.
[1982]   Generalized Banach-Mazur functionals in the topos of recursive sets, *J. Pure
         Appl. Alg.* 26 (1982) 71–83.

Mundici, D.
[1981]   Irreversibility, uncertainty, relativity and computer limitations, *Nuovo Cimento*
         61 (1981) 297–305.

Mycielski, J., and Steinhaus, H.
[1962] A mathematical axiom contradicting the axiom of choice, *Bull. Acad. Pol. Sci.* 10 (1962) 1–3.

Myhill, J.
[1953] Criteria of constructibility for real numbers, *J. Symb. Log.* 18 (1953) 7–10.
[1955] Creative sets, *Zeit. Math. Log. Grund. Math.* 1 (1955) 97–108.
[1956] The lattice of recursively enumerable sets, *J. Symb. Log.* 21 (1956) 220.
[1958] Recursive equivalence types and combinatorial functions, *Bull. Am. Math. Soc.* 64 (1958) 373–376.
[1959] Recursive diagraphs, splinters and cylinders, *Math. Ann.* 138 (1959) 211–218.
[1961] Category methods in recursion theory, *Pac. J. Math.* 11 (1961) 1479–1486.
[1961a] Note on degrees of partial functions, *Proc. Am. Math. Soc.* 12 (1961) 519–521.
[1963] The converse of Moore's Garden-of-Eden theorem, *Proc. Am. Math. Soc.* 14 (1963) 685–686.

Myhill, J., and Shepherdson, J.C.
[1955] Effective operations on partial recursive functions, *Zeit. Math. Log. Grund. Math.* 1 (1955) 310–317.

Nerode, A.
[1957] General topology and partial recursive functionals, *Talks Cornell Summ. Inst. Symb. Log.*, Cornell, 1957, pp. 247–251.
[1961] Extensions to isols, *Ann. Math.* 73 (1961) 362–403.
[1962] Extensions to isolic integers, *Ann. Math.* 75 (1962) 419–448.
[1966] Diophantine correct non-standard models in the isols, *Ann. Math.* 84 (1966) 421–432.

Nerode, A., and Manaster, A.B.
[1971] The degree of the theory of addition of isols, *Notices Am. Math. Soc.* 18 (1971) 83.

Nerode, A., and Remmel, J.
[1985] A survey of lattices of r.e. substructures, *Proc. Symp. Pure Math.* 42 (1985) 323–375.

Nerode, A., and Shore, R.A.
[1980] Second-order logic and first-order theories of reducibility orderings, in *Kleene symposium*, Barwise et al. eds., North Holland, 1980, pp. 181–200.
[1980a] Reducibility orderings: theories, definability and automorphisms, *Ann. Math. Log.* 18 (1980) 61–89.

Newton, I.
[1669] De analysi per aequationes numero terminorum infinitas, in *The mathematical work of Isaac Newton*, Whiteside ed., Cambridge University Press, 1964, vol. II, pp. 206–247.
[1687] *Philosophiae naturalis principia mathematica*, London, 1687.

Nies, A.
[1996] Undecidable fragments of elementary theories, *Alg. Univ.* 35 (1996) 8–33.

Nies, A., Shore, R.A., and Slaman, T.A.
[1998] Interpretability and definability in the recursively enumerable degrees, *Proc. Lond. Math. Soc.* 77 (1998) 241–291.

Nogina, E.J.
[1966] On effectively topological spaces, *Dokl. Acad. Nauk* 169 (1966) 28–31, transl. 7 (1966) 865–868.
[1978] Numbered topological spaces, *Zeit. Mat. Log. Grund. Math.* 24 (1978) 141–176.

Normann, D.
- [1980]   *Recursion on the countable functionals*, Springer Lect. Not. Math. 811, 1980.
- [1981]   The continuous functionals: computations, recursions and degrees, *Ann. Math. Log.* 21 (1981) 1–26.

Novikov, P.S.
- [1954]   On the unsolvability of the word problem in group theory, *Isv. Acad. Nauk* 18 (1954) 485–524, A.M.S. transl. 9 (1958) 1–124.

Odifreddi, P.G.
- [1981]   Strong reducibilities, *Bull. Am. Math. Soc.* 4 (1981) 37–86.
- [1981a] Trees and degrees, *Springer Lect. Not. Math.* 839 (1981) 235–271.
- [1985]   The structure of m-degrees, *Springer Lect. Not. Math.* 1141 (1985) 315–332.

Omanadze, R.
- [1976]   On completeness for recursively enumerable sets, *Bull. Acad. Sci. Georgia* 81 (1976) 529–532.
- [1976a] On one kind of reducibility, *Bull. Acad. Sci. Georgia* 83 (1976) 281–284.
- [1978]   On some generalizations of the notion of productive set, *Sov. Math.* 196 (1978) 84–88, transl. 196 (1978) 65–68.
- [1978a] On reducibilities for the class of recursively enumerable sets, *Bull. Acad. Sci. Georgia* 91 (1978) 549–552.
- [1980]   On bounded Q-reducibility, *Bull. Acad. Sci. Georgia* 100 (1980) 57–60.

Ord-Smith, R.J., and Stephenson, J.
- [1975]   *Computer simulation of continuous systems*, Cambridge University Press, 1975.

Ostrowski, A.
- [1920]   Über Dirichletesche Reihen und algebraische Differentialgleichungen, *Math. Zeit.* 8 (1920) 241–298.

Owings, J.C.
- [1973]   Diagonalization and the recursion theorem, *Notre Dame J. Form. Log.* 14 (1973) 95–99.

Oxtoby, J.C.
- [1957]   The Banach-Mazur game and Banach Category theorem, *Ann. Math. Stud.* 39 (1957) 159–163.

Paliutin, E.
- [1975]   Addendum to the paper of Ershov [1975], *Alg. Log.* 14 (1975) 284–287, transl. 14 (1975) 176–178.

Paris, J.B.
- [1977]   Measure and minimal degrees, *Ann. Math. Log.* 11 (1977) 203–216.

Paris, J.B., and Harrington, L.
- [1977]   A mathematical incompleteness in $\mathcal{PA}$, in *Handbook of Mathematical Logic*, Barwise ed., North Holland, 1977, pp. 1133–1142.

Park, D.M.
- [1970]   The $\mathcal{Y}$ combinator in Scott's lambda-calculus, *Memo*, University of Warwick, 1970.

Paterson, M.S.
- [1968]   Program schemata, *Mach. Intell.* 3 (1968) 19–31.

Peano, G.
- [1889]   *Arithmetices principia, novo methodo exposita*, Torino, 1889, transl. in Van Heijenoort [1967], pp. 85–97.
- [1891]   Sul concetto di numero, *Riv. Mat.* 1 (1891) 87–102.

Pearson, K.
[1892]   *The grammar of science*, Scott, 1892.

Peitgen, H.O., and Richter, P.H.
[1986]   *The beauty of fractals*, Springer, 1986.

Peter, R.
[1934]   Über den Zusammenhang der verschiedenen Begriffe der rekursiven Funktion, *Math. Ann.* 110 (1934) 612–632.
[1935]   Konstruktion nichtrekursiver Funktionen, *Math. Ann.* 111 (1935) 42–60.
[1951]   *Recursive Funktionen*, Akadémiai Kiadó, 1951, transl. Academic Press, 1967.
[1958]   Graphschemata und rekursive Funktionen, *Dial.* 12 (1958) 373–393.
[1959]   Über die Partiell-Rekursivität der durch Graphschemata definierten Zahlentheoretischen Funktionen, *Ann. Univ. Sci. Budap.* 2 (1959) 41–48.
[1963]   Programmierung und partiell-rekursive funktionen, *Acta Mat. Acad. Sci. Hung.* 14 (1963) 373–401.

Peterson, J.L.
[1981]   *Petri net theory and the modeling of systems*, Prentice Hall, 1981.

Petri, C.
[1962]   *Kommunication mit Automaten*, Ph.D. Thesis, University of Bonn, 1962.

Platek, R.A.
[1966]   *Foundations of Recursion Theory*, Ph.D. Thesis, Stanford, 1966.
[1969]   Eliminating the continuum hypothesis, *J. Symb. Log.* 34 (1969) 219–225.

Plotkin, G.D.
[1972]   A set-theoretical definition of application, *Memo*, Univ. of Edinburgh, 1972.

Poincaré H.
[1903]   *La science et l'hypothèse*, Paris, 1903.
[1906]   Les mathématiques et la logique, *Rev. Métaph. Mor.* 14 (1906) 294–317.
[1913]   *Dernieres pensées*, Paris, 1913.

Poliakov, E.A.
[1964]   The algebra of recursive functions, *Alg. Log.* 3.1 (1964) 41–56.
[1964a] On some properties of the algebra of recursive functions, *Alg. Log.* 3.3 (1964) 39–57.

Popper, K.R., and Eccles, J.C.
[1977]   *The self and the brain*, Springer, 1977.

Posner, D.
[1981]   A survey of non r.e. degrees below $0'$, *Lond. Math. Soc. Lect. Not.* 45 (1981) 52–109.

Posner, D., and Epstein, R.L.
[1978]   Diagonalization in degree constructions, *J. Symb. Log.* 43 (1978) 280–283.

Post, E.L.
[1921]   Introduction to a general theory of elementary propositions, *Am. J. Math.* 43 (1921) 163–185, also in Van Heijenoort [1967], pp. 265–283.
[1922]   Absolutely unsolvable problems and relatively undecidable propositions. Account of an anticipation, in Davis [1965], pp. 340– 433.
[1936]   Finite combinatory processes. Formulation I, *J. Symb. Log.* 1 (1936) 103–105, also in Davis [1965], pp. 289–291.
[1943]   Formal reductions of the general combinatorial decision problem, *Am. J. Math.* 65 (1943) 197–215.
[1944]   Recursively enumerable sets of positive integers and their decision problems, *Bull. Am. Math. Soc.* 50 (1944) 284–316, also in Davis [1965], pp. 305–337.

[1947] Recursive unsolvability of a problem of Thue, *J. Symb. Log.* 12 (1947) 1–11, also
        in Davis [1965], pp. 293–303.
[1948] Degrees of recursive unsolvability, *Bull. Am. Math. Soc.* 54 (1948) 641–642.

Poundstone, W.
[1985] *The recursive universe*, Contemporary Books, 1985.

Pour El, M.B.
[1960] A comparison of five 'computable' operators, *Zeit. Math. Log. Grund. Math.* 6
        (1960) 325–340.
[1964] Gödel numberings versus Friedberg numberings, *Proc. Am. Math. Soc.* 15 (1964)
        252–256.
[1968] Effectively extensible theories, *J. Symb. Log.* 33 (1968) 56–68.
[1968a] Independent axiomatization and its relation to the hypersimple set, *Zeit. Math.
        Log. Grund. Math.* 14 (1968) 449–456.
[1974] Abstract computability and its relations to the general purpose analog computer,
        Trans. Am. Math. Soc. 199 (1974) 1–28.

Pour El, M.B., and Howard, W.A.
[1964] A structural criterion for recursive enumeration without repetitions, *Zeit. Math.
        Log. Grund. Math.* 10 (1964) 105–114.

Pour El, M.B., and Kripke, S.
[1967] Deduction-preserving 'recursive isomorphisms' between theories, *Bull. Am.
        Math. Soc.* 73 (1967) 145–148.

Pour El, M.B., and Putnam, H.
[1965] Recursively enumerable classes and their application to recursive sequences of
        formal theories, *Arch. Math. Log. Grund.* 8 (1965) 104–121.

Pour El, M.B., and Richards, I.
[1979] A computable ordinary differential equation which possesses no computable
        solutions, *Ann. Math. Log.* 17 (1979) 61–90.
[1981] The wave equation with computable initial data such that its unique solution is
        not computable, *Adv. Math.* 39 (1981) 215–239.
[1983] Non computability in analysis and physics: a complete determination of the class
        of non computable linear operators, *Adv. Math.* 48 (1983) 44–74.
[1983a] Computability and non computability in classical analysis, *Trans. Am. Math.
        Soc.* 275 (1983) 539–560.

Preston, K., and Duff, M.J.
[1984] *Modern cellular automata*, Plenum Press, 1984.

Priese, L.
[1979] Towards a precise characterization of the complexity of universal and non uni-
        versal Turing machines, *S.I.A.M. J. Comp.* 8 (1979) 508–523.

Putnam, H.
[1957] Decidability and essential undecidability, *J. Symb. Log.* 22 (1957) 39–54.
[1963] A note on constructible sets of integers, *Notre Dame J. Form. Log.* 4 (1963) 270–
        273.
[1964] On hierarchies and systems of notations, *Proc. Am. Math. Soc.* 15 (1964) 44–50.
[1965] Trial and error predicates and the solution to a problem of Mostowski, *J. Symb.
        Log.* 30 (1965) 49–57.

Putnam, H., and Smullyan, R.M.
[1960] Exact separation of recursively enumerable sets within theories, *Proc. Am. Math.
        Soc.* 11 (1960) 574–577.

Quine, W.V.
[1946] Concatenation as basis for arithmetic, *J. Symb. Log.* 11 (1946) 105–114.

Rabin, M.O.
[1960] Computable algebra, general theory and theory of computable fields, *Trans. Am. Math. Soc.* 95 (1960) 341–360.

Rakic, P.
[1975] Local circuit neurons, *Neur. Res. Prog. Bull.* 13 (1975) 291–446.

Rice, H.G.
[1953] Classes of recursively enumerable sets and their decision problems, *Trans. Am. Math. Soc* 74 (1953) 358–366.
[1954] Recursive real numbers, *Proc. Am. Math. Soc.* 5 (1954) 784–790.
[1956] On completely r.e. classes and their key arrays, *J. Symb. Log.* 21 (1956) 304–308.
[1956a] Recursive and recursively enumerable orders, *Trans. Am. Math. Soc.* 83 (1956) 277–300.

Richardson, D.
[1968] Some undecidable problems involving elementary functions of a real variable, *J. Symb. Log.* 33 (1968) 514–520.

Richter, L.
[1979] On automorphisms of the degrees that preserve jumps, *Isr. J. Math.* 32 (1979) 27–31.

Roberts, A., and Bush, M.H., eds.
[1981] *Neurones without impulses,* Cambridge University Press, 1981.

Robinson, J.
[1950] General recursive functions, *Proc. Am. Math. Soc.* 1 (1950) 703–718.
[1955] A note on primitive recursive functions, *Proc. Am. Math. Soc.* 6 (1955) 667–670.
[1968] Recursive functions of one variable, *Proc. Am. Math. Soc.* 19 (1968) 815–820.

Robinson, R.M.
[1947] Primitive recursive functions, *Bull. Am. Math. Soc.* 53 (1947) 925–942.
[1950] An essentially undecidable axiom system, *Proc. Int. Congr. Math.* (1950) 729–730.
[1951] Review of Peter [1951], *J. Symb. Log.* 16 (1951) 282.
[1955] Primitive recursive functions II, *Proc. Am. Math. Soc.* 6 (1955) 663–666.

Robinson, R.W.
[1967] Two theorems on hyperhypersimple sets, *Trans. Am. Math. Soc.* 128 (1967) 531–538.
[1967a] Simplicity of recursively enumerable sets, *J. Symb. Log.* 32 (1967) 162–172.

Rogers, H.
[1958] Gödel numberings of partial recursive functions, *J. Symb. Log.* 23 (1958) 331–341.
[1965] On universal functions, *Proc. Am. Math. Soc.* 16 (1965) 39–44.
[1967] *Theory of recursive functions and effective computability,* McGraw Hill, 1967.

Rose, G.F., and Ullian, J.S.
[1963] Approximations of functions on the integers, *Pac. J. Math.* 13 (1963) 693–701.

Rosenbloom, P.C.
[1950] *The elements of mathematical logic,* Dover Press, 1950.

Rosolini, G.
[1986] *Continuity and effectiveness in topoi,* Ph.D. Thesis, Oxford University, 1986.

Rosser, B.J.
[1935] A mathematical logic without variables, *Ann. Math.* 36 (1935) 127–150.
[1936] Extensions of some theorems of Gödel and Church, *J. Symb. Log.* 1 (1936) 87–91, also in Davis [1965], pp. 230–235.

[1984]   Highlights of the history of the lambda calculus, *Ann. Hist. Comp.* 6 (1984) 337–349.

Rowbottom, F.
[1971]   Some strong axioms of infinity incompatible with the axiom of constructibility, *Ann. Math. Log.* 3 (1971) 1–44.

Rubel, L.A.
[1981]   A universal differential equation, *Bull. Am. Math. Soc.* 4 (1981) 345–349.
[1982]   An elimination theorem for systems of algebraic differential equations, *Houst. J. Math.* 8 (1982) 289–295.
[1985]   The brain as an analog computer, *J. Theor. Neur.* 4 (1985) 73–81.

Rubel, L.A., and Singer, M.F.
[1985]   A differential algebraic elimination theorem with application to analog computability in the calculus of variations, *Proc. Am. Math. Soc.* 94 (1985) 653–658.

Russell, B.
[1903]   *The principles of mathematics*, Cambridge University Press, 1903.
[1906]   On some difficulties in the theory of transfinite numbers and order types, *Proc. Lond. Math. Soc.* 4 (1906) 29–53.
[1908]   Mathematical logic as based on the theory of types, *Am. J. Math.* 30 (1908) 222–262, also in Van Heijenoort [1967], pp. 150–182.

Ryll-Nardzewski, C.
[1952]   The role of the axiom of induction in elementary arithmetic, *Fund. Math.* 39 (1952) 239–263.

Sacks, G.E.
[1961]   On suborderings of degrees of recursive unsolvability, *Zeit. Math. Log. Grund. Math.* 7 (1961) 46–56.
[1963]   *Degrees of unsolvability*, Princeton University Press, 1963, 2nd ed. 1966.
[1964]   A simple set which is not effectively simple, *Proc. Am. Math. Soc.* 15 (1964) 51–55.
[1971]   Forcing with perfect closed sets, *Proc. Symp. Pure Math.* 17 (1971) 331–355.

Sacks, G.E., and Simpson, S.G.
[1972]   The α-finite injury method, *Ann. Math. Log.* 4 (1972) 323–367.

Sammett, J.E.
[1969]   *Programming languages: history and fundamentals*, Prentice Hall, 1969.

Sasso, L.P.
[1971]   *Degrees of unsolvability of partial functions*, Ph.D. Thesis, University of California, Berkeley, 1971.
[1974]   A minimal degree not realizing the least possible jump, *J. Symb. Log.* 39 (1974) 571–573.
[1975]   A survey of partial degrees, *J. Symb. Log.* 40 (1975) 130–140.

Sčedrov, A.
[1984]   Differential equations in constructive analysis and in recursive realizability topos, *J. Pure Appl. Alg.* 33 (1984) 69–80.
[1987]   Some aspects of categorical semantics: sheaves and glueing, in *Logic Colloquium '85*, Jervell et al. eds., North Holland, 1987, pp. 281–301.

Schinzel, B.
[1977]   Decomposition of Gödel-numberings into Friedberg-numberings, *Zeit. Math. Log. Grund. Math.* 23 (1977) 393–399.

Schnorr, C.P.
[1975] Optimal enumerations and optimal Gödel numberings, *Math. Syst. Th.* 8 (1975) 182–191.

Schönfinkel, M.
[1924] Über die Bausteine der mathematischen Logik, *Math. Ann.* 92 (1924) 305–316, transl. in Van Heijenoort [1967], pp. 357–366.

Scott, D.
[1962] Algebras of sets binumerable in complete extension of arithmetic, *Proc. Symp. Pure Appl. Math.* 5 (1962) 117–121.
[1972] Continuous lattices, *Springer Lect. Not. Math.* 274 (1972) 97–136.
[1973] Models for various type-free calculi, *Log. Phil. Meth. Sci.* 4 (1973) 157–187.
[1975] Lambda Calculus and Recursion Theory, in *Proceedings of the Third Scandinavian Logic Symposium*, Kanger ed., North Holland, 1975, pp. 154–193.
[1975a] Some philosophical issues concerning theories of combinators, *Springer Lect. Not. Comp. Sci.* 37 (1975) 346–366.
[1976] Data types as lattices, *S.I.A.M. J. Comp.* 5 (1976) 522–587.
[1977] Logic and programming languages, *Comm. Ass. Comp. Mach.* 20 (1977) 634–641.
[1980] λ-calculus: some models, some philosophy, in *The Kleene Symposium*, Barwise et al. eds., North Holland, 1980, pp. 223–266.
[1980a] Relating theories of the λ-calculus, in *To H.B. Curry: essays on combinatory logic, lambda calculus and formalism*, Seldin et al. eds., Academic Press, 1980, pp. 403–450.
[1982] Domains for denotational semantics, *Springer Lect. Not. Comp. Sci.* 140 (1982) 577–613.

Scott, D., and Tennenbaum, S.
[1960] On the degrees of complete extensions of arithmetic, *Notices Am. Math. Soc.* 7 (1960) 242–243.

Selivanov, V.L.
[1982] On one class of reducibilities in the theory of recursive functions, *Probl. Math. Cyb.* 18 (1982) 83–100.

Selverston, A.I., ed.
[1985] *Model neural network and behavior*, Plenum Press, 1985.

Shanker, S.G.
[1985] Wittgenstein versus Turing on the nature of Church's Thesis, *Notre Dame J. Form. Log.* 28 (1987) 615–649.

Shannon, C.E.
[1938] A symbolic analysis of relay and switching circuits, *Trans. Am. Inst. Electr. Eng.* 57 (1938) 713–723.
[1941] Mathematical theory of the differential analyzer, *J. Math. Phys.* 20 (1941) 337–354.
[1948] A mathematical theory of communication, *Bell Syst. Tech. J.* 27 (1948) 379–423, 623–656.
[1956] A universal Turing machine with two internal states, in Shannon and McCarthy [1956], pp. 157–165.

Shannon, C.E., and McCarthy, J., eds.
[1956] *Automata studies*, Princeton University Press, 1956.

Shepherd, G.M.
[1979] *The synaptic organization of the brain*, Oxford University Press, 1979.

Shepherdson, J.C.
[1951] Inner models for set theory, *J. Symb. Log.* 16 (1951) 161–190.
[1960] Representability of recursively enumerable sets in formal theories, *Arch. Math. Log. Grund.* 5 (1963) 119–127.
[1975] Computation over abstract structures: serial and parallel procedures in Friedman's effective definitional schemes, in *Logic Colloquium '73*, Rose et al. eds., North Holland, 1975, pp. 445–513.
[1985] Algorithmic procedures, generalized Turing algorithms, and elementary Recursion Theory, in Harrington, Morley, Ščedrov, and Simpson [1985], pp. 285–308.

Shepherdson, J.C., and Sturgis, H.E.
[1963] Computability of recursive functions, *J. Ass. Comp. Mach.* 10 (1963) 217–255.

Shoenfield, J.R.
[1957] Quasicreative sets, *Proc. Am. Math. Soc.* 8 (1957) 964–967.
[1958] Degrees of formal systems, *J. Symb. Log.* 23 (1958) 389–392.
[1959] On degrees of unsolvability, *Ann. Math.* 69 (1959) 644–653.
[1960] An uncountable set of incomparable degrees, *Proc. Am. Math. Soc.* 11 (1960) 61–62.
[1960a] Degrees of models, *J. Symb. Log.* 25 (1960) 233–237.
[1961] Undecidable and creative theories, *Fund. Math.* 49 (1961) 171–179.
[1961a] The problem of predicativity, in *Essays on the foundations of mathematics*, Bar Hillel et al. eds., Magnes Press, 1961, pp. 132–142.
[1962] The form of the negation of a predicate, *Proc. Symp. Pure Appl. Math.* 5 (1962) 131–134.
[1966] A theorem on minimal degrees, *J. Symb. Log.* 31 (1966) 539–544.

Shore, R.A.
[1978] On the ∀∃ sentences of α-recursion theory, in *Generalized recursion theory*, Fenstad et al. eds., North Holland, 1978, pp. 331–353.
[1979] The homogeneity conjecture, *Proc. Nat. Acad. Sci.* 76 (1979) 4218–4219.
[1981] The degrees of unsolvability: global results, *Springer Lect. Not. Math.* 859 (1981) 283–301.
[1982] On homogeneity and definability in the first-order theory of the Turing degrees, *J. Symb. Log.* 47 (1982) 8–16.
[1982a] The theories of the truth-table and Turing degrees are not elementarily equivalent, in *Logic Colloquium '80*, Van Dalen et al. eds., North Holland, 1982, pp. 231–237.

Sierpinski, W.
[1924] Sur une propriété des ensembles ambigus, *Fund. Math.* 6 (1924) 1–5.
[1925] Sur une classe d'ensembles, *Fund. Math.* 7 (1925) 237–243.
[1950] *Les ensembles projectifs at analytiques*, Gauthier-Villars, 1950.

Simpson, S.G.
[1974] Post's problem for admissible sets, in *Generalized Recursion Theory*, Fenstad et al. eds., North Holland, 1974, pp. 437–441.
[1977] First-order theory of the degrees of recursive unsolvability, *Ann. Math.* 105 (1977) 121–139.

Skolem, T.
[1920] Logisch-kombinatorische Untersuchungen über die Erfüllbarkeit oder Beweisbarkeit mathematischer Sätze nebst einem Theoreme über dichte Mengen, *Vidensk. Skrifter*, no. 4, transl. in Van Heijenoort [1967], pp. 254–263.
[1923] Begründung der elementaren Arithmetik durch die rekurriende Denkweise ohne Anwendung scheinbarer Veränderlichen mit unendlichem Ausdehnungsbereich, *Vidensk. Skrifter*, no. 6, transl. in Van Heijenoort [1967], pp. 303–333.

[1934] Über die Nicht-charakterisierbarkeit der Zahlenreihe mittels endlich oder abzähl-
bar unendlich vieler Aussagen mit ausschliesslich Zahlenvariablen, *Fund. Math.*
23 (1934) 150–161.

Slaman, T.A., and Woodin, H.W.
[1986] Definability in the Turing degrees, *Ill. J. Math.* 30 (1986) 320–334.

Smith, A.R.
[1972] Simple computation-universal cellular spaces, *J. Ass. Comp. Mach.* 18 (1972)
339–353.

Smorynski, C.
[1977] The incompleteness theorems, in *Handbook of Mathematical Logic*, Barwise ed.,
North Holland, 1977, pp. 821–865.

Smullyan, R.M.
[1958] Undecidability and recursive inseparability, *Zeit. Math. Log. Grund. Math.* 4
(1958) 143–147.
[1961] *Theory of formal systems*, Princeton University Press, 1961.
[1964] Effectively simple sets, *Proc. Am. Math. Soc.* 15 (1964) 893–894.

Soare, R.I.
[1972] The Friedberg-Muchnik theorem re-examined, *Can. J. Math.* 24 (1972) 1070–
1078.
[1987] *Recursively enumerable sets and degrees*, Springer, 1987.

Solomonov, R.
[1964] A formal theory of inductive inference, *Inf. Contr.* 7 (1964) 1–2.

Solovay, R.M.
[1967] A non-constructible Δ_3^1 set of integers, *Trans. Am. Math. Soc.* 127 (1967) 58–75.
[1969] On the cardinality of Σ_2^1 sets of reals, in *Foundations of mathematics*, Bullof et al.
eds., Springer, 1969, pp. 58–73.

Soloviev, V.D.
[1974] Q-reducibility and hyperhypersimple sets, *Probl. Meth. Cyb.* 10 (1974) 121–128.
[1976] Some generalization of the notion of reducibility and creativity, *Math. Univ.*
News 166 (1976) 65–72.
[1976a] Superhypersimple sets, *Math. Univ. News* 165 (1976) 108–110.

Specker, E.
[1949] Nicht konstruktiv beweisbare Sätze der Analysis, *J. Symb. Log.* 14 (1949) 145–
158.
[1959] Der Sätz vom Maximum in der rekursiven Analysis, in Heyting [1959], pp.
254–265.

Spector, C.
[1955] Recursive well-orderings, *J. Symb. Log.* 20 (1955) 151–163.
[1956] On degrees of recursive unsolvability, *Ann. Math.* 64 (1956) 581–592.

Spreen, D., and Young, P.
[1984] Effective operators in a topological setting, *Springer Lect. Not. Math.* 1104
(1984) 437–451.

Statman, R.
[1977] Herbrand's theorem and Gentzen's notion of a direct proof, in *Handbook of Math-*
ematical Logic, Barwise ed., North Holland, 1977, pp. 897–912.

Stillwell, J.
[1972] Decidability of the 'almost all' theory of degrees, *J. Symb. Log.* 37 (1972) 501–
506.

Stoy, J.E.
[1977] *Denotational semantics*, M.I.T. Press, 1977.

Strong, H.R.
[1968] Algebraically generalized recursive function theory, *I.B.M. J. Res. Dev.* 12 (1968) 465–475.
[1971] Translating recursion equations into flowcharts, *J. Comp. Syst. Sci.* 5 (1971) 254–285.

Sudan, G.
[1927] Sur le nombre transfini ω^ω, *Bull. Soc. Roum. Sci.* 30 (1927) 11-30.

Suslin, M.
[1917] Sur une définition des ensembles mesurables, *Compt. Rend. Acad. Sci.* 164 (1917) 88–91.

Suzuki, Y.
[1959] Enumerations of recursive sets, *J. Symb. Log.* 24 (1959) 311.

Takeuti, G.
[1957] On the theory of ordinals numbers, *J. Math. Soc. Japan* 9 (1957) 93–113.
[1960] On the recursive functions of ordinal numbers, *J. Math. Soc. Japan* 12 (1960) 119–127.

Takeuti, G., and Kino, A.
[1962] On hierarchies of predicates of ordinal numbers, *J. Math. Soc. Japan* 14 (1962) 199–232.

Tarski, A.
[1931] Sur les ensembles définissables de nombres réels, *Fund. Math.* 17 (1931) 210–239.
[1933] Einige Betrachtungen über die Begriffe ω-Widerspruchsfreiheit und der ω-Vollständigkeit, *Monash. Math. Phys.* 40 (1933) 97–112.
[1936] Der Wahrheitsbegriff in der formalisierten Sprachen, *Studia Phil.* 1 (1936) 261–405, transl. in [1956], pp. 152–278.
[1949] On essential undecidability, *J. Symb. Log.* 14 (1949) 75–76.
[1955] A lattice-theoretical fixed-point theorem and its applications, *Pac. J. Math.* 5 (1955) 285–309.
[1956] *Logic, semantics, metamathematics*, Oxford University Press, 1956.

Tarski, A., Mostowski, A., and Robinson, R.M.
[1953] *Undecidable theories*, North Holland, 1953.

Tennenbaum, S.
[1959] Non archimedian models for arithmetic, *Notices Am. Math. Soc.* 6 (1959) 270.
[1961] Degrees of unsolvability and the rate of growth of functions, *Notices Am. Math. Soc.* 8 (1961) 608.
[1961a] Inseparable sets and reducibility, *Technical note*, Univ. of Michigan, 1961.

Thatcher, T.W.
[1963] The construction of a self-describing Turing machine, *Proc. Symp. Math. Th. Aut.* (1963) 165–171.

Thomason, S.K.
[1970] A theorem on initial segments of degrees, *J. Symb. Log.* 35 (1970) 41–45.

Thue, A.
[1914] Probleme über Veränderungen von Zeichenreihen nach gegebenen Regeln, *Skrif. Viden. Krist.* 10 (1914).

Titgemeyer, D.
[1962] Unterseuchen über die Struktur des Kleene-Postchen Halbverbandes der Grade der rekursiven Unlösbarkeit, *Arch. Math. Log. Grund.* 8 (1962) 45–62.

Toffoli, T.

[1977] Computation and construction universality of reversible cellular automata, *J. Comp. Syst. Sci.* 15 (1977) 213–231.

[1981] Bicontinuous extensions of invertible combinatorial functions, *Math. Syst. Th.* 14 (1981) 13–23.

[1984] Cellular automata as an alternative to (rather than an approximation of) differential equations in modelling physics, in Farmer, Toffoli and Wolfram [1984], pp. 117–127.

Toffoli, T., and Margolus, N.

[1987] *Cellular automata machines*, M.I.T. Press, 1987.

Trakhtenbrot, B.A.

[1953] On recursive separability, *Dokl. Acad. Nauk* 88 (1953) 953–956.

[1955] Tabular representation of recursive operators, *Dokl. Acad. Nauk* 101 (1955) 417–420.

[1970] On autoreducibility, *Dokl. Acad. Nauk.* 192 (1970) 1224–1227, transl. 11 (1972) 814–817.

Trakhtenbrot, B.A., and Bardzin, Y.M.

[1973] *Finite automata: behavior and synthesis*, North Holland, 1973.

Troelstra, A.S.

[1973] *Metamathematical investigation of intuitionistic arithmetic and analysis*, Springer Lect. Not. Math. 344, 1973.

Turing, A.M.

[1936] On computable numbers with an application to the Entscheidungsproblem, *Proc. London Math. Soc.* 42 (1936) 230–265, corrections ibid. 43 (1937) 544–546, also in Davis [1965], pp. 116–154.

[1937] Computability and λ-definability, *J. Symb. Log.* 2 (1937) 153–163.

[1937a] The p-function in λK-conversion, *J. Symb. Log.* 2 (1937) 164.

[1939] Systems of logic based on ordinals, *Proc. Lond. Math. Soc.* 45 (1939) 161–228, also in Davis [1965], pp. 155–222.

[1950] Computing machinery and intelligence, *Mind* 59 (1950) 433–460.

Turner, D.A.

[1979] A new implementation technique for applicative languages, *Softw. Pract. Exp.* 9 (1979) 31–49.

Ullian, J.S.

[1960] Splinters of recursive functions, *J. Symb. Log.* 25 (1960) 33–38.

Uspenskii, V.A.

[1953] Gödel's theorem and the theory of algorithms, *Dokl. Acad. Nauk* 91 (1953) 737–740, A.M.S. transl. 23 (1963) 103–107.

[1955] On enumeration operators, *Dokl. Acad. Nauk* 103 (1955) 773–776.

[1955a] Systems of enumerable sets and their enumerations, *Dokl. Acad. Nauk* 105 (1955) 1155–1158.

[1956] Enumeration operators and the concept of program, *Usp. Mat. Nauk* 11 (1956) 172–176.

[1957] Some remarks on r.e. sets, *Zeit. Math. Log. Grund. Math.* 3 (1957) 157–170, A.M.S. transl. 23 (1963) 89–101.

Van Heijenoort, J., ed.

[1967] *From Frege to Gödel*, Harvard University Press, 1967.

Vaught, R.L.

[1960] On a theorem of Cobham concerning undecidable theories, *Log. Phil. Meth. Sci.* 1 (1960) 14–25.

Vichniac, G.Y.
    [1984]   Simulating physics with cellular automata, in Farmer, Toffoli and Wolfram [1984], pp. 96–116.

Von Neumann, J.
    [1923]   Zur Einführung der transfiniten Zahlen, *Acta Sci. Math.* 1 (1923) 199–208, transl. in Van Heijenoort [1967], pp. 347–354.
    [1928]   Über die Definition durch transfinite Induktion und verwandte Fragen der allgemeinen Mengenlehre, *Math. Ann.* 99 (1928) 373–391.
    [1932]   *Mathematische grundlagen der Quantum-mechanik*, Berlin, 1932.
    [1951]   The general and logical theory of automata, in *The Hixon Symposium*, Jeffress ed., Wiley, 1948, pp. 1–31.
    [1954]   The role of mathematics in the sciences and society, in *Collected works* 6 (1963) 477–498.
    [1956]   Probabilistic logic and the synthesis of reliable organisms from unreliable components, in Shannon and McCarthy [1956], pp. 43–98.
    [1958]   *The computer and the brain*, Yale University Press, 1958.
    [1966]   *The theory of self-reproducing automata*, University of Illinois Press, 1966.

Vučkovich, V.D.
    [1967]   Creative and weakly creative sequences of r.e. sets, *Proc. Am. Math. Soc.* 17 (1967) 478–483.

Wagner, E.G.
    [1969]   Uniformly reflexive structures: on the nature of Gödelizations and relative computability, *Trans. Am. Math. Soc.* 144 (1969) 1–41.

Wang, H.
    [1957]   A variant of Turing's theory of calculating machines, *J. Ass. Comp. Mach.* 4 (1957) 63–92.
    [1957a]   Universal Turing machines: an exercise in coding, *Zeit. Math. Log. Grund. Math.* 3 (1957) 69–80.
    [1974]   *From mathematics to philosophy*, Routledge and Kegan, 1974.

Wang, P.
    [1974]   The undecidability of the existence of zeros of real elementary functions, *J. Ass. Comp. Mach.* 21 (1974) 586–589.

Watson, J.D.
    [1970]   *Molecular biology of the gene*, Benjamin, 1970.

Webb, J.C.
    [1980]   *Mechanism, mentalism and metamathematics*, Reidel, 1980.

Wexelblat, R.L., ed.
    [1981]   *History of programming languages*, Academic Press, 1981.

Weyl, H.
    [1918]   *Das Kontinuum*, Veit, 1918.

Wiener, N.
    [1948]   *Cybernetics*, Wiley, 1948.

Winograd, S., and Cowan, J.D.
    [1963]   *Reliable computation in the presence of noise*, M.I.T. Press, 1963.

Wirth, N.
    [1971]   The programming language PASCAL, *Acta Inf.* 1 (1971) 35–63.

Wittgenstein, L.
    [1921]   Logisch-philosophische Abhandlung, *Ann. Naturphil.* 14 (1921) 185–262.

Wolfram, S., ed.
[1986]   *Theory and applications of cellular automata*, World Scientific, 1986.

Yates, C.E.M.
[1962]   Recursively enumerable sets and retracing functions, *Zeit. Mat. Log. Grund. Math.* 8 (1962) 331–345.
[1965]   Three theorems on the degrees of r.e. sets, *Duke Math. J.* 32 (1965) 461–468.
[1969]   On the degrees of index sets II, *Trans. Am. Math. Soc.* 135 (1969) 249–266.
[1970]   Initial segments of the degrees of unsolvability, part I, in *Mathematical logic and foundations of set theory*, Bar Hillel ed., North Holland, 1970, pp. 63–83.
[1972]   Initial segments and implications for the structure of degrees, *Springer Lect. Not. Math.* 255 (1972) 305–335.
[1976]   Banach-Mazur games, comeager sets and degrees of unsolvability, *Math. Proc. Cambr. Phil. Soc.* 79 (1976) 195–220.

Young, J.Z.
[1978]   *Programs of the brain*, Oxford University Press, 1978.

Young, P.R.
[1964]   On reducibility by recursive functions, *Proc. Am. Math. Soc.* 15 (1964) 889–892.
[1964a]   A note on pseudo-creative sets and cylinders, *Pac. J. Math.* 14 (1964) 749–753.
[1965]   On semi-cylinders, splinters and bounded truth-table reducibility, *Trans. Am. Math. Soc.* 115 (1965) 329–339.
[1966]   A theorem on recursively enumerable classes and splinters, *Proc. Am. Math. Soc.* 1050–1056.
[1966a]   Linear orderings under one-one reducibility, *J. Symb. Log.* 31 (1966) 70–85.
[1967]   On pseudo-creative sets, splinters and bounded truth-table reducibility, *Zeit. Mat. Log. Grund. Math.* 13 (1967) 25–31.
[1968]   An effective operator, continuous but not partial recursive, *Proc. Am. Math. Soc.* 19 (1968) 103–108.
[1969]   Toward a theory of enumerations, *J. Ass. Comp. Mach.* 16 (1969) 328–348.

Zakharov, S.D.
[1984]   On *e* and *s*-degrees, *Alg. Log.* 23 (1984) 395–406, transl. 273–281.
[1986]   Degrees of denumerability reducibilities, *Alg. Log.* 25 (1986) 121–135, transl. 25 (1986) 75–85.

Zbierski, P.
[1971]   Models for higher order arithmetics, *Bull. Acad. Polon. Sci.* 19 (1971) 557–562.

Zermelo, E.
[1908]   Untersuchungen über die Grundlagen der Mengenlehre I, *Math. Ann.* 65 (1908) 261–281, transl. in Van Heijenoort [1967], pp. 200–215.

Notation Index

Introduction

ω	set of natural numbers	13
$\mathcal{P}(\omega)$	power set of ω (set of all subsets of ω)	13
ω^ω	set of total functions from ω to ω	13
$\mathcal{P}$	set of partial functions from ω to ω	13
$|A|$	cardinality of A	14
$A \oplus B$	disjoint union of A and B	14
$A \cdot B$	recursive product of A and B	14
c_A	characteristic function of A	14
$(\exists x \leq y)$	bounded existential quantifier	15
$(\forall x \leq y)$	bounded universal quantifier	15

Chapter I

$\mathcal{S}$	successor function	19
μ	least number operator	21
$\mathcal{O}$	constant zero function	22
$\mathcal{I}_i^n$	i-th projection of n arguments	22
$\mathcal{Q}$	Robinson Arithmetic	23
$\mathcal{PA}$	Peano Arithmetic	24
$x - y$	integer difference	25
$\sum_{y \leq z}$	bounded sum	25
$\prod_{y \leq z}$	bounded product	25
$x|y$	x divides y	25
$Pr(x)$	x is a prime	25
p_x	the x-th prime number	26
$\mu y_{\leq z}$	bounded μ-operator	26
$exp(y, k)$	exponent of k in the decomposition of y	26
$\mathcal{J}$	pairing function	27
$\mathcal{R}, \mathcal{L}$	right and left inverses of $\mathcal{J}$	27
β	Gödel's β-function	28
$\overline{n}$	numeral for n	32
$\mathcal{E}(f_1, \ldots, f_n; \vec{z})$	system of equations	32
$\mathcal{R}$	Tarski, Mostowski and Robinson arithmetic	44
q_i	state of a Turing machine	48
s_i	symbol of a Turing machine	48
I_i	instruction of a Turing machine	48

$:=$	assignment	62
λ	abstraction operator	76
$\mathcal{Y}$	fixed-point paradoxical combinator	79
$\mathbf{S}, \mathbf{K}, \mathbf{I}$	combinators	83
$\langle x_1, \ldots, x_n \rangle$	sequence number coding $x_1, \ldots, x_n$	88
$(x)_n$	n-th component of x	88
$ln(x)$	length of x	88
$Seq(x)$	x is a sequence number	88
$x * y$	concatenation of x and y	89
$\sqsubseteq, \sqsubset$	subsequence predicates	89
$\bar{f}$	history function (course-of-values) of f	89
T_n	normal form predicate	90
$\mathcal{U}$	normal form function	90

Chapter II

$\simeq$	equality for partial functions	127
$\varphi(x)\downarrow$	φ is defined (converges) at x	127
$\varphi(x)\uparrow$	φ is undefined (diverges) at x	127
$\alpha \subseteq \beta$	β extends α as a partial function	127
$\varphi_e^n, \varphi_{e,s}^n$	n-ary partial recursive function of index e	130
$\{e\}^n, \{e\}_s^n$	n-ary partial recursive function of index e	130
S_n^m	parametrization function	131
$\mathcal{W}_e^n, \mathcal{W}_{e,s}^n$	n-ary r.e. relation of index e	134
Tot	set of indices of total recursive functions	146
$\mathcal{K}$	diagonal r.e. set	147
$\mathcal{K}_0$	master r.e. set	150
$\theta\mathcal{A}$	index set of $\mathcal{A}$	150
$K(x)$	Kolmogorov complexity of x	151
$\leq_T, \equiv_T$	Turing reducibility	176
$\mathcal{D}$	structure of Turing degrees	176
$F(\alpha, x)$	functional	177
$T_{m,n}$	normal form predicate for functionals	179, 180
$\varphi_e^A, \varphi_{e,s}^A$	partial recursive function with oracle A of index e	181
$\{e\}^A, \{e\}_s^A$	partial recursive function with oracle A of index e	181
$\mathcal{P}$	set of partial functions from ω to ω	186
$\hat{u}$	basic open set determined by u	186
$[D \to D]$	set of continuous functions from D to D	194
$\trianglelefteq$	order relation on c.p.o.'s	195
D_∞	inverse limit of a chain of c.p.o.'s	195
$\leq_e$	enumeration reducibility	197
$\leq_{wT}$	weak Turing reducibility	198
$\mathcal{PR}$	set of partial recursive functions	205
$\mathcal{R}$	set of total recursive functions	208
$\{\psi_e^n\}_{e \in \omega}$	system of indices	215
$Char$	set of characteristic indices of recursive sets	226
Rec	set of r.e. indices of recursive sets	226
D_e	finite set of canonical index e	226
Fin	set of r.e. indices of finite sets	228
$\{\mathcal{W}_{f(x)}\}_{x \in \omega}$	r.e. class of r.e. sets	228
$\nu_0 \leq \nu_1$	reducibility for enumerations	236
$\nu_0 \equiv \nu_1$	equivalence for enumerations	236

$\mathcal{L}^{\circ}(\mathcal{A})$ structure of equivalence classes of r.e. enumerations of $\mathcal{A}$ 236
$\mathcal{L}(\mathcal{A})$ structure of equivalence classes of enumerations of $\mathcal{A}$ 238

Chapter III

$\mathbf{0}$	T-degree of recursive sets	252
$\mathbf{0}'$	T-degree of $\mathcal{K}$	252
$\leq_m, \equiv_m$	m-reducibility	257
$\mathcal{D}_m$	structure of m-degrees	257
$\mathbf{0}_m$	m-degree of nontrivial recursive sets	257
$\mathbf{0}'_m$	m-degree of $\mathcal{K}$	257
$\leq_c$	conjunctive reducibility	268
$\leq_d$	disjunctive reducibility	268
$\leq_p$	positive reducibility	268
σ_n	truth-table condition	268
$\leq_{tt}, \equiv_{tt}$	truth-table reducibility	268
$\leq_l$	linear reducibility	269
$\mathcal{D}_{tt}$	structure of tt-degrees	270
$\mathbf{0}_{tt}$	tt-degree of recursive sets	270
$\mathbf{0}'_{tt}$	tt-degree of $\mathcal{K}$	270
$\leq_Q$	Q-reducibility	281
$\sigma(e, x, s)$	e-state of x at stage s	291
$\eta, \eta_A, \eta_\varphi, \eta_\varphi^i$	positive equivalence relations	300
$[A]_\eta$	η-closure of A	300
$f \leq_m A$	m-reducibility for functions	308
$\leq_1, \equiv_1$	1-reducibility	320
$\mathcal{D}_1$	structure of 1-degrees	320
$\mathbf{0}_1$	1-degree of infinite coinfinite recursive sets	320
$\mathbf{0}'_1$	1-degree of $\mathcal{K}$	320
$\equiv$	recursive isomorphism	324
$\cong$	recursive equivalence	328
Λ	set of isols	328
f_Λ, R_Λ	extensions to isols of f and R	329
$\leq_{btt}, \equiv_{btt}$	bounded truth-table equivalence	331
$\mathcal{D}_{btt}$	structure of btt-degrees	331
$\mathbf{0}_{btt}$	btt-degree of recursive sets	331
$\mathbf{0}'_{btt}$	btt-degree of $\mathcal{K}$	331
$\leq_{wtt}, \equiv_{wtt}$	weak truth-table equivalence	337
$\mathcal{D}_{wtt}$	structure of wtt-degrees	337
$\mathbf{0}_{wtt}$	wtt-degree of recursive sets	337
$\mathbf{0}'_{wtt}$	wtt-degree of $\mathcal{K}$	337
$\leq_s$	s-reducibility	340
$\leq_{bs}$	bounded search reducibility	340

Chapter IV

$\mathcal{L}, \mathcal{L}^*$	languages for first-order arithmetic	364
$\mathcal{A}, \mathcal{A}^*$	structures for first-order arithmetic	364
$\mathcal{A} \models \varphi$	φ is true in $\mathcal{A}$	364
$\Sigma_n^0, \Pi_n^0, \Delta_n^0$	arithmetical hierarchy classes	367
Δ_ω^0	arithmetical relations	367
$\Sigma_{0,n}^0, \Pi_{0,n}^0, \Delta_{0,n}^0$	bounded arithmetical hierarchy classes	368

$\Sigma_n^{-1}, \Pi_n^{-1}, \Delta_n^{-1}$	Boolean hierarchy classes	373
$\Sigma_n^{0,X}, \Pi_n^{0,X}, \Delta_n^{0,X}$	relativized arithmetical hierarchy classes	374
$\leq_{\Delta_n^0}$	Δ_n^0-reducibility	375
$\leq_a, \equiv_a$	arithmetical reducibility	375
$\mathcal{D}_a$	structure of arithmetical degrees	375
$\mathcal{L}_2, \mathcal{L}_2^*$	languages for second-order arithmetic	376
$\mathcal{A}_2, \mathcal{A}_2^*$	structures for second-order arithmetic	376
$\mathcal{A}_2 \models \varphi$	φ is true in $\mathcal{A}_2$	376
$\Sigma_n^1, \Pi_n^1, \Delta_n^1$	analytical hierarchy classes	379
Δ_ω^1	analytical relations	379
T	set of characteristic indices of recursive well-founded trees	383
$ord_T(u)$	ordinal of the node u on the tree T	383
$ord(T)$	ordinal of the tree T	383
ω_1^{ck}	Church-Kleene ordinal	384
$T_\prec$	tree associated to the ordering $\prec$	385
$\prec_T$	ordering associated to the tree T	386
$\mathcal{W}$	set of characteristic indices of recursive well-orderings of ω	386
$\Sigma_n^{1,X}, \Pi_n^{1,X}, \Delta_n^{1,X}$	relativized analytical hierarchy classes	394
$\leq_{\Delta_n^1}, \equiv_{\Delta_n^1}$	Δ_n^1-reducibility	394
$\mathcal{D}_n$	structure of Δ_n^1-degrees	394
ZF, ZFC	Zermelo-Fraenkel set theory	399
GKP	generalized Kripke-Platek set theory	399
ZF^-	ZF minus the power set axiom	400
V_α	cumulative hierarchy classes	400
V	universe of set theory	400
$\langle A, \varepsilon \rangle$	structure for set theory	400
$\mathcal{A} \models \varphi$	φ is true in $\mathcal{A}$	401
$\Sigma_n, \Pi_n, \Delta_n$	set-theoretical language hierarchy classes	406
$\Sigma_n^T, \Pi_n^T, \Delta_n^T$	set-theoretical theory hierarchy classes	406
$\Sigma_n^\mathcal{A}, \Pi_n^\mathcal{A}, \Delta_n^\mathcal{A}$	set-theoretical structure hierarchy classes	406
$Tc(x)$	transitive closure of x	409
$\mathcal{HF}$	hereditarily finite sets	414
L_α	constructible hierarchy classes	423
L	constructible sets	423
NGB	Von Neumann-Gödel-Bernays set theory	423
$V = L$	axiom of constructibility	423
$\leq_\alpha, \leq_L$	well-ordering of L	430
$\mathcal{P}(\omega) \cap L$	constructible sets of natural numbers	432
$\mathcal{HC}$	hereditarily countable sets	441
$L[A], L(A)$	relative constructibility	444
$\leq_L, \equiv_L$	constructibility reducibility	444
$\mathcal{D}_L$	structure of constructibility degrees	444

Chapter V

$a|b$	incomparable degrees a and b	448
$a \cup b$	least upper bound (join) of a and b	449
$a \cap b$	greatest lower of bound of a and b	449
$\oplus_{n \in I} A_n$	infinite join	449
a'	jump of a	450
$\mathcal{D}'$	structure of degrees with jump	450
$A^{(n)}, a^{(n)}$	n-th jump	451

$A^{(\omega)}$, $a^{(\omega)}$	ω-jump	451
$\mathcal{D}(\geq a)$	cone above a	452
AD	axiom of determinacy	453
σ	a string	456
$|\sigma|$	length of σ	456
$\sigma * \tau$	concatenation of σ and τ	457
θ	coinfinite condition	484
$\mathcal{D}(\leq a)$	principal ideal below a	487
T	a tree	493
$T(\sigma)$, $T(\sigma * i)$	nodes of T	493
f_L, f_R	left and right functions of an admissible triple	516

Chapter VI

$A \cdot N$	cylinder of A	592
A^{tt}	tt-cylinder of A	593

Index

0′ 252, 459, 461, 462, 464, 468, 469, 470,
    488, 495, 498, 500, 501, 503,
    508, 513, 570, 589, 594, 595
0″ 464, 497, 498, 500, 501, 523
0$^{(3)}$ 551
0$^{(\omega)}$ 543, 544, 548, 552, 553
1-generic 468, 484
1-l.u.b. 553
1-tree 520
2-l.u.b. 552, 553

α-rule 77
β-rule 77
Δ_n^{-1} 374
$\Delta_{0,n}^0$ 368, 369
Δ_2^0 285, 341, 373, 507, 514, 515
Δ_n^0 367, 369, 372–375, 514, 515
Δ_ω^0 367
Δ_1^1 387–392, 395
Δ_2^1 396, 437
Δ_n^1 5, 379–381, 391, 395, 438, 540, 543
Δ_ω^1 379
Δ_1^{GKP} 406–410, 419
Δ_n^{GKP} 406–411
$\Delta_1^{ZF^-}$ 426, 430, 431, 443
$\in$-induction 399
η-
    closed 300
    finite 300
    (hyper)hypersimple 300
    infinite 300
    maximal 280, 301, 302
    rule 82
    simple 300
λ-
    abstraction 76
    algebra 224
    calculus 76–87, 194–196, 223–225
    definability 84, 105, 132
    model 224, 225
μ-recursion 21, 128, 129

Π_n^{-1} 374
$\Pi_{0,n}^0$ 368, 369
Π_1^0 classes 505–512
Π_n^0 367, 369–372, 514
Π_1^1 380–387, 419, 420
Π_n^1 379–381
Π_n^{GKP} 411–413
Σ_n^{-1} 373, 374
$\Sigma_{0,n}^0$ 368, 369
Σ_1^0 368
Σ_n^0 367, 369–372, 514
Σ_2^1 361, 420, 437–440, 442
Σ_n^1 379–381
Σ_n^{GKP} 411–413
$\Sigma_1^{ZF^-}$ 426, 430
ω-$BRFT$ 222–224
ω-consistency 160, 163, 167
ω-jump 451, 543, 552, 553
ω_1^{ck} 384, 385, 387, 443
ω_1^L 432

Abel 7
Aberth 9, 214
Abian 189
Abraham 529, 571, 589
absoluteness 418–420, 426, 442, 537, 575
acceptable system of indices 215–221, 236,
    271, 272, 292, 308, 346
Ackermann 165, 415, 417
AD 453–456
Addison 10, 373, 391, 393, 395, 440
adjacent strings 520
Adler 9
admissible
    ordinal 9, 443, 444
    set 421
    triple 516, 520
Agerwala 74
algebra 7, 8
    combinatory 223–225

649

algebraic c.p.o. 192, 201
ALGOL 70, 186
Ambos-Spies x
analogue machine 109–113, 117
analysis 2, 8, 9, 213, 214, 376, 422
    see also second-order arithmetic
analytic set 392, 393
analytical
    hierarchy 375–397, 418, 419, 420,
        438, 441, 442
    engine 134
    set 3, 361, 376, 377, 540, 543
Anderson 115
antichain 464, 466, 530, 597, 599, 601
Appel 242, 243, 295
Arbib 50, 53, 116, 117, 133, 172, 173
Aristotle 113, 127, 363
arithmetic 11
    first-order 24, 363–365
    intuitionistic 10
    Peano 23, 28, 46, 164, 169, 417, 418,
        442, 510–515
    primitive recursive 23
    $\mathcal{R}$ 44, 99, 101, 159, 162, 165–168,
        307, 353, 355, 356, 359, 368,
        513
    Robinson 23, 44, 165, 169, 359
    second-order 24, 376, 430, 442, 536–
        539, 541–543, 545, 546, 550,
        551, 579, 581, 582, 584, 589
arithmetical (a-)
    degree 375, 541–545, 547
    hierarchy 362–375, 381, 392, 393,
        415–417, 438
    reducibility 375
    set 3, 5, 361, 364, 371, 375, 390, 415,
        419, 451, 514, 541–545, 548,
        550, 577
arithmetization 11, 87–90
array 228
Arslanov ix, 255, 256, 277, 308, 338
Asanuma 117
Ashby 115
assembly programs 60
associate 200
atomism 51, 113
automaton
    cellular 109, 170, 173, 174
    finite 52, 53, 116
    Life 173, 174
automorphisms 546–551, 576, 577, 590
    basis 548, 550

autoreducible 501, 502, 523, 528
axiom
    choice 399, 430–432, 442, 444, 453–
        455
    collection 398, 421, 427, 442
    comprehension 397, 398, 442
    constructibility 393, 423, 429–432,
        434, 440–442, 544
    determinacy 393, 453–456, 544, 546
    extensionality 397, 400–405
    foundation 399, 422
    infinity 398, 400, 415, 417, 418, 426,
        442
    Martin 475
    pairing 398
    power set 398, 400, 442
    replacement 398
    separation 398, 421, 427
    union 398
axiomatic Recursion Theory 221
axiomatizability 350, 352, 357–360, 427,
    510, 537, 581
    finite 357, 510
    independent 357–360

Babbage 113, 134
Backus 64, 68
Baire vii, 4, 392, 448, 471, 473–475, 477,
    478, 484, 495
    category 471–484, 495
        theorem 475, 477, 495
    property 477
    space 471, 473
Baker 221
Banach 193, 475, 478
    fixed-point theorem 193, 194
    -Mazur game 475, 476, 478
Bardzin 53
Barendregt 83, 196, 223, 225
Bar-Hillel 397, 423, 427
Barna 193
bar-recursion 382
Barwise 421
basis 507–512
    automorphism 548, 550
    Kreisel 507
    low 508
    Scott 510
Beeson 11, 196, 210, 214, 223–225
Beigel 298
Bell 112
Benioff 52

Bennett 51, 52
Berlekamp 173
Bernays 1, 72, 90, 169, 352, 423
Bernstein 324
Berry (paradox) 262
Bessel 111
Beth 118
Bezboruah 169
Bienestock 117
bi-immune 266, 267, 498
biology 172
Birkoff 112
Bishop 214
Blackwell 393
Blum 220, 314
Böhm 70, 81
Boole 113
Boolean hierarchy 373, 374
Boolos 436, 437
Boone 8
Borel 8, 213, 392, 456
Born 107
bounded
    μ-operator 26
    arithmetical hierarchy 368, 369
    product 25
    quantifier 15, 25, 365, 369, 405, 411
    reducibility 340
    sum 25
    truth-table (btt-)
        completeness 331, 333–335, 337,
            341–343, 347, 349
        degree 331
        reducibility 331, 332, 373, 591
Bourbaki 113
brain 115–118, 122
branch 381, 494, 505–510
Bremermann 51
BRFT 222
Brouwer 119, 122, 127
Brown 189
Bulitko 269, 591
Burks 172–174
Bush 117
Byerly 223, 225

canonical
    index 226
    system 105, 143, 144, 253, 254
Cantor 1, 23, 26, 146, 324, 398, 402, 471,
            473, 474, 494
    space 471, 473

theorem 146, 402
cardinal 9
    measurable 393, 456
Cardone x
Carnap 165
Carpentier 319
cartesian closed category 196, 218
Casalegno 199
category 196, 218, 223
    Baire 471–484, 495
Cauchy 211, 213
Cayley 193
Čeitin 209, 211, 213
cellular automaton 109, 170, 173, 174
chain of degrees 466, 489, 505, 523
Chaitin 151, 263
characteristic
    function 14, 15
    index 225–227
chinese remainder theorem 29
choice
    axiom 399, 430–432, 442, 444, 453–
        455
    functions 137, 138, 229, 230, 232
Chomsky 144
Chong 421, 443
Church 1, 37, 78, 82–84, 98, 100–123, 147,
            148, 152, 162, 164–166, 254,
            385
    -Kleene ordinal 384, 385, 387, 443
    rule 121
    theorem 164, 165
    thesis 101–123, 254
class 400, 423, 427
Cleave 319
Cobham 44
Codd 173
coding 26, 27, 31, 88
    procedure 279
Cohen P.J. 2, 164, 392, 397, 432, 437, 468
Cohen P.F. 266, 273, 294, 316
cohesive 288–290, 498
coinfinite
    condition 484
    extension 484, 493, 500, 520–523
collapsing lemma 402–405, 421, 434, 435,
        439, 442
collection axiom 398, 421, 427, 442
Colmerauer 39
combinator 81, 83, 87, 223
combinatorial (set) function 329
combinatory

algebra 223–225
 logic 223, 224
comeager 473–477, 479, 483, 489, 495,
  503, 527, 546, 549–550
compactness 181, 190, 192, 199, 206, 209,
  211, 473
compiler 61
complementary index 225
complete
  Π_1^1 set 383, 386
  $\Sigma_n^{-1}, \Pi_n^{-1}, \Delta_n^{-1}$ set 374
  Σ_n^0, Π_n^0 set 370, 372, 451
  Σ_n^1, Π_n^1 set 381
  chain- 189, 190
  enumeration 237
  extension 510–515
  formal system 350
  partial ordering 189, 192
complete r.e. set 306, 341–349
  1- 320, 321, 348, 349
  bs- 340, 341
  btt- 331, 333–335, 337, 341–343, 347,
   349
  c- 275, 306, 311, 313, 341–343, 348,
   349
  d- 306, 311, 312, 335, 341–344, 347–
   349
  m- 258, 259, 305, 307, 308, 319, 341,
   343, 347–349
  p- 295, 341, 343, 344, 347, 349
  Q- 282, 286, 287, 294, 295, 297, 299,
   301, 306, 314, 316, 341–343,
   349
  T- 253, 255, 264, 265, 277, 293–295,
   304, 309, 339, 341, 342, 344,
   349, 353
  tt- 270–272, 274, 334, 341, 344, 346,
   347, 349
  wtt- 338–342, 344, 347, 349
completely
  autoreducible 502, 523, 528
  productive 304, 322
component 89
comprehension axiom 397, 398, 442
computability 6, 17, 101, 102, 104, 125–
  126, 197, 202–205, 361, 386,
  391
  flowchart 61–70, 99, 101, 132, 197–
   198
  Herbrand-Gödel 36–38, 98, 101, 105,
   132, 198
  prime 203–205, 222, 396

 search 204, 205, 396, 421
 Turing 53–54, 99, 101, 132, 197
computation
  deterministic 18, 50, 107–109, 197
  (ir)reversible 51, 52, 174
  nondeterministic 18, 50, 74, 198
  theory 223
  tree 91
computers 6, 64, 104, 110, 112, 115–117,
  133
concatenation 89
cone 452, 454, 503, 526, 544, 545, 548,
  550, 551, 576, 584, 590, 591
configuration of a Turing machine 49
conjunctive (c-)
  completeness 275, 306, 311, 313,
   341–343, 348, 349
  reducibility 268, 282, 591
consistent
  extension 491, 492, 510–515
  formal system 114, 115, 168–170
  with ZFC 431, 432, 434, 435, 438,
   440, 445, 453, 454, 529
constructibility axiom 393, 423, 429–432,
  434, 440–442, 544
constructible 4, 5, 361, 422–445, 543
constructive mathematics 7–10
constructively immune 267
contiguous degree 279
continuous function(al) 187–192, 205,
  206, 211, 214, 393
  effectively 188, 189, 191, 207–210,
   212, 213
continuum hypothesis 164, 392, 422, 430,
  434, 459, 468
contraction of quantifiers 365, 377, 405
contraproductive 310, 311, 313, 319, 322
conversion rule 77, 82
Conway 173
Cook 197
Cooper 285, 498, 501, 523
countable functional 200, 201
course-of-value 89, 410
Cowan 116
Craig 357
creative 306–312, 314, 318, 319, 321, 322,
  323, 327, 348, 356, 593
  pseudo- 323, 324, 593
  quasi- 311–313, 336, 348
  semi- 313
  sub- 314, 315, 349
Crick 117

Crossley 8, 330
cumulative hierarchy 400
Curry H.B. 79, 83
Curry J. 193
cut 295
cylinder 583, 592–594, 597
  tt- 593, 594

$\mathcal{D}$ 176, 449, 451–455, 457, 459, 462, 463, 483, 488, 490–492, 493, 516, 526, 528–530, 534, 536–550, 587, 591
$\mathcal{D}'$ 450, 467, 469, 541, 550–553
$\mathcal{D}_1$ 320, 582–584, 591
$\mathcal{D}_m$ 257, 555, 556, 558, 561, 562, 565–568, 570–577, 581, 582, 591
$\mathcal{D}_{tt}$ 584–591
$\mathcal{D}_{wtt}$ 584, 589–591
Dahl 68
Dante 100, 152, 162, 170, 304, 575
Darwin 152
Davis 53, 70, 106, 133, 135, 179–181, 197, 200, 368, 391
De Broglie 107
De Leeuw 50
De Witt 107
decidability 253, 350, 462, 492, 581
Dedekind 1, 7, 19–24, 213, 328, 490
deficiency
  set 263, 265, 266, 287, 349
  stage 246, 277
definability
  absolute 540–543
  arithmetical
   first-order 361, 364, 415
   second-order 361, 376, 436, 437, 542, 543, 545, 551
  invariant 44, 45
  of functions 45, 46
  over $\mathcal{D}$ 449, 450, 530, 534, 540–543
  over $\mathcal{D}'$ 550, 551, 553
  over $\mathcal{D}_1$ 584
  over $\mathcal{D}_m$ 576, 577
  over $\mathcal{D}_{tt}$ 590
  over $\mathcal{HC}$ 442
  over $\mathcal{HF}$ 415
  over L 438–440, 443
  set-theoretical 361, 401, 423
degrees (notions of) 2, 4
  1- 320, 325, 582–584, 590, 591, 594, 597, 600
  Δ_n^1- 394, 440, 543

a- 375, 541–545, 547
btt- 331
e- 340
hyper- 395
L- 444, 445, 543
m- 238, 257, 352, 502, 555–582, 590, 591, 594–598, 600
partial 197, 199
Q- 282
T- 175, 176, 263, 276, 278, 287, 297, 299, 304, 310, 313, 316, 323, 328, 351, 447–553, 568, 570, 574, 575, 583, 589–591, 595
tt- 270, 276, 296, 317, 338, 352, 584–591, 595, 598–601
wtt- 337, 338, 584–591
degrees (types of)
  below $\mathbf{0}'$ 459, 461, 462, 464, 487, 488, 498, 500, 501, 589
  completely autoreducible 502, 523, 528
  hyperimmune-free 495, 496, 497, 500, 501, 505, 509, 520, 523, 588, 589, 600
  low 508, 513
  minimal 465, 479, 483, 484, 498–502, 513–523, 527, 528, 549, 577, 583, 588
  r.e. 252, 257, 263, 270, 276, 278, 287, 297, 299, 304, 313, 316, 317, 320, 323, 328, 331, 337, 338, 351, 352, 508, 509, 513, 577, 595
Degtev x, 244, 248, 298, 302, 328, 340, 581, 591, 595, 601
Dehn 8
Dekker 150, 205, 228, 234, 238, 239, 241, 242, 245, 247–249, 259, 263, 275–277, 289, 304, 306, 308–311, 313, 322, 323, 327, 328, 330, 498, 597
Democritus 113
Demongeot 174
Denisov ix, 317, 596
Dennett 115
dense
  immune 273
  (open) set 474, 475, 495
  simple 273, 289
Descartes 113, 115, 170
descriptive set theory 10, 392–394
determinacy axiom 393, 453–456, 544, 546

projective 393, 544
deterministic computation 18, 50, 107–109, 197
Detlovs 145
Devlin , 423, 424
diagonalization 11, 145, 146, 152–154, 162, 468, 496, 500
    lemma 496, 517
diamond 527, 528
Dijksterhuis 106
Dijkstra 68
Di Paola 218, 223
directed 192
Dirichelet 75
disjoint array 228
disjunctive $(d-)$
    completeness 306, 311, 312, 335, 341–344, 347–349
    reducibility 268, 339, 591
distributive 529, 555–561, 565, 567, 568, 572, 579–581, 583, 584
DNA 172
domain of a partial recursive function 16, 192
domination 273, 274
dovetailing 26
Downey 360, 595
Drake 398
Dreben 165
Du Bois Reymond 146
Duff 174

e-splitting 463, 498
e-state 291, 292
Eccles 115, 116
effective
    algebraic c.p.o. 192, 210
    definitional schemata 203
    metric space 211
    operation 205–213
        on partial recursive functions 205–208
        on total recursive functions 208–210
        weak 213
    topological space 211
    topos 214
effectively
    bi-immune 267
    extensible 356
    hyperhypersimple 288
    hyperimmune 277

hypersimple 277, 280
immune 267
inseparable 318–319, 327, 356
maximal 293
non-recursive 280, 304–306, 318
not m-reducible 305
simple 263–266, 271, 280, 294, 338, 349
effectivity principle 280
Ehrenfeucht 159
Eilenberg 53, 218, 223
Elgot 64, 197, 218, 223
Einstein 107, 112
elementary
    (in)equivalence 540, 544–546, 551, 587, 590, 591
    map 547
Ellentuck 329, 330
embeddings 459, 461, 462, 490–493, 526, 528, 529, 537
Enderton 552
energy (consumption) 51–52
Engels 152
Entscheidungsproblem 165
enumeration of
    Σ_n^0 or Π_n^0 relations 370
    Σ_n^1 or Π_n^1 relations 380
    Σ_n^{GKP} or Π_n^{GKP} relations 411
    classes of r.e. sets 228–236, 248
    higher type recursive objects 200
    partial recursive
        functions 130, 215, 216
        operators 197
    primitive recursive functions 96
    r.e. set 138–140
    recursive
        functions 146
        operators 197
        set 139–140
enumeration (types of)
    complete 237
    minimal 236
    one-one 140, 229
    principal 236
    reducibility $(e-)$ 197
enumerations (theory of) 236–238
Epimenides 166
Epstein ix, 5, 468, 500, 501, 516, 529, 538, 551
equations 31–38
equivalent
    1- 320

a- 375
btt- 331
m- 257
L- 444
T- 176
tt- 268
wtt- 337
Ershov A.P. 65, 202
Ershov Y.L. ix, 8, 192, 201, 205, 211, 237,
           238, 300, 319, 341, 353, 373,
           555, 559, 570, 572, 573, 575,
           596
     topology 205, 211
essential undecidability 353–357, 513
Euclid 2, 26
Euler 111
exact pair 485–489, 584
expansionary 190, 193
extension
     coinfinite 484, 493, 500, 520–523
     complete 510–515
     consistent 491, 492, 510–515
     finite 456, 457, 477–479, 487–490,
           493, 496, 500
extension of
     embedding 490–493
     formal system 354, 510
     Peano arithmetic 510–515
     string 456, 494
extensional
     λ-calculus 82, 194–196, 224–225
     combinatory algebra 224–225
     operator 155, 205
extensionality axiom 397, 400–405

f_0-space 192
fap 202, 204
Farber 145
Farmer 172, 174
Feferman x, 6, 120, 159, 201, 351
Feiner 551
Fenstad 200, 223, 421
Feynman 52, 106, 112
Feys 83
filter space 201
finite 6
     automaton 52–53
     Dedekind- 9, 328
     definability 33–36, 97, 101
     extension 456, 457, 477–479, 487–
           490, 493, 496, 500
finitely

axiomatizable 357, 510
branching tree 509, 510
strongly hypersimple 285, 286, 298
finitism 23, 119
first-order
     arithmetic 24, 363–365
     formal system 350–352, 593
     theory of degrees 530, 536–539, 546,
           579–582, 584, 589
Fischer 332
fixed-point 157, 182, 189, 192–194, 255,
           256, 308, 338
     least 182–185, 189–191
     operator 79
     theorem 79, 152–158, 165, 184, 185,
           217, 235, 237, 255, 413
        Banach 193
        double 155
        with parameters 155
Florence 235
flowcharts 58, 61–70, 99, 101, 132, 197,
           198
Fogelman Soulié 117
follower 230
'for' 70, 71
forcing method 468
formal systems 11, 39, 45, 46, 98, 99, 114,
           159–170, 254, 349–360
formalism 119, 158–165
FORTRAN 64
foundation axiom 399, 422
fractals 193
Fraenkel 2, 397, 399, 423, 427
Fredkin 52
Frege 119, 397
Freivald 213
Freud 152
Freyd 225
Friedberg 213, 217, 230, 232, 277, 288,
           290, 306, 311, 337, 338, 468
Friedman H. 164, 202–204, 222, 223
Friedman S. 421
full subtree 494
function
     choice 137, 138, 229, 230, 232
     coding 27, 31, 89
     combinatorial 329
     continuous 187–192, 214, 393
     expansionary 190, 193
     history 89
     monotone 181, 189, 190

partial recursive 125–132, 157, 158,
    222, 223
primitive recursive 4, 20, 22, 24–28,
    70–74, 88–90, 96, 147, 407
recursive 3, 22, 28, 34, 37, 43, 54,
    65, 84, 97–102
functional 177
    countable 200, 201
    partial recursive 3, 178–181, 188,
        196, 197, 269
    restricted 177, 179

Gale 453, 454
Galilei 106
Galois 7
game 453, 456, 475, 476
Gandy 107, 108, 110, 200, 201, 442
gap 436, 437
garden-of-Eden 174
Garnett 193
Gasarch x, 298
Gauss 7
genetics 172
Georgieva 74
Gill 286, 298, 308, 314, 316, 341
GKP 399, 405–421, 426, 441
Gladstone 73, 74
g.l.b. of degrees 449, 465, 488, 489, 582
global results 458, 530–549, 574–582
'go to' 64
God 165, 175
Gödel 1, 2, 5, 9, 18, 22, 27, 28, 31, 34,
    36–39, 43, 44, 88, 98, 102, 105,
    106, 113–115, 120, 132, 147,
    159, 160, 162, 164–167, 169,
    170, 198, 254, 350, 356, 365,
    392, 393, 397, 398, 401, 402,
    418, 422–427, 430–432, 440,
    443
    theorem
        first 114, 115, 162–164, 167, 254
        second 168–170, 401, 427
    theory 356
Gold 226, 373
Goldfarb 165
Goldstine 62
Golès 174
Goodstein 214
Gordon C.E. 203, 204
Gordon G. 63, 421
GPSS 63
Graham 107

grammar 144
graph 135–137, 538, 579, 580
    model 194
Grassmann 20, 22
Greenspan 109
Greibach 144
Grilliott 200, 204
Griswold 145
Groszek 467, 529, 574
Grzegorczyk 33, 72, 213, 214
Guy 173

Hack 74
Haddon 116
Hajnal 444
halting problem 150, 263
Hamilton 108, 109
Hanf 351, 510
Hanson 8, 353
Hardy 146
Harrington x, 164, 421, 438, 542, 544, 545,
    547, 548
Harrow 368
Hartmanis x, 50, 53, 221
Hausdorff 186, 393
$\mathcal{HC}$ 441–443
head of a Turing machine 47, 50
Hebb 116
Heisenberg 107
Heller 218, 223
Helm 212, 213
Herbrand 18, 33, 34, 36–39, 132, 198, 350
Herbrand-Gödel computability 36–38, 98,
    101, 105, 132, 198
hereditarily
    countable 441–443
    finite 414–418, 424, 438, 441
hereditary set 298
Hermes 53, 133
hidden-variable 112
hierarchy
    analytical 375–397, 418, 419, 420,
        438, 441, 442
    arithmetical 362–375, 381, 392, 393,
        415–417, 438
    Boolean 373, 374
    bounded arithmetical 368, 369
    constructible 423–445
    cumulative 400
    Grzegorczyck 72
    hyperarithmetical 362, 391
    Jensen 423

Levy 397–421, 438
projective 377, 392, 393
ramified analytical 436, 437
set-theoretical 397–421, 438
theorems 371, 381, 413, 425
higher type 199–201, 422
Higman 8
Hilbert 1, 7, 23, 88, 90, 105, 119, 135, 169, 422
tenth problem 135
Hindley 83, 196, 225
Hinman 5, 200, 396
history function 89
$\mathcal{HF}$ 414–418, 424, 438, 441
Hoare 68
Hobbes 113
Hodges 134, 164
Hofstadter 18, 115, 170
homogeneity 452, 543–546, 551, 576, 589
Hopcroft 50, 53, 133, 144
Hopfield 117
Horn 39, 330
Horowitz 322
Howard 232, 236
Hugill 527, 528
Hyland 201, 214
hyper-
arithmetical 5, 33, 361, 391, 510
degree 395, 510
hyperimmune 284, 285, 498
hypersimple 282–288, 294, 297, 316, 349
immune 272, 273, 276, 277, 495, 498
effectively 277
-free 495, 496, 497, 500, 501, 505, 509, 520, 523, 588, 589, 600
simple 272–280, 285, 287, 295, 297, 317, 338, 339, 349, 357, 358, 514
strongly 285, 286

Ianov 198, 202
ideal 487, 488, 530, 538–540, 546, 558, 561, 572, 573, 575, 579, 580, 582
immune 141, 239, 259, 263, 266, 267, 273, 321, 328, 498, 579, 583
hyper- 272, 273, 276, 277, 495, 498
hyperhyper- 284, 285, 498
incomparable degrees 457, 459, 463, 464, 468, 479, 483, 600
incompatible strings 456

incompleteness 11, 114, 115, 162–164, 167, 254, 262
indecomposable 557
independent
degrees 462, 464, 466–468
of ZFC 392, 467, 468, 475, 540, 574
sets 460
independently axiomatizable 357–360
index of
finite sets 226, 227
partial recursive functions 129, 150
r.e. sets 134
recursive functions 90, 91
recursive sets 225, 226
index set 150, 370
index system
acceptable 215–221, 236, 271, 272, 292, 308, 346
standard 215, 220, 221
index (types of)
canonical 226
characteristic 225, 226
complementary 225
r.e. 225–228
induction 18–21, 46, 399, 402, 510
infinity axiom 398, 400, 415, 417, 418, 426, 442
initial segment of
$\mathcal{D}$ 487, 492, 516, 526–529, 537, 538
$\mathcal{D}_1$ 583, 584
$\mathcal{D}_m$ 561–574
$\mathcal{D}_{tt}$ 589
string 456, 457
injury 261, 292
inseparable
Δ^0_n- 372
effectively 318, 319, 327, 356
recursively 148, 316–319, 334, 353–355
interpretable 353
interpreter 61
introreducible 241, 502, 534, 579
intuitionism 10, 22, 119–122, 210, 214
invariant
choice function 137
definability 44, 45
irreducible 594–596, 600, 601
irreversible computation 51
isols 328–330
isomorphism
of cones 545, 551, 576, 590, 591
theorem 325

type 324, 325, 536, 538
iteration 72, 73, 192–194
    theorem 132

Jacopini 70
Janiczak 350
Jech 397, 475
Jensen 423, 436, 438
Jeroslow 169
Jockusch ix, 241, 256, 266, 268, 271, 273,
    276, 279, 280, 294–297, 316,
    337, 338, 340, 341, 469, 489,
    498, 501–503, 506–510, 513–
    515, 523, 528, 537, 540–542,
    544–551, 598, 600
John 363
join 448, 449, 452, 523
Jones 44, 135
Joyce 16
Julia 193
jump 450–452, 467–471, 501, 509, 546,
    550–553, 556, 584, 586, 588,
    594
    inversion theorem 468, 556, 584,
    586, 588

$\mathcal{K}$ 147, 148, 164, 252, 254, 258, 259, 265,
    266, 268, 280, 298, 299, 305,
    307, 310, 317, 332, 337, 349,
    353, 373, 374, 450, 579, 594
$\mathcal{K}_0$ 150, 254, 307
Kallibekov 587
Kalmar 33, 103
Kandel 116
Kanoui 39
Kanovich 277, 338, 339
Karp 310, 410, 425, 430, 443
Kechris 200
Kelley 186, 201, 471
Kfoury 202, 204
Khutorezkii 232, 237
Kino 438, 442, 443
Kleene vii, ix, 1, 5, 10, 21, 22, 28, 34, 37,
    79, 83, 84, 86, 90, 98, 100, 118,
    119, 121, 122, 128–132, 134,
    137, 138, 140, 142, 146–148,
    152, 156–158, 178–182, 184,
    198–201, 318, 328, 350, 354,
    364, 367, 369–372, 376, 378–
    382, 385–387, 391, 393, 395,
    436, 449–452, 457, 460, 462,
    463, 485, 488, 490–492, 506,
    552

Kleene-Brouwer ordering 386
Klein 327
Klop 224
Knaster 189, 190
Knorr 2
Kobzev x, 334, 335, 337, 594
Kolmogorov 10, 107, 151, 236, 261, 263
    complexity 151, 152, 261–263
König 214, 270, 473, 505–507
    lemma 505–507
Kowalski 39
Koymans 196
Kozmnikh 74
KP 421, 443
Kreisel ix, x, 6, 10, 33, 38, 45, 97, 102–104,
    107, 109–112, 119–123, 179,
    200, 202, 208, 211, 213, 224,
    357, 359, 379, 397, 430, 442,
    459, 507, 508
    basis lemma 507
Kreitz 214
Kripke 122, 356, 399, 411, 421, 443
Kripke-Platek set theory 399, 421
Krivine 397
Kronecker 7, 22
Kučera 256, 515
Kunen 397
Kuratowski 142, 366, 378, 393, 406, 476,
    481
Kuratowski-Ulam theorem 476, 477
Kurtz 484
Kušner 214
Kuznekov 272

L 423–440, 444, 445
La Budde 109
La Mettrie 113, 116, 118
Lachlan 213, 221, 232, 234, 235, 241, 255,
    265, 272, 293, 309, 311, 313,
    319, 335, 341, 344, 346, 347,
    520, 522, 523, 528, 529, 537,
    555, 558, 561, 562, 565, 566–
    569, 583, 590
Lacombe 9, 204, 208, 211, 213, 214, 237,
    238, 485, 500
Ladner 339
Lagrange 7
lambda-
    abstraction 76
    algebra 224
    calculus 76–87, 194–196, 223–225
    definability 84, 105, 132

model 224, 225
Lambek 64
Lamola 116
Landauer 51, 52
Landin 87
Langton 173
lattices 15, 488, 489, 528, 529, 537, 538,
        557, 558, 561, 565, 567, 568,
        579–581, 583, 584
Laventrieff 149
Lavrov 8, 74, 298, 353, 579
leaf 381
least
        fixed-point 182–185, 189–191
        number principle 21, 399
        possible jump 469, 479, 501
        upper bound
            of degrees 449, 479, 489, 490, 552,
                582
            principle 214
Lebesgue 22, 381, 392, 484
Lebeuf 529
Lee 165
Leeds 437
left-narrow 402
Leibniz 87, 88, 140, 164, 397
Leighton 106
length of
        sequence number 89
        string 456
Lerman x, 5, 256, 462, 490, 492, 501, 516,
        529, 538, 543, 581
Leucippus 113
Levy 397, 406, 411, 421, 423, 426, 427,
        442–444, 475
        absoluteness 442, 443
        hierarchy 397–421, 438
Lewis 165
Li Xiang 267
liar paradox 166
Life 173, 174
limit
        lemma 373
        sets 4, 361, 373
limitation of size 398
linear (l-)
        ordering 385, 527
        reducibility 269, 591
Lipschitz 110
Lipton x
LISP 7, 87, 186
Löb 169, 170, 213

theorem 169, 170
local results 458
logic 11, 165
Lolli x
Longo 201
low
        basis theorem 508
        degree 508, 513
Löwenheim 432, 433, 435, 439, 440, 442,
        537
Löwenheim-Skolem theorem 432, 433,
        435, 439, 442, 537
Luckham 202
Lucretius 113
Lullus 87
Lusin 148, 376, 379–382, 385, 386, 392,
        393
Lynch 308

Mac Lane 218
machine
        analogue 109–113, 117
        nondeterministic 50, 51
        probabilistic 50, 51
        random access 64
        SECD 87
        SK 87
        Turing 46–61, 116, 132, 203, 204,
            254
Machover 443
Machtey 217, 221
Malc'ev A.A. 574, 591
Malc'ev A.I. 237, 238, 300, 319
Manaster 330
Mandelbrot 193
Manna 186
Mansfield 240
many-one (m-)
        completeness 258, 259, 305, 307,
            308, 319, 341, 343, 347–349
        degree 238, 257, 352, 502, 555–582,
            590, 591, 594–598, 600
        reducibility 251, 257, 331, 591
Mapertuis 109
Marchenkov 232, 248, 295, 301
Marek 436
Margolus 174
Markov 8, 10, 145, 210, 213, 214
        algorithms 145
        chains 112, 122
Marques 314
Martin D.A. ix, 255, 264, 266, 273, 286,
        287, 289, 296, 297, 360, 393,

454–456, 475, 481, 484, 495–
497, 500, 501, 503, 505, 517,
523, 528, 544, 546, 588, 600
axiom 475
Martin R.L. 166
Marx 152
materialism 118
Matijasevich 7, 135, 164, 368
maximal 279, 288–294, 316
η- 280, 301, 302
maximum degree principle 280
Mazur 475, 478
McCarthy 87, 186, 203
McCarty 122, 328
McClelland 117
McCulloch 52, 116
McLaughlin 242, 243, 249, 266, 267, 287,
295, 296, 310, 311, 319, 328,
330
McNaughton 206, 229, 427
meager 474–477, 479, 481, 495, 502, 507,
549
measurable cardinal 393, 456
measure 455, 456, 484
mechanics 106–113
mechanism 113–115
mechanisms 150–152
Medvedev 272
Melzak 64
memory 47, 52, 53, 64, 117
Mermin 112
Metakides 7
method
    Baire category 471–484, 495
    coinfinite extension 484, 493, 500,
        520–523
    e-state 291, 292
    finite extension 456, 457, 477–479,
        487–490, 493, 496, 500
    forcing 468
    permitting 277–280, 338
    priority 232, 261, 292
    splitting 463, 464
    tree 493–495, 500
Meyer 70, 71, 194, 224
mezoic 322, 323
Miller 495–497, 500, 501, 505, 517, 523
mind 113–118
minimal
    cover 502, 503, 523, 527, 569, 570,
        573, 587, 588, 589
    degree 465, 479, 483, 484, 498–502,

513–523, 527, 528, 549, 577,
583, 588
    enumeration 236
    pair 464–466, 479, 490, 499
    predecessor 483, 528
    upper bound 502–505, 528, 552, 553
minimality lemma 499, 504, 518, 520, 526
Minsky 27, 50, 53, 133, 144, 253
Mirimanoff 399
Mitchell 309, 311
model of
    λ-calculus 194–196, 223–225
        graph model 194, 225
        D_∞ 195, 225
    arithmetic 536–539, 541–543
    set theory 401
modulus of continuity 188, 209
Moggi 201
Mohrherr 584, 587, 590
Moldestad 204
Monk 397
monotonicity 181, 189, 190, 206
Montague 24, 426, 427
Moore 50, 174
Morley x, 164
Morris 286, 308, 314, 316, 341
Moschovakis ix, 5, 200, 203, 204, 211, 213,
222, 223, 393, 394, 396
Moses 165
Mostowski 8, 33, 39, 40, 42–45, 134, 140,
142, 159, 162, 352–354, 364,
367, 369–371, 391, 393, 402,
419, 421, 424, 437, 440, 461
Muchnik 155, 213, 233, 258, 277, 310, 318,
319, 327
Mulry 214
Mundici 51
Mycielski 453
Myhill 11, 174, 196, 198, 206, 207, 211,
213, 229, 234, 239, 241, 242,
244, 245, 249, 276, 288, 289,
305, 307–310, 321, 323, 325,
329, 330, 459, 477, 498, 592,
593, 594, 597

Nerode x, 7, 8, 186–188, 191, 269, 330,
538, 540, 547, 548, 551, 579,
581, 582, 584, 589, 590
neuronic net 52, 116, 117
Newton 106, 108, 192, 193
NGB 423, 427
Nies 551, 581

n-l.u.b. 552
node 381
Nogina 211
non-
    deficiency 246, 248
    deterministic computations 18, 50,
        74, 198
    standard integer 330
norm 331
normal
    canonical system 144
    object 200
normal form of
    λ-terms 78
    Π^1_1 sets 380
    analytical relations 378
    arithmetical relations 366, 367
    higher type recursive objects 200
    partial recursive
        functions 129
        functionals 179, 180
    r.e. relations 134
    recursive functions 90
    set-theoretical relations 406
Normann 201
Novikov 8, 148
number theory 7, 10, 29
numeral 32, 39, 84

Odifreddi 341
Omanadze 311, 340
one-one (1-)
    completeness 320, 321, 348, 349
    degree 320, 325, 582–584, 590, 591,
        594, 597, 600
    enumeration 140, 229
    reducibility 320
open set 186, 205, 472, 474
    dense 474, 475
    effectively 187, 207
oracle 175, 251
ordinal of
    node 383
    structure 402
    tree 383
ordinal (types of)
    admissible 9, 443, 444
    Church-Kleene 384, 385, 387, 443
    countable 384
    gap 436, 437
    recursive 384–386
    stable 443

Ord-Smith 109
Ostrowski
Owings 154, 298
Oxtoby 476

padding lemma 131, 218, 237
pairing
    axiom 398
    function 27
Paliutin 573, 575
paradoxes 23
    Berry 262
    liar 166
    Russell 76, 81, 148, 159, 397
paradoxical combinator 81, 155, 185
parallelism 74, 197
parametrization theorem 132, 215, 221
Paris 164, 484
Park 194, 202
partial
    degree 197, 199
    function 127
    ordering 15, 385, 461
        algebraic 192, 210
        chain-complete 189, 190
        complete 192
        effective 192, 210
        linear 385, 527
        reflexive 195
    recursive
        function 125–132, 157, 158, 222,
            223
        functional 178–181, 201
        operator 196–198
        tree 494
        uniformly 178, 199
Pascal 113
PASCAL 4, 7, 70
Pasero 39
Paterson 202, 502, 523, 528
Paul 166
Peano 23, 24, 28, 39, 45, 84, 442, 510
    arithmetic 23, 28, 46, 164, 169, 417,
        418, 442, 510–515
Pearson 115
Peitgen 193
perfect set 494, 596
permitting method 277–280, 338
permutation
    of quantifiers 365, 377, 405
    recursive 324, 327
Peter 1, 64, 65, 74, 89, 96, 99

Peterson 75
Petri (net) 74, 75
Picard 9
Pilate 363
Pitts 52, 116
Plank 106
Platek 203, 399, 411, 421, 430, 443
Plato 113, 175
platonism 119
Plotkin 194
Podolski 112
Poincaré 22, 399, 422
pointed tree 503, 504
Poliakov 74
Polonsky 145
Popper 115
positive (p-)
    completeness 295, 341, 343, 344,
        347, 349
    equivalence relation 300
    information topology 186
    reducibility 268, 306, 340, 591
Posner 465, 468, 500, 501, 549, 550
Post 1, 8, 18, 46, 48, 53, 63, 64, 103, 105,
        106, 114, 118, 130, 132, 134,
        140, 142–144, 147, 162, 176,
        178, 251–254, 256–259, 267,
        268, 270–272, 274, 275, 279–
        283, 288, 294, 296, 304–307,
        320, 324, 327, 331, 333, 339,
        341, 350, 353, 372, 395, 396,
        421, 444, 449–452, 457, 460,
        462, 463, 485, 488, 490–492,
        506, 552
    hypersimple set 271
    problem 251–254, 256, 258, 259, 263,
        267, 268, 270, 282, 288, 293,
        294, 304, 305, 321, 333, 349,
        353, 421, 446
    simple set 259, 265, 266, 271, 272,
        275, 278, 279, 294, 339, 346,
        349, 506
    theorem 140, 372, 373, 395–397
Pound v
Poundstone 174
Pour El 9, 110, 111, 211, 213, 214, 232,
        234–237, 356, 357, 359, 360
power set axiom 398, 400, 442
precomputation theory 222
predicate calculus 165
predicativity 6, 422
prenex normal form 366, 367, 378, 406

Preston 174
Previale x
Priese 50, 133
prime computability 203–205, 222, 396
primitive recursive
    arithmetic 23
    function 4, 20, 22, 24–28, 70–74, 88–
        90, 96, 147, 407
principal
    enumeration 236
    ideal of degrees 487, 488
Principia Mathematica 253, 254
priority method 232, 261, 292
probabilistic machine 50, 51
probability 51, 111, 112
productions 143, 144
productive 306–311, 322
    completely 304, 322
    contra- 310, 311, 313, 319, 322
program schemata 202
programming languages 4, 6, 59–61, 63,
        64, 68–70, 86, 87, 144, 145, 186
    ALGOL 70, 186
    FORTRAN 64
    GPSS 63
    LISP 7, 87, 186
    PASCAL 4, 7, 70
    PROLOG 7, 38, 39, 186
    SNOBOL 7, 145
projective
    determinacy 393, 544
    hierarchy 377, 392
    set 393
PROLOG 7, 38, 39, 186
provability 166–170
pseudo-
    creative 323, 324, 593
    simple 323, 334
Putnam 232, 234, 235, 352, 354, 355, 373,
        435–437, 552

Q-
    completeness 282, 286, 287, 294,
        295, 297, 299, 301, 306, 314,
        316, 341–343, 349
    degree 282
    reducibility 281, 282
Q 23, 24, 44, 165, 169, 359
quantifier 15, 25, 365, 369, 377–381, 405,
        406, 411, 462, 490, 492, 530,
        538, 581
    contraction 365, 377, 405

one- 462
permutation 365, 377, 405
two- 490, 492, 581
quasicreative 311–313, 336, 348
Quine 31

$\mathcal{R}$ 44, 99, 101, 159, 162, 165–168, 307, 353, 355, 356, 359, 368, 513
Rabin 7, 579
Rakic 117
ramified analytical hierarchy 436, 437
Ramsey theorem 164
random
    access machine 64
    object 116, 152, 261–263, 265
range of (partial) recursive functions 138–140
realizability 10
recursion theorem
    first 181–186, 189, 192
    second 156, 157, 165, 184, 185
Recursion Theory
    axiomatic 221–223
    basic 222
    classical 1
    generalized 1, 421
Recursion Theory on
    abstract domains 158, 202–205, 222, 396
    admissible sets 421
    continuum 395–397
    higher type objects 158, 199–201
    ordinals 423, 443, 444
recursive
    admissible triple 517
    analysis 213, 214
    branch 506
    completely 150
    enumeration 236
    equivalence 328
    function 3, 22, 28, 34, 37, 43, 54, 65, 84, 97–102
    graph 136, 137
    in E 391
    in $\mathcal{K}$ 373
    isomorphism type 324, 325
    linear ordering 527
    operator 197, 198
    ordinal 9, 384–386
    partial
        function 125–132, 157, 158, 222, 223

functional 3, 178–181, 188, 196, 197, 269
permutation 324, 327
potentially 147
predicate 22
product 14
program 185, 186
real 8, 213
schemata 203, 204
set 3, 139–142, 159, 233, 239–241, 257, 258, 270, 298, 308, 320, 323, 327, 331, 353, 355, 367, 368, 507, 514, 593
topos 214
tree 382, 494, 505–510
uniformly 178, 199, 233
recursively
    bounded tree 509
    independent 460
    inseparable 148, 316–319, 334, 353–355
    invariant 327, 328
    isomorphic 324
    pointed tree 503, 504
recursively enumerable
    class of
        partial recursive functions 150, 205, 206, 228, 232
        r.e. sets 228–236
    completely 205, 228, 229, 236
    degree 252, 257, 263, 270, 276, 278, 287, 297, 299, 304, 313, 316, 317, 320, 323, 328, 331, 337, 338, 351, 352, 508, 509, 513, 577, 595
    index 225–228
    set 3, 134–143, 147–149, 159, 242–249, 252, 254, 255, 257, 268, 281, 288, 304, 331–333, 355, 368, 376, 515
    without repetitions 229–236
reducibilities
    1- 320
    Δ_n^1- 394
    a- 375
    bounded 340
    btt- 331, 332, 373, 591
    c- 268, 282, 591
    d- 268, 339, 591
    e- 197, 340
    l- 269, 591
    L- 444

m- 251, 257, 331, 591
p- 268, 306, 340, 591
partial 341
Q- 281,282
s- 340
T- 175, 176, 254, 280, 294, 306, 338, 340
tt- 251, 268, 306, 337, 591
wT- 198
wtt- 337
reduction
property 142, 372
rule 77, 82
reflection 413, 426, 427
reflexive c.p.o. 195
regressive 238, 242–245, 249, 276, 284, 285, 297, 327, 328
regular predicate 21
relative recursion for
functionals 199
partial functions 176
total functions 175–177, 251
relativity 51
relativization 177, 374, 375, 394, 444
Remmel 8
replacement axiom 398
representability 39–44, 98, 99, 101, 159–166, 198, 307, 352, 353, 355, 513–515
strong 40, 42–44, 99
weak 40, 41, 43, 98, 99, 159–163, 165, 166, 307, 352, 353, 513–515
representation theorems for Π_1^1 sets 382, 385
requirements 232, 261, 292, 477, 478
restricted functional 177, 179
retraceable 239–241, 245–249, 276, 284–286, 297, 327, 328
reversible computation 51, 52, 174
Rice 150, 151, 205, 213, 228, 229, 272
theorem 150, 151,237
Richards 9, 111, 214
Richardson 9
Richter L. 551
Richter P.H. 193
Riemann 111
rigid structure 547
Ritchie 70, 71
r-maximal 285
RNA 172
Roberts 117

Robinson A. 64, 197
Robinson J. 31, 74, 135
Robinson R.M. 8, 23, 39, 40, 42–44, 72–74, 159, 162, 165, 213, 352–354
arithmetic 23, 44, 165, 169, 359
Robinson R.W. 285, 286
Rogers ix, 5, 133, 156, 184, 196, 197, 210, 215, 216, 218, 219, 221, 227, 254, 305, 306, 308, 311, 322, 324, 327, 337, 338, 544, 592, 593, 597, 600
root 381
Rose 288
Rosen 112
Rosenbloom 79
Rosolini 214
Rosser 78, 83, 84, 134, 140, 142, 148, 161, 162, 167, 355, 356
theory 356
trick 161
Roussel 39
Rowbottom 432, 435
Rubel 110, 111, 117
rudimentary predicates 368
Rumelhart 117
Russell 76, 81, 119, 148, 159, 262, 397–399, 422
paradox 76, 81, 148, 159, 397
Rutledge 197
Ryll-Nardzewski 24, 33

S_n^m-theorem 131, 132
Sacks vii, 264, 421, 436, 444, 461, 462, 466, 467, 477, 484, 492, 493, 503–505, 527, 552, 553, 587
Sammett 63, 64, 70, 87, 145
Sands 106
Sasso 178, 179, 188, 198, 199, 501
Ščedrov 164, 210, 214, 225
Schmerl 492
Schnorr 221
Schönfinkel 83
Schröder 324
Schrödinger 107
Schwartz 116
Scott 192, 194, 196, 211, 224, 510, 511, 513, 579
basis theorem 510
topology 192, 211
search computability 204, 205, 396, 421
SECD machine 87
second-order

arithmetic 24, 376, 430, 442, 536–
539, 541–543, 545, 546, 550,
551, 579, 581, 582, 584, 589
definability 361, 376, 436, 437, 542,
543, 545, 551
Seldin 83, 196, 225
selection theorem 200
self-
application 76
membership 76
reference 146, 153, 157, 162, 165–
170
reproduction 170–174
Selivanov ix, 269, 591
Selverston 117
semi-
creative 313
intuitionists 22
recursive 140, 294–299, 302, 337, 349
separable 354, 355, 506
separated space 473
separation axiom 398, 421, 427
sequence number 89, 381
Set Theory 2, 9
descriptive 10, 392–394
GKP 399, 405–421, 426, 441
KP 421, 443
NGB 423, 427
ZF^- 400, 419, 425–427, 430, 431,
459
$ZF(C)$ 2, 399, 401, 417, 420, 427,
430–432, 434–438, 440–445,
453, 467, 475, 529, 537, 540,
574
Shanker
Shannon 50, 52, 110, 133
Shapiro 50, 206, 229
Shepherd 117
Shepherdson 44, 64, 159, 169, 197, 202,
206, 207, 211, 213, 229, 426
Shoenfield 208, 211, 213, 311–313, 316,
318, 334, 351, 353, 373, 374,
394, 395, 420, 430, 437, 464,
471, 493, 498, 503, 508, 529
Shore x, 156, 490, 492, 529, 538, 540–545,
547, 548, 551, 571, 577, 579,
581, 582, 584, 589, 590
Sierpinski 149, 376, 379, 381, 382, 385,
386, 392, 393
simple
hyper- 272–280, 285, 287, 295, 297,
317, 338, 339, 349, 357, 358,
514
hyperhyper- 282–288, 294, 297, 316,
349
effectively 263–266, 271, 280, 294,
338, 349
pseudo- 323, 334
set 259–267, 270, 276, 278, 280, 284,
295, 301, 308, 313, 317, 321,
323, 333, 334, 338, 339, 349
Simpson M. x
Simpson S. 164, 421, 444, 498, 527, 536–
538, 546, 551
theorem 536–538, 546
simultaneous
iteration 72
recursion 89
Singer 111
SK machine 87
Skolem 1, 7, 22–24, 89, 432, 433, 435, 439,
440, 442, 537
Slaman vii, 467, 529, 530, 534, 538, 547,
551, 574
Smith 174
Smorynski 170
Smullyan 31, 155, 258, 263, 319, 327, 354–
356, 360, 368
SNOBOL 7, 145
Soare x, 5, 256, 271, 279, 280, 328, 469,
506–510, 513–515, 550
Socrates 113
Solomonov 151
Solovay 256, 393, 438, 456, 475, 511, 551
Soloviev 277, 281, 286, 311, 339, 340
space
Baire 471, 473
Cantor 471, 473
countable T_0- 211
f_0- 192
filter 201
metric 211
separated 473
topological 211
Specker 9, 213, 214
Spector 120, 385, 386, 467, 470, 471, 485,
487–489, 491, 498–500, 503,
517–519, 530, 537, 581, 584,
589
theorem 485–487, 530, 537, 584, 589
splinter 244, 323, 324, 593
splitting method 463, 464
Spreen 211
Srebrny 436

stable ordinal 443
standard
    class 235
    model of arithmetic 536–539, 541–543
    part of a model 44
    structure 401–405, 418–421
    system of indices 215, 220, 221
state of a Turing machine 47, 50
Statman
Stearns 50, 53
Steel ix
Steinhaus 453
Steinitz 7, 351
Stephan 599
Stephenson 109
Sterling 329
Stewart 453, 454
Stillwell 479, 484
Stoltenberg-Hansen 204
Stoy 196
string 456
Strong 185, 222
strong
    array 228, 229, 235
    homogeneity 544, 545, 576
    minimal cover 503, 527, 569, 570, 573
    normalization 78
strongly
    effectively
        immune 267
        simple 266, 273, 294, 316, 339, 349
    hyperhyperimmune 284, 285
    hypersimple 285, 286
    representable 40, 42–44, 99
    uniform tree 520–523, 568, 583
Sturgis 64
sub-
    computation 217, 223
    creative 314, 315, 349
    sequence 89
    tree 494
substitution property 178, 179
Sudan 147
Sullivan 193
superthesis 103, 115–118
Suslin 380, 397, 392
Suslin-Kleene theorem 387, 391
Suzuki 225, 233
system of indices 215–221

acceptable 215–221, 236, 271, 272, 292, 308, 346
    standard 215, 220, 221

tag 253
Taimanov 8, 353
Taislin 8, 353
Tait 33, 38, 97, 201
Takeuti vii, 425, 438, 442, 443
tape of a Turing machine 46, 50
Tarski 8, 24, 39, 40, 42–44, 88, 114, 159, 160, 162, 166, 189, 190, 352–354, 363, 364, 366, 371, 376, 378, 400, 406
    theorem 114, 166, 376, 413
Tchuente 174
Tennenbaum 24, 238, 241, 274, 281, 289, 318, 513
terse 298, 299
Thatcher 165
Thomason 529, 537
three-
    element chain 523–528
    quantifier theory 492, 581
Thue 8, 144
Titgemeyer 524–526
Toffoli 52, 109, 174
topology 186–192, 205–208, 472, 473, 494, 495, 527
    Ershov 205, 211
    positive information 186
    Scott 192, 211
total
    functional 177
    recursive function 129
totality lemma 496, 517, 520, 552
Trakhtenbrot 53, 148, 269, 502
transitive
    closure 409
    structure 401–405
tree 381, 493
    1- 520
    binary 505
    e-splitting 498, 500
    finitely branching 509, 510
    identity 494
    method 493–495, 500
    of trees 497
    pointed 503, 504
    recursive 494, 505–510
    recursively bounded 509
    strongly uniform 520–523, 568, 583

total 494
uniform 516, 527, 583
Troelstra 10, 121, 122
true stage 246
truth 114, 166, 363, 364, 370, 376, 400,
    401, 468
truth-table (*tt-*) 268, 473
    completeness 270–272, 274, 334,
        341, 344, 346, 347, 349
    cylinder 593, 594
    degree 270, 276, 296, 317, 338, 352,
        584–591, 595, 598–601
    reducibility 251, 268, 306, 337, 591
    weak 337–342, 344, 347, 349, 584–
        591
Tucker 204
Turing vii, 1, 6, 46, 48, 53, 54, 64, 79, 99,
        102, 106, 115, 117, 118, 130,
        132–134, 146, 150, 164, 165,
        175, 176, 197, 213, 254, 337,
        538, 581, 582, 589, 590, 595,
        600, 601
    completeness 253, 255, 264, 265,
        277, 293–295, 304, 309, 339,
        341, 342, 344, 349, 353
    computability 53–54, 99, 101, 132,
        197
    degree 175, 176, 263, 276, 278, 287,
        297, 299, 304, 310, 313, 316,
        323, 328, 351, 447–553, 568,
        570, 574, 575, 583, 589–591,
        595
    machine 46–61, 116, 132, 203, 204,
        254
    reducibility 175, 176, 254, 280, 294,
        306, 338, 340
Turner 87
two-quantifier theory 490, 492, 581
type
    isomorphism 324, 325
    equivalence 328

Ulam 476, 481
Ullian 244, 267, 288, 323
Ullman 50, 53, 133, 144
undecidability 11, 103, 114, 147, 148, 150,
        151, 162–166, 253, 254, 263,
        350, 352–357, 492, 513, 537,
        550, 581
    essential 353–357, 513
uniform tree 516, 527, 583
    strongly 520–523, 568, 583

uniformization 137
uniformly recursive 178, 199, 233
union axiom 398
universal
    canonical system 254
    constructor 172, 173
    differential algebraic function 111
    isol 330
    partial function 132, 217
    Turing machine 132, 133, 172
unsolvability 103, 148, 150, 151, 254
uppersemilattice 15, 451, 529, 556, 558,
        559, 561, 572, 573, 575, 579,
        583, 589
urelement 399, 400
Uspenskii 107, 137, 186–188, 191, 207,
        215, 234, 236, 272, 318, 323

Van Der Mey 81
Van Emden 39
Van Heijenoort
Vaught 44, 352
verbose 298, 299
vertex of a tree 381
Vichniac 109
vicious circle principle 399, 422
Von Neumann 6, 62, 109, 112, 115, 116,
        151, 171, 173, 407, 410, 423
Von Neumann-Gödel-Bernays set theory
        423, 427
Vučkovich 319

Wagner 222
Wang H. 50, 64, 65, 99, 118, 120, 133
Wang P. 9
Watson 172
weak
    array 228, 229
    effective operation 213
    partial ordering 14
    Turing (*wT-*) 198
    truth-table (*wtt-*)
        completeness 338–342, 344, 347,
            349
        degree 337, 338, 584–591
        reducibility 337
weakly representable 40, 41, 43, 98, 99,
        159–163, 165, 166, 307, 352,
        353, 513–515
Webb 18
well-
    founded

relation 407, 421
structure 402
tree 381–384, 386
ordering 395, 386
of *L* 430, 431
principle 399
Weierstrass 8
Weihrauch 214
Weisbuch 117
Wexelblat 63, 64, 70, 87, 145
Weyl 422
'while' 68–70, 185, 186
Wiener 117
Winklmann 217, 221
winning strategy 453
Winograd 116
Wirth 70
Wittgenstein 84
Wolfram 174
Woodin 530, 534, 538, 547
word problem 8

Xiang Li 267

$\mathcal{Y}$ 81, 155, 185
Yates 212, 233, 247, 248, 277, 278, 283–
        285, 287, 292, 293, 313, 453,
        478, 495, 528, 544, 551
Young J.Z. 116
Young, P.R. 211–213, 217, 221, 235, 244,
        285, 324, 341, 582, 590, 592,
        593, 597

Zakharov 340
Zbierski 442
Zermelo 2, 399
Zermelo-Fraenkel set theory 399
ZF^- 400, 419, 425–427, 430, 431, 459
$ZF(C)$ 2, 399, 401, 417, 420, 427, 430–
        432, 434–438, 440–445, 453,
        467, 475, 529, 537, 540, 574
Zorn 464